AF356207

Animal Models of Human Pathology: Revision, Relevance and Refinements

Animal Models of Human Pathology: Revision, Relevance and Refinements

Guest Editor

Martina Perše

Basel • Beijing • Wuhan • Barcelona • Belgrade • Novi Sad • Cluj • Manchester

Guest Editor
Martina Perše
University of Ljubljana
Ljubljana
Slovenia

Editorial Office
MDPI AG
Grosspeteranlage 5
4052 Basel, Switzerland

This is a reprint of the Special Issue, published open access by the journal *Biomedicines* (ISSN 2227-9059), freely accessible at: https://www.mdpi.com/journal/biomedicines/special_issues/animal_models_pathology.

For citation purposes, cite each article independently as indicated on the article page online and as indicated below:

Lastname, A.A.; Lastname, B.B. Article Title. *Journal Name* **Year**, *Volume Number*, Page Range.

ISBN 978-3-7258-2691-9 (Hbk)
ISBN 978-3-7258-2692-6 (PDF)
https://doi.org/10.3390/books978-3-7258-2692-6

Contents

Preface

Animal models play a crucial role in enhancing our understanding of complex human pathologies. They not only provide insights into the underlying mechanisms of diseases but are also used for testing potential treatments and interventions.

Recent advancements in science and technology have ushered in new approaches for characterizing animal disease models. Research has increasingly focused on a deeper exploration of the molecular mechanisms driving disease and the various factors implicated in it. This shift underscores the necessity of critically re-evaluating existing models, a fundamental process for advancing scientific research.

While animal models are essential for unraveling the biological mechanisms of human pathology, the growing depth of knowledge regarding these mechanisms—along with the intricate interplay of factors contributing to human diseases—demands ongoing characterization and refinement of existing models, as well as the development of improved disease models to better meet the needs of scientific inquiry.

This reprint addresses the underlying mechanisms, risk factors, opportunities, hurdles, and challenges associated with a wide range of animal models. These include models of perinatal asphyxia, developmental hypertension, cisplatin toxicity, interstitial cystitis/urinary bladder pain syndrome, autism spectrum disorder, brain aging and neurodegeneration, peripheral neuropathy, tauopathies, iatrogenic chronic hypercortisolism, cancer, early-onset glaucoma, mucosal irritation studies, testing cardioplegic solutions for cardiopulmonary bypass surgery, and skin healing.

We thank the authors for their contributions and invite readers to explore the diverse topics presented in this issue. This reprint serves as an invaluable resource for anyone interested in the forefront of biomedical research and the understanding of animal models in complex diseases.

Martina Perše
Guest Editor

biomedicines

Review

Pathophysiology of Perinatal Asphyxia in Humans and Animal Models

Daniel Mota-Rojas [1,*,†], Dina Villanueva-García [2,*,†], Alfonso Solimano [3], Ramon Muns [4], Daniel Ibarra-Ríos [2] and Andrea Mota-Reyes [5]

[1] Neurophysiology, Behavior and Animal Welfare Assessment, Universidad Autónoma Metropolitana (UAM), Mexico City 04960, Mexico

[2] Division of Neonatology, National Institute of Health Hospital Infantil de México Federico Gómez, Mexico City 06720, Mexico; ibarraneonato@hotmail.com

[3] Department of Pediatrics, University of British Columbia, Vancouver, BC V6H 3V4, Canada; asolimano@cw.bc.ca

[4] Livestock Production Sciences Unit, Agri-Food and Biosciences Institute, Hillsborough BT26 6DR, UK; rmunsvila@gmail.com

[5] School of Medicine and Health Sciences, TecSalud, Instituto Tecnológico y de Estudios Superiores de Monterrey (ITESM), Monterrey 64849, Mexico; andreamreyes2020@gmail.com

* Correspondence: dmota100@yahoo.com.mx (D.M.-R.); dinavg21@yahoo.com (D.V.-G.)

† These authors contributed equally to this work.

Citation: Mota-Rojas, D.; Villanueva-García, D.; Solimano, A.; Muns, R.; Ibarra-Ríos, D.; Mota-Reyes, A. Pathophysiology of Perinatal Asphyxia in Humans and Animal Models. *Biomedicines* **2022**, *10*, 347. https://doi.org/10.3390/biomedicines10020347

Academic Editor: Martina Perše

Received: 31 December 2021
Accepted: 28 January 2022
Published: 1 February 2022

Publisher's Note: MDPI stays neutral with regard to jurisdictional claims in published maps and institutional affiliations.

Abstract: Perinatal asphyxia is caused by lack of oxygen delivery (hypoxia) to end organs due to an hypoxemic or ischemic insult occurring in temporal proximity to labor (peripartum) or delivery (intrapartum). Hypoxic–ischemic encephalopathy is the clinical manifestation of hypoxic injury to the brain and is usually graded as mild, moderate, or severe. The search for useful biomarkers to precisely predict the severity of lesions in perinatal asphyxia and hypoxic–ischemic encephalopathy (HIE) is a field of increasing interest. As pathophysiology is not fully comprehended, the gold standard for treatment remains an active area of research. Hypothermia has proven to be an effective neuroprotective strategy and has been implemented in clinical routine. Current studies are exploring various add-on therapies, including erythropoietin, xenon, topiramate, melatonin, and stem cells. This review aims to perform an updated integration of the pathophysiological processes after perinatal asphyxia in humans and animal models to allow us to answer some questions and provide an interim update on progress in this field.

Keywords: brain injury; hypoxic–ischemic encephalopathy; human and animal models; meconium aspiration syndrome; perinatal asphyxia

1. Introduction

"Globally 2.5 million children died in the first month of life in 2018, approximately 7000 newborn deaths every day, with about one third dying on the day of birth and close to three quarters dying within the first week of life" [1]. Newborn mortality differs, depending on the country. With 27 deaths per 1000 live births in 2019, sub-Saharan Africa had the highest newborn mortality rate, followed by 24 deaths per 1000 live births in central and southern Asia. An infant born in sub-Saharan Africa or southern Asia is 10 times more likely than an infant born in a high-income country (HIC) to die in the first month of life [1].

"Preterm birth, intrapartum-related complications (birth asphyxia or lack of breathing at birth), infections and congenital disabilities cause most neonatal deaths" in 2017 [1].

Progress in newborn survival has been slow, and the reduction in stillbirths has been even slower. An accelerated scaleup of care strategies targeting the major causes of death is needed in order to meet Every Newborn targets of 10 or fewer neonatal deaths and 10 or fewer stillbirths per 1000 births in every country by 2035. The most effective strategy to decrease perinatal and neonatal deaths is through interventions delivered during labor and birth, with 41% dying due to obstetric complications, and 30% followed by the care of small and ill newborn babies [2].

The incidence of neonatal hypoxic–ischemic (HI) brain injury is higher in preterm newborns than in the term newborns. Although the risk factors of perinatal asphyxia in preterm newborns are similar to those observed in the term newborns, the immature brain of preterm newborns, particularly those born before 32 weeks gestational age, is highly vulnerable to HI injury. Two primary reasons for this have been documented: (1) preterm newborns are at higher risk of hypoperfusion during transition, especially when transition is impaired; (2) their immature brains possess reduced autoregulatory capacity [3–6].

The increased survival of extremely immature infants poses an additional challenge. Data analysis on 4274 infants born after just 22–24 weeks of gestation in three periods (2000–2003, 2004–2007, 2008–2011) at the Neonatal Research Network Centers of the National Institute of Child Health and Human Development (NICHD) in the United States produced results that indicated an overall increase in survival from 30 to 36% and a rate of survival free of neurodevelopmental impairment that rose from 16 to 20% between period 1 and period 3. The frequency of cases of cerebral palsy of moderate-to-severe degree, however, did not show a significant decrease from period 1 to periods 2 and 3 (15% in period 1 vs. 11% in periods 2 and 3) [7–10]. Although evidence of this kind suggests that improvement has occurred in the affected population, indices of death and neurodevelopmental impairment, among other undesirable outcomes, are still high and raise concern that the evident decrease in mortality rates means that more infants with neurodevelopmental problems will survive. Hence, it is important to gather data on these two outcomes—death and impairment—so that physicians and family members can make informed decisions regarding early care for these high-risk children. The goal of this article is to clarify the major issues involved in the obstetric management of these cases, as well as aspects of the pathophysiology of the birth process, while stressing that it is unacceptable to decrease blood flow to the brain of fetuses and neonates in this condition [7–10]. This review aims to perform an updated integration of the pathophysiological processes after perinatal asphyxia, as well as recent investigations in different animal models, to allow us to answer some questions.

2. Defining Birth Asphyxia

The term birth asphyxia was introduced by the World Health Organization (WHO) in 1997 to describe the clinical condition of a newborn who either fails to establish or sustain regular breathing at birth [11,12]. In that context, birth asphyxia, implies a condition or a state in which the newborn requires immediate assistance to establish breathing. However, this makes the term imprecise and non-diagnostic of a causal pathology, which may vary from an intrapartum-related hypoxic event to a physiologic condition, such as prematurity, perinatal sedation, congenital structural abnormality of the brain or other maternal conditions.

Asphyxia has also been defined as a condition characterized by the impairment of gas exchange that can generate distinct degrees of hypoxia, hypercarbia and acidosis according to the duration and severity of airflow interruption. Asphyxia at birth—that is, impeded perinatal gas exchange—has no exact biochemical criteria, and we lack a gold standard that could ensure reliable diagnoses. For this reason, researchers have proposed numerous clinical and biochemical markers to predict, confirm or determine the condition called intrapartum asphyxia. These include verifying the pH of the umbilical cord, calculating Apgar scores, and evaluating the presence of acidosis, as well as signs of fetal distress. This is complicated, however, because intrapartum physiology can be very dynamic. These markers have limitations and remain controversial [13]. As such, caution must be exercised in labeling a neonate with asphyxia. Unfortunately, this term is often inappropriately linked with the poor neurodevelopmental outcome commonly referred to as cerebral palsy [14,15].

3. Circulatory Changes during Labor and Neonatal Transition

Human fetuses develop in an hypoxicemic state—but not hypoxic—where various vital mechanisms allow them to thrive. By binding to high-affinity fetal hemoglobin, for example, oxygen easily diffuses into the fetus' bloodstream from the mother's circulatory system. The blood that enters the placenta in this way is returned to the fetus via the umbilical vein, most of it entering the ductus venosus. The PO_2 level of this blood ranges from 32 to 35 mmHg, but en route to the right atrium, it meets less oxygenated blood from the inferior vena cava. In the right atrium, the blood with greater oxygenation from the umbilical vein flows into the left atrium through the foramen ovale before exiting through the left ventricle to supply two arteries, the carotid and coronary. After that, it streams into the aorta with blood from the right ventricle via the ductus arteriosus [12,16]. Thus, the fetus preferentially supplies more oxygenated blood to the brain (PO_2 of approximately 28 mmHg) and heart. The less oxygenated blood from the inferior and superior vena cava also mixes with placental blood and then exits the right side of the heart via the pulmonary trunk. Most of this blood (~90%) then bypasses the lungs via the ductus arteriosus and enters the aorta distal to the carotid and coronary arteries. The right ventricular blood has a PO_2 of 15 to 25 mmHg; however, distal oxygen delivery is supported by the fact that a large proportion of the cardiac output of both ventricles, pumping in parallel, is systemic in the fetus [12,16]. Then, a portion of this combined ventricular output (30% during gestational weeks 20 to 30 and 20% or less at or near term) enters the placental circulation. During labor, uterine contractions lead to intermittently decreased uterine arterial blood flow and decreased flow into the intervillous spaces. Transplacental gas exchange is also impaired intermittently, but this is generally inconsequential during normal labor. When the fetal side of the circulation is examined, uterine contractions do not seem to affect umbilical blood flow. At the time when normal birth occurs, several simultaneous circulatory changes favor the adaptation of the fetus to extrauterine conditions [12,16–19].

Because circulation switches from in-parallel to in-series, there is a need to equilibrate the outflows from the left and right ventricles (LVO, RVO), but completing this process takes several days—sometimes weeks—after birth, particularly in cases of preterm infants. This may occur due to a delay in closing of the fetal channels. When infants begin to cry right after birth, pulmonary vascular resistance decreases, and the newborns' lungs expand rapidly. Another important factor in systemic circulation is the removal of the (low-resistance) placenta by clamping the umbilical cord [20,21]. Pulmonary blood flow increases significantly, and so does the pulmonary venous return to the left atrium, closing the foramen ovale. Right-to-left shunting at the ductus arteriosus then decreases and eventually reverses as pressure of the pulmonary artery decreases below and the systemic blood pressure increases, aiding in the reversal of the ductal shunt. Increases in PaO_2 stimulate ductal closure. Finally, especially in some very preterm neonates, the inability of the immature myocardium to pump against the suddenly increased systemic vascular resistance (SVR) might lead to a transient decrease in systemic blood flow, which in turn could also contribute to a decrease in cerebral blood flow (CBF). Thus, the transition from intrauterine to extrauterine life involves fast, complex, and well-organized steps to guarantee neonatal survival [20,21]. When all of these changes are completed, an adult circulation pattern is established [14,22].

4. Pathophysiology of Birth Asphyxia

Delay in starting pulmonary ventilation at birth brings about a reduction in oxygen saturation in the blood and decreased oxygen delivery to the brain, which depends on aerobic metabolism to sustain the mitochondrial respiratory chain and adenosine triphosphate (ATP) ATPase activity. When hypoxia persists, there is a metabolic switch to glycolysis, which, for neurons, is a poor metabolic option because of low stores of glucose in brain tissue and the deficient ATP output by the glycolysis pathway. Glycolysis culminates in the generation of lactate, which then accumulates in extracellular compartments, causing acidosis, although it has also been suggested that lactate is an energy source for neurons [23].

Zheng and Wang [24] investigated the regulatory mechanisms of energy metabolism in neurons and astrocytes in the basal ganglia of a neonatal hypoxic–ischemic brain injury piglet model. Their results showed that lactate levels had peaked at 2–6 h after hypoxic–ischemic injury, and glucose peaked at 6–12 h. The expression levels of monocarboxylic acid and glucose transporter proteins (MCTs and GLUTs) increased after HI (peak at 12–24 h) and then decreased. Astrocytes and neuronal damage after HI were not synchronized. These results indicate that lactate and glucose transporters have a synergistic effect on the energy metabolism of neurons and astrocytes following hypoxic–ischemic reperfusion brain injury.

Reoxygenation is another process necessary for survival, but this involves uneven metabolism with certain organs that are metabolically privileged (adrenal medulla, brain, heart), others that are less privileged in this regard (the body in general, kidneys, muscles), and brain regions with irregular metabolism. While reoxygenation occurs, there is an increase in the levels of extracellular glutamate, which enhances activation of Na^+/K^+ ATPase to further increase the consumption of ATP. These high levels of extracellular glutamate exceed the buffering ability of astrocytes and produce a continuous overactivation of glutamate receptors. The receptors most severely affected are of the N-methyl-D-aspartate (NMDA) subtype. These reactions intensify Ca^{2+} conductance, leading to improper homeostasis and a condition of metabolic crisis marked by acidosis and an accumulation of lactate, which is reflected differentially during development in distinct areas of the brain. Under these conditions, lactate levels in the neostriatum remain high (>2-fold); however, this does not occur in other zones of the basal ganglia of rats subjected to severe asphyxia when compared to a group of control rats evaluated on postpartum day 8 [25]. This condition can arise in still-immature brains, causing a lesion that can alter the initial plasticity that is required to establish synapses and circuits. In fact, the extra energy the organism consumes in order to reestablish homeostasis could affect the levels it needs to consolidate synapses and circuits. Research based on the Swedish experimental model has identified various alterations that may be associated with perinatal asphyxia (PA). These include thicker pre- and post-synaptic densities, ubiquitination, errors in protein folding, altered levels of synapsin and neurotrophic factors, aggregation, interrupted postnatal neurogenesis, decreased length and branching of neurites, and deficits in myelination. Some studies suggest that plastic changes may occur in an effort to compensate for such neuronal loss. However, compensatory mechanisms, such as over-plasticity, for example, are not necessarily functional and may even aggravate cognitive impairment. Unfortunately, behavioral deficits inevitably and irreparably accompany this insufficiency of neural circuits [26].

Perinatal brain injury can affect infants born at any gestational age; however, preterm fetuses (born < 32 weeks of gestation) are less equipped to adapt to perinatal insults as term infants, making them more predisposed to brain injury [27].

Perinatal asphyxia remains a significant cause of neurological morbidity and mortality in the newborns [14], comprising a decrease in gas exchange, leading to a deficit of O_2 (hypoxemia) and excess CO_2 (hypercapnia), with consequent metabolic acidosis. If the episode is prolonged, blood flow and oxygen delivery to tissues are reduced (ischemia) [28]. This condition can have significant repercussions for the neonate, mainly at the level of the central nervous system [29].

Fetuses can suffer asphyxia at any point before, during, or after parturition [14]. In terms of frequency, studies indicate that around 50% of cases occur prepartum, 40% during birthing, and the other 10% immediately postpartum [30–33]. The fact that asphyxia has also been associated with various risk factors during gestation makes the pathophysiology of this condition exceedingly complicated. Risk factors identified to date include chronic diseases in the mother (preeclampsia, hypertension, diabetes) and her age, as these can restrict fetal blood flow and alter placental vasculature [14] (Figure 1), two processes associated with parturition and prolonged labor. Risk factors related to the placenta, meanwhile, include detachment, fetal–maternal hemorrhaging, inflammation, and insufficiency. In addition, the umbilical cord may become occluded, the fetus may develop an anomaly or malformation, intrauterine growth may be retarded, and neurological disorders or spinal cord injuries, among other conditions, may occur [14,34].

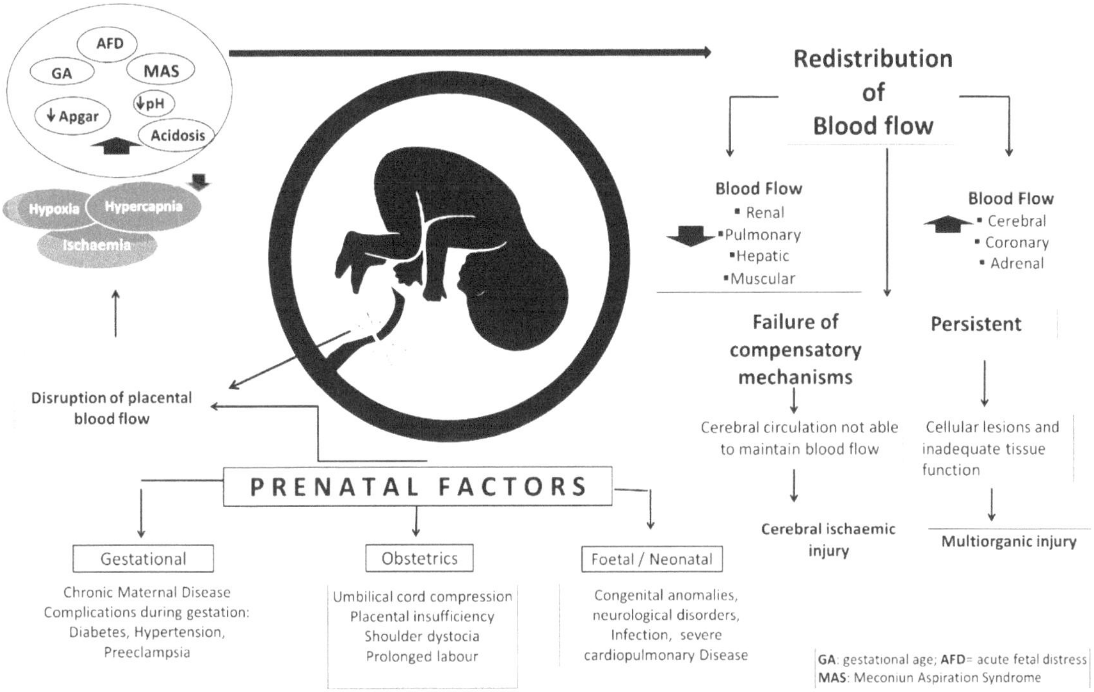

Figure 1. Prenatal risk factors and effect on cerebral ischemic injury.

Birth in all mammal species is accompanied by a period of obligatory or transient asphyxia in the newborn [30–32]. However, the severity of hypoxia–ischemia during these events can be affected by different factors related to the dam, placenta and/or fetus, or the neonate. Asphyxia can occur during fetal life, before, during, or after birth [14]. Approximately 50% of asphyxia occurs before delivery, 40% during parturition, and the remaining 10% during the neonatal period [33].

The pathophysiology of asphyxia is extraordinarily complex and related to several gestational risk factors, such as the dam's age and chronic maternal diseases, including diabetes, hypertension, or preeclampsia, that can affect the placental vasculature and decrease fetal blood flow [14] (Figure 1) associated with the process of parturition and prolonged labor. Placental risk factors are detachment, fetal–maternal hemorrhage, placental insufficiency, or inflammation. Other fetal neonatal factors involve occlusion of the umbilical cord, fetal anomalies, malformations, and intrauterine growth retardation, as well as spinal cord injuries and neurological disorders, among others [14,34].

4.1. Physiological Changes during Birth Asphyxia

Asphyxia markedly alters the physiology of the transition from the intrauterine to extrauterine life. Under conditions of placental hypoxia, circulatory and non-circulatory changes occur in the fetus. From the circulatory changes, the most important mechanism is the centralization of blood flow, or diving reflex, which consists of concentrating blood flow in vital organs, such as the brain, myocardium, and adrenal glands, reducing flow to less essential organs, such as skin, kidney, intestines, and muscle [35]. It is proposed that this mechanism is originated by the chemoreceptor of the carotid body that produces the release of catecholamines that are responsible for centralizing the blood flow; in turn, there is also a decrease in cerebral vascular resistance in order to improve cerebral perfusion. This mechanism of redistribution of flow has limits, and when its severity or duration exceeds these limits, a compromise of the blood flow begins to the encephalon and myocardium, predisposing neuronal damage. Although this mechanism of centralization can ideally preserve fetal physiology in hypoxic conditions, there is evidence that in some neonates, these mechanisms do not develop normally, which can favor brain damage [35].

At the respiratory level, after 30 s of absolute hypoxia, there is a brief period of rhythmic and rapid breaths, which leads to a brief stage of primary apnea (30–60 s). During this phase, there is also bradycardia followed again by panting breaths lasting 4 min after which there is again a secondary apnea, and in the lack of resuscitation, death follows [14,35].

Other mechanisms that attempt to compensate for this oxygen deficiency occur at the level of fetal hemoglobin, which increases its transport capacity and dissociation of oxygen. Additionally, the brain can use other energetic substrates under hypoxia to remain functional, such as lactate and ketone bodies. This is because neuroglial glycogen stores are depleted very quickly (<5 min) [14]. In some cases, neonates successfully recover from hypoxic episodes; however, in prolonged hypoxia, the redistribution of cardiac output causes dysfunction and multiorgan injury [36], leading to hypoxic–ischemic encephalopathy (HIE). Survivors of moderate-to-severe HIE show a high incidence of permanent neurologic and psychiatric disorders, including seizures, cerebral palsy, cognitive delays, motor disabilities, visual loss, hearing loss, behavioral alterations, schizophrenia, and epilepsy [30,37,38].

Basic and clinical research on asphyxia at birth has been supported by animal models, which contribute to the understanding of pathophysiology and lead to the discovery of biomarkers to accurately detect rapid alterations of biochemical pathways [39–42] in response to the hypoxic environment, as well as the severity of the injury by HIE, allowing for intervention and timely therapy. It is important to understand that the physiology of the fetus differs in fundamental ways from that of the newborn, and those distinctions are both structural and functional in nature. Ensuring the survival of neonates requires several well-orchestrated steps during the passage from intra- to extrauterine life, but these are complex and occur rapidly. Therefore, all caregivers handling newborns need to be well-versed in the physiology of this crucial transition; otherwise, they may fail to perceive deviations from normal physiology or to respond adequately when such scenarios arise. Precisely because asphyxia alters the physiology of transition, managing affected newborns must be based on a thoughtful approach [21].

Human and animal newborns are extremely vulnerable to hypoxia during and immediately after labor as a result of different anatomical, physiological, and biochemical factors. Several mechanisms can be responsible for the lack of fetal or newborn oxygenation: (a) fetal asphyxia due to obstruction of the blood supply in the umbilical cord, for example, torsion or tight circular cord; (b) imbalances of the placental exchange of oxygen, for example, premature placental detachment or retroplacental hematomas; (c) placental perfusion imbalances, for example, maternal hypotension, dehydration, cardiopathies, or severe maternal anemia; (d) failure of the expansion or ventilation or in the pulmonary perfusion of the newborn (asphyxia after birth is usually produced by the latter mechanism, since pulmonary ventilation does not adequately occur) [40–49].

Current clinical criteria of perinatal hypoxia require the presence of four conditions: (1) evidence of metabolic acidosis detected in a sample of umbilical cord arterial blood at birth (pH less than 7.00 and base deficit higher than 12 mmol/L); (2) early onset of moderate or severe encephalopathy in newborns of 34 weeks or more from gestation; (3) cerebral palsy of the spastic or dyskinetic type; and (4) exclusion of other etiologies, such as trauma, sepsis, and coagulopathy, among others [14]. During perinatal asphyxia, several physiologic changes occur in an attempt to compensate for the deficiency of oxygen supply in the newborn (Figure 2).

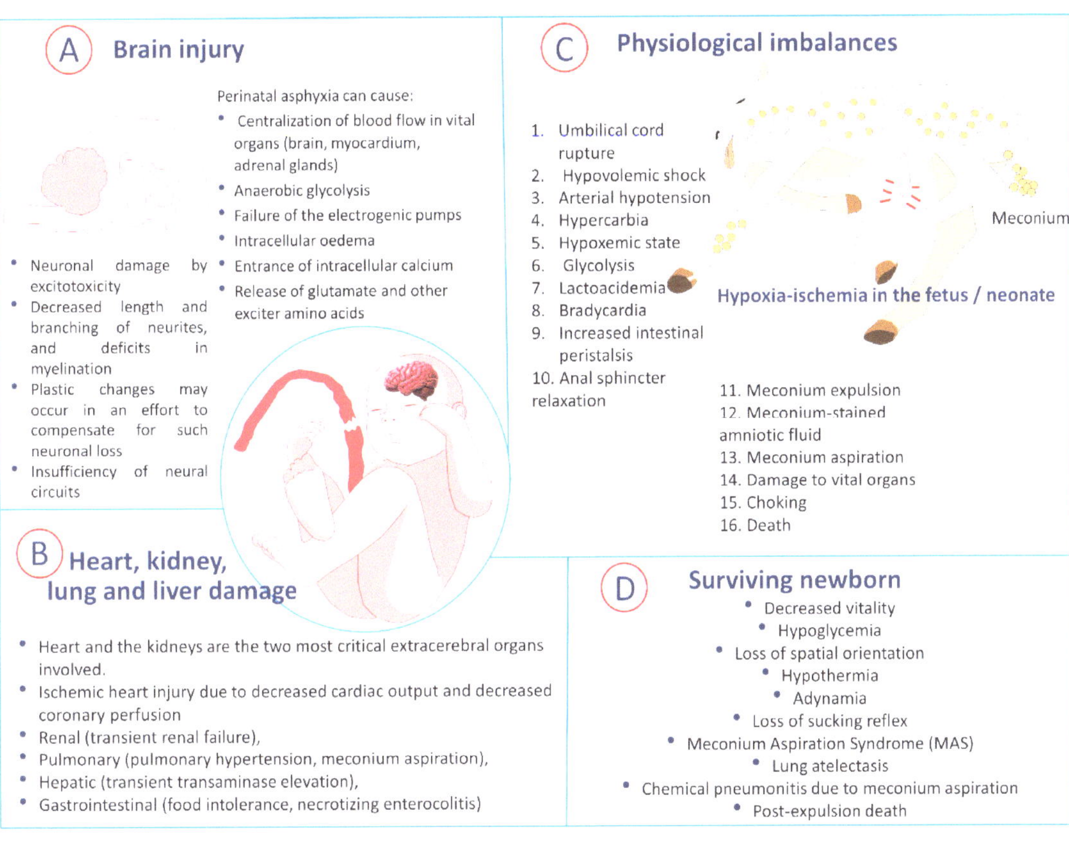

Figure 2. Fetal asphyxia in human and non-human animals, organ injury and physiological imbalances in response to hypoxia–ischemia. Image (**A**) summarizes the neuronal damage at a biochemical and pathological level. Image (**B**) shows damage in other vital organs. In image (**C**), the physiological imbalances are appreciated due to the drastic reduction of oxygen to the fetus, and promote the expulsion of meconium. In letter (**D**) some adverse effects if the newborn survives.

4.2. Mechanisms of Neuronal Injury in Perinatal Asphyxia

Once the compensation mechanisms mentioned above are overcome, a cascade of biochemical events begins in the brain, which eventually leads to different degrees of neuronal and cerebral injury.

As soon as the demands of cerebral O_2 are not adequately satisfied, the cerebral metabolism begins to perform anaerobic glycolysis with an increase in lactate levels (favoring the process of acidosis) and therefore a reduction in the formation of ATP. This energy deficit affects, in the first instance, the sodium/potassium membrane ATPase pumps, are involved in the maintenance of the membrane potential at neuronal rest [45,46,50,51].

This failure of the electrogenic pumps leads to a secondary accumulation of intracellular sodium, which induces intracellular oedema and loss of the membrane potential at rest, leading to the entrance of intracellular calcium, which favors the release of glutamate and other exciter amino acids, which, in turn, could be involved in the dispersion of neuronal damage caused by excitotoxicity [50]. The entrance of neuronal calcium also activates diverse calcium-dependent enzymatic systems (lipases, proteases, caspases), which are responsible for carrying out proteolysis of cytoplasmic components, also inducing mitochondrial damage. This mitochondrial damage, in turn, produces an increase in oxygen free radicals, which also enhance damage to macromolecules, such as the peroxidation of lipids, proteins, carbohydrates, and nucleic acids [52]. Additionally, all these processes of damage produce the release of inflammatory mediators by the microglial cells, leading to the recruitment of inflammatory cells from the periphery, interstitial oedema and alterations in the permeability of the blood–brain barrier [52].

All of these mechanisms are closely related, and all contribute, to a greater or lesser extent, to primary and secondary neuronal and brain injury. Additionally, it is also widely described that delayed restoration of oxygenation and perfusion to brain tissue can generate even more damage due to previously developed alterations in the permeability of the blood–brain barrier, as well as biochemical changes. This late reperfusion can increase inflammation, the generation of oxygen free radicals, and the interstitial oedema of brain tissue, which can increase final brain damage [14].

Despite the fact that at birth, the immune system has not fully matured, it does have the capacity to react to certain kinds of stimuli. When the condition called neonatal hypoxia–ischemia occurs, immune cells in the blood, brain, and peripheral organs are activated and trigger complex, dynamic interactions among these corporal regions. During or after HI, innate and adaptive immune cells in the bloodstream are activated and proliferate, and some extravasation into the parenchyma of the brain occurs. In addition, peripheral immune organs, such as the spleen, may be activated. This seems to exacerbate the insult, since performing splenectomies has been shown to provide some neuroprotection. However, we still do not know exactly how or to what extent specific innate or adaptive immune cells participate in neonatal cerebral lesions. We do know that excessive local cerebral inflammation has detrimental effects, but recent evidence suggests that microglia activation post-lesion may have a protective effect. Another important factor to be considered is the complex interaction that occurs between two systems: the central nervous (CNS) and the peripheral immune system. Responses by the CNS can be modulated by peripheral immune cells and mediators that penetrate the brain; however, at the same time, induction of tolerance and immunity can be orchestrated by brain-derived antigens that drain to peripheral lymph nodes [53].

Finally, because of the activity from all these mechanisms of neuronal damage, diverse strategies of neuroprotection have been implemented with different therapeutic targets (antiexcitotoxic, antioxidant, anti-inflammatory) to limit the neuronal damage associated with perinatal hypoxia and to improve the functional prognosis of the newborns. A complete review of these topics may be found in several recent publications [54,55].

5. Cardiovascular Alterations and Multiorgan Dysfunction

As a consequence of asphyxia and ischemia, multiple biochemical mechanisms are responsible for the deterioration of different organs and systems (central nervous system, 28%; cardiovascular system, 25%; kidneys, 50%; and lungs, 23%) [56]. However, the cardiovascular system and hemodynamic instability due to hypoxia, either in the uterus or during the transition of the newborn, have received considerable research attention [36].

During oxygen deprivation in the event of fetal hypoxia–ischemia, compensatory mechanisms are responsible for redistributing cardiac output, centralization of blood flow to vital organs, and reducing oxygen consumption [57]. Although hypoxia–ischemia can affect other tissues and systems, such as the renal (transient renal failure), pulmonary (pulmonary hypertension, meconium aspiration), hepatic (transient transaminase elevation), and gastrointestinal (food intolerance, necrotizing enterocolitis) systems, the heart and the kidneys are the two most critical extracerebral organs involved [56,58]. In a study carried out by Hankins et al. [59], they observed that low oxygen level was not the cause of hypoxic–ischemic encephalopathy (HIE) but that it was a condition developed secondarily to renal, hepatic, and cardiac dysfunction after asphyxia at birth (Figure 2).

5.1. Cardiovascular Response

Cardiovascular response consists of transient fetal bradycardia, systemic hypertension, peripheral vasoconstriction, and centralization of blood flow triggered by the release of catecholamines derived from the stimulation of chemoreceptors of the carotid sinus, which detect hypoxemia and transmit afferent impulses to the cardiovascular center in the medulla, then send parasympathetic efferent discharges and peripheral α- and β-adrenergic to the heart, peripheral vasculature, and adrenal glands [14,60,61]. If hypoxia persists, redistribution of blood flow to vital organs is maintained by peripheral vasoconstriction mediated by circulant vasoconstrictors norepinephrine, arginine, vasopressin, neuropeptide Y, and angiotensin II [62].

Hypoxia affects myocardial contractility and relaxation, causing hypotension; thus, babies require aminergic support [63]. Hypotension is a common problem found by neonatologists [64]. Asphyxiated newborns are known to have an increased risk of ischemic heart injury due to decreased cardiac output and decreased coronary perfusion [65]. The presence of high levels of cardiac enzymes diagnoses cardiac injury; however, the immediate and long-term structural consequences are not well known, since reported histological findings are minimal [66].

Other pathologies, such as cardiomegaly, have been described [67], including electrocardiogram abnormalities, signs of myocardial ischemia, arrhythmias, atrioventricular valve dysfunction, sustained sinus bradycardia, and decreased ventricular contractility [68,69].

Ikeda et al. [70] reported evidence of necrosis with phagocytosis in hearts of lambs with severe asphyxiation [71].

However, although blood supply to the heart is prioritized during an hypoxic event, studies reported deficits in the cardiac function after asphyxia at birth [72]. Sehgal et al. [65], indicate that when redistribution of cardiac output fails to maintain myocardial oxygenation, depletion of cardiac glycogen, anaerobic respiration, and metabolic acidosis occur, and without intervention, continuous circulatory deterioration will occur, eventually leading to myocardial dysfunction, circulatory shock, right and left ventricular insufficiency, tricuspid regurgitation, hypotension, and eventual cardiac arrest. In newborns surviving an intrapartum asphyxia event, the multiorgan damage that may result represents a high risk for the development of severe morbidities throughout life [66].

5.2. Renal, Hepatic, Pulmonary, and Gastrointestinal Injury

The reduction in blood supply to the kidney due to the blood redistribution to the peripheral organs to maintain the cerebral, cardiac, and adrenal perfusion during an hypoxic episode allowed for the recognition of Acute Kidney Injury (AKI) as an inevitable consequence of intrapartum asphyxia [66]. It is estimated that between 50 and 72% of asphyxiated newborns with an Apgar score $\leq$ 6 at 5 min show signs of renal compromise [73]. Cells of renal parenchyma have a limited capacity under anaerobic conditions and a high susceptibility to injury by reperfusion. Saikumar and Venkatachalam [74] suggest that in hypoxia/reoxygenation in vitro models, apoptotic cell death can occur during reoxygenation as a consequence of the mitochondrial release of cytochrome c during hypoxia.

The decreased excretory function of the kidney, which leads to a reduced capacity to filter blood and maintain blood volume, electrolyte levels, and acid-base homeostasis [75], correlates positively with risk of morbidity and mortality in the asphyxiated newborn [68]. If the kidney injury exacerbates, other organs, particularly the brain, will be damaged [76]. Regarding liver injury, Beath [77], points out that it is probably the cause of hypoperfusion secondary to hypoxia, rather than asphyxia itself. It has been seen that an increase in transaminase levels has some correlation with the severity of perinatal asphyxia [78]. However, Gluckman et al. [79] reported a weak relationship in the transaminase levels in a model of HIE in piglets.

The lungs are organs also commonly affected by asphyxia, so many babies require mechanical ventilation at birth [36]. Pulmonary hypertension, hemorrhage, and significant coagulopathy are frequently observed complications [80]. Concerning gastrointestinal complications, such as necrotizing enterocolitis, are described in infants with asphyxia; however, they are not common. The study of brain injury after intrapartum asphyxia has been the focus of much fundamental scientific and clinical research. When compensatory mechanisms are exceeded and cerebral blood flow can no longer meet demand, a cascade of molecular reactions and mechanisms related to calcium flow, free radical formation, free iron accumulation, and nitric oxide production begins, which leads to cell death [81]. The degree and location of brain injury may vary depending on the type and duration of injury, gestational age, and whether the baby was treated with hypothermia. Studies in both humans and animals have suggested that intermittent asphyxia for less than one hour is not likely to cause brain injury, but severe total asphyxiations can cause brain injury much earlier [82,83]. Brain injury after asphyxia is hypoxic–ischemic in nature, and the neurology injury patterns include selective neuronal necrosis, parasagittal cerebral injury, periventricular leukomalacia, and focal ischemic necrosis [14]. The adverse effects of an hypoxic–ischemic process in the fetus are summarized in Figure 2: both organic damage (A,B) and physiological imbalances (C), as well as some adverse effects if the newborn survives (D).

6. Meconium Aspiration Syndrome

Meconium-stained amniotic fluid (MSAF) results from the passage of fetal intestinal content to the amniotic fluid, which occurs with an incidence of 15 to 20% in term pregnancies and 30 to 40% in post-term pregnancies [84]. Among MASF newborns, 3 to 12% may develop meconium aspiration syndrome (MAS) [85]. When a neonate aspirates meconium during intrauterine panting or with the first breaths at birth, MAS develops [86]. MAS is characterized by respiratory distress (from mild tachypnea to severe respiratory failure), which occurs in newborns as a result of bronchoalveolar aspiration of meconium [85]. Among perinatal dysfunctions, MAS is considered a consequence of fetal distress and one of the predominant causes of morbidity and mortality in term infants [86], with important sequelae in lungs and neurologic development at short and long term [44,87]. Its association with asphyxia and pulmonary hypertension is well recognized [85]. It is worth considering that the incidence of MAS, pathophysiology, prevention, and management have changed in the last 20 years.

Meconium is mainly composed of water (70–80%), as well as intestinal epithelial cells, squamous cells, lanugo, amniotic fluid, bile acids and salts, phospholipase A2, interleukin-8, mucous glycoproteins, lipids, and proteases [48] (Figure 3). Meconium is present for the first time in the fetal intestinal tract between 70 to 85 days of gestation. It is recognized that the passage from meconium to the amniotic fluid is due to fetal gastrointestinal maturity or fetal stress secondary to hypoxia and infection. However, the pathophysiology of MAS is very complex and not yet fully understood. Fundamental questions arise, as some neonates exposed to MASF develop MAS, while others do not [88].

It is known that several mechanisms are implicated in the pathophysiology of MAS, including acute airway obstruction (considered the primary mechanism of MAS in the past), surfactant dysfunction, chemical pneumonitis and the direct toxic effect, and persistent pulmonary hypertension of the newborn (PPHN) with a right-to-left shunt and secondary infection. Other authors suggest that fetal systemic inflammation is also a risk factor for the development of MAS in newborns with MSAF [89–91].

Despite the complexity of the passage of meconium to the amniotic fluid, what is certain is the damage it generates when it is aspirated, such as airway obstruction, air trapping, hyperinflation of the lungs and subsequent pneumomediastinum or pneumothorax, partial alveolar collapse, and complete obstruction of the smaller airways, which may cause atelectasis [44,51,91–93]. Additionally, pneumonitis can develop due to the direct toxic effect of meconium components and the infiltration of a large number of polymorphonuclear leukocytes and macrophages. A decrease in lung capacity and oxygenation has been reported as a result of a reduction in surface tension of the surfactant due to meconium [84].

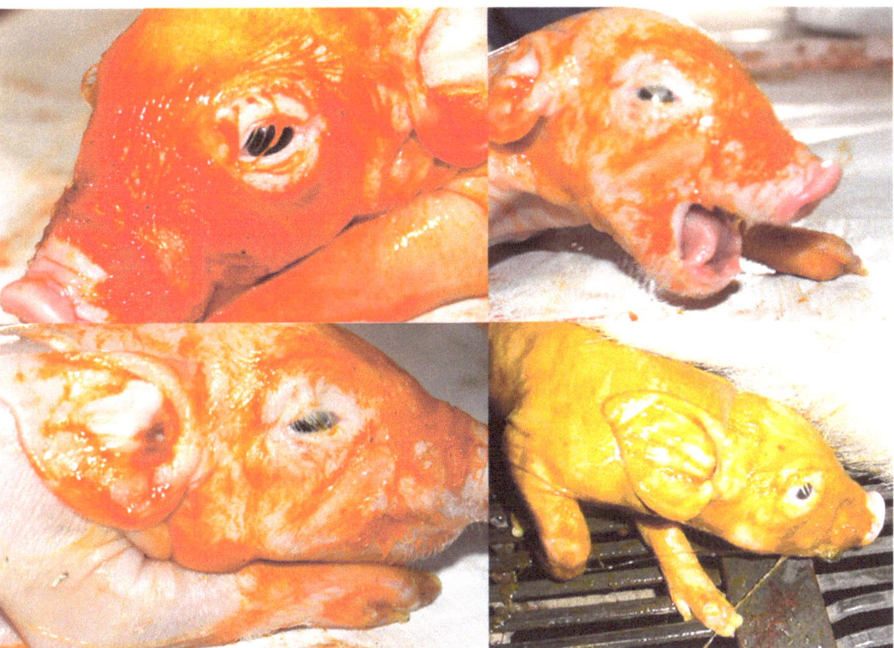

Figure 3. Neonatal piglets with perinatal hypoxia syndrome. When the neonate goes through a severe hypoxia process, it displays bradycardia, yellow to greenish meconium staining, tachypnea, lactic acidemia, hypothermia, hypoglycemia, adynamia, and flaccid muscle tone. Neonates with this perinatal syndrome do not recover, since they do not connect with the teat and are sluggish, which is why they die in the following postpartum hours. There is no neonatal intensive therapy area on pig farms.

7. Criteria for Diagnosis of Hypoxia–Ischemia

Animal studies in newborns are crucial for our understanding of transitional physiology and current resuscitation practice; as Dawes showed in the sixties [93], sustained hypoxia in utero, during labor or postnatally, manifests as an increased or absent respiratory effort, followed by a period of apnea (primary apnea). During this period, heart rate falls, but blood pressure is maintained. If asphyxia continues the infant gasp and the heart rate and blood pressure fall, and secondary apnea appears with anaerobic metabolism, lactic acidosis and myocardial performance compromise [94]. Ventilation is required to recover positive pressure, and if the insult lasted long enough, cardiac compressions are needed. Hypoxic–ischemic encephalopathy is the clinical manifestation of generalized disordered neurologic function due to hypoxia. In HIE, as opposed to other etiologies of neonatal encephalopathy, criteria for its diagnosis include (in neonates $\geq$ 35 weeks gestational age) [95]: evidence of intrapartum hypoxia (significant hypoxic or ischemic event immediately before or during labor or history consistent with fetal heart-rate compromise) and two or more of the following:

a. Apgar score of <5 at 5 and 10 min;
b. Need for mechanical ventilation or resuscitation at 10 min;
c. Acidemia documented in fetal umbilical artery (pH < 7.0 or base deficit $\geq$ 12 mmol/L);
d. Multisystem organ failure;
e. Evidence of moderate or severe encephalopathy staging, often supported by neuroimaging with evidence of acute brain injury consistent with hypoxia–ischemia.

Clinical Assessment

Clinical assessment and disease staging are based on the scale developed by Harvey and Margaret Sarnat [96]. Frequently, a modified (simplified) score, such as Thompson's, is used [97]. Moderate and severe grades of encephalopathy have a poor prognosis, as opposed to mild. Therefore, two combinations of signs have been used as criteria [79,98], including:

1. Reduced responsiveness with hypotonia or incomplete reflexes (including weak suck) or clinical seizures.
2. At least three signs from the following categories:
   (a) Reduced responsiveness;
   (b) Reduced activity;
   (c) Abnormal posture;
   (d) Abnormal tone;
   (e) Incomplete reflexes;
   (f) Abnormal pupil response, heart rate, or respiration.

8. Neonatal Hypoxic–Ischemic Encephalopathy: Clinical Aspects

Neonatal hypoxic–ischemic encephalopathy (NHIE) in babies born at term or preterm may occur at different periods of the newborn's life. These periods are antepartum, intrapartum, and postpartum. NHIE puts the neonate at a higher risk of suffering visible neurological alterations [99].

Neonatal hypoxic–ischemic encephalopathy, as a clinical syndrome, was described in the second half of the twentieth century and included: (1) evidence of fetal stress based on records of heart rhythm and amniotic fluid stained with meconium; (2) signs of depression in the neonate at birth; (3) neurological syndrome progressing in the first hours/days of life [100]. The clinical features of NHIE from childbirth and after 12 h include low levels of consciousness, for example, stupor or coma; a respiratory abnormality, such as periodic breathing or failure; intact pupil response; hypo or hypertonia; and seizures.

Clinical characteristics from 12 to 24 h of life are stupor–coma or some other alert impairment, more seizures, apnea events, and weakness (in proximal limbs or hemiparesis in full-term infants; in lower limbs in premature infants) [101].

From 24 to 72 h, infants show stupor–coma, respiratory arrest, abnormalities of the oculomotor brain stem and pupil, deterioration of consciousness after a hemorrhagic infringement in full-term infants, or interventricular hemorrhage in preterm infants [101].

After 72 h, the stupor–coma remains persistent, as well as abnormal movements of suckling and swallowing, hypo and hypertonia, and weakness [101].

Some pathological brain changes involve [102] selective neuronal necrosis of the cortex, basal ganglia, thalamus, brain stem (reticular formation and other nuclei), cerebellum, and anterior horns from the spinal cord.

- Parasagittal injury of the cerebral cortex, in subcortical white matter in the lateral convection of the superior-medial orientation, in the posterior–anterior direction.
- Periventricular leukomalacia with necrosis in the subcortical white matter of the hemisphere, including descending motor fibers, optical radiations, and association fibers.
- Focal and multifocal necrotic ischemia in the cerebral cortex and subcortical necrosis in white matter, mainly unilateral with a vascular distribution.
- Unilateral brain lesion with the presence of minor injury in the contralateral hemisphere. The most frequent lesion is in the area of irrigation of the medial cerebral artery: cerebral cortex, white matter, basal ganglia, and posterior limb of the internal capsule (Figures 4 and 5).

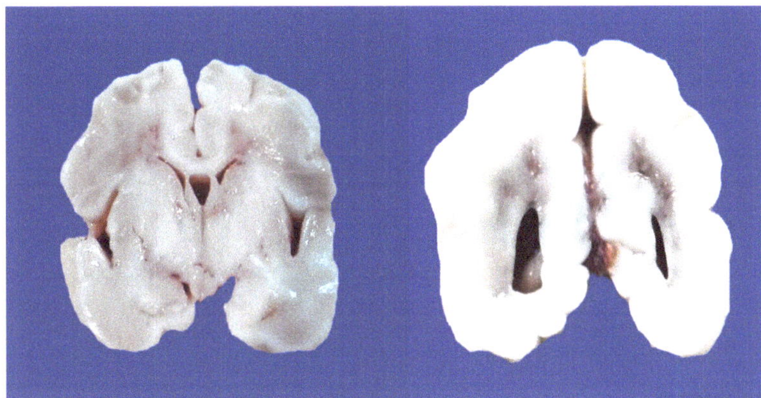

Figure 4. Periventricular white matter greyish discoloration and edema in a male, 32 weeks of gestational age old, with HIE. (Courtesy: María de Lourdes Cabrera-Muñoz, MD, Department of Pathology, Hospital Infantil de México Federico Gómez).

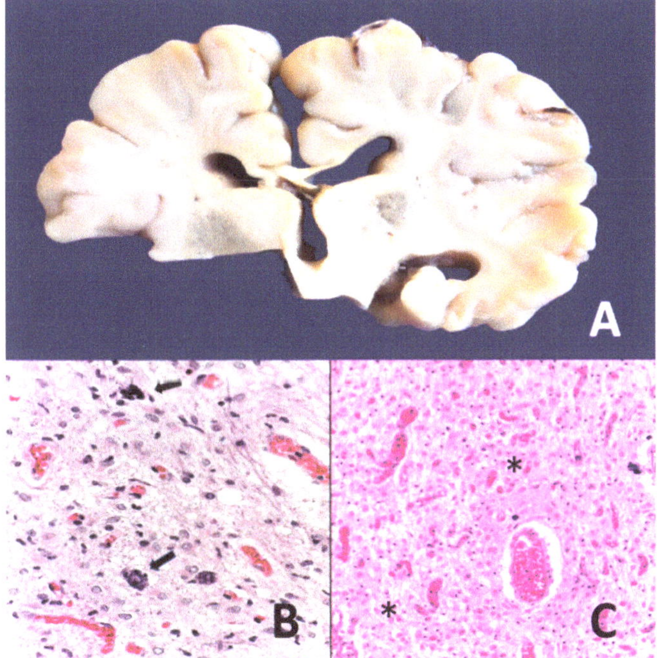

Figure 5. (**A**). Bilateral basal ganglia necrosis in a male at 39.6 weeks gestation with severe perinatal asphyxia secondary to congenital heart disease. (**B**). Neuronal necrosis and calcification (arrows). (**C**) Withe matter infarct (*) HE 40X. (Courtesy: María de Lourdes Cabrera-Muñoz MD. Department of Pathology, Hospital Infantil de México Federico Gómez).

However, the prevention of brain damage or the use of neuroprotective drugs in animal models was not successful as expected. More research is necessary for the benefit of newborns suffering NHIE in order to avoid neurological sequelae and development deviations.

8.1. Inflammatory Biomarkers of Birth Asphyxia

Understanding the causes of asphyxia and the intermediate steps between hypoxia and fetal death can allow for the identification of biomarkers that enable prediction and prevention of fetal death, mainly in women at risk of this clinical complication [103].

Several biomarkers have been studied. Creatinine, liver enzymes, cardiac troponin I, prolonged coagulation times, and thrombocytopenia help to address multiorgan dysfunction. In one study, lactic dehydrogenase (LDH) sampled within 6 h of birth was found to provide prognostic information: <2085 U/L survived without disability vs. 3555 U/L (interquartile range 3003 to 8705) who were disabled [104].

In the last decade, the search for useful biomarkers to accurately predict the severity of the lesion in perinatal asphyxia and HIE has become an area of growing interest in neonatal research [38]. However, despite the potential of many promising markers, few studies have been validated or used in clinical practice [105].

8.2. Placental Inflammatory Biomarkers

It is well known that the placenta mediates the interactions between the mother and the fetus during pregnancy. Thus, in a clinical study with a population of neonates with high-risk evidence, with acidosis and encephalopathy, placentas were analyzed, and 95% were found to have abnormalities. Additionally, 65% met the criteria for diagnosis of an important placental pathology, with a high incidence of inflammatory processes (chorioamnionitis and chronic patchy/diffuse villitis), which were significantly associated with abnormal outcomes of neurological development two years after treatment with hypothermia. Therefore, these findings suggest an important contribution of placental inflammatory mechanisms in the severity of asphyxia [106,107].

8.3. Serum Brain Biomarkers

Within the neuronal biomarkers detected in the serum of neonates with HIE we can find glial fibrillary acid protein (GFAP), which is an intermediate filament protein released from astrocytes [108]; ubiquitin carboxyl-terminal hydrolase (UCH-L1), a specific cytoplasmic neuron enzyme and a marker for neuronal apoptosis that is concentrated in dendrites [108]; S-100B, a calcium-binding protein released in astrocytes; neuron-specific enolase (NSE), a glycolytic enzyme in neurons [109]; ubiquitin carboxy-terminal hydrolase L1 (UCH-L1); or total tau protein, which indicates astrocytic and neuronal damage. These proteins are released into the blood through the increased permeability of the blood–brain barrier [81].

Another characteristic of traumatic brain injury is an imbalance between oxygen delivery to the brain and consumption. Post-insult alterations in the flow rate and volume of oxygen in the blood and arteries can immediately compromise the delivery of oxygen to the brain. When this occurs, the area around the lesion suffers an hypoxic condition that affects brain resident cells and infiltrated neutrophils. Various published reports mention that under hypoxic conditions, microglia can release nitric oxide and a series of cytokines. This sustains neutrophil activation, triggering persistent hypoxia and neuronal death [110]. Sorokina et al. [111] suggested the involvement of nitric oxide and its products in immune responses and that traumatic brain injuries (TBI) may be accompanied by nitrosative stress. Regarding the first two days of mild-to-moderate vs. severe TBI, these authors suggested that the opposed natural levels of NR2 (NMDA) antibodies could be related to a key mechanism that operates to protect neurons from Glu excitotoxicity.

In theory, biomarkers can identify at-risk pregnancies and allow a proper intervention before an irreversible lesion occurs in the fetus due to hypoxia; however, many studies, although novel, are small pilot tests that have no power to predict long-term results. Therefore, it is necessary to continue researching this topic [81,112]. Several biomarkers have recently been isolated that may aid in identifying neonatal neurologic injuries in the immediate postpartum period. Trials with 10-day-old mice showed that the HI condition significantly increases osteopontin significantly, but that this did not occur after administering LPS. The mRNA of osteopontin was induced in the brain but not the blood. Since immunostaining showed the expression of osteopontin by the microglia/macrophages in brains lesioned by HI, the authors suggested osteopontin as a possible prognostic blood biomarker in cases of HIE related to the activation of microglia in the brain. They further posited that an increase in osteopontin in the bloodstream may be indicative of a perinatal event that occurred up to 24 h earlier. Hence, osteopontin might be an effective marker of previous HI stress in the uterus and of an inadequate response to treatment for hypothermia [112].

The most reliable approach for identifying and evaluating the seriousness, timing, and pattern of these kinds of insults may well consist in measuring an array of inflammatory and neuronal biomarkers at a precise point of care at various intervals. Full pathway analysis employing various omic strategies could lead to the identification of new therapeutic targets for the treatment of neonatal HI cerebral lesions. However, research into neonatal brain biomarkers is still at its outset, so we may have to wait some time for major advances in this area [113].

8.4. Electrophysiology

Amplitude-integrated electroencephalography (aEEG) has been widely used in neonatal intensive care units (NICU) and can help to diagnose encephalopathy by the bedside. It provides an objective record that can be reviewed by a pediatric neurologist if needed. In mild forms of encephalopathy, the background pattern might change from continuous to discontinuous. This is called discontinuous normal voltage and does not represent an important abnormality. If the injury is more severe, a low-voltage background with periods of normal amplitude might be seen; this is called burst suppression. On more severe disturbance is when there is no burst, only continuous low-voltage activity that can progress to a flat trace. Burst suppression, continuous low voltage, and flat trace are severely abnormal. Spitzmiller et al. [114] found in metanalysis (8 studies) that aEEG has an overall sensitivity of 91% (95% CI 87–95%) and a negative likelihood ratio of 0.09 (95% CI 0.06–0.15) in prediction of poor outcomes [114].

8.5. Near-Infrared Spectroscopy

Regional cerebral saturation ($rScO_2$) and fractional tissue oxygen extraction (cFTOE) have also been used to monitor oxygen delivery to the brain by near-infrared spectroscopy (NIRS). High cerebral oxygenation on NIRS with a low electrical activity aEEG background in severely HIE neonates on hypothermia treatment at 12 h of age has a 91%, positive predictive value for long-term adverse neurological outcomes (magnetic resonance imaging and neurodevelopmental assessment at 18 months of age), and the absence of these results in a negative predictive value of 100% [115] (See Figure 6).

8.6. Neuroimaging

Cranial ultrasound is useful to document antenatal injury. After 24 h, cerebral vasodilatation can be documented in a low cerebral resistance index (cRI). Low RI is not predictive of poor outcome during hypothermia, but in normothermic infants or after rewarming, a value below 0.55 has an 84% positive predictive value for death or disability [116] (See Figure 7).

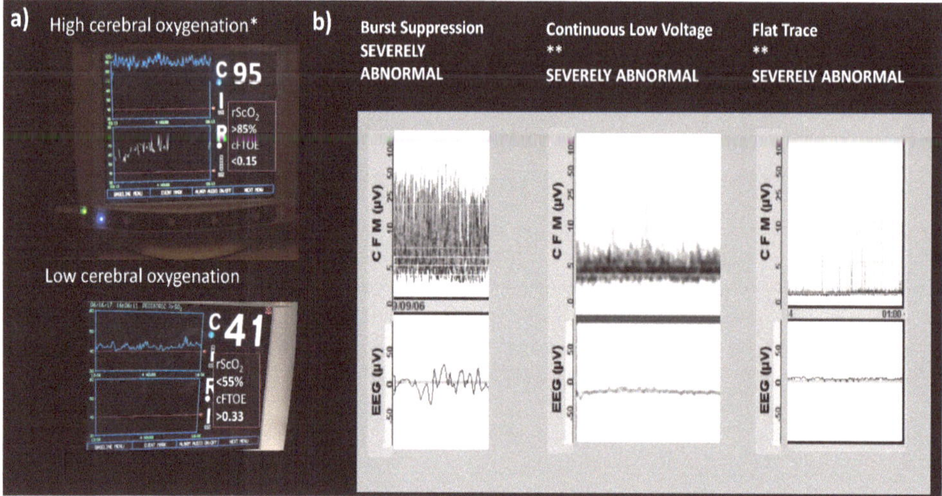

Figure 6. (**a**) High cerebral oxygenation on NIRS * with (**b**) a low-electrical-activity aEEG background ** in severe HIE neonates on hypothermia treatment at 12 h of age has a 91%positive predictive value for long-term adverse neurological outcome (magnetic resonance imaging and neurodevelopmental assessment at 18 months of age), and the absence of these results in a negative predictive value of 100%. NIRS: near infrared spectroscopy; aEEG: amplitude-integrated electroencephalography; $rScO_2$: regional cerebral oxygenation; cFTOE: cerebral fractional tissue oxygen extraction. Courtesy: Daniel Ibarra-Ríos, MD, Department of Neonatology, Hospital Infantil de México Federico Gómez.

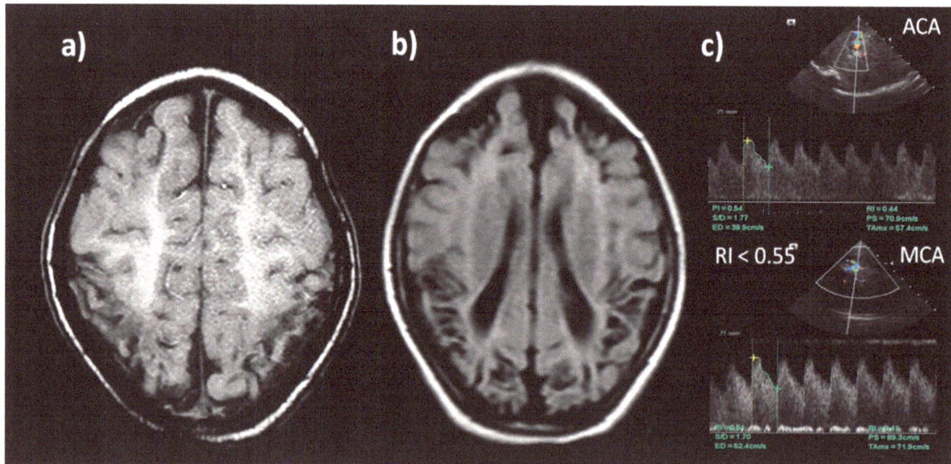

Figure 7. Magnetic resonance: (**a**) axial image of sequence enhanced in T1 towards convexity and (**b**) enhanced in FLAIR, where hyperintensity of the bilateral semioval centers is observed, as well as a decrease in volume in bordering territory in parietal regions and in the smaller frontal portion with retraction of the lateral ventricles associated with hyperintensity of the periventricular white matter. Courtesy: Eduardo M. Flores Armas, MD, Department of Medical Imaging. Cranial ultrasound: (**c**) low RI (<0.55) in normothermic infants or after rewarming has an 84% positive predictive value for death or disability. Courtesy: Daniel Ibarra-Ríos, MD, Department of Neonatology, Hospital Infantil de México Federico Gómez. ACA: anterior cerebral artery; MCA: medial cerebral artery; RI: resistive index.

Magnetic resonance is the imaging choice for prognosis. Diffusion-weighted imaging helps to detect abnormalities within the first week. A study of between 7 and 21 days is essential to document injury to basal ganglia, the internal capsule, white matter, brainstem and cortex. In conjunction with cranial ultrasound, it can help to detect complications or comorbidities such as infarction, hemorrhage, and malformations. The evolution of diffusion abnormalities on MRI following HIE in term infants on therapeutic hypothermia (current treatment) has been studied [117]. The above led to the development an MRI injury scoring system. Higher MRI injury grades were significantly associated with worse outcomes in the cognitive, motor, and language domains of the Bayley-III [118,119].

8.7. Hemodynamic Management

Echocardiography is an important diagnostic tool to delineate two different clinical scenarios [120]:

1. Low systemic blood flow with normal oxygenation: with echocardiographic findings consistent of left ventricle (LV)/right ventricle (RV) dysfunction in which management must include positive inotropes.
2. Low systemic blood flow with impaired oxygenation. In this scenario, the echocardiographic finding can show:

    (a) Persistent pulmonary hypertension, where management should include pulmonary vasodilation, subsequentially augmenting systemic blood flow after pulmonary venous return improves;
    (b) LV dysfunction with PPHN, where management must include positive inotropy and maintenance of right-to-left ductal shunt to support systolic blood flow;
    (c) RV dysfunction with PPHN, where management must include positive inotropy, reduced RV afterload (pulmonary vasodilation and consideration of prostaglandin E1 if the ductus arteriosus is restrictive) and maintenance of adequate RV preload.

Nitric oxide remains the vasodilator of choice. Low-dose vasopressin is a physiologically suited drug in HIE newborns with systemic hypotension and oxygenation failure, as it can produce systemic constriction and pulmonary vasodilation, as well as theoretical stimulation the endogenous production of nitric oxide, as noted in animal models [121]. Milrinone, a widely used inotropic/dilator in neonatology, is not recommended, as drug clearance is lower (especially in patients on therapeutic hypothermia), and might cause extreme hypotension [122].

Hochwald et al. [123] found that neonates with brain injury on magnetic resonance imaging had higher superior vena cava (SVC) flow pre-rewarming compared with newborns without brain injury, possibly reflecting a lack of cerebral vascular adaptation in newborns with more severe brain injury [123]. Is important to note that LVO is lower in patients receiving therapeutic hypothermia (mean ± SD 126 ± 38 mL/kg/min), so one must be cautious interpreting it. Sakhuja et al. [124] found that celiac artery and superior mesenteric artery blood flow velocity and LVO did not vary during hypothermia but rose after rewarming, suggesting a protective effect of therapeutic hypothermia on the gastrointestinal system [124].

Hemodynamic assessment might have prognostic value, as Giesinger et al. [125] found studying 53 patients with HIE undergoing hypothermia. RV dysfunction was associated with risk of adverse outcome; high brain-regional oxygen saturation and low middle-cerebral artery resistive index were associated with RV dysfunction in post hoc analysis [125].

9. Targets for Neuroprotection

Potential Interventions for Birth Asphyxia: A Window for Reducing Further Brain Injury

Studies of animals and humans show that both the stage of development and the seriousness of HI lesions impact the brain's selective regional susceptibility to damage and, as a result, the clinical manifestations that ensue from such damage. This approach, however, has one serious drawback: the fact that of every six or seven neonates with HI treated with therapeutic hypothermia (TH), only one is saved from death or serious disability by the age of 18–22 months. The HELIX trial, which looked into hypothermia for encephalopathy in economically developing countries with low and middle incomes, was published in 2021, adding complexity. It was a multicenter, open trial performed with a randomized control method. The study, which included 408 newborns, was conducted with a thoroughly strict procedure. For the group with hypothermia, 202 subjects were included, and for the control group, 206. The study concluded that hypothermia in these developing countries did not reduce mortality or moderate-to-severe impairment at 18 months, but it remarkably increased the number of deaths [126]. The lack of hypothermic neuroprotection in developing countries could potentially be associated with the subacute nature of brain injury in these countries, as demonstrated by the pathway of partial prolonged hypoxia (intermittent hypoxic injury) presented in economically developing countries vs. acute hypoxia manifested in high-income countries, as proven by MRI spectroscopy and gene-expression studies [127]. These results indicate that alternative techniques must be developed to increase the efficiency of TH or replace it [128].

Based on different animal models on newborn swine, rat pups, and fetal sheep [129,130], clinical trials of head cooling combined with body cooling and whole-body cooling alone were conducted [131]. As previously mentioned, current evidence from 11 randomized control trials on asphyxiated newborns with 36 weeks of gestation or more found that moderate hypothermia (33.5 °C for the whole body and 34.5 °C for head combined with body cooling) initiated less than 6 h after birth for 72 h with a rate of rewarming of 0.5 °C/h reduces death and neurodevelopmental disability among moderate and severe cases of HIE (64). A recent trial demonstrated that among infants of at least 36 weeks of gestation with HIE, more extended cooling was not superior to 72 h of cooling, and more profound cooling was not superior to cooling to 33.5 °C [132]. Gunn et al. [130] studied a fetal sheep model of hypothermia, cooling at 1.5, 5.5, and 8 h. They found the highest levels of neuroprotection (neuronal loss score) at 1.5 h, less (but favorable) neuroprotection at 5.5 h, and no neuroprotection at 8 h [133]. These findings constitute the basis for the conclusion that hypothermic neuroprotection is time-sensitive. A question that has not been answered is whether, on humans, a wider window of opportunity exists beyond 6 h, which is very important in limited-resource settings where transport to third-level care might take longer. At present, a clinical trial is addressing that question [8]. It is crucial to prevent hyperthermia, as one of the clinical trials showed that the odds of death or disability were quadrupled for each 1 degree C increase in the highest quartile of the skin or esophageal temperatures [134]. Some groups use the term therapeutic normothermia to designate all the measures taken at the bedside to avoid iatrogenic hyperthermia. Another critical question to be addressed is whether there is a benefit of cooling late preterm infants. So far, case reports have shown mixed results, but a lack of safety is an issue. Currently, a clinical trial is recruiting preterm infants of 33 to 35 weeks gestational age who present at < 6 h postnatal age with moderate to severe neonatal encephalopathy [134]. Supportive management for neonatal encephalopathy includes adequate resuscitation, avoiding hyperthermia, maintaining $PaCO_2$ in the normal range, adequate hemodynamic management according to clinical scenario, avoiding hypertension, and cautious volume management (initial volume restriction) following serum sodium and fluid balance and serial glucose, with prompt correction when abnormal [120,135].

In addition to hypothermia, the most promising adjuvant therapies include Xenon and erythropoietin [135]. Xenon is an anesthetic gas that acts as a non-competitive NMDA antagonist. The largest trial in humans is the TOBY-Xe study, wherein one arm was treated with hypothermia and xenon (administered within 12 h of life for a duration of 24 h vs. hypothermia alone). No differences were found in biomarkers of cerebral damage, and this is believed to be influenced by the late initiation (median 10 h) of the drug [136]. Erythropoietin (EPO) is assumed to work as an antioxidant, an anti-inflammatory, and an inducer of antiapoptotic factors [137]. Protection has been found in stroke and hypoxic–

ischemic animal models [138–140]. Razak et al. [141], in metanalysis (6 studies 5 with EPO and one with darbepoetin, N; 454), found a reduced risk of brain injury identified in EPO-treated infants, as well as a trend toward a lower risk of death [10].

To date, EPO is the only approach that has been tested in primates, accompanied by follow-up that allows for the calculation of anticipated effect sizes for the combined outcomes of cerebral palsy and death.

With the possible exception of melatonin, none of the therapeutical approaches examined herein necessarily requires intensive care measures. However, additional pre-clinical research on the processes of infection and inflammation are needed before trials can be contemplated [142].

Adverse neurodevelopmental results due to HIE can be reduced by treating (subclinical) seizures and routine EEG neuromonitoring. The large number of patients who suffer undesirable outcomes, however, indicates the need to focus research on complementary therapies and interventions that can improve results. There is an urgent need for funding to carry out studies of this kind. The evidence available at this time supports beginning TH as soon as indicated to obtain optimal results in terms of neuroprotection [143]. Topiramate is a sulfamate-substituted monosaccharide and carbonic anhydrase inhibitor. A substance similar to acetazolamide, it was first analyzed as an anticonvulsant with potential for neuronal repair. It also seemed promising for repairing post-ischemic myocardial damage. The best protection seems to come from treatment with melatonin, a pineal hormone with circadian secretion and maximum nocturnal values. Melatonin has several important abilities, as it can diminish oxidative stress, scavenge free radicals, and improve cellular physiology. Recent stem cell research has raised expectations regarding regenerative medicine, where the functionality of cells damaged by hypoxic insult could be improved or replaced by progenitor cells. The paracrine role of progenitor cells and stimulation of myocardial angiogenesis/repair are reflected in current data [144].

10. Neurodevelopment of Babies with Asphyxia

Deviations in the neurodevelopment of infants after an NHIE event have recently been the focus on many investigations. Researchers have found a large percentage of child survivors of NHIE with significant neurodevelopment sequelae [145] (Figure 8). There is a close relationship between the degree of NHIE and the outcome. When the risk of death was 12.5%, neurological disabilities had a frequency of 14.3%—the more severe the NHIE event, the greater the number and severity of sequelae. Therefore, clinical classification of NHIE has prognostic value [146]. More severe symptoms of neurological dysfunction in the first days of life are associated with a more severe outcome [145].

The main neurological alterations after an NHIE event are cerebral palsy, blindness, deafness, epilepsy, mental retardation, delayed motor control, speech delay, attention deficit hyperactivity disorder, reading–writing disability, and psychiatric alterations. At-risk Infants must be followed by a long interval of time, at least during the first years of the school-age period. If the necessary resources are available, it is recommended to continue monitoring until adolescence and youth [146]. The observation of the findings related to the language of children with NHIE indicates that deficits may become evident only with advancing age and stages of childhood neurodevelopment [147]. The aim is to prevent the development of neurological disabilities, control seizures and abnormal movements, aid in orthopedic development, and promote the development of cognitive functions and brain plasticity. The rational use of therapeutic measures should produce normal or almost normal neurologic development of these at-risk infants.

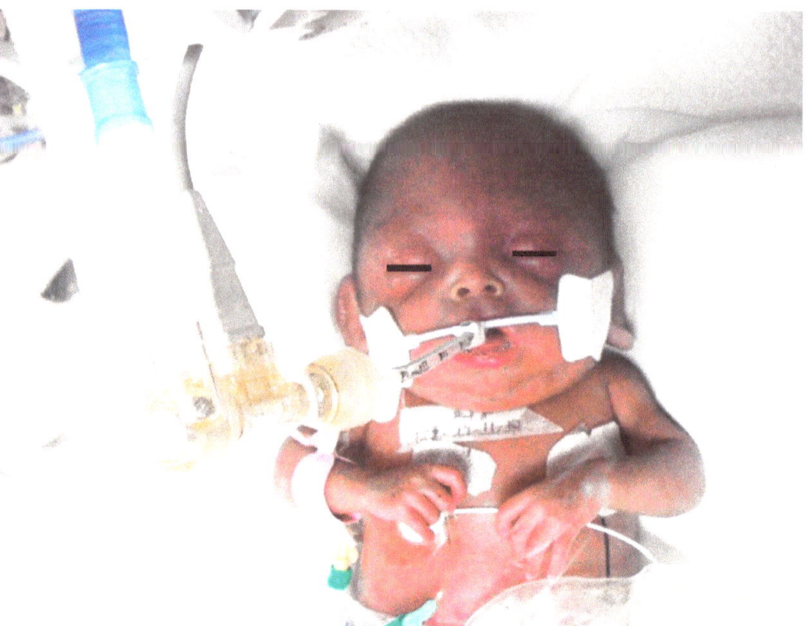

Figure 8. Male of 34 weeks of gestational with multiorgan damage (heart, lungs, liver, gut, and kidneys) after severe perinatal asphyxia. (Courtesy: Dina Villanueva-García, MD, Department of Neonatology, Hospital Infantil de México Federico Gómez).

Prognoses of the clinical course, severity, and results of any disease are important. In this regard, studies have analyzed organ-specific proteins as potential plasma-based biomarkers of insults. To date, they have been found to be of poor specificity and sensitivity for routine clinical use, but research to identify biomarkers that can characterize brain injury in critically ill children is ongoing [148].

11. Animal Models of Perinatal Asphyxia

The use of animal models has significantly contributed to the understanding of perinatal asphyxia. Several studies have used animal models for the induction of fetal asphyxia, including maternal hypoxemia, infrared thermography (IRT), reduction of in uteroplacental blood flow, umbilical cord occlusion, and embolization, among others [46–48,51,92,147–153] (Figures 3 and 9). Infrared thermography (IRT) is a technique used in both veterinary and human medicine to quantify the surface temperature of the skin based on visualizations of thermographic changes [154]. The use of IRT is essential to understanding the changes in the vascular microcirculation in the study of hypothermia newborns with asphyxia (Figure 9). In veterinary medicine, thermography has brought several benefits for animal models in terms of evaluating lesions, diseases, and surgical procedures [150,155–158].

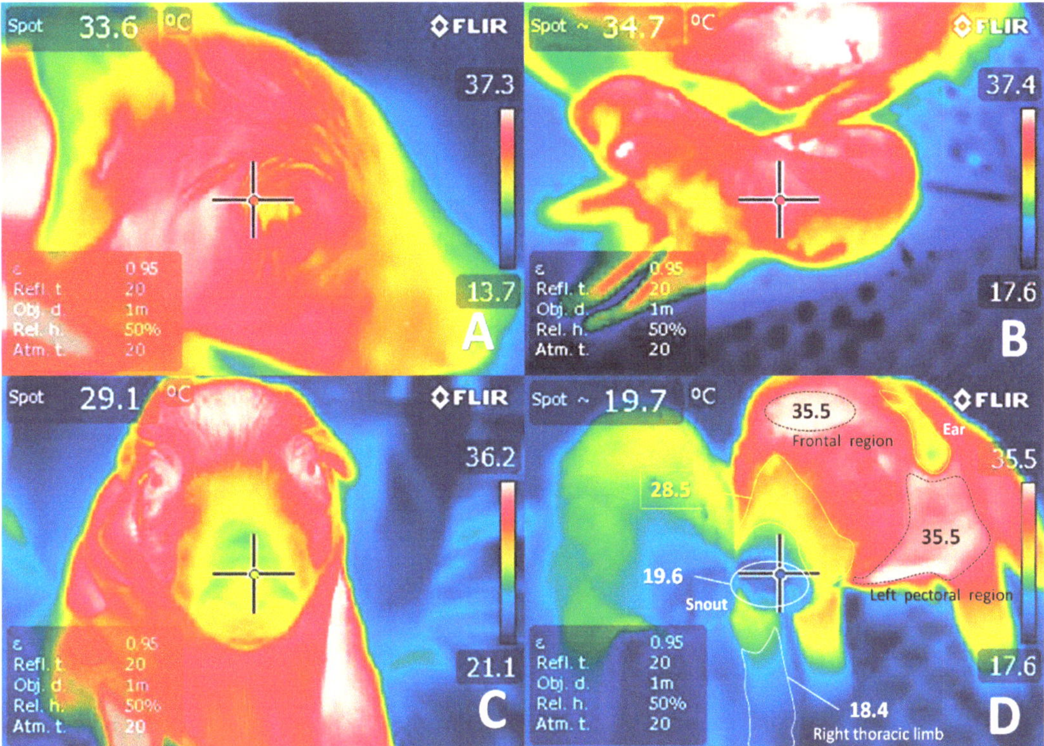

Figure 9. Thermograms of newborn piglets with severe hypoxia and meconium staining on the skin to a severe degree. It is important to dry the newborn immediately, since the humidity of the amniotic fluid favors a rapid drop in body temperature. The areas marked in yellow on the thermograms indicate temperatures between 28 and 32 °C, especially in images (**A–C**), where a marked drop in temperature is seen in peripheral areas of the auricular pavilion, thoracic limbs, and especially the face, particularly the snout. In thermographic image (**D**), we can see a piglet stained with severe meconium grade, with umbilical-cord rupture, a failing vitality score, and severe hypothermia, despite having been dried. It is important to note that different temperatures are seen depending on the body region. In the frontal area of the head and left pectoral region, the highest surface temperatures are observed (35.5 °C). In yellow (image (**D**)), the proximal region of the snout and the auricular pavilion (28–29 °C) are distinguished, and in the distal region of the thoracic limbs and the distal snout area (in blue), the lowest temperature ranges of 18–19 °C are shown. The use of infrared thermography is essential to understanding the changes in the vascular microcirculation in the study of hypothermia in newborns with asphyxia.

During the last six decades of the study of perinatal asphyxia, various animal species have been included from non-human primates, sheep, pigs, canine, and rodents, among others.

While it is true that models of small animals with mouse, rats, and rabbits to analyze lesions for neonatal asphyxia/hypoxia have provided relevant data, models of larger animals that use pigs, sheep, and non-human primates have provided additional essential information necessary to initiate clinical trials [159]. For example, piglets have been frequently used for studies of oxidative-stress conditions as a result of the hyperoxic challenge due to the transition from the hypoxic intrauterine environment to extrauterine life [160] (Figure 3). The ovine fetus has been widely used for physiological and pathophysiological studies of the brain [161]. The non-human primate model is unique because of the physiological and anatomical similarities to humans and because neurological development tests adapted from those of humans can be used [159].

However, a benefit of neonatal models with rodents is the advantage that pups exhibit postnatal cerebral development analogous to human development in the third trimester, and litters are easily produced and easy to handle. The main disadvantages of neonatal rodent models are related to their lissencephalic brain organization, a lower proportion of white and grey matter compared to primates, and the limited behavioral repertory. Compared to rodents, the development and complexity of the brain in non-human primates are similar to those in humans, allowing for applications of sophisticated neurological tests over time [162].

12. Scientific Findings of Perinatal Asphyxia in Animal Models: Advantages and Limitations

In general, animal models have significantly contributed to the field of neonatal medicine. Specifically, research on animal models has provided the evidence base for the therapeutic treatment of newborns with NHIE [163].

Some animal models that have been proven crucial for perinatal brain research are non-human primate fetuses and neonates, as well as pregnant sheep, lambs, puppies, piglets, and immature rodents. Although no model is perfectly ideal in terms of capturing the diversity and complexity of human brain pathology, the investigator must evaluate the strengths and limitations of each model in the context of the research questions [164].

Clinically, it is challenging to estimate the exact time or duration of hypoxia damage, and even more, neurologic lesions as hypoxic–ischemic injury. However, animal models give us the ability to control the time of the injury, allowing for programmed detection of the alteration of metabolites and clarification of the pathophysiological mechanisms [38].

In the case of sheep, they have been used for a variety of purposes, helping to elucidate, for example, how subjection to moderate hyperglycemia before ischemia for a pro-longed time and during reperfusion does not influence the length of brain injury. On the contrary, immense damage to the fetal brain can result from exposure to an additional acute increase in plasma glucose concentration before ischemia [161].

Other relevant contributions in lambs include the fact that renal tubular injury occurs with all degrees of asphyxia, despite varying degrees of brain injury. Lamb models have also helped to establish a correlation between morphological changes in the myocardium and liver, which were previously only associated with severe brain damage. Therefore, oxidative stress appears to play a role in liver damage pathogenesis (Table 1) [70].

Additionally, there have been successful studies to determine variations in the regional blood flow of fetal sheep under severe asphyxia and presenting neurological impairment (presence of seizures) (Table 1) [165].

Table 1. Relevant findings of studies related to perinatal asphyxia in sheep.

Species/ Models	Objective	Contribution	Authors
Ovine (fetuses)	Effects of dexamethasone on brain injury due to asphyxia using one dose and a clinically relevant form of administration (12 mg of maternal IM)	It highlights the possible adverse neural effects of glucocorticoid treatment before perinatal asphyxia	[166]
Ovine (lambs)	1. To determine the effects on the survival and the behavior of the lamb of a brief asphyxial attack induced by occlusion of the umbilical cord at 132 days of gestation 2. To report the type and distribution of brain injury present in the newborn after an asphyxia event at 132 days of gestation.	It was shown that brief fetal asphyxia in the uterus in late pregnancy, increase the probability of premature delivery, and the lambs have significant behavioral deficits after birth that appear to arise from the underlying neuropathology caused by asphyxia, and not from premature delivery *per se*. They identified specific areas of the brain vulnerable to hypoxic damage in late pregnancy.	[167]
Ovine (fetuses)	To determine the changes in the regional blood flow of the fetal sheep during severe asphyxia, and with neurological damage (presence of seizures)	The pattern of redistribution of the blood flow of the ovine fetus exposed to severe asphyxia is comparable to the response of the mild asphyxia, except that a significant increase in total cerebral blood flow does not occur, a relevant finding in the likely association with the development of long-term neurological damage	[165]
Ovine (fetuses)	To evaluate the consequences of acute hypoxia on arterial and central venous pressures, carotid and femoral blood flows and HR in intact and carotid denervated fetal sheep.	The initial cardiovascular responses to hypoxia in the near-term sheep fetus have a strong carotid chemoreflex component. Moreover, fetal survival during hypoxia is dependent on this chemoreflex and the release of catecholamines from the adrenal medulla.	[61]
Ovine (Fetuses)	To evaluate the role of oxidative stress in asphyxia induced perinatal brain injury in near-term fetal lambs subjected to umbilical cord occlusion	Authors suggest that the developing telencephalic white matter seems to be most vulnerable to the effects of intrauterine fetal asphyxia and that oxidative stress may be a significant contributing factor in the pathogenesis of perinatal HIE	[71]

On the other hand, preterm piglets present inadequate ventilation, along with clinical risks comparable to the development of respiratory distress syndrome, thus representing a viable alternative to animal models of sheep and non-human primates [168].

Other studies in piglets have aimed to investigate in detail the prospective implications of the success of cardiopulmonary resuscitation and short-time survival, as well as the presence of non-perfusing cardiac rhythms in asphyxiated newborns [169]. Interesting research has helped in the development and validation of a model of birth asphyxia by performing umbilical cord clamping in term piglets during caesarean sections under general anesthesia as an imitation of the progress of birth asphyxia during natural parturition [170]. Other relevant studies and their contributions in piglets are shown in Table 2.

Table 2. Relevant findings of studies related to perinatal asphyxia in piglets.

Species/ Models	Objective	Contribution	Authors
Porcine (Piglets)	To examine the relationship between isovolumic relaxation time constant (IVR Tau), functional heart parameters. and heart rate (HR) during normoxia and hypoxia–asphyxia (HA) in newborn piglets.	It was demonstrated that HR and IVR Tau significantly accoupled in normoxia; however, they uncoupled during hypoxia–asphyxia (HA) in a piglet model of asphyxia.	[171]
Porcine (Piglets)	To establish methods for free DNA evaluation from circulant cells (cfDNA) and to investigate the temporary changes of cfDNA in blood for a clinically relevant piglet model of hypoxia–reoxygenation.	First methodological study for the extraction and evaluation of cfADN using a piglet model of hypoxia–reoxygenation. cfADN could be an early indicator of the damage caused by perinatal asphyxia.	[160]
Porcine (Piglets)	To investigate whether different metabolomic profiles are produced according to the oxygen administered during resuscitation.	The results indicated that the use of 21% oxygen seems to be better for resuscitation in piglets with normocapnic hypoxia.	[172]
Porcine (Piglets)	To evaluate the effects of asphyxia and resuscitation with different concentrations of oxygen on plasma metabolites in newborn piglets.	Identification of a set of markers with good correlation with the duration of hypoxia. Plasma metabolites indicated an earlier recovery of mitochondrial function when 21% oxygen is used for resuscitation compared to 100% oxygen.	[173]
Porcine (Piglets)	To develop an hypoxic-preconditioning (PC) model of ischemic tolerance in newborn piglets that imitates relevant clinical similarities to humans with birth asphyxia and to characterize some of the molecular mechanisms implicated in PC-induced neuroprotection in rodent models.	Results confirm, for the first time, the protective efficacy of PC against hypoxic–ischemic injury in a newborn piglet model, which reiterates many pathophysiological features of asphyxiated human neonates. PC-induced protection in neonatal piglets may involve upregulation of VEGF.	[174]

Term rodent animal models are used in basic research on the mechanisms of specific diseases [175]. Baboons and premature lambs provide an idea for clinical applications, and rabbits seem to be the most useful animal models due to their possible application in both situations [176,177], in addition to the fact that their medium size greatly facilitates handling in many investigations, reducing costs [168].

Other studies in rodents have contributed to the study and comprehension of the conditions caused by the model of subchronic perinatal asphyxia, which has proven to be effective for the study of conditions of asphyxia during pregnancy. It is non-invasive, reliable, and easy to replicate, and it also allows for control of the conditions of asphyxia. It appears to be adequate for screening and research on asphyxia indicators in the dam and fetus [178]. Other relevant studies and their contributions in rodents are shown in Table 3.

Table 3. Relevant findings of studies related to perinatal asphyxia in rodents.

Species/ Models	Objective	Contribution	Authors
Murine (Rats)	To study the neuroprotector role of palmitoylethanolamide (PEA) on the hippocampus of a 30 day-old rat after perinatal asphyxia.	Treatment with PEA (10 mg/kg) during the first hour of life could attenuate the alterations induced by perinatal asphyxia in the CA1 hippocampus neurons. Hence, PEA represents a recognized protective agent for hippocampal disorders.	[26]
Murine (Fetal Rats)	To investigate the acute changes that occur in the sphingomyelin/ceramide pathway after sublethal fetal asphyxia injury. To identify relevant molecules for brain tolerance.	Acute and persistent prenatal and postnatal changes in the metabolism of ceramide were found in rat brain under asphyxia, leading to positive regulation of ceramide and an increase in apoptosis.	[179]
Murine (Rats)	To evaluate the kinetics of arginine–vasopressin (AVP)/copeptin release during asphyxia and validate the use of the current rodent model in preclinical work on asphyxia at birth	Demonstrated that the proposed rat model meets the standard acid–base criteria for the diagnosis of asphyxia at birth and identified the production of a massive wave of AVP.	[30]
Murine (Rats)	To assess whether the ifetime exposure to an enriched environment (EE) (18 months) could counteract the cognitive anomalies observed in middle-aged rats that suffered 19 minutes of asphyxia at birth.	Lifelong EE was able to counteract cognitive anomalies and improved the performance of spatial learning. Results support the relevance of EE across the lifespan to prevent cognitive deficits induced by perinatal asphyxia.	[180]
Murine (Rats)	To evaluate the effects of both the physiological body temperature (33 oC) and excessive body temperature (37 and 39 oC) in neonatal rats exposed to a severe anoxia and of post-anoxic chelation of iron in neonatal rats exposed to both critical anoxia and hyperthermia on stress responses of the animals at the age of 4 months.	Authors concluded that permanent post-anoxic behavioral disorders are caused by iron-dependent oxidative brain injury, which can be prevented by reducing neonatal body temperature.	[181]
Guinea Pigs	To determine whether sildenafil increased fetal weight and favored fetal tolerance to induced asphyxia at birth.	Low doses of sildenafil administered from day 35 to the end of pregnancy favored fetal tolerability of intrapartum-induced asphyxia. High doses of sildenafil increased fetal weight.	[182]

On the other hand, baboons have been used as animal models since 1980 in research on infectious and cardiovascular diseases, obesity, and hypertension, among others, since the stages of intrauterine development of the lungs, brain, kidneys, and adrenals are very similar to those in human fetuses. However, non-human primates have high economic handling costs, in addition to the fact that, for ethical reasons, their use as animal models for the study of bronchopulmonary dysplasia has been discontinued [183–185].

Research in primates has been relevant in the study of perinatal asphyxia, such as the first study in a modified model that aimed to detect thalamic lesions with magnetic resonance, which allowed for the opportune detection of biomarkers related to specific patterns of newborn brain injury that could be useful for the validation of possible treatments of neonatal hypoxic–ischemic encephalopathy [162]. Measurements of the resistive indices at the thalamus level have also been important; they could potentially supplement other measures for anticipating results from the population of infants with hypoxic–ischemic encephalopathy [186]. For more details of the contributions of experiments in non-human primates, consult Table 4.

Due to rapid advances in technology, in vitro studies are widely used. Furthermore, for ethical reasons, it is not currently possible to test on totally healthy animals without a highly relevant justification [187].

Table 4. Relevant findings of studies related to perinatal asphyxia in primates.

Species/ Models	Objective	Contribution	Authors
Primates (*Macaca nemestrina*)	To investigate the sse of metabolomic and analysis tools to detect potential biomarkers of perinatal asphyxia. To evaluate a model of asphyxia by clamping the umbilical cord and to evaluate the differences between pre- and post-asphyxia.	Through metabolomic analyses, a profile of metabolites was identified with a significant elevation in response to asphyxia at birth (succinic acid, lactate, glucose, malate, arachidonic acid, glutamate, and butanoic acid, among others).	[188]
Primates (*Macaca nemestrina*)	To evaluate the safety and efficacy of erythropoietin (EPO) plus hypothermia for the treatment of perinatal HIE in a non-human primate model. To characterize the acute and chronic consequences of perinatal asphyxia with diagnostic imaging tools to correlate brain injury and neurodevelopmental tests to evaluate early motor and cognitive outcomes.	Occlusion of the umbilical cord for between 15 and 18 minutes can induce severe asphyxia at birth. Asphyxiated neonates developed long-term physical and cognitive deficits.	[159]
Primates (*Macaca nemestrina*)	To establish a non-human primate model of perinatal asphyxia suitable for preclinical evaluation of neuroprotective treatment strategies in conditions resembling human neonatal emergencies and testing erythropoietin neuroprotective treatment.	The model demonstrated changes in magnetic resonance/spectroscopy images consistent with hypoxia, significant motor and behavioral anomalies, and evidence of brain gliosis and was found to be an appropriate model of moderate-to-severe perinatal hypoxic–ischemic injury	[189]

Other benefits of working with animal models include the ability to manipulate the genetic variety, ensuring that the phenotype is representative of the lesion, a small but significant statistical sample, and the ability to control environmental conditions [190]. Ethical problems derived from the use of control groups (with no treatment) and informed consent are eliminated. However, the complexity of mechanisms and organic interactions of perinatal asphyxia, as well as the presence of multiple comorbidities, particularly those of neurological development, cause animal studies to vary in terms of species, injury methods, metabolic approach, and biospecific sample, among other factors, as can be seen in the following table of revised animal models [26,30,61,70,71,159,162,165–167,169–174,178–182,186,188,189,191–197].

13. Conclusions

Perinatal asphyxia remains a significant cause of morbidity and neurological mortality in newborns. Hypoxic–ischemic encephalopathy is the clinical manifestation of generalized disordered neurologic function due to hypoxia. The pathophysiology of asphyxia is extraordinarily complex and related to several gestational, obstetric, and fetal risk factors. Mechanisms are closely related, and all contribute, to a greater or lesser extent, to primary and secondary neuronal and brain injury. Many clinical and biochemical markers have been used to evaluate intrapartum injury, but controversies remain. Recent studies have identified several biomarkers that may improve the identification of neonatal neurologic damage in the period shortly after birth. The most reliable approach to identify and evaluate the seriousness, timing, and pattern of these kinds of insults may consist in measuring an array of inflammatory and neuronal biomarkers at a precise point of care at various intervals. Diverse strategies of neuroprotection have been performed with different therapeutic targets to limit the neuronal damage associated with perinatal hypoxia and to improve the functional prognosis of newborns. Current evidence indicates that therapeutic hypothermia should start as soon as indicated to obtain the best neuroprotective results. Research in animal models has significantly contributed and provided the evidence base for the therapeutic treatment of newborns with NHIE. However, the complexity of mechanisms and interactions of perinatal asphyxia cause animal studies to vary in terms of species, injury methods, metabolic approach, and biospecific sample, among other factors. Various experimental approaches to treat neonatal HIE have advanced to the stage of clinical applications in recent years, but more optimal animal models, additional support/sponsorship from industry, and greater utilization of juvenile toxicology are all urgently required. These aspects could be complemented by dose-ranging studies based on pharmacokinetic–pharmacodynamic modeling and carefully programmed clinical trials that do not expose subjects to harmful medications or result in abandonment of potential treatments. Further investigation should focus on add-on treatment modalities or interventions to further improve outcomes.

Author Contributions: Conceptualization, D.M.-R., D.V.-G. and R.M.; investigation, D.M.-R., D.V.-G., A.M.-R., R.M. and D.I.-R.; writing—original draft preparation, D.M.-R., and D.V.-G.; writing—review and editing, D.V.-G., A.M.-R., R.M., D.I.-R., A.S., and D.M.-R.; project administration, D.M.-R., and D.V.-G. All authors have read and agreed to the published version of the manuscript.

Funding: This research received no external funding.

Institutional Review Board Statement: Not applicable.

Conflicts of Interest: All the authors declare that there is no conflict of interest or ethical concern in this article.

References

1. WHO. Newborns: Reducing Mortality. Available online: https://www.who.int/news-room/fact-sheets/detail/newborns-reducing-mortality (accessed on 11 November 2021).
2. Bhutta, Z.A.; Das, J.K.; Bahl, R.; Lawn, J.E.; Salam, R.A.; Paul, V.K.; Sankar, M.J.; Blencowe, H.; Rizvi, A.; Chou, V.B.; et al. Can available interventions end preventable deaths in mothers, newborn babies, and stillbirths, and at what cost? *Lancet* **2014**, *384*, 347–370. [CrossRef]
3. Lawn, J.E.; Cousens, S.; Zupan, J.; Lancet Neonatal Survival Steering Team. 4 million neonatal deaths: When? Where? Why? *Lancet* **2005**, *365*, 891–900. [CrossRef]
4. Liu, L.; Johnson, H.L.; Cousens, S.; Perin, J.; Scott, S.; Lawn, J.E.; Rudan, I.; Campbell, H.; Cibulskis, R.; Li, M.; et al. Global, regional, and national causes of child mortality: An updated systematic analysis for 2010 with time trends since 2000. *Lancet* **2012**, *379*, 2151–2161. [CrossRef]
5. Perez, A.; Ritter, S.; Brotschi, B.; Werner, H.; Caflisch, J.; Martin, E.; Latal, B. Long-Term Neurodevelopmental Outcome with Hypoxic-Ischemic Encephalopathy. *J. Pediatr.* **2013**, *163*, 454–459.e1. [CrossRef] [PubMed]
6. Li, B.; Concepcion, K.; Meng, X.; Zhang, L. Brain-immune interactions in perinatal hypoxic-ischemic brain injury. *Prog. Neurobiol.* **2017**, *159*, 50–68. [CrossRef] [PubMed]
7. Younge, N.; Goldstein, R.F.; Bann, C.M.; Hintz, S.R.; Patel, R.M.; Smith, P.B.; Bell, E.F.; Rysavy, M.A.; Duncan, A.F.; Vohr, B.R.; et al. Survival and Neurodevelopmental Outcomes among Periviable Infants. *N. Engl. J. Med.* **2017**, *376*, 617–628. [CrossRef] [PubMed]
8. National Institute of Health. Late Hypothermia for Hypoxic-Ischemic Encephalopathy. Available online: https://clinicaltrials.gov/ct2/show/NCT00614744 (accessed on 20 July 2021).
9. National Institute of Health. Preemie Hypothermia for Neonatal Encephalopathy. Available online: https://clinicaltrials.gov/ct2/show/NCT01793129 (accessed on 11 August 2021).
10. National Institute of Health. Erythropoietin for Hypoxic Ischemic Encephalopathy in Newborns (PAEAN). Available online: https://clinicaltrials.gov/ct2/show/NCT03079167 (accessed on 11 November 2021).
11. Lincetto, O. *Birth Asphyxia Summary of the Previous Meeting and Protocol Overview*; World Health Organization: Gevena, Switzerland, 2007; pp. 1–34.
12. Kamath-Rayne, B.D.; Hobe, A. Birth asphyxia. In *Clinics in Perinatology*; Elsevier: Philadelphia, PA, USA, 2016; p. 621.
13. American Academy of Pediatrics; Committee on Fetus and Newborn; American College of Obstetricians and Gynecologists; Committee on Obstetric Practice. The Apgar score. *Adv. Neonatal Care Off. J. Natl. Assoc. Neonatal Nurses* **2006**, *6*, 220–223. [CrossRef] [PubMed]
14. Rainaldi, M.A.; Perlman, J.M. Pathophysiology of Birth Asphyxia. *Clin. Perinatol.* **2016**, *43*, 409–422. [CrossRef]

15. Pacora, P.; Romero, R.; Jaiman, S.; Erez, O.; Bhatti, G.; Panaitescu, B.; Benshalom-Tirosh, N.; Jung, E.J.; Hsu, C.-D.; Hassan, S.S.; et al. Mechanisms of death in structurally normal stillbirths. *J. Perinat. Med.* **2019**, *47*, 222–240. [CrossRef]
16. Yli, B.M.; Kjellmer, I. Pathophysiology of fetal oxygenation and cell damage during labor. *Best Pract. Res. Clin. Obstet. Gynaecol.* **2016**, *30*, 9–21. [CrossRef]
17. Richardson, B.S. Fetal Adaptive Responses to Asphyxia. *Clin. Perinatol.* **1989**, *16*, 595–611. [CrossRef]
18. Jensen, A.; Roman, C.; Rudolph, A.M. Effects of reducing uterine blood flow on fetal blood flow distribution and oxygen delivery. *J. Dev. Physiol.* **1991**, *15*, 309–323. [PubMed]
19. Low, J.A. Determining the contribution of asphyxia to brain damage in the neonate. *J. Obstet. Gynaecol. Res.* **2004**, *30*, 276–286. [CrossRef] [PubMed]
20. Britton, J.R. The transition to extrauterine life and disorders of transition. *Clin. Perinatol.* **1998**, *25*, 271–294. [CrossRef]
21. Morton, S.U.; Brodsky, D. Fetal Physiology and the Transition to Extrauterine Life. *Clin. Perinatol.* **2016**, *43*, 395–407. [CrossRef]
22. Noori, S.; Wlodaver, A.; Gottipati, V.; McCoy, M.; Schultz, D.; Escobedo, M. Transitional Changes in Cardiac and Cerebral Hemodynamics in Term Neonates at Birth. *J. Pediatrics* **2012**, *160*, 943–948. [CrossRef]
23. Wyss, M.T.; Jolivet, R.; Buck, A.; Magistretti, P.J.; Weber, B. In Vivo Evidence for Lactate as a Neuronal Energy Source. *J. Neurosci.* **2011**, *31*, 7477–7485. [CrossRef]
24. Zheng, Y.; Wang, X.-M. Expression Changes in Lactate and Glucose Metabolism and Associated Transporters in Basal Ganglia following Hypoxic-Ischemic Reperfusion Injury in Piglets. *Am. J. Neuroradiol.* **2018**, *39*, 569–576. [CrossRef]
25. Chen, Y.; Engidawork, E.; Loidl, F.; Dell'Anna, E.; Goiny, M.; Lubec, G.; Andersson, K.; Herrera-Marschitz, M. Short- and long-term effects of perinatal asphyxia on monoamine, amino acid and glycolysis product levels measured in the basal ganglia of the rat. *Brain Res. Dev. Brain Res.* **1997**, *104*, 19–30. [CrossRef]
26. Herrera, M.I.; Otero-Losada, M.; Udovin, L.D.; Kusnier, C.; Kölliker-Frers, R.; De Souza, W.; Capani, F. Could perinatal asphyxia induce a synaptopathy? New highlights from an experimental model. *Neural Plast.* **2017**, *2017*, 3436943. [CrossRef]
27. Larroque, B.; Ancel, P.-Y.; Marret, S.; Marchand, L.; André, M.; Arnaud, C.; Pierrat, V.; Rozé, J.-C.; Messer, J.; Thiriez, G.; et al. Neurodevelopmental disabilities and special care of 5-year-old children born before 33 weeks of gestation (the EPIPAGE study): A longitudinal cohort study. *Lancet* **2008**, *371*, 813–820. [CrossRef]
28. Laptook, A.R. Birth Asphyxia and Hypoxic-Ischemic Brain Injury in the Preterm Infant. *Clin. Perinatol.* **2016**, *43*, 529–545. [CrossRef] [PubMed]
29. Antonucci, R.; Porcella, A.P.M. Perinatal asphyxia in the term newborn. *J. Pediatr. Neonatal Individ. Med.* **2014**, *3*, e030269.
30. Summanen, M.; Bäck, S.; Voipio, J.; Kaila, K. Surge of Peripheral Arginine Vasopressin in a Rat Model of Birth Asphyxia. *Front. Cell. Neurosci.* **2018**, *12*, 2. [CrossRef] [PubMed]
31. VanWoudenberg, C.D.; Wills, C.A.; Rubarth, L.B. Newborn Transition to Extrauterine Life. *Neonatal Netw.* **2012**, *31*, 317–322. [CrossRef] [PubMed]
32. Evers, K.S.; Wellmann, S. Arginine vasopressin and copeptin in perinatology. *Front. Pediatr.* **2016**, *4*, 2–11. [CrossRef] [PubMed]
33. Dilenge, M.E.; Majnemer, A.; Shevell, M.I. Long-term developmental outcome of asphyxiated term neonates. *J. Child Neurol.* **2001**, *16*, 781–792. [CrossRef]
34. Milsom, I.; Ladfors, L.; Thiringer, K.; Niklasson, A.; Odeback, A.; Thornberg, E. Influence of maternal, obstetric and fetal risk factors on the prevalence of birth asphyxia at term in a Swedish urban population. *Acta Obstet. Gynecol. Scand.* **2002**, *81*, 909–917. [CrossRef]
35. Giussani, D.A. The fetal brain sparing response to hypoxia: Physiological mechanisms. *J. Physiol.* **2016**, *594*, 1215–1230. [CrossRef]
36. Polglase, G.R.; Ong, T.; Hillman, N.H. Cardiovascular Alterations and Multiorgan Dysfunction After Birth Asphyxia. *Clin. Perinatol.* **2016**, *43*, 469–483. [CrossRef]
37. Fattuoni, C.; Palmas, F.; Noto, A.; Fanos, V.; Barberini, L. Perinatal asphyxia: A review from a metabolomics perspective. *Molecules* **2015**, *20*, 7000–7016. [CrossRef] [PubMed]
38. Denihan, N.M.; Boylan, G.B.; Murray, D.M. Metabolomic profiling in perinatal asphyxia: A promising new field. *BioMed Res. Int.* **2015**, *2015*, 254076. [CrossRef] [PubMed]
39. Mota-Rojas, D.; Martínez-Burnes, J.; Trujillo, M.E.; López, A.; Rosales, A.M.; Ramírez, R.; Orozco, H.; Merino, A.; Alonso-Spilsbury, M. Uterine and fetal asphyxia monitoring in parturient sows treated with oxytocin. *Anim. Reprod. Sci.* **2005**, *86*, 131–141. [CrossRef] [PubMed]
40. Mota-Rojas, D.; Rosales, A.M.; Trujillo, M.E.; Orozco, H.; Ramírez, R.; Alonso-Spilsbury, M. The effects of vetrabutin chlorhydrate and oxytocin on stillbirth rate and asphyxia in swine. *Theriogenology* **2005**, *64*, 1889–1897. [CrossRef] [PubMed]
41. Mota-Rojas, D.; Nava-Ocampo, A.A.; Trujillo, M.E.; Velázquez-Armenta, Y.; Ramírez-Necoechea, R.; Martínez-Burnes, J.; Alonso-Spilsbury, M. Dose minimization study of oxytocin in early labor in sows: Uterine activity and fetal outcome. *Reprod. Toxicol.* **2005**, *20*, 255–259. [CrossRef] [PubMed]
42. Mota-Rojas, D.; Martínez-Burnes, J.; Napolitano, F.; Domínguez-Muñoz, M.; Guerrero-Legarreta, I.; Mora-Medina, P.; Ramírez-Necoechea, R.; Lezama-García, K.; González-Lozano, M.; Mota-Rojas, D.; et al. Dystocia: Factors affecting parturition in domestic animals. *CAB Rev.* **2020**, *15*, 1–16. [CrossRef]
43. Alonso-Spilsbury, M.; Mota-Rojas, D.; Villanueva-García, D.; Martínez-Burnes, J.; Orozco, H.; Ramírez-Necoechea, R.; Mayagoitia, A.L.; Trujillo, M.E. Perinatal asphyxia pathophysiology in pig and human: A review. *Anim. Reprod. Sci.* **2005**, *90*, 1–30. [CrossRef] [PubMed]

44. Mota-Rojas, D.; Martinez-Burnes, J.; Alonso-Spilsbury, M.L.; Lopez, A.; Ramirez-Necoechea, R.; Trujillo-Ortega, M.E.; Medina-Hernandez, F.J.; de la Cruz, N.I.; Albores-Torres, V.; Loredo-Osti, J. Meconium staining of the skin and meconium aspiration in porcine intrapartum stillbirths. *Livest. Sci.* **2006**, *102*, 155–162. [CrossRef]

45. Mota, D.; Orozco, H.; Villanueva, D.; Bonilla, H.; Suárez, X.; Hernandez, R.; Roldan, P.; Trujillo, M. Fetal and neonatal energy metabolism in pigs and humans: A review. *Vet. Med.* **2011**, *56*, 215–225. [CrossRef]

46. Mota-Rojas, D.; Velarde, A.; Maris-Huertas, S.; Cajiao, M.N. *Animal Welfare, a Global Vision in Ibero-America*, 3rd ed.; Mota-Rojas, D., Velarde, A., Maris-Huertas, S., Cajiao, M.N., Eds.; Elsevier: Barcelona, Spain, 2016; ISBN 978-84-9113-026-0.

47. Mota-Rojas, D.; López, A.; Martínez-Burnes, J.; Muns, R.; Villanueva-García, D.; Mora-Medina, P.; González-Lozano, M.; Olmos-Hernández, A.; Ramírez-Necoechea, R. Is vitality assessment important in neonatal animals? *CAB Rev.* **2018**, *13*, 1–13. [CrossRef]

48. Martínez-Burnes, J.; Mota-Rojas, D.; Villanueva-García, D.; Ibarra-Rios, D.; Lezama-García, K.; Barrios-García, H.; Lopez, A. Meconium aspiration syndrome in mammals. *CAB Rev.* **2019**, *14*, 1–11. [CrossRef]

49. González-Lozano, M.; Mota-Rojas, D.; Orihuela, A.; Martínez-Burnes, J.; Di Francia, A.; Braghieri, A.; Berdugo-Gutiérrez, J.; Mora-Medina, P.; Ramírez-Necoechea, R.; Napolitano, F. Review: Behavioral, physiological, and reproductive performance of buffalo cows during eutocic and dystocic parturitions. *Appl. Anim. Sci.* **2020**, *36*, 407–422. [CrossRef]

50. Placha, K.; Luptakova, D.; Baciak, L.; Ujhazy, E.; Juranek, I. Neonatal brain injury as a consequence of insufficient cerebral oxygenation. *Neuro Endocrinol.* **2016**, *37*, 79–96.

51. Mota-Rojas, D.; Martinez-Burnes, J.; Villanueva-Garcia, D.; Roldan-Santiago, P.; Trujillo-Ortega, M.E.; Orozco-Gregorio, H.; Bonilla-Jaime, H.; Lopez-Mayagoitia, A. Animal welfare in the newborn piglet: A review. *Vet. Med.* **2012**, *57*, 338–349. [CrossRef]

52. Riljak, V.; Kraf, J.; Daryanani, A.; Jiruška, P.; Otáhal, J. Pathophysiology of perinatal hypoxic-ischemic encephalopathy—Biomarkers, animal models and treatment perspectives. *Physiol. Res.* **2016**, *65*, S533–S545. [CrossRef]

53. Engelhardt, B.; Carare, R.O.; Bechmann, I.; Flügel, A.; Laman, J.D.; Weller, R.O. Vascular, glial, and lymphatic immune gateways of the central nervous system. *Acta Neuropathol.* **2016**, *132*, 317–338. [CrossRef]

54. Van Bel, F.V.; Groenendaal, F. Drugs for neuroprotection after birth asphyxia: Pharmacologic adjuncts to hypothermia. *Semin. Perinatol.* **2016**, *40*, 152–159.

55. Hassell, K.J.; Ezzati, M.; Alonso-Alconada, D.; Hausenloy, D.J.; Robertson, N.J. New horizons for newborn brain protection: Enhancing endogenous neuroprotection. *Arch. Dis. Child. Fetal Neonatal Ed.* **2015**, *100*, F541–F551. [CrossRef]

56. Bhatti, A.; Kumar, P. Systemic effects of perinatal asphyxia. *Indian J. Pediatrics* **2014**, *81*, 231–233. [CrossRef]

57. Jensen, A.; Garnier, Y.; Berger, R. Dynamics of fetal circulatory responses to hypoxia and asphyxia. *Eur. J. Obstet. Gynecol. Reprod. Biol.* **1999**, *84*, 155–172. [CrossRef]

58. Pertierra Cortada, À.; Figueras Aloy, J.; Sebastiani, G.; Rovira Girabal, N.; Krauel Vidal, X. Asfixia perinatal: Relación entre afectación cardiovascular, neurológica y multisistémica. *Acta Pediatr. Esp.* **2008**, *66*, 494–501.

59. Hankins, G.D.V.; Koen, S.; Gei, A.F.; Lopez, S.M.; Van Hook, J.W.; Anderson, G.D. Neonatal organ system injury in acute birth asphyxia sufficient to result in neonatal encephalopathy. *Obstet. Gynecol.* **2002**, *99*, 688–691. [PubMed]

60. Kara, T.; Narkiewicz, K.; Somers, V.K. Chemoreflexes—Physiology and clinical implications. *Acta Physiol. Scand.* **2003**, *177*, 377–384. [CrossRef] [PubMed]

61. Giussani, D.A.; Spencer, J.A.; Moore, P.J.; Bennet, L.; Hanson, M.A. Afferent and efferent components of the cardiovascular reflex responses to acute hypoxia in term fetal sheep. *J. Physiol.* **1993**, *461*, 431–449. [CrossRef]

62. Giussani, D.A.; Spencer, J.A.; Hanson, M.A. Fetal cardiovascular reflex responses to hypoxaemia. *Fetal Matern. Med. Rev.* **1994**, *6*, 17–37. [CrossRef]

63. Cullen, P.; Salgado, E. Conceptos básicos para el manejo de la asfixia perinatal y la encefalopatía hipóxico-isquémica en el neonato. *Rev. Mex. Pediatr.* **2009**, *76*, 174–180.

64. Jacobs, S.E.; Berg, M.; Hunt, R.; Tarnow-Mordi, W.O.; Inder, T.E.; Davis, P.G. Cooling for newborns with hypoxic ischemic encephalopathy. *Cochrane Database Syst. Rev.* **2013**, *2013*, CD003311.

65. Sehgal, A.; Wong, F.; Mehta, S. Reduced cardiac output and its correlation with coronary blood flow and troponin in asphyxiated infants treated with therapeutic hypothermia. *Eur. J. Pediatr.* **2012**, *171*, 1511–1517. [CrossRef]

66. LaRosa, D.A.; Ellery, S.J.; Walker, D.W.; Dickinson, H. Understanding the Full Spectrum of organ injury following intrapartum asphyxia. *Front. Pediatr.* **2017**, *5*, 1–11. [CrossRef] [PubMed]

67. Dattilo, G.; Tulino, V.; Tulino, D.; Lamari, A.; Falanga, G.; Marte, F.; Patanè, S. Perinatal asphyxia and cardiac abnormalities. *Int. J. Cardiol.* **2011**, *147*, e39–e40. [CrossRef]

68. Perlman, J.M. Acute Systemic Organ Injury in Term Infants After Asphyxia. *Arch. Pediatr. Adolesc. Med.* **1989**, *143*, 617. [CrossRef] [PubMed]

69. Gunn, A.J.; Bennet, L. Fetal Hypoxia Insults and Patterns of Brain Injury: Insights from Animal Models. *Clin. Perinatol.* **2009**, *36*, 579–593. [CrossRef] [PubMed]

70. Ikeda, T.; Murata, Y.; Quilligan, E.J.; Parer, J.T.; Murayama, T.; Koono, M. Histologic and biochemical study of the brain, heart, kidney, and liver in asphyxia caused by occlusion of the umbilical cord in near-term fetal lambs. *Am. J. Obstet. Gynecol.* **2000**, *182*, 449–457. [CrossRef]

71. Ikeda, T.; Choi, B.H.; Yee, S.; Murata, Y.; Quilligan, E.J. Oxidative stress, brain white matter damage and intrauterine asphyxia in fetal lambs. *Int. J. Dev. Neurosci.* **1999**, *17*, 1–14. [CrossRef]

72. Shankaran, S.; Woldt, E.; Koepke, T.; Bedard, M.P.; Nandyal, R. Acute neonatal morbidity and long-term central nervous system sequelae of perinatal asphyxia in term infants. *Early Hum. Dev.* **1991**, *25*, 135–148. [CrossRef]

73. Aggarwal, A.; Kumar, P.; Chowdhary, G.; Majumdar, S.; Narang, A. Evaluation of Renal Functions in Asphyxiated Newborns. *J. Trop. Pediatr.* **2005**, *51*, 295–299. [CrossRef]

74. Saikumar, P.; Venkatachalam, M.A. Role of apoptosis in hypoxic/ischemic damage in the kidney. *Semin. Nephrol.* **2003**, *23*, 511–521. [CrossRef]

75. Askenazi, D.J.; Feig, D.I.; Graham, N.M.; Hui-Stickle, S.; Goldstein, S.L. 3–5 year longitudinal follow-up of pediatric patients after acute renal failure. *Kidney Int.* **2006**, *69*, 184–189. [CrossRef]

76. Jetton, J.G.; Askenazi, D.J. Acute Kidney Injury in the Neonate. *Clin. Perinatol.* **2014**, *41*, 487–502. [CrossRef]

77. Beath, S.V. Hepatic function and physiology in the newborn. *Semin. Neonatol.* **2003**, *8*, 337–346. [CrossRef]

78. Islam, M.T.; Islam, M.N.; Mollah, A.H.; Hoque, M.A.; Hossain, M.A.; Nazir, F.; Ahsan, M.M. Status of liver enzymes in babies with perinatal asphyxia. *Mymensingh Med. J. MMJ* **2011**, *20*, 446–449. [PubMed]

79. Gluckman, P.D.; Wyatt, J.S.; Azzopardi, D.; Ballard, R.; Edwards, A.D.; Ferriero, D.M.; Polin, R.A.; Robertson, C.M.; Thoresen, M.; Whitelaw, A.; et al. Selective head cooling with mild systemic hypothermia after neonatal encephalopathy: Multicentre randomised trial. *Lancet* **2005**, *365*, 663–670. [CrossRef]

80. Forman, K.R.; Diab, Y.; Wong, E.C.C.; Baumgart, S.; Luban, N.L.C.; Massaro, A.N. Coagulopathy in newborns with hypoxic ischemic encephalopathy (HIE) treated with therapeutic hypothermia: A retrospective case-control study. *BMC Pediatr.* **2014**, *14*, 277. [CrossRef]

81. Paprocka, J.; Kijonka, M.; Rzepka, B.; Sokół, M. Melatonin in Hypoxic-Ischemic Brain Injury in Term and Preterm Babies. *Int. J. Endocrinol.* **2019**, *2019*, 9626715. [CrossRef] [PubMed]

82. American College of Obstetricians and Gynecologists Committee on Obstetric Practice. Inappropriate use of the terms fetal distress and birth asphyxia. *Obstet. Gynecol.* **2005**, *106*, 1469–1470. [CrossRef] [PubMed]

83. Committee on Obstetric Practice, ACOG; American Academy of Pediatrics; Committee on Fetus and Newborn, ACOG. ACOG Committee Opinion No. 333: The Apgar Score. *Obstet. Gynecol.* **2006**, *107*, 1209. [CrossRef] [PubMed]

84. Rahman, S.; Unsworth, J.; Vause, S. Meconium in labor. *Obstet. Gynaecol. Reprod. Med.* **2013**, *23*, 247–252. [CrossRef]

85. Vain, N.E.; Batton, D.G. Meconium "aspiration" (or respiratory distress associated with meconium-stained amniotic fluid?). *Semin. Fetal Neonatal Med.* **2017**, *22*, 214–219. [CrossRef]

86. Chettri, S.; Bhat, B.V.; Adhisivam, B. Current Concepts in the Management of Meconium Aspiration Syndrome. *Indian J. Pediatr.* **2016**, *83*, 1125–1130. [CrossRef]

87. Yurdakök, M. Meconium aspiration syndrome: Do we know? *Turk. J. Pediatr.* **2011**, *53*, 121–129.

88. Lee, J.; Romero, R.; Lee, K.A.; Kim, E.N.; Korzeniewski, S.J.; Chaemsaithong, P.; Yoon, B.H. Meconium aspiration syndrome: A role for fetal systemic inflammation. *Am. J. Obstet. Gynecol.* **2016**, *214*, 366.e1–366.e9. [CrossRef] [PubMed]

89. Mendoza, M.M.; Martinez-Burnes, J.; Calderon, R.; Hayen, S.; Mota-Rojas, D.; Medellin, J.A. Meconium staining of foals at birth. In *Animal Perinatology: Clinical and Experimental Approaches*; Mota-Rojas, D., Nava-Ocampo, A.A., Villanueva-Garcia, D.A.-S.M., Eds.; BM Editores: Ciudad de México, Mexico, 2008; pp. 351–362.

90. Villanueva-García, D. Hipertensión arterial pulmonar. In *Neonatología, Esencia, Arte y Praxis*; Murguía, T., Villanueva, D.L.G., Eds.; Mc Graw Hill: Ciudad de México, Mexico, 2011; pp. 88–94.

91. Ibarra-Ríos, D.; Villanueva-García, D. Síndrome de Aspiración de Meconio. Programa de Actualización Continua. In *PAC®Neonatología-4*; Libro 2; Insuficiencia Respiratoria Neonatal, Ed.; Intersistemas Editores: Ciudad de México, Mexico, 2016; pp. 59–67. ISBN 978-607-443-552-8.

92. Mota-Rojas, D.; Villanueva-García, D.; Hernández, R.; Martínez-Rodríguez, R.; Mora-Medina, P.; Gonzalez, B.; Sánchez, M. Assessment of the vitality of the newborn: An overview. *Sci. Res. Essays* **2012**, *7*, 712–718. [CrossRef]

93. Dawes, G.S. The natural history of asphyxia and resuscitation at birth. In *Fetal and Neonatal Physiology: A Comparative Study of the Changes at Birth*; Yearbook Medical Publishers: Chicago, IL, USA, 1968.

94. Dawes, G.S.; Jacobson, H.N.; Mott, J.C.; Shelley, H.J.; Stafford, A. The treatment of asphyxiated, mature fetal lambs and rhesus monkeys with intravenous glucose and sodium carbonate. *J. Physiol.* **1963**, *169*, 167–184. [CrossRef] [PubMed]

95. Jacobs, S.E.; Morley, C.J.; Inder, T.E.; Stewart, M.J.; Smith, K.R.; McNamara, P.J.; Wright, I.M.R.; Kirpalani, H.M.; Darlow, B.A.; Doyle, L.W. Whole-body hypothermia for term and near-term newborns with hypoxic-ischemic encephalopathy: A randomized controlled trial. *Arch. Pediatr. Adolesc. Med.* **2011**, *165*, 692–700. [CrossRef] [PubMed]

96. Sarnat, H.B. Neonatal Encephalopathy Following Fetal Distress. *Arch. Neurol.* **1976**, *33*, 696–705. [CrossRef] [PubMed]

97. Thompson, C.; Puterman, A.; Linley, L.; Hann, F.; Elst, C.; Molteno, C.; Malan, A. The value of a scoring system for hypoxic ischemic encephalopathy in predicting neurodevelopmental outcome. *Acta Paediatr.* **1997**, *86*, 757–761. [CrossRef] [PubMed]

98. Shankaran, S.; Laptook, A.R.; Ehrenkranz, R.A.; Tyson, J.F.; McDonald, S.A.; Donovan, E.F.; Fanaroff, A.A.; Poole, W.K.; Wright, L.L.; Higgins, R.D.; et al. Whole-Body Hypothermia for Neonates with Hypoxic-Ischemic Encephalopathy. *N. Engl. J. Med.* **2005**, *353*, 1574–1584. [CrossRef]

99. Ducsay, C.A.; Goyal, R.; Pearce, W.J.; Wilson, S.; Hu, X.Q.; Zhang, L. Gestational hypoxia and developmental plasticity. *Physiol. Rev.* **2018**, *98*, 1241–1334. [CrossRef]

100. Dos Riesgo, R.S.; Becker, M.M.; Ranzan, J.; Winckler, M.I.B.; Ohlweiler, L. Avances en el abordaje de la hipoxia neonatal. *Rev. Neurol.* **2013**, *57*, S17–S21. [CrossRef]

101. Negro, S.; Benders, M.J.N.L.; Tataranno, M.L.; Coviello, C.; de Vries, L.S.; van Bel, F.; Groenendaal, F.; Longini, M.; Proietti, F.; Belvisi, E.; et al. Early prediction of hypoxic-ischemic brain injury by a new panel of biomarkers in a population of term newborns. *Oxidative Med. Cell. Longev.* **2018**, *2018*, 1–10. [CrossRef] [PubMed]
102. Volpe, J.J.; Inder, T.; Darras, B.; Vries, L.; Plessis, A.; Neil, J.; Perlman, J. *Volpe 's Neurology of the Newborn. Philadelphia*; Saunders: Philadelphia, PA, USA, 2001; pp. 1–1224.
103. Chaiworapongsa, T.; Romero, R.; Erez, O.; Tarca, A.L.; Conde-Agudelo, A.; Chaemsaithong, P.; Kim, C.J.; Kim, Y.M.; Kim, J.-S.; Yoon, B.H.; et al. The prediction of fetal death with a simple maternal blood test at 24–28 weeks: A role for angiogenic index-1 (PlGF/sVEGFR-1 ratio). *Am. J. Obstet. Gynecol.* **2017**, *217*, 682.e1–682.e13. [CrossRef] [PubMed]
104. Thoresen, M.; Liu, X.; Jary, S.; Brown, E.; Sabir, H.; Stone, J.; Cowan, F.; Karlsson, M. Lactate dehydrogenase in hypothermia-treated newborn infants with hypoxic-ischemic encephalopathy. *Acta Paediatr.* **2012**, *101*, 1038–1044. [CrossRef] [PubMed]
105. Ramaswamy, V.; Horton, J.; Vandermeer, B.; Buscemi, N.; Miller, S.; Yager, J. Systematic review of biomarkers of brain injury in term neonatal encephalopathy. *Pediatr. Neurol.* **2009**, *40*, 215–226. [CrossRef] [PubMed]
106. Chalak, L.F. Inflammatory Biomarkers of Birth Asphyxia. *Clin. Perinatol.* **2016**, *43*, 501–510. [CrossRef]
107. Mir, I.N.; Johnson-Welch, S.F.; Nelson, D.B.; Brown, L.S.; Rosenfeld, C.R.; Chalak, L.F. Placental pathology is associated with severity of neonatal encephalopathy and adverse developmental outcomes following hypothermia. *Am. J. Obstet. Gynecol.* **2015**, *213*, 849.e1–849.e7. [CrossRef]
108. Chalak, L.F.; Sánchez, P.J.; Adams-Huet, B.; Laptook, A.R.; Heyne, R.J.; Rosenfeld, C.R. Biomarkers for severity of neonatal hypoxic-ischemic encephalopathy and outcomes in newborns receiving hypothermia therapy. *J. Pediatr.* **2014**, *164*, 1–8. [CrossRef]
109. Massaro, A.N.; Chang, T.; Kadom, N.; Tsuchida, T.; Scafidi, J.; Glass, P.; McCarter, R.; Baumgart, S.; Vezina, G.; Nelson, K.B. Biomarkers of brain injury in neonatal encephalopathy treated with hypothermia. *J. Pediatr.* **2012**, *161*, 434–440. [CrossRef]
110. Liu, Y.-W.; Li, S.; Dai, S.-S. Neutrophils in traumatic brain injury (TBI): Friend or foe? *J. Neuroinflamm.* **2018**, *15*, 146. [CrossRef]
111. Sorokina, E.G.; Semenova, Z.B.; Reutov, V.P.; Arsenieva, E.N.; Karaseva, O.V.; Fisenko, A.P.; Roshal, L.M.; Pinelis, V.G. Brain Biomarkers in children after mild and severe traumatic brain injury. In *Intracranial Pressure and Neuromonitoring XVII*; Depreitere, B., Meyfroidt, G., Güiza, F., Eds.; Springer Nature Switzerland AG: Cham, Switzerland, 2021; pp. 103–107.
112. Li, Y.; Dammer, E.B.; Zhang-Brotzge, X.; Chen, S.; Duong, D.M.; Seyfried, N.T.; Kuan, C.-Y.; Sun, Y.-Y. Osteopontin is a blood biomarker for microglial activation and brain injury in experimental hypoxic-ischemic encephalopathy. *Eneuro* **2017**, *4*. [CrossRef]
113. Graham, E.M.; Everett, A.D.; Delpech, J.C.; Northington, F.J. Blood biomarkers for evaluation of perinatal encephalopathy: State of the art. *Curr. Opin. Pediatr.* **2018**, *30*, 199–203. [CrossRef]
114. Spitzmiller, R.E.; Phillips, T.; Meinzen-Derr, J.; Hoath, S.B. Amplitude-integrated EEG is useful in predicting neurodevelopmental outcome in full-term infants with hypoxic-ischemic encephalopathy: A meta-analysis. *J. Child Neurol.* **2007**, *22*, 1069–1078. [CrossRef] [PubMed]
115. Lemmers, P.M.A.; Zwanenburg, R.J.; Benders, M.J.N.L.; de Vries, L.S.; Groenendaal, F.; van Bel, F.; Toet, M.C. Cerebral oxygenation and brain activity after perinatal asphyxia: Does hypothermia change their prognostic value? *Pediatr. Res.* **2013**, *74*, 180–185. [CrossRef] [PubMed]
116. Archer, L.N.; Levene, M.; Evans, D. Cerebral artery doppler ultrasonography for prediction of outcome after perinatal asphyxia. *Lancet* **1986**, *328*, 1116–1118. [CrossRef]
117. Bednarek, N.; Mathur, A.; Inder, T.; Wilkinson, J.; Neil, J.; Shimony, J. Impact of therapeutic hypothermia on MRI diffusion changes in neonatal encephalopathy. *Neurology* **2012**, *78*, 1420–1427. [CrossRef] [PubMed]
118. Trivedi, S.B.; Vesoulis, Z.A.; Rao, R.; Liao, S.M.; Shimony, J.S.; McKinstry, R.C.; Mathur, A.M. A validated clinical MRI injury scoring system in neonatal hypoxic-ischemic encephalopathy. *Pediatr. Radiol.* **2017**, *47*, 1491–1499. [CrossRef] [PubMed]
119. Groenendaal, F.; De Vries, L.S. Fifty years of brain imaging in neonatal encephalopathy following perinatal asphyxia. *Pediatr. Res.* **2017**, *81*, 150–155. [CrossRef] [PubMed]
120. Giesinger, R.E.; Bailey, L.J.; Deshpande, P.; McNamara, P.J. Hypoxic-Ischemic Encephalopathy and Therapeutic Hypothermia: The Hemodynamic Perspective. *J. Pediatr.* **2017**, *180*, 22–30.e2. [CrossRef]
121. Pelletier, J.-S.; LaBossiere, J.; Dicken, B.; Gill, R.S.; Sergi, C.; Tahbaz, N.; Bigam, D.; Cheung, P.-Y. Low-Dose vasopressin improves cardiac function in newborn piglets with acute hypoxia-reoxygenation. *Shock* **2013**, *40*, 320–326. [CrossRef]
122. Bischoff, A.R.; Habib, S.; McNamara, P.J.; Giesinger, R.E. Hemodynamic response to milrinone for refractory hypoxemia during therapeutic hypothermia for neonatal hypoxic ischemic encephalopathy. *J. Perinatol.* **2021**, *41*, 2345–2354. [CrossRef]
123. Hochwald, O.; Jabr, M.; Osiovich, H.; Miller, S.P.; McNamara, P.J.; Lavoie, P.M. Preferential cephalic redistribution of left ventricular cardiac output during therapeutic hypothermia for perinatal hypoxic-ischemic encephalopathy. *J. Pediatr.* **2014**, *164*, 999–1004.e1. [CrossRef]
124. Sakhuja, P.; More, K.; Ting, J.Y.; Sheth, J.; Lapointe, A.; Jain, A.; McNamara, P.J.; Moore, A.M. Gastrointestinal hemodynamic changes during therapeutic hypothermia and after rewarming in neonatal hypoxic-Ischemic encephalopathy. *Pediatr. Neonatol.* **2019**, *60*, 669–675. [CrossRef] [PubMed]
125. Giesinger, R.E.; El Shahed, A.I.; Castaldo, M.P.; Breatnach, C.R.; Chau, V.; Whyte, H.E.; El-Khuffash, A.F.; Mertens, L.; McNamara, P.J. Impaired right ventricular performance is associated with adverse outcome after hypoxic ischemic encephalopathy. *Am. J. Respir. Crit. Care Med.* **2019**, *200*, 1294–1305. [CrossRef] [PubMed]

126. Thayyil, S.; Pant, S.; Montaldo, P.; Shukla, D.; Oliveira, V.; Ivain, P.; Bassett, P.; Swamy, R.; Mendoza, J.; Moreno-Morales, M.; et al. Hypothermia for moderate or severe neonatal encephalopathy in low-income and middle-income countries (HE-LIX): A randomised controlled trial in India, Sri Lanka, and Bangladesh. *Lancet Glob. Health* **2021**, *9*, e1273–e1285. [CrossRef]

127. Montaldo, P.; Cunnington, A.; Oliveira, V.; Swamy, R.; Bandya, P.; Pant, S.; Lally, P.J.; Ivain, P.; Mendoza, J.; Atreja, G.; et al. Transcriptomic profile of adverse neurodevelopmental outcomes after neonatal encephalopathy. *Sci. Rep.* **2020**, *10*, 13100. [CrossRef]

128. Tetorou, K.; Sisa, C.; Iqbal, A.; Dhillon, K.; Hristova, M. Current therapies for neonatal hypoxic-ischemic and infection-sensitised hypoxic-ischemic brain damage. *Front. Synaptic Neurosci.* **2021**, *13*, 1–30. [CrossRef]

129. Trescher, W.H.; Ishiwa, S.; Johnston, M.V. Brief post-hypoxic-ischemic hypothermia markedly delays neonatal brain injury. *Brain Dev.* **1997**, *19*, 326–338. [CrossRef]

130. Gunn, A.J.; Gunn, T.R.; De Haan, H.H.; Williams, C.E.; Gluckman, P.D. Dramatic neuronal rescue with prolonged selective head cooling after ischemia in fetal lambs. *J. Clin. Investig.* **1997**, *99*, 248–256. [CrossRef]

131. Bingham, A.L.A. Hypothermia for Neonatal Hypoxic-Ischemic Encephalopathy: Different Cooling Regimens and Infants not Included in Prior Trials. In *Neonatology Questions and Controversies*; Perlman, J.C.M., Ed.; Elsevier: Philadelphia, PA, USA, 2019; pp. 77–91.

132. Shankaran, S.; Laptook, A.R.; Pappas, A.; McDonald, S.A.; Das, A.; Tyson, J.E.; Poindexter, B.B.; Schibler, K.; Bell, E.F.; Heyne, R.J.; et al. Effect of Depth and Duration of Cooling on Deaths in the NICU Among Neonates With Hypoxic Ischemic Encephalopathy. *JAMA* **2014**, *312*, 2629. [CrossRef]

133. Gunn, A.J.; Gunn, T.R.; Gunning, M.I.; Williams, C.E.; Gluckman, P.D. Neuroprotection with prolonged head cooling started before postischemic seizures in fetal sheep. *Pediatrics* **1998**, *102*, 1098–1106. [CrossRef]

134. Laptook, A.; Tyson, J.; Shankaran, S.; McDonald, S.; Ehrenkranz, R.; Fanaroff, A.; Donovan, E.; Goldberg, R.; O'Shea, T.M.; Higgins, R.D.; et al. Elevated temperature after hypoxic-ischemic encephalopathy: Risk factor for adverse outcomes. *Pediatrics* **2008**, *122*, 491–499. [CrossRef]

135. Kasdorf, E.; Perlman, J. General supportive management of the term Infant with neonatal encephalopathy following intrapartum hypoxia-ischemia. In *Neonatology Questions and Controversies*; Perlman, J.C.M., Ed.; Elsevier: Philadelphia, PA, USA, 2019; pp. 77–91.

136. Azzopardi, D.; Robertson, N.J.; Bainbridge, A.; Cady, E.; Charles-Edwards, G.; Deierl, A.; Fagiolo, G.; Franks, N.P.; Griffiths, J.; Hajnal, J.; et al. Moderate hypothermia within 6 h of birth plus inhaled xenon versus moderate hypothermia alone after birth asphyxia (TOBY-Xe): A proof-of-concept, open-label, randomised controlled trial. *Lancet Neurol.* **2016**, *15*, 145–153. [CrossRef]

137. Kelen, D.; Robertson, N.J. Experimental treatments for hypoxic ischemic encephalopathy. *Early Hum. Dev.* **2010**, *86*, 369–377. [CrossRef] [PubMed]

138. Gonzalez, F.F.; Abel, R.; Almli, C.R.; Mu, D.; Wendland, M.; Ferriero, D.M. Erythropoietin sustains cognitive function and brain volume after neonatal stroke. *Dev. Neurosci.* **2009**, *31*, 403–411. [CrossRef] [PubMed]

139. Aydin, A.; Genç, K.; Akhisaroglu, M.; Yorukoglu, K.; Gokmen, N.; Gonullu, E. Erythropoietin exerts neuroprotective effect in neonatal rat model of hypoxic-ischemic brain injury. *Brain Dev.* **2003**, *25*, 494–498. [CrossRef]

140. Kellert, B.A.; McPherson, R.J.; Juul, S.E. A comparison of high-dose recombinant erythropoietin treatment regimens in brain-injured neonatal rats. *Pediatr. Res.* **2007**, *61*, 451–455. [CrossRef]

141. Razak, A.; Hussain, A. Erythropoietin in perinatal hypoxic-ischemic encephalopathy: A systematic review and meta-analysis. *J. Perinat. Med.* **2019**, *47*, 478–489. [CrossRef]

142. Victor, S.; Rocha-Ferreira, E.; Rahim, A.; Hagberg, H.; Edwards, D. New possibilities for neuroprotection in neonatal hypoxic-ischemic encephalopathy. *Eur. J. Pediatr.* **2021**, 1–13. [CrossRef]

143. Groenendaal, F. Time to start hypothermia after perinatal asphyxia: Does it matter? *BMJ Paediatr. Open* **2019**, *14*, e000494. [CrossRef] [PubMed]

144. Popescu, M.R.; Panaitescu, A.M.; Pavel, B.; Zagrean, L.; Peltecu, G.; Zagrean, A.-M. Getting an early start in understanding perinatal asphyxia impact on the cardiovascular system. *Front. Pediatr.* **2020**, *8*, 1–18. [CrossRef]

145. Shah, P.; Anvekar, A.; McMichael, J.; Rao, S. Outcomes of infants with Apgar score of zero at 10 min: The West Australian experience. *Arch. Dis. Child. Fetal Neonatal Ed.* **2015**, *100*, F492–F494. [CrossRef]

146. Milner, K.M.; Neal, E.F.G.; Roberts, G.; Steer, A.C.; Duke, T. Long-term neurodevelopmental outcome in high-risk newborns in resource-limited settings: A systematic review of the literature. *Paediatr. Int. Child Health* **2015**, *35*, 227–242. [CrossRef] [PubMed]

147. Martinez, C.; Carneiro, L.; Vernier, L.; Cesa, C.; Guardiola, A.; Vidor, D. Language in children with Neonatal Hypoxic-Ischemic Encephalopathy. *Int. Arch. Otorhinolaryngol.* **2014**, *18*, 255–259. [CrossRef] [PubMed]

148. Fisher, P.G. In search of biomarkers for HIE. *J. Pediatr.* **2018**, *194*, 3. [CrossRef] [PubMed]

149. Mota-Rojas, D.; Flerro, R.; Roldan-Santiago, P.; Orozco-Gregorio, H.; González-Lozano, M.; Bonilla, H.; Martínez-Rodríguez, R.; García-Herrera, R.; Mora-Medina, P.; Flores-Peinado, S.; et al. Outcomes of gestation length in relation to farrowing performance in sows and daily weight gain and metabolic profiles in piglets. *Anim. Prod. Sci.* **2015**, *55*, 93–100. [CrossRef]

150. Mota-Rojas, D.; Olmos-Hernández, A.; Verduzco-Mendoza, A.; Lecona-Butrón, H.; Martínez-Burnes, J.; Mora-Medina, P.; Gómez-Prado, J.; Orihuela, A. Infrared thermal imaging associated with pain in laboratory animals. *Exp. Anim.* **2021**, *70*, 1–12. [CrossRef]

151. Martínez-Burnes, J.; Muns, R.; Barrios-García, H.; Villanueva-García, D.; Domínguez-Oliva, A.; Mota-Rojas, D. Parturition in mammals: Animal models, pain and distress. *Animals* **2021**, *11*, 2960. [CrossRef]
152. Villanueva-García, D.; Mota-Rojas, D.; Miranda-Cortés, A.; Ibarra-Ríos, D.; Casas-Alvarado, A.; Mora-Medina, P.; Martínez-Burnes, J.; Olmos-Hernández, A.; Hernández-avalos, I. Caffeine: Cardiorespiratory effects and tissue protection in animal models. *Exp. Anim.* **2021**, *70*, 20–0185. [CrossRef]
153. Villanueva-García, D.; Mota-Rojas, D.; Miranda-Cortés, A.; Mora-Medina, P.; Hernández-Avalos, I.; Casas-Alvarado, A.; Olmos-Hernández, A.; Martínez-Burnes, J. Neurobehavioral and neuroprotector effects of caffeine in animal models. *J. Anim. Behav. Biometeorol.* **2020**, *8*, 298–307. [CrossRef]
154. Redaelli, V.; Papa, S.; Marsella, G.; Grignaschi, G.; Bosi, A.; Ludwig, N.; Luzi, F.; Vismara, I.; Rimondo, S.; Veglianese, P.; et al. A refinement approach in a mouse model of rehabilitation research. Analgesia strategy, reduction approach and infrared thermography in spinal cord injury. *PLoS ONE* **2019**, *14*, e0224337. [CrossRef]
155. Casas-Alvarado, A.; Mota-Rojas, D.; Hernández-Avalos, I.; Mora-Medina, P.; Olmos-Hernández, A.; Verduzco-Mendoza, A.; Reyes-Sotelo, B.; Martínez-Burnes, J. Advances in infrared thermography: Surgical aspects, vascular changes, and pain monitoring in veterinary medicine. *J. Therm. Biol.* **2020**, *92*, 102664. [CrossRef]
156. Yáñez, A.; Mota-Rojas, D.; Ramírez-Necoechea, R.; Castillo-Rivera, M.; Roldán, P.; Mora-Medina, P.; González-Lozano, M. Application of infrared thermography to assess the effect of different types of environmental enrichment on the ocular, auricular pavilion and nose area temperatures of weaned piglets. *Comput. Electron. Agric.* **2019**, *156*, 33–42. [CrossRef]
157. Mota-Rojas, D.; Wang, D.; Titto, C.G.; Gómez-Prado, J.; Carvajal-de la Fuente, V.; Ghezzi, M.; Boscato-Funes, L.; Barrios-García, H.; Torres-Bernal, F.; Casas-Alvarado, A.; et al. Pathophysiology of fever and application of infrared thermography (IRT) in the detection of sick domestic animals: Recent advances. *Animals* **2021**, *11*, 2316. [CrossRef] [PubMed]
158. Mota-Rojas, D.; Titto, C.G.; Orihuela, A.; Martínez-Burnes, J.; Gómez-Prado, J.; Torres-Bernal, F.; Flores-Padilla, K.; Carvajal-de la Fuente, V.; Wang, D. Physiological and Behavioral Mechanisms of Thermoregulation in Mammals. *Animals* **2021**, *11*, 1733. [CrossRef] [PubMed]
159. Misbe, E.N.J.; Richards, T.L.; McPherson, R.J.; Burbacher, T.M.; Juul, S.E. Perinatal asphyxia in a nonhuman primate model. *Dev. Neurosci.* **2011**, *33*, 210–221. [CrossRef]
160. Manueldas, S.; Benterud, T.; Rueegg, C.S.; Garberg, H.T.; Huun, M.U.; Pankratov, L.; Åsegg-Atneosen, M.; Solberg, R.; Escobar, J.; Saugstad, O.D.; et al. Temporal patterns of circulating cell-free DNA (cfDNA) in a newborn piglet model of perinatal asphyxia. *PLoS ONE* **2018**, *13*, e0206601. [CrossRef]
161. Petersson, K.H.; Pinar, H.; Stopa, E.G.; Sadowska, G.B.; Hanumara, R.C.; Stonestreet, B.S. Effects of exogenous glucose on brain ischemia in ovine fetuses. *Pediatr. Res.* **2004**, *56*, 621–629. [CrossRef]
162. McAdams, R.M.; McPherson, R.J.; Kapur, R.P.; Juul, S.E. Focal brain injury associated with a model of severe hypoxic-ischemic encephalopathy in nonhuman primates. *Dev. Neurosci.* **2017**, *39*, 107–123. [CrossRef]
163. Whitelaw, A.; Thoresen, M. Animal research has been essential to saving babies' lives. *BMJ* **2014**, *348*, g4174. [CrossRef]
164. Raju, T.N. Some animal models for the study of perinatal asphyxia. *Neonatology* **1992**, *62*, 202–214. [CrossRef]
165. Ball, R.H.; Espinoza, M.I.; Parer, J.T.; Alon, E.; Vertommen, J.; Johnson, J. Regional blood flow in asphyxiated fetuses with seizures. *Am. J. Obstet. Gynecol.* **1994**, *170*, 156–161. [CrossRef]
166. Lear, C.A.; Davidson, J.O.; Mackay, G.R.; Drury, P.P.; Galinsky, R.; Quaedackers, J.S.; Gunn, A.J.; Bennet, L. Antenatal dexamethasone before asphyxia promotes cystic neural injury in preterm fetal sheep by inducing hyperglycemia. *J. Cereb. Blood Flow Metab.* **2018**, *38*, 706–718. [CrossRef] [PubMed]
167. Castillo-Melendez, M.; Baburamani, A.A.; Cabalag, C.; Yawno, T.; Witjaksono, A.; Miller, S.L.; Walker, D.W. Experimental modelling of the consequences of brief late gestation asphyxia on newborn lamb behaviour and brain structure. *PLoS ONE* **2013**, *8*, e77377. [CrossRef] [PubMed]
168. Nardiello, C.; Mižíková, I.; Morty, R.E. Looking ahead: Where to next for animal models of bronchopulmonary dysplasia? *Cell Tissue Res.* **2017**, *367*, 457–468. [CrossRef] [PubMed]
169. Solevåg, A.L.; Luong, D.; Lee, T.-F.; O'Reilly, M.; Cheung, P.-Y.; Schmölzer, G.M. Non-perfusing cardiac rhythms in as-phyxiated newborn piglets. *PLoS ONE* **2019**, *14*, e0214506.
170. van Dijk, A.J.; van Loon, J.P.A.M.; Taverne, M.A.M.; Jonker, F.H. Umbilical cord clamping in term piglets: A useful model to study perinatal asphyxia? *Theriogenology* **2008**, *70*, 662–674. [CrossRef] [PubMed]
171. Shen, W.; Xu, X.; Lee, T.-F.; Schmölzer, G.; Cheung, P.-Y. The relationship between heart rate and left ventricular isovolumic relaxation during rormoxia and hypoxia-asphyxia in newborn piglets. *Front. Physiol.* **2019**, *10*, 1–6. [CrossRef] [PubMed]
172. Fanos, V.; Noto, A.; Xanthos, T.; Lussu, M.; Murgia, F.; Barberini, L.; Finco, G.; D'Aloja, E.; Papalois, A.; Iacovidou, N.; et al. Metabolomics network characterization of resuscitation after normocapnic hypoxia in a newborn piglet model supports the hypothesis that room air is better. *Biomed. Res. Int.* **2014**, *2014*, 1–7. [CrossRef] [PubMed]
173. Solberg, R.; Enot, D.; Deigner, H.-P.; Koal, T.; Scholl-Bürgi, S.; Saugstad, O.D.; Keller, M. metabolomic analyses of plasma reveals new insights into asphyxia and resuscitation in pigs. *PLoS ONE* **2010**, *5*, e9606. [CrossRef] [PubMed]
174. Ara, J.; Fekete, S.; Frank, M.; Golden, J.A.; Pleasure, D.; Valencia, I. Hypoxic-preconditioning induces neuroprotection against hypoxia–ischemia in newborn piglet brain. *Neurobiol. Dis.* **2011**, *43*, 473–485. [CrossRef]
175. Villanueva-García, D.; Wang, K.T.; Nielsen, H.C.; Ramadurai, S.M. Expression of specific protein kinase c (PKC) isoforms and ligand-specific activation of pkca in late gestation fetal lung. *Exp. Lung Res.* **2007**, *33*, 185–196. [CrossRef]

176. Boynton, B.R.; Villanueva, D.; Hammond, M.D.; Vreeland, P.N.; Buckley, B.; Frantz, I.D. Effect of mean airway pressure on gas exchange during high-frequency oscillatory ventilation. *J. Appl. Physiol.* **1991**, *70*, 701–707. [CrossRef] [PubMed]
177. Salaets, T.; Gie, A.; Tack, B.; Deprest, J.; Toelen, J. modelling bronchopulmonary dysplasia in animals: Arguments for the preterm rabbit model. *Curr. Pharm. Des.* **2017**, *23*, 5887–5901. [CrossRef] [PubMed]
178. Ujhazy, E.; Dubovicky, M.; Navarova, J.; Sedlackova, N.; Danihel, L.; Brucknerova, I.; Mach, M. Subchronic perinatal asphyxia in rats: Embryo–foetal assessment of a new model of oxidative stress during critical period of development. *Food Chem. Toxicol.* **2013**, *61*, 233–239. [CrossRef] [PubMed]
179. Vlassaks, E.; Mencarelli, C.; Nikiforou, M.; Strackx, E.; Ferraz, M.J.; Aerts, J.M.; De Baets, M.H.; Martinez-Martinez, P.; Gavilanes, A.W.D. Fetal asphyxia induces acute and persisting changes in the ceramide metabolism in rat brain. *J. Lipid Res.* **2013**, *54*, 1825–1833. [CrossRef]
180. Galeano, P.; Blanco, E.; Logica Tornatore, T.M.A.; Romero, J.I.; Holubiec, M.I.; RodríGuez De Fonseca, F.; Capani, F. Life-long environmental enrichment counteracts spatial learning, reference and working memory deficits in middle-aged rats subjected to Perinatal asphyxia. *Front. Behav. Neurosci.* **2015**, *8*, 1–12. [CrossRef]
181. Caputa, M.; Rogalska, J.; Wentowska, K.; Nowakowska, A. Perinatal asphyxia, hyperthermia and hyperferremia as factors inducing behavioural disturbances in adulthood: A rat model. *Behav. Brain Res.* **2005**, *163*, 246–256. [CrossRef]
182. Sánchez-Aparicio, P.; Mota-Rojas, D.; Nava-Ocampo, A.A.; Trujillo-Ortega, M.E.; Alfaro-Rodríguez, A.; Arch, E.; Alonso-Spilsbury, M. Effects of sildenafil on the fetal growth of guinea pigs and their ability to survive induced intrapar-tum asphyxia. *Am. J. Obstet. Gynecol.* **2008**, *198*, 127.e1–127.e6. [CrossRef]
183. Coalson, J.J.; Kuehl, T.J.; Escobedo, M.B.; Leonard Hilliard, J.; Smith, F.; Meredith, K.; Null, D.M.; Walsh, W.; Johnson, D.; Robotham, J.L. A baboon model of bronchopulmonary dysplasia. *Exp. Mol. Pathol.* **1982**, *37*, 335–350. [CrossRef]
184. Escobedo, M.B.; Leonard Hilliard, J.; Smith, F.; Meredith, K.; Walsh, W.; Johnson, D.; Coalson, J.J.; Kuehl, T.J.; Null, D.M.; Robotham, J.L. A baboon model of bronchopulmonary dysplasia. I. Clinical features. *Exp. Mol. Pathol.* **1982**, *37*, 323–334. [CrossRef]
185. Yoder, B.A.; Coalson, J.J. Animal models of bronchopulmonary dysplasia. The preterm baboon models. *Am. J. Physiol.-Lung Cell Mol. Physiol.* **2014**, *307*, L970–L977. [CrossRef]
186. Peeples, E.S.; Ezeokeke, C.K.; Juul, S.E.; Mourad, P.D. Evaluating a targeted bedside measure of cerebral perfusion in a nonhuman primate model of neonatal hypoxic-ischemic encephalopathy. *J. Ultrasound Med.* **2018**, *37*, 913–920. [CrossRef] [PubMed]
187. Lewis, K.J.R.; Tibbitt, M.W.; Zhao, Y.; Branchfield, K.; Sun, X.; Balasubramaniam, V.; Anseth, K.S. In vitro model alveoli from photodegradable microsphere templates. *Biomater. Sci.* **2015**, *3*, 821–832. [CrossRef] [PubMed]
188. Beckstrom, A.C.; Humston, E.M.; Snyder, L.R.; Synovec, R.E.; Juul, S.E. Application of comprehensive two-dimensional gas chromatography with time-of-flight mass spectrometry method to identify potential biomarkers of perinatal asphyxia in a non-human primate model. *J. Chromatogr. A* **2011**, *1218*, 1899–1906. [CrossRef] [PubMed]
189. Juul, S.E.; Aylward, E.; Richards, T.; McPherson, R.J.; Kuratani, J.; Burbacher, T.M. Prenatal cord clamping in newborn Macaca nemestrina: A model of perinatal asphyxia. *Dev. Neurosci.* **2007**, *29*, 311–320. [CrossRef] [PubMed]
190. Dunn, W.B.; Wilson, I.D.; Nicholls, A.W.; Broadhurst, D. The importance of experimental design and QC samples in large-scale and MS-driven untargeted metabolomic studies of humans. *Bioanalysis* **2012**, *4*, 2249–2264. [CrossRef]
191. Pomfy, M.; Franko, J. Validation of a four-vessel occlusion model for transient global cerebral ischemia in dogs. *J. Hirnforsch.* **1999**, *39*, 465–471. [PubMed]
192. Uleanya, N.D.; Aniwada, E.C.; Ekwochi, U.; Uleanya, N.D. Short term outcome and predictors of survival among birth asphyxiated babies at a tertiary academic hospital in Enugu, South East, Nigeria. *Afr. Health Sci.* **2019**, *19*, 1554. [CrossRef]
193. Ment, L.R.; Stewart, W.B.; Gore, J.C.; Duncan, C.C. Beagle puppy model of perinatal asphyxia: Alterations in cerebral blood flow and metabolism. *Pediatr. Neurol.* **1988**, *4*, 98–104. [CrossRef]
194. Walsh, W.F.; Butler, D.; Schmidt, J.W. Report of a pilot study of Cooling four preterm infants 32–35 weeks gestation with HIE. *J. Neonatal Perinatal Med.* **2015**, *8*, 47–51. [CrossRef]
195. Choudhary, M.; Sharma, D.; Dabi, D.; Lamba, M.; Pandita, A.; Shastri, S. Hepatic dysfunction in asphyxiated neonates: Prospective case-controlled study. *Clin. Med. Insights Pediatr.* **2015**, *9*, 1–6. [CrossRef]
196. Liang, C.; Hong, Q.; Jiang, T.-T.; Gao, Y.; Yao, X.-F.; Luo, X.-X.; Zhuo, X.-H.; Shinn, J.B.; Jones, R.O.; Zhao, H.-B.; et al. The effects and outcomes of electrolyte disturbances and asphyxia on newborns hearing. *Int. J. Pediatr. Otorhinolaryngol.* **2013**, *77*, 1072–1076. [CrossRef] [PubMed]
197. Aly, H.; Hassanein, S.; Nada, A.; Mohamed, M.H.; Atef, S.H.; Atiea, W. Vascular endothelial growth factor in neonates with perinatal asphyxia. *Brain Dev.* **2009**, *31*, 600–604. [CrossRef]

Review

Pathophysiological Studies of Monoaminergic Neurotransmission Systems in Valproic Acid-Induced Model of Autism Spectrum Disorder

Hsiao-Ying Kuo [1],* and Fu-Chin Liu [2],*

[1] Institute of Anatomy and Cell Biology, National Yang Ming Chiao Tung University, Taipei 11221, Taiwan
[2] Institute of Neuroscience, National Yang Ming Chiao Tung University, Taipei 11221, Taiwan
* Correspondence: kuohy@nycu.edu.tw (H.-Y.K.); fuchin@nycu.edu.tw (F.-C.L.);
Tel.: +886-2-2826-7075 (H.-Y.K.); +886-2-2826-7216 (F.-C.L.); Fax: +886-2-2821-2884 (H.-Y.K.);
+886-2-2820-0259 (F.-C.L.)

Abstract: Autism spectrum disorder (ASD) is a neurodevelopmental disorder with complex etiology. The core syndromes of ASD are deficits in social communication and self-restricted interests and repetitive behaviors. Social communication relies on the proper integration of sensory and motor functions, which is tightly interwoven with the limbic function of reward, motivation, and emotion in the brain. Monoamine neurotransmitters, including serotonin, dopamine, and norepinephrine, are key players in the modulation of neuronal activity. Owing to their broad distribution, the monoamine neurotransmitter systems are well suited to modulate social communication by coordinating sensory, motor, and limbic systems in different brain regions. The complex and diverse functions of monoamine neurotransmission thus render themselves as primary targets of pathophysiological investigation of the etiology of ASD. Clinical studies have reported that children with maternal exposure to valproic acid (VPA) have an increased risk of developing ASD. Extensive animal studies have confirmed that maternal treatments of VPA include ASD-like phenotypes, including impaired social communication and repetitive behavior. Here, given that ASD is a neurodevelopmental disorder, we begin with an overview of the neural development of monoaminergic systems with their neurochemical properties in the brain. We then review and discuss the evidence of human clinical and animal model studies of ASD with a focus on the VPA-induced pathophysiology of monoamine neurotransmitter systems. We also review the potential interactions of microbiota and monoamine neurotransmitter systems in ASD pathophysiology. Widespread and complex changes in monoamine neurotransmitters are detected in the brains of human patients with ASD and validated in animal models. ASD animal models are not only essential to the characterization of pathogenic mechanisms, but also provide a preclinical platform for developing therapeutic approaches to ASD.

Keywords: valproic acid; autism spectrum disorder; serotonin; dopamine; norepinephrine; histamine; neurodevelopment

Citation: Kuo, H.-Y.; Liu, F.-C. Pathophysiological Studies of Monoaminergic Neurotransmission Systems in Valproic Acid-Induced Model of Autism Spectrum Disorder. *Biomedicines* **2022**, *10*, 560. https://doi.org/10.3390/biomedicines10030560

Academic Editor: Martina Perše

Received: 7 January 2022
Accepted: 22 February 2022
Published: 27 February 2022

Publisher's Note: MDPI stays neutral with regard to jurisdictional claims in published maps and institutional affiliations.

1. Introduction

Autism spectrum disorder (ASD) is a devasting neurodevelopmental disease with an increasing prevalence of ~18.5 per 1000 people [1]. The etiology of ASD is highly heterogeneous with genetic and environmental roots. The core syndromes of ASD patients are deficits in social communication/interaction and restrictive/repetitive behavior [2]. Because of its complicated and heterogeneous etiology, the pathological mechanisms are not yet fully characterized, which prevents the development of effective therapy for ASD.

1.1. Animal Models of ASD

ASD is a complex and heterogeneous neurodevelopmental disorder. Genetic mutations of genes involved in synaptic wiring and neurotransmission in the development and function of neural circuits have been linked to ASD pathogenesis [3,4]. Genetic animal models of ASD provide research platforms to investigate ASD pathophysiology caused by mutations of ASD-risk genes, including *MECP2, SHANK3, NLGN3*, etc. [3–5]. As developing brains are vulnerable to environmental insults, environmental factors also play a significant role in the pathogenesis of ASD. It has been estimated up to 40–50% of the variance in ASD pathogenesis is influenced by environmental factors [6]. Maternal exposure to pharmacological reagents, heavy metals, and vaccination against virus infection has been implicated in the etiology of ASD [6].

1.2. VPA-Induced ASD-Like Animal Models

Valproic acid (VPA) is prescribed medicine for treating epilepsy, migraine, and mood disorders [7]. However, clinical studies indicate that children with maternal exposure to VPA have a higher chance of developing ASD [8]. Subsequent animal studies confirm that maternally VPA-treated offspring develop prominent ASD-like phenotypes resembling the core syndromes of human ASD patients.

Most VPA-induced animal models are established in the rodent owing to the easy accessibility to rodents in the laboratory setting. VPA-induced models have also been reported in non-human primates [9,10]. The time windows of VPA administration in producing offspring with ASD-like phenotypes are from neural tube closure to neurogenesis, i.e., embryonic (E) day 9-E12.5 in pregnant rats (Tables 1 and 2) [11–13]. Time windows of VPA exposure appear to critically influence the developmental consequences in the VPA models. Previous studies have compared the teratogenic effects of VPA in different time windows of exposure. For example, Rodier et al. have found that VPA exposure at E11.5, E12, or E12.5 results in differential vulnerabilities of cranial nerves in the brains of offspring [13]. Animals exposed to VPA at different gestational time windows also show distinct developmental trajectories and abnormalities in cortical and brainstem regions [14]. At the behavioral level, Kim et al. have reported that VPA exposure at E12, but not at E7, E9.5, and E15, causes significant ASD-like social abnormalities, suggesting that E12 is a critical period for inducing ASD-like phenotypes [15].

Intriguingly, postnatal VPA administration during the first two weeks after birth causes ASD-like behavioral phenotypes in rodents as well [16–18], suggesting that postnatal developmental processes, including synaptogenesis, myelination, and synaptic pruning are also vulnerable to VPA exposure. Therefore, ASD-like phenotypes could be induced by VPA administrated at either gestational or early postnatal stages.

Regarding the issue of gender, the prevalence of human ASD patients among males is ~4.3 times higher than among females [1]. Most studies use male animals in VPA-induced ASD model studies. However, some groups also include females and found ASD phenotypes in female animals as well (Tables 1 and 2). To investigate sexual dimorphism of VPA effects, several studies have compared gender-specific alterations in maternally VPA-treated offspring [19–23]. Evidence shows gender-specific changes in the immune responses [19,20,22,24], cell density of cerebellum [23,25], network connectivity [26], anandamide signaling [27], attentional processing [28], and composition of gut microbiota [29] in maternally VPA-treated animal offspring. These sexual dimorphisms may be related to histone deacetylase (HDAC) inhibition by VPA, as Konopko et al. have reported differential epigenetic modification of *Bdnf* genes in maternally VPA-treated fetal brains of different genders [24]. In general, maternally VPA-treated male offspring have more severe ASD-like phenotypes compared to female offspring.

With respect to the dosage of VPA administration, maternal VPA exposure affects the nervous systems in a dose-dependent manner. Magnetic resonance imaging (MRI) studies have shown dose-dependent changes in networks of the primary motor cortex-premotor cortex and the primary motor cortex-supplementary motor area in humans, suggesting that VPA induces changes in specific neural circuits [30,31]. The higher dose of the expecting mother taking VPA, the higher risk of the offspring being diagnosed with ASD [32]. Frisch et al. have reported that offspring of rat dams receiving a high dose of VPA (720 mg/kg) daily during the entire pregnancy show the most severe phenotype of impaired learning and memory. In the same study, MRI data show that compared to a medium dose of VPA (470 mg/kg), the higher dose of maternal treatment of VPA (720 mg/kg) induced more significant reductions in cortical and brainstem regions of offspring [33]. Prenatal administration of a lower dose of VPA (200 mg/kg) did not induce ASD-like behavioral phenotypes [16]. Therefore, for the experimental induction of ASD-like behavioral phenotypes without severe developmental retardation in the offspring, VPA administered at the dosage of 300–600 mg/kg in pregnant rodents and non-human primates are mostly used (Tables 1 and 2) [9,34–37].

1.3. Potential Mechanisms Underlying VPA-Induced ASD-Like Pathophysiology

It is yet unclear how maternal VPA exposure induces ASD pathophysiology. VPA is known to have multiple pharmacological properties. Firstly, VPA is a simple branched-chain fatty acid (2-propylpentanoic acid) that can directly inhibit voltage-gated sodium channels and suppress the high-frequency firing activity of neurons. VPA also can indirectly inhibit GABA transferase resulting in increased GABA neurotransmission in the brain [38]. Note that the imbalanced excitatory/inhibitory (E/I) ratio in neural circuits has been proposed as a prevailing pathogenic mechanism of ASD [39,40]. Hence, VPA may reduce the excitability of neural circuits by inhibiting GABA uptake, leading to disrupted E/I balance, which may account for ASD phenotypes. In supporting this hypothesis, previous studies show that abnormal E/I balance and aberrant synaptic transmission have been found in the medial prefrontal cortex (mPFC), primary somatosensory cortex, amygdala, and dorsal raphe nucleus of maternally VPA-treated rodent offspring [41–46]. Furthermore, drugs that regulate neuronal excitability can partially rescue ASD-like phenotypes in VPA-treated animals [37,47], suggesting that E/I imbalance resulting from VPA treatment may be a key mechanism of ASD pathogenesis.

Secondly, VPA functions as a non-specific histone deacetylase (HDAC) inhibitor that can affect neural development by chromosome remodeling [6–8]. VPA-induced changes in neural development have been reported in many studies and discussed by several review articles [12,35,48]. The epigenetic effects of VPA have been confirmed by VPA-induced site-specific epigenetic alterations in rodents [49–51] and non-human primates [9], suggesting that VPA-mediated epigenetic modifications may contribute to VPA-induced ASD-like pathophysiology. Several genome-wide screening studies have identified the alteration of genetic networks by VPA. In non-human primates, prenatal VPA administration changes the expression of human ASD-associated genes and the gene expression patterns that are related to synaptogenesis, critical period, neural plasticity, and E/I balance, including *CADM1*, *LRRTM4*, *GRIN1*, and *GRIN2A* in marmoset infant brains [52]. Another RNA sequencing study has investigated transcriptome profiles in maternal VPA-treated marmoset embryos and found marked alterations in the gene expression pattern of neuronal development, including axon guidance and calcium signaling [9]. In rodent models, maternal VPA treatment induces changes in circadian rhythm- and extracellular matrix-related genes in the mPFC [53]. Altered gene expression profiles in neural development and function, cellular development and function, cell death, and immune system are also detected in the amygdala of maternally VPA-treated offspring brains [54]. Proteomics study has further found changes in protein expression profiles in synaptic function, energy metabolism, cytoskeleton and neuropsychiatric disorders in the cerebral cortex of maternally VPA-treated rat offspring [55]. Note that molecular and cellular changes in the VPA-induced animal models are similar, but not identical to those in idiopathic ASD patients.

On the other hand, VPA-induced changes in the immune system are of particular interest, because maternal immune activation during embryonic stages has been proposed as a pathogenic mechanism of ASD [56]. Inflammatory responses are detected in human ASD brains, including elevated levels of inflammatory biomarkers and activation of astrocytes and microglia [57]. Evidence suggests abnormalities in the interaction of the microbiota–gut–brain axis and the immune system in ASD brains [58], highlighting the multifaceted and potential pathogenic role of inflammation in ASD pathogenesis. In VPA-treated animals, marked signs of neuroinflammation have been detected, including elevated levels of reactive oxygen species, NFκB, and pro-inflammatory cytokines [59]. Moreover, anti-inflammatory drugs could alleviate ASD-like phenotypes in maternally VPA-treated animal offspring [37,47]. In light of these findings, we will discuss how monoamines interplay with the microbiota–gut–brain axis and the immune system later.

1.4. Advantages and Limitations of VPA-Induced ASD-Like Animal Models

Since the first rodent model in the 1970s, VPA exposure in experimental animals has been widely used as an ASD model to study the ASD etiology of environmental risk factors. The major advantage of studying VPA models is that the VPA-exposed animals develop robust ASD-like phenotypes resembling the core symptoms of ASD patients [7,11,12,35]. Its simple experimental protocols also allow it to be widely adopted as an animal model for studying ASD pathophysiology from molecular, cellular to behavioral levels for decades [12,35,48,60,61]. Moreover, as VPA is an HDAC inhibitor that can influence chromosome remodeling, VPA animal models may help understand the epigenetic basis of ASD etiology.

The limitations of VPA models are, however, that VPA exposure is one of many environmental risk factors of ASD and maternally VPA-exposed ASD patients represent a small population of ASD. The VPA-induced pathophysiology may thus account for part of ASD pathophysiology. Nonetheless, because VPA-induced pathophysiology shares some similarities with ASD genetic mutation models, e.g., synaptopathy [4,62], the easy accessibility and the extensive knowledge built on this ASD model have been leveraged to screen for pharmacological reagents for treating ASD patients [37,47,63].

1.5. Dysfunction of Monoaminergic Neurotransmission in VPA-Induced ASD-Like Animal Models of ASD

Alterations of monoaminergic neurotransmitter systems have been detected in many brain regions as well as in the peripheral system in ASD pathophysiology (Figure 1; Tables 1 and 2). Unlike the classical neurotransmitters of glutamate and GABA that directly dictate neuronal activity, monoamines are neuromodulators that modulate the excitability of neurons by regulating synaptic neurotransmission (Figure 2; Tables 1 and 2).

Monoamine-modulating drugs are so far the available medications for treating ASD-associated syndromes. The atypical antipsychotics, risperidone and aripiprazole, are U.S. Food and Drug Administration (FDA)-approved clinical drugs for ASD. Risperidone and aripiprazole pharmacologically act on dopamine (DA) and serotonin of the monoaminergic systems. In supporting the importance of the monoaminergic systems for ASD, several studies have shown that antipsychotics targeting monoaminergic systems are effective in improving maternal VPA-induced ASD-like abnormalities [64–67].

The neural development of monoaminergic systems protracts a long time from embryonic to postnatal stages. The prolonged developmental time frame makes it vulnerable to pathological alterations that are induced by genetic and epigenetic challenges related to ASD. In the present review, we focus on the VPA-induced ASD model, because it has been widely adopted to investigate the neuropathophysiology of ASD and used as an experimental platform to develop therapeutic reagents [35,37,68]. We begin with an overview of the neural development and neurochemical property of each monoamine neurotransmitter, we then discuss the abnormalities of ASD in human patients and VPA-induced animal models. Schematic summary of the monoaminergic neurotransmission-related endophenotypes and other pathophysiological changes in the central nervous system and peripheral system of VPA-treated animals are illustrated in Figures 1 and 2 and listed in Tables 1 and 2. Schematic drawings of synaptic transmission of monoaminergic neurotransmissions and VPA-induced dysfunction of monoaminergic neurotransmission are illustrated in Figure 2.

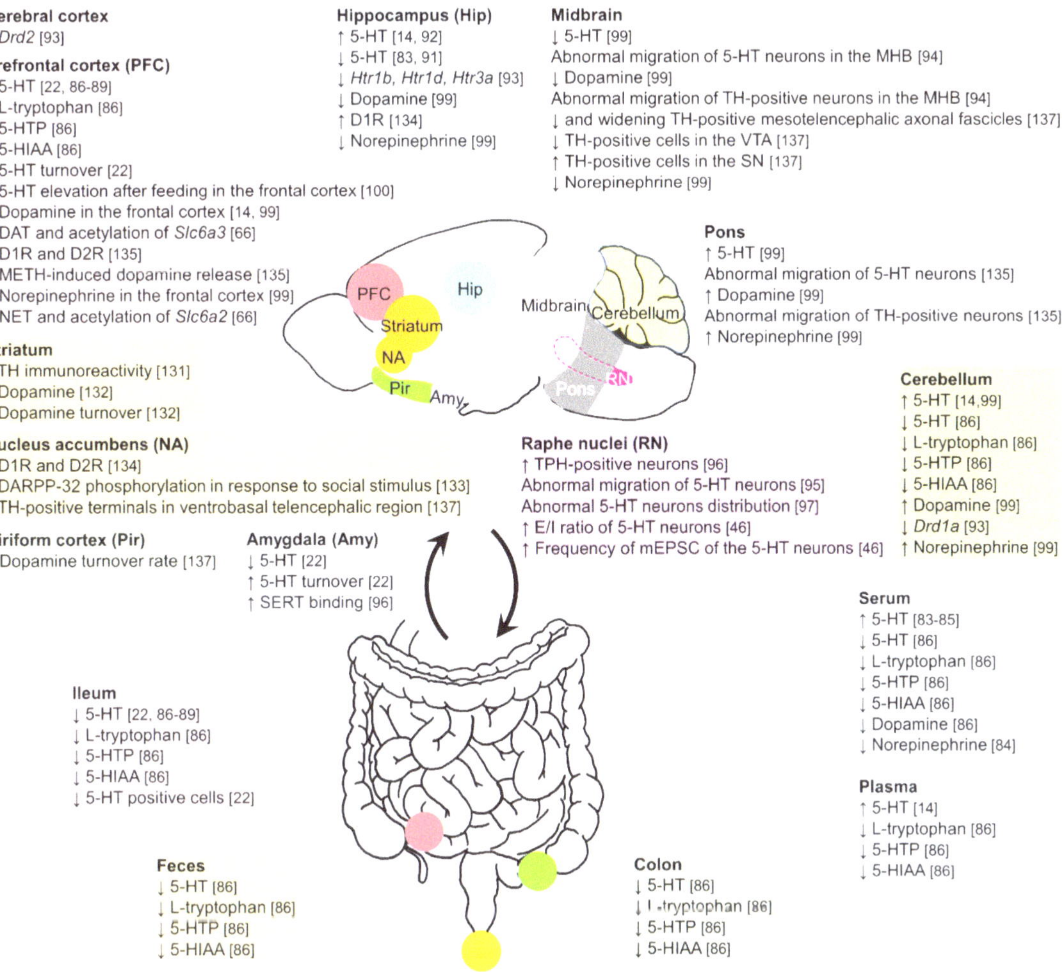

Figure 1. Alterations of monoaminergic systems in the brain and peripheral system of maternally VPA-treated rodent offspring. Schematic drawings illustrate pathophysiological changes in monoaminergic systems in the central nervous system and peripheral system of maternally VPA-treated rodent offspring. The bi-directional arrows indicate cross-talk between the central and peripheral systems. Abbreviations: ↑, increased; ↓, decreased; 5-HIAA, 5-hydroxyindoleacetic acid; 5-HT, 5-hydroxytryptamine (serotonin); 5-HTP, 5-hydroxytryptophan; D1R, dopamine D1 receptor; D2R, dopamine D2 receptor; DAT, dopamine transporter; DOPAC, 3,4-dihydroxyphenylacetic acid; E/I, excitatory/inhibitory; Htr, 5-hydroxytryptamine receptors; mEPSCs, miniature excitatory postsynaptic currents; METH, methamphetamine; MHB, midbrain-hindbrain boundary; NET, norepinephrine transporter; SERT, serotonin transporter; SN, substantia nigra; TH, tyrosine hydroxylase; TPH, tryptophan hydroxylase; VTA, ventral tegmental area.

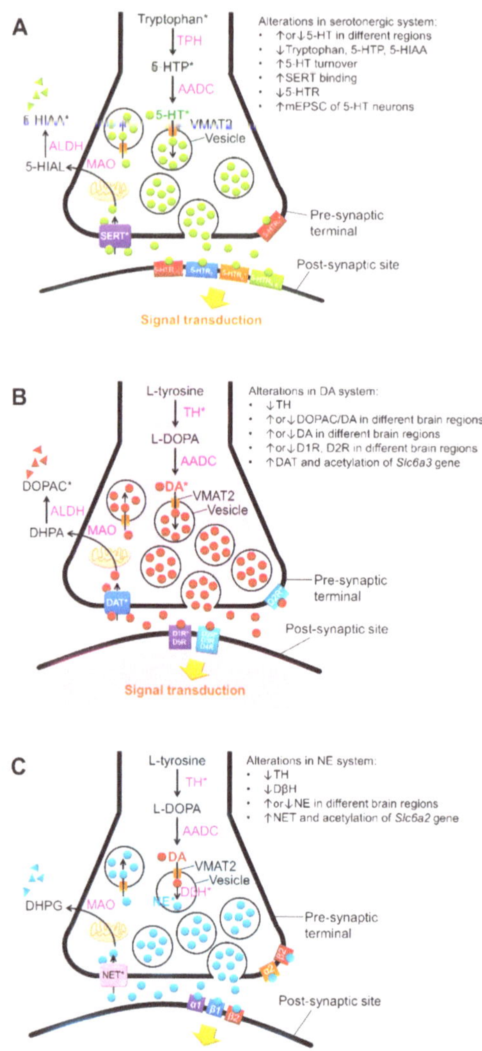

Figure 2. Schematic drawings illustrate monoaminergic neurotransmissions and their dysregulation in response to maternal VPA challenge. (**A**) Serotonergic neurotransmission. (**B**) Dopaminergic neurotransmission. (**C**) Norepinephrinergic neurotransmission. The neurotransmission molecules that are altered by maternal VPA treatments are indicated by asterisk*. Abbreviations: ↑, increased; ↓, decreased; 5-HIAA, 5-hydroxyindoleacetic acid; 5-HIAL: 5-hydroxyindole acetaldehyde; 5-HT, 5-hydroxytryptamine (serotonin); 5-HTP, 5-hydroxytryptophan; 5-HTR, 5-hydroxytryptamine receptor; AADC, aromatic L-amino acid decarboxylase; ALDH, aldehyde dehydrogenase; D1R, dopamine D1 receptor; D2R, dopamine D2 receptor; D3R, dopamine D3 receptor; D4R, dopamine D4 receptor; D5R, dopamine D5 receptor; DA, dopamine; DAT, dopamine transporter; DβH, dopamine-β-hydroxylase; DHPG, dihydroxyphenylglycol; DOPAC, 3,4-dihydroxyphenylacetic acid; MAO, monoamine oxidase; mEPSCs, miniature excitatory postsynaptic currents; METH, methamphetamine; NE, norepinephrine; NET, norepinephrine transporter; SERT, serotonin transporter; TH, tyrosine hydroxylase; TPH, tryptophan hydroxylase; VMAT, vesicular monoamine transporter.

2. Alterations of Monoaminergic Systems in VPA-Induced ASD Models

2.1. Serotoninergic Systems

2.1.1. Neural Development of Serotonergic Neurons

In the central nervous system, the majority of serotonergic neurons are localized in the raphe nuclei. All the serotonergic neurons are derived from the developing rhombencephalon. Serotonergic progenitors are generated in the isthmus and p3 domain of the ventral hindbrain. Serotonergic neurons undergo neurogenesis at E 9.5–E12 in mice and E10.5–E13 in rats. Shortly after cell mitosis, differentiating serotonergic neurons begin their migration as early as E10 [69]. The early-born neurons migrate to form the anterior cluster of serotonergic neurons (B5–B9) that extensively innervate the forebrain, midbrain, cerebellum, and raphe nuclei themselves. The late-born neurons migrate caudally to form the posterior cluster of serotonergic neurons (B1–B4) that innervate the brainstem and spinal cord [70]. Upon the end of differentiation, serotonergic neurons produce 5-hydroxytryptamine (5-HT, i.e., serotonin) not only for neurotransmission, but also for facilitating the maturation of themselves. Along with axonal outgrowth and navigation, serotonergic neurons start to release 5-HT during axonal outgrowth as early as E16.5. The extensive and prolonged axonal innervations of serotonergic neurons are not completed until several weeks after birth [71].

2.1.2. Neurochemical Properties of Serotonin

Tryptophan is the precursor for the synthesis of serotonin. Tryptophan hydroxylase (TPH), the rate-limiting enzyme found only in serotonergic neurons, converts tryptophan into 5-hydroxytryptophan (5-HTP). Then, aromatic L-amino acid decarboxylase (AADC) converts 5-HTP into 5-HT. Cytoplasmic 5-HT is packed into synaptic vesicles through vesicular monoamine transporter 2 (VMAT2), where it is stored until being released into the synaptic cleft by exocytosis. The neurotransmission activity of 5-HT in the synaptic cleft is terminated by the serotonin transporter (SERT) that reuptakes 5-HT back to presynaptic terminals [70]. Because SERT is critical to 5-HT activity, the kinetics of 5-HT neurotransmission is governed by the level of SERT. 5-HT is metabolically degraded by monoamine oxidase (MAO) and converted into 5-hydrozy-indoleacetaldehyde, which is subsequently oxidized by aldehyde dehydrogenase to derive 5-hydroxyindoleacetic acid (5-HIAA). See Figure 2 for the summary of serotoninergic neurotransmission.

2.1.3. Clinical Evidence of Abnormalities of Serotoninergic Systems Related to ASD

Clinical evidence of dysregulated serotoninergic systems in ASD patients is yet disputed. The first clinical study has reported hyperserotonemia in early-onset ASD patients [72]. Consistent with this report, a meta-analysis study has revealed elevated levels of 5-HT in the blood of 25.4% of ASD patients [73]. Another group has also reported an increased number of dystrophic 5-HT axons in different regions of human ASD brains [74,75]. Despite the evidence of hyperserotonemia as reported in some clinical studies, postmortem studies indicate different aspects of serotonin dysfunction in ASD pathophysiology. Significant reductions in receptor-binding density of two 5-HT receptors, G_i-coupled 5-HT$_{1A}$ and G_q-coupled 5-HT$_2$, were found in the posterior cingulate cortex and fusiform gyrus of ASD patients [76]. Moreover, abnormally increased and decreased levels of serotonin receptors were, respectively, observed in ASD patients with and without seizure histories [76]. Age-dependent alterations of the serotonergic systems have recently been identified, especially in the anterior cingulate cortex [77]. In addition to neuroanatomical alterations, genetic linkage studies have observed that SERT variants are correlated with platelet hyperserotonemia and impaired social communication in ASD patients [78–80]. Animal studies have found abnormal serotonergic neurotransmission in ASD-like pathophysiology as well. Normal serotonin levels are critical to the maintenance of E/I balance in cortical neurons, and restoration of serotonin levels can alleviate ASD-like phenotypes [81]. The causal relationship between the ASD-associated SLC6A4/SERT Ala56 coding variant and hyperserotonemia has been demonstrated in SLC6A4/SERT Ala56 gain-of-function transgenic mice that exhibited ASD-like behaviors [82].

2.1.4. Abnormalities of Serotonergic Systems in VPA-Induced ASD Models

In the peripheral system, previous studies have found elevated 5-HT levels in serum, plasma or gastrointestinal tract of VPA-treated rodents, which support the general hyperserotonemia theory of ASD etiology [14,83–85]. However, other studies have reported the reduction in 5-HT-positive cells in the ileum, 5-HT levels and its precursors (tryptophan and 5-HTP), and metabolites (5-HIAA) in both the peripheral system and brain [22,86–89]. Note that short-term depletion of 5-HT precursor tryptophan deteriorates repetitive behavior and elevates anxiety status in ASD patients [90]. Given the complexity of clinical case studies, e.g., variations in the genetic, comorbidity, medication, and diet, it remains to be clarified how dysfunction of peripheral 5-HT takes part in ASD pathophysiology.

In the central nervous system, alterations of 5-HT levels have been detected in several brain regions of VPA-treated offspring. Serotonergic neurons of the median raphe nucleus primarily project to the hippocampus, whereas serotonergic neurons of the dorsal raphe nuclei preferentially project to the prefrontal cortex, basal ganglia, and amygdala [70]. In contrast to the more consistent results of VPA-induced 5-HT levels in the regions that are innervated by the dorsal raphe nucleus, the VPA-induced alterations of 5-HT levels are varied in the hippocampus [14,18,83,91,92]. The dorsal and median raphe nuclei may differ in vulnerabilities to VPA treatments. Reduced serotonin receptors have also been found in the hippocampus [93]. In other brain regions, alterations in 5-HT levels are more consistent. In VPA-treated rodents, increased 5-HT levels are detected in the cerebellum and pons; however, reduced 5-HT levels are detected in the prefrontal cortex, midbrain, and amygdala. Considering developmental changes in VPA-treated rodents, administration of VPA during E9-E12.5 is at the time window of neurogenesis, differentiation, and migration of serotonergic neurons in the raphe nuclei. Kuwagata et al. have reported differential vulnerabilities of 5-HT-positive neurons to different time windows of maternal VPA administration. They found abnormal migration of 5-HT-positive neurons in rats receiving VPA at E11, but not in rats receiving VPA at E9 [94]. Wang et al. found an increased number of 5-HT-positive neurons in the raphe magnus nucleus (caudal part of raphe nuclei), but Miyazaki et al. report no significant changes in the total number of 5-HT-positive cells in the raphe nuclei [95,96]. These two studies are different at VPA dose and the timing of VPA injections. The discrepancy of these findings may be partly explained by VPA-induced abnormal migration in the raphe nuclei as reported by other groups [94,95,97]. Abnormal differentiation of serotonergic neurons has been reported in the embryos of VPA-treated zebrafish [98]. Because VPA is an HDAC inhibitor that can regulate chromatin remodeling, it would be of interest to study how VPA-induced epigenetic changes in chromatin remodeling affect the development of serotonergic neurons.

2.1.5. Behavioral Phenotypes Related to Abnormal Serotonergic Systems in VPA-Induced ASD Models

At the functional level, Wang et al. (2018) reported abnormal excitatory/inhibitory balance and mEPSCs of 5-HT neurons in the dorsal raphe nucleus of VPA-treated brains, implicating a role of 5-HT in the modulation of excitatory/inhibitory balance of neuronal activity [46]. With respect to behavioral phenotypes, most of the studies have found abnormal social behaviors along with abnormalities in serotonergic systems [18,22,83,84,86–89,91,92,96,99]. Impaired social communication is a hallmark of ASD symptoms, and defective social interaction also occurs in VPA-induced ASD animals. Other ASD-like behavioral phenotypes in serotonergic deficient VPA mice are increased repetitive behaviors and anxiety levels, abnormal sensation, delayed developmental milestones, and impaired memory [18,83,84,86–89,92,96,99]. By contrast, the findings of locomotion are inconsistent in that some studies have reported reduced locomotion [18,84,86], but other studies found elevated locomotion [87–89,91,92,100]. This inconsistency cannot be explained by species of animals or dosage of VPA. An interesting finding is that abnormal circadian rhythm and activity were found in rats that maternally received VPA at E9.5. Note that a high proportion of patients with autism show sleep disturbance, especially insomnia [101,102]. Regarding the causality of abnormal serotonergic systems and behavioral abnormalities, Wang et al. (2013) have treated VPA-induced ASD rats with 5-HT$_{1A}$ receptor agonists during behavioral tests. They found that the treatment of 5-HT$_{1A}$ receptor agonists could restore social impairments and improve extinction of fear memory, suggesting that abnormal serotonergic signaling may contribute to abnormal social behaviors and memory deficits in VPA-treated rodents [96]. In the future, more studies are required to further clarify the causal relationship between behavioral phenotypes and serotonergic abnormalities (Table 1).

Table 1. Summary of valproic acid (VPA)-induced teratogenicity in serotonergic systems.

Monoamines-Related Endophenotypes (Age of Animals during Analysis)	Other Molecular, Cellular and Physiological Phenotypes (Age of Animals during Analysis)	Behavioral Phenotypes (Age of Animals during Analysis)	Animals/ Gender	VPA Exposure	References
		Serotonergic System			
↓ 5-HT in the hippocampus ↑ 5-HT in serum (> 3 m/o)	↑ Inflammation in colonic tissues ↓ Tight junction protein (claudin1 and occluding) in the colonic tissues (> 3 m/o)	↓ Social play behavior Abnormal olfactory habituation/dishabituation (11 wks–3 m/o)	Wistar rats/male	600 mg/kg, E12.5, i.p.	[83]
↑ 5-HT in serum (P28)	↑ Proinflammatory cytokines: IL-17A, TNF-α, IL-6 ↓ GABA in serum Abnormal gut microbiota	Delayed developmental milestones (P2-21) ↓ Locomotion ↓ Social behaviors ↑ Repetitive behavior (P28)	Wistar rats/male	600 mg/kg, E12.5, i.p.	[84]
↑ 5-HT in serum (N/A)	Abnormal gut microbiota Altered several metabolic pathways caused by microbial dysbiosis (P21, 28, 35)	↑ Anxiety ↓ Social behaviors (6 wks)	C57BL/6 mice/male	500 mg/kg, E12.5, i.p.	[85]
↑ 5-HT in the hippocampus, cerebellum, and plasma (P50)			SD rats/male	800 mg/kg, E9, oral feeding	[14]
↓ 5-HT and its precursors (L-tryptophan and 5-HTP) and final metabolites (5-HIAA) in the colon, feces, serum, cerebellum, and PFC (8 wks)	↓ GABA in serum, cerebellum, and PFC ↑ Glutamate in serum, cerebellum, and PFC ↓ Acetylcholine in serum and PFC Abnormal gut microbiota (8 wks)	↓ Novel object recognition ↓ Social behavior ↓ Locomotion ↑ Depression ↑ Repetitive behavior (8 wks)	Wistar rats/male	500 mg/kg, E12.5, i.p.	[86]
↓ 5-HT in the PFC and amygdala ↑ 5-HT turnover (5-HIAA/5-HT) in the PFC and amygdala ↓ 5-HT levels and numbers of 5-HT positive cells in the ileum (P28)	↑ Epithelial loss and intestinal inflammation in the ileum ↑ Neuroinflammation in the dorsal hippocampus (P28)	↓ Social interaction (P28)	BALB/c mice/both gender	500 or 600 mg/kg, E11.5, s.c.	[22]

Table 1. *Cont.*

Monoamines-Related Endophenotypes (Age of Animals during Analysis)	Other Molecular, Cellular and Physiological Phenotypes (Age of Animals during Analysis)	Behavioral Phenotypes (Age of Animals during Analysis)	Animals/ Gender	VPA Exposure	References
↓ 5-HT in the PFC and ileum (P50)	↓ Gastrointestinal tract motility ↓ Activity of mitochondrial enzyme complex-I, II, V in PFC ↑ BBB permeability ↑ Oxidative stress ↑ Nitrosative stress (nitrate/nitrite) ↑ Inflammation in the brain and ileum ↑ Calcium (P50)	↓ Social behaviors (P44, 45) ↓ Exploratory behaviors (P47) ↑ Locomotion (P49) ↑ Repetitive behaviors (P46) ↑ Anxiety (P47, 48)	Wistar rats/male	500 mg/kg, E12.5, s.c.	[87–89]
↑ 5-HT in the hippocampus (P110)	↑ Inflammation: ↓Glutathione and catalase; ↑Total nitrite ↓ Cerebellar Purkinje cells ↑ Neurodegenerative chromatolysis in the cerebellum (P110)	Developmental delay (P9–12) Impaired nociceptive sensation ↑ Locomotion ↓ Rearing and hole poking ↑ Anxiety ↓ Social behaviors (P90–110)	Rats/male	600 mg/kg, E12.5, i.p.	[92]
↑ 5-HT in the hippocampus (P40)	↑ Oxidative stress ↓ Number of Purkinje cell layer in the cerebellum (P40)	↓ Locomotion ↓ Motor coordination ↓ Social behavior ↓ Spatial learning and memory ↑ Anxiety Impaired nociceptive sensation Delayed negative geotaxis (P40)	BALB/c mice/both gender	400 mg/kg, P14, s.c.	[18]
↓ 5-HT in the midbrain ↑ 5-HT in the cerebellum and pons (P22)	↑ Free amino acid in the frontal cortex (P22)	↓ Social behaviors ↓ Spatial memory (P21)	Albino rats/male	800 mg/kg, E12.5, oral feeding	[99]
↓ 5-HT in the hippocampus n.s. SERT and 5-HIAA among a variety of brain regions (P50)		↑ Locomotion ↓ Social behaviors (P31)	Wistar rats/male	500 mg/kg, E9, i.p.	[91]
↑ TPH-positive neurons in the raphe magnus nucleus (P28) ↑ SERT binding in the amygdala (P60)	↑ Amplitude of mEPSC in the lateral amygdala ↓ Paired pulse facilitation ratio in the thalamo-amygdaloid synapses (P35)	↓ Social behaviors Impaired fear memory extinction (P28–35)	SD rats/male	500 mg/kg, E12.5, i.p.	[96]
↑ Elevation of 5-HT after feeding in the frontal cortex (P56–105)		↑ Locomotion (P18) ↓ Body weight (4 wks) Abnormal circadian rhythm and activity (P28–43)	Wistar rats/both gender	800 mg/kg, E9.5. oral feeding	[100]
↓ Serotonin receptor, *Htr1b, Htr1d, Htr3a* mRNA in the hippocampus (P35)	Altered expression in genes encoding cholinergic and adrenergic receptors in the cortex and cerebellum (P35)	N/A	Wistar rats/both gender	800 mg/kg, E11, oral feeding	[93]
↓ or absent 5-HT neuronal differentiation in embryos (48 and 72 hpf)	Lack of Mauthner neurons ↓ Proneural gene *ascl1b* (28 and 32 hpf)	N/A	Zebrafish/N/A	0.625 mM, from 50% epiboly to 27 hpf, incubated in normal medium 0.625 mM, from 24 to 48 hpf, incubated in normal medium	[98]

Table 1. *Cont.*

Monoamines-Related Endophenotypes (Age of Animals during Analysis)	Other Molecular, Cellular and Physiological Phenotypes (Age of Animals during Analysis)	Behavioral Phenotypes (Age of Animals during Analysis)	Animals/ Gender	VPA Exposure	References
Abnormal migration of 5-HT in adult dorsal raphe nucleus n.s. in number of 5-HT positive neurons (P50)	↓ Shh mRNA level at E9	N/A	Wistar rats/male	800 mg/kg, E9, oral feeding	[95]
Abnormal migration of 5-HT-positive neurons in the pons Abnormal navigation of 5-HT-positive neurons in the boundary of midbrain and hindbrain (E16)	Disorganization of cortical lamination (E16)	N/A	SD rats/both gender	800 mg/kg, E9 or E11, oral feeding	[94]
Abnormal 5-HT neurons distribution (dorsal tangential migration) in the rostral raphe nucleus (E15.5)	↓ *Shh* mRNA expression around the isthmus (E11.5)	N/A	Wistar rats/both gender	800 mg/kg, E9.5, oral feeding	[97]
↑ The excitation/inhibition ratio by enhancing glutamatergic synaptic transmission of the 5-HT neurons in the dorsal raphe nucleus ↑ Frequency of mEPSCs of the 5-HT neurons in the dorsal raphe nucleus (6–8 wks)	↓ Spike-timing-dependent long-term potentiation in the dorsal raphe nucleus (6–8 wks)	↑ Anxiety (8 wks) ↓ Body weight (P1)	Long Evans/male	400 mg/kg, E12.5, s.c.	[46]

All the animals in the studies listed in this table received a single administration of VPA. Abbreviations: ↑, increased; ↓, decreased; 5-HIAA, 5-hydroxyindoleacetic acid; 5-HT, 5-hydroxytryptamine (serotonin); 5-HTP, 5-hydroxytryptophan; BBB, blood–brain barrier; dpf, days post-fertilization; E, embryonic day; GABA, γ-Aminobutyric acid; hpf, hours post-fertilization; Htr, 5-hydroxytryptamine receptors; IL, interleukin; m/o, months old, i.p., intraperitoneal injection; mEPSCs, miniature excitatory postsynaptic currents; N/A, not applicable; P, postnatal day; PFC, prefrontal cortex; s.c., subcutaneous injection; SERT, serotonin transporter; Shh, sonic hedgehog; TH, tyrosine hydroxylase; TNF, tumor necrosis factor; TPH, tryptophan hydroxylase; VPA, valproic acid; wks, weeks old.

2.1.6. Microbiota–Gut–Brain Axis and Serotoninergic Systems in VPA-Induced ASD Models

Peripheral 5-HT is mainly produced by enterochromaffin cells from dietary tryptophan in the gastrointestinal tract, and is taken up by platelets or SERT on serotoninergic presynaptic terminals. Peripheral 5-HT regulates gastrointestinal motility and is involved in immune function [103]. An intriguing issue is how peripheral 5-HT affects the central nervous system, given that 5-HT in the central and peripheral nervous systems is separated by the blood–brain barrier. 5-HT may influence enteric nerves, vagal afferent activity, or inflammatory responses, and then modulate the central nervous system indirectly [58]. Another potential mechanism of peripheral 5-HT influence neurodevelopmental process is mediated by the microbiota–gut–brain axis. Previous studies have demonstrated the causal relationship of the microbiota–gut–brain axis and the pathogenesis of ASD, and ~70% of ASD patients show comorbid gastrointestinal disturbances [58]. 5-HT has been shown to modulate bacterial motility and gene expression profile of bacteria in vivo [58,104], indicating complex feedback interactions between 5-HT and microbiota in the gastrointestinal tract. Conversely, the microbiota is known not only to synthesize 5-HT, but also to modulate the serotonergic system in the gastrointestinal tract of the host, suggesting that 5-HT may be a communication molecule of gut–brain interactions. In germ-free mice, TPH (rate-limiting enzyme of 5-HT) is upregulated, and the 5-HT level in the hippocampus is reduced with enhanced anxiety [105,106]. Animal studies have found the abnormal composition of microbiota in ASD-like models, including VPA-treated rodents [84–86]. The VPA-treated rodents also showed altered 5-HT, L-tryptophan, 5-HTP, and 5-HIAA levels in serum, colon, feces, cerebellum, and prefrontal cortex [84–86]. Given the important role of the microbiota–gut–brain axis in ASD pathophysiology, it would be of interest to see if the modulation of the gut–brain axis by serotonin may be involved in the etiology of ASD.

2.2. Catecholamines

2.2.1. Neural Development of Catecholaminergic Neurons

DA and norepinephrine (NE) are the primary catecholamines that are released from groups of catecholamine neurons in different brain regions (A1–A17). An additional three adrenaline-containing groups (C1–C3) are distributed in the telencephalon, mesencephalon, and rhombencephalon [107]. Catecholamines are important neurotransmitters that are essential to brain function. Dysfunction of catecholamines is involved in the pathophysiology of neurodevelopmental and neurodegenerative disorders, including ASD and Parkinson's disease.

Although dopaminergic neurons comprise a small population of neurons in the brain, they profoundly control brain function through the extensive innervations of their network [107]. Among the different groups of dopaminergic neurons, substantia nigra pars compacta (SNc, A9) and ventral tegmental area (VTA, A10) play important roles in the control of motor, reward, motivation, and emotion. Progenitors of dopaminergic neurons are originated from the ventral midline of the floor/basal plate that is located near to the midbrain–hindbrain boundary (MHB), or isthmus. In mice, SNc dopaminergic neurons are generated before E11 and their neurogenesis is peaked at E11–E12, whereas the peak of neurogenesis of VTA dopaminergic neurons occurs at E12–E13 [108]. After exiting the cell cycle progression, midbrain dopaminergic neurons migrate radially and/or tangentially from the ventricular zone toward the pial surface, and this migratory process continues until the first week after birth. [109,110].

On the other hand, cell bodies of norepinephrinergic neurons are clustered in the medulla oblongata, pons, and midbrain. The locus coeruleus (LC) is known as the major source of NE in the brain, which is located near the floor of the fourth ventricle [70]. Progenitors of LC neurons are mainly derived from rhombomere 1, and LC neurons are born early and migrate toward the basal plate before E10.5 [111]. Neurogenesis of norepinephrinergic neurons occurs predominantly between E10.5–E12.5. Norepinephrinergic neurons are functionally active by birth [112], and they gradually mature during postnatal periods [113].

2.2.2. Neurochemical Properties of Catecholamines

All the catecholamines are derived from L-tyrosine. The rate-limiting enzyme, tyrosine hydroxylase (TH), converts L-tyrosine to L-DOPA, and AADC subsequently converts L-DOPA to DA. For neurons that synthesize NE or epinephrine, dopamine-β-hydroxylase catabolizes DA into NE, and phenylethanolamine N-methyltransferase further converts NE into epinephrine. The degrative metabolism of catecholamines is mediated by MAO and catechol-O-methyltransferase (COMT). DA and NE are transported into synaptic vesicles by VMAT2. After being released into the synaptic cleft, NE and DA are recycled back into the presynaptic terminal, respectively, through the uptake by norepinephrine transporter (NET) and dopamine transporter (DAT). The DA receptor family consists of two members, D_1-like and D_2-like receptors. Activation of D_1-like receptor increases cAMP levels, whereas activation of D2-like receptor decreases cAMP levels in DA responsive cells. As for the NE receptor, NE binds to three NE receptor families (α1, α2, β) [70]. See Figure 2 for the summary of catecholaminergic neurotransmission.

2.2.3. Clinical Evidence of Abnormalities of Catecholaminergic Systems Related to ASD

An early study documented dysregulation of catecholamines in ASD patients whose plasma level of NE is increased, but dopamine-β-hydroxylase is decreased [114]. Clinical study has shown reduced DA level in the mPFC of medication-free ASD patients, which suggests an aberrant function of the dopaminergic systems in ASD [115]. Genetic linkage studies have revealed that mutations of many dopaminergic systems regulators are associated with ASD, including dopamine receptor *DRD1* [116], *DRD2* [117], *DRD3* [118], *DRD4* [119], and *DAT* [120,121]. The dopaminergic systems are of particular interest with respect to ASD pathophysiology, because it participates in the reward system that is involved in social motivation and social interaction [122,123]. Because of the multifaceted roles of the dopaminergic systems in reward prediction, decision making, and motivation [124–126] that are affected in ASD patients, it is imperative to study the pathophysiological changes in dopaminergic systems in ASD.

As for the norepinephrinergic systems, previous studies have reported abnormalities of resting-state functional connectivity in LC and elevated tone of NE activity in ASD children [127,128]. Another positron emission tomography study found a correlation between the binding signals of D1R and NET and ASD symptoms [129]. Moreover, antipsychotic drugs targeting NE receptors have been shown to have beneficial effects on ASD symptoms [130].

2.2.4. Abnormalities of Catecholaminergic Systems in VPA-Induced ASD Models

Alterations of dopaminergic systems have been extensively reported in studies of VPA-induced ASD animal models. Reduction in the rate-limiting enzyme of TH has been found in the striatum of VPA-treated rats [131]. The reduced TH level presumably would result in decreased DA level in the striatum. However, another study has reported an elevated DA level in the striatum of VPA-treated mice [132]. In fact, the increased DA levels were found not only in the striatum, but also in the frontal cortex, cerebellum, and pons [14,99,132]. By contrast, decreased DA levels were found in the hippocampus, midbrain, and serum of VPA-treated animals [86,99]. The variations in the changes in DA levels in different brain regions may result from the differences in DA turnover and/or the expression of DA-related signaling molecules in different brain regions, including DA receptor, acetylation of DAT, and phosphorylation of DARPP-32 in response to social stimulus [66,93,133–135]. Along this line, an interesting question is whether the dopaminergic systems may be affected by the external environment. Indeed, elevated DA levels and reduced DA turnover were found in VPA mice raised in different social environments [132].

In the norepinephrinergic systems, reduced NE levels are detected in the hippocampus, midbrain, and serum; whereas increased NE levels are found in the frontal cortex, cerebellum, and pons [84,99]. Furthermore, increased NET expression and acetylation levels have been reported in VPA-treated rats [66]. Regarding VPA-induced changes in neuronal development, abnormal migration of TH-positive neurons have been found in the pons, and the boundary of the midbrain and hindbrain [94], which implicates that pathological changes in the catecholamine systems may result from VPA-induced developmental abnormalities.

2.2.5. Neuroanatomical Phenotypes Related to Altered Catecholamine Systems in VPA-Induced ASD Models

In addition to altered metabolic, abnormal migration, and distribution of TH-positive DA neurons in the SNc and VTA have been found in VPA-treated rats and chickens [94,136]. Regarding dopaminergic projections, a recent study has used the tissue clearing iDISCO method combined with a laser light-sheet confocal microscope to investigate whole-brain dopaminergic axonal projections. Ádám et. al. (2020) have found reduced ventrobasal telencephalic projections from the VTA along with a reduced number of DA neurons in the VTA of VPA-treated mice. By contrast, an increased number of DA neurons in the substantia nigra was observed in VPA-treated mice. These findings suggest region- and cell-type specificity in VPA-induced pathophysiology and further suggest potential compensatory mechanisms among different dopaminergic neuronal populations in the midbrain [137].

2.2.6. Behavioral Phenotypes Related to Altered Catecholamine Systems in VPA-Induced ASD Models

At the behavioral level, consistent behavioral phenotypes have been found along with catecholaminergic abnormalities in VPA-treated rodents, including reduced social behaviors, increased repetitive, depressive and anxious behaviors, and impaired memory [14,66,84,86,93,99,131,133,134,138]. Impaired cognitive flexibility and temporal processing have also been observed in VPA-treated rodents, implicating abnormal functions of the prefrontal cortex [131,132]. Reduced levels of hyperlocomotion in response to methamphetamine further support the abnormalities of dopaminergic systems in VPA-treated mice [135]. Risperidone and aripirazole are two atypical antipsychotics for alleviating ASD syndromes in VPA-treated mice, including impairments in social behaviors, vocalization, and recognition memory [64,65]. Note that risperidone inhibits D2R and 5-HT$_{2A}$ receptors, whereas aripirazole acts as an agonist of 5-HT$_{2A}$ receptor and also as a partial agonist of D2R. The causality between dopaminergic and behavioral phenotypes awaits further investigation. On the other hand, hyperactivity, repetitive behaviors, impaired social interaction, and recognition memory can be alleviated by atomoxetine, a blocker of NET. The neurochemical mechanisms by which altered catecholaminergic systems contribute to ASD pathogenesis require further studies.

2.2.7. Altered Reciprocal Interactions of Microbiota and Catecholamine Systems in VPA-Induced ASD Models

Decreased levels of DA and NE in serum have been found in VPA-treated rats along with the abnormal composition of gut microbiota [84,86]. NE has been shown to regulate various aspects of bacteria, from bacterial growth, migration to the gene expression profile of bacteria [58]. Some bacteria are found to express monoamine transporter on their plasma membrane [139]. Other bacteria can convert host-derived inactive forms of NE and DA into biologically active forms [140], suggesting a role of microbiota in modulating the activity of catecholaminergic systems. The mechanisms underlying reciprocal interactions between peripheral catecholaminergic systems and microbiota–gut–brain axis and how dysfunctional microbiota–gut–brain axis affect central catecholaminergic systems await future studies.

2.3. Histamine

2.3.1. Neural Development of Histaminergic Neurons

In the postnatal stages, histaminergic neurons are mainly distributed in the tuberomammillary nucleus of the hypothalamus, and they extensively innervate a variety of brain regions, including the cerebral cortex, hippocampus, preoptic area, striatum, and thalamus [141]. Histidine decarboxylase (HDC) is an enzyme that catalyzes the decarboxylation of histidine to form histamine. Neurogenesis of HDC-positive neurons in the tuberomammillary nucleus begins from E13-18 and is peaked at E16 in the rat brain [142]. Notably, there is a transient histaminergic system during prenatal brain development. Transient histamine-immunoreactive neurons are located in the mesencephalon, metencephalon, and rhombencephalon, and they are no longer detected at the end of embryonic stages in the rat brain. The peak of neurogenesis in the brain and the highest level of histamine coincide at ~E14, implicating a potential role of histamine in regulating neurogenesis in the rat brain [143]. Later studies have revealed a regulatory role of histamine in promoting neural proliferation, cell-type specific differentiation, and axogenesis [143]. It has also been reported that histamine systems can cross-talk with other neurotransmitter systems by forming heteroreceptors (see below) and by other unknown mechanisms. For example, the expression levels of dopamine, D2R, and D3R are increased in *HDC* knockout mice [144,145].

2.3.2. Neurochemical Properties of Histamine

The histaminergic systems play important roles in modulating the sleep–wake cycle, reward, neuroinflammation, emotion, and learning and memory [146]. Histamine is converted from L-histidine by histidine decarboxylase (HDC), and is packaged into synaptic vesicles by VMAT2 before releasing. Although there are no high-affinity reuptake mechanisms for uptaking histamine, histamine N-methyltransferase (HNMT) is responsible for the clearance of histamine in the synaptic cleft [70,141]. In the brain, neuronal expression of histamine receptor H1 (H1R), histamine receptor H2 (H2R), and histamine receptor H3 (H3R) are engaged in the regulation of different neurological functions. For example, H1R and H3R signalings are important for regulating the sleep–wake cycle, whereas H2R signaling is involved in aggression [147].

2.3.3. Clinical Evidence of Abnormalities of Histaminergic Systems Related to ASD

Clinical studies have linked abnormal histaminergic systems to ASD pathogenesis. Human postmortem brain study has found increased HNMT levels and histamine receptors in the dorsolateral prefrontal cortex of ASD patients [148], suggestive of altered histaminergic systems in ASD pathophysiology. The histaminergic system transiently but profoundly affects the developmental processes of other neural systems, including serotonergic and dopaminergic systems [144,149]. Furthermore, histamine not only modulates the proliferation and differentiation of neural stem cells, but also regulates the biological functions of astrocytes and microglia that are involved in neuroinflammation [144,149]. Taken together, histaminergic dysfunction may contribute to autistic abnormalities either directly as a neuromodulator or indirectly through affecting neuronal development (Table 2).

Table 2. Summary of valproic acid (VPA)-induced teratogenicity in catecholaminergic and histaminergic systems.

Monoamines-Related Endophenotypes (Age of Animals during Analysis)	Other Molecular, Cellular, and Physiological Phenotypes (Age of Animals during Analysis)	Behavioral Phenotypes (Age of Animals during Analysis)	Animals	VPA Exposure	References
Catecholaminergic Systems: Dopamine and Norepinephrine					
↓ TH immunoreactivity in the striatum (P30)	N/A	↓ Vocalization (P11) ↓ Cognitive flexibility (P29) ↑ Repetitive behavior (P29) ↓ Play behavior (P30)	Wistar rats/male	400 mg/kg, E12.5, i.p.	[131]
↓ TH-positive neurons (5 dpf) ↓ Dopamine β-Hydroxylase (5 dpf and 6 mpf) ↓ DOPAC (5 dpf and 6 mpf) ↓ NE (5 dpf and 6 mpf)	N/A	↓ Locomotion in larvae (5 dpf) Abnormal dark-flash response Abnormal social behaviors (6 mpf)	Zebrafish/male	25 μM, from 10 hpf to 24 hpf, incubated in embryonic medium	[150]
↓ DA in serum (8 wks)	↓ GABA in serum, cerebellum and PFC ↑ Glutamate in serum, cerebellum and PFC ↓ Acetylcholine in serum and PFC Abnormal gut microbiota (8 wks)	↓ Novel object recognition ↓ Social behavior ↓ Locomotion ↑ Depression ↑ Repetitive behavior (8 wks)	Wistar rats/male	500 mg/kg, E12.5, i.p.	[86]

Table 2. *Cont.*

Monoamines-Related Endophenotypes (Age of Animals during Analysis)	Other Molecular, Cellular, and Physiological Phenotypes (Age of Animals during Analysis)	Behavioral Phenotypes (Age of Animals during Analysis)	Animals	VPA Exposure	References
↑ DA in the frontal cortex (P50)	N/A	N/A	SD rats/male	800 mg/kg, E9, oral feeding	[14]
↑ DA level in the dorsal striatum ↓ DA turnover (DOPAC/DA) in the dorsal striatum (1 day after behavioral test)	N/A	Abnormal temporal processing (P60)	CrlFcen:CF1 mice/both gender	600 mg/kg, E12.5, s.c.	[132]
↓ DA in the hippocampus and midbrain ↑ DA in the frontal cortex, cerebellum and pons ↓ NE in the hippocampus and the midbrain ↑ NE in frontal cortex, cerebellum and pons (P22)	↑ Free amino acid in the frontal cortex (P22)	↓ Social behaviors ↓ Spatial memory (P21)	Albino rats/male	800 mg/kg, E12.5, oral feeding	[99]
↑ DA turnover (DOPAC/DA) in the piriform cortex (P60)	↑ c-fos immunoreactivity in the piriform cortex Altered pattern of brain glucose metabolism (P60)	↓ Social behaviors ↑ Repetitive behavior ↑ Anxiety (8 wks)	CrlFcen:CF1 mice/male	600 mg/kg, E12.5, s.c.	[138]
↓ DARPP-32 phosphorylation in response to social stimulus in the nucleus accumbens (P60)	↑ PPARα in the VTA ↑ Vglut and Vglut/Vgat ratio in the caudate-putamen of male mice ↓ PSD95 expression in the caudate-putamen of male mice ↑ NR2B in the caudate-putamen of male mice (P60)	Delayed negative geotaxis ↓ Social behaviors ↑ Repetitive behavior ↑ Depression in male mice ↑ Anxiety (P48–53)	SD rats/both gender	500 mg/kg, E12.5, i.p.	[133]
↑ D1R in the nucleus accumbens and hippocampus ↑ D2R in the nucleus accumbens (P35–40 and P90–95)	↓ Resting potential of medium spiny neurons in the striatum ↓ Excitability ↓ Inwardly rectifying potassium currents density (P30–35)	↓ Social behaviors n.s. amphetamine-induced hyperlocomotion (P35–40)	Wistar rats/male	500 mg/kg, E12.5, i.p.	[134]
↓ *Drd1a* mRNA in the cerebellum ↓ *Drd2* mRNA in the cerebral cortex (P35)	Altered expression in genes encoding cholinergic and adrenergic receptors in the cortex and cerebellum	N/A	Wistar rats/both gender	800 mg/kg, E11, oral feeding	[93]

Table 2. *Cont.*

Monoamines-Related Endophenotypes (Age of Animals during Analysis)	Other Molecular, Cellular, and Physiological Phenotypes (Age of Animals during Analysis)	Behavioral Phenotypes (Age of Animals during Analysis)	Animals	VPA Exposure	References
↓ NE in serum (P28)	↑ Proinflammatory cytokines: IL-17A, TNF-α, IL-6 ↓ GABA in serum Abnormal gut microbiota	Delayed developmental milestones (P2–21) ↓ Locomotion ↓ Social behaviors ↑ Repetitive behavior (P28)	Wistar rats/male	600 mg/kg, E12.5, i.p.	[84]
↑ DAT expression and acetylation of histone H3 bound to *Slc6a3* gene in the PFC ↑ NET expression and acetylation of histone H3 bound to *Slc6a2* gene in the PFC (4 wks)	N/A	Hyperactivity ↑ Repetitive rearing (4 wks)	SD rats/male	400 mg/kg, E12, s.c.	[66]
↓ D1R and D2R in the prefrontal cortex ↓ METH-induced DA release in the prefrontal cortex (8 wks)	↓ c-fos positive neurons in the prefrontal cortex after METH administration (8 wks)	↓ METH-induced hyperlocomotion (8 wks)	ICR mice/male	500 mg/kg, E12.5, i.p.	[135]
Abnormal migration of TH-positive neurons in the pons Abnormal navigation of TH-positive neurons in the boundary of midbrain and hindbrain (E16)	Disorganization of cortical lamination (E16)	N/A	SD rats/both gender	800 mg/kg, E9 or E11, oral feeding	[94]
Abnormal distribution of TH-positive neurons in substantia nigra and VTA ↓ DRD1 and GRIN2A expression in the septum (P2)	N/A	N/A	Chicken (*Gallus gallus*) (N/A)	35 μmoles, E14, dropping VPA solution into the air sac	[136]
↓ Widening TH-positive mesotelecephalic axonal fascicles ↓ TH-positive cells in the VTA ↓ TH-positive terminals in ventrobasal telencephalic region ↑ TH-positive cells in the substantia nigra (P7)	N/A	N/A	C57BL/6/both gender	400 mg/kg, E13.5, s.c.	[137]

Table 2. *Cont.*

Monoamines-Related Endophenotypes (Age of Animals during Analysis)	Other Molecular, Cellular, and Physiological Phenotypes (Age of Animals during Analysis)	Behavioral Phenotypes (Age of Animals during Analysis)	Animals	VPA Exposure	References
		Histamine			
↓ Histamine receptor H3 ↓ Histidine decarboxylase ↓ Histaminergic neurons (5 dpf and 6 mpf)	N/A	↓ Locomotion in larvae (5 dpf) Abnormal dark-flash response Abnormal social behaviors (6 mpf)	Zebrafish/male	25 μM, from 10 hpf to 24 hpf, incubated in embryonic medium	[150]

All the animals in the studies listed in this table received a single administration of VPA. Abbreviations: ↑, increased; ↓, decreased; D1R, dopamine D1 receptor; D2R, dopamine D2 receptor; DAT, dopamine transporter; DOPAC, 3,4-dihydroxyphenylacetic acid; dpf, days post fertilization; E, embryonic day; GABA, γ-Aminobutyric acid; hpf, hours post-fertilization; IL, interleukin; i.p., intraperitoneal injection; METH, methamphetamine; mpf, months post-fertilization; m/o, months old; n.s., no significant difference; N/A, not available; NR2B, N-methyl D-aspartate receptor subtype 2B; P, postnatal day; PFC, prefrontal cortex; PPARα, peroxisome proliferator-activated receptor alpha; PSD95, postsynaptic density protein 95; s.c., subcutaneous injection; TH, tyrosine hydroxylase; TNF, tumor necrosis factor; VPA, valproic acid; VTA, ventral tegmental area; wks, weeks old.

2.3.4. Abnormalities in the Histaminergic System in VPA-Induced ASD Models

Similar to what is observed in clinical studies, abnormalities in the histaminergic system have been found in VPA-induced ASD-like animal models as well. Reduced levels of histamine receptor H3 (H3R), HDC and histaminergic neurons have been found in VPA-treated zebrafish [150]. Notably, H3R antagonists have therapeutic effects on VPA-treated mice in reducing inflammation and ASD-like behaviors [151–153]. H3R antagonists also alleviate ASD-like symptoms in other ASD animal models [154,155]. Given that Gi-coupled H3R can indirectly modulate other neurotransmitter systems by forming heteroreceptors, the involvement of histaminergic systems in ASD pathophysiology may have been underestimated. Until now, there is no study revealing alterations of histaminergic systems except VPA-treated zebrafish. No information is yet available for potential changes in the central and peripheral systems in mammals.

Histamine is not only synthesized in the central nervous system, but also in the peripheral mast cells and gastric enterochromaffin-like cells that play important roles in the microbiota–gut–brain axis [58]. Moreover, some microbiota in the gastrointestinal tract of the host also synthesize histamine. Although histamine hardly penetrates the blood–brain barrier, in addition to the microbiota–gut–brain axis, histamine may indirectly affect the nervous system via the immune system and the vagal system [58,156]. It is worthwhile to note that histamine is a key player in many immune responses, because maternal immune activation is a risk factor for offspring to develop ASD [56]. It is imperative to decipher the mechanisms by which histamine by itself and/or through regulating other neurotransmission systems contribute to the pathophysiology of ASD.

3. Conclusions

Over the decades of studies, the VPA-induced ASD-like animal models have provided important information about the pathophysiology of ASD. The monoaminergic systems, key players of neuromodulation in the brain, are extensively affected in the pathophysiology of ASD as demonstrated in the VPA-induced ASD model. In this review, we summarize and discuss the pathological changes in monoaminergic systems related to ASD etiology. Given the pharmacological nature of VPA that can regulate cellular signaling as well as chromatin remodeling as an HDAC inhibitor, the information gained from VPA model studies with respect to monoaminergic systems would not only help clarify the pathogenic trajectories of

ASD at the cellular and epigenetic levels, but it may also help develop therapeutic reagents for ASD [37].

Author Contributions: H.-Y.K. and F.-C.L. wrote the manuscript. All authors have read and agreed to the published version of the manuscript.

Funding: This review was supported by the Ministry of Science and Technology-Taiwan grants MOST110-2636-B-A49-001, MOST111-2636-B-A49-007 (H.-Y.K.), MOST107-2320-B-010-041-MY3, MOST110-2320-B-A49A-532-MY3, MOST111-2321-B-001-011, MOST110-2326-B-A49A-504, and Featured Areas Research Center Program Grant from the Ministry of Education-Taiwan (F.-C.L.).

Institutional Review Board Statement: Not applicable.

Informed Consent Statement: Not applicable.

Conflicts of Interest: The authors declare no conflict of interest.

References

1. Maenner, M.J.; Shaw, K.A.; Baio, J.; Washington, A.; Patrick, M.; DiRienzo, M.; Christensen, D.L.; Wiggins, L.D.; Pettygrove, S.; Andrews, J.G.; et al. Prevalence of Autism Spectrum Disorder Among Children Aged 8 Years - Autism and Developmental Disabilities Monitoring Network, 11 Sites, United States, 2016. *MMWR. Surveill. Summ.* **2020**, *69*, 1–12. [CrossRef]
2. Association, A.P. *Diagnostic and Statistical Manual of Mental Disorders: DSM-5*, 5th ed.; American Psychiatric Publishing: Arlington, VA, USA, 2013.
3. Bourgeron, T. From the genetic architecture to synaptic plasticity in autism spectrum disorder. *Nat. Rev. Neurosci.* **2015**, *16*, 551–563. [CrossRef] [PubMed]
4. Guang, S.; Pang, N.; Deng, X.; Yang, L.; He, F.; Wu, L.; Chen, C.; Yin, F.; Peng, J. Synaptopathology involved in autism spectrum disorder. *Front. Cell. Neurosci.* **2018**, *12*, 470. [CrossRef] [PubMed]
5. Möhrle, D.; Fernández, M.; Peñagarikano, O.; Frick, A.; Allman, B.; Schmid, S. What we can learn from a genetic rodent model about autism. *Neurosci. Biobehav. Rev.* **2020**, *109*, 29–53. [CrossRef]
6. Modabbernia, A.; Velthorst, E.; Reichenberg, A. Environmental risk factors for autism: An evidence-based review of systematic reviews and meta-analyses. *Mol. Autism* **2017**, *8*, 13. [CrossRef]
7. Ximenes, J.C.M.; Verde, E.C.L.; da Graça Naffah-Mazzacoratti, M.; de Barros Viana, G.S. Valproic acid, a drug with multiple molecular targets related to its potential neuroprotective action. *Neurosci. Med.* **2012**, *3*, 107–123. [CrossRef]
8. Christensen, J.; Grønborg, T.K.; Sørensen, M.J.; Schendel, D.; Parner, E.T.; Pedersen, L.H.; Vestergaard, M. Prenatal valproate exposure and risk of autism spectrum disorders and childhood autism. *JAMA* **2013**, *309*, 1696–1703. [CrossRef] [PubMed]
9. Zhao, H.; Wang, Q.; Yan, T.; Zhang, Y.; Xu, H.J.; Yu, H.P.; Tu, Z.; Guo, X.; Jiang, Y.H.; Li, X.J.; et al. Maternal valproic acid exposure leads to neurogenesis defects and autism-like behaviors in non-human primates. *Transl. Psychiatry* **2019**, *9*, 267. [CrossRef]
10 Mimura, K.; Oga, T.; Sasaki, T.; Nakagaki, K.; Sato, C.; Sumida, K.; Hoshino, K.; Saito, K.; Miyawaki, I.; Suhara, T.; et al. Abnormal axon guidance signals and reduced interhemispheric connection via anterior commissure in neonates of marmoset ASD model. *Neuroimage* **2019**, *195*, 243–251. [CrossRef]
11. Binkerd, P.E.; Rowland, J.M.; Nau, H.; Hendrickx, A.G. Evaluation of valproic acid (VPA) developmental toxicity and pharmacokinetics in Sprague-Dawley rats. *Fundam. Appl. Toxicol.* **1988**, *11*, 485–493. [CrossRef]
12. Roullet, F.I.; Lai, J.K.; Foster, J.A. In utero exposure to valproic acid and autism–a current review of clinical and animal studies. *Neurotoxicol. Teratol.* **2013**, *36*, 47–56. [CrossRef] [PubMed]
13. Rodier, P.M.; Ingram, J.L.; Tisdale, B.; Nelson, S.; Romano, J. Embryological origin for autism: Developmental anomalies of the cranial nerve motor nuclei. *J. Comp. Neurol.* **1996**, *370*, 247–261. [CrossRef]
14. Narita, N.; Kato, M.; Tazoe, M.; Miyazaki, K.; Narita, M.; Okado, N. Increased monoamine concentration in the brain and blood of fetal thalidomide- and valproic acid-exposed rat: Putative animal models for autism. *Pediatr. Res.* **2002**, *52*, 576–579. [CrossRef] [PubMed]
15. Kim, K.C.; Kim, P.; Go, H.S.; Choi, C.S.; Yang, S.I.; Cheong, J.H.; Shin, C.Y.; Ko, K.H. The critical period of valproate exposure to induce autistic symptoms in Sprague-Dawley rats. *Toxicol. Lett.* **2011**, *201*, 137–142. [CrossRef]
16. Wagner, G.C.; Reuhl, K.R.; Cheh, M.; McRae, P.; Halladay, A.K. A new neurobehavioral model of autism in mice: Pre- and postnatal exposure to sodium valproate. *J. Autism Dev. Disord.* **2006**, *36*, 779–793. [CrossRef] [PubMed]
17. Yochum, C.L.; Dowling, P.; Reuhl, K.R.; Wagner, G.C.; Ming, X. VPA-induced apoptosis and behavioral deficits in neonatal mice. *Brain Res.* **2008**, *1203*, 126–132. [CrossRef] [PubMed]
18. Pragnya, B.; Kameshwari, J.S.; Veeresh, B. Ameliorating effect of piperine on behavioral abnormalities and oxidative markers in sodium valproate induced autism in BALB/C mice. *Behav. Brain Res.* **2014**, *270*, 86–94. [CrossRef]
19. Schneider, T.; Roman, A.; Basta-Kaim, A.; Kubera, M.; Budziszewska, B.; Schneider, K.; Przewlocki, R. Gender-specific behavioral and immunological alterations in an animal model of autism induced by prenatal exposure to valproic acid. *Psychoneuroendocrinology* **2008**, *33*, 728–740. [CrossRef]

20. Kazlauskas, N.; Seiffe, A.; Campolongo, M.; Zappala, C.; Depino, A.M. Sex-specific effects of prenatal valproic acid exposure on sociability and neuroinflammation: Relevance for susceptibility and resilience in autism. *Psychoneuroendocrinology* **2019**, *110*, 104441. [CrossRef]

21. Thornton, A.M.; Humphrey, R.M.; Kerr, D.M.; Finn, D.P.; Roche, M. Increasing endocannabinoid tone alters anxiety-like and stress coping behaviour in female rats prenatally exposed to valproic acid. *Molecules* **2021**, *26*, 3720. [CrossRef]

22. De Theije, C.G.; Koelink, P.J.; Korte-Bouws, G.A.; Lopes da Silva, S.; Korte, S.M.; Olivier, B.; Garssen, J.; Kraneveld, A.D. Intestinal inflammation in a murine model of autism spectrum disorders. *Brain. Behav. Immun.* **2014**, *37*, 240–247. [CrossRef]

23. Mowery, T.M.; Wilson, S.M.; Kostylev, P.V.; Dina, B.; Buchholz, J.B.; Prieto, A.L.; Garraghty, P.E. Embryological exposure to valproic acid disrupts morphology of the deep cerebellar nuclei in a sexually dimorphic way. *Int. J. Dev. Neurosci.* **2015**, *40*, 15–23. [CrossRef] [PubMed]

24. Konopko, M.A.; Densmore, A.L.; Krueger, B.K. Sexually dimorphic epigenetic regulation of brain-derived neurotrophic factor in fetal brain in the valproic acid model of autism spectrum disorder. *Dev. Neurosci.* **2017**, *39*, 507–518. [CrossRef] [PubMed]

25. Roux, S.; Bailly, Y.; Bossu, J.L. Regional and sex-dependent alterations in Purkinje cell density in the valproate mouse model of autism. *Neuroreport* **2019**, *30*, 82–88. [CrossRef]

26. Cho, H.; Kim, C.H.; Knight, E.Q.; Oh, H.W.; Park, B.; Kim, D.G.; Park, H.J. Changes in brain metabolic connectivity underlie autistic-like social deficits in a rat model of autism spectrum disorder. *Sci. Rep.* **2017**, *7*, 13213. [CrossRef]

27. Melancia, F.; Schiavi, S.; Servadio, M.; Cartocci, V.; Campolongo, P.; Palmery, M.; Pallottini, V.; Trezza, V. Sex-specific autistic endophenotypes induced by prenatal exposure to valproic acid involve anandamide signalling. *Br. J. Pharmacol.* **2018**, *175*, 3699–3712. [CrossRef]

28. Anshu, K.; Nair, A.K.; Kumaresan, U.D.; Kutty, B.M.; Srinath, S.; Laxmi, T.R. Altered attentional processing in male and female rats in a prenatal valproic acid exposure model of autism spectrum disorder. *Autism Res.* **2017**, *10*, 1929–1944. [CrossRef]

29. Gu, Y.Y.; Han, Y.; Liang, J.J.; Cui, Y.N.; Zhang, B.; Zhang, Y.; Zhang, S.B.; Qin, J. Sex-specific differences in the gut microbiota and fecal metabolites in an adolescent valproic acid-induced rat autism model. *Front. Biosci. (Landmark Ed.)* **2021**, *26*, 1585–1598. [CrossRef] [PubMed]

30. Li, X.; Large, C.H.; Ricci, R.; Taylor, J.J.; Nahas, Z.; Bohning, D.E.; Morgan, P.; George, M.S. Using interleaved transcranial magnetic stimulation/functional magnetic resonance imaging (fMRI) and dynamic causal modeling to understand the discrete circuit specific changes of medications: Lamotrigine and valproic acid changes in motor or prefrontal effective connectivity. *Psychiatry Res. Neuroimaging* **2011**, *194*, 141–148. [CrossRef]

31. Wandschneider, B.; Koepp, M.J. Pharmaco fMRI: Determining the functional anatomy of the effects of medication. *Neuroimage Clin.* **2016**, *12*, 691–697. [CrossRef]

32. Meador, K.; Reynolds, M.W.; Crean, S.; Fahrbach, K.; Probst, C. Pregnancy outcomes in women with epilepsy: A systematic review and meta-analysis of published pregnancy registries and cohorts. *Epilepsy Res.* **2008**, *81*, 1–13. [CrossRef] [PubMed]

33. Frisch, C.; Hüsch, K.; Angenstein, F.; Kudin, A.; Kunz, W.; Elger, C.E.; Helmstaedter, C. Dose-dependent memory effects and cerebral volume changes after in utero exposure to valproate in the rat. *Epilepsia* **2009**, *50*, 1432–1441. [CrossRef] [PubMed]

34. Favre, M.; Barkat, T.; Mendola, D.; Khazen, G.; Markram, H.; Markram, K. General developmental health in the VPA-rat model of autism. *Front. Behav. Neurosci.* **2013**, *7*, 88. [CrossRef]

35. Nicolini, C.; Fahnestock, M. The valproic acid-induced rodent model of autism. *Exp. Neurol.* **2018**, *299*, 217–227. [CrossRef] [PubMed]

36. Mast, T.J.; Cukierski, M.A.; Nau, H.; Hendrickx, A.G. Predicting the human teratogenic potential of the anticonvulsant, valproic acid, from a non-human primate model. *Toxicology* **1986**, *39*, 111–119. [CrossRef]

37. Kuo, H.Y.; Liu, F.C. Molecular pathology and pharmacological treatment of autism spectrum disorder-like phenotypes using rodent models. *Front. Cell. Neurosci.* **2018**, *12*, 422. [CrossRef] [PubMed]

38. Rosenberg, G. The mechanisms of action of valproate in neuropsychiatric disorders: Can we see the forest for the trees? *Cell. Mol. Life Sci.* **2007**, *64*, 2090–2103. [CrossRef]

39. Hussman, J.P. Letters to the editor: Suppressed GABAergic inhibition as a common factor in suspected etiologies of autism. *J. Autism Dev. Disord.* **2001**, *31*, 247–248. [CrossRef]

40. Rubenstein, J.L.; Merzenich, M.M. Model of autism: Increased ratio of excitation/inhibition in key neural systems. *Genes Brain Behav.* **2003**, *2*, 255–267. [CrossRef]

41. Kim, J.W.; Park, K.; Kang, R.J.; Gonzales, E.L.T.; Kim, D.G.; Oh, H.A.; Seung, H.; Ko, M.J.; Kwon, K.J.; Kim, K.C.; et al. Pharmacological modulation of AMPA receptor rescues social impairments in animal models of autism. *Neuropsychopharmacology* **2018**, *44*, 314–323. [CrossRef]

42. Wu, H.F.; Chen, Y.J.; Chu, M.C.; Hsu, Y.T.; Lu, T.Y.; Chen, I.T.; Chen, P.S.; Lin, H.C. Deep brain stimulation modified autism-like deficits via the serotonin system in a valproic acid-induced rat model. *Int. J. Mol. Sci.* **2018**, *19*, 2840. [CrossRef]

43. Rinaldi, T.; Silberberg, G.; Markram, H. Hyperconnectivity of local neocortical microcircuitry induced by prenatal exposure to valproic acid. *Cerebral. Cortex.* **2008**, *18*, 763–770. [CrossRef]

44. Rinaldi, T.; Perrodin, C.; Markram, H. Hyper-connectivity and hyper-plasticity in the medial prefrontal cortex in the valproic Acid animal model of autism. *Front. Neural Circ.* **2008**, *2*, 4. [CrossRef]

45. Rinaldi, T.; Kulangara, K.; Antoniello, K.; Markram, H. Elevated NMDA receptor levels and enhanced postsynaptic long-term potentiation induced by prenatal exposure to valproic acid. *Proc. Natl. Acad. Sci. USA* **2007**, *104*, 13501–13506. [CrossRef] [PubMed]
46. Wang, R.; Hausknecht, K.; Shen, R.Y.; Haj-Dahmane, S. Potentiation of glutamatergic synaptic transmission onto dorsal raphe serotonergic neurons in the valproic acid model of autism. *Front. Pharmacol.* **2018**, *9*, 1185. [CrossRef] [PubMed]
47. Taleb, A.; Lin, W.; Xu, X.; Zhang, G.; Zhou, Q.G.; Naveed, M.; Meng, F.; Fukunaga, K.; Han, F. Emerging mechanisms of valproic acid-induced neurotoxic events in autism and its implications for pharmacological treatment. *Biomed. Pharmacother.* **2021**, *137*, 111322. [CrossRef] [PubMed]
48. Chomiak, T.; Hu, B. Alterations of neocortical development and maturation in autism: Insight from valproic acid exposure and animal models of autism. *Neurotoxicol. Teratol.* **2013**, *36*, 57–66. [CrossRef] [PubMed]
49. Ryu, Y.K.; Park, H.Y.; Go, J.; Choi, D.H.; Choi, Y.K.; Rhee, M.; Lee, C.H.; Kim, K.S. Sodium phenylbutyrate reduces repetitive self-grooming behavior and rescues social and cognitive deficits in mouse models of autism. *Psychopharmacology* **2021**, *238*, 1833–1845. [CrossRef] [PubMed]
50. Kataoka, S.; Takuma, K.; Hara, Y.; Maeda, Y.; Ago, Y.; Matsuda, T. Autism-like behaviours with transient histone hyperacetylation in mice treated prenatally with valproic acid. *Int. J. Neuropsychopharmacol.* **2011**, *16*, 91–103. [CrossRef]
51. Moldrich, R.X.; Leanage, G.; She, D.; Dolan-Evans, E.; Nelson, M.; Reza, N.; Reutens, D.C. Inhibition of histone deacetylase in utero causes sociability deficits in postnatal mice. *Behav. Brain Res.* **2013**, *257*, 253–264. [CrossRef]
52. Watanabe, S.; Kurotani, T.; Oga, T.; Noguchi, J.; Isoda, R.; Nakagami, A.; Sakai, K.; Nakagaki, K.; Sumida, K.; Hoshino, K.; et al. Functional and molecular characterization of a non-human primate model of autism spectrum disorder shows similarity with the human disease. *Nat. Commun.* **2021**, *12*, 5388. [CrossRef]
53. Olde Loohuis, N.F.M.; Martens, G.J.M.; van Bokhoven, H.; Kaplan, B.B.; Homberg, J.R.; Aschrafi, A. Altered expression of circadian rhythm and extracellular matrix genes in the medial prefrontal cortex of a valproic acid rat model of autism. *Prog. Neuropsychopharmacol. Biol. Psychiatry* **2017**, *77*, 128–132. [CrossRef]
54. Barrett, C.E.; Hennessey, T.M.; Gordon, K.M.; Ryan, S.J.; McNair, M.L.; Ressler, K.J.; Rainnie, D.G. Developmental disruption of amygdala transcriptome and socioemotional behavior in rats exposed to valproic acid prenatally. *Mol. Autism* **2017**, *8*, 42. [CrossRef]
55. Lin, J.; Zhang, K.; Cao, X.; Zhao, Y.; Ullah Khan, N.; Liu, X.; Tang, X.; Chen, M.; Zhang, H.; Shen, L. iTRAQ-based proteomics analysis of rat cerebral cortex exposed to valproic acid before delivery. *ACS Chem. Neurosci.* **2022**. [CrossRef]
56. Careaga, M.; Murai, T.; Bauman, M.D. Maternal immune activation and autism spectrum disorder: From rodents to nonhuman and human primates. *Biol. Psychiatry* **2017**, *81*, 391–401. [CrossRef] [PubMed]
57. Kern, J.K.; Geier, D.A.; Sykes, L.K.; Geier, M.R. Relevance of neuroinflammation and encephalitis in autism. *Front. Cell. Neurosci.* **2015**, *9*, 519. [CrossRef]
58. Cryan, J.F.; O'Riordan, K.J.; Cowan, C.S.M.; Sandhu, K.V.; Bastiaanssen, T.F.S.; Boehme, M.; Codagnone, M.G.; Cussotto, S.; Fulling, C.; Golubeva, A.V.; et al. The microbiota-gut-brain axis. *Physiol. Rev.* **2019**, *99*, 1877–2013. [CrossRef] [PubMed]
59. Tung, E.W.; Winn, L.M. Valproic acid increases formation of reactive oxygen species and induces apoptosis in postimplantation embryos: A role for oxidative stress in valproic acid-induced neural tube defects. *Mol. Pharmacol.* **2011**, *80*, 979–987. [CrossRef] [PubMed]
60. Mabunga, D.F.; Gonzales, E.L.; Kim, J.W.; Kim, K.C.; Shin, C.Y. Exploring the validity of valproic acid animal model of autism. *Exp. Neurobiol.* **2015**, *24*, 285–300. [CrossRef]
61. Ergaz, Z.; Weinstein-Fudim, L.; Ornoy, A. Genetic and non-genetic animal models for autism spectrum disorders (ASD). *Reprod. Toxicol.* **2016**, *64*, 116–140. [CrossRef]
62. Spooren, W.; Lindemann, L.; Ghosh, A.; Santarelli, L. Synapse dysfunction in autism: A molecular medicine approach to drug discovery in neurodevelopmental disorders. *Trends Pharmacol. Sci.* **2012**, *33*, 669–684. [CrossRef] [PubMed]
63. Ornoy, A.; Weinstein-Fudim, L.; Ergaz, Z. Prevention or amelioration of autism-like symptoms in animal models: Will it bring us closer to treating human ASD? *Int. J. Mol. Sci.* **2019**, *20*, 1074. [CrossRef] [PubMed]
64. Kuo, H.Y.; Liu, F.C. Valproic acid induces aberrant development of striatal compartments and corticostriatal pathways in a mouse model of autism spectrum disorder. *FASEB J.* **2017**, *31*, 4458–4471. [CrossRef]
65. Hara, Y.; Ago, Y.; Taruta, A.; Hasebe, S.; Kawase, H.; Tanabe, W.; Tsukada, S.; Nakazawa, T.; Hashimoto, H.; Matsuda, T.; et al. Risperidone and aripiprazole alleviate prenatal valproic acid-induced abnormalities in behaviors and dendritic spine density in mice. *Psychopharmacology* **2017**, *234*, 3217–3228. [CrossRef]
66. Choi, C.S.; Hong, M.; Kim, K.C.; Kim, J.W.; Yang, S.M.; Seung, H.; Ko, M.J.; Choi, D.H.; You, J.S.; Shin, C.Y.; et al. Effects of atomoxetine on hyper-locomotive activity of the prenatally valproate-exposed rat offspring. *Biomol. Ther.* **2014**, *22*, 406–413. [CrossRef]
67. Hara, Y.; Ago, Y.; Taruta, A.; Katashiba, K.; Hasebe, S.; Takano, E.; Onaka, Y.; Hashimoto, H.; Matsuda, T.; Takuma, K. Improvement by methylphenidate and atomoxetine of social interaction deficits and recognition memory impairment in a mouse model of valproic acid-induced autism. *Autism Res.* **2016**, *9*, 926–939. [CrossRef]
68. Chaliha, D.; Albrecht, M.; Vaccarezza, M.; Takechi, R.; Lam, V.; Al-Salami, H.; Mamo, J. A systematic review of the valproic-acid-induced rodent model of autism. *Dev. Neurosci.* **2020**, *42*, 12–48. [CrossRef] [PubMed]

69. Hawthorne, A.L.; Wylie, C.J.; Landmesser, L.T.; Deneris, E.S.; Silver, J. Serotonergic neurons migrate radially through the neuroepithelium by dynamin-mediated somal translocation. *J. Neurosci.* **2010**, *30*, 420. [CrossRef]
70. Brady, S. *Basic Neurochemistry: Molecular, Cellular and Medical Aspects*; Elsevier: Amsterdam, The Netherlands, 2005.
71. Deneris, E.; Gaspar, P. Serotonin neuron development: Shaping molecular and structural identities. *Wiley Interdiscip. Rev. Dev. Biol.* **2018**, *7*, e301. [CrossRef]
72. Schain, R.J.; Freedman, D.X. Studies on 5-hydroxyindole metabolism in autistic and other mentally retarded children. *J. Pediatr.* **1961**, *58*, 315–320. [CrossRef]
73. Gabriele, S.; Sacco, R.; Persico, A.M. Blood serotonin levels in autism spectrum disorder: A systematic review and meta-analysis. *Eur. Neuropsychopharmacol.* **2014**, *24*, 919–929. [CrossRef]
74. Azmitia, E.C.; Singh, J.S.; Whitaker-Azmitia, P.M. Increased serotonin axons (immunoreactive to 5-HT transporter) in postmortem brains from young autism donors. *Neuropharmacology* **2011**, *60*, 1347–1354. [CrossRef] [PubMed]
75. Azmitia, E.C.; Singh, J.S.; Hou, X.P.; Wegiel, J. Dystrophic serotonin axons in postmortem brains from young autism patients. *Anat. Rec.* **2011**, *294*, 1653–1662. [CrossRef]
76. Oblak, A.; Gibbs, T.T.; Blatt, G.J. Reduced serotonin receptor subtypes in a limbic and a neocortical region in autism. *Autism Res.* **2013**, *6*, 571–583. [CrossRef]
77. Brandenburg, C.; Blatt, G.J. Differential serotonin transporter (5-HTT) and 5-HT(2) receptor density in limbic and neocortical areas of adults and children with autism spectrum disorders: Implications for selective serotonin reuptake inhibitor efficacy. *J. Neurochem.* **2019**, *151*, 642–655. [CrossRef]
78. Cook, E.H., Jr.; Courchesne, R.; Lord, C.; Cox, N.J.; Yan, S.; Lincoln, A.; Haas, R.; Courchesne, E.; Leventhal, B.L. Evidence of linkage between the serotonin transporter and autistic disorder. *Mol. Psychiatry* **1997**, *2*, 247–250. [PubMed]
79. Tordjman, S.; Gutknecht, L.; Carlier, M.; Spitz, E.; Antoine, C.; Slama, F.; Carsalade, V.; Cohen, D.J.; Ferrari, P.; Roubertoux, P.L.; et al. Role of the serotonin transporter gene in the behavioral expression of autism. *Mol. Psychiatry* **2001**, *6*, 434–439. [CrossRef]
80. Coutinho, A.M.; Oliveira, G.; Morgadinho, T.; Fesel, C.; Macedo, T.R.; Bento, C.; Marques, C.; Ataide, A.; Miguel, T.; Borges, L.; et al. Variants of the serotonin transporter gene (SLC6A4) significantly contribute to hyperserotonemia in autism. *Mol. Psychiatry* **2004**, *9*, 264–271. [CrossRef]
81. Nakai, N.; Nagano, M.; Saitow, F.; Watanabe, Y.; Kawamura, Y.; Kawamoto, A.; Tamada, K.; Mizuma, H.; Onoe, H.; Watanabe, Y.; et al. Serotonin rebalances cortical tuning and behavior linked to autism symptoms in 15q11-13 CNV mice. *Sci. Adv.* **2017**, *3*, e1603001. [CrossRef] [PubMed]
82. Veenstra-VanderWeele, J.; Muller, C.L.; Iwamoto, H.; Sauer, J.E.; Owens, W.A.; Shah, C.R.; Cohen, J.; Mannangatti, P.; Jessen, T.; Thompson, B.J.; et al. Autism gene variant causes hyperserotonemia, serotonin receptor hypersensitivity, social impairment and repetitive behavior. *Proc. Natl. Acad. Sci. USA* **2012**, *109*, 5469–5474. [CrossRef]
83. Wang, J.; Zheng, B.; Zhou, D.; Xing, J.; Li, H.; Li, J.; Zhang, Z.; Zhang, B.; Li, P. Supplementation of diet with different n-3/n-6 PUFA ratios ameliorates autistic behavior, reduces serotonin, and improves intestinal barrier impairments in a valproic acid rat model of autism. *Front. Psychiatry* **2020**, *11*, 552345. [CrossRef] [PubMed]
84. Qi, Z.; Lyu, M.; Yang, L.; Yuan, H.; Cao, Y.; Zhai, L.; Dang, W.; Liu, J.; Yang, F.; Li, Y. A novel and reliable rat model of autism. *Front. Psychiatry* **2021**, *12*, 549810. [CrossRef] [PubMed]
85. Lim, J.S.; Lim, M.Y.; Choi, Y.; Ko, G. Modeling environmental risk factors of autism in mice induces IBD-related gut microbial dysbiosis and hyperserotonemia. *Mol. Brain* **2017**, *10*, 14. [CrossRef] [PubMed]
86. Kong, Q.; Wang, B.; Tian, P.; Li, X.; Zhao, J.; Zhang, H.; Chen, W.; Wang, G. Daily intake of Lactobacillus alleviates autistic-like behaviors by ameliorating the 5-hydroxytryptamine metabolic disorder in VPA-treated rats during weaning and sexual maturation. *Food Funct.* **2021**, *12*, 2591–2604. [CrossRef]
87. Kumar, H.; Sharma, B. Memantine ameliorates autistic behavior, biochemistry & blood brain barrier impairments in rats. *Brain Res. Bull.* **2016**, *124*, 27–39. [CrossRef]
88. Kumar, H.; Sharma, B. Minocycline ameliorates prenatal valproic acid induced autistic behaviour, biochemistry and blood brain barrier impairments in rats. *Brain Res.* **2016**, *1630*, 83–97. [CrossRef]
89. Kumar, H.; Sharma, B.M.; Sharma, B. Benefits of agomelatine in behavioral, neurochemical and blood brain barrier alterations in prenatal valproic acid induced autism spectrum disorder. *Neurochem. Int.* **2015**, *91*, 34–45. [CrossRef]
90. McDougle, C.J.; Naylor, S.T.; Cohen, D.J.; Aghajanian, G.K.; Heninger, G.R.; Price, L.H. Effects of tryptophan depletion in drug-free adults with autistic disorder. *Arch. Gen. Psychiatry* **1996**, *53*, 993–1000. [CrossRef]
91. Dufour-Rainfray, D.; Vourc'h, P.; Le Guisquet, A.M.; Garreau, L.; Ternant, D.; Bodard, S.; Jaumain, E.; Gulhan, Z.; Belzung, C.; Andres, C.R.; et al. Behavior and serotonergic disorders in rats exposed prenatally to valproate: A model for autism. *Neurosci. Lett.* **2010**, *470*, 55–59. [CrossRef]
92. Sandhya, T.; Sowjanya, J.; Veeresh, B. Bacopa monniera (L.) Wettst ameliorates behavioral alterations and oxidative markers in sodium valproate induced autism in rats. *Neurochem. Res.* **2012**, *37*, 1121–1131. [CrossRef]
93. Zieminska, E.; Ruszczynska, A.; Augustyniak, J.; Toczylowska, B.; Lazarewicz, J.W. Zinc and copper brain levels and expression of neurotransmitter receptors in two rat ASD models. *Front. Mol. Neurosci.* **2021**, *14*, 656740. [CrossRef]
94. Kuwagata, M.; Ogawa, T.; Shioda, S.; Nagata, T. Observation of fetal brain in a rat valproate-induced autism model: A developmental neurotoxicity study. *Int. J. Dev. Neurosci.* **2009**, *27*, 399–405. [CrossRef] [PubMed]

95. Miyazaki, K.; Narita, N.; Narita, M. Maternal administration of thalidomide or valproic acid causes abnormal serotonergic neurons in the offspring: Implication for pathogenesis of autism. *Int. J. Dev. Neurosci.* **2005**, *23*, 287–297. [CrossRef] [PubMed]

96. Wang, C.C.; Lin, H.C.; Chan, Y.H.; Gean, P.W.; Yang, Y.K.; Chen, P.S. 5-HT1A-receptor agonist modified amygdala activity and amygdala-associated social behavior in a valproate-induced rat autism model. *Int. J. Neuropsychopharmacol.* **2013**, *16*, 2027–2039. [CrossRef] [PubMed]

97. Oyabu, A.; Narita, M.; Tashiro, Y. The effects of prenatal exposure to valproic acid on the initial development of serotonergic neurons. *Int. J. Dev. Neurosci.* **2013**, *31*, 202–208. [CrossRef]

98. Jacob, J.; Ribes, V.; Moore, S.; Constable, S.C.; Sasai, N.; Gerety, S.S.; Martin, D.J.; Sergeant, C.P.; Wilkinson, D.G.; Briscoe, J. Valproic acid silencing of ascl1b/Ascl1 results in the failure of serotonergic differentiation in a zebrafish model of fetal valproate syndrome. *Dis. Model. Mech.* **2014**, *7*, 107–117. [CrossRef]

99. Ali, E.H.; Elgoly, A.H. Combined prenatal and postnatal butyl paraben exposure produces autism-like symptoms in offspring: Comparison with valproic acid autistic model. *Pharmacol. Biochem. Behav.* **2013**, *111*, 102–110. [CrossRef]

100. Tsujino, N.; Nakatani, Y.; Seki, Y.; Nakasato, A.; Nakamura, M.; Sugawara, M.; Arita, H. Abnormality of circadian rhythm accompanied by an increase in frontal cortex serotonin in animal model of autism. *Neurosci. Res.* **2007**, *57*, 289–295. [CrossRef]

101. Lai, M.C.; Lombardo, M.V.; Baron-Cohen, S. Autism. *Lancet* **2014**, *383*, 896–910. [CrossRef]

102. Lord, C.; Elsabbagh, M.; Baird, G.; Veenstra-Vanderweele, J. Autism spectrum disorder. *Lancet* **2018**, *392*, 508–520. [CrossRef]

103. Namkung, J.; Kim, H.; Park, S. Peripheral serotonin: A new player in systemic energy homeostasis. *Mol. Cells* **2015**, *38*, 1023–1028. [CrossRef]

104. Knecht, L.D.; O'Connor, G.; Mittal, R.; Liu, X.Z.; Daftarian, P.; Deo, S.K.; Daunert, S. Serotonin activates bacterial quorum sensing and enhances the virulence of pseudomonas aeruginosa in the host. *EBioMedicine* **2016**, *9*, 161–169. [CrossRef]

105. Yano, J.M.; Yu, K.; Donaldson, G.P.; Shastri, G.G.; Ann, P.; Ma, L.; Nagler, C.R.; Ismagilov, R.F.; Mazmanian, S.K.; Hsiao, E.Y. Indigenous bacteria from the gut microbiota regulate host serotonin biosynthesis. *Cell* **2015**, *161*, 264–276. [CrossRef]

106. Crumeyrolle-Arias, M.; Jaglin, M.; Bruneau, A.; Vancassel, S.; Cardona, A.; Daugé, V.; Naudon, L.; Rabot, S. Absence of the gut microbiota enhances anxiety-like behavior and neuroendocrine response to acute stress in rats. *Psychoneuroendocrinology* **2014**, *42*, 207–217. [CrossRef]

107. Bjorklund, A.; Dunnett, S.B. Dopamine neuron systems in the brain: An update. *Trends Neurosci.* **2007**, *30*, 194–202. [CrossRef] [PubMed]

108. Bayer, S.A.; Wills, K.V.; Triarhou, L.C.; Ghetti, B. Time of neuron origin and gradients of neurogenesis in midbrain dopaminergic neurons in the mouse. *Exp. Brain Res.* **1995**, *105*, 191–199. [CrossRef] [PubMed]

109. Prakash, N.; Wurst, W. Development of dopaminergic neurons in the mammalian brain. *Cell. Mol. Life Sci.* **2006**, *63*, 187–206. [CrossRef]

110. Smidt, M.P.; Burbach, J.P.H. How to make a mesodiencephalic dopaminergic neuron. *Nat. Rev. Neurosci.* **2007**, *8*, 21–32. [CrossRef]

111. Aroca, P.; Lorente-Cánovas, B.; Mateos, F.R.; Puelles, L. Locus coeruleus neurons originate in alar rhombomere 1 and migrate into the basal plate: Studies in chick and mouse embryos. *J. Comp. Neurol.* **2006**, *496*, 802–818. [CrossRef]

112. Poe, G.R.; Foote, S.; Eschenko, O.; Johansen, J.P.; Bouret, S.; Aston-Jones, G.; Harley, C.W.; Manahan-Vaughan, D.; Weinshenker, D.; Valentino, R.; et al. Locus coeruleus: A new look at the blue spot. *Nat. Rev. Neurosci.* **2020**, *21*, 644–659. [CrossRef]

113. Nakamura, S.; Kimura, F.; Sakaguchi, T. Postnatal development of electrical activity in the locus ceruleus. *J. Neurophysiol.* **1987**, *58*, 510–524. [CrossRef] [PubMed]

114. Lake, C.R.; Ziegler, M.G.; Murphy, D.L. Increased norepinephrine levels and decreased dopamine-beta-hydroxylase activity in primary autism. *Arch. Gen. Psychiatry* **1977**, *34*, 553–556. [CrossRef] [PubMed]

115. Ernst, M.; Zametkin, A.J.; Matochik, J.A.; Pascualvaca, D.; Cohen, R.M. Low medial prefrontal dopaminergic activity in autistic children. *Lancet* **1997**, *350*, 638. [CrossRef]

116. Hettinger, J.A.; Liu, X.; Schwartz, C.E.; Michaelis, R.C.; Holden, J.J. A DRD1 haplotype is associated with risk for autism spectrum disorders in male-only affected sib-pair families. *Am. J. Med. Genet. B Neuropsychiatr. Genet.* **2008**, *147*, 628–636. [CrossRef]

117. Hettinger, J.A.; Liu, X.; Hudson, M.L.; Lee, A.; Cohen, I.L.; Michaelis, R.C.; Schwartz, C.E.; Lewis, S.M.; Holden, J.J. DRD2 and PPP1R1B (DARPP-32) polymorphisms independently confer increased risk for autism spectrum disorders and additively predict affected status in male-only affected sib-pair families. *Behav. Brain Funct.* **2012**, *8*, 19. [CrossRef]

118. De Krom, M.; Staal, W.G.; Ophoff, R.A.; Hendriks, J.; Buitelaar, J.; Franke, B.; de Jonge, M.V.; Bolton, P.; Collier, D.; Curran, S. A common variant in DRD3 receptor is associated with autism spectrum disorder. *Biol. Psychiatry* **2009**, *65*, 625–630. [CrossRef] [PubMed]

119. Gadow, K.D.; Devincent, C.J.; Olvet, D.M.; Pisarevskaya, V.; Hatchwell, E. Association of DRD4 polymorphism with severity of oppositional defiant disorder, separation anxiety disorder and repetitive behaviors in children with autism spectrum disorder. *Eur. J. Neurosci.* **2010**, *32*, 1058–1065. [CrossRef]

120. Hamilton, P.J.; Campbell, N.G.; Sharma, S.; Erreger, K.; Herborg Hansen, F.; Saunders, C.; Belovich, A.N.; Consortium, N.A.A.S.; Sahai, M.A.; Cook, E.H.; et al. De novo mutation in the dopamine transporter gene associates dopamine dysfunction with autism spectrum disorder. *Mol. Psychiatry* **2013**, *18*, 1315–1323. [CrossRef]

121. Nguyen, M.; Roth, A.; Kyzar, E.J.; Poudel, M.K.; Wong, K.; Stewart, A.M.; Kalueff, A.V. Decoding the contribution of dopaminergic genes and pathways to autism spectrum disorder (ASD). *Neurochem. Int.* **2014**, *66*, 15–26. [CrossRef]

122. Dichter, G.S.; Damiano, C.A.; Allen, J.A. Reward circuitry dysfunction in psychiatric and neurodevelopmental disorders and genetic syndromes: Animal models and clinical findings. *J. Neurodev. Disord.* **2012**, *4*, 19. [CrossRef]
123. Gunaydin, L.A.; Deisseroth, K. Dopaminergic dynamics contributing to social behavior. *Cold Spring Harb. Symp. Quant. Biol.* **2014**, *79*, 221–227. [CrossRef] [PubMed]
124. Matsumoto, M.; Hikosaka, O. Two types of dopamine neuron distinctly convey positive and negative motivational signals. *Nature* **2009**, *459*, 837–841. [CrossRef] [PubMed]
125. Ilango, A.; Kesner, A.J.; Keller, K.L.; Stuber, G.D.; Bonci, A.; Ikemoto, S. Similar roles of substantia nigra and ventral tegmental dopamine neurons in reward and aversion. *J. Neurosci.* **2014**, *34*, 817–822. [CrossRef] [PubMed]
126. Lerner, T.N.; Shilyansky, C.; Davidson, T.J.; Evans, K.E.; Beier, K.T.; Zalocusky, K.A.; Crow, A.K.; Malenka, R.C.; Luo, L.; Tomer, R.; et al. Intact-brain analyses reveal distinct information carried by SNc dopamine subcircuits. *Cell* **2015**, *162*, 635–647. [CrossRef] [PubMed]
127. Keehn, B.; Kadlaskar, G.; Bergmann, S.; McNally Keehn, R.; Francis, A. Attentional disengagement and the locus coeruleus - norepinephrine system in children with autism spectrum disorder. *Front. Integr. Neurosci.* **2021**, *15*, 716447. [CrossRef]
128. Huang, Y.; Yu, S.; Wilson, G.; Park, J.; Cheng, M.; Kong, X.; Lu, T.; Kong, J. Altered extended locus coeruleus and ventral tegmental area networks in boys with autism spectrum disorders: A resting-state functional connectivity study. *Neuropsychiatr. Dis. Treat.* **2021**, *17*, 1207–1216. [CrossRef]
129. Kubota, M.; Fujino, J.; Tei, S.; Takahata, K.; Matsuoka, K.; Tagai, K.; Sano, Y.; Yamamoto, Y.; Shimada, H.; Takado, Y.; et al. Binding of dopamine D1 receptor and noradrenaline transporter in individuals with autism spectrum disorder: A PET study. *Cereb. Cortex* **2020**, *30*, 6458–6468. [CrossRef]
130. Beversdorf, D.Q. The role of the noradrenergic system in autism spectrum disorders, implications for treatment. *Semin. Pediatr. Neurol.* **2020**, *35*, 100834. [CrossRef]
131. Cezar, L.C.; Kirsten, T.B.; da Fonseca, C.C.N.; de Lima, A.P.N.; Bernardi, M.M.; Felicio, L.F. Zinc as a therapy in a rat model of autism prenatally induced by valproic acid. *Prog. Neuropsychopharmacol. Biol. Psychiatry* **2018**, *84*, 173–180. [CrossRef]
132. Acosta, J.; Campolongo, M.A.; Hocht, C.; Depino, A.M.; Golombek, D.A.; Agostino, P.V. Deficits in temporal processing in mice prenatally exposed to valproic acid. *Eur. J. Neurosci.* **2018**, *47*, 619–630. [CrossRef]
133. Scheggi, S.; Guzzi, F.; Braccagni, G.; De Montis, M.G.; Parenti, M.; Gambarana, C. Targeting PPARα in the rat valproic acid model of autism: Focus on social motivational impairment and sex-related differences. *Mol. Autism* **2020**, *11*, 62. [CrossRef]
134. Schiavi, S.; Iezzi, D.; Manduca, A.; Leone, S.; Melancia, F.; Carbone, C.; Petrella, M.; Mannaioni, G.; Masi, A.; Trezza, V. Reward-related behavioral, neurochemical and electrophysiological changes in a rat model of autism based on prenatal exposure to valproic acid. *Front. Cell. Neurosci.* **2019**, *13*, 479. [CrossRef] [PubMed]
135. Hara, Y.; Takuma, K.; Takano, E.; Katashiba, K.; Taruta, A.; Higashino, K.; Hashimoto, H.; Ago, Y.; Matsuda, T. Reduced prefrontal dopaminergic activity in valproic acid-treated mouse autism model. *Behav. Brain Res.* **2015**, *289*, 39–47. [CrossRef] [PubMed]
136. Adiletta, A.; Pross, A.; Taricco, N.; Sgadò, P. Embryonic valproate exposure alters mesencephalic dopaminergic neurons distribution and septal dopaminergic gene expression in domestic chicks. *bioRxiv* **2021**. [CrossRef]
137. Ádám, Á.; Kemecsei, R.; Company, V.; Murcia-Ramón, R.; Juarez, I.; Gerecsei, L.I.; Zachar, G.; Echevarría, D.; Puelles, E.; Martínez, S.; et al. Gestational exposure to sodium valproate disrupts fasciculation of the mesotelencephalic dopaminergic tract, with a selective reduction of dopaminergic output from the ventral tegmental area. *Front. Neuroanat.* **2020**, *14*, 29. [CrossRef] [PubMed]
138. Campolongo, M.; Kazlauskas, N.; Falasco, G.; Urrutia, L.; Salgueiro, N.; Hocht, C.; Depino, A.M. Sociability deficits after prenatal exposure to valproic acid are rescued by early social enrichment. *Mol. Autism* **2018**, *9*, 36. [CrossRef] [PubMed]
139. Lyte, M.; Brown, D.R. Evidence for PMAT- and OCT-like biogenic amine transporters in a probiotic strain of Lactobacillus: Implications for interkingdom communication within the microbiota-gut-brain axis. *PLoS ONE* **2018**, *13*, e0191037. [CrossRef] [PubMed]
140. Asano, Y.; Hiramoto, T.; Nishino, R.; Aiba, Y.; Kimura, T.; Yoshihara, K.; Koga, Y.; Sudo, N. Critical role of gut microbiota in the production of biologically active, free catecholamines in the gut lumen of mice. *Am. J. Physiol. Gastrointest. Liver Physiol.* **2012**, *303*, G1288–G1295. [CrossRef]
141. Scammell, T.E.; Jackson, A.C.; Franks, N.P.; Wisden, W.; Dauvilliers, Y. Histamine: Neural circuits and new medications. *Sleep* **2019**, *42*, zsy183. [CrossRef]
142. Reiner, P.B.; Semba, K.; Fibiger, H.C.; McGeer, E.G. Ontogeny of histidine-decarboxylase-immunoreactive neurons in the tuberomammillary nucleus of the rat hypothalamus: Time of origin and development of transmitter phenotype. *J. Comp. Neurol.* **1988**, *276*, 304–311. [CrossRef]
143. Molina-Hernández, A.; Díaz, N.F.; Arias-Montaño, J.A. Histamine in brain development. *J. Neurochem.* **2012**, *122*, 872–882. [CrossRef] [PubMed]
144. Carthy, E.; Ellender, T. Histamine, neuroinflammation and neurodevelopment: A review. *Front. Neurosci.* **2021**, *15*, 870. [CrossRef] [PubMed]
145. Baldan, L.C.; Williams, K.A.; Gallezot, J.D.; Pogorelov, V.; Rapanelli, M.; Crowley, M.; Anderson, G.M.; Loring, E.; Gorczyca, R.; Billingslea, E.; et al. Histidine decarboxylase deficiency causes tourette syndrome: Parallel findings in humans and mice. *Neuron* **2014**, *81*, 77–90. [CrossRef] [PubMed]
146. Haas, H.L.; Sergeeva, O.A.; Selbach, O. Histamine in the nervous system. *Physiol. Rev.* **2008**, *88*, 1183–1241. [CrossRef]

147. Naganuma, F.; Nakamura, T.; Yoshikawa, T.; Iida, T.; Miura, Y.; Kárpáti, A.; Matsuzawa, T.; Yanai, A.; Mogi, A.; Mochizuki, T.; et al. Histamine N-methyltransferase regulates aggression and the sleep-wake cycle. *Sci. Rep.* **2017**, *7*, 15899. [CrossRef]
148. Wright, C.; Shin, J.H.; Rajpurohit, A.; Deep-Soboslay, A.; Collado-Torres, L.; Brandon, N.J.; Hyde, T.M.; Kleinman, J.E.; Jaffe, A.E.; Cross, A.J.; et al. Altered expression of histamine signaling genes in autism spectrum disorder. *Transl. Psychiatry* **2017**, *7*, e1126. [CrossRef]
149. Panula, P.; Sundvik, M.; Karlstedt, K. Developmental roles of brain histamine. *Trends Neurosci.* **2014**, *37*, 159–168. [CrossRef]
150. Baronio, D.; Puttonen, H.A.J.; Sundvik, M.; Semenova, S.; Lehtonen, E.; Panula, P. Embryonic exposure to valproic acid affects the histaminergic system and the social behaviour of adult zebrafish (Danio rerio). *Br. J. Pharmacol.* **2018**, *175*, 797–809. [CrossRef]
151. Eissa, N.; Jayaprakash, P.; Azimullah, S.; Ojha, S.K.; Al-Houqani, M.; Jalal, F.Y.; Lazewska, D.; Kiec-Kononowicz, K.; Sadek, B. The histamine H3R antagonist DL77 attenuates autistic behaviors in a prenatal valproic acid-induced mouse model of autism. *Sci. Rep.* **2018**, *8*, 13077. [CrossRef]
152. Baronio, D.; Gonchoroski, T.; Castro, K.; Zanatta, G.; Gottfried, C.; Riesgo, R. Histaminergic system in brain disorders: Lessons from the translational approach and future perspectives. *Ann. Gen. Psychiatry* **2014**, *13*, 34. [CrossRef]
153. Eissa, N.; Azimullah, S.; Jayaprakash, P.; Jayaraj, R.L.; Reiner, D.; Ojha, S.K.; Beiram, R.; Stark, H.; Łażewska, D.; Kieć-Kononowicz, K.; et al. The dual-active histamine H3 receptor antagonist and acetylcholine esterase inhibitor E100 alleviates autistic-like behaviors and oxidative stress in valproic acid induced autism in mice. *Int. J. Mol. Sci.* **2020**, *21*, 3996. [CrossRef] [PubMed]
154. Eissa, N.; Jayaprakash, P.; Stark, H.; Łażewska, D.; Kieć-Kononowicz, K.; Sadek, B. Simultaneous blockade of histamine H3 receptors and inhibition of acetylcholine esterase alleviate autistic-like behaviors in BTBR T+ tf/J mouse model of autism. *Biomolecules* **2020**, *10*, 1251. [CrossRef] [PubMed]
155. Molenhuis, R.T.; Hutten, L.; Kas, M.J.H. Histamine H3 receptor antagonism modulates autism-like hyperactivity but not repetitive behaviors in BTBR T+Itpr3tf/J inbred mice. *Pharmacol. Biochem. Behav.* **2022**, *212*, 173304. [CrossRef] [PubMed]
156. Alstadhaug, K.B. Histamine in migraine and brain. *Headache* **2014**, *54*, 246–259. [CrossRef]

Review

Animal Models for DOHaD Research: Focus on Hypertension of Developmental Origins

Chien-Ning Hsu [1,2] and You-Lin Tain [3,4,*]

[1] Department of Pharmacy, Kaohsiung Chang Gung Memorial Hospital, Kaohsiung 833, Taiwan; cnhsu@cgmh.org.tw
[2] School of Pharmacy, Kaohsiung Medical University, Kaohsiung 807, Taiwan
[3] Department of Pediatrics, Kaohsiung Chang Gung Memorial Hospital and Chang Gung University College of Medicine, Kaohsiung 833, Taiwan
[4] Institute for Translational Research in Biomedicine, Kaohsiung Chang Gung Memorial Hospital, Chang Gung University College of Medicine, Kaohsiung 833, Taiwan
* Correspondence: tainyl@cgmh.org.tw; Tel.: +886-975-056-995; Fax: +886-7733-8009

Abstract: Increasing evidence suggests that fetal programming through environmental exposure during a critical window of early life leads to long-term detrimental outcomes, by so-called developmental origins of health and disease (DOHaD). Hypertension can originate in early life. Animal models are essential for providing convincing evidence of a causal relationship between diverse early-life insults and the developmental programming of hypertension in later life. These insults include nutritional imbalances, maternal illnesses, exposure to environmental chemicals, and medication use. In addition to reviewing the various insults that contribute to hypertension of developmental origins, this review focuses on the benefits of animal models in addressing the underlying mechanisms by which early-life interventions can reprogram disease processes and prevent the development of hypertension. Our understanding of hypertension of developmental origins has been enhanced by each of these animal models, narrowing the knowledge gap between animal models and future clinical translation.

Keywords: animal model; developmental origins of health and disease (DOHaD); hypertension; oxidative stress; pregnancy; renin-angiotensin system; gut microbiota; reprogramming

Citation: Hsu, C.-N.; Tain, Y.-L. Animal Models for DOHaD Research: Focus on Hypertension of Developmental Origins. *Biomedicines* **2021**, *9*, 623. https://doi.org/10.3390/biomedicines9060623

Academic Editor: Martina Perše

Received: 4 May 2021
Accepted: 28 May 2021
Published: 31 May 2021

Publisher's Note: MDPI stays neutral with regard to jurisdictional claims in published maps and institutional affiliations.

1. Introduction

The association between fetal development and the increased risk of adult disease has attracted a great deal of attention to the concept of developmental programming or developmental origins of health and disease (DOHaD) [1,2]. The DOHaD hypothesis gained momentum after the emergence of observational studies from the 1944–1945 Dutch famine cohort, illustrating that maternal starvation is associated with an increased risk of metabolic and cardiovascular diseases in adult offspring [3]. These findings, combined with numerous subsequent epidemiologic investigations, indicate that the perinatal period, a critical window of organogenesis, is a vulnerable time in terms of the impact of adverse environmental insults [4]. Several hypotheses, such as thrifty phenotype [5], maternal capital [6], and predictive adaptive responses [7], have been developed to explain the epidemiological observations of an association between early life insults and diseases in adulthood. However, these hypotheses do not propose mechanistic pathways by which disease proceeds or suggest potential interventions for the prevention of adult diseases. Accordingly, animal models that have been developed and characterized have been instrumental in indicating the biological plausibility of the associations observed in epidemiological research, providing proof of causality. Emerging evidence indicates that animal models are valuable tools for understanding the pathogenesis of developmental programming and developing therapeutic interventions for DOHaD-related diseases [8–10]. A variety of small (e.g., rats, mice, and guinea pigs) and large (e.g., sheep and pigs) animals have been used to test aspects of the DOHaD hypothesis, and each offers different advantages.

Hypertension and related cardiovascular diseases are leading causes of mortality worldwide [11]. The WHO reported that 1 in 4 men and 1 in 5 women have hypertension [12]. Due to the multifactorial nature of hypertension, the use of various animal models, which induce hypertension by various mechanisms and produce the same end result, is advantageous [13,14]. In the past decades, novel drug classes and interventional strategies for the treatment of hypertension have been developed using hypertensive animal models [15]. However, the prevalence of hypertension remains high and continues to increase globally [16]. All this raises the question of how to prevent and not just treat hypertension based on the DOHaD concept.

A broad range of early-life insults can induce developmental programming, resulting in hypertension. These include maternal undernutrition or overnutrition, maternal disease states, lifestyle changes, substance abuse, environmental exposure to toxins/chemicals, and medication use during pregnancy [10,17–20]. Hypertension, diabetes, kidney disease, and inflammation are common maternal diseases that complicate pregnancy. On the other hand, programming processes geared toward disease could be reversed by shifting therapy from adulthood to the perinatal period, that is to say, by reprogramming [21]. Although the pathogenesis behind hypertension of developmental origins is poorly understood at present, our understanding of animal models used to study common mechanistic pathways has advanced greatly in recent years, which helps in developing efficient strategies to reprogram hypertension and prevent it from happening.

This review summarizes the contributions of animal models to DOHaD research with a focus on hypertension. It is proposed that integrating evidence from diverse animal models is essential in order to advance our understanding of hypertension of developmental origins and develop novel reprogramming strategies to alleviate the global burden of hypertension.

We retrieved related literature from all articles indexed in PubMed/MEDLINE. Search terms were as follows: "blood pressure", "developmental programming", "DOHaD", "animal model", "mother", "maternal", "pregnancy", "gestation", "offspring", "progeny", "prenatal", "perinatal", "reprogramming", and "hypertension". Additional studies were then selected and assessed based on appropriate references in eligible papers. The last search was conducted on 20 April 2021.

2. Choice of Animal Models

A broad range of animal models have been established to validate that the associations found in human observational studies can be replicated under experimental conditions. Animal models can be categorized in many different ways. First, models for DOHaD research can be categorized by types of environmental insult. For example, global caloric restriction and protein restriction in animals can mimic the starvation associated with famine in human cohorts [8,9]. Second, animal models can be classified according to molecular mechanisms. Since different environmental insults during pregnancy and lactation produce similar outcomes with respect to hypertension in adult offspring, there might be common mechanisms behind the developmental programming of hypertension. To date, hypertension of developmental origins has been attributed to mechanisms [10,17–21] including reduced nephron number, oxidative stress, an aberrant renin–angiotensin system (RAS), gut microbiota dysbiosis, and sex differences, among others. Animal models have been developed to test such proposed mechanisms. Finally, various small- and large-animal models have been established for DOHaD research, each with its own natural advantages and disadvantages [8]. Although non-human primates have long been regarded as the gold standard because of their high genetic and biological similarity to humans, the most commonly used species in the DOHaD field are rodents [22]. Rat and mouse models provide a low-cost option with a short life cycle that is easy to handle. Mice also provide ample access that allows for genetic modification. Depending on the experimental approach, other species such as rabbits, sheep, and pigs have also been used to evaluate developmental programming related to offspring outcomes [22]. Rabbits are useful for studies as their lipid metabolism and placental structure are similar to those in humans [23]. Pigs are considered to be a suitable model for evaluating the early stages of fertilization and development. Sheep have a long gestation period, and their fetal size and developmental rate are close to those in humans [24]. Cows are large, monotocous animals with a long gestation period, as in humans [24]. Thus, many aspects of animal models have to be taken into consideration when choosing one species over another, such as genetic background, anatomy, physiology, length of gestation, litter size, life cycle, and application to the clinical context. A summary of the selection of animal models for the study of hypertension of developmental origins is depicted in Figure 1.

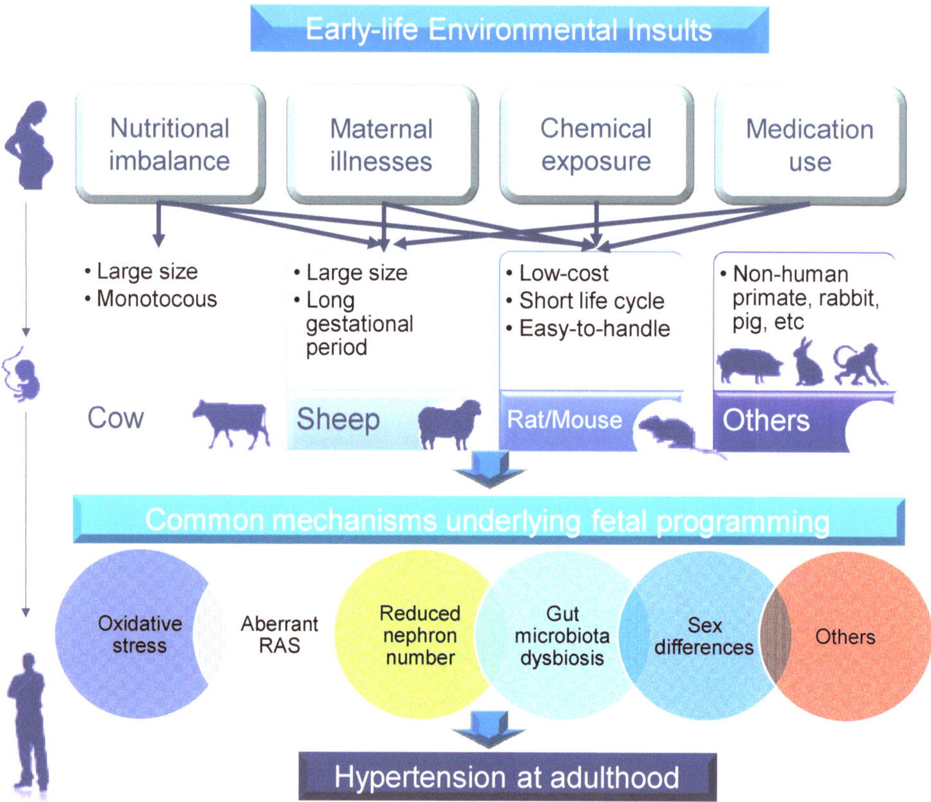

Figure 1. Schematic illustration of the selection of animal models for studying hypertension of developmental origins in adulthood according to early-life environmental insults, animal species, and common mechanisms. Lines with arrows (top section) indicate types of early-life insults produced in particular species of animals to induce hypertension in adult offspring. The study of other animals in DOHaD research (non-human primates, rabbits, pigs, etc.) is limited.

3. Hypertension of Developmental Origins: Early-Life Insults

Several suboptimal environmental conditions during fetal development are relevant to hypertension in adult offspring, including maternal nutritional imbalance, maternal illnesses and conditions, exposure to environmental chemicals, and medication use during pregnancy and lactation [10,17–20]. Each category is discussed in turn.

3.1. Maternal Nutritional Imbalance

Within the context of DOHaD research, studies of nutritional programming using small animal models have been ongoing since the early 1990s [8]. Nutritional interventions during critical developmental phases can have long-lasting effects on blood pressure (BP) in adult offspring [17]. Excessive or insufficient consumption of a specific nutrient has been used to induce hypertension of developmental origins in animal models, as shown in Figure 2 [25].

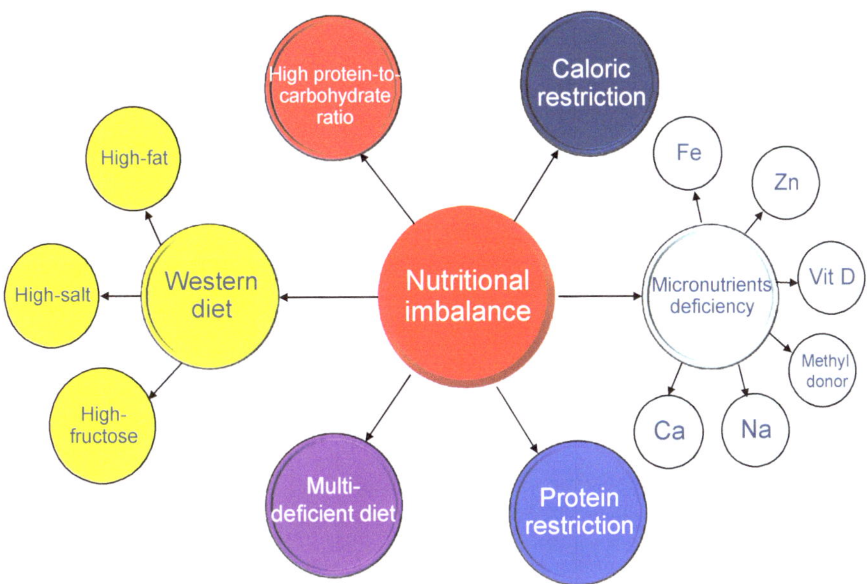

Figure 2. Overview of nutritional interventions used to modulate nutritional status during pregnancy and/or lactation to study hypertension of developmental origins in animal models. Fe = iron; Zn = zinc; Vit D = vitamin D; Na = sodium; Ca = calcium.

Caloric restriction refers to an overall reduction of energy and nutrient intake without incurring malnutrition. Caloric restriction in a range of 30–70% in pregnant rats has been reported to induce elevated BP in their adult offspring [26–28]. Hypertension programmed by maternal caloric restriction has also been observed in other species, including sheep [29,30] and cows [31]. In general, more severe caloric restriction resulted in earlier development of hypertension in adult offspring [25]. The protein restriction model has also been widely used to explore the mechanisms of nutritional programming [32]. As in the caloric restriction model, when pregnant rats were exposed to a greater degree of protein restriction, their adult offspring were likely to have high BP earlier [33–35]. Moreover, deficiencies in micronutrients, including iron [36], zinc [37], vitamin D [38], methyl donor nutrients (folic acid; choline; methionine; and vitamins B2, B6, and B12) [39], sodium [40], and calcium [41] in pregnant rats were associated with hypertension in their offspring. In a Brazilian study, when dams were fed with a multi-deficient diet developed from a basic regional diet, this was also shown to induce hypertension in adult rat offspring [42,43]. On the other hand, the excessive intake of certain nutrients can result in programmed hypertension in male adult offspring [25]. The Western diet is a modern dietary pattern characterized by the high intake of high-fat products, high-sugar drinks, and excess salt. In animal models of maternal diets containing key components based on the human Western diet, synergistic effects of fat, sugar, and salt on the rise of BP in adult progeny were observed [44–46]. The most frequently used model to induce obesity-related disorders is a high-fat diet [47]. The BP of adult offspring exposed to a maternal high-fat diet varies according to age, sex, diverse fatty acid composition, and strain [48–50]. Similarly, the intake of solely a high-fructose diet by rodent mothers results in BP elevation in the offspring [51–53]. A maternal high-fructose diet was developed into an animal model frequently used for studying hypertension and metabolic syndrome of developmental origins [54]. Male rat offspring exposed to a high protein-to-carbohydrate ratio in the maternal diet were also characterized by elevated BP [55]. In addition, high salt consumption during gestation and lactation has also been associated with hypertension in the offspring in a rat model [40]. However, little is known about the use of large animals to evaluate nutritional programming-induced hypertension.

Worthy of note is that nutritional programming can also be advantageous. Several nutritional interventions have proven to be effective in preventing the development of many adult diseases, including hypertension, with the use of animal models [56]. Since all nutrients during pregnancy play a crucial role in fetal growth and development, studies utilizing animal models of nutritional programming will lead to a better understanding of the timing, optimal dose, and intake duration of nutritional interventions for clinical practice.

3.2. Maternal Illnesses and Conditions

Maternal illnesses and complications during pregnancy can cause fetal programming and increase the risk of developing hypertension in offspring. Thus, animal models that mimic chronic illnesses and pregnancy complications have been established to study hypertension of developmental origins. Table 1 shows that rats are the most commonly used animal species. Diverse animal models resembling human illnesses and pregnancy complications have been evaluated, such as hypertensive disorders of pregnancy [57,58], preeclampsia [59–61], chronic kidney disease [62], diabetes [63,64], polycystic ovary syndrome [65], maternal inflammation [66,67], maternal hypoxia [68,69], and sleep disorder [70,71].

Hypertensive disorders affect around 10% of pregnancies, which includes the 3–5% of all pregnancies complicated by preeclampsia [72]. A previous cohort study showed that there is an association between maternal hypertension and adverse cardiometabolic outcomes in offspring at 40 years of age, including a 67% increased risk of hypertension [73]. Studies in two animal models—spontaneously hypertensive rat (SHR) and renovascular hypertensive rat—support an association between maternal hypertension and rising BP in the offspring during young adulthood [57,58]. Several animal models have been established that mimic changes in maternal preeclampsia. For example, pregnant rats were administered suramin [59] or N^G-nitro-L-arginine-methyl ester (L-NAME, an inhibitor of nitric oxide synthase) [60], or underwent a reduced uterine perfusion procedure [61], resulting in elevated BP in their adult offspring. Pregnant women with chronic kidney disease (CKD) are at risk of adverse outcomes for themselves and their offspring [74]. An adenine-induced maternal CKD model was used to study uremia-related adverse outcomes in pregnancy and offspring, including hypertension of developmental origins [59].

Epidemiological observations have established that exposure to gestational diabetes mellitus in utero leads to a high risk of high BP in childhood [75,76]. Hypertension in offspring induced by maternal diabetes is also demonstrable in animal models [63,64]. Although many models have been used for diabetes research [77], only streptozotocin (STZ)-induced diabetes has been modelled for hypertension of developmental origins [63,64]. Both type 1 and type 2 diabetes can be induced by STZ when given to adult [63,64] or neonate rats [63]. Another common pregnancy complication is iron-deficiency anemia. A previous report demonstrated that adult offspring of both sexes in a rat model of maternal iron deficiency had hypertension at 16 weeks of age [36].

Additionally, polycystic ovary syndrome (PCOS), inflammatory disorders, and hypoxia are associated with an increased risk of maternal pregnancy complications [78,79]. In the case of PCOS, the fetus is exposed to high levels of testosterone from the maternal circulation [80]. Thus, a model of maternal hyperandrogenemia by testosterone cypionate administration in pregnant rats in late gestation was developed to study BP in adult offspring [65]. As a result, female offspring exposed to prenatal androgen developed hypertension at 120 days of age [65]. Prenatal exposure to two pyrogens, LPS and zymosan, has been used to mimic maternal inflammation, and both models showed elevated BP in adult offspring [66,67]. Likewise, hypertension can be programmed by prenatal hypoxia in rats [68] or sheep [69]. Moreover, sleep disorders or chronodisruption in pregnant women could have harmful consequences for their offspring, as we reviewed elsewhere [81]. Table 1 shows that adult rat offspring exposed to maternal sleep restriction or constant light prenatally were found to develop hypertension [70,71]. Based on evidence gathered from the above-mentioned studies, various maternal illnesses and conditions indeed impact the offspring's BP and validate the epidemiological observations. However, whether other maternal conditions such as depression are relevant to the developmental programming of hypertension has not yet been adequately addressed.

It is noteworthy that most animal models employ rats and may evaluate short-term but not long-term outcomes in offspring. Research on DOHaD should now be intensified to validate the observed effects, with long-term follow-up studies using different species to identify the underlying common mechanisms.

3.3. Chemical and Medication Exposure

In addition to maternal conditions, early-life chemical and medication exposure has been associated with the developmental programming of hypertension. Table 2 illustrates that prenatal exposure to 2,3,7,8-tetrachlorodibenzo-p-dioxin (TCDD) or bisphenol A leads to increased BP in adult rat offspring [82–84]. These findings support the epidemiological data indicating that exposure to environmental chemicals such as endocrine-disrupting chemicals (EDCs) during critical developmental stages can increase the risk of cardiovascular disease later in life [85].

Substance abuse is also a major maternal insult; about 6–16% of pregnant women in the United States are alcohol users, cigarette smokers, or illicit drug users [86]. Previous reports on animal models demonstrated that maternal nicotine, alcohol, or caffeine exposure caused elevated BP in rat offspring [87–89]. However, similar models using large animals are not applied at the present time.

Additionally, medication use during pregnancy is also involved in the pathogenesis of programmed hypertension. As shown in Table 2, cyclosporine [90], gentamicin [91], minocycline [92], tenofovir [93], or glucocorticoid [94–98] administration in critical periods of development has been reported to induce hypertension of developmental origins in offspring. Unlike in humans, renal development in rodents continues up to postnatal week 1–2. Thus, adverse events during gestation and the early lactation period can impair nephrogenesis and reduce nephron numbers, resulting in hypertension in later life [99]. Cyclosporine, gentamicin, and glucocorticoid have been related to renal programming and reduced nephron numbers in various animal models [99]. Particularly noteworthy is glucocorticoid, the most extensively studied medication in animal models of programmed hypertension. A developing fetus is prone to being exposed to excessive glucocorticoids through excess maternal corticosteroid use (e.g., due to a stressed pregnancy) or through exogenous administration (e.g., during preterm birth). In rats, both maternal and neonatal administration of dexamethasone induced hypertension in adult offspring [94–96]. Likewise, prenatal glucocorticoid administration in a sheep model caused increased BP in the offspring [97,98]. Moreover, the use of minocycline, a tetracycline antibiotic, during pregnancy and lactation was shown to induce programmed hypertension in rat offspring, coinciding with alterations of the gut microbiota and its derived metabolites [92]. Tenofovir, an antiviral drug, can also program hypertension in a rat model [93]. To sum up, different classes of medications contribute to developmental programming of hypertension. It is possible that various insults can cause similar adult phenotypes that converge on common mechanisms, culminating in the development of hypertension.

Table 1. Summary of animal models of the developmental programming of hypertension, categorized according to maternal illness and pregnancy complications.

Maternal Illnesses and Conditions	Animal Models	Species/Gender	Age at Hypertension Development	Ref.
Hypertensive disorders of pregnancy	Genetic hypertension model	SHR/M	12 weeks	[57]
	2-kidney, 1-clip renovascular hypertension model	SD rat/M,F	16 weeks	[58]
Preeclampsia	Intraperitoneal administration of 60 mg/kg suramin on gestational days 10 and 11	SD rat/M	12 weeks	[59]
	Subcutaneous administration of 60 mg/kg L-NAME during pregnancy	SD rat/M	12 weeks	[60]
	Reduced uterine perfusion	SD rat/M	16 weeks	[61]
Chronic kidney disease	0.5% adenine supplementation from 3 weeks before pregnancy until 3 weeks after delivery	SD rat/M	12 weeks	[62]
Type 1 diabetes	Single intraperitoneal injection of 45 mg/kg STZ on gestational day 0	SD rat/M	12 weeks	[63]
	Single intraperitoneal injection of 35 mg/kg STZ on gestational day 0	SD rat/M	6 months	[64]
Type 2 diabetes	Mother rat received single intraperitoneal injection of 50 mg/kg STZ at newborn stage	SD rat/M	12 weeks	[63]
Anemia	Iron-deficiency diet from 4 weeks before pregnancy until delivery	Rowett hooded Lister rat/M & F	16 weeks	[36]
Polycystic ovary syndrome	Subcutaneous injection of 5 mg/kg testosterone cypionate on gestational day 20	Wistar rat/F	120 days	[65]
Maternal inflammation	Intraperitoneal administration of 0.79 mg/kg LPS on gestational days 8, 10, and 12	SD rat/M & F	12 weeks	[66]
	Intraperitoneal injection of 2.37 mg/kg zymosan on gestation days 8, 10, and 12	SD rat/M	66 weeks	[67]
Maternal hypoxia	Hypoxia maintained at constant inspired fraction of oxygen of 13% from gestational day 6 to 20	Wistar rat/M	4 months	[68]
Sleep disorder	Hypoxia maintained at 10% oxygen from gestational day 105 to 145	Sheep/F	9 months	[69]
	Sleep restriction	Wistar rat/M	3 months	[70]
	24 h constant light exposure during pregnancy	SD rat/M	12 weeks	[71]

Studies tabulated according to types of maternal illnesses and conditions, animal model, and age at evaluation. L$-$NAME = N^G-nitro-L-arginine-methyl ester; STZ = streptozotocin; LPS = lipopolysaccharide; SHR = spontaneously hypertensive rat; SD = Sprague-Dawley.

Table 2. Summary of animal models of the developmental programming of hypertension, categorized according to chemical and medication exposure.

Chemical or Medication	Animal Models	Species/Gender	Age at Hypertension Development	Ref.
TCDD	Oral administration of 200 ng/kg TCDD on gestational days 14 and 21 and postnatal days 7 and 14	SD rat/M	12 weeks	[82]
	Oral administration of 200 ng/kg TCDD on gestational days 14 and 21 and postnatal days 7 and 14	SD rat/M	16 weeks	[83]
Bisphenol A	Oral administration of 50 μg/kg/day bisphenol A during pregnancy and lactation	SD rat/M	16 weeks	[84]
Nicotine	Nicotine administration via osmotic mini-pump at 4 μg/kg/min from gestational day 4 to postnatal day 10	SD rat/M	8 months	[87]
Alcohol	Ethanol 1 g/kg by oral gavage on gestational days 13.5 and 14.5	SD rat/M,F	6 months	[88]
Caffeine	Subcutaneous injection of 20 mg/kg caffeine daily during pregnancy	C57BL/6 mouse/M	3 months	[89]
Cyclosporine	Cyclosporine 3.3 mg/kg from gestational day 10 to postnatal day 7	SD rat/M	11 weeks	[90]
Gentamicin	Subcutaneous injection of 110 mg/kg gentamicin from gestational day 10 to 15 or 15 to 20	SD rat/F	1 year	[91]
Minocycline	Minocycline 50 mg/kg via oral gavage during pregnancy and lactation	SD rat/M	12 weeks	[92]
Tenofovir	Tenofovir 100 mg/kg diet from 1 week before mating and during pregnancy	Wistar rat/M	6 months	[93]
Glucocorticoid	Intraperitoneal injection of 0.2 mg/kg dexamethasone on gestational days 15 and 16	SD rat/M	12 weeks	[94]
	Intraperitoneal injection of 0.1 mg/kg dexamethasone from gestational day 16 to 22	SD rat/M	12 weeks	[95]
	Intraperitoneal injection of 0.5 mg/kg dexamethasone on postnatal day 1, 0.3 mg/kg on day 2, and 0.1 mg/kg on day 3.	SD rat/M	12 weeks	[96]
	Intramuscular injection of 0.17 mg/kg betamethasone on gestational days 80 and 81	Sheep/M,F	18 months	[97]
	Intravenous treatment with 0.48 mg/h dexamethasone for 48 h on gestational day 27	Sheep/M,F	16 months	[98]

Studies tabulated according to type of chemical or medication, animal model, and age at evaluation. TCDD = 2,3,7,8-tetrachlorodibenzo-p-dioxin; SD = Sprague-Dawley.

Emerging evidence supports a "two-hit" hypothesis that explains the developmental programming of adult diseases [8]. Hypertension can develop with two sequential hits, the first hit being the response to a prenatal insult, followed by the second hit in response to ongoing programming induced by the first hit. During fetal development, the first hit can lead to morphological changes and functional adaption of vital organ systems, which alone in not sufficient to alter the adult phenotype. Another type of insult may act as a second hit, during which the same mechanism is targeted and could unmask or amplify the underlying defects culminating in a disease state. Accordingly, a number of two-hit models have been used to evaluate whether two distinct hits affect offspring outcomes synergistically or differently when combined as compared to either hit alone. For example, models of a high-fructose diet and TCDD exposure [82], TCDD plus dexamethasone exposure [83], combined bisphenol A and a high-fat diet [84], and a high-fructose diet plus a post-weaning high-fat diet [100] have been established to study hypertension of developmental origins. Together, these animal models have provided evidence of a number of common mechanisms behind hypertension of developmental origins, which will be discussed in turn.

4. Common Mechanisms Underlying Hypertension of Developmental Origins

In view of the fact that diverse early-life insults create very similar outcomes in adult offspring, there might be some common mechanistic pathways contributing to the pathogenesis of hypertension of developmental origins. So far, the proposed mechanisms include oxidative stress, aberrant RAS, reduced nephron numbers, gut microbiota dysbiosis, and sex differences [10,18–21].

4.1. Oxidative Stress

During fetal development, overproduction of reactive oxygen species (ROS) under adverse conditions in utero prevails over the defensive antioxidant system, resulting in oxidative stress damage [101]. There are several types of early-life insults linked to oxidative stress in mediating hypertension of developmental origins, including maternal caloric restriction [28,29], a zinc-deficient diet [37], a methyl-donor diet [39], high fat intake [50], high-fructose consumption [51], preeclampsia [60,61], maternal CKD [62], gestational diabetes [63], maternal hypoxia [68,69], TCDD exposure [83], bisphenol A exposure [84], nicotine exposure [87], and glucocorticoid use [94].

Reported mechanisms behind oxidative stress-induced hypertension of developmental origins consist of increased ROS generation [61], decreased antioxidant capacity [35], impaired nitric oxide (NO) signaling pathway [33,59,62,94], and increased oxidative damage [29,82,84,94]. Markers of lipid peroxidation such as malondialdehyde (MDA) and F2-isoprostanes were proven to be elevated in animal models of programmed hypertension induced by a maternal low-protein diet [35], maternal L-NAME administration [60], and reduced uterine perfusion [61]. Additionally, the expression of 8-hydroxydeoxyguanosine (8-OHdG), an oxidative DNA damage marker, was increased in animal models of hypertension programmed by a maternal methyl-donor diet [39], prenatal dexamethasone plus TCDD exposure [82], combined high-fat diet and bisphenol A exposure [84], prenatal dexamethasone exposure [95], and a maternal high-fructose diet [102].

Conversely, many natural and synthetic antioxidants have been used as a reprogramming strategy to prevent hypertension of developmental origins in diverse of animal models [20,103]. These observations suggest the notion that the developmental programming of hypertension might be driven by oxidative stress.

4.2. Aberrant Renin-Angiotensin System

Blood pressure is tightly controlled by the renin-angiotensin system (RAS) [104]. The blockade of the RAS provides the rationale for current antihypertensive therapies. The kidney is a major target for all components of the RAS. During kidney development, constituents of the RAS are highly expressed and play key roles in mediating proper renal morphology and physiological function [105]. In humans, RAS blockers have been avoided for pregnant women due to fetopathy and renal maldevelopment [106]. The adult progeny of animals that are transgenic for RAS genes or received angiotensin receptor blocker (ARB) during the nephrogenesis stage to block the RAS have a concurrent reduction in nephron numbers and hypertension [107,108].

An increasing number of animal models related to aberrant RAS are now being developed to evaluate hypertension of developmental programming [109]. Various nutritional insults can program the kidney and RAS concurrently—protein restriction [34], calorie restriction [30], a high-fructose diet [51], and a high-fat diet [110]—resulting in hypertension in adult offspring.

Adult rat offspring of diabetic mothers developed hypertension coinciding with increased angiotensin-converting enzyme (ACE) activity [111]. Other maternal illnesses and conditions such as hypertension [58], CKD [62], chronodisruption [78], and preeclampsia [60] also interfere with aberrant RAS and programmed hypertension. Moreover, programmed hypertension coinciding with dysregulated RAS can be triggered by maternal exposure to TCDD [83], caffeine [89], minocycline [92], or glucocorticoid [94,98].

On the other hand, reprogramming strategies targeting the RAS to prevent hypertension of developmental origins have been employed in various animal models [109]. So far, several early-life interventions have been demonstrated, including renin inhibitor [112], ACE inhibitor [113], ARB [114], and ACE2 activator [115]. Overall, the findings suggest that the interplay between the RAS and other mechanisms in early life is implicated in renal programming and consequently, hypertension in adulthood.

4.3. Reduced Nephron Numbers

A nephron is the basic unit of the kidney; however, there are large individual differences in the number of nephrons, ranging from 0.25 to 1.1 million per human kidney [116]. Epidemiologic studies have associated low birth weight and prematurity with low nephron numbers as risk factors for hypertension in later life [117]. Reduced nephron numbers can cause compensatory glomerular hyperfiltration and glomerular hypertension, consequently leading to further nephron loss later in life. Therefore, reduced nephron number has been considered as a key mechanism behind renal programming [118]. Likewise, animal studies have indicated that there are vulnerable periods during kidney development that could lead to a reduced nephron endowment.

In rats, adult offspring develop hypertension coinciding with reduced nephron numbers in response to diverse environmental insults during kidney development. These animal models of renal programming involved maternal exposure to cyclosporine [90], gentamicin [91], or glucocorticoid [94], or maternal diabetes [63], a low-protein diet [119], inflammation [120], or hypoxia [121]. However, reduced nephron numbers per se would not be essential for hypertension of developmental origins and renal programming [118]. The role of altering the nephron number in hypertension of developmental origins is still awaiting discovery but is certainly a subject of great interest.

4.4. Gut Microbiota Dysbiosis

Recent evidence suggests that early development of the gut microbiota may impact the programming of adult diseases, including hypertension [122,123]. During gestation, diet-gut microbiota interactions can alter global histone acetylation and methylation, not only in the mother but also in the fetus via contact with her metabolites [124]. Several mechanisms that link gut microbiota dysbiosis to hypertension have been proposed, including increased sympathetic activity, NO inhibition, aberrant RAS, and altered microbial metabolites, such as short-chain fatty acids (SCFAs) [125].

Data from many animal models indicate that gut microbiota dysbiosis may be involved in the developmental programming of hypertension. Various rat models of maternal insults such as hypertension [57], CKD [62], PCOS [65], TCDD exposure [82], minocycline use [92], a high-fructose diet [102], and a high-fat diet [126] have been examined with regard to the impact of gut microbiota dysbiosis on hypertension of developmental origins.

Worth noting is the consumption of probiotics or prebiotics, which has become one dietary strategy for modulating the gut microbiota. Our prior studies reported that maternal consumption of a high-fructose or high-fat diet induced hypertension in adult offspring, which can be prevented by modulating the gut microbiota through the intake of prebiotic inulin or probiotic *Lactobacillus casei* [127,128]. Despite recent studies showing that microbiota-targeted therapies can be applied to various diseases [129], their role in hypertension of developmental origins, especially their use in gestation, awaits further exploration.

4.5. Sex Differences

There is a considerable body of literature indicating that sex-dependent differences exist in hypertension of developmental origins [130,131]. It has long been observed that male offspring are more prone to hypertension than female offspring [25,130,131]. Additionally, several mechanisms mentioned above, such as oxidative stress [132], the RAS [133], and gut microbiota [134], are known to respond to environmental stimuli in a sex-specific manner.

Some early-life insults, such as maternal caloric restriction [27], low-protein diet [55], high-fat diet [110], or high-fructose diet [112], or prenatal dexamethasone exposure [135], have been reported to induce hypertension in male but not female offspring. This difference has led many researchers to investigate only males instead of both sexes, as listed in Table 2.

In a prenatal dexamethasone exposure model [135], we found that glucocorticoid-programmed hypertension developed in male but not in female adult offspring. We also observed the absence of hypertension in female offspring coinciding with lower *Agt* mRNA expression, suggesting that sex-dependent renal programming within the RAS may underlie the pathogenesis of programmed hypertension. Additionally, we found that the renal transcriptome is sex-specific in hypertension in offspring programmed by a maternal high-fructose diet [112]. One possible protective mechanism of females being refractory to high-fructose-diet-induced programmed hypertension is related to sex differences in the renal transcriptome. However, whether the increased female sensitivity to insult is beneficial or harmful to the developmental programming of various organs in female fetuses remains unclear. Thus, a better understanding of the sex-dependent mechanisms that underlie hypertension of developmental origins will aid in developing a novel sex-specific strategy to prevent programmed hypertension across genders.

4.6. Others

Other molecular mechanisms relevant to the developmental programming of hypertension are evaluated in different animal models, such as impaired sodium transport [10], dysregulated nutrient-sensing signaling [136], increased sympathetic nerve activity [137], and epigenetic regulation [138].

These observations suggest that there might be considerable interplay among the common mechanisms behind the pathogenesis of hypertension of developmental origins, even though this remains speculative. Although numerous mechanisms are outlined above, attention will need to be focused on exploring other potential mechanisms and validating them in different types of animal models. A better understanding of the mechanisms behind hypertension of developmental origins is the key to developing novel reprogramming interventions for further clinical translation.

5. Moving Forward: Promising Prospects of Early-Life Interventions

Given the advances in our understanding of the DOHaD research field, it has become apparent that early-life interventions can reprogram molecular mechanisms behind hypertension of developmental origins to prevent the development of hypertension in adulthood. Animal models have been essential in providing ideal reprogramming strategies. As described earlier, many antioxidants have been used as reprogramming strategies to prevent hypertension in offspring in a number of animal models [20,103]: L-arginine [139], L-taurine [140], L-citrulline [60], vitamin C [69], vitamin E [28], folic acid [141], selenium [28], melatonin [39,60,71], resveratrol [83,84,102], and N-acetylcysteine [57,59,60,95].

Additionally, several lines of evidence support the idea that early-life interventions targeting specific signaling pathways are of benefit in the prevention of developmental hypertension. First, reprogramming strategies targeting the NO pathway in early life have been employed in various animal models to prevent the development of hypertension in adult progeny. These interventions include supplementation of NO substrate [142], agents that lower asymmetric dimethylarginine (ADMA, an inhibitor of NOS) [95], NO donors [143], and enhancement of NOS expression [144], as reviewed elsewhere [142]. Second, several RAS-based interventions have also shown benefits in protecting against programmed hypertension, such as renin inhibitor, ACEI, ARB, and ACE2 activator [109]. Third, the reprogramming effects of hydrogen sulfide (H_2S)-based interventions have been shown in diverse animal models [145]. Currently available reprogramming interventions targeting the H_2S pathway are L-cysteine [146], D-cysteine [146], NAC [147], sodium hydrosulfide [148], and garlic [126]. Finally, the targeting of nutrient-sensing signals such as cyclic adenosine monophosphate-activated protein kinase (AMPK) or peroxisome proliferator-activated receptor (PPAR) has been noted to regulate downstream target genes, thereby reprogramming hypertension induced by various maternal insults [149–154]. This review provides a general overview of the various early-life interventions that show benefits with regard to hypertension of developmental origins. Despite the tremendous advances made from animal research, their clinical translation is still a long way off.

6. Selection of Appropriate Animal Models to Study Hypertension of Developmental Origins

6.1. Important Issues for Consideration

Even though significant advances have been made in developing diverse animal models to study hypertension of developmental origins, the need for meaningful clinical translation remains a research priority. The following conditions should be taken into consideration when we select animal models. First, the timing of the animal's organogenesis is similar to that of humans. Second, the gestation period and litter size should be comparable to those of humans. Third, it is crucial that animal models share similar features of adverse outcomes to those seen in human studies, which can be measured. Finally, any effective therapeutic intervention must be evaluated and validated.

6.2. Timing of Organogenesis

Across different species, critical development periods for major organ systems are not uniform. Blood pressure is tightly controlled by coordination among the kidney, heart, brain, and other organ systems. As such, the translatability of studies performed in animals should be approached with caution, as many key stages of BP-controlled organ development that occur before birth in humans occur after birth in some species [103,136,137,155].

Many animal studies on hypertension of developmental origins focus on renal programming [118,137]. Kidney development starts at week 3 and ceases at approximately 36 weeks of gestation in humans [156]. Unlike in humans, rat kidneys continue to develop after birth and complete at 1 to 2 weeks postnatally [157]. Accordingly, adverse environmental conditions during pregnancy as well as lactation can impair kidney development, consequently resulting in hypertension in rodents [118]. For example, repeated dexamethasone administration on embryonic days 15 and 16 [94], from gestational days 16 to 22 [95], or from postnatal days 1 to 3 [96] was associated with developmental programming of hypertension in adult rat offspring.

Another unsolved problem is that almost no studies have taken a comprehensive approach to simultaneously evaluating every BP-controlled organ system in response to in utero exposure at specific developmental stages to assess their relative vulnerability in an experiment. Due to the complex nature of the interplay between organogenesis and environmental insults, the programming effect on various organs might be dissimilar in different animal species. Hence, it is apparent that the selected animal paradigm should mirror the timing of human organ development as closely as possible so that the effects of early-life insults can be fully assessed.

6.3. Gestation Period and Litter Size

The advantages of a shorter gestation period and higher offspring yield compared to large animal models make rodent models the most commonly used in DOHaD research. There is a large set of studies on hypertension of developmental origins that were carried out in rats (Tables 1 and 2). The average gestation period for rats is within 23 days, compared to 280 days for humans [158]. If an early-life insult is induced by surgical manipulation or if delivery requires repeated procedures, short gestation in rodents could become disadvantageous. In addition, the short gestation time may not allow for the permanent resolution of developmental plasticity and the identification of critical time periods that are vulnerable to insults.

Unlike humans, rodents generally have more than one offspring, and litter sizes of 8–12 pups are usually seen. Such a large litter size is also a disadvantage when compared to singleton births common in humans and large animal models. Accordingly, normalizing the size of each litter after birth should be considered to control for differences in offspring food intake, maternal care, and pup growth [159]. Since these limitations exist, the complete translation of findings in rodents to human medicine is seriously compromised.

On the other hand, gestational length in sheep is around 150 days, during which the fetal size and development rate are similar to those of humans [160]. With the use of ewe models, maternal caloric restriction [30], maternal hypoxia [69], and prenatal glucocorticoid exposure [97,98] have been shown to cause hypertension in adult progeny. Although these early insults have shown the same adverse effects on offspring BP in sheep and rats, whether different gestation periods and litter sizes differentially impact hypertension of developmental origins in rats and large animals awaits further evaluation.

6.4. Outcome Measurements

As we mentioned earlier, rats are the most commonly used species for the developmental programming of hypertension. However, a critical assessment of the data show that this phenomenon is mostly observed when BP is typically measured by the tail cuff method; in contrast, hypertension is not detected in telemetrically instrumented animals [161]. Although BP data obtained from the tail cuff method are reported to correlate well with findings of direct arterial catheter and telemetry methods [162], part of the increased BP in offspring found after early-life insults may be due to an increased stress response related to sympathetic nerve activity.

In adulthood, one rat month is roughly equivalent to three human years [158]. Accordingly, Table 1 lists the timing of hypertension development measured in rats from 12 weeks to 8 months of age, which is equivalent to humans of a specific age group ranging from childhood to early adulthood. Thus, there remain gaps in our knowledge regarding the long-term adverse effects of maternal insults on BP in older adult offspring.

Several species have been studied for cardiovascular outcomes programmed by maternal adverse exposure, including guinea pigs [163], swine [164], and non-human primates [165]. However, none of them have been used to study hypertension of developmental origins. It is important to remember that large animals should not be neglected, as they are generally more physiologically suitable models with regard to human conditions.

In the current review, the wide range of early-life insults certainly influenced the outcomes, resulting in the reported heterogeneity. The results depended strongly on the applied measurement technique and animal model. Methodological heterogeneity is another reason for the observed heterogeneity. A huge percentage of studies employed male-only small animal models with small sample sizes. Future animal studies should improve the methodological quality by applying randomization, blinding, and sample size calculation techniques in order to avoid bias and collect data of better quality.

6.5. Effective Interventions

Currently, reprogramming strategies could be categorized as nutritional intervention, lifestyle modification, or pharmacological therapy. It stands to reason that avoiding in utero exposure to adverse conditions is the most effective strategy for preventing hypertension of developmental origins. Another approach is the use of nutritional intervention during pregnancy and lactation [56]. Although the targeting of specific nutrients as a reprogramming strategy opens a new avenue for prevention [25], there remains a lack of accurate dietary recommendations for specific nutritional requirements for pregnant women in case of deficiencies [166,167].

Research on short-lived rodent models has provided significant results, revealing potential pharmacological therapies for preventing hypertension of developmental origins. However, disparities in the therapeutic doses, timing and duration, and animal models used are among the major concerns. The standardization of animal experiments will improve the comparability of such studies. During the preparation of the current review, we found that almost no studies tested different doses or the use of different species. Additionally, the follow-up period after the cessation of interventions in most cited studies was rather short.

The efficacy of the intervention can be influenced by its duration with respect to organ development in a dose- and species-specific manner. Thus, further translational research into the pharmacokinetics and metabolism of pharmacological intervention is required to validate and compare its safety and therapeutic potential between humans and other species.

7. Conclusions and Future Perspectives

Various small (e.g., rat and mouse) and large (e.g., cow and sheep) animal models have made important contributions to the DOHaD research field, giving rise to convincing evidence of a causal relationship between various early-life insults and the risk of developing hypertension in later life. Our review highlights that animal models are not only used to investigate the mechanisms behind hypertension of developmental origins, but also have an impact on the development of early-life interventions as a reprogramming strategy to prevent the development of hypertension in adulthood.

There are still several questions that need to be answered. In the last decades, many insults have been identified by epidemiological and animal studies. Nevertheless, there is a growing need to identify all factors that can adversely impact the BP of offspring. Additionally, this review did not consider the potential for the programming of hypertension by paternal factors that clearly exist in the DOHaD field [168]. Moreover, little reliable information currently exists with regard to optimal doses and durations of pharmacological interventions for pregnant women and the long-term effects on their offspring. Currently, preventive strategies should focus on avoiding exposure to theoretically harmful agents perinatally and promoting a healthy lifestyle.

Each of the abovementioned animal models was used to study a specific hypothesis and neither can be considered superior with regard to all aspects of research on hypertension of developmental origins. Therefore, further research is needed to gain a better understanding of the types of early-life insults, other mechanisms behind hypertension of developmental origins, the ideal therapeutic dose and duration of early intervention, and the appropriate animal species. It is proposed that taking a DOHaD approach with maximum use of the animal evidence should be of benefit in reducing the global burden of hypertension.

Author Contributions: C.-N.H.: contributed to data interpretation, drafting of the manuscript, critical revision of the manuscript, concept generation, and approval of the article; Y.-L.T.: drafting of the manuscript, data interpretation, contributed to concept generation, critical revision of the manuscript, and approval of the article. All authors have read and agreed to the published version of the manuscript.

Funding: This research was funded by Chang Gung Memorial Hospital, Kaohsiung, Taiwan, grants CMRPG8J0253, CORPG8J0121, CORPG8L0121, CORPG8L0261, and CORPG8L0301.

Institutional Review Board Statement: Not applicable.

Informed Consent Statement: Not applicable.

Data Availability Statement: Data will be available upon request.

Conflicts of Interest: The authors declare no conflict of interest.

References

1. Barker, D.J. The origins of the developmental origins theory. *J. Intern. Med.* **2007**, *261*, 412–417. [CrossRef] [PubMed]
2. Hanson, M. The birth and future health of DOHaD. *J. Dev. Orig. Health Dis.* **2015**, *6*, 434–437. [CrossRef] [PubMed]
3. Roseboom, T.; de Rooij, S.; Painter, R. The Dutch famine and its long-term consequences for adult health. *Early Hum. Dev.* **2006**, *82*, 485–491. [CrossRef] [PubMed]
4. Hanson, M.; Gluckman, P. Developmental origins of noncommunicable disease: Population and public health implications. *Am. J. Clin. Nutr.* **2011**, *94*, 1754S–1758S. [CrossRef]
5. Hales, C.N.; Barker, D.J. The thrifty phenotype hypothesis. *Br. Med. Bull.* **2001**, *60*, 5–20. [CrossRef] [PubMed]
6. Wells, J.C. Maternal capital and the metabolic ghetto: An evolutionary perspective on the transgenerational basis of health inequalities. *Am. J. Hum. Biol.* **2010**, *22*, 1–17. [CrossRef] [PubMed]
7. Gluckman, P.D.; Hanson, M.A. Living with the past: Evolution, development, and patterns of disease. *Science* **2004**, *305*, 1733–1736. [CrossRef]
8. McMullen, S.; Mostyn, A. Animal models for the study of the developmental origins of health and disease. *Proc. Nutr. Soc.* **2009**, *68*, 306–320. [CrossRef]
9. McMillen, I.C.; Robinson, J.S. Developmental origins of the metabolic syndrome: Prediction, plasticity, and programming. *Physiol. Rev.* **2005**, *85*, 571–633. [CrossRef] [PubMed]
10. Ojeda, N.B.; Grigore, D.; Alexander, B.T. Developmental programming of hypertension: Insight from animal models of nutritional manipulation. *Hypertension* **2008**, *52*, 44–50. [CrossRef] [PubMed]
11. Bromfield, S.; Muntner, P. High blood pressure: The leading global burden of disease risk factor and the need for worldwide prevention programs. *Curr. Hypertens. Rep.* **2013**, *15*, 134–136. [CrossRef] [PubMed]
12. World Health Organization. Hypertension. 2019. Available online: https://www.who.int/news-room/fact-sheets/detail/hypertension (accessed on 13 April 2021).
13. Lerman, L.O.; Kurtz, T.W.; Touyz, R.M.; Ellison, D.H.; Chade, A.R.; Crowley, S.D.; Mattson, D.L.; Mullins, J.J.; Osborn, J.; Eirin, A.; et al. Animal Models of Hypertension: A Scientific Statement From the American Heart Association. *Hypertension* **2019**, *73*, e87–e120. [CrossRef] [PubMed]
14. Pinto, Y.M.; Paul, M.; Ganten, D. Lessons from rat models of hypertension: From Goldblatt to genetic engineering. *Cardiovasc. Res.* **1998**, *39*, 77–88. [CrossRef]
15. Oparil, S.; Schmieder, R.E. New approaches in the treatment of hypertension. *Circ. Res.* **2015**, *116*, 1074–1095. [CrossRef] [PubMed]
16. Mills, K.T.; Bundy, J.D.; Kelly, T.N.; Reed, J.E.; Kearney, P.M.; Reynolds, K.; Chen, J.; He, J. Global Disparities of Hypertension Prevalence and Control: A Systematic Analysis of Population-Based Studies From 90 Countries. *Circulation* **2016**, *134*, 441–450. [CrossRef] [PubMed]
17. Bagby, S.P. Maternal nutrition, low nephron number, and hypertension in later life: Pathways of nutritional programming. *J. Nutr.* **2007**, *137*, 1066–1072. [CrossRef] [PubMed]
18. Chong, E.; Yosypiv, I.V. Developmental programming of hypertension and kidney disease. *Int. J. Nephrol.* **2012**, *2012*, 760580. [CrossRef] [PubMed]
19. Paixão, A.D.; Alexander, B.T. How the kidney is impacted by the perinatal maternal environment to develop hypertension. *Biol. Reprod.* **2013**, *89*, 144. [CrossRef]
20. Hsu, C.N.; Tain, Y.L. Early Origins of Hypertension: Should Prevention Start Before Birth Using Natural Antioxidants? *Antioxidants* **2020**, *9*, 1034. [CrossRef] [PubMed]
21. Tain, Y.L.; Joles, J.A. Reprogramming: A Preventive Strategy in Hypertension Focusing on the Kidney. *Int. J. Mol. Sci.* **2015**, *17*, 23. [CrossRef]
22. Chavatte-Palmer, P.; Tarrade, A.; Rousseau-Ralliard, D. Diet before and during Pregnancy and Offspring Health: The Importance of Animal Models and What Can Be Learned from Them. *Int. J. Environ. Res. Public Health* **2016**, *13*, 586. [CrossRef]
23. Furukawa, S.; Kuroda, Y.; Sugiyama, A. A comparison of the histological structure of the placenta in experimental animals. *J. Toxicol. Pathol.* **2014**, *27*, 11–18. [CrossRef]
24. Morel, O.; Laporte-Broux, B.; Tarrade, A.; Chavatte-Palmer, P. The use of ruminant models in biomedical perinatal research. *Theriogenology* **2012**, *78*, 1763–1773. [CrossRef] [PubMed]
25. Hsu, C.N.; Tain, Y.L. The Double-Edged Sword Effects of Maternal Nutrition in the Developmental Programming of Hypertension. *Nutrients* **2018**, *10*, 1917. [CrossRef] [PubMed]

26. Ozaki, T.; Nishina, H.; Hanson, M.A.; Poston, L. Dietary restriction in pregnant rats causes gender-related hypertension and vascular dysfunction in offspring. *J. Physiol.* **2001**, *530*, 141–152. [CrossRef] [PubMed]
27. Franco Mdo, C.; Ponzio, B.F.; Gomes, G.N.; Gil, F.Z.; Tostes, R.; Carvalho, M.H.; Fortes, Z.B. Micronutrient prenatal supplementation prevents the development of hypertension and vascular endothelial damage induced by intrauterine malnutrition. *Life Sci.* **2009**, *85*, 327–333. [CrossRef] [PubMed]
28. Tain, Y.L.; Hsieh, C.S.; Lin, I.C.; Chen, C.C.; Sheen, J.M.; Huang, L.T. Effects of maternal L-citrulline supplementation on renal function and blood pressure in offspring exposed to maternal caloric restriction: The impact of nitric oxide pathway. *Nitric. Oxide.* **2010**, *23*, 34–41. [CrossRef] [PubMed]
29. Gilbert, J.S.; Lang, A.L.; Grant, A.R.; Nijland, M.J. Maternal nutrient restriction in sheep: Hypertension and decreased nephron number in offspring at 9 months of age. *J. Physiol.* **2005**, *565*, 137–147. [CrossRef] [PubMed]
30. Gopalakrishnan, G.S.; Gardner, D.S.; Rhind, S.M.; Rae, M.T.; Kyle, C.E.; Brooks, A.N.; Walker, R.M.; Ramsay, M.M.; Keisler, D.H.; Stephenson, T.; et al. Programming of adult cardiovascular function after early maternal undernutrition in sheep. *Am. J. Physiol. Regul. Integr. Comp. Physiol.* **2004**, *287*, R12–R20. [CrossRef] [PubMed]
31. Mossa, F.; Carter, F.; Walsh, S.W.; Kenny, D.A.; Smith, G.W.; Ireland, J.L.; Hildebrandt, T.B.; Lonergan, P.; Ireland, J.J.; Evans, A.C. Maternal undernutrition in cows impairs ovarian and cardiovascular systems in their offspring. *Biol. Reprod.* **2013**, *88*, 92. [CrossRef] [PubMed]
32. Zohdi, V.; Lim, K.; Pearson, J.T.; Black, M.J. Developmental programming of cardiovascular disease following intrauterine growth restriction: Findings utilising a rat model of maternal protein restriction. *Nutrients* **2014**, *7*, 119–152. [CrossRef] [PubMed]
33. Sathishkumar, K.; Elkins, R.; Yallampalli, U.; Yallampalli, C. Protein restriction during pregnancy induces hypertension and impairs endothelium-dependent vascular function in adult female offspring. *J. Vasc. Res.* **2009**, *46*, 229–239. [CrossRef] [PubMed]
34. Woods, L.L.; Ingelfinger, J.R.; Nyengaard, J.R.; Rasch, R. Maternal protein restriction suppresses the newborn renin-angiotensin system and programs adult hypertension in rats. *Pediatr. Res.* **2001**, *49*, 460–467. [CrossRef] [PubMed]
35. Cambonie, G.; Comte, B.; Yzydorczyk, C.; Ntimbane, T.; Germain, N.; Lê, N.L.; Pladys, P.; Gauthier, C.; Lahaie, I.; Abran, D.; et al. Antenatal antioxidant prevents adult hypertension, vascular dysfunction, and microvascular rarefaction associated with in utero exposure to a low-protein diet. *Am. J. Physiol. Regul. Integr. Comp. Physiol.* **2007**, *292*, R1236–R1245. [CrossRef] [PubMed]
36. Gambling, L.; Dunford, S.; Wallace, D.I.; Zuur, G.; Solanky, N.; Srai, K.S.; McArdle, H.J. Iron deficiency during pregnancy affects post-natal blood pressure in the rat. *J. Physiol.* **2003**, *552*, 603–610. [CrossRef] [PubMed]
37. Tomat, A.; Elesgaray, R.; Zago, V.; Fasoli, H.; Fellet, A.; Balaszczuk, A.M.; Schreier, L.; Costa, M.A.; Arranz, C. Exposure to zinc deficiency in fetal and postnatal life determines nitric oxide system activity and arterial blood pressure levels in adult rats. *Br. J. Nutr.* **2010**, *104*, 382–389. [CrossRef] [PubMed]
38. Tare, M.; Emmett, S.J.; Coleman, H.A.; Skordilis, C.; Eyles, D.W.; Morley, R.; Parkington, H.C. Vitamin D insufficiency is associated with impaired vascular endothelial and smooth muscle function and hypertension in young rats. *J. Physiol.* **2011**, *589*, 4777–4786. [CrossRef] [PubMed]
39. Tain, Y.L.; Chan, J.Y.H.; Lee, C.T.; Hsu, C.N. Maternal melatonin therapy attenuates methyl-donor diet-induced programmed hypertension in male adult rat offspring. *Nutrients* **2018**, *10*, 1407. [CrossRef]
40. Koleganova, N.; Piecha, G.; Ritz, E.; Becker, L.E.; Müller, A.; Weckbach, M.; Nyengaard, J.R.; Schirmacher, P.; Gross-Weissmann, M.L. Both high and low maternal salt intake in pregnancy alter kidney development in the offspring. *Am. J. Physiol. Renal Physiol.* **2011**, *301*, F344–F354. [CrossRef] [PubMed]
41. Bergel, E.; Belizán, J.M. A deficient maternal calcium intake during pregnancy increases blood pressure of the offspring in adult rats. *BJOG* **2002**, *109*, 540–545. [CrossRef]
42. Paixão, A.D.; Maciel, C.R.; Teles, M.B.; Figueiredo-Silva, J. Regional Brazilian diet-induced low birth weight is correlated with changes in renal hemodynamics and glomerular morphometry in adult age. *Biol. Neonate* **2001**, *80*, 239–246. [CrossRef] [PubMed]
43. Vieira-Filho, L.D.; Cabral, E.V.; Farias, J.S.; Silva, P.A.; Muzi-Filho, H.; Vieyra, A.; Paixão, A.D. Renal molecular mechanisms underlying altered Na+ handling and genesis of hypertension during adulthood in prenatally undernourished rats. *Br. J. Nutr.* **2014**, *111*, 1932–1944. [CrossRef] [PubMed]
44. Yamada-Obara, N.; Yamagishi, S.I.; Taguchi, K.; Kaida, Y.; Yokoro, M.; Nakayama, Y.; Ando, R.; Asanuma, K.; Matsui, T.; Ueda, S.; et al. Maternal exposure to high-fat and high-fructose diet evokes hypoadiponectinemia and kidney injury in rat offspring. *Clin. Exp. Nephrol.* **2016**, *20*, 853–886. [CrossRef] [PubMed]
45. Tain, Y.L.; Lee, W.C.; Leu, S.; Wu, K.; Chan, J. High salt exacerbates programmed hypertension in maternal fructose-fed male offspring. *Nutr. Metab. Cardiovasc. Dis.* **2015**, *25*, 1146–1151. [CrossRef]
46. Tain, Y.L.; Wu, K.L.H.; Lee, W.C.; Leu, S.; Chan, J.Y.H. Prenatal Metformin Therapy Attenuates Hypertension of Developmental Origin in Male Adult Offspring Exposed to Maternal High-Fructose and Post-Weaning High-Fat Diets. *Int. J. Mol. Sci.* **2018**, *19*, 1066. [CrossRef] [PubMed]
47. Buettner, R.; Schölmerich, J.; Bollheimer, L.C. High-fat diets: Modeling the metabolic disorders of human obesity in rodents. *Obesity* **2007**, *15*, 798–808. [CrossRef] [PubMed]
48. Williams, L.; Seki, Y.; Vuguin, P.M.; Charron, M.J. Animal models of in utero exposure to a high fat diet: A review. *Biochim. Biophys. Acta* **2014**, *1842*, 507–519. [CrossRef] [PubMed]

49. Tain, Y.L.; Lin, Y.J.; Sheen, J.M.; Yu, H.R.; Tiao, M.M.; Chen, C.C.; Tsai, C.C.; Huang, L.T.; Hsu, C.N. High fat diets sex-specifically affect the renal transcriptome and program obesity, kidney injury, and hypertension in the offspring. *Nutrients* **2017**, *9*, 357. [CrossRef] [PubMed]

50. Resende, A.C.; Emiliano, A.F.; Cordeiro, V.S.; de Bem, G.F.; de Cavalho, L.C.; de Oliveira, P.R.; Neto, M.L.; Costa, C.A.; Boaventura, G.T.; de Moura, R.S. Grape skin extract protects against programmed changes in the adult rat offspring caused by maternal high-fat diet during lactation. *J. Nutr. Biochem.* **2013**, *24*, 2119–2120. [CrossRef]

51. Tain, Y.L.; Wu, K.L.; Lee, W.C.; Leu, S.; Chan, J.Y. Maternal fructose-intake-induced renal programming in adult male offspring. *J. Nutr. Biochem.* **2015**, *26*, 642–650. [CrossRef] [PubMed]

52. Tain, Y.L.; Chan, J.Y.; Hsu, C.N. Maternal Fructose Intake Affects Transcriptome Changes and Programmed Hypertension in Offspring in Later Life. *Nutrients* **2016**, *8*, 757. [CrossRef]

53. Seong, H.Y.; Cho, H.M.; Kim, M.; Kim, I. Maternal High-Fructose Intake Induces Multigenerational Activation of the Renin-Angiotensin-Aldosterone System. *Hypertension* **2019**, *74*, 518–525. [CrossRef] [PubMed]

54. Lee, W.C.; Wu, K.L.H.; Leu, S.; Tain, Y.L. Translational insights on developmental origins of metabolic syndrome: Focus on fructose consumption. *Biomed. J.* **2018**, *41*, 96–101. [CrossRef] [PubMed]

55. Thone-Reineke, C.; Kalk, P.; Dorn, M.; Klaus, S.; Simon, K.; Pfab, T.; Godes, M.; Persson, P.; Unger, T.; Hocher, B. High-protein nutrition during pregnancy and lactation programs blood pressure, food efficiency, and body weight of the offspring in a sex-dependent manner. *Am. J. Physiol. Regul. Integr. Comp. Physiol.* **2006**, *291*, R1025–R1030. [CrossRef] [PubMed]

56. Hsu, C.N.; Tain, Y.L. The Good, the Bad, and the Ugly of Pregnancy Nutrients and Developmental Programming of Adult Disease. *Nutrients* **2019**, *11*, 894. [CrossRef] [PubMed]

57. Hsu, C.N.; Hou, C.Y.; Chang-Chien, G.P.; Lin, S.; Tain, Y.L. Maternal N-Acetylcysteine Therapy Prevents Hypertension in Spontaneously Hypertensive Rat Offspring: Implications of Hydrogen Sulfide-Generating Pathway and Gut Microbiota. *Antioxidants* **2020**, *9*, 856. [CrossRef] [PubMed]

58. Guo, Q.; Feng, X.; Xue, H.; Jin, S.; Teng, X.; Duan, X.; Xiao, L.; Wu, Y. Parental Renovascular Hypertension-Induced Autonomic Dysfunction in Male Offspring Is Improved by Prenatal or Postnatal Treatment With Hydrogen Sulfide. *Front. Physiol.* **2019**, *10*, 1184. [CrossRef] [PubMed]

59. Tain, Y.L.; Hsu, C.N.; Lee, C.T.; Lin, Y.J.; Tsai, C.C. N-Acetylcysteine Prevents Programmed Hypertension in Male Rat Offspring Born to Suramin-Treated Mothers. *Biol. Reprod.* **2016**, *95*, 8. [CrossRef] [PubMed]

60. Tain, Y.L.; Lee, C.T.; Chan, J.Y.; Hsu, C.N. Maternal melatonin or N-acetylcysteine therapy regulates hydrogen sulfide-generating pathway and renal transcriptome to prevent prenatal N(G)-Nitro-L-arginine methyl ester (L-NAME)-induced fetal programming of hypertension in adult male offspring. *Am. J. Obstet. Gynecol.* **2016**, *215*, 636. [CrossRef]

61. Ojeda, N.B.; Hennington, B.S.; Williamson, D.T.; Hill, M.L.; Betson, N.E.; Sartori-Valinotti, J.C.; Reckelhoff, J.F.; Royals, T.P.; Alexander, B.T. Oxidative stress contributes to sex differences in blood pressure in adult growth-restricted offspring. *Hypertension* **2012**, *60*, 114–122. [CrossRef] [PubMed]

62. Hsu, C.N.; Yang, H.W.; Hou, C.Y.; Chang-Chien, G.P.; Lin, S.; Tain, Y.L. Maternal Adenine-Induced Chronic Kidney Disease Programs Hypertension in Adult Male Rat Offspring: Implications of Nitric Oxide and Gut Microbiome Derived Metabolites. *Int. J. Mol. Sci.* **2020**, *21*, 7237. [CrossRef] [PubMed]

63. Tain, Y.L.; Lee, W.C.; Hsu, C.N.; Lee, W.C.; Huang, L.T.; Lee, C.T.; Lin, C.Y. Asymmetric dimethylarginine is associated with developmental programming of adult kidney disease and hypertension in offspring of streptozotocin-treated mothers. *PLoS ONE* **2013**, *8*, e55420. [CrossRef] [PubMed]

64. Dib, A.; Payen, C.; Bourreau, J.; Munier, M.; Grimaud, L.; Fajloun, Z.; Loufrani, L.; Henrion, D.; Fassot, C. In Utero Exposure to Maternal Diabetes Is Associated With Early Abnormal Vascular Structure in Offspring. *Front. Physiol.* **2018**, *9*, 350. [CrossRef]

65. Sherman, S.B.; Sarsour, N.; Salehi, M.; Schroering, A.; Mell, B.; Joe, B.; Hill, J.W. Prenatal androgen exposure causes hypertension and gut microbiota dysbiosis. *Gut Microbes* **2018**, *9*, 400–421. [CrossRef] [PubMed]

66. Wang, J.; Yin, N.; Deng, Y.; Wei, Y.; Huang, Y.; Pu, X.; Li, L.; Zheng, Y.; Guo, J.; Yu, J.; et al. Ascorbic Acid Protects against Hypertension through Downregulation of ACE1 Gene Expression Mediated by Histone Deacetylation in Prenatal Inflammation-Induced Offspring. *Sci. Rep.* **2016**, *6*, 39469. [CrossRef] [PubMed]

67. Liao, W.; Wei, Y.; Yu, C.; Zhou, J.; Li, S.; Pang, Y.; Li, G.; Li, X. Prenatal exposure to zymosan results in hypertension in adult offspring rats. *Clin. Exp. Pharmacol. Physiol.* **2008**, *35*, 1413–1418. [CrossRef]

68. Giussani, D.A.; Camm, E.J.; Niu, Y.; Richter, H.G.; Blanco, C.E.; Gottschalk, R.; Blake, E.Z.; Horder, K.A.; Thakor, A.S.; Hansell, J.A.; et al. Developmental programming of cardiovascular dysfunction by prenatal hypoxia and oxidative stress. *PLoS ONE* **2012**, *7*, e31017. [CrossRef] [PubMed]

69. Brain, K.L.; Allison, B.J.; Niu, Y.; Cross, C.M.; Itani, N.; Kane, A.D.; Herrera, E.A.; Skeffington, K.L.; Botting, K.J.; Giussani, D.A. Intervention against hypertension in the next generation programmed by developmental hypoxia. *PLoS Biol.* **2019**, *17*, e2006552. [CrossRef] [PubMed]

70. Thomal, J.T.; Palma, B.D.; Ponzio, B.F.; Franco Mdo, C.; Zaladek-Gil, F.; Fortes, Z.B.; Tufik, S.; Gomes, G.N. Sleep restriction during pregnancy: Hypertension and renal abnormalities in young offspring rats. *Sleep* **2010**, *33*, 1357–1362. [CrossRef] [PubMed]

71. Tain, Y.L.; Lin, Y.J.; Chan, J.Y.H.; Lee, C.T.; Hsu, C.N. Maternal melatonin or agomelatine therapy prevents programmed hypertension in male offspring of mother exposed to continuous light. *Biol. Reprod.* **2017**, *97*, 636–643. [CrossRef] [PubMed]

72. Fox, R.; Kitt, J.; Leeson, P.; Aye, C.Y.L.; Lewandowski, A.J. Preeclampsia: Risk Factors, Diagnosis, Management, and the Cardiovascular Impact on the Offspring. *J. Clin. Med.* **2019**, *8*, 1625. [CrossRef] [PubMed]
73. Kurbasic, A.; Fraser, A.; Mogren, I.; Hallmans, G.; Franks, P.W.; Rich-Edwards, J.W.; Timpka, S. Maternal Hypertensive Disorders of Pregnancy and Offspring Risk of Hypertension: A Population-Based Cohort and Sibling Study. *Am. J. Hypertens.* **2019**, *32*, 331–334. [CrossRef] [PubMed]
74. Hladunewich, M.A. Chronic Kidney Disease and Pregnancy. *Semin. Nephrol.* **2017**, *37*, 337–346. [CrossRef] [PubMed]
75. Wright, C.S.; Rifas-Shiman, S.L.; Rich-Edwards, J.W.; Taveras, E.M.; Gillman, M.W.; Oken, E. Intrauterine exposure to gestational diabetes, child adiposity, and blood pressure. *Am. J. Hypertens.* **2009**, *22*, 215–220. [CrossRef] [PubMed]
76. Lu, J.; Zhang, S.; Li, W.; Leng, J.; Wang, L.; Liu, H.; Li, W.; Zhang, C.; Qi, L.; Tuomilehto, J.; et al. Maternal Gestational Diabetes Is Associated With Offspring's Hypertension. *Am. J. Hypertens.* **2019**, *32*, 335–342. [CrossRef] [PubMed]
77. King, A.J. The use of animal models in diabetes research. *Br. J. Pharmacol.* **2012**, *166*, 877–894. [CrossRef] [PubMed]
78. Bahri Khomami, M.; Joham, A.E.; Boyle, J.A.; Piltonen, T.; Silagy, M.; Arora, C.; Misso, M.L.; Teede, H.J.; Moran, L.J. Increased maternal pregnancy complications in polycystic ovary syndrome appear to be independent of obesity-A systematic review, meta-analysis, and meta-regression. *Obes. Rev.* **2019**, *20*, 659–674. [CrossRef] [PubMed]
79. Kalagiri, R.R.; Carder, T.; Choudhury, S.; Vora, N.; Ballard, A.R.; Govande, V.; Drever, N.; Beeram, M.R.; Uddin, M.N. Inflammation in Complicated Pregnancy and Its Outcome. *Am. J. Perinatol.* **2016**, *33*, 1337–1356. [CrossRef]
80. Carlsen, S.M.; Romundstad, P.; Jacobsen, G. Early second-trimester maternal hyperandrogenemia and subsequent preeclampsia: A prospective study. *Acta Obstet. Gynecol. Scand.* **2005**, *84*, 117–121. [CrossRef]
81. Hsu, C.N.; Tain, Y.L. Light and Circadian Signaling Pathway in Pregnancy: Programming of Adult Health and Disease. *Int. J. Mol. Sci.* **2020**, *21*, 2232. [CrossRef]
82. Hsu, C.N.; Chan, J.Y.H.; Yu, H.R.; Lee, W.C.; Wu, K.L.H.; Chang-Chien, G.P.; Lin, S.; Hou, C.Y.; Tain, Y.L. Targeting on Gut Microbiota-Derived Metabolite Trimethylamine to Protect Adult Male Rat Offspring against Hypertension Programmed by Combined Maternal High-Fructose Intake and Dioxin Exposure. *Int. J. Mol. Sci.* **2020**, *21*, 5488. [CrossRef]
83. Hsu, C.N.; Lin, Y.J.; Lu, P.C.; Tain, Y.L. Maternal resveratrol therapy protects male rat offspring against programmed hypertension induced by TCDD and dexamethasone exposures: Is it relevant to aryl hydrocarbon receptor? *Int. J. Mol. Sci.* **2018**, *19*, 2459. [CrossRef] [PubMed]
84. Hsu, C.N.; Lin, Y.J.; Tain, Y.L. Maternal exposure to bisphenol A combined with high-fat diet-induced programmed hypertension in adult male rat offspring: Effects of resveratrol. *Int. J. Mol. Sci.* **2019**, *20*, 4382. [CrossRef] [PubMed]
85. Kirkley, A.G.; Sargis, R.M. Environmental endocrine disruption of energy metabolism and cardiovascular risk. *Curr. Diab. Rep.* **2014**, *14*, 494. [CrossRef] [PubMed]
86. Slotkin, T.A. Cholinergic systems in brain development and disruption by neurotoxicants: Nicotine, environmental tobacco smoke, organophosphates. *Toxicol. Appl. Pharmacol.* **2004**, *198*, 132–151. [CrossRef] [PubMed]
87. Xiao, D.; Huang, X.; Li, Y.; Dasgupta, C.; Wang, L.; Zhang, L. Antenatal Antioxidant Prevents Nicotine-Mediated Hypertensive Response in Rat Adult Offspring. *Biol. Reprod.* **2015**, *93*, 66. [CrossRef] [PubMed]
88. Gray, S.P.; Denton, K.M.; Cullen-McEwen, L.; Bertram, J.F.; Moritz, K.M. Prenatal exposure to alcohol reduces nephron number and raises blood pressure in progeny. *J. Am. Soc. Nephrol.* **2010**, *21*, 1891–1902. [CrossRef] [PubMed]
89. Serapiao Moraes, D.F.; Souza-Mello, V.; Aguila, M.B.; Mandarim-de-Lacerda, C.A.; Faria, T.S. Maternal caffeine administration leads to adverse effects on adult mice offspring. *Eur. J. Nutr.* **2013**, *52*, 1891–1900. [CrossRef] [PubMed]
90. Slabiak-Blaz, N.; Adamczak, M.; Gut, N.; Grajoszek, A.; Nyengaard, J.R.; Ritz, E.; Wiecek, A. Administration of cyclosporine a in pregnant rats—The effect on blood pressure and on the glomerular number in their offspring. *Kidney Blood Press. Res.* **2015**, *40*, 413–423. [CrossRef] [PubMed]
91. Chahoud, I.; Stahlmann, R.; Merker, H.J.; Neubert, D. Hypertension and nephrotoxic lesions in rats 1 year after prenatal expo-sure to gentamicin. *Arch. Toxicol.* **1988**, *62*, 274–284. [CrossRef] [PubMed]
92. Hsu, C.N.; Chan, J.Y.H.; Wu, K.L.H.; Yu, H.R.; Lee, W.C.; Hou, C.Y.; Tain, Y.L. Altered Gut Microbiota and Its Metabolites in Hypertension of Developmental Origins: Exploring Differences between Fructose and Antibiotics Exposure. *Int. J. Mol. Sci.* **2021**, *22*, 2674. [CrossRef] [PubMed]
93. Gois, P.H.; Canale, D.; Luchi, W.M.; Volpini, R.A.; Veras, M.M.; Costa Nde, S.; Shimizu, M.H.; Seguro, A.C. Tenofovir during pregnancy in rats: A novel pathway for programmed hypertension in the offspring. *J. Antimicrob. Chemother.* **2015**, *70*, 1094–1105. [CrossRef] [PubMed]
94. Tain, Y.L.; Sheen, J.M.; Chen, C.C.; Yu, H.R.; Tiao, M.M.; Kuo, H.C.; Huang, L.T. Maternal citrulline supplementation prevents prenatal dexamethasone-induced programmed hypertension. *Free Radic. Res.* **2014**, *48*, 580–586. [CrossRef] [PubMed]
95. Tai, I.H.; Sheen, J.M.; Lin, Y.J.; Yu, H.R.; Tiao, M.M.; Chen, C.C.; Huang, L.T.; Tain, Y.L. Maternal N-acetylcysteine therapy regulates hydrogen sulfide-generating pathway and prevents programmed hypertension in male offspring exposed to prenatal dexamethasone and postnatal high-fat diet. *Nitric Oxide* **2016**, *53*, 6–12. [CrossRef] [PubMed]
96. Chang, H.Y.; Tain, Y.L. Postnatal dexamethasone-induced programmed hypertension is related to the regulation of melatonin and its receptors. *Steroids* **2016**, *108*, 1–6. [CrossRef] [PubMed]
97. Gwathmey, T.M.; Shaltout, H.A.; Rose, J.C.; Diz, D.I.; Chappell, M.C. Glucocorticoid-induced fetal programming alters the functional complement of angiotensin receptor subtypes within the kidney. *Hypertension* **2011**, *57*, 620–626. [CrossRef] [PubMed]

98. Dodic, M.; Abouantoun, T.; O'Connor, A.; Wintour, E.M.; Moritz, K.M. Programming effects of short prenatal exposure to dexamethasone in sheep. *Hypertension* **2002**, *40*, 729–734. [CrossRef]

99. Tain, Y.L.; Hsu, C.N. Developmental Origins of Chronic Kidney Disease: Should We Focus on Early Life? *Int. J. Mol. Sci.* **2017**, *18*, 381. [CrossRef] [PubMed]

100. Tain, Y.L.; Lee, W.C.; Wu, K.; Leu, S.; Chan, J.Y.H. Maternal high fructose intake increases the vulnerability to post-weaning high-fat diet induced programmed hypertension in male offspring. *Nutrients* **2018**, *10*, 56. [CrossRef]

101. Dennery, P.A. Oxidative stress in development: Nature or nurture? *Free Radic. Biol. Med.* **2010**, *49*, 1147–1151. [CrossRef]

102. Tain, Y.L.; Lee, W.C.; Wu, K.L.H.; Leu, S.; Chan, J.Y.H. Resveratrol Prevents the Development of Hypertension Programmed by Maternal Plus Post-Weaning High-Fructose Consumption through Modulation of Oxidative Stress, Nutrient-Sensing Signals, and Gut Microbiota. *Mol. Nutr. Food Res.* **2018**, *30*, e1800066. [CrossRef] [PubMed]

103. Hsu, C.N.; Tain, Y.L. Developmental Origins of Kidney Disease: Why Oxidative Stress Matters? *Antioxidants* **2021**, *10*, 33. [CrossRef]

104. Te Riet, L.; van Esch, J.H.; Roks, A.J.; van den Meiracker, A.H.; Danser, A.H. Hypertension: Renin-angiotensin-aldosterone system alterations. *Circ. Res.* **2015**, *116*, 960–975. [CrossRef] [PubMed]

105. Gubler, M.C.; Antignac, C. Renin-angiotensin system in kidney development: Renal tubular dysgenesis. *Kidney Int.* **2010**, *77*, 400–406. [CrossRef] [PubMed]

106. Cragan, J.D.; Young, B.A.; Correa, A. Renin-Angiotensin System Blocker Fetopathy. *J. Pediatr.* **2015**, *167*, 792–794. [CrossRef] [PubMed]

107. Mullins, J.J.; Peters, J.; Ganten, D. Fulminant hypertension in transgenic rats harbouring the mouse Ren-2 gene. *Nature* **1990**, *344*, 541–544. [CrossRef] [PubMed]

108. Woods, L.L.; Rasch, R. Perinatal ANG II programs adult blood pressure, glomerular number and renal function in rats. *Am. J. Physiol. Regul. Integr. Comp. Physiol.* **1998**, *275*, R1593–R1599. [CrossRef] [PubMed]

109. Hsu, C.N.; Tain, Y.L. Targeting the Renin–Angiotensin–Aldosterone System to Prevent Hypertension and Kidney Disease of Developmental Origins. *Int. J. Mol. Sci.* **2021**, *22*, 2298. [CrossRef] [PubMed]

110. Lin, Y.J.; Huang, L.T.; Tsai, C.C.; Sheen, J.M.; Tiao, M.M.; Yu, H.R.; Lin, I.C.; Tain, Y.L. Maternal high-fat diet sex-specifically alters placental morphology and transcriptome in rats: Assessment by next-generation sequencing. *Placenta* **2019**, *78*, 44–53. [CrossRef] [PubMed]

111. Wichi, R.B.; Souza, S.B.; Casarini, D.E.; Morris, M.; Barreto-Chaves, M.L.; Irigoyen, M.C. Increased blood pressure in the offspring of diabetic mothers. *Am. J. Physiol. Regul. Integr. Comp. Physiol.* **2005**, *288*, R1129–R1133. [CrossRef] [PubMed]

112. Hsu, C.N.; Wu, K.L.; Lee, W.C.; Leu, S.; Chan, J.Y.; Tain, Y.L. Aliskiren Administration during Early Postnatal Life Sex-Specifically Alleviates Hypertension Programmed by Maternal High Fructose Consumption. *Front. Physiol.* **2016**, *7*, 299. [CrossRef] [PubMed]

113. Manning, J.; Vehaskari, V.M. Postnatal modulation of prenatally programmed hypertension by dietary Na and ACE inhibition. *Am. J. Physiol. Regul. Integr. Comp. Physiol.* **2005**, *288*, R80–R84. [CrossRef] [PubMed]

114. Sherman, R.C.; Langley-Evans, S.C. Early administration of angiotensin-converting enzyme inhibitor captopril, prevents the development of hypertension programmed by intrauterine exposure to a maternal low-protein diet in the rat. *Clin. Sci.* **1998**, *94*, 373–381. [CrossRef]

115. Bessa, A.S.M.; Jesus, É.F.; Nunes, A.D.C.; Pontes, C.N.R.; Lacerda, I.S.; Costa, J.M.; Souza, E.J.; Lino-Júnior, R.S.; Biancardi, M.F.; Dos Santos, F.C.A.; et al. Stimulation of the ACE2/Ang-(1-7)/Mas axis in hypertensive pregnant rats attenuates cardiovascular dysfunction in adult male offspring. *Hypertens. Res.* **2019**, *42*, 1883–1893. [CrossRef] [PubMed]

116. Bertram, J.F.; Douglas-Denton, R.N.; Diouf, B.; Hughson, M.D.; Hoy, W.E. Human nephron number: Implications for health and disease. *Pediatr. Nephrol.* **2011**, *26*, 1529–1533. [CrossRef] [PubMed]

117. Luyckx, V.A.; Brenner, B.M. The clinical importance of nephron mass. *J. Am. Soc. Nephrol.* **2010**, *21*, 898–910. [CrossRef] [PubMed]

118. Tain, Y.L.; Chan, S.H.H.; Chan, J.Y.H. Biochemical basis for pharmacological intervention as a reprogramming strategy against hypertension and kidney disease of developmental origin. *Biochem. Pharmacol.* **2018**, *153*, 82–90. [CrossRef] [PubMed]

119. Vehaskari, V.M.; Stewart, T.; Lafont, D.; Soyez, C.; Seth, D.; Manning, J. Kidney angiotensin and angiotensin receptor expression in prenatally programmed hypertension. *Am. J. Physiol. Ren. Physiol.* **2004**, *287*, F262–F267. [CrossRef]

120. Hao, X.Q.; Zhang, H.G.; Yuan, Z.B.; Yang, D.L.; Hao, L.Y.; Li, X.H. Prenatal exposure to lipopolysaccharide alters the intrarenal renin-angiotensin system and renal damage in offspring rats. *Hypertens. Res.* **2010**, *33*, 76–82. [CrossRef] [PubMed]

121. Walton, S.L.; Bielefeldt-Ohmann, H.; Singh, R.R.; Li, J.; Paravicini, T.M.; Little, M.H.; Moritz, K.M. Prenatal hypoxia leads to hypertension, renal renin-angiotensin system activation and exacerbates salt-induced pathology in a sex-specific manner. *Sci. Rep.* **2017**, *7*, 8241. [CrossRef] [PubMed]

122. Tamburini, S.; Shen, N.; Wu, H.C.; Clemente, J.C. The microbiome in early life: Implications for health outcomes. *Nat. Med.* **2016**, *22*, 713–722. [CrossRef]

123. Stiemsma, L.T.; Michels, K.B. The role of the microbiome in the developmental origins of health and disease. *Pediatrics* **2018**, *141*, e20172437. [CrossRef] [PubMed]

124. Krautkramer, K.A.; Kreznar, J.H.; Romano, K.A.; Vivas, E.I.; Barrett-Wilt, G.A.; Rabaglia, M.E.; Keller, M.P.; Attie, A.D.; Rey, F.E.; Denu, J.M. Diet-Microbiota Interactions Mediate Global Epigenetic Programming in Multiple Host Tissues. *Mol. Cell* **2016**, *64*, 982–992. [CrossRef] [PubMed]

125. Khodor, S.A.; Reichert, B.; Shatat, I.F. The Microbiome and Blood Pressure: Can Microbes Regulate Our Blood Pressure? *Front. Pediatr.* **2017**, *5*, 138. [CrossRef] [PubMed]
126. Hsu, C.N.; Hou, C.Y.; Chang-Chien, G.P.; Lin, S.; Tain, Y.L. Maternal Garlic Oil Supplementation Prevents High-Fat Diet-induced Hypertension in Adult Rat Offspring: Implications of H2S-generating Pathway in the Gut and Kidneys. *Mol. Nutr. Food Res.* **2021**, e2001116. [CrossRef] [PubMed]
127. Hsu, C.N.; Lin, Y.J.; Hou, C.Y.; Tain, Y.L. Maternal Administration of Probiotic or Prebiotic Prevents Male Adult Rat Offspring against Developmental Programming of Hypertension Induced by High Fructose Consumption in Pregnancy and Lactation. *Nutrients* **2018**, *10*, 1229. [CrossRef]
128. Hsu, C.N.; Hou, C.Y.; Chan, J.Y.H.; Lee, C.T.; Tain, Y.L. Hypertension Programmed by Perinatal High-Fat Diet: Effect of Maternal Gut Microbiota-Targeted Therapy. *Nutrients* **2019**, *11*, 2908. [CrossRef] [PubMed]
129. Lankelma, J.M.; Nieuwdorp, M.; de Vos, W.M.; Wiersinga, W.J. The gut microbiota in internal medicine: Implications for health and disease. *Neth. J. Med.* **2015**, *73*, 61–68. [PubMed]
130. Ojeda, N.B.; Intapad, S.; Alexander, B.T. Sex differences in the developmental programming of hypertension. *Acta Physiol.* **2014**, *210*, 307–316. [CrossRef] [PubMed]
131. Tomat, A.L.; Salazar, F.J. Mechanisms involved in developmental programming of hypertension and renal diseases. Gender differences. *Horm. Mol. Biol. Clin. Investig.* **2014**, *18*, 63–77. [CrossRef]
132. Vina, J.; Gambini, J.; Lopez-Grueso, R.; Abdelaziz, K.M.; Jove, M.; Borras, C. Females live longer than males: Role of oxidative stress. *Curr. Pharm. Des.* **2011**, *17*, 3959–3965. [CrossRef] [PubMed]
133. Hilliard, L.M.; Sampson, A.K.; Brown, R.D.; Denton, K.M. The "his and hers" of the renin-angiotensin system. *Curr. Hypertens. Rep.* **2013**, *15*, 71–79. [CrossRef]
134. Beale, A.L.; Kaye, D.M.; Marques, F.Z. The role of the gut microbiome in sex differences in arterial pressure. *Biol. Sex. Differ.* **2019**, *10*, 22. [CrossRef]
135. Tain, Y.L.; Wu, M.S.; Lin, Y.J. Sex differences in renal transcriptome and programmed hypertension in offspring exposed to prenatal dexamethasone. *Steroids* **2016**, *115*, 40–46. [CrossRef] [PubMed]
136. Tain, Y.L.; Hsu, C.N. Interplay between oxidative stress and nutrient sensing signaling in the developmental origins of cardiovascular disease. *Int. J. Mol. Sci.* **2017**, *18*, 841. [CrossRef] [PubMed]
137. Kett, M.M.; Denton, K.M. Renal programming: Cause for concern? *Am. J. Physiol. Regul. Integr. Comp. Physiol.* **2011**, *300*, R791–R803. [CrossRef] [PubMed]
138. Bogdarina, I.; Welham, S.; King, P.J.; Burns, S.P.; Clark, A.J. Epigenetic modification of the renin-angiotensin system in the fetal programming of hypertension. *Circ. Res.* **2007**, *100*, 520–526. [CrossRef] [PubMed]
139. Cavanal Mde, F.; Gomes, G.N.; Forti, A.L.; Rocha, S.O.; Franco Mdo, C.; Fortes, Z.B.; Gil, F.Z. The influence of L-arginine on blood pressure, vascular nitric oxide and renal morphometry in the offspring from diabetic mothers. *Pediatr. Res.* **2007**, *62*, 145–150. [CrossRef]
140. Thaeomor, A.; Teangphuck, P.; Chaisakul, J.; Seanthaweesuk, S.; Somparn, N.; Roysommuti, S. Perinatal Taurine Supplementation Prevents Metabolic and Cardiovascular Effects of Maternal Diabetes in Adult Rat Offspring. *Adv. Exp. Med. Biol.* **2017**, *975*, 295–305. [PubMed]
141. Torrens, C.; Brawley, L.; Anthony, F.W.; Dance, C.S.; Dunn, R.; Jackson, A.A.; Poston, L.; Hanson, M.A. Folate supplementation during pregnancy improves offspring cardiovascular dysfunction induced by protein restriction. *Hypertension* **2006**, *47*, 982–987. [CrossRef] [PubMed]
142. Hsu, C.N.; Tain, Y.L. Regulation of Nitric Oxide Production in the Developmental Programming of Hypertension and Kidney Disease. *Int. J. Mol. Sci.* **2019**, *20*, 681. [CrossRef]
143. Wesseling, S.; Essers, P.B.; Koeners, M.P.; Pereboom, T.C.; Braam, B.; van Faassen, E.E.; Macinnes, A.W.; Joles, J.A. Perinatal exogenous nitric oxide in fawn-hooded hypertensive rats reduces renal ribosomal biogenesis in early life. *Front. Genet.* **2011**, *2*, 52. [CrossRef]
144. Uson-Lopez, R.A.; Kataoka, S.; Mukai, Y.; Sato, S.; Kurasaki, M. Melinjo (*Gnetum gnemon*) Seed Extract Consumption during Lactation Improved Vasodilation and Attenuated the Development of Hypertension in Female Offspring of Fructose-Fed Pregnant Rats. *Birth Defects Res.* **2018**, *110*, 27–34. [CrossRef] [PubMed]
145. Hsu, C.N.; Tain, Y.L. Preventing Developmental Origins of Cardiovascular Disease: Hydrogen Sulfide as a Potential Target? *Antioxidants* **2021**, *10*, 247. [CrossRef] [PubMed]
146. Hsu, C.N.; Lin, Y.J.; Lu, P.C.; Tain, Y.L. Early supplementation of D-cysteine or L-cysteine prevents hypertension and kidney damage in spontaneously hypertensive rats exposed to high-salt intake. *Mol. Nutr. Food Res.* **2018**, *62*, 2. [CrossRef] [PubMed]
147. Xiao, D.; Wang, L.; Huang, X.; Li, Y.; Dasgupta, C.; Zhang, L. Protective Effect of Antenatal Antioxidant on Nicotine-Induced Heart Ischemia-Sensitive Phenotype in Rat Offspring. *PLoS ONE* **2016**, *11*, e0150557. [CrossRef] [PubMed]
148. Feng, X.; Guo, Q.; Xue, H.; Duan, X.; Jin, S.; Wu, Y. Hydrogen Sulfide Attenuated Angiotensin II-Induced Sympathetic Excitation in Offspring of Renovascular Hypertensive Rats. *Front. Pharmacol.* **2020**, *11*, 565726. [CrossRef] [PubMed]
149. Tain, Y.L.; Hsu, C.N. AMP-Activated Protein Kinase as a Reprogramming Strategy for Hypertension and Kidney Disease of Developmental Origin. *Int. J. Mol. Sci.* **2018**, *19*, 1744. [CrossRef] [PubMed]

150. Torres, T.S.; D'Oliveira Silva, G.; Aguila, M.B.; de Carvalho, J.J.; Mandarim-De-Lacerda, C.A. Effects of rosiglitazone (a peroxysome proliferator-activated receptor γ agonist) on the blood pressure and aortic structure in metabolically programmed (perinatal low protein) rats. *Hypertens. Res.* **2008**, *31*, 965–975. [CrossRef] [PubMed]

151. Gray, C.; Vickers, M.H.; Segovia, S.A.; Zhang, X.D.; Reynolds, C.M. A maternal high fat diet programmes endothelial function and cardiovascular status in adult male offspring independent of body weight, which is reversed by maternal conjugated linoleic acid (CLA) supplementation. *PLoS ONE* **2015**, *10*, e0115994.

152. Yousefipour, Z.; Newaz, M. PPARα ligand clofibrate ameliorates blood pressure and vascular reactivity in spontaneously hypertensive rats. *Acta Pharmacol. Sin.* **2014**, *35*, 476–482. [CrossRef] [PubMed]

153. Wu, L.; Wang, R.; de Champlain, J.; Wilson, T.W. Beneficial and deleterious effects of rosiglitazone on hypertension development in spontaneously hypertensive rats. *Am. J. Hypertens.* **2004**, *17*, 749–756. [CrossRef] [PubMed]

154. Tain, Y.L.; Hsu, C.N.; Chan, J.Y. PPARs link early life nutritional insults to later programmed hypertension and metabolic syndrome. *Int. J. Mol. Sci.* **2015**, *17*, 20. [CrossRef] [PubMed]

155. Hunter, D.S.; Hazel, S.J.; Kind, K.L.; Owens, J.A.; Pitcher, J.B.; Gatford, K.L. Programming the brain: Common outcomes and gaps in knowledge from animal studies of IUGR. *Physiol. Behav.* **2016**, *164*, 233–248. [CrossRef] [PubMed]

156. Rosenblum, S.; Pal, A.; Reidy, K. Renal development in the fetus and premature infant. *Semin. Fetal Neonatal Med.* **2017**, *22*, 58–66. [CrossRef] [PubMed]

157. Hartman, H.A.; Lai, H.L.; Patterson, L.T. Cessation of renal morphogenesis in mice. *Dev. Biol.* **2007**, *310*, 379–387. [CrossRef] [PubMed]

158. Sengupta, P. The Laboratory Rat: Relating Its Age with Human's. *Int. J. Prev. Med.* **2013**, *4*, 624–630. [PubMed]

159. Chahoud, I.; Paumgartten, F.J.R. Influence of litter size on the postnatal growth of rat pups: Is there a rationale for litter-size standardization in toxicity studies? *Environ. Res.* **2009**, *109*, 1021–1027. [CrossRef] [PubMed]

160. Barry, J.S.; Anthony, R.V. The pregnant sheep as a model for human pregnancy. *Theriogenology* **2008**, *69*, 55–67. [CrossRef] [PubMed]

161. Van Abeelen, A.F.; Veenendaal, M.V.; Painter, R.C.; de Rooij, S.R.; Thangaratinam, S.; van der Post, J.A.; Bossuyt, P.M.; Elias, S.G.; Uiterwaal, C.S.; Grobbee, D.E.; et al. The fetal origins of hypertension: A systematic review and meta-analysis of the evidence from animal experiments of maternal undernutrition. *J. Hypertens.* **2012**, *30*, 2255–2267. [CrossRef]

162. Krege, J.H.; Hodgin, J.B.; Hagaman, J.R.; Smithies, O. A noninvasive computerized tail-cuff system for measuring blood pressure in mice. *Hypertension* **1995**, *25*, 1111–1115. [CrossRef] [PubMed]

163. Morrison, J.L.; Botting, K.J.; Darby, J.R.T.; David, A.L.; Dyson, R.M.; Gatford, K.L.; Gray, C.; Herrera, E.A.; Hirst, J.J.; Kim, B.; et al. Guinea pig models for translation of the developmental origins of health and disease hypothesis into the clinic. *J. Physiol.* **2018**, *596*, 5535–5569. [CrossRef] [PubMed]

164. Gonzalez-Bulnes, A.; Astiz, S.; Ovilo, C.; Lopez-Bote, C.J.; Torres-Rovira, L.; Barbero, A.; Ayuso, M.; Garcia-Contreras, C.; Vazquez-Gomez, M. Developmental Origins of Health and Disease in swine: Implications for animal production and biomedical research. *Theriogenology* **2016**, *86*, 110–119. [CrossRef] [PubMed]

165. Kuo, A.H.; Li, C.; Li, J.; Huber, H.F.; Nathanielsz, P.W.; Clarke, G.D. Cardiac remodeling in a baboon model of intrauterine growth restriction mimics accelerated ageing. *J. Physiol.* **2017**, *595*, 1093–1110. [CrossRef] [PubMed]

166. Haider, B.A.; Bhutta, Z.A. Multiple-micronutrient supplementation for women during pregnancy. *Cochrane Database Syst. Rev.* **2017**, *4*, CD004905. [CrossRef] [PubMed]

167. Schwarzenberg, S.J.; Georgieff, M.K.; Committee on Nutrition. Advocacy for Improving Nutrition in the First 1000 Days to Support Childhood Development and Adult Health. *Pediatrics* **2018**, *141*, e20173716. [CrossRef] [PubMed]

168. Safi-Stibler, S.; Gabory, A. Epigenetics and the Developmental Origins of Health and Disease: Parental environment signalling to the epigenome, critical time windows and sculpting the adult phenotype. *Semin. Cell Dev. Biol.* **2020**, *97*, 172–180. [CrossRef]

Review

Cisplatin Mouse Models: Treatment, Toxicity and Translatability

Martina Perše

Medical Experimental Centre, Institute of Pathology, Faculty of Medicine, University of Ljubljana, 1000 Ljubljana, Slovenia; martina.perse@mf.uni-lj.si; Tel.: +386-4047-4675

Abstract: Cisplatin is one of the most widely used chemotherapeutic drugs in the treatment of a wide range of pediatric and adult malignances. However, it has various side effects which limit its use. Cisplatin mouse models are widely used in studies investigating cisplatin therapeutic and toxic effects. However, despite numerous promising results, no significant improvement in treatment outcome has been achieved in humans. There are many drawbacks in the currently used cisplatin protocols in mice. In the paper, the most characterized cisplatin protocols are summarized together with weaknesses that need to be improved in future studies, including hydration and supportive care. As demonstrated, mice respond to cisplatin treatment in similar ways to humans. The paper thus aims to illustrate the complexity of cisplatin side effects (nephrotoxicity, gastrointestinal toxicity, neurotoxicity, ototoxicity and myelotoxicity) and the interconnectedness and interdependence of pathomechanisms among tissues and organs in a dose- and time-dependent manner. The paper offers knowledge that can help design future studies more efficiently and interpret study outcomes more critically. If we want to understand molecular mechanisms and find therapeutic agents that would have a potential benefit in clinics, we need to change our approach and start to treat animals as patients and not as tools.

Keywords: cisplatin; toxicity; kidney; gut; nerve system; mouse; treatment; tumors; mouse model

Citation: Perše, M. Cisplatin Mouse Models: Treatment, Toxicity and Translatability. *Biomedicines* **2021**, 9, 1406. https://doi.org/10.3390/biomedicines9101406

Academic Editor: Julie Chan

Received: 1 September 2021
Accepted: 5 October 2021
Published: 7 October 2021

Publisher's Note: MDPI stays neutral with regard to jurisdictional claims in published maps and institutional affiliations.

1. Introduction

In 2018, 18.1 million new cases and 9.5 million cancer-related deaths were diagnosed worldwide. It is estimated that by 2040 the number of new cancer cases per year will rise to 29.5 million and the number of cancer-related deaths to 16.4 millions [1].

Despite intensive research and progress in cancer therapy, chemotherapeutic drugs are still the basis of systemic therapy for many cancers. Cisplatin is one of the most widely prescribed chemotherapeutic drugs, used to treat a wide range of pediatric and adult malignances such as ovarian, testicular, bladder, head, neck, breast and lung [2,3]. It is prescribed in nearly 50% of all tumor chemotherapies [4]. However, it has limited use in clinical practice due to various deleterious side effects. Currently, around 40 side effects of cisplatin have been reported [5]. Extensive supportive medical care of cisplatin treated cancer patients enables the use of very high-dose cisplatin regimens [3,6–8]. However, with the use of high-dose treatment regimens acute kidney injury, persistent diarrhea, neurological disorders and loss of hearing became major hurdles of cisplatin therapy. These unwanted effects result in reduction or cessation of therapy or have a major impact on patients' quality of life, leading to higher levels of negative states such as depression and anxiety. There is no effective therapy for the prevention of these side effects; the current treatment strategy is symptomatic with limited effectiveness.

Based on extensive research, and after 40 years of cisplatin use, the anti-cancer effects of cisplatin are well understood, while the mechanisms of cisplatin toxicity remain unclear [9–11]. Therefore, increasing emphasis is being placed on various strategies to reveal the mechanisms responsible for toxicities and to overcome cisplatin side effects.

Cisplatin mouse models are a promising strategy; however, despite intensive investigation and numerous promising results, no significant improvement in the treatment outcomes has been reached in clinical practice [12–14].

The aim of the present paper is firstly to illustrate the complexity of cisplatin mouse models. The review summarizes the data to demonstrating that mice respond to cisplatin treatment in a similar way as humans. Mice develop all the same cisplatin side effects that humans do. However, in contrast to cisplatin treated cancer patients, in which all cisplatin side effects are monitored and treated, in animal studies usually one cisplatin toxicity is under investigation, while other side effects of cisplatin are mostly neglected or ignored. For instance, in cisplatin nephrotoxicity or neurotoxicity studies gastrointestinal injury is usually neglected. Secondly, to encourage researchers to take into consideration all events that are taking place in a mouse and to reconsider the severity and the time course of toxicity in accordance with other interdependent and interconnected mechanisms and toxicities in the body. Understanding the complexity of cisplatin side effects in a dose- and time-dependent manner as well as the interconnectedness and interplay of pathomechanisms among tissues and organs can help design future studies more efficiently and interpret study outcomes more critically. Thirdly, the review aims to expose limitations and weaknesses of current cisplatin protocols together with suggestions for future studies. It is important to recognize that not only the lack of complex knowledge and approaches but also a lack of robust and validated cisplatin mouse models are important factors that contributed to poor translatability. Since the literature is extremely numerous, only the most relevant articles are included.

2. Cisplatin Mouse Models

Cisplatin mouse models have been used to investigate pharmacokinetics and tissue distribution of cisplatin [15–17], the repair capacity of cisplatin-DNA adducts [18], the molecular mechanisms of cisplatin toxicity [19–23] and to test a new generation of platinum-based chemotherapy drugs or adjunctive therapies [24–35], or other potential agents or strategies to prevent or treat cisplatin toxicities [36–41].

Table 1 shows cisplatin protocols in mice reported in publications in 2020/2021. Nephrotoxicity [42–44] was by far the most frequently studied toxicity, followed by neurotoxicity [45–47], ototoxicity [16,48,49], gonadotoxicity [50–53], gastrointestinal toxicity [54–56], muscle wasting [57–59] and anemia [60].

Table 1. Examples of mouse cisplatin toxicity protocols used in studies published from April 2020 to February 2021.

Cisplatin Toxicity (100%)	Cisplatin Protocols	Endpoint
Nephrotoxicity (57.1%)	8, 10, 15, 20, 25, 30, 40 mg/kg single ip (most frequently used 20 mg/kg single ip)	d3–d4
Gastrointestinal toxicity (3.6%)	20 mg/kg single ip	d3
Ototoxicity (10.7%)	30 mg/kg single ip (FVB; hydration 1mL 2xdaily; 50% mortality)	d21
	3 cycles: 3 × (3–3.5 mg/kg/daily for 4 days followed by 10 days recovery) Cumulative dose = 36 or 42 mg/kg	d42
Neurotoxicity (10.7%)	2 cycles: 2 × (2.3 mg/kg/daily for 5 days followed by 5 days recovery) Cumulative dose = 23 mg/kg	d15, d30, d65
Gonadotoxicity (10.7%)	3 × 5 mg/kg ip	d4
	5 × 3mg/kg ip	d14
Muscle (3.6%)	4 × 3 mg/kg/daily	d4
Anemia (3.6%)	4 × 7mg/kg/week ip	2 months

A MEDILINE/PubMed search, using keywords "cisplatin","mouse","toxicity" was conducted in February 2021. Due to a huge number of publications (more than 10.000 results), the search was limited to studies from April 2020 to February 2021. One hundred full text articles were retrieved and examined. Special attention was paid to the cisplatin treatment protocol, duration of the study, model (healthy or tumor bearing), humane intervention and endpoints used, clinical markers of toxicity, necropsy or histology findings, randomization (group allocation). To better evaluate cisplatin treatment protocols in mouse studies, publications were divided in two groups. In the first group, cisplatin treatment was used mostly as a positive control to evaluate the antitumor activity of a novel agent or treatment strategy in tumor bearing mice. Due to huge variability in cisplatin protocols, results are presented in Table S1. In the second group cisplatin treatment was used to investigate cisplatin toxicity in healthy mice. Most frequently used cisplatin protocols for specific toxicity are presented in Table 1. In toxicity studies only the toxicity under investigation was examined, while other side effects of cisplatin were not reported. Hydration was rarely used, supportive care never. Legend: ip—intraperitonealy, d—day.

3. Cisplatin Nephrotoxicity

Experimental nephrotoxicity, first reported in 1971 [61], is the most frequently studied cisplatin side effect. Over the past decades, researchers have demonstrated that cisplatin can cause nephrotoxicity or acute kidney injury (AKI) of varying severity in a dose-dependent manner [62]. Depending on the dose (single or cumulative) rodents may develop acute (early) or chronic (advanced) kidney injury. However, AKI evolves slowly and predictably after initial and repeated exposure. Unlike other drug toxicities, clinical evidence of cisplatin nephrotoxicity develops within a few days after administration. In clinical practice, nephrotoxicity typically presents approximately 10 days after cisplatin treatment [3]. In mice, clinical evidence of nephrotoxicity develops 4–6 days after a single sub-lethal dose of cisplatin (Figure 1). To better understand development of the nephrotoxicity after a single sub-lethal nephrotoxic dose of cisplatin, the development over time of morphological, functional, and clinical changes is schematically presented in Figure 1.

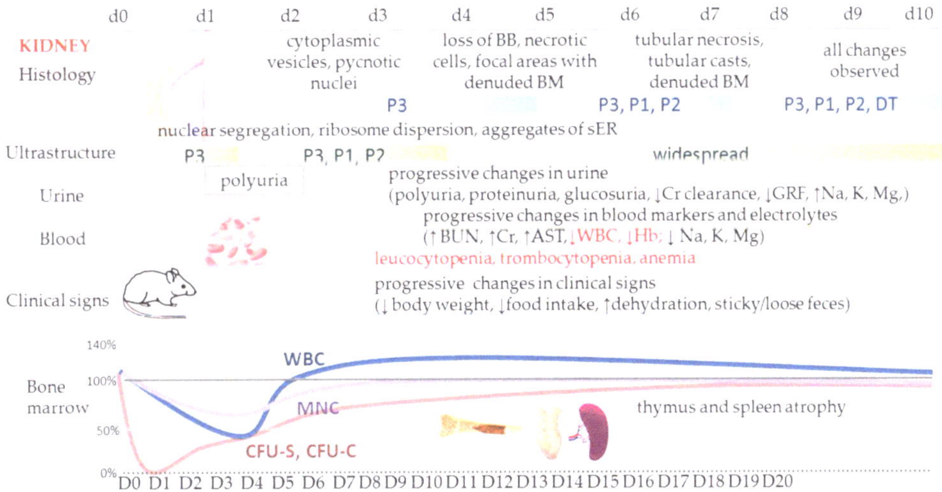

Figure 1. Schematic presentation of clinical signs and kidney function in mice after a nephrotoxic dose of cisplatin. First, two days after a single high dose of cisplatin (10–13 mg/kg; ip), minimal structural changes in the proximal tubules (P3) can be detected (i.e., mitochondria alterations, focal loss of the microvillus brush border, pycnotic nuclei, increased cytoplasmic vesicles) [63,64]. More obvious changes such as loss of the brush border or necrotic cells sloughing into the tubular lumen are usually seen 3–4 days after injection and changes are located in all parts of the proximal tubules (P1–3) [63–65]. Depending on the dose, increased BUN/Cr are usually observed 3–7 days after cisplatin injection [66–69],

and if nephrotoxicity is reversible, BUN/Cr return to the baseline levels within 14 days [70]. In such cases, the first signs of structural regeneration can be observed 7 days after cisplatin injection [64,71]. A single high dose of cisplatin (B6D2F1: 8 mg/kg, 10 mg/kg, 12 mg/kg, 14 mg/kg; ip) induces dose-dependent weight loss (11–26%), reticulocytopenia with the lowest levels of body weight and reticulocytes observed 6 days after cisplatin injection. Necrosis in kidney tubular cells can be seen up to 10–22 days post-treatment [72]. When a lethal dose is used, death may occur within 10 days [73] and the time course of AKI development or mortality can occur slightly faster, but still 1–2 days after cisplatin injection. Cisplatin (F1 CBAxC57BL, 12 mg/kg, ip) induces lymphocytopenia, thrombocytopenia and anemia. Cisplatin exhibits cytotoxicity to spleen (CFU-S), granulocyte–macrophage (CFU-C) colony-forming units and mononuclear cells (MNC) in bone marrow and white blood cells (WBC) (adapted and modified from Nowrousian et al. [74]) Legend: P1–3 denotes kidney proximal tubules parts 1–3, DT—distal tubules; BB—brush border; BM—basal membrane; BUN- blood urea nitrogen; Cr—serum creatinine; AST—aspartate aminotransferase; WBC—white blood cells; Hb—hemoglobin; GFR—glomerular filtration rate; CFU—colony-forming unit; MNC—mononuclear cells; ER—endoplasmic reticulum.

In patients, cisplatin treatment is usually administered in cycles with 1- or 3-week intervals and a cycle consisting of a single high dose of cisplatin or multiple lower daily doses (see Table 2). In mice or rats, nephrotoxicity is mostly induced by a single cisplatin administration. Repeated cisplatin protocols for nephrotoxicity are extremely rare [44], which is also the main critique of AKI models [13]. Moreover, in mice, a wide variation of cisplatin dosage is used to induce renal toxicity, i.e., from low sub-therapeutic (5 mg/kg), sub-lethal nephrotoxic (10–12 mg/kg) to lethal dosage (14–18 mg/kg) or even higher (>20 mg/kg) [44]. The use of a different dosage of cisplatin can be useful when the time course and/or the severity of nephrotoxicity and its functional, morphological or molecular abnormalities are under systematic investigation. However, for testing potential agents or treatment strategies we need a robust and validated cisplatin mouse model. Currently, there is no standardized, robust or validated cisplatin mouse model of AKI that is clinically or physiologically relevant to patients [44].

Recently, Siskind and coworkers [75] established a mouse model of repeated administration of cisplatin (FVD, 7 mg/kg per week for 4 weeks) [76]. However, their aim was to obtain a model for chronic kidney disease [76–78]. In the past, it has already been demonstrated that cisplatin can have long term effects on kidney morphology and function after single [64,79,80] or repeated [81–83] cisplatin administration. However, the long-term toxic effects of cisplatin became the subject of investigation recently, when it was realized that even a mild and reversible AKI can have long term effects in patients [84–86] or that chronic kidney disease may develop undetected [87,88].

Table 2. Examples of cisplatin regimes used in the clinics and incidence of AKI complications.

Cisplatin Clinical Dose in Humans [1] (iv)		AKI Incidence, Severity	ref	MED??
50–75 mg/m^2#	1.35–2.03 mg/kg	25–33%, mild-moderate	[62]	16.7–25.0 mg/kg
15–20 mg/m^2 daily for 5 days#	0.41–0.54 mg/kg	50–75%, mild-moderate	[62]	5.0–6.7 mg/kg
100 mg/m^2#	2.7 mg/kg	severe to irreversible	[62]	33.4 mg/kg
75 mg/m^2 every 3 weeks up to 6 cycles *	2.03 mg/kg	53%, mild-moderate	[89]	25 mg/kg
100 mg/m^2 with concurrent radiation **	2.7 mg/kg **	47–60% of patients discontinued therapy	[90]	33.4 mg/kg
80 mg/m^2 1 h iv infusion	2.2 mg/kg	H#	[91]	27 mg/kg

[1] dose of cisplatin in humans measured as mg per skin area (mg/m^2) was translated in mg/kg using the correction factor for human body weight of 60 kg and the body surface area 1.62 m^2 (K_m = 37) [92]. #data from 1978 when supportive care measures were not established; * therapy cisplatin/docetaxel (lung cancer); cisplatin 2 h infusion every 3 weeks, antiemetic prophylaxis, hydration with up to 3000 mL of normal saline; cumulative dose = 340 mg/m^2 [89]; ** 2–3 cycles every 3 weeks; doses of cisplatin for subsequent cycles were adjusted at the discretion of the physician; (squamous cell cancer of the head and neck); cisplatin 2 h infusion diluted in 1 L of 0.9% saline and 1–2 h hydration with 1 L of saline pre and post cisplatin infusion; antiemetic premedication (dexamethasone, 5-HT$_3$ antagonist, neurokinin-1 receptor antagonist) [90]; H#—the highest dose recommended as a single administration [91]. MED—mouse equivalent dose, needs to be treated with caution (see warning in Section 3.3). Calculation was done according to guide for dose conversion using correction factor for 20 g mouse (K_m = 0.081) [92].

3.1. Pathopysiological Mechanisms

The severity of kidney injury (mild, moderate, severe) and consequently functional, morphological, molecular, and inflammatory alterations in the kidney as well as morbidity and mortality depends on the cisplatin dosage and the time of study termination [44]. The pathophysiological mechanisms of cisplatin AKI in rodents involve cellular uptake, damage of proteins, lipids and mitochondria, oxidative stress, disruption of the cytoskeletal integrity of the cell polarity, alterations in membrane proteins and water channels, leading to damage of epithelial cells of renal tubules, loss of brush border, activation of cytokines, receptors and inflammatory cells, and finally reduced reabsorptive capacity, which reflects clinically as polyuria, proteinuria, glycosuria, electrolyte wasting (hyponatremia, hypomagnesemia, hypokalemia, hypocalcemia), reduced creatinine (Cr) clearance and glomerular filtration rate (GFR) and failure to clear nitrogenous wastes from the blood. As a result, blood urea nitrogen (BUN) and uric acid accumulate in the blood (described in detail in [44]). Intensive investigation of molecular mechanisms of cisplatin nephrotoxicity resulted in a plethora of information, including contradictory ones. The latter is the consequence of above-mentioned heterogeneity of cisplatin protocols. Namely, the course and the signature of underlying mechanisms (i.e., severity of oxidative stress, intensity of inflammation, activation of particular immune cell types, inflammatory and molecular crosstalk and response, type of cell death, etc.) strongly depend on the severity of AKI (mild, moderate–reversible, severe, irreversible) or cisplatin dose (subtherapeutic, therapeutic, lethal, intoxication). An update on molecular mechanisms involved in the pathogenesis of cisplatin nephrotoxicity can be found elsewhere [93–96].

3.2. Weaknesses and Translatability

To study AKI, researchers use a single dose of cisplatin in a dosage above LD100 and terminate study 48–96 h after cisplatin treatment (Table 1). To avoid the inevitable death of animals, they use shorter endpoints. Differences in AKI severity are then confirmed by the histology report, BUN, Cr or other molecular markers. If differences in AKI severity are significant, the testing agent or strategy is evaluated as beneficial and promising [37,44]. The problems of such studies are numerous. First, the lethal dosage of cisplatin causes systemic toxicity and multi-organ failure, which clinically represent different pathology (i.e., intoxication) and treatment (i.e., detoxication). Second, evaluation of potential treatment strategies based on the significant differences in BUN, Cr, renal histology or survival in such protocols does not have any clinical useful value. At the end, animals die despite significant improvements in some markers (i.e., BUN, Cr, severity of tubular necrosis, time of mortality), which is very likely an effect of biological variability (inter-individual variability is high in both, mice and humans). In humans, severity of AKI is graded according to the levels of serum creatinine: grade 1: increased serum Cr levels to 1.5–1.9 times baseline, grade 2: increased Cr levels to 2.0–2.9 times baseline, grade 3: increased Cr levels to 3.0 times baseline or >4.0 mg/dl or initiation of renal replacement therapy [12]. Third, at autopsy and when interpreting results scientists often forget or ignore other pathologies or side effects (i.e., gastrointestinal, myelotoxicity, anemia, vasculitis, etc,) which significantly affect the study outcomes and hamper comparison of results and development of valid therapeutic strategies. Forth, cisplatin treated mice usually do not receive any supportive care. Moreover, some studies even use water deprivation prior cisplatin treatment [44]. Hydration and supportive care affect nephrotoxicity and mortality enormously and also influence Cr and BUN levels. Dehydration, degradation due to starvation or loss of body weight, gastric or intestinal bleeding [97–99], all of which are usually seen in cisplatin models (see gastrointestinal toxicity section) affect the levels of BUN/Cr resulting in misinterpretation of the actual degree of renal damage [65,97,99].

In humans, treatment with cisplatin consists of repeated cycles of as high a dose as possible. The dose in humans is balanced between antitumor efficacy and toxicity to avoid unacceptable toxic side effects. However, the patient is constantly monitored and provided with extensive supportive medical care (i.e intravenous hydration, diuretics, slow infusion of the drug, anti-emetics) [3,12,100]. Despite all care measures, severe kidney injury is observed in one third of cisplatin treated patients (inter-individual variability; Table 2) [3,12]. In addition, 16–40% of the patients treated with cisplatin develop myelotoxicities, of which leukocytopenia and neutropenia have the highest incidence [89]. Depending on a dose (single or cumulative) cisplatin can cause leucocytopenia, thrombocytopenia, and anemia also in mice (toxicity on hemopoietic cells, Figure 1). Myelotoxicity can be observed already a day after cisplatin administration and is more toxic for earlier hemopoietic progenitor cells than for the mature cells [74].

To improve cisplatin protocols, it is thus important to understand why selection of certain cisplatin dosage, the time-point of measurement or observation (i.e., scientific endpoints) and the use of supportive care are the key variables that directly affect the measured outcomes of a study and the translatability. Not only the severity of cisplatin nephrotoxicity but also the incidence of nephro-, myelo-, neuro-, oto-toxicity are dose-related in both, humans and animals (explained in the following sections). The dose of cisplatin (single and/or cumulative) is thus important not only from the animal welfare point of view but mostly from the scientific and clinical point of view. When evaluating potential treatment agents both, sub-therapeutic and lethal dosages result in a lack of translatability, unnecessary suffering of animals and time and money costs.

3.3. Mouse Equivalent Dose–Simplistic Pharmacological Guides

Some research papers refer to mouse cisplatin dose, which was calculated from the human clinical dose using simplistic calculation. Table 2 includes examples of such calculations. Based on numerous studies investigating maximum tolerated dose (MTD) or lethal dose (LD100) of cisplatin in mice (Table 3) it is obvious that such simplistic guides for dose conversion between animals and humans [92] (or other similar papers) can do more harm than benefit. "This overly simplistic conversion neglects discussion of interspecies differences in drug absorbance, metabolism, clearance, etc. These differences in pharmacokinetics greatly affect the resulting peak plasma concentration (C_{max}) values and exposure derived from area under curve (AUC), which influence the dose response relationship of potential therapeutics" [101]. In addition, in humans cisplatin is given as intravenous 1–2 h infusion treatment with pre and post hydration with up to 3 L of saline, while in mice intraperitoneal administration without any hydration is usually used. Both, route of administration and particularly hydration have profound effects on the distribution and elimination rates and consequently LD100 or MTD [91]. LD100 and MTD doses for intraperitoneal administration of cisplatin in mice without hydration are summarized in Table 3. However, it is important to emphasize that currently used MTD dose in mice does not necessary represent a clinically relevant dose of cisplatin for therapeutic efficacy [15]. Concentration may vary between a MTD dose in mice and concentration achieved in humans (due to above explained reasons) which means that study outcomes can have limited value from translational perspective of a drug. Since doses of cisplatin in published animal studies vary widely (from 1–40mg/kg, Table 1, Table S1) a publication with a clinically relevant cisplatin doses in mice, like published for some other drugs [102], is more than needed.

As explained above, the dose (single/cumulative) and the timepoint are two variables that directly relate to severity and the incidence of cisplatin nephrotoxicity. However, in the following sections, it will be shown that cisplatin induces also gastrointestinal toxicities in the body whose severity is also dose dependent and can affect cisplatin nephrotoxicity significantly. The kidney participates in the control of fluid osmolality, acid-base balance and electrolyte concentrations (i.e., Mg, K, Na, Cl), and is the main organ responsible for filtration and detoxification of the blood and is thus directly confronted with all toxins, cytokines, detrimental waste products, or microorganisms that are flushed or penetrate into the circulation from or though other organs.

Table 3. Cisplatin acute toxicity and single or repeated maximum tolerated dose given intraperitonealy varies among mouse strains.

Strain, Sex, Age	Single Dose LD100	Endpoint	Ref.
BALB/c, female, (N = 8)	14.5mg/kg; ip	d7	[69,103]
C57BL/6, male, 11–15wk; (N = 5)	15 mg/kg; ip	d10	[104]
CBA; female, 24 months, (N = 3)	16 mg/kg; ip	d7	[105]
Single MTD			
BALB/c, female, 8–10 wk; (N = 3)	6 mg/kg; ip	d10	[106]
C57BL/6J, female, 8–10 wk; (N = 3)	6 mg/kg; ip	d10	[106]
Repeated MTD			
C57BL/6J, female, 8–10 wk; (N = 3)	3 × 4 mg/kg; ip	d21	[106]

Repeated administration: once mice had recovered to 100% of their starting weight or a clinical score of 0, a second MDT was given (d0, d8, d16). MTD is defined as a dose as high as possible that causes no unacceptable toxicity such as no clinical evidence of toxicity, no reduction in mean body weight >10% to 15% and, no mortality [106]. Legend: N: number of animals; ip: intraperitonealy; iv: intravenously; LD: lethal dose; LD100: dose of cisplatin that results in 100% mortality in animals (without hydration or supportive care); d: day, MTD: maximum tolerated dose.

4. Cisplatin Gastrointestinal Toxicity

Cisplatin is one of the most emetogenic drugs in the clinic [107] causing profound and long lasting gastrointestinal symptoms such as nausea, vomiting, bloating, diarrhea, constipation [108,109]. Gastrointestinal side effects can occur in up to 40% patients receiving standard dose chemotherapy or 100% patients receiving high dose chemotherapy. Gastrointestinal problems can persist up to 10 years after the treatment cessation (late/chronic toxicity). Despite guidelines to navigate management of gastrointestinal side effects, diarrhea is responsible for about 5% of early deaths during chemotherapy [110].

Cisplatin can cause acute (within 24 h) and delayed vomiting/pica (24 h after cisplatin) in both, humans and rodents [111–113]. In rodents, acute vomiting reflects as a reduction in food intake, an increase in non-nutritive substance intake, and a delay in gastric emptying (so-called pica behavior; rodents do not have vomiting reflex). Acute pica occurs after low and high cisplatin doses, while delayed pica, including gastric stasis and stomach distension is dose-dependent (single or cumulative) [56,112–114] and worsens after repeated cisplatin administration [113].

Cisplatin causes damage to the gastrointestinal mucosa along the whole gastrointestinal tract (the stomach, small intestine and colon), however in the colon mucosal lesions appear later and are less severe [72]. Alterations are seen in the morphology [72], kinetics, secretory and digestive function and nutrition uptake [108,115]. Changes are similar to those observed in humans. Mucosal damage after single cisplatin injection can persist up to 10 days [72]. The severity of mucosal damage along the gastrointestinal tract is dose-dependent [72]. Mucosal damage with inflammation, digestive dysfunction, disruption of water and electrolyte balance are responsible for dehydration, malnutrition, and changes in feces consistency [110]. However, an initial increase in gastrointestinal transit, associated with acute intestinal inflammation, is followed by a slowing in transit. Recent studies have shown that cisplatin can cause morphological and functional alterations in the enteric neurons in a dose-dependent manner [116]. Partial loss of enteric neurons and gial cells [55] was suggested to be responsible for reduced gut motility (Figure 2).

4.1. Weaknesses and Translatability

Like humans also cisplatin treated mice suffer from nausea/pica, stomach and gut inflammation, abdominal pain, and have reduced food and water intake and altered feces consistency (from sticky, loose to diarrhea), all of which by itself is a risk factor for kidney impairment. Therefore, extensive hydration and treatment of gastrointestinal symptoms are routinely applied into clinical settings to reduce intravascular depletion of fluid and electrolytes (Mg, K, Na, Cl) and consequently the incidence and the severity of renal injury in cisplatin treated patients [3]. An example of an incidence and severity of gastrointestinal symptoms in cisplatin treated cancer patients is shown in Table 4. However, in cancer patients, age, co-morbidities (diabetes, hypertension), and concomitant nephrotoxic medications (antibiotics-infections, NSAID, etc) can increase the risk of cisplatin-induced kidney injury [3] Many cisplatin treated patients have thus AKI with mixed renal etiology, while in cisplatin animal studies AKI is mostly result of severe dehydration, malnutrition, electrolyte wasting, systemic toxicity, and cisplatin nephrotoxicity [117]. In addition, up to 100 % of patients develop Mg depletion, which has been associated with increased cisplatin transport to the kidney and enhanced cisplatin nephrotoxicity [3,12]. Thus, the development of AKI in mice and in cancer patients differs in the etiology, underlying mechanisms and importantly, the treatment [117].

Table 4. An example of an incidence and severity of cisplatin acute toxicities in cancer patients.

Severity (Grade)	Any (1–4)	Severe (3–4)		Any (1–4)	Severe (3–4)
Nausea	90.7%	23.6%	Anaemia	76.7%	2.3%
Vomiting	58.1%	14%	Leukopenia	83.7%	44.2%
Diarrhea	65.1%	18.6%	Neutropenia	72.1%	55.8%
Constipation	27.9%	0%	Thrombocytopenia	32.6%	9.3%
Stomatitis	55.8%	9.3%	Creatinine	55.8%	2.3%
Neurosensory	53.5%	2.3%	Infection	41.9%	25.6%
Fatigue	81.4%	20.1%	Fever	23.6%	0%
Weight loss	41.9%	2.3%			

Cisplatin 75 mg/m^2 every 3 weeks up to 6 cycles or until cessation (cumulative = 340 mg/m^2). Therapy cisplatin/docetacel; cisplatin 2h infusion every 3 weeks, antiemetic prophylaxis, pre and post cisplatin hydration with up to 3000 mL of normal saline [89].

4.2. Mechanisms

Investigation of the molecular mechanisms involved in gastrointestinal toxicity has not been paid much attention; thus, the literature is very scarce. The mechanism by which cisplatin induces damage to epithelial cells, neurons or glia cells is not known. Inflammation and oxidative stress involving NF-κB and TNF-α pathways have been proposed as key players (for more information see [118]). However, although enteric neurons have control over the intestinal movement [119], as shown in Hirschsprung disease in humans where loss of intrinsic enteric nervous system results in reduced or absent gut motility, other factors are also important for normal gut motility. Interestingly, cisplatin can have long-term effects in the gastrointestinal tract also in mice (Figure 3).

4.3. Gastrointestinal Toxicity Can Impair Kidney and Brain Function and Vice Versa

Motility in the gastrointestinal tract is regulated by the autonomic nervous system composed of extrinsic (i.e., parasympathetic, vagal nerve (the rest-and-digest), sympathetic (fight-or-flight)) and intrinsic enteric nervous systems (ENS) [119–121]. The primary regulator of gut motility is intrinsic ENS, followed by extrinsic ENS (parasympathetic, symphatetic) and the central nerve system. However, the gut microbiota, immune system and gut secretions also interact and modulate gut motility [122]. The gut microbiota can affect intestinal transit by modulating the anatomy of the adult ENS (in a serotonin (5-HT)-dependent fashion) [123] and activity of gut-extrinsic sympathetic neurons [124].

In addition, gut barrier dysfunction (i.e., leaky gut) is associated with various kidney disorders. Recent animal studies have demonstrated a direct link between gut inflammation and structural alterations in the kidneys [125], suggesting that persistent gastrointestinal problems of cisplatin treated patients could be involved in the pathogenesis of long-term kidney pathology. Interestingly, renal complications develop in up to 23% of patients with inflammatory bowel disease [125]. On the other hand, impaired kidney function may contribute to long-term gastrointestinal problems in cancer survivors (uremia, cytokines, etc.) [126]. Furthermore, recent studies have demonstrated multiple complex pathways between the gut and the brain [119], linking chemotherapy induced gut–brain axis dysregulation with cognitive impairment, depression and fatigue [127]. The gut microbiota has also been linked with various neurological disorders [128]. In fact, cisplatin causes gut microbiota dysbiosis directly (i.e., cisplatin affects microbiota [129,130]) and indirectly (injury of epithelial cells and inflammation; mucositis [131]), which in the long term can contribute to chronic kidney disease and cognitive impairment [127,132], all of which are frequent complications of cisplatin therapy in cancer survivors. To date, no work has been undertaken to investigate the effects of cisplatin on the submucosal plexus, smooth muscle cells of the muscle layer in the gut wall, extrinsic nerves (i.e., parasympathetic and sympathetic), or the gut–kidney–brain axis dysfunction.

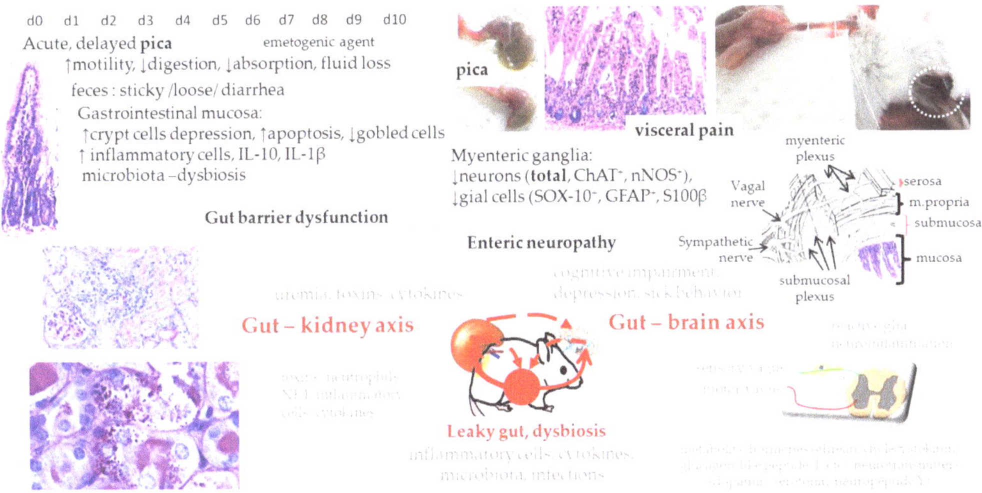

Figure 2. Cisplatin causes acute and chronic effects in the gastrointestinal tract. A single injection of cisplatin causes pica, a rodent-specific behavior of nausea, which reflects as a progressive reduction in food intake, increase in non-nutritive material intake (for instance bedding) and decreased gastric motility [107]. As a result the stomach is full of bedding and markedly enlarged/distended (white arrow) [114]. Reduction in food (68%) and water intake (45%) and an increase in stomach content (threefold) is evident from day 2 on (C57BL/6; 6 mg/kg ip) [114]. First morphological changes in the small intestinal mucosa (i.e., apoptosis, necrosis, decreased number of goblet cells, shortened villi and inflammatory cell infiltration) can be seen 1 day after a single cisplatin injection (B6D2F1: 8 mg/kg, 10 mg/kg, 12 mg/kg, 14 mg/kg; ip; d1,3,6,10,14) followed by reduced mucosal digestive function (depletion in maltase, sucrose, disaccharidase activity and reduced absorption) [108,115]. Depression in crypt cell production is already evident 2h after cisplatin and is maximal between 12 and 24 h post-treatment (CBA: 10 mg/kg, ip). Cisplatin causes lesions also in the colon mucosa, however, they appear later and are less severe [72]. The severity of gastrointestinal damage and mucosal dysfunction is dose-dependent and can persist up to 10 days after a single sub-lethal dose of cisplatin (B6D2F1: 8 mg/kg, 10 mg/kg, 12 mg/kg, 14 mg/kg; ip; d1,3,6,10,14) [72]. Mucosal recovery is slow, first signs of recovery can be observed 7 days post-treatment [72]. Repeated cisplatin administration (C57BL/6; 4 mg/kg/week for 4 weeks, ip; ↓20% BW) besides gut lesions (↑IL-1β and IL-10) also causes delayed pica, [55] and alterations in the ENS seen as loss of neurons in the myenteric ganglia of mouse gastric fundus (total and nNOS⁺) [56] and colon (neurons (total, ChAT⁺, nNOS⁺) and gial cells (SOX-10⁺, GFAP⁺, S100β⁺) [55]. Circulation and the nervous system are the main pathways for communication between the gut, the kidney and the brain in health or disease (the brain–gut–kidney axis). Legend: BW—body weight; ChAT—choline acetyltransferase; ENS—enteric nerve system; GFAP—glial fibrillary acidic protein; NET—neutrophil extracellular traps; nNOS—neuronal nitric oxide synthase; ip—intraperitonealy.

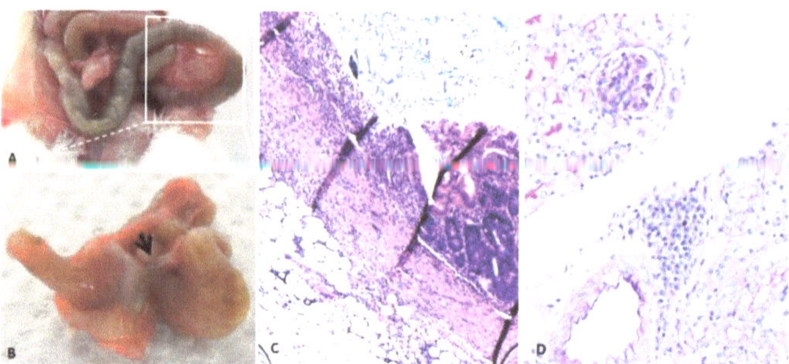

Figure 3. Cisplatin can have long-term effects in the gastrointestinal tract (**A**). A case of penetrating ulcer (**B**, arrow and **C**) in a mouse that survived a single lethal dose of cisplatin (17 mg/kg). Three months after cisplatin recovery, body weight started to decrease, and the mouse was killed and autopsy performed. Inflammatory cells found in the kidney (**D**).

5. Cisplatin Neurotoxicity

Cisplatin causes dose related, cumulative toxic effects on the peripheral and central nervous systems (i.e., peripheral neuropathy, chemo brain). Peripheral neuropathy is characterized by sensory loss, often accompanied by pain, starting in the distal extremities [8,133,134]. Chemobrain is characterized by subtle to moderate cognitive deficits such as a decrease in processing speed, memory, executive functioning, and attention [11]. In humans, 49% to 100% of cisplatin treated patients develop some symptoms of neuropathy [135]. The incidence and the severity increase with higher cumulative dose and longer exposure time to cisplatin. Peripheral neuropathy generally develops after a cumulative dose of 250 to 350 mg/m^2 [136], usually as mild neuropathy in a few patients. When cumulative dose reaches 350–420 mg/m^2, neuropathy occurs in up to 50% of patients and after 600 mg/m^2, neuropathy occurs in almost all patients, however, 30–40% of them develop moderate neuropathy, and 10% of them severe and disabling neuropathy [7,8,135,137].

In mice, serial testing at different cumulative doses of cisplatin showed that neuropathy develops progressively with higher cumulative doses [18,138]. Declines in sensory nerve conduction velocity (SNCV) and sudomotor responses were found from cumulative doses of 10 mg/kg, while reduction in the intensity of the nociceptive response to pinprick painful stimuli occurred at cumulative doses of 40 mg/kg (5 or 10 mg/kg/week up to cumulative doses of 40 mg/kg) [138]. In another study SNCV occur at cumulative dose 16 mg/kg (0.5 mg/kg twice per week up to cumulative doses of 32 mg/kg) [18]. There are many protocols of cisplatin induced mouse neurotoxicity [139]. They differ in the dosage, frequency of administration, cumulative dose and consequently in the severity of neurotoxicity and measured outcomes, i.e., the mortality, intensity and the incidence. The most characterized protocol for cisplatin neurotoxicity in mice is administration of cisplatin in two cycles, where one cycle is composed of daily intraperitoneal injection of cisplatin at a dose of 2.3 mg/kg for 5 days, followed by 5 days of recovery (cumulative dose 23 mg/kg; see Figure 4). This protocol induces structural, functional and molecular changes in the peripheral sensory neurons, dorsal root ganglia (DRG), spinal cord, and the brain. Changes can be observed 3–5 weeks after first cisplatin injection. Mice show altered behavioral responses to thermal and mechanical stimuli and impaired performance in the novel object and place recognition tasks. However, although the induced neuropathy is mild and reversible [46], no study reported how many mice develop peripheral neuropathy (the incidence and severity of neuropathy is dose dependent). It has been recognized that models of mild neuropathy have higher inter-individual differences, which requires a higher number of animals per group [140]. In the literature, we can find cisplatin protocols for peripheral neuropathy with an even lower cumulative dose of cisplatin and/or a shorter time point of testing. Considering that neurotoxicity is dose- and time-dependent, such cisplatin protocols do not induce all characteristics of peripheral neuropathy and need to be taken with caution.

Research on cisplatin toxicity of the central system started recently. Advanced neuroimaging techniques in cancer patients have revealed that chemotherapy causes structural alterations in white and gray matter, alterations in the activation of the fronto-parietal attentional network in cancer patients [141], and changes in structural brain networks [142,143]. Cisplatin can cross the blood–brain barrier and penetrate into the brain in low concentrations [15,133] and causes alteration in various parts of the brain in humans and rodents [144]. Structural abnormalities in cerebral white matter [145,146], reduction in myelin density [147], and cerebral neurogenesis [146], changes in synaptic integrity in the prefrontal cortex [148] and decrease in global functional neuronal connectivity in the brain were found also in mice [147]. Cisplatin induced mitochondrial dysfunction and structural abnormalities in brain synaptosomes in the hippocampus [147]. Mice with higher cumulative dose and longer exposure time to cisplatin developed even more severe impairment of mitochondrial transport and mitochondrial dysfunction [149], showing dose dependent toxicity.

5.1. Behavioral Tests and Their Weakness

Various behavioral tests have been used to evaluate mice wellbeing, motor activity behavioral responses to mechanical and thermal stimuli, and cognitive performance. It was consistently reported that this cisplatin protocol (Figure 4) induces changes in mice response to radiant heat-paw, tail immersion, adhesive removal test and the von Frey test. Alterations were interpreted as heat hyperalgesia and mechanical allodynia [150–153]. The pattern of onset and progression of the heat hyperalgesia was similar to the mechanical allodynia and persisted for up to 5 weeks post treatment [150]. No difference was observed in the open field test, motor coordination or signs of paresis (the rotarod test) [153], cold plate test, locomotor activity, grip strength (muscle strength) [150,151]. However, activity patterns of cisplatin treated mice did alter moderately [153], the exploratory activity and body weight of mice were reduced and recovered after cessation of the cisplatin treatment [151]. It was claimed that this cisplatin protocol does not cause significant deterioration in the general health of mice. However, two independent research groups reported body weight decrease (10% after the first cycle and 17% after the second cycle) and sudden death of a mouse during the study [150,151,153].

Why is all this information important? In humans, peripheral neuropathy is characterized by sensory loss and pain. Patients describe a range of predominantly sensory, bilateral symptoms in both hands and feet (i.e., a stocking and glove distribution) such as numbness, tingling, spontaneous pain, and hypersensitivity to mechanical and/or cold stimuli [8,14,133,134]. Loss of cognitive abilities of concentration, attention, learning and memory, and executive functions are characteristics of chemotherapy induced cognitive impairment [154].

The pain, sensory abnormalities and cognitive abilities are difficult to evaluate without verbal communication. In animals, therefore, various behavioral tests are used. However, the behavioral tests have many drawbacks. A major shortcoming is that they are all evoked responses. Mice and rats are prey species and when distressed they will mask their spontaneous behavior, sensations and signs of pain. There are many factors that can influence and confound behavioral tests, for instance, aggression (males are prone to aggression) [155], gender of the experimenter (exposure to male experimenters causes in mice stress that results in stress-induced analgesia) [156], anxiety and/or agitation (caused by over-handling or repeated testing) [153] and health states like kidney injury and visceral pain. We have explained that cisplatin causes pica and dose related injuries and inflammation along the gastrointestinal tract, all of which result in visceral pain. Mice suffering from visceral pain of lower abdomen respond to mechanical and thermal stimulation of the hind-paw or tail in the same manner as mice with peripheral neuropathy [157]. It was demonstrated that inflammation in the gastrointestinal tract activates satellite glial cells in DRG and cause excitation of those DRG neurons that innervate particular parts of the gut [158,159].

Major sensory nerves that arise from the L4–L6 DRG neurons innervate the colon [120,160]. These DRG neurons are examined in cisplatin neuropathy studies. Accordingly, visceral pain can be mistakenly diagnosed as peripheral neuropathy. In addition, repeated cisplatin treatment worsens gut toxicity and induces delayed pica, thus, conditioned place preference test used to test the analgesics for cisplatin neuropathy [161] might also be mistakenly interpreted as peripheral neuropathy treatment. None of the neurotoxicity studies evaluated kidney or gut damage.

All the above demonstrates the need for understanding the characteristics and the complexity of cisplatin mouse models to correctly design and interpret the study outcomes. It also demonstrates that outcomes of behavioral tests alone are not sufficient to characterize the model or to evaluate the role of a particular gene or therapeutic agent in the model.

To evaluate and confirm cisplatin induced neuropathy in rodents it is recommended to use behavioral, electrophysiological and histological tests [162]. Electrophysiological tests have limitations. The most significant drawback of the conduction velocity changes is that nerve conduction velocities do not correlate with symptoms [162]. In addition, results of the electrophysiologic tests can vary among studies and even within the laboratory, due to many factors including mice's body temperature during the recording (the tests are done under anesthesia) [162]. The most reliable is histological assessment, light and electron microscopy. A relevant indicator of small-diameter sensory nerve fiber status in neurotoxicity studies is analysis of intra-epidermal nerve fibers, a method also used for evaluation of peripheral neuropathy in patients, which has yet to become a routine end point in nonclinical safety testing [163]. However, we also need to perform autopsies and analyze all vital organs, particularly the gut and the kidney to evaluate the severity of the injury and inflammation and correctly report and interpret the study outcomes.

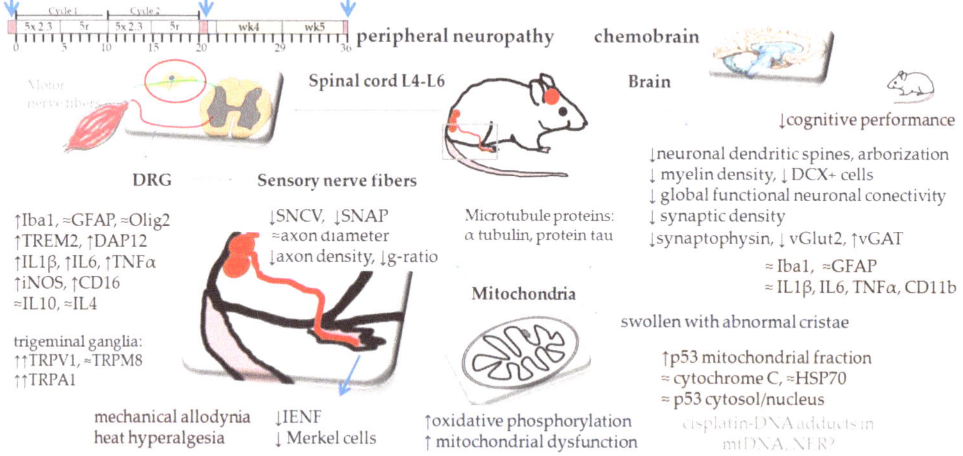

Figure 4. Cisplatin neurotoxicity. In mice, two cycles of cisplatin (2.3 mg/kg/daily for 5 days followed by 5 days recovery; 5d+5r/5d+5r; cumulative dose 23 mg/kg) resulted in reduced density of intraepidermal nerve fibers (IENF) (wk3, and wk5) [152,164] and epidermal Merkel cells [152] in the mouse plantar footpad. Merkel cells, mechanosensory cells actively involved in touch reception (tactile sensation), [165–167] are proposed to underlie sensory dysfunction in diabetic patients and animals [168]. In sensory nerves (sciatic, caudal, tibial) mild hypomyelination with few degenerating axons (reduced density of myelinated fibers without alterations in axon diameter) can be observed together with a slight decrease in the sensory nerve conduction velocity (SNCV; indication of demyelination) and the sensory nerve action potential (SNAP) [153].

In sensory neurons (trigeminal ganglia) cisplatin activated the transient receptor potential (TRP) channels (TRPA1, TRPV1) [151], a non-selective cation channels involved in chemical and thermal evoked pain sensation [169]. In the spinal cord (L4-L6) cisplatin activated microglia (Iba1), induced pro-inflammatory cytokines (IL-1β, IL-6, TNFα, iNOS, CD16, a marker of pro-inflammatory microglia (wk3) and increased protein levels of triggering receptor expressed on myeloid cells 2 (TREM2) and DNAX activating protein of 12 kDa (DAP12) (wk3) [152]. TREM2/DNAX is a receptor complex predominantly expressed on microglia in the central nervous system associated with neurodegenerative diseases and inflammatory response of microglia [152]. Cisplatin induced structural abnormalities in cerebral white matter (loss of neuronal dendritic spines and arborizations) [145,146] and reduced myelin density in the cingulated cortex [147]. It also [145] decreased cerebral neurogenesis (DCX$^+$ cells) [146] but did not cause inflammation (IL1β, IL6, TNFα, GFAP, CD11b) [146] or microglia (Iba1$^-$, GFAP$^-$) activation [145]. However, decreased synaptic integrity (synaptophysin, vGlut2, vGAT) in the prefrontal cortex [148] and global functional neuronal connectivity in the mouse brain was found (fMRI) [147]. Cisplatin induced mitochondrial dysfunction and structural abnormalities in brain synaptosomes [147]. Mice treated with three cycles of cisplatin (protocol 2.3 mg/kg 5d + 5r/5d + 5r/5d + 5r; cumulative dose 34.5mg/kg) developed more severe impairment of mitochondrial transport and mitochondrial dysfunction in the hippocampus [149] (43% decrease in cytochrome C activity, ATP production, 96% increase in ROS, 29% decrease in mitochondrial membrane potential, impaired mitochondrial transport, reduced α-tubulin acetylation in the hippocampus, decrease in dendritic spine and synaptic density (vGlut1 and PSD95) [149]. Legend: DAP12—DNAX activating protein of 12 kDa; DRG—dorsal root ganglia; GFAP—glial fibrillary acidic protein; IENF—intraepidermal nerve fibers; IL—interleukine; Iba1—ionized calcium-binding adaptor molecule 1; iNOS—inducible nitric oxide synthase; L4-L6—lumbal vertebra; mtDNA—mitochondrial DNA; NER - nucleotide excision repair; Olig-2—oligodendrocyte lineage gene 2; ROS—reactive oxidative species; SNAP—sensory nerve action potential; SNCV—sensory nerve conduction velocity; TNFα—tumor necrosis factor alpha; TRP—transient receptor potential channels (TRPA1, TRPV1); TREM2—triggering receptor expressed on myeloid cells 2; vGlut2—vesicular glutamate transporter 2; vGAT—vesicular GABA transporter.

5.2. Cisplatin Mechanisms

Cisplatin exerts its antitumor activity by binding to guanine and adenine residues, forming cisplatin–DNA adducts that bend and unwind the DNA helix (i.e., distorting its structure by intra- and inter-strand DNA cross-linkage), thus interfering with DNA replication and/or transcription which results in DNA damage, induction of cell cycle arrest, inhibition of DNA synthesis and repair, senescence or cell death (by activating necrotic and apoptotic pathways). While these effects of cisplatin on cancer cells are desired, the same process in normal tissue causes varying degrees of toxicity [5,129,130]. In dividing stem or progenitor cells (myelotoxicity, gut stem cells, etc) cisplatin induces different types of cell death, while in non-dividing cells transcription and translation are more affected leading to senescence, degeneration or dysfunction. Similar to the effect in kidneys (tubular cells), cisplatin was reported to cause DNA damage, activation of apoptotic pathways like p53 activation, Bax translocation, mitochondrial cytochrome C release, activation of caspase-3 and caspase-9 and cell death also in DRG sensory neurons [170].

DRC neurons are non-dividing cells that need a high level of active transcription to sustain their large size, high metabolism, and long axons [171]. Repeated cisplatin administration results in accumulation of cisplatin–DNA adducts in DRG neurons, which is subsequently removed and repaired by nucleotide excision repair (NER) [18]. NER is one of the major DNA repair pathways particularly relevant for cisplatin–DNA adduct repair. Serial testing with increasing cumulative doses of cisplatin showed that mice with NER dysfunctions accumulated higher numbers of cisplatin–DNA adducts in DRG neurons and developed higher severity of peripheral neuropathy [18].

In 2011, Podratz and coworkers demonstrated that cisplatin binds not only nuclear DNA but also mitochondrial DNA (mtDNA), both with the same binding affinity [170]. However, in contrast to nuclear DNA, in mitochondrial DNA isplatin–DNA adducts inhibited mtDNA replication and transcription of mitochondrial genes which resulted in mitochondrial vacuolization and degradation. It was proposed that mitochondrial dysfunction is very likely the consequence of reduced repair of cisplatin adducts in mtDNA, particularly NER [170]. Until recently it was believed that mitochondria do not possess NER. However, extensive investigation in the last decade has shown that mitochondrial

DNA repair is very diverse and complex. Mitochondria have an NER mechanism, but it differs from the nuclear one. The proteins that participate in the NER mechanism are imported into the mitochondria in response to oxidative stress [172]. Thus, it is possible, that the NER mechanism is indeed involved in mitochondrial dysfunction (not only in DRG neurons but also in other tissues with high amounts of mitochondria like proximal tubular cells [19] and the brain [41]). However, the contribution of NER in mitochondrial dysfunction remains to be determined.

Loss of mitochondrial number was found in axons of the sensory nerve (tibial) [164], while in the DRG neurons and the brain, mostly alterations in mitochondrial morphology [170] and gene expression were observed [146,164], which suggests that mitochondria were injured but still able to cope and maintain basal functions. However, with higher cumulative dose and longer exposure time to cisplatin more severe impairment of mitochondrial transport and function occurs [149], showing dose dependent toxicity. Interestingly, mitochondrial damage has been investigated and linked with cisplatin toxicity in renal cells of proximal tubules already in the 1980s [63]. Nevertheless, the main cause of cisplatin toxicity remains unknown. We must recall that cisplatin can affect a wide variety of molecules and mechanisms in the cell, it binds not only to DNA but also to various proteins and affects their numerous functions, influences the transport in the cells, etc., [173] (Figure 5).

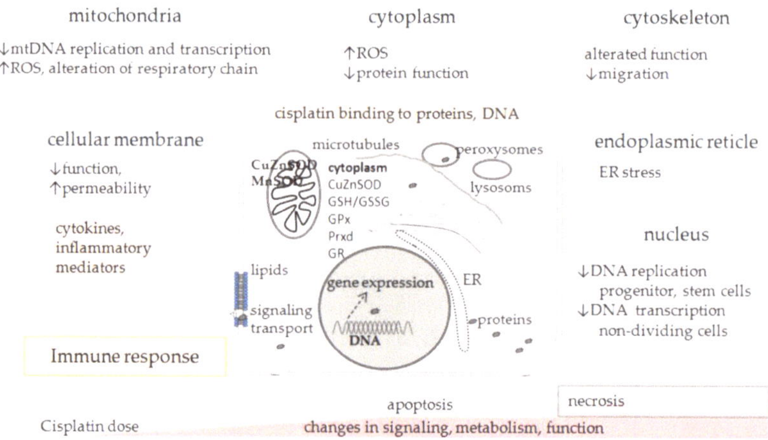

Figure 5. Schematic presentation of cisplatin toxicity in non-tumor cells in the body. Extent and intensity of oxidative stress, changes in signaling, metabolism, function, intensity of inflammation, activation of certain immune cell types, inflammatory and molecular crosstalk and response, type of cell death, etc., depend on cisplatin dose (single or cumulative) and severity of toxicity. ER—endoplasmic reticulum; mtDNA—mitochondrial DNA; ROS—reactive oxygen species.

6. Cisplatin Ototoxicity

Cisplatin ototoxicity is a common cisplatin side effect. Cisplatin treated cancer patients experience progressive, bilateral, primarily high-frequency sensorineural hearing loss. Ototoxicity is dose-dependent cisplatin side effect which can start at doses from 60 mg/m^2/cycle and affects approximately 62% patients. However, in high dose treatment schedules (150–225 mg/m^2/cycle) up to 100% of patients can be affected [174]. It is reported that 40–80% of adults [16] and 60% of children develop permanent hearing loss [175]. A recent study reported that young children (<5 years) are more susceptible than older children (>5 years). Since young children develop hearing loss at lower cumulative dose and early during cisplatin therapy, audiological monitoring is recommended at each cisplatin cycle [176]. The exact mechanism responsible for hearing loss is not fully understood [174,177,178], but data suggest that cisplatin directly stimulates the production of cytokines leading to inflammation, oxidative stress, endoplasmic reticulum stress and, finally, to various forms of cell death [179]. Currently, there is no treatment to reduce cisplatin ototoxicity [174,177,178]. However, sodium thiosulfate, a thiol-containing antioxidant, has shown promising results in a phase III clinical trial [180].

In the past, a wide variety of cisplatin protocols has been used to model ototoxicity. Most frequently a single, high dose of cisplatin has been used and effects were evaluated a few days later (due to high mortality rate, similarly to nephrotoxicity studies) [178]. Repeated administration of low dosage was also used, but frequently resulted in high mortality or inconsistent and small changes in hearing sensitivity [48]. Protocols were recently summarized and can be found elsewhere [178].

Mouse model of ototoxicity is included in this review mostly as an example of an animal study that aimed to establish a clinically relevant and reproducible mouse model [16,48]. Specifically, it is the first study that reported extensive supportive care for mice during cisplatin treatment. The study [16,48] is summarized with hope that supportive care becomes a part of every cisplatin protocol in animal studies.

To get clinically relevant model of cisplatin ototoxicity, Cunningham and coworkers [16] used an already-established cisplatin protocol. However, to optimize the protocol, firstly a pharmacokinetic study was done to get information on cisplatin distribution and elimination rates from various tissues (kidney, liver, inner ear, brain and heart) before and after each cycle (protocol composed of three cycles of a daily ip injection of cisplatin at a dose 3.5 mg/kg for 4 days, followed by 10 days of recovery; cumulative dose 42 mg/kg) [16]. An auditory function and the three doses were evaluated after each cycle and finally the protocol with clinically relevant and reproducible hearing loss with the lowest suffering of the animals was established [48]. Briefly, CBA/CaJ male and female mice were used and treated with above stated cisplatin protocol but three different doses of cisplatin (2.5, 3.0, and 3.5 mg/kg, cumulative dose 30, 36 and 42 mg/kg, respectively) were evaluated. After the second cycle, minimal hearing loss was observed (at 3.5 mg/kg/day) but without significant threshold shifts across frequencies [16]. Mice developed a dose-dependent loss of cochlear outer hair cell function (distortion product otoacoustic emissions; DPOAEs) and hearing sensitivity (auditory brainstem response; ABR). No significant difference was found between male and female mice. A cisplatin dose of 3.0 mg/kg/day showed better health state of mice than 3.5 mg/kg/day but similarly robust hearing loss across all frequencies, most severe at the high frequencies [48]. This dose (3.0 mg/kg) was thus selected for further characterization of cochleotoxicity and vestibulotoxicity. Assessment of auditory function follows 42 days after the first cisplatin injection. It was found that after cessation of cisplatin administration hearing loss in mice even progresses over time [16,48], similar to cisplatin ototoxicity in humans [174].

Hydration and Supportive Care in Cisplatin Protocols

From the first day of the study, all cisplatin treated mice received intensive supportive care (twice daily). Supportive care was composed of hydration (1 mL of 0.9%NaCl and 1ml of Normasol injected subcutenously) and supplemental nutrition (0.3 mL high calorie liquid supplement, DietGel Recovery cups and pellets on the floor cage). Body weight, overall health, activity and body condition scoring [181] was used to monitor the overall condition of each mouse on daily basis (muscular tone, body fat content, coat maintenance, overall energy level. Using supportive care protocol, all mice in the study survived although their body weight progressively decreased during each cycle and at the end of the experiment reached significant loss of their initial weight (21% at dose 2.5 mg/kg and 27% at dose 3.0, 3.5 mg/kg). The only drawback of this study is that the kidney function and gastrointestinal damage in mice were not examined. Inflammation has significant effects on health and disease. Treatment of inflammation (in the gut and kidney) could improve the mice's health state. Particularly because during the auditory testing mice need to be anesthetized, and diseased animals are at higher risk of death during the anesthesia. Thus, during anesthesia special care is needed to avoid additional hypothermia, hypoxia, acidosis, and death.

We must recall that mice treated with cisplatin suffer from acute and delayed pica, gastric distension (delay in gastric emptying, stomach filled with bedding), reduced food intake, inflammation in intestine, polyuria (malnutrition, dehydration and electrolyte waste). In addition, mice treated with cisplatin are hypothermic [182]. Mice that are ill and suffer abdominal pain (intestine inflammation, nausea/pica/full stomach, kidney injury) are less active and vital, do not rear/climb up after water and food, and do not care for their nests. Well-structured nests are important for their body temperature maintenance. Supportive care is mandatory, to prevent agonistic death from dehydration, malnutrition and hypothermia. Vitamin C and sodium bicarbonate pretreatments has been show to improve mice's health and reduce cisplatin nephrotoxicity [182], while dexamethasone, a corticosteroid used in humans and/or ondansetron, a serotonin 5-HT3 receptor antagonist, showed confounding results [106].

It is interesting that in the 1980s the effects of hydration and cisplatin vehicle (Table 5) on nephrotoxicity and tumor burden were tested in cisplatin treated mice and rats. Although both hydration [183] and the vehicle in which cisplatin was dissolved [184,185] markedly reduced mortality and nephrotoxicity, hydration became a routinely used method of nephrotoxicity prevention only in clinics but not in preclinical models. Intravenous hydration using isotonic saline solution significantly reduces cisplatin half-life, urinary cisplatin concentrations and proximal tubule transit time [3,12], which reduces nephrotoxicity and allows higher doses of cisplatin for the cancer treatment. Thus, hydration affects the MTD dose and, consequently, also the therapeutic effect of cisplatin (dose dependent) in preclinical studies (see Section 3.3). In addition, not only the incidence but also the severity of cisplatin toxicity is dose dependent. Based on the cisplatin protocol mice thus can develop (Figure 6): changes in molecular mechanisms without structural damage (process is in the range of the physiological limits and does not affect the clinical picture, although molecular markers can show significant increases; MTD); changes in molecular mechanisms with structural damage (although structural damage is histologically confirmed and clinical signs are present, damage is still in the range where regression and restitution or repair is possible; mild, moderate); clinical signs are present and structural damage is obvious, regression and repair is possible only if properly treated (severe-systemic inflammation) and intoxication (irreversible).

Table 5. Effect of cisplatin vehicle on cisplatin toxicity/mortality [184,185].

Cisplatin Vehicle	LD50
distilled water	10.8 ± 1.0 mg/kg
0.9% NaCl	15.3 ± 1.6 mg/kg
4.5% NaCl	24.5 ± 0.7 mg/kg

LD50—dose of cisplatin that results in 50% mortality in animals.

Various factors can affect response to cisplatin treatment such as strain, substrain [44,106], age [105,186], hydration [183], circadian rhythms [23,187–189]. However, there is high interindividual variability also among mice within the same inbred strain (genetically uniform) showing that environmental and phenotypic factors like physical state play important roles in cisplatin toxicity. Since there are many factors that can influence cisplatin effects (therapeutic or toxic) scientists are encouraged to thoroughly report all details in their study and follow the ARRIVE guidelines [190] or the Gold Standard Publication Checklist [191], FELASA recommendations [192,193] and standardized genetic nomenclature of rodents (http://www.informatics.jax.org/nomen/strains.shtml) 6th October 2021.

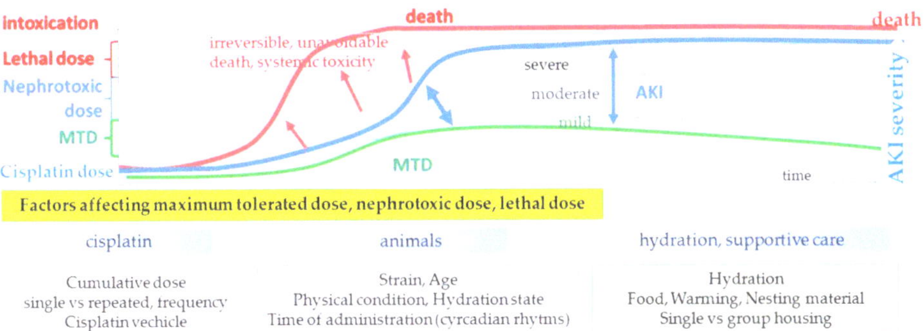

Figure 6. Dose-dependent toxicity of cisplatin and factors affecting maximum tolerated dose (MTD), nephrotoxic and lethal dose.

7. Cisplatin Distribution and Elimination

To better understand the complexity of cisplatin toxicity in various organs, basic knowledge about cisplatin distribution, elimination and accumulation is briefly summarized. Cisplatin reaches systemic circulation within 10 min after systemic administration (ip, iv) [15,16,194] and within 1 h cisplatin is already distributed in almost all tissues studied (kidney, liver, lung, inner ear, heart and brain), with the highest concentration in the kidney [16]. There is a linear correlation between cisplatin dose (3.75, 7.5 or 15 mg/kg) and cisplatin concentration in the blood or tissues (kidney, liver, tumor, brain and testis) 1 h after ip administration [15]. Free cisplatin eliminates from the blood predominantly by the kidney, much less by biliary [194] or intestinal excretion [184].

It appears that the rate of cisplatin clearance in repeated treatment depends on the dose (cumulative) and the frequency interval (daily vs weekly). Repeated administration of low dose of cisplatin (16 mg/m^2 or 2.5 mg/kg) did not affect the elimination rates of cisplatin until the fifth cycle (ip; five cycles with 3-week intervals between each cycle). After the fifth cycle elimination of cisplatin significantly decreased (cumulative dose reached 12.5 mg/kg) [195]. In contrast, repeated administration of higher doses of cisplatin (5 mg/kg iv; three cycles with 3 weeks between each cycle) resulted in decreased renal clearance and increased accumulation of cisplatin in the kidney by each cycle (cumulative dose at the second cycle reached 10 mg/kg) [196], suggesting a longer elimination half-life of cisplatin and an impaired elimination/detoxification mechanisms when reaching critical levels of cisplatin (Table 3). A similar situation occurred in the case of cisplatin protocol for ototoxicity (three cycles of 14 mg/kg (3.5 mg/kg/daily) with 10-day intervals between each cycle). After each cycle the elimination of cisplatin decreased, resulting in gradual retention of cisplatin in tissues. After the third cycle (42 days after the start) cisplatin in all examined tissues reached levels twofold higher than after the first cycle. The highest concentration of cisplatin was detected in the liver, followed by spleen, femur, kidney, inner ear, lung, heart, skeletal muscle, small intestine and brain. Two months later (60 days recovery) marked decline was observed in all tissues except femur and inner ear. However, in all examined tissues cisplatin was still present at the detected levels [16].

While elimination of cisplatin from the blood is very rapid (mostly within 1 h), elimination from the tissues is a longer process lasting weeks or even years. In tissues cisplatin accumulates in all cell compartments, the mitochondria, nucleus, cytoplasm, microsomes [195,197]. In general, the larger decline of cisplatin concentration in tissues occurs within the first 24 h [15,16], followed by slower elimination rates during the first 30 days [15,79,195] reaching an almost steady state 3 months after a single nephrotoxic dose of cisplatin [195]. Recently it was found that elimination rates from the inner ear are much lower than in other organs. Cisplatin retains in the inner ear for months in both mice and humans (at least 18 months after patients last cycle) [16]. It was also found that the highest levels of cisplatin in the inner ear accumulate in the stria vascularis (the region of the inner ear that maintains the ionic composition of endolymph), while cisplatin accumulation in mechanosensory hair cells is more limited. Similar cisplatin distribution was also found in humans. Long-term retention of cisplatin was associated with progressive hearing loss in mice [16].

Cisplatin retention in the tissues can be evaluated also by detecting cisplatin–DNA adducts, a method usually used in the nervous system (see section neurotoxicity). However, cisplatin–DNA adducts can be found also in other tissues (kidney, liver, testis and brain). Nevertheless, the formation of cisplatin–DNA adducts is a slower process; depending on the tissue it can take up to 4 h or more [15,198]. After single dose of cisplatin (7.5 mg/kg) the highest levels were observed in the kidney cortex, particularly tubules. The levels persisted for 24 h (liver, kidney), followed by a slow decline, while in other tissues (tumor, testis) decline was observed within first 12 h. Formation of cisplatin–DNA adducts was dose dependent with large inter-individual variations, particularly for kidney and tumor [15]. Cisplatin–DNA adducts can be detected in various tissues in patients treated with cisplatin for many months after therapy [198].

8. Discussion

As shown in the paper, there are many similarities between mice and humans. Mice develop all cisplatin side effects in a dose- and time-dependent manner. Just as humans, mice also develop cisplatin side effects of varying severity from mild to multi-organ failure, each pathology with its own time course and pathophysiological response or molecular signature. Despite all the similarities, there is an apparent gap between the results in animal models and human clinical trials.

As described, there are many drawbacks in the currently used cisplatin protocols. Besides a wide variability in protocols [44], most of cisplatin protocols have no similarities to the treatment schedules used in cancer patients. In humans, cisplatin is given in cycles with extensive hydration and supportive care to provide the highest possible dose of cisplatin to improve the success of therapy, while in tumor bearing mice a wide variety of cisplatin protocols with no hydration or supportive care are used. In mice, cisplatin treatment ranges from a single to repeated (multiple) administration, where cumulative doses range from sub-therapeutic to lethal doses or even higher (see Table 1 and Table S1). To evaluate potential beneficial effects of therapy or toxicity, in mice studies, most frequently only the size or the volume of the tumor is used as a measure of successful treatment and the body weight is used as a marker of systemic toxicity. No examination of gut toxicity, myelotoxicity or neurotoxicity is performed. Rarely, a few blood parameters are examined. Body condition of the animals and mortality rate are rarely reported and necropsy and histology of all vital organs are rarely performed (Table S1). Importantly, cisplatin protocol, hydration and supportive care all together affect not only the MTD or lethal dose but also the therapeutic dose of cisplatin and its side effects (dose-dependent). Higher doses of cisplatin result in higher cisplatin tissue retention (see section cisplatin distribution and elimination).

As demonstrated in the article, mice respond to cisplatin therapy in a similar way to humans. Importantly, mouse response to cisplatin is highly dependent on cisplatin protocols. Thus, we can say that we get what we design. If we want to understand molecular mechanisms and find therapeutic agents that would have a potential benefit in clinics, we need to use similar cisplatin treatment protocols as are used in cancer patients.

In this paper, only the most characterized cisplatin protocols were presented together with weaknesses that need to be improved in future studies. An example of hydration and supportive care in repeated cisplatin protocol is summarized with the hope that in the future hydration and supportive care become a part of cisplatin protocols. The use of the same cisplatin protocol by various research groups around the world could help evaluate, optimize and validate particular cisplatin protocols. Investigating cisplatin effects in all organs of a currently established model and gaining insight into complex cisplatin toxicology would help understand the underlying mechanisms of cisplatin toxicity in a time-dependent manner. It would enable the use of optimal markers of a certain toxicity at a given time period/point in the development of the toxicity. Optimized and validated models can then be used to test potential treatment strategies for cisplatin toxicity. However, first optimization with hydration and supportive care is needed. This may affect the dose adjustment in cisplatin protocols. Then protocols need to be tested and optimized in tumor-bearing animals.

Research on mice enables systematic and controlled investigation of complex mechanisms involved in the development of cisplatin therapeutic or toxic effects. In addition, it enables investigation of pathogenesis of cisplatin toxicity in a time- and dose-dependent manner. However, it is important that we change our approach to animal studies and start to treat animals in research as patients and not as a tool. Otherwise we must ask ourselves *"what have we chosen to ignore in this model, and at what cost?"* [199].

Supplementary Materials: The following are available online at https://www.mdpi.com/article/10.3390/biomedicines9101406/s1, Table S1: Publications reporting cisplatin protocols in tumor-bearing mice published in the period from April 2020 to February 2021.

Funding: This work has in part been supported by ARRS (Slovenian Research Agency, P3-054).

Institutional Review Board Statement: Not applicable.

Informed Consent Statement: Not applicable.

Data Availability Statement: Not applicable.

Conflicts of Interest: The author declares no conflict of interest.

Abbreviations

AKI	acute kidney injury
AST	aspartate aminotransferase
BB	brush border
BM	basal membrane
BUN	blood urea nitrogen
BW	body weight
ChAT	choline acetyltransferase
Cr	serum creatinine
CFU	colony-forming unit
d	day
DAP12	DNAX activating protein of 12 kDa
DRG	dorsal root ganglia
DT	distal tubules
ENS	enteric nerve system
ER	endoplasmic reticulum
GFAP	glial fibrillary acidic protein
GFR	glomerular filtration rate
Hb	hemoglobin
IENF	intraepidermal nerve fibers
IL	interleukine
Iba1	ionized calcium-binding adaptor molecule 1
iNOS	inducible nitric oxide synthase
L4-L6	lumbal vertebra
LD	lethal dose
LD100 MNC	dose of cisplatin that results in 100% mortality in animalsmononuclear cells
MTD	maximum tolerated dose
mtDNA	mitochondrial DNA
N	number of animals
NET	neutrophil extracellular traps

nNOS	neuronal nitric oxide synthase
NER	nucleotide excision repair
Olig-2	oligodendrocyte lineage gene 2
P1–3	proximal tubules pars 1–3
ROS	reactive oxidative species
SNAP	sensory nerve action potential
SNCV	sensory nerve conduction velocity
TNFα	tumor necrosis factor alpha
TRP	transient receptor potential channels (TRPA1, TRPV1)
TREM2	triggering receptor expressed on myeloid cells 2
vGlut2	vesicular glutamate transporter 2
vGAT	vesicular GABA transporter
WBC	white blood cells
ip	intraperitonealy
iv	intravenously

References

1. Siegel, R.L.; Miller, K.D.; Fuchs, H.E.; Jemal, A. Cancer Statistics, 2021. *CA Cancer J. Clin.* **2021**, *71*, 7–33. [CrossRef]
2. Casanova, A.G.; Hernández-Sánchez, M.T.; Martínez-Salgado, C.; Morales, A.I.; Vicente-Vicente, L.; López-Hernández, F.J. A meta-analysis of preclinical studies using antioxidants for the prevention of cisplatin nephrotoxicity: Implications for clinical application. *Crit. Rev. Toxicol.* **2020**, *50*, 780–800. [CrossRef] [PubMed]
3. Crona, D.J.; Faso, A.; Nishijima, T.F.; McGraw, K.A.; Galsky, M.D.; Milowsky, M.I. A Systematic Review of Strategies to Prevent Cisplatin-Induced Nephrotoxicity. *Oncologist* **2017**, *22*, 609–619. [CrossRef] [PubMed]
4. Galanski, M.; Jakupec, M.A.; Keppler, B.K. Update of the preclinical situation of anticancer platinum complexes: Novel design strategies and innovative analytical approaches. *Curr. Med. Chem.* **2005**, *12*, 2075–2094. [CrossRef] [PubMed]
5. Qi, L.; Luo, Q.; Zhang, Y.; Jia, F.; Zhao, Y.; Wang, F. Advances in Toxicological Research of the Anticancer Drug Cisplatin. *Chem. Res. Toxicol.* **2019**, *32*, 1469–1486. [CrossRef] [PubMed]
6. Di Maio, M.; Perrone, F.; Gallo, C.; Iaffaioli, R.V.; Manzione, L.; Piantedosi, F.V.; Cigolari, S.; Illiano, A.; Barbera, S.; Robbiati, S.F.; et al. Supportive care in patients with advanced non-small-cell lung cancer. *Br. J. Cancer* **2003**, *89*, 1013–1021. [CrossRef] [PubMed]
7. Grunberg, S.M.; Sonka, S.; Stevenson, L.L.; Muggia, F.M. Progressive paresthesias after cessation of therapy with very high-dose cisplatin. *Cancer Chemother. Pharmacol.* **1989**, *25*, 62–64. [CrossRef]
8. Van der Hoop, R.G.; van der Burg, M.E.; ten Bokkel Huinink, W.W.; van Houwelingen, C.; Neijt, J.P. Incidence of neuropathy in 395 patients with ovarian cancer treated with or without cisplatin. *Cancer* **1990**, *66*, 1697–1702. [CrossRef]
9. Banach, M.; Juranek, J.K.; Zygulska, A.L. Chemotherapy-induced neuropathies-a growing problem for patients and health care providers. *Brain Behav.* **2017**, *7*, e00558. [CrossRef]
10. McWhinney, S.R.; Goldberg, R.M.; McLeod, H.L. Platinum neurotoxicity pharmacogenetics. *Mol. Cancer Ther.* **2009**, *8*, 10–16. [CrossRef]
11. Vichaya, E.G.; Chiu, G.S.; Krukowski, K.; Lacourt, T.E.; Kavelaars, A.; Dantzer, R.; Heijnen, C.J.; Walker, A.K. Mechanisms of chemotherapy-induced behavioral toxicities. *Front. Neurosci.* **2015**, *9*, 131. [CrossRef] [PubMed]
12. Hamroun, A.; Lenain, R.; Bigna, J.J.; Speyer, E.; Bui, L.; Chamley, P.; Pottier, N.; Cauffiez, C.; Dewaeles, E.; Dhalluin, X.; et al. Prevention of Cisplatin-Induced Acute Kidney Injury: A Systematic Review and Meta-Analysis. *Drugs* **2019**, *79*, 1567–1582. [CrossRef] [PubMed]
13. Skrypnyk, N.I.; Siskind, L.J.; Faubel, S.; de Caestecker, M.P. Bridging translation for acute kidney injury with better preclinical modeling of human disease. *Am. J. Physiol. Renal Physiol.* **2016**, *310*, F972–F984. [CrossRef] [PubMed]
14. Acklin, S.; Xia, F. The Role of Nucleotide Excision Repair in Cisplatin-Induced Peripheral Neuropathy: Mechanism, Prevention, and Treatment. *Int. J. Mol. Sci.* **2021**, *22*, 1975. [CrossRef]
15. Johnsson, A.; Olsson, C.; Nygren, O.; Nilsson, M.; Seiving, B.; Cavallin-Stahl, E. Pharmacokinetics and tissue distribution of cisplatin in nude mice: Platinum levels and cisplatin-DNA adducts. *Cancer Chemother. Pharmacol.* **1995**, *37*, 23–31. [CrossRef] [PubMed]
16. Breglio, A.M.; Rusheen, A.E.; Shide, E.D.; Fernandez, K.A.; Spielbauer, K.K.; McLachlin, K.M.; Hall, M.D.; Amable, L.; Cunningham, L.L. Cisplatin is retained in the cochlea indefinitely following chemotherapy. *Nat. Commun.* **2017**, *8*, 1654. [CrossRef]
17. Rix, A.; Drude, N.I.; Mrugalla, A.; Baskaya, F.; Pak, K.Y.; Gray, B.; Kaiser, H.J.; Tolba, R.H.; Fiegle, E.; Lederle, W.; et al. Assessment of Chemotherapy-Induced Organ Damage with Ga-68 Labeled Duramycin. *Mol. Imaging Biol.* **2020**, *22*, 623–633. [CrossRef]
18. Dzagnidze, A.; Katsarava, Z.; Makhalova, J.; Liedert, B.; Yoon, M.S.; Kaube, H.; Limmroth, V.; Thomale, J. Repair capacity for platinum-DNA adducts determines the severity of cisplatin-induced peripheral neuropathy. *J. Neurosci.* **2007**, *27*, 9451–9457. [CrossRef]

19. Hu, Y.; Yang, C.; Amorim, T.; Maqbool, M.; Lin, J.; Li, C.; Fang, C.; Xue, L.; Kwart, A.; Fang, H.; et al. Cisplatin-Mediated Upregulation of APE2 Binding to MYH9 Provokes Mitochondrial Fragmentation and Acute Kidney Injury. *Cancer Res.* **2021**, *81*, 713–723. [CrossRef]

20. Zager, R.A.; Johnson, A.C.M.; Therapeutics, R. Iron sucrose ('RBT-3') activates the hepatic and renal HAMP1 gene, evoking renal hepcidin loading and resistance to cisplatin nephrotoxicity. *Nephrol. Dial. Transplant.* **2021**, *36*, 465–474. [CrossRef]

21. Freitas-Lima, L.C.; Budu, A.; Arruda, A.C.; Perilhão, M.S.; Barrera-Chimal, J.; Araujo, R.C.; Estrela, G.R. PPAR-α Deletion Attenuates Cisplatin Nephrotoxicity by Modulating Renal Organic Transporters MATE-1 and OCT-2. *Int. J. Mol. Sci.* **2020**, *21*, 7416. [CrossRef]

22. Chiou, Y.Y.; Jiang, S.T.; Ding, Y.S.; Cheng, Y.H. Kidney-based in vivo model for drug-induced nephrotoxicity testing. *Sci. Rep.* **2020**, *10*, 13640. [CrossRef] [PubMed]

23. Zha, M.; Tian, T.; Xu, W.; Liu, S.; Jia, J.; Wang, L.; Yan, Q.; Li, N.; Yu, J.; Huang, L. The circadian clock gene Bmal1 facilitates cisplatin-induced renal injury and hepatization. *Cell Death Dis.* **2020**, *11*, 446. [CrossRef] [PubMed]

24. Cepeda, V.; Fuertes, M.A.; Castilla, J.; Alonso, C.; Quevedo, C.; Pérez, J.M. Biochemical mechanisms of cisplatin cytotoxicity. *Anticancer Agents Med. Chem.* **2007**, *7*, 3–18. [CrossRef] [PubMed]

25. Zhu, S.; Pabla, N.; Tang, C.; He, L.; Dong, Z. DNA damage response in cisplatin-induced nephrotoxicity. *Arch. Toxicol.* **2015**, *89*, 2197–2205. [CrossRef]

26. Zhang, M.; Hagan, C.T.; Foley, H.; Tian, X.; Yang, F.; Au, K.M.; Mi, Y.; Medik, Y.; Roche, K.; Wagner, K.; et al. Co-delivery of etoposide and cisplatin in dual-drug loaded nanoparticles synergistically improves chemoradiotherapy in non-small cell lung cancer models. *Acta Biomater.* **2021**, *124*, 327–335. [CrossRef] [PubMed]

27. Zhang, H.; Almuqbil, R.M.; Alhudaithi, S.S.; Sunbul, F.S.; da Rocha, S.R.P. Pulmonary administration of a CSF-1R inhibitor alters the balance of tumor-associated macrophages and supports first-line chemotherapy in a lung cancer model. *Int. J. Pharm.* **2021**, *598*, 120350. [CrossRef] [PubMed]

28. Wang, B.; Hu, W.; Yan, H.; Chen, G.; Zhang, Y.; Mao, J.; Wang, L. Lung cancer chemotherapy using nanoparticles: Enhanced target ability of redox-responsive and pH-sensitive cisplatin prodrug and paclitaxel. *Biomed. Pharmacother.* **2021**, *136*, 111249. [CrossRef]

29. Zhang, Y.; Dong, Y.; Fu, H.; Huang, H.; Wu, Z.; Zhao, M.; Yang, X.; Guo, Q.; Duan, Y.; Sun, Y. Multifunctional tumor-targeted PLGA nanoparticles delivering Pt(IV)/siBIRC5 for US/MRI imaging and overcoming ovarian cancer resistance. *Biomaterials* **2021**, *269*, 120478. [CrossRef]

30. Zhang, M.; Liu, Q.; Cao, C.; Liu, X.; Li, G.; Xu, C.; Zhang, X. Enhanced antitumor effects of follicle-stimulating hormone receptor-mediated hexokinase-2 depletion on ovarian cancer mediated by a shift in glucose metabolism. *J. Nanobiotechnol.* **2020**, *18*, 161. [CrossRef]

31. Jin, S.; Muhammad, N.; Sun, Y.; Tan, Y.; Yuan, H.; Song, D.; Guo, Z.; Wang, X. Multispecific Platinum(IV) Complex Deters Breast Cancer via Interposing Inflammation and Immunosuppression as an Inhibitor of COX-2 and PD-L1. *Angew. Chem. Int. Ed. Engl.* **2020**, *59*, 23313–23321. [CrossRef]

32. Fei, L.; Huimei, H.; Dongmin, C. Pivalopril improves anti-cancer efficiency of cDDP in breast cancer through inhibiting proliferation, angiogenesis and metastasis. *Biochem. Biophys. Res. Commun.* **2020**, *533*, 853–860. [CrossRef]

33. Ledezma-Gallegos, F.; Jurado, R.; Mir, R.; Medina, L.A.; Mondragon-Fuentes, L.; Garcia-Lopez, P. Liposomes Co-Encapsulating Cisplatin/Mifepristone Improve the Effect on Cervical Cancer: In Vitro and In Vivo Assessment. *Pharmaceutics* **2020**, *12*. [CrossRef] [PubMed]

34. Zeng, L.; Boggs, D.H.; Xing, C.; Zhang, Z.; Anderson, J.C.; Wajapeyee, N.; Veale, C.; Bredel, M.; Shi, L.Z.; Bonner, J.A.; et al. Combining PARP and DNA-PK Inhibitors with Irradiation Inhibits HPV-Negative Head and Neck Cancer Squamous Carcinoma Growth. *Front. Genet.* **2020**, *11*, 1036. [CrossRef] [PubMed]

35. Iannelli, F.; Zotti, A.I.; Roca, M.S.; Grumetti, L.; Lombardi, R.; Moccia, T.; Vitagliano, C.; Milone, M.R.; Ciardiello, C.; Bruzzese, F.; et al. Valproic Acid Synergizes with Cisplatin and Cetuximab. *Front. Cell Dev. Biol.* **2020**, *8*, 732. [CrossRef] [PubMed]

36. Ali, B.H.; Al Moundhri, M.S. Agents ameliorating or augmenting the nephrotoxicity of cisplatin and other platinum compounds: A review of some recent research. *Food Chem. Toxicol.* **2006**, *44*, 1173–1183. [CrossRef]

37. Večerić-Haler, Ž.; Cerar, A.; Perše, M. (Mesenchymal) Stem Cell-Based Therapy in Cisplatin-Induced Acute Kidney Injury Animal Model: Risk of Immunogenicity and Tumorigenicity. *Stem Cells Int.* **2017**, *2017*, 17. [CrossRef] [PubMed]

38. Wu, W.J.; Tang, Y.F.; Dong, S.; Zhang, J. Ginsenoside Rb3 Alleviates the Toxic Effect of Cisplatin on the Kidney during Its Treatment to Oral Cancer via TGF-. *Evid.-Based Complement. Alternat. Med.* **2021**, *2021*, 6640714. [CrossRef] [PubMed]

39. Arita, M.; Watanabe, S.; Aoki, N.; Kuwahara, S.; Suzuki, R.; Goto, S.; Abe, Y.; Takahashi, M.; Sato, M.; Hokari, S.; et al. Combination therapy of cisplatin with cilastatin enables an increased dose of cisplatin, enhancing its antitumor effect by suppression of nephrotoxicity. *Sci. Rep.* **2021**, *11*, 750. [CrossRef] [PubMed]

40. Zhang, Q.; Lu, Q.B. New combination chemotherapy of cisplatin with an electron-donating compound for treatment of multiple cancers. *Sci. Rep.* **2021**, *11*, 788. [CrossRef]

41. Alexander, J.F.; Seua, A.V.; Arroyo, L.D.; Ray, P.R.; Wangzhou, A.; Heiß-Lückemann, L.; Schedlowski, M.; Price, T.J.; Kavelaars, A.; Heijnen, C.J. Nasal administration of mitochondria reverses chemotherapy-induced cognitive deficits. *Theranostics* **2021**, *11*, 3109–3130. [CrossRef] [PubMed]

42. Ozkok, A.; Edelstein, C.L. Pathophysiology of cisplatin-induced acute kidney injury. *Biomed. Res. Int.* **2014**, *2014*, 967826. [CrossRef] [PubMed]

43. Zhang, J.; Goering, P.L.; Espandiari, P.; Shaw, M.; Bonventre, J.V.; Vaidya, V.S.; Brown, R.P.; Keenan, J.; Kilty, C.G.; Sadrieh, N.; et al. Differences in immunolocalization of Kim-1, RPA-1, and RPA-2 in kidneys of gentamicin-, cisplatin-, and valproic acid-treated rats: Potential role of iNOS and nitrotyrosine. *Toxicol. Pathol.* **2009**, *37*, 629–643. [CrossRef]

44. Perše, M.; Večerić-Haler, Ž. Cisplatin-Induced Rodent Model of Kidney Injury: Characteristics and Challenges. *Biomed. Res. Int.* **2018**, *2018*, 1462802. [CrossRef] [PubMed]

45. Acklin, S.; Zhang, M.; Du, W.; Zhao, X.; Plotkin, M.; Chang, J.; Campisi, J.; Zhou, D.; Xia, F. Depletion of senescent-like neuronal cells alleviates cisplatin-induced peripheral neuropathy in mice. *Sci. Rep.* **2020**, *10*, 14170. [CrossRef]

46. Zhang, M.; Du, W.; Acklin, S.; Jin, S.; Xia, F. SIRT2 protects peripheral neurons from cisplatin-induced injury by enhancing nucleotide excision repair. *J. Clin. Investig.* **2020**, *130*, 2953–2965. [CrossRef]

47. Podratz, J.L.; Tang, J.J.; Polzin, M.J.; Schmeichel, A.M.; Nesbitt, J.J.; Windebank, A.J.; Madigan, N.N. Mechano growth factor interacts with nucleolin to protect against cisplatin-induced neurotoxicity. *Exp. Neurol.* **2020**, *331*, 113376. [CrossRef]

48. Fernandez, K.; Wafa, T.; Fitzgerald, T.S.; Cunningham, L.L. An optimized, clinically relevant mouse model of cisplatin-induced ototoxicity. *Hear. Res.* **2019**, *375*, 66–74. [CrossRef]

49. Gersten, B.K.; Fitzgerald, T.S.; Fernandez, K.A.; Cunningham, L.L. Ototoxicity and Platinum Uptake Following Cyclic Administration of Platinum-Based Chemotherapeutic Agents. *J. Assoc. Res. Otolaryngol.* **2020**, *21*, 303–321. [CrossRef]

50. Huang, J.; Shan, W.; Li, N.; Zhou, B.; Guo, E.; Xia, M.; Lu, H.; Wu, Y.; Chen, J.; Wang, B.; et al. Melatonin provides protection against cisplatin-induced ovarian damage and loss of fertility in mice. *Reprod. Biomed. Online* **2021**, *42*, 505–519. [CrossRef]

51. Zhang, K.; Sun, C.; Hu, Y.; Yang, J.; Wu, C. Network pharmacology reveals pharmacological effect and mechanism of Panax notoginseng (Burk.) F. H. Chen on reproductive and genetic toxicity in male mice. *J. Ethnopharmacol.* **2021**, *270*, 113792. [CrossRef] [PubMed]

52. Eldani, M.; Luan, Y.; Xu, P.C.; Bargar, T.; Kim, S.Y. Continuous treatment with cisplatin induces the oocyte death of primordial follicles without activation. *FASEB J.* **2020**, *34*, 13885–13899. [CrossRef] [PubMed]

53. Gouveia, B.B.; Barberino, R.S.; Dos Santos Silva, R.L.; Lins, T.L.B.G.; da Silva Guimarães, V.; do Monte, A.P.O.; Palheta, R.C.; de Matos, M.H.T. Involvement of PTEN and FOXO3a Proteins in the Protective Activity of Protocatechuic Acid against Cisplatin-Induced Ovarian Toxicity in Mice. *Reprod. Sci.* **2021**, *28*, 865–876. [CrossRef] [PubMed]

54. Hu, J.N.; Yang, J.Y.; Jiang, S.; Zhang, J.; Liu, Z.; Hou, J.G.; Gong, X.J.; Wang, Y.P.; Wang, Z.; Li, W. Panax quinquefolium saponins protect against cisplatin evoked intestinal injury via ROS-mediated multiple mechanisms. *Phytomedicine* **2021**, *82*, 153446. [CrossRef] [PubMed]

55. Nardini, P.; Pini, A.; Bessard, A.; Duchalais, E.; Niccolai, E.; Neunlist, M.; Vannucchi, M.G. GLP-2 Prevents Neuronal and Glial Changes in the Distal Colon of Mice Chronically Treated with Cisplatin. *Int. J. Mol. Sci.* **2020**, *21*, 8875. [CrossRef] [PubMed]

56. Pini, A.; Garella, R.; Idrizaj, E.; Calosi, L.; Baccari, M.C.; Vannucchi, M.G. Glucagon-like peptide 2 counteracts the mucosal damage and the neuropathy induced by chronic treatment with cisplatin in the mouse gastric fundus. *Neurogastroenterol. Motil.* **2016**, *28*, 206–216. [CrossRef] [PubMed]

57. Sakai, H.; Ikeno, Y.; Tsukimura, Y.; Inomata, M.; Suzuki, Y.; Kon, R.; Ikarashi, N.; Chiba, Y.; Yamada, T.; Kamei, J. Upregulation of ubiquitinated proteins and their degradation pathway in muscle atrophy induced by cisplatin in mice. *Toxicol. Appl. Pharmacol.* **2020**, *403*, 115165. [CrossRef] [PubMed]

58. Damrauer, J.S.; Stadler, M.E.; Acharyya, S.; Baldwin, A.S.; Couch, M.E.; Guttridge, D.C. Chemotherapy-induced muscle wasting: Association with NF-κB and cancer cachexia. *Eur. J. Transl. Myol.* **2018**, *28*, 7590. [CrossRef]

59. Moreira-Pais, A.; Ferreira, R.; Gil da Costa, R. Platinum-induced muscle wasting in cancer chemotherapy: Mechanisms and potential targets for therapeutic intervention. *Life Sci.* **2018**, *208*, 1–9. [CrossRef]

60. Zhang, X.; Lei, Y.; Hu, T.; Wu, Y.; Li, Z.; Jiang, Z.; Yang, C.; Zhang, L.; You, Q. Discovery of Clinical Candidate (5-(3-(4-Chlorophenoxy)prop-1-yn-1-yl)-3-hydroxypicolinoyl)glycine, an Orally Bioavailable Prolyl Hydroxylase Inhibitor for the Treatment of Anemia. *J. Med. Chem.* **2020**, *63*, 10045–10060. [CrossRef]

61. Kociba, R.J.; Sleight, S.D. Acute toxicologic and pathologic effects of cis-diamminedichloroplatinum (NSC-119875) in the male rat. *Cancer Chemother. Rep.* **1971**, *55*, 1–8.

62. Madias, N.E.; Harrington, J.T. Platinum nephrotoxicity. *Am. J. Med.* **1978**, *65*, 307–314. [CrossRef]

63. Singh, G. A possible cellular mechanism of cisplatin-induced nephrotoxicity. *Toxicology* **1989**, *58*, 71–80. [CrossRef]

64. Dobyan, D.C.; Levi, J.; Jacobs, C.; Kosek, J.; Weiner, M.W. Mechanism of cis-platinum nephrotoxicity: II. Morphologic observations. *J. Pharmacol. Exp. Ther.* **1980**, *213*, 551–556. [PubMed]

65. McKeage, M.J.; Morgan, S.E.; Boxall, F.E.; Murrer, B.A.; Hard, G.C.; Harrap, K.R. Lack of nephrotoxicity of oral ammine/amine platinum (IV) dicarboxylate complexes in rodents. *Br. J. Cancer* **1993**, *67*, 996–1000. [CrossRef] [PubMed]

66. Bodenner, D.L.; Dedon, P.C.; Keng, P.C.; Katz, J.C.; Borch, R.F. Selective protection against cis-diamminedichloroplatinum(II)-induced toxicity in kidney, gut, and bone marrow by diethyldithiocarbamate. *Cancer Res.* **1986**, *46*, 2751–2755.

67. Kramer, R.A. Protection against cisplatin nephrotoxicity by prochlorperazine. *Cancer Chemother. Pharmacol.* **1989**, *25*, 156–160. [CrossRef]

68. Naganuma, A.; Satoh, M.; Imura, N. Prevention of lethal and renal toxicity of cis-diamminedichloroplatinum(II) by induction of metallothionein synthesis without compromising its antitumor activity in mice. *Cancer Res.* **1987**, *47*, 983–987.

69. Baldew, G.S.; van den Hamer, C.J.; Los, G.; Vermeulen, N.P.; de Goeij, J.J.; McVie, J.G. Selenium-induced protection against cis-diamminedichloroplatinum(II) nephrotoxicity in mice and rats. *Cancer Res.* **1989**, *49*, 3020–3023. [PubMed]

70. Mizushima, Y.; Nagahama, H.; Yokoyama, A.; Morikage, T.; Yano, S. Studies on nephrotoxicity following a single and repeated administration of cis-diamminedichloroplatinum (CDDP) in rats. *Tohoku J. Exp. Med.* **1987**, *151*, 129–135. [CrossRef] [PubMed]
71. Yamate, J.; Tatsumi, M.; Nakatsuji, S.; Kuwamura, M.; Kotani, T.; Sakuma, S. Immunohistochemical observations on the kinetics of macrophages and myofibroblasts in rat renal interstitial fibrosis induced by cis-diamminedichloroplatinum. *J. Comp. Pathol.* **1995**, *112*, 27–39. [CrossRef]
72. Harrison, S.D. Toxicologic evaluation of cis-diamminedichloroplatinum II in B6D2F1 mice. *Fundam. Appl. Toxicol.* **1981**, *1*, 382–385. [CrossRef]
73. Wagner, T.; Kreft, B.; Bohlmann, G.; Schwieder, G. Effects of fosfomycin, mesna, and sodium thiosulfate on the toxicity and antitumor activity of cisplatin. *J. Cancer Res. Clin. Oncol.* **1988**, *114*, 497–501. [CrossRef]
74. Nowrousian, M.R.; Schmidt, C.G. Effects of cisplatin on different haemopoietic progenitor cells in mice. *Br. J. Cancer* **1982**, *46*, 397–402. [CrossRef] [PubMed]
75. Sharp, C.N.; Siskind, L.J. Developing better mouse models to study cisplatin-induced kidney injury. *Am. J. Physiol. Renal Physiol.* **2017**, *313*, F835–F841. [CrossRef] [PubMed]
76. Sharp, C.N.; Doll, M.A.; Dupre, T.V.; Shah, P.P.; Subathra, M.; Siow, D.; Arteel, G.E.; Megyesi, J.; Beverly, L.J.; Siskind, L.J. Repeated administration of low-dose cisplatin in mice induces fibrosis. *Am. J. Physiol. Renal Physiol.* **2016**, *310*, F560–F568. [CrossRef] [PubMed]
77. Sears, S.M.; Sharp, C.N.; Krueger, A.; Oropilla, G.B.; Saforo, D.; Doll, M.A.; Megyesi, J.; Beverly, L.J.; Siskind, L.J. C57BL/6 mice require a higher dose of cisplatin to induce renal fibrosis and CCL2 correlates with cisplatin-induced kidney injury. *Am. J. Physiol. Renal Physiol.* **2020**, *319*, F674–F685. [CrossRef] [PubMed]
78. Sharp, C.N.; Doll, M.; Dupre, T.V.; Beverly, L.J.; Siskind, L.J. Moderate aging does not exacerbate cisplatin-induced kidney injury or fibrosis despite altered inflammatory cytokine expression and immune cell infiltration. *Am. J. Physiol. Renal Physiol.* **2019**, *316*, F162–F172. [CrossRef]
79. Dobyan, D.C. Long-term consequences of cis-platinum-induced renal injury: A structural and functional study. *Anat. Rec.* **1985**, *212*, 239–245. [CrossRef]
80. Behling, E.B.; Sendão, M.C.; Francescato, H.D.; Antunes, L.M.; Costa, R.S.; Bianchi, M.e.L. Comparative study of multiple dosage of quercetin against cisplatin-induced nephrotoxicity and oxidative stress in rat kidneys. *Pharmacol. Rep.* **2006**, *58*, 526–532. [PubMed]
81. Yamate, J.; Ishida, A.; Tsujino, K.; Tatsumi, M.; Nakatsuji, S.; Kuwamura, M.; Kotani, T.; Sakuma, S. Immunohistochemical study of rat renal interstitial fibrosis induced by repeated injection of cisplatin, with special reference to the kinetics of macrophages and myofibroblasts. *Toxicol. Pathol.* **1996**, *24*, 199–206. [CrossRef]
82. Yamate, J.; Sato, K.; Ide, M.; Nakanishi, M.; Kuwamura, M.; Sakuma, S.; Nakatsuji, S. Participation of different macrophage populations and myofibroblastic cells in chronically developed renal interstitial fibrosis after cisplatin-induced renal injury in rats. *Vet. Pathol.* **2002**, *39*, 322–333. [CrossRef]
83. Yamate, J.; Machida, Y.; Ide, M.; Kuwamura, M.; Kotani, T.; Sawamoto, O.; LaMarre, J. Cisplatin-induced renal interstitial fibrosis in neonatal rats, developing as solitary nephron unit lesions. *Toxicol. Pathol.* **2005**, *33*, 207–217. [CrossRef]
84. Palant, C.E.; Amdur, R.L.; Chawla, L.S. The Acute Kidney Injury to Chronic Kidney Disease Transition: A Potential Opportunity to Improve Care in Acute Kidney Injury. *Contrib. Nephrol.* **2016**, *187*, 55–72. [CrossRef]
85. Ferenbach, D.A.; Bonventre, J.V. Mechanisms of maladaptive repair after AKI leading to accelerated kidney ageing and CKD. *Nat. Rev. Nephrol.* **2015**, *11*, 264–276. [CrossRef]
86. Ferenbach, D.A.; Bonventre, J.V. Acute kidney injury and chronic kidney disease: From the laboratory to the clinic. *Nephrol. Ther.* **2016**, *12* (Suppl. 1), S41–S48. [CrossRef]
87. Chawla, L.S.; Eggers, P.W.; Star, R.A.; Kimmel, P.L. Acute kidney injury and chronic kidney disease as interconnected syndromes. *N. Engl. J. Med.* **2014**, *371*, 58–66. [CrossRef] [PubMed]
88. Palant, C.E.; Chawla, L.S.; Faselis, C.; Li, P.; Pallone, T.L.; Kimmel, P.L.; Amdur, R.L. High serum creatinine nonlinearity: A renal vital sign? *Am. J. Physiol. Renal Physiol.* **2016**, *311*, F305–F309. [CrossRef] [PubMed]
89. Atmaca, A.; Al-Batran, S.E.; Werner, D.; Pauligk, C.; Güner, T.; Koepke, A.; Bernhard, H.; Wenzel, T.; Banat, A.G.; Brueck, P.; et al. A randomised multicentre phase II study with cisplatin/docetaxel vs oxaliplatin/docetaxel as first-line therapy in patients with advanced or metastatic non-small cell lung cancer. *Br. J. Cancer* **2013**, *108*, 265–270. [CrossRef] [PubMed]
90. McKibbin, T.; Cheng, L.L.; Kim, S.; Steuer, C.E.; Owonikoko, T.K.; Khuri, F.R.; Shin, D.M.; Saba, N.F. Mannitol to prevent cisplatin-induced nephrotoxicity in patients with squamous cell cancer of the head and neck (SCCHN) receiving concurrent therapy. *Support Care Cancer* **2016**, *24*, 1789–1793. [CrossRef]
91. Liston, D.R.; Davis, M. Clinically Relevant Concentrations of Anticancer Drugs: A Guide for Nonclinical Studies. *Clin. Cancer Res.* **2017**, *23*, 3489–3498. [CrossRef]
92. Nair, A.B.; Jacob, S. A simple practice guide for dose conversion between animals and human. *J. Basic Clin. Pharm.* **2016**, *7*, 27–31. [CrossRef] [PubMed]
93. McSweeney, K.R.; Gadanec, L.K.; Qaradakhi, T.; Ali, B.A.; Zulli, A.; Apostolopoulos, V. Mechanisms of Cisplatin-Induced Acute Kidney Injury: Pathological Mechanisms, Pharmacological Interventions, and Genetic Mitigations. *Cancers* **2021**, *13*, 1572. [CrossRef]

94. Holditch, S.J.; Brown, C.N.; Lombardi, A.M.; Nguyen, K.N.; Edelstein, C.L. Recent Advances in Models, Mechanisms, Biomarkers, and Interventions in Cisplatin-Induced Acute Kidney Injury. *Int. J. Mol. Sci.* **2019**, *20*, 3011. [CrossRef] [PubMed]

95. Loren, P.; Saavedra, N.; Saavedra, K.; Zambrano, T.; Moriel, P.; Salazar, L.A. Epigenetic Mechanisms Involved in Cisplatin-Induced Nephrotoxicity: An Update. *Pharmaceuticals* **2021**, *14*, 491. [CrossRef] [PubMed]

96. Tanase, D.M.; Gosav, E.M.; Radu, S.; Costea, C.F.; Ciocoiu, M.; Carauleanu, A.; Lacatusu, C.M.; Maranduca, M.A.; Floria, M.; Rezus, C. The Predictive Role of the Biomarker Kidney Molecule-1 (KIM-1) in Acute Kidney Injury (AKI) Cisplatin-Induced Nephrotoxicity. *Int. J. Mol. Sci.* **2019**, *20*, 5238. [CrossRef]

97. Gautier, J.C.; Riefke, B.; Walter, J.; Kurth, P.; Mylecraine, L.; Guilpin, V.; Barlow, N.; Gury, T.; Hoffman, D.; Ennulat, D.; et al. Evaluation of novel biomarkers of nephrotoxicity in two strains of rat treated with Cisplatin. *Toxicol. Pathol.* **2010**, *38*, 943–956. [CrossRef]

98. Espandiari, P.; Rosenzweig, B.; Zhang, J.; Zhou, Y.; Schnackenberg, L.; Vaidya, V.S.; Goering, P.L.; Brown, R.P.; Bonventre, J.V.; Mahjoob, K.; et al. Age-related differences in susceptibility to cisplatin-induced renal toxicity. *J. Appl. Toxicol.* **2010**, *30*, 172–182. [CrossRef]

99. Jodrell, D.I.; Newell, D.R.; Morgan, S.E.; Clinton, S.; Bensted, J.P.; Hughes, L.R.; Calvert, A.H. The renal effects of N10-propargyl-5,8-dideazafolic acid (CB3717) and a non-nephrotoxic analogue ICI D1694, in mice. *Br. J. Cancer* **1991**, *64*, 833–838. [CrossRef]

100. Launay-Vacher, V.; Rey, J.B.; Isnard-Bagnis, C.; Deray, G.; Daouphars, M. Prevention of cisplatin nephrotoxicity: State of the art and recommendations from the European Society of Clinical Pharmacy Special Interest Group on Cancer Care. *Cancer Chemother. Pharmacol.* **2008**, *61*, 903–909. [CrossRef] [PubMed]

101. Blanchard, O.L.; Smoliga, J.M. Translating dosages from animal models to human clinical trials-revisiting body surface area scaling. *FASEB J.* **2015**, *29*, 1629–1634. [CrossRef] [PubMed]

102. Spilker, M.E.; Chen, X.; Visswanathan, R.; Vage, C.; Yamazaki, S.; Li, G.; Lucas, J.; Bradshaw-Pierce, E.L.; Vicini, P. Found in Translation: Maximizing the Clinical Relevance of Nonclinical Oncology Studies. *Clin. Cancer Res.* **2017**, *23*, 1080–1090. [CrossRef] [PubMed]

103. Baldew, G.S.; McVie, J.G.; van der Valk, M.A.; Los, G.; de Goeij, J.J.; Vermeulen, N.P. Selective reduction of cis-diamminedichloroplatinum(II) nephrotoxicity by ebselen. *Cancer Res.* **1990**, *50*, 7031–7036. [PubMed]

104. Nakamura, T.; Yonezawa, A.; Hashimoto, S.; Katsura, T.; Inui, K. Disruption of multidrug and toxin extrusion MATE1 potentiates cisplatin-induced nephrotoxicity. *Biochem. Pharmacol.* **2010**, *80*, 1762–1767. [CrossRef]

105. Parham, K. Can intratympanic dexamethasone protect against cisplatin ototoxicity in mice with age-related hearing loss? *Otolaryngol. Head Neck Surg.* **2011**, *145*, 635–640. [CrossRef] [PubMed]

106. Aston, W.J.; Hope, D.E.; Nowak, A.K.; Robinson, B.W.; Lake, R.A.; Lesterhuis, W.J. A systematic investigation of the maximum tolerated dose of cytotoxic chemotherapy with and without supportive care in mice. *BMC Cancer* **2017**, *17*, 684. [CrossRef]

107. Vera, G.; Chiarlone, A.; Martín, M.I.; Abalo, R. Altered feeding behaviour induced by long-term cisplatin in rats. *Auton. Neurosci.* **2006**, *126–127*, 81–92. [CrossRef] [PubMed]

108. Allan, S.G.; Smyth, J.F.; Hay, F.G.; Leonard, R.C.; Wolf, C.R. Protective effect of sodium-2-mercaptoethanesulfonate on the gastrointestinal toxicity and lethality of cis-diamminedichloroplatinum. *Cancer Res.* **1986**, *46*, 3569–3573.

109. Stavraka, C.; Ford, A.; Ghaem-Maghami, S.; Crook, T.; Agarwal, R.; Gabra, H.; Blagden, S. A study of symptoms described by ovarian cancer survivors. *Gynecol. Oncol.* **2012**, *125*, 59–64. [CrossRef]

110. McQuade, R.M.; Stojanovska, V.; Abalo, R.; Bornstein, J.C.; Nurgali, K. Chemotherapy-Induced Constipation and Diarrhea: Pathophysiology, Current and Emerging Treatments. *Front. Pharmacol.* **2016**, *7*, 414. [CrossRef]

111. Cubeddu, L.X. Serotonin mechanisms in chemotherapy-induced emesis in cancer patients. *Oncology* **1996**, *53* (Suppl. 1), 18–25. [CrossRef]

112. Cabezos, P.A.; Vera, G.; Castillo, M.; Fernández-Pujol, R.; Martín, M.I.; Abalo, R. Radiological study of gastrointestinal motor activity after acute cisplatin in the rat. Temporal relationship with pica. *Auton. Neurosci.* **2008**, *141*, 54–65. [CrossRef]

113. Cabezos, P.A.; Vera, G.; Martín-Fontelles, M.I.; Fernández-Pujol, R.; Abalo, R. Cisplatin-induced gastrointestinal dysmotility is aggravated after chronic administration in the rat. Comparison with pica. *Neurogastroenterol. Motil.* **2010**, *22*, 797–805. [CrossRef]

114. Liu, Y.L.; Malik, N.; Sanger, G.J.; Friedman, M.I.; Andrews, P.L. Pica—A model of nausea? Species differences in response to cisplatin. *Physiol. Behav.* **2005**, *85*, 271–277. [CrossRef]

115. Allan, S.G.; Smyth, J.F. Small intestinal mucosal toxicity of cis-platinum-comparison of toxicity with platinum analogues and dexamethasone. *Br. J. Cancer* **1986**, *53*, 355–360. [CrossRef] [PubMed]

116. Vera, G.; Castillo, M.; Cabezos, P.A.; Chiarlone, A.; Martín, M.I.; Gori, A.; Pasquinelli, G.; Barbara, G.; Stanghellini, V.; Corinaldesi, R.; et al. Enteric neuropathy evoked by repeated cisplatin in the rat. *Neurogastroenterol. Motil.* **2011**, *23*, 370–378. [CrossRef] [PubMed]

117. Makris, K.; Spanou, L. Acute Kidney Injury: Definition, Pathophysiology and Clinical Phenotypes. *Clin. Biochem. Rev.* **2016**, *37*, 85–98. [PubMed]

118. Shahid, F.; Farooqui, Z.; Khan, F. Cisplatin-induced gastrointestinal toxicity: An update on possible mechanisms and on available gastroprotective strategies. *Eur. J. Pharmacol.* **2018**, *827*, 49–57. [CrossRef] [PubMed]

119. Jakob, M.O.; Kofoed-Branzk, M.; Deshpande, D.; Murugan, S.; Klose, C.S.N. An Integrated View on Neuronal Subsets in the Peripheral Nervous System and Their Role in Immunoregulation. *Front. Immunol.* **2021**, *12*, 679055. [CrossRef] [PubMed]

120. Spencer, N.J.; Hu, H. Enteric nervous system: Sensory transduction, neural circuits and gastrointestinal motility. *Nat. Rev. Gastroenterol. Hepatol.* **2020**, *17*, 338–351. [CrossRef]
121. Rao, M.; Gershon, M.D. The bowel and beyond: The enteric nervous system in neurological disorders. *Nat. Rev. Gastroenterol. Hepatol.* **2016**, *13*, 517–528. [CrossRef]
122. Dimidi, E.; Christodoulides, S.; Scott, S.M.; Whelan, K. Mechanisms of Action of Probiotics and the Gastrointestinal Microbiota on Gut Motility and Constipation. *Adv. Nutr.* **2017**, *8*, 484–494. [CrossRef] [PubMed]
123. De Vadder, F.; Grasset, E.; Mannerås Holm, L.; Karsenty, G.; Macpherson, A.J.; Olofsson, L.E.; Bäckhed, F. Gut microbiota regulates maturation of the adult enteric nervous system via enteric serotonin networks. *Proc. Natl. Acad. Sci. USA* **2018**, *115*, 6458–6463. [CrossRef] [PubMed]
124. Muller, P.A.; Schneeberger, M.; Matheis, F.; Wang, P.; Kerner, Z.; Ilanges, A.; Pellegrino, K.; Del Mármol, J.; Castro, T.B.R.; Furuichi, M.; et al. Microbiota modulate sympathetic neurons via a gut-brain circuit. *Nature* **2020**, *583*, 441–446. [CrossRef] [PubMed]
125. Chang, C.J.; Wang, P.C.; Huang, T.C.; Taniguchi, A. Change in Renal Glomerular Collagens and Glomerular Filtration Barrier-Related Proteins in a Dextran Sulfate Sodium-Induced Colitis Mouse Model. *Int. J. Mol. Sci.* **2019**, *20*. [CrossRef] [PubMed]
126. Vaziri, N.D.; Zhao, Y.Y.; Pahl, M.V. Altered intestinal microbial flora and impaired epithelial barrier structure and function in CKD: The nature, mechanisms, consequences and potential treatment. *Nephrol. Dial. Transplant.* **2016**, *31*, 737–746. [CrossRef] [PubMed]
127. Yang, T.; Richards, E.M.; Pepine, C.J.; Raizada, M.K. The gut microbiota and the brain-gut-kidney axis in hypertension and chronic kidney disease. *Nat. Rev. Nephrol.* **2018**, *14*, 442–456. [CrossRef]
128. Chen, Y.; Xu, J. Regulation of Neurotransmitters by the Gut Microbiota and Effects on Cognition in Neurological Disorders. *Nutrients* **2021**, *13*, 2099. [CrossRef] [PubMed]
129. Dilruba, S.; Kalayda, G.V. Platinum-based drugs: Past, present and future. *Cancer Chemother. Pharmacol.* **2016**, *77*, 1103–1124. [CrossRef]
130. Raudenska, M.; Balvan, J.; Fojtu, M.; Gumulec, J.; Masarik, M. Unexpected therapeutic effects of cisplatin. *Metallomics* **2019**, *11*, 1182–1199. [CrossRef]
131. Stojanovska, V.; Sakkal, S.; Nurgali, K. Platinum-based chemotherapy: Gastrointestinal immunomodulation and enteric nervous system toxicity. *Am. J. Physiol. Gastrointest. Liver Physiol.* **2015**, *308*, G223–G232. [CrossRef]
132. Bajic, J.E.; Johnston, I.N.; Howarth, G.S.; Hutchinson, M.R. From the Bottom-Up: Chemotherapy and Gut-Brain Axis Dysregulation. *Front. Behav. Neurosci.* **2018**, *12*, 104. [CrossRef]
133. Thompson, S.W.; Davis, L.E.; Kornfeld, M.; Hilgers, R.D.; Standefer, J.C. Cisplatin neuropathy. Clinical, electrophysiologic, morphologic, and toxicologic studies. *Cancer* **1984**, *54*, 1269–1275. [CrossRef]
134. Roelofs, R.I.; Hrushesky, W.; Rogin, J.; Rosenberg, L. Peripheral sensory neuropathy and cisplatin chemotherapy. *Neurology* **1984**, *34*, 934–938. [CrossRef]
135. Zajączkowska, R.; Kocot-Kępska, M.; Leppert, W.; Wrzosek, A.; Mika, J.; Wordliczek, J. Mechanisms of Chemotherapy-Induced Peripheral Neuropathy. *Int. J. Mol. Sci.* **2019**, *20*, 1451. [CrossRef]
136. Cavaletti, G.; Marzorati, L.; Bogliun, G.; Colombo, N.; Marzola, M.; Pittelli, M.R.; Tredici, G. Cisplatin-induced peripheral neurotoxicity is dependent on total-dose intensity and single-dose intensity. *Cancer* **1992**, *69*, 203–207. [CrossRef]
137. Van den Bent, M.J.; van Putten, W.L.; Hilkens, P.H.; de Wit, R.; van der Burg, M.E. Retreatment with dose-dense weekly cisplatin after previous cisplatin chemotherapy is not complicated by significant neuro-toxicity. *Eur. J. Cancer* **2002**, *38*, 387–391. [CrossRef]
138. Verdú, E.; Vilches, J.J.; Rodríguez, F.J.; Ceballos, D.; Valero, A.; Navarro, X. Physiological and immunohistochemical characterization of cisplatin-induced neuropathy in mice. *Muscle Nerve* **1999**, *22*, 329–340. [CrossRef]
139. Hopkins, H.L.; Duggett, N.A.; Flatters, S.J.L. Chemotherapy-induced painful neuropathy: Pain-like behaviours in rodent models and their response to commonly used analgesics. *Curr. Opin. Support Palliat. Care* **2016**, *10*, 119–128. [CrossRef] [PubMed]
140. Screnci, D.; McKeage, M.J. Platinum neurotoxicity: Clinical profiles, experimental models and neuroprotective approaches. *J. Inorg. Biochem.* **1999**, *77*, 105–110. [CrossRef]
141. Simó, M.; Rifà-Ros, X.; Rodriguez-Fornells, A.; Bruna, J. Chemobrain: A systematic review of structural and functional neuroimaging studies. *Neurosci. Biobehav. Rev.* **2013**, *37*, 1311–1321. [CrossRef]
142. Amidi, A.; Hosseini, S.M.H.; Leemans, A.; Kesler, S.R.; Agerbæk, M.; Wu, L.M.; Zachariae, R. Changes in Brain Structural Networks and Cognitive Functions in Testicular Cancer Patients Receiving Cisplatin-Based Chemotherapy. *J. Natl. Cancer Inst.* **2017**, *109*. [CrossRef]
143. Stouten-Kemperman, M.M.; de Ruiter, M.B.; Boogerd, W.; Kerst, J.M.; Kirschbaum, C.; Reneman, L.; Schagen, S.B. Brain Hyperconnectivity >10 Years after Cisplatin-Based Chemotherapy for Testicular Cancer. *Brain Connect.* **2018**, *8*, 398–406. [CrossRef]
144. Seigers, R.; Schagen, S.B.; Van Tellingen, O.; Dietrich, J. Chemotherapy-related cognitive dysfunction: Current animal studies and future directions. *Brain Imaging Behav.* **2013**, *7*, 453–459. [CrossRef]
145. Zhou, W.; Kavelaars, A.; Heijnen, C.J. Metformin Prevents Cisplatin-Induced Cognitive Impairment and Brain Damage in Mice. *PLoS ONE* **2016**, *11*, e0151890. [CrossRef]
146. Chiu, G.S.; Maj, M.A.; Rizvi, S.; Dantzer, R.; Vichaya, E.G.; Laumet, G.; Kavelaars, A.; Heijnen, C.J. Pifithrin-μ Prevents Cisplatin-Induced Chemobrain by Preserving Neuronal Mitochondrial Function. *Cancer Res.* **2017**, *77*, 742–752. [CrossRef]

147. Chiu, G.S.; Boukelmoune, N.; Chiang, A.C.A.; Peng, B.; Rao, V.; Kingsley, C.; Liu, H.L.; Kavelaars, A.; Kesler, S.R.; Heijnen, C.J. Nasal administration of mesenchymal stem cells restores cisplatin-induced cognitive impairment and brain damage in mice. *Oncotarget* **2018**, *9*, 35581–35597. [CrossRef] [PubMed]

148. Huo, X.; Reyes, T.M.; Heijnen, C.J.; Kavelaars, A. Cisplatin treatment induces attention deficits and impairs synaptic integrity in the prefrontal cortex in mice. *Sci. Rep.* **2018**, *8*, 17400. [CrossRef] [PubMed]

149. Wang, D.; Wang, B.; Liu, Y.; Dong, X.; Su, Y.; Li, S. Protective Effects of ACY-1215 against Chemotherapy-Related Cognitive Impairment and Brain Damage in Mice. *Neurochem. Res.* **2019**, *44*, 2460–2469. [CrossRef] [PubMed]

150. Ta, L.E.; Low, P.A.; Windebank, A.J. Mice with cisplatin and oxaliplatin-induced painful neuropathy develop distinct early responses to thermal stimuli. *Mol. Pain* **2009**, *5*, 9. [CrossRef]

151. Ta, L.E.; Bieber, A.J.; Carlton, S.M.; Loprinzi, C.L.; Low, P.A.; Windebank, A.J. Transient Receptor Potential Vanilloid 1 is essential for cisplatin-induced heat hyperalgesia in mice. *Mol. Pain* **2010**, *6*, 15. [CrossRef]

152. Hu, L.Y.; Zhou, Y.; Cui, W.Q.; Hu, X.M.; Du, L.X.; Mi, W.L.; Chu, Y.X.; Wu, G.C.; Wang, Y.Q.; Mao-Ying, Q.L. Triggering receptor expressed on myeloid cells 2 (TREM2) dependent microglial activation promotes cisplatin-induced peripheral neuropathy in mice. *Brain Behav. Immun.* **2018**, *68*, 132–145. [CrossRef] [PubMed]

153. Boehmerle, W.; Huehnchen, P.; Peruzzaro, S.; Balkaya, M.; Endres, M. Electrophysiological, behavioral and histological characterization of paclitaxel, cisplatin, vincristine and bortezomib-induced neuropathy in C57Bl/6 mice. *Sci. Rep.* **2014**, *4*, 6370. [CrossRef] [PubMed]

154. Joly, F.; Lange, M.; Dos Santos, M.; Vaz-Luis, I.; Di Meglio, A. Long-Term Fatigue and Cognitive Disorders in Breast Cancer Survivors. *Cancers* **2019**, *11*, 1896. [CrossRef]

155. Larson, C.M.; Wilcox, G.L.; Fairbanks, C.A. The Study of Pain in Rats and Mice. *Comp. Med.* **2019**, *69*, 555–570. [CrossRef]

156. Sorge, R.E.; Martin, L.J.; Isbester, K.A.; Sotocinal, S.G.; Rosen, S.; Tuttle, A.H.; Wieskopf, J.S.; Acland, E.L.; Dokova, A.; Kadoura, B.; et al. Olfactory exposure to males, including men, causes stress and related analgesia in rodents. *Nat. Methods* **2014**, *11*, 629–632. [CrossRef]

157. Laird, J.M.A.; Martinez-Caro, L.; Garcia-Nicas, E.; Cervero, F. A new model of visceral pain and referred hyperalgesia in the mouse. *Pain* **2001**, *92*, 335–342. [CrossRef]

158. Huang, T.Y.; Belzer, V.; Hanani, M. Gap junctions in dorsal root ganglia: Possible contribution to visceral pain. *Eur. J. Pain* **2010**, *14*, 49–e41. [CrossRef]

159. Hanani, M. Role of satellite glial cells in gastrointestinal pain. *Front. Cell Neurosci.* **2015**, *9*, 412. [CrossRef] [PubMed]

160. Spencer, N.J.; Kyloh, M.A.; Travis, L.; Dodds, K.N. Identification of spinal afferent nerve endings in the colonic mucosa and submucosa that communicate directly with the spinal cord: The gut-brain axis. *J. Comp. Neurol.* **2020**, *528*, 1742–1753. [CrossRef] [PubMed]

161. Park, H.J.; Stokes, J.A.; Pirie, E.; Skahen, J.; Shtaerman, Y.; Yaksh, T.L. Persistent hyperalgesia in the cisplatin-treated mouse as defined by threshold measures, the conditioned place preference paradigm, and changes in dorsal root ganglia activated transcription factor 3: The effects of gabapentin, ketorolac, and etanercept. *Anesth. Analg.* **2013**, *116*, 224–231. [CrossRef]

162. Höke, A.; Ray, M. Rodent models of chemotherapy-induced peripheral neuropathy. *ILAR J.* **2014**, *54*, 273–281. [CrossRef]

163. Mangus, L.M.; Rao, D.B.; Ebenezer, G.J. Intraepidermal Nerve Fiber Analysis in Human Patients and Animal Models of Peripheral Neuropathy: A Comparative Review. *Toxicol. Pathol.* **2020**, *48*, 59–70. [CrossRef] [PubMed]

164. Ma, J.; Trinh, R.T.; Mahant, I.D.; Peng, B.; Matthias, P.; Heijnen, C.J.; Kavelaars, A. Cell-specific role of histone deacetylase 6 in chemotherapy-induced mechanical allodynia and loss of intraepidermal nerve fibers. *Pain* **2019**, *160*, 2877–2890. [CrossRef] [PubMed]

165. Maricich, S.M.; Wellnitz, S.A.; Nelson, A.M.; Lesniak, D.R.; Gerling, G.J.; Lumpkin, E.A.; Zoghbi, H.Y. Merkel cells are essential for light-touch responses. *Science* **2009**, *324*, 1580–1582. [CrossRef] [PubMed]

166. Maksimovic, S.; Nakatani, M.; Baba, Y.; Nelson, A.M.; Marshall, K.L.; Wellnitz, S.A.; Firozi, P.; Woo, S.H.; Ranade, S.; Patapoutian, A.; et al. Epidermal Merkel cells are mechanosensory cells that tune mammalian touch receptors. *Nature* **2014**, *509*, 617–621. [CrossRef]

167. Abraham, J.; Mathew, S. Merkel Cells: A Collective Review of Current Concepts. *Int. J. Appl. Basic Med. Res.* **2019**, *9*, 9–13. [CrossRef] [PubMed]

168. Christianson, J.A.; Ryals, J.M.; Johnson, M.S.; Dobrowsky, R.T.; Wright, D.E. Neurotrophic modulation of myelinated cutaneous innervation and mechanical sensory loss in diabetic mice. *Neuroscience* **2007**, *145*, 303–313. [CrossRef] [PubMed]

169. Samanta, A.; Hughes, T.E.T.; Moiseenkova-Bell, V.Y. Transient Receptor Potential (TRP) Channels. *Subcell. Biochem.* **2018**, *87*, 141–165. [CrossRef] [PubMed]

170. Podratz, J.L.; Knight, A.M.; Ta, L.E.; Staff, N.P.; Gass, J.M.; Genelin, K.; Schlattau, A.; Lathroum, L.; Windebank, A.J. Cisplatin induced mitochondrial DNA damage in dorsal root ganglion neurons. *Neurobiol. Dis.* **2011**, *41*, 661–668. [CrossRef]

171. Yan, F.; Liu, J.J.; Ip, V.; Jamieson, S.M.; McKeage, M.J. Role of platinum DNA damage-induced transcriptional inhibition in chemotherapy-induced neuronal atrophy and peripheral neurotoxicity. *J. Neurochem.* **2015**, *135*, 1099–1112. [CrossRef]

172. Zinovkina, L.A. Mechanisms of Mitochondrial DNA Repair in Mammals. *Biochemistry* **2018**, *83*, 233–249. [CrossRef]

173. Ma, J.; Huo, X.; Jarpe, M.B.; Kavelaars, A.; Heijnen, C.J. Pharmacological inhibition of HDAC6 reverses cognitive impairment and tau pathology as a result of cisplatin treatment. *Acta Neuropathol. Commun.* **2018**, *6*, 103. [CrossRef]

174. Callejo, A.; Sedó-Cabezón, L.; Juan, I.D.; Llorens, J. Cisplatin-Induced Ototoxicity: Effects, Mechanisms and Protection Strategies. *Toxics* **2015**, *3*, 268–293. [CrossRef]
175. Brock, P.R.; Knight, K.R.; Freyer, D.R.; Campbell, K.C.; Steyger, P.S.; Blakley, B.W.; Rassekh, S.R.; Chang, K.W.; Fligor, B.J.; Rajput, K.; et al. Platinum-induced ototoxicity in children: A consensus review on mechanisms, predisposition, and protection, including a new International Society of Pediatric Oncology Boston ototoxicity scale. *J. Clin. Oncol.* **2012**, *30*, 2408–2417. [CrossRef] [PubMed]
176. Meijer, A.J.M.; Li, K.H.; Brooks, B.; Clemens, E.; Ross, C.J.; Rassekh, S.R.; Hoetink, A.E.; van Grotel, M.; van den Heuvel-Eibrink, M.M.; Carleton, B.C. The cumulative incidence of cisplatin-induced hearing loss in young children is higher and develops at an early stage during therapy compared with older children based on 2052 audiological assessments. *Cancer* **2021**. [CrossRef] [PubMed]
177. Tang, Q.; Wang, X.; Jin, H.; Mi, Y.; Liu, L.; Dong, M.; Chen, Y.; Zou, Z. Cisplatin-induced ototoxicity: Updates on molecular mechanisms and otoprotective strategies. *Eur. J. Pharm. Biopharm.* **2021**, *163*, 60–71. [CrossRef] [PubMed]
178. Yu, D.; Gu, J.; Chen, Y.; Kang, W.; Wang, X.; Wu, H. Current Strategies to Combat Cisplatin-Induced Ototoxicity. *Front. Pharmacol.* **2020**, *11*, 999. [CrossRef]
179. Gentilin, E.; Simoni, E.; Candito, M.; Cazzador, D.; Astolfi, L. Cisplatin-Induced Ototoxicity: Updates on Molecular Targets. *Trends Mol. Med.* **2019**, *25*, 1123–1132. [CrossRef]
180. Freyer, D.R.; Chen, L.; Krailo, M.D.; Knight, K.; Villaluna, D.; Bliss, B.; Pollock, B.H.; Ramdas, J.; Lange, B.; Van Hoff, D.; et al. Effects of sodium thiosulfate versus observation on development of cisplatin-induced hearing loss in children with cancer (ACCL0431): A multicentre, randomised, controlled, open-label, phase 3 trial. *Lancet Oncol.* **2017**, *18*, 63–74. [CrossRef]
181. Ullman-Culleré, M.H.; Foltz, C.J. Body condition scoring: A rapid and accurate method for assessing health status in mice. *Lab. Anim. Sci.* **1999**, *49*, 319–323.
182. Guindon, J.; Deng, L.; Fan, B.; Wager-Miller, J.; Hohmann, A.G. Optimization of a cisplatin model of chemotherapy-induced peripheral neuropathy in mice: Use of vitamin C and sodium bicarbonate pretreatments to reduce nephrotoxicity and improve animal health status. *Mol. Pain* **2014**, *10*, 56. [CrossRef] [PubMed]
183. Hrushesky, W.J.; Levi, F.A.; Halberg, F.; Kennedy, B.J. Circadian stage dependence of cis-diamminedichloroplatinum lethal toxicity in rats. *Cancer Res.* **1982**, *42*, 945–949. [PubMed]
184. Litterst, C.L. Alterations in the toxicity of cis-dichlorodiammineplatinum-II and in tissue localization of platinum as a function of NaCl concentration in the vehicle of administration. *Toxicol. Appl. Pharmacol.* **1981**, *61*, 99–108. [CrossRef]
185. Mannel, R.S.; Stratton, J.A.; Moran, G.; Rettenmaier, M.A.; Liao, S.Y.; DiSaia, P.J. Intraperitoneal cisplatin: Comparison of antitumor activity and toxicity as a function of solvent saline concentration. *Gynecol. Oncol.* **1989**, *34*, 50–53. [CrossRef]
186. Hill, G.W.; Morest, D.K.; Parham, K. Cisplatin-induced ototoxicity: Effect of intratympanic dexamethasone injections. *Otol. Neurotol.* **2008**, *29*, 1005–1011. [CrossRef]
187. Levi, F.A.; Hrushesky, W.J.; Halberg, F.; Langevin, T.R.; Haus, E.; Kennedy, B.J. Lethal nephrotoxicity and hematologic toxicity of cis-diamminedichloroplatinum ameliorated by optimal circadian timing and hydration. *Eur. J. Cancer Clin. Oncol.* **1982**, *18*, 471–477. [CrossRef]
188. Levi, F.A.; Hrushesky, W.J.; Blomquist, C.H.; Lakatua, D.J.; Haus, E.; Halberg, F.; Kennedy, B.J. Reduction of cis-diamminedichloroplatinum nephrotoxicity in rats by optimal circadian drug timing. *Cancer Res.* **1982**, *42*, 950–955.
189. To, H.; Kikuchi, A.; Tsuruoka, S.; Sugimoto, K.; Fujimura, A.; Higuchi, S.; Kayama, F.; Hara, K.; Matsuno, K.; Kobayashi, E. Time-dependent nephrotoxicity associated with daily administration of cisplatin in mice. *J. Pharm. Pharmacol.* **2000**, *52*, 1499–1504. [CrossRef]
190. Kilkenny, C.; Browne, W.J.; Cuthill, I.C.; Emerson, M.; Altman, D.G. Improving bioscience research reporting: The ARRIVE guidelines for reporting animal research. *PLoS Biol.* **2010**, *8*, e1000412. [CrossRef]
191. Hooijmans, C.R.; Leenaars, M.; Ritskes-Hoitinga, M. A gold standard publication checklist to improve the quality of animal studies, to fully integrate the Three Rs, and to make systematic reviews more feasible. *Altern. Lab. Anim.* **2010**, *38*, 167–182. [CrossRef] [PubMed]
192. Mähler Convenor, M.; Berard, M.; Feinstein, R.; Gallagher, A.; Illgen-Wilcke, B.; Pritchett-Corning, K.; Raspa, M.; FELASA working group on revision of guidelines for health monitoring of rodents and rabbits. FELASA recommendations for the health monitoring of mouse, rat, hamster, guinea pig and rabbit colonies in breeding and experimental units. *Lab. Anim.* **2014**, *48*, 178–192. [CrossRef]
193. Rülicke, T.; Montagutelli, X.; Pintado, B.; Thon, R.; Hedrich, H.J.; FELASA Working Group. FELASA guidelines for the production and nomenclature of transgenic rodents. *Lab. Anim.* **2007**, *41*, 301–311. [CrossRef] [PubMed]
194. Siddik, Z.H.; Newell, D.R.; Boxall, F.E.; Harrap, K.R. The comparative pharmacokinetics of carboplatin and cisplatin in mice and rats. *Biochem. Pharmacol.* **1987**, *36*, 1925–1932. [CrossRef]
195. Esteban-Fernández, D.; Verdaguer, J.M.; Ramírez-Camacho, R.; Palacios, M.A.; Gómez-Gómez, M.M. Accumulation, fractionation, and analysis of platinum in toxicologically affected tissues after cisplatin, oxaliplatin, and carboplatin administration. *J. Anal. Toxicol.* **2008**, *32*, 140–146. [CrossRef]
196. Okada, A.; Fukushima, K.; Fujita, M.; Nakanishi, M.; Hamori, M.; Nishimura, A.; Shibata, N.; Sugioka, N. Alterations in Cisplatin Pharmacokinetics and Its Acute/Sub-chronic Kidney Injury over Multiple Cycles of Cisplatin Treatment in Rats. *Biol. Pharm. Bull.* **2017**, *40*, 1948–1955. [CrossRef]

197. Levi, J.; Jacobs, C.; Kalman, S.M.; McTigue, M.; Weiner, M.W. Mechanism of cis-platinum nephrotoxicity: I. Effects of sulfhydryl groups in rat kidneys. *J. Pharmacol. Exp. Ther.* **1980**, *213*, 545–550.
198. Poirier, M.C.; Shamkhani, H.; Reed, E.; Tarone, R.E.; Gupta-Burt, S. DNA adducts induced by platinum drug chemotherapeutic agents in human tissues. *Prog. Clin. Biol. Res.* **1992**, *374*, 197–212. [PubMed]
199. Garner, J.P.; Gaskill, B.N.; Weber, E.M.; Ahloy-Dallaire, J.; Pritchett-Corning, K.R. Introducing Therioepistemology: The study of how knowledge is gained from animal research. *Lab. Anim.* **2017**, *46*, 103–113. [CrossRef] [PubMed]

 *biomedicines*

Review

Relevance and Recommendations for the Application of Cardioplegic Solutions in Cardiopulmonary Bypass Surgery in Pigs

Anna Glöckner, Susann Ossmann, Andre Ginther, Jagdip Kang, Michael A. Borger, Alexandro Hoyer [†] and Maja-Theresa Dieterlen *,[†]

University Department for Cardiac Surgery, Heart Center Leipzig, HELIOS Clinic, 04289 Leipzig, Germany;
Anna.Gloeckner@gmx.de (A.G.); Susann.Ossmann@helios-gesundheit.de (S.O.);
Andre.Ginther@helios-gesundheit.de (A.G.); jagdip.kang@helios-gesundheit.de (J.K.);
Michael.Borger@helios-gesundheit.de (M.A.B.); Alexandro.Hoyer@helios-gesundheit.de (A.H.)
* Correspondence: mdieterlen@web.de; Tel.: +49-341-865-256-144
† These authors equally contributed to the work.

Abstract: Cardioplegic solutions play a major role in cardiac surgery due to the fact that they create a silent operating field and protect the myocardium against ischemia and reperfusion injury. For studies on cardioplegic solutions, it is important to compare their effects and to have a valid platform for preclinical testing of new cardioplegic solutions and their additives. Due to the strong anatomical and physiological cardiovascular similarities between pigs and humans, porcine models are suitable for investigating the effects of cardioplegic solutions. This review provides an overview of the results of the application of cardioplegic solutions in adult or pediatric pig models over the past 25 years. The advantages, disadvantages, limitations, and refinement strategies of these models are discussed.

Keywords: pig model; animal model; cardioplegia; refinement; cardiopulmonary bypass; cardiac surgery

Citation: Glöckner, A.; Ossmann, S.; Ginther, A.; Kang, J.; Borger, M.A.; Hoyer, A.; Dieterlen, M.-T. Relevance and Recommendations for the Application of Cardioplegic Solutions in Cardiopulmonary Bypass Surgery in Pigs. *Biomedicines* **2021**, *9*, 1279. https://doi.org/10.3390/biomedicines9091279

Academic Editor: Martina Perše

Received: 30 July 2021
Accepted: 15 September 2021
Published: 21 September 2021

Publisher's Note: MDPI stays neutral with regard to jurisdictional claims in published maps and institutional affiliations.

1. Introduction

Cardioplegic solutions are essential in cardiac surgery since they create a silent operating field and protect the myocardium against extensive ischemic damage and ischemia-reperfusion injury (IRI). Cardioplegia is defined as controlled-induced cardiac arrest [1,2]. A cardioplegic solution induces cardioplegia leading to reversible cardiac arrest. To create a bloodless surgical field, the heart should be excluded from circulation by aortic clamping. This induces whole-organ ischemia of the heart, which can be tolerated for only a few minutes without additional protection [1,2]. The application of a cardioplegic solution increases the time of ischemic tolerance in the heart for up to several hours. Furthermore, cardiopulmonary bypass (CPB) compensates for the pump function of the heart, provides oxygen and nutrients to organs and tissues, and removes metabolites.

The basic principle of any cardioplegic solution is electromechanical decoupling, which influences the extracellular and intracellular ion concentrations. With this, the energy consumption of the myocardium is significantly reduced and ischemia tolerance is increased [2]. In recent years, several different cardioplegic solutions have been developed [3,4], which are based on either a crystalloid electrolyte solution or patients' blood with added electrolytes. However, there are no national or international guidelines or recommendations for choosing cardioplegic solutions for different cardiac surgery procedures [5]. Hence, the selection is largely based on the personal preferences of the surgeon. Furthermore, this must be constantly adapted to new conditions.

Due to demographic changes, patients undergoing cardiac surgery are becoming older and sicker, which necessitates complex cardiac procedures, such as combined heart valve/coronary bypass surgery [6]. Cardioplegic solutions must be adapted to this patient population to provide sufficient myocardial protection and to minimize cardiac damage.

2. Clinical Relevance of Data Analyzing Cardioplegic Solutions in Pig Models

Due to the changing characteristics of the patient cohort and more complex surgical interventions in cardiac patients, it is essential to investigate the effects of cardioplegic solutions. Furthermore, it is necessary to perform structured comparisons and to identify the advantages of different cardioplegic solutions. Especially for new compositions of cardioplegic solutions, adequate tests for their safety and efficiency are necessary. Also, translational research requires testing new drugs in two independent species to fulfill the criteria of the application for ethical and regulatory approval [7].

The pathophysiological processes that are induced by cardioplegic arrest of the heart and CPB are very complex and also affect the kidneys, brain, gut, and lungs. If the effects of cardiac cardioplegia and CPB need to be further investigated, invasive procedures, such as biopsy withdrawal, are necessary. For ethical reasons, it is not possible to conduct studies, including extensive biopsy withdrawal, directly in humans. Hence, animal models are used. Alternative methods to investigate the effects of cardioplegic solutions, such as cell cultures or isolated organs, are not able to fully display the effects of surgical intervention and CPB, such as surgical trauma, blood loss, blood contact with foreign surfaces and shear stress during CPB, inflammatory response to CPB, and changes in the coagulation system [8–10]. The animal model is therefore of particular clinical relevance. However, the ethical consideration of the risk-benefit balance in animal experiments is significant. The benefit of information from a study should always be greater than the expected risks and suffering of the animals. Throughout the entire study, the focus must be on animal welfare along with the achieved results. Therefore, good experimental planning is necessary, and the requirements of the study must be precisely defined to achieve satisfactory validity of the results. Owing to the reproducibility of the study, it is important to investigate meaningful parameters in a targeted manner [11,12]. The first step is the selection of a suitable animal model that produces transferable results for future human clinical applications. In many cardiac surgery studies, pig models have been established due to their special anatomical and physiological similarities to the human heart [13]. Thus, not only heart valves and coronary care are comparable, but also the hemodynamics of the circulatory system. Furthermore, the responses to certain events, such as the lack of volume, are very similar in pigs and humans [14,15]. Thus, the results obtained from pig models can be transferred to humans [14].

This review provides an overview of the in vivo application of cardioplegic solutions in adult and pediatric pig models over the past 25 years. This review focuses on the induction of cardioplegic arrest in CPB procedures, except for the preservation strategies necessary for heart transplantation. Investigations in isolated pig hearts and in vitro studies were excluded from the analysis. The advantages, disadvantages, limitations, and refinement strategies of the pig models are discussed.

3. Comparability of the Heart Anatomy and Physiology in Pig Models and Humans

The pig model has useful biometric conditions regarding the size and anatomy of the cardiovascular system (heart, atria, aorta, femorales and jugulars, coronary vessels, and coronary sinus) [16]. Furthermore, several physiological and hemodynamic similarities exist between the cardiovascular system of pigs and humans [17]. The receptor profiles, ion channels, sympathoadrenal innervation, coronary circulation, and electrophysiology of the pig heart are comparable to those in humans [18]. A lack of volume or loss of blood induced a comparable response in pigs and humans [18]. In response to cardiac arrest and CPB support, it is necessary that the left ventricle ends at the apex, which simplifies physiological measurements such as pressure-volume loops. The pig heart shows limited collateral blood flow, which is analogous to humans and makes it ideal for ischemia studies [19]. However, there is higher cardiac output in pigs, which results from higher heart rate and stroke volume. Furthermore, both parameters resulted from lower hematocrit and oxygen transport capacity [20]. Despite these similarities, long-term follow-up in pig models is considered problematic [19]. If juvenile animals are utilized, changes in animal weight result in alterations in basic cardiac physiology. For example, the heart/body weight ratio is approximately 5 g/kg in healthy humans and 25–30 kg in juvenile farm pigs. However, this ratio decreases up to 50% in farm pigs exceeding 100 kg [19]. Interpreting data obtained in pigs exceeding 100 kg is difficult and not comparable to the human setting. Thus, studies investigating the effects of cardioplegic solutions in pig models have good conditions for a high translation into the human setting. However, handling these animals for longer follow-up periods requires either the use of special requirements for pig husbandry, personnel staff, and institutional facilities.

4. Investigations in Adult Pig Models

A total of 42 studies reported the application of cardioplegic solutions in inducing cardiac arrest during cardiac surgery of adult pig models (Supplementary Table S1).

The St. Thomas-based cardioplegia was investigated in 27 studies (St. Thomas I: $n = 7$; St. Thomas II: $n = 20$), followed by 10 on blood cardioplegia, 9 on histidine-tryptophan-ketoglutarate (HTK)-based cardioplegia, and 8 that did not specify crystalloid cardioplegia. Cardioplegia induced by HTK-N ($n = 2$), Buckberg's solution ($n = 3$), Del Nido ($n = 1$) and Braile ($n = 1$) were investigated to a lesser extent.

Cardioplegia studies aimed to identify the solution with the best properties for the human application. Therefore, direct comparisons of different cardioplegic solutions were performed. Comparisons between St. Thomas I and II cardioplegic solutions have shown enhanced functional recovery, better contractile efficiency, and improved energy status with St. Thomas I cardioplegia [21–24]. The novel HTK-N solution stabilizes hemoglobin and blood calcium levels, which can potentially increase kidney function [25]. Furthermore, HTK-N-induced cardioplegia resulted in fewer cerebral effects and inflammation during CPB surgery than HTK and appeared to exert protective effects in the brain [26]. The comparison of HTK and St. Thomas II cardioplegic solution showed better preservation of post-ischemic mechanoenergetic function and lower troponin T release with St. Thomas II-induced cardiopelgia [27]. Additives such as adenosine [28–30], pentazocine [28], lidocaine [28], procaine [29], cyclosporine A (CsA) [26,31], pyruvate [32,33], amrinone [34], cariporide [35], eniporide [36], aprotinin [37,38], nicorandil [39], H_2S [40], zink-bis-histidinate [41], germinated brown rice extract (GBR) [42], and diazoxide [43] have been added to improve cardioplegic solutions. The addition of diaxozide, adenosine, and nicorandil to cardioplegic solutions preserved ventricular function [29,39,43,44]. Meanwhile, treatment with H_2S, pyruvate, zink-bis-histidinate, or a combination of adenosine/lidocaine/pentazocine improved myocardial protection [33,40–42]. Pyruvate supplementation in cardioplegic solutions also decreased CPB-induced myocardial inflammation [32]. Aprotinin is able to reduce IRI and myocardial tissue edema, and preserve the vascular endothelial barrier [37,38]. A promoting effect on the coronary microcirculation was reported for a Mg^{2+}-enriched crystalloid cardioplegic solution when compared with a potassium-enriched crystalloid cardioplegic solution [45]. GBR was reported to reduce the lactate production in CPB surgery [42]. The phosphodiesterase III inhibitor amrinone and CsA, which inhibits the mitochondrial permeability transition pore, promote cardiac function during cardioplegia [31,34]. While amrinone promotes rapid and sustained cardiac functional recovery by replenishing myocardial cyclic adenosine monophosphate [34], low-dose CsA supplementation enhanced basal mitochondrial respiration and preserved mitochondrial function, thereby diminishing the effects of IRI [31]. Inhibition of the Na^+/H^+ exchanger by eniporide and cariporide failed to show an effect on ventricular function or myocardial damage [35,36]. The majority of the investigations have been performed with an ischemic period ranging between 60–120 min (Supplementary Table S1) which correlates with the duration of ischemic periods in human surgery. Only four studies on St. Thomas cardioplegic solutions defined an ischemia duration of 30 min [30,46–48]. The on-pump reperfusion time varied between 10–180 min. A reperfusion period with a disconnection from CPB device, called off-pump reperfusion, ranged between 30–300 min.

5. Investigations in Pediatric Models

Cardioplegic solutions have been used in pediatric surgery. Thus, pediatric pig models were used to investigate the effects of the cardioplegic solutions. Sixteen studies reported the effects of different cardioplegic solutions, including St. Thomas cardioplegia ($n = 9$), HTK ($n = 4$), Del Nido ($n = 2$), and Calafiore/blood cardioplegia ($n = 4$) (Supplementary Table S2).

Direct comparisons between different cardioplegic solutions revealed that the modified Calafiore cardioplegia had a superior contractility after CBP surgery when compared to HTK [49]. For adult pig models, additives such as ebselene [50], olprinone [51], diazoxide [52], and sivelestat [53] have been investigated. A reduction in myocardial IRI with the antioxidants ebselene and olprinone has been proven [50,51]. Diazoxide protected the integrity of the mitochondrial structure when applied to a cardioplegic solution [52]. The neutrophil elastase inhibitor sivelestat reduced neutrophilic activation in the lungs and improved oxygenation after CPB in 7 to 14-week-old pigs [53]. The pediatric pig models investigated short-lasting ischemic periods of 10–45 min [54,55] as well as longer ischemic periods of 60–120 min (Supplementary Table S2). The on-pump and off-pump reperfusion periods ranged from 10 to 120 min and 30 min to 48 h, respectively.

6. Impact of Breeds, Strains, Age, and Sex

The species *Sus scrofa domestica* comprises several breeds that may vary in size and appearance, and can be classified into farm pigs and minipigs [17]. Farm pigs include breeds such as Yorkshire, Landrace, and Duroc. Minipig strains such as Yucatan, Göttingen, and Hanford are attractive due to low body weight at birth, early sexual maturity, and adult age. Their tissue properties are more mature and more resistant to surgical procedures [17]. The majority of investigations of cardioplegic solutions in infant and adult pig models have been performed in farm pigs. Only the groups of Sayk et al. [56] and Wu et al. [28] performed experimental investigations on minipigs. Farm pig breeds differ in their susceptibility to stress, growth rate, and fat content. In particular, susceptibility to stress during the preoperative period could influence the outcome of cardiovascular studies. Furthermore, their core body temperature and metabolism could differ slightly, which results in a bias on outcome parameters between different breeds. The age of the pigs had an indirect impact wherein the body weight of farm pigs rapidly increases with age. Consequently, the heart/body weight ratio decreased as described in the section on "Comparability of the heart anatomy and physiology in pig models and humans" and leads to alterations in basic cardiac physiology.

The impact of the sex of the pig on the outcome of cardioplegia-induced effects is unknown. Several studies have used pigs of both sexes to balance possible gender differences. However, there are also studies that exclusively used either male [40,41,49,57,58] or female pigs [21,48,59–62]. This can be influenced by additional experimental factors. For example, the withdrawal of urine in studies investigating kidney function during cardiac cardioplegia is easier in female pigs due to their anatomical features. Therefore, these investigations were performed only in female pigs. Thus, careful selection is necessary to determine the suitable breeds, strains, age, and sex for this study.

7. Refinement Strategies

Several aspects could be considered to refine preclinical investigations of cardioplegic solutions in pig models. Due to a special susceptibility to distress, it is necessary to avoid each conscious perceived stressful moment, such as a noise or any painful handling. Transportation to the operating room should be kept as short as possible. Intramuscular premedication consisting of midazolam, atropine, and ketamine is recommended.

To avoid the perception of any procedure-related pain, premedication of the sedated pigs (e.g., metamizole, fentanyl, or sufentanil) and anesthesia maintained by propofol and fentanyl, respectively, sufentanil is indicated.

Excessive hemodilution can have a significant impact on the outcome. Therefore, the total volume of the cardioplegic solution should be completely filtered. If blood donor pigs are included in the study, experiments should be planned such that the blood of one donor pig could be provided to several surgically-treated pigs. Furthermore, arterial and venous cannulas should be removed at the end of the CPB period, and the remaining blood in the tubes of the perfusion system should be re-transferred to the pig.

Intraischemic temperature could have a critical impact on the development of IRI. Therefore, monitoring of the cardiac temperature is recommended in the septum and left and right ventricles.

Sample withdrawal should comprise all organ systems that could be affected by IRI or supplements added to cardioplegic solutions. This allows for further investigation of this research field.

8. Limitations of Pig Models

Numerous limitations have been reported in studies investigating cardioplegic solutions for cardiac arrest. The restrictions relate to the small number of animals included in the studies, time limits, randomization and blinding, comparability with the clinical setting, study endpoints, and missing data and measurements (Table 1).

The most frequently mentioned restriction was the limited number of animals. In principle, the number of cases for an animal study should be determined according to the statistical calculation of the power and sample size, and is dependent on the primary and secondary endpoints of the study. Financial or human resources should not influence the sample size of the studies. Another limiting factor is the choice of the duration of the aortic cross-clamp, reperfusion, and recovery/observation period. A sufficiently long period of reperfusion is required for physiological weaning from CPB. However, pig models are known to deteriorate over time. In addition, statements about molecular changes in the organism can only be made with an appropriate duration of the reperfusion period, since some parameters requires hours to change. This led to the trend that parameters or markers are used for the analysis of cardioplegic effects that respond early and in a sensitive way to cardioplegia-induced ischemia or in the early reperfusion period (e.g., translocation of hypoxia-inducible factor 1α for oxidative stress or troponin T release into the blood). Additionally, a short reperfusion or observation period increased the risk that the study endpoints were not fully reached.

Implementation problems may arise when conducting the study in a blinded manner. For example, while experimental observers may be blinded to the study groups, it may be difficult to blind the surgeons or perfusionists when comparing crystalloid and blood cardioplegic solutions. However, blinding of the experimenters is strongly recommended to avoid biased results. Furthermore, simple or adaptive randomization is sufficient along with the learning curves of the surgical team (veterinarians, surgeons, and perfusionists).

Another key limitation is the use of young and healthy animals that do not have relevant clinical pathologies. Patients who undergo CPB surgery are usually older and have several comorbidities. Multimorbid patients may be more sensitive to CPB surgery. However, some of these effects can be reproduced in healthy animal models.

Table 1. Limitations in pig models for cardioplegic arrest.

Limitation	References
Animal number	
limited number of animals	[21,27–29,31,35,36,49,57–64]
Time limits	
short cross clamping time	[23,24,49,58]
short reperfusion/recovery time	[23,25,27,28,31–33,38,60–62,64–66]
short observation period/no long-term follow up	[26,32,59,63,65]
Randomization & blinding	
no randomization	[60,63]
surgeon/observer not blinded	[37,60,63]
Comparability with clinical settings	
use of young, healthy animals without clinically relevant pathology	[21,23–25,29,31–33,44,57,61,64]
results not fully comparable with humans	[26,29,44,49,58,67,68]
the use of neonatal piglets not allowed (animal protection requirements)	[49,58]
standardization of interventions/no individual treatment	[21,23–26,31,32,46,49,64]
model restricted to mild ischemia	[35,48,66]
reperfusion phase departed from clinical normality	[66]
Study endpoints	
effect on study endpoint not fully reached	[36,69]
endpoint not suited	[56,62]
lack of measurement of end-point related parameters	[26,50,68]
use of surrogate markers for endpoint measurement	[26]
Missing data and measurements	
missing control	[32,50]
the number of tested factors in one study limited	[24,57,65]
missing measurement/correlation with cardiac function	[56,61,69]
myocardial temperature not monitored	[55]
missing dose-response relationship for tested supplement	[50]
missing pressure-volume measurements	[21]
missing histological examination	[62]
wrong time point of blood/biopsy withdrawal	[44,60]

A further limitation results from the reduced oxygen transport capacity and hematocrit of pigs, leading to increased blood flow. This may result in a significantly stronger left ventricular wall, since it occurs in patients with pathological heart disease. The potassium serum concentration in healthy pigs ranges between 4.6 and 5.8 mmol/L [70]. According to Seutter et al., breed has no influence on potassium content in serum [71]. Despite the great similarity between pigs and humans, the results of these studies cannot be fully adapted to human medicine.

Particularly for pediatric applications, difficulties arise due to conflicts between newborn animal models and animal welfare [49,58]. Hence, this study aimed to investigate the following: (I) the choice of the appropriate animal model and duration of ischemia and reperfusion, (II) a detailed study planning including the consideration of all relevant factors and a statistical calculation of the sample size, and (III) a standardized operational process to ensure good reproducibility.

9. Conclusions

Porcine models for testing cardioplegic solutions in cardiac surgery have been used for the last 25 years, which generated information on cellular effects that could not be obtained from human trials. These investigations comprised results for cardioplegia in CPB procedures using adult and in infant porcine models in vivo. Different cardioplegic solutions have been compared or supplemented with drugs or additives that promote cell stability and protection to diminish the effects of IRI. Furthermore, the major limitations of pig models for investigating cardioplegic solutions are known. However, experimenters and preclinical investigator teams are encouraged to reduce these limitations within an experimental setting to achieve the best possible translation into the clinic.

Supplementary Materials: The following are available online at https://www.mdpi.com/article/10.3390/biomedicines9091279/s1, Table S1: Overview about investigations of cardioplegic solutions in adult pig models, Table S2: Overview about investigations of cardioplegic solutions in pediatric pig models.

Author Contributions: Conceptualization, A.G. (Anna Glöckner), A.H. and M.-T.D.; formal analysis, A.G. (Anna Glöckner), M.-T.D.; interpretation, A.G. (Anna Glöckner), S.O., A.G. (Andre Ginther), J.K., M.A.B., A.H. and M.-T.D.; resources, M.A.B.; data curation, A.G. (Anna Glöckner), M.-T.D.; writing—original draft preparation, A.G. (Anna Glöckner), M.-T.D.; writing—review and editing, A.G. (Anna Glöckner), S.O., A.G. (Andre Ginther), J.K., A.H. and M.-T.D.; supervision, M.A.B., A.H. All authors have read and agreed to the published version of the manuscript.

Funding: This research received no external funding.

Institutional Review Board Statement: Not applicable.

Informed Consent Statement: Not applicable.

Data Availability Statement: No new data were created or analyzed in this study. Data sharing is not applicable to this article.

Conflicts of Interest: The authors declare no conflict of interest.

References

1. Seyboldt-Epting, W. *Kardioplegie: Myokardschutz Während Extrakorporaler Zirkulation*; Springer: Berlin/Heidelberg, Germany, 2013.
2. Gravlee, G.P. *Cardiopulmonary Bypass: Principles and Practice*; Wolters Kluwer Health/Lippincott Williams and Wilkins: Philadelphia, PA, USA, 2008.
3. Donnelly, A.J.; Djuric, M. Cardioplegia solutions. *Am. J. Hosp. Pharm.* **1991**, *48*, 2444–2460. [CrossRef]
4. Hoyer, A.; Kiefer, P.; Borger, M. Cardioplegia and myocardial protection: Time for a reassessment? *J. Thorac. Dis.* **2019**, *11*, e76–e78. [CrossRef] [PubMed]
5. Ferguson, Z.G.; Yarborough, D.E.; Jarvis, B.L.; Sistino, J.J. Evidence-based medicine and myocardial protection-where is the evidence? *Perfusion* **2015**, *30*, 415–422. [CrossRef] [PubMed]
6. Wegscheider, K. *Deutscher Herzbericht 2016*; Deutsche Herzstiftung: Frankfurt am Main, Germany, 2016; ISBN 978-3-9817032-5-2.
7. Holers, V.M.; Thurman, J.M. The alternative pathway of complement in disease: Opportunities for therapeutic targeting. *Mol. Immunol.* **2004**, *41*, 147–152. [CrossRef] [PubMed]
8. Cavarocchi, N.C.; England, M.D.; Schaff, H.; Russo, P.; Orszulak, T.A.; Schnell, W.A.; O'Brien, J.F.; Pluth, J.R. Oxygen free radical generation during cardiopulmonary bypass: Correlation with complement activation. *Circulation* **1986**, *74*, 130–133.
9. Janeway, C.A.; Travers, P.; Walport, M.; Shlomchik, M. *Immunologie*; Spektrum Akademischer Verlag: Heidelberg, Germany, 2002.
10. Cheluvappa, R.; Scowen, P.; Eri, R. Ethics of animal research in human disease remediation, its institutional teaching; and alternatives to animal experimentation. *Pharmacol. Res. Perspect.* **2017**, *5*, e00332. [CrossRef]
11. Hooijmans, C.R.; De Vries, R.; Leenaars, M.; Curfs, J.; Ritskes-Hoitinga, M. Improving planning, design, reporting and scientific quality of animal experiments by using the Gold Standard Publication Checklist, in addition to the ARRIVE guidelines. *Br. J. Pharmacol.* **2011**, *162*, 1259–1260. [CrossRef] [PubMed]
12. Crick, S.J.; Sheppard, M.N.; Ho, S.Y.; Gebstein, L.; Anderson, R.H. Anatomy of the pig heart: Comparisons with normal human cardiac structure. *J. Anat.* **1998**, *193*, 105–119. [CrossRef]
13. Nguyen, P.K.; Wu, J.C. Large Animal Models of Ischemic Cardiomyopathy: Are They Enough to Bridge the Translational Gap? *J. Nucl. Cardiol.* **2015**, *22*, 666–672. [CrossRef] [PubMed]
14. Leonhardt, H. *Anatomie des Menschen. Band II: Innere Organe*; Georg Thieme Verlag: Stuttgart, Germany, 1987.
15. Sim, E.K.; Muskawad, S.; Lim, C.-S.; Yeo, J.H.; Lim, K.H.; Grignani, R.T.; Durrani, A.; Lau, G.; Duran, C. Comparison of human and porcine aortic valves. *Clin. Anat.* **2003**, *16*, 193–196. [CrossRef]
16. Garg, S.; Singh, P.; Sharma, A.; Gupta, G. A Gross Comparative Anatomical Study of Hearts in Human Cadavers and Pigs. *Int. J. Med. Dent. Sci.* **2013**, *2*, 170–176. [CrossRef]
17. Lelovas, P.P.; Kostomitsopoulos, N.; Xanthos, T.T. A Comparative Anatomic and Physiologic Overview of the Porcine Heart. *J. Am. Assoc. Lab. Anim. Sci.* **2014**, *53*, 432–438.
18. Hannon, J.P.; Bossone, C.A.; Wade, C.E. Normal physiological values for conscious pigs used in biomedical research. *Lab. Anim. Sci.* **1990**, *40*, 293–298. [PubMed]

19. Gallegos, R.P.; Rivard, A.L.; Bianco, R.W. Animal models for cardiac research. In *Handbbok of Cardiac Anatomy, Physiology and Devices*; Springer: Berlin, Germany, 2005.

20. Hiebl, B.; Mrowietz, C.; Ploetze, K.; Matschke, K.; Jung, F. Critical hematocrit and oxygen partial pressure in the beating heart of pigs. *Microvasc. Res.* **2010**, *80*, 389–393. [CrossRef]

21. Santer, D.; Kramer, A.; Kiss, A.; Aumayr, K.; Hackl, M.; Heber, S.; Chambers, D.J.; Hallström, S.; Podesser, B.K. St Thomas' Hospital polarizing blood cardioplegia improves hemodynamic recovery in a porcine model of cardiopulmonary bypass. *J. Thorac. Cardiovasc. Surg.* **2019**, *158*, 1543–1554. [CrossRef] [PubMed]

22. Aass, T.; Stangeland, L.; Moen, C.A.; Solholm, A.; Dahle, G.O.; Chambers, D.J.; Urban, M.; Nesheim, K.; Haaverstad, R.; Matre, K.; et al. Left ventricular dysfunction after two hours of polarizing or depolarizing cardioplegic arrest in a porcine model. *Perfusion* **2019**, *34*, 67–75. [CrossRef] [PubMed]

23. Aass, T.; Stangeland, L.; Chambers, D.J.; Hallström, S.; Rossmann, C.; Podesser, B.K.; Urban, M.; Nesheim, K.; Haaverstad, R.; Matre, K.; et al. Myocardial energy metabolism and ultrastructure with polarizing and depolarizing cardioplegia in a porcine model. *Eur. J. Cardio-Thorac. Surg.* **2017**, *52*, 180–188. [CrossRef]

24. Aass, T.; Stangeland, L.; Moen, C.A.; Salminen, P.-R.; Dahle, G.O.; Chambers, D.J.; Markou, T.; Eliassen, F.; Urban, M.; Haaverstad, R.; et al. Myocardial function after polarizing versus depolarizing cardiac arrest with blood cardioplegia in a porcine model of cardiopulmonary bypass. *Eur. J. Cardio-Thorac. Surg.* **2016**, *50*, 130–139. [CrossRef]

25. Feirer, N.; Dieterlen, M.T.; Klaeske, K.; Kiefer, P.; Oßmann, S.; Salameh, A.; Borger, M.A.; Hoyer, A. Impact of Custodiol-N cardioplegia on acute kidney injury after cardiopulmonary bypass. *Clin. Exp. Pharmacol. Physiol.* **2020**, *47*, 640–649. [CrossRef] [PubMed]

26. Hoyer, A.; Bergh, F.T.; Klaeske, K.; Lehmann, S.; Misfeld, M.; Borger, M.; Dieterlen, M.T. Custodiol-N™ cardioplegia lowers cerebral inflammation and activation of hypoxia-inducible factor-1α. *Interact. Cardiovasc. Thorac. Surg.* **2019**, *28*, 884–892. [CrossRef]

27. Aarsaether, E.; Stenberg, T.A.; Jakobsen, Ø.; Busund, R. Mechanoenergetic function and troponin T release following cardioplegic arrest induced by St Thomas' and histidine-tryptophan-ketoglutarate cardioplegia-an experimental comparative study in pigs. *Interact. Cardiovasc. Thorac. Surg.* **2009**, *9*, 635–639. [CrossRef] [PubMed]

28. Wu, T.; Dong, P.; Chen, C.; Yang, J.; Hou, X. The myocardial protection of polarizing cardioplegia combined with delta-opioid receptor agonist in swine. *Ann. Thorac. Surg.* **2011**, *91*, 1914–1920. [CrossRef]

29. Jakobsen, Ø.; Muller, S.; Aarsæther, E.; Steensrud, T.; Sørlie, D.G. Adenosine instead of supranormal potassium in cardioplegic solution improves cardioprotection. *Eur. J. Cardio-Thorac. Surg.* **2007**, *32*, 493–500. [CrossRef] [PubMed]

30. Vähäsilta, T.; Virtanen, J.; Saraste, A.; Luotolahti, M.; Pulkki, K.; Valtonen, M.; Voipio-Pulkki, L.M.; Savunen, T. Adenosine in myocardial protection given through three windows of opportunity. An experimental study with pigs. *Scand. Cardiovasc. J.* **2001**, *35*, 409–414. [CrossRef]

31. Hoyer, A.A.; Klaeske, K.; Garnham, J.; Kiefer, P.; Salameh, A.; Witte, K.; Borger, M.; Dieterlen, M.T. Cyclosporine A-enhanced cardioplegia preserves mitochondrial basal respiration after ischemic arrest. *Perfusion* **2021**. [CrossRef]

32. Ryou, M.-G.; Flaherty, D.C.; Hoxha, B.; Gurji, H.; Sun, J.; Hodge, L.M.; Olivencia-Yurvati, A.H.; Mallet, R.T. Pyruvate-enriched cardioplegia suppresses cardiopulmonary bypass-induced myocardial inflammation. *Ann. Thorac. Surg.* **2010**, *90*, 1529–1535. [CrossRef]

33. Ryou, M.-G.; Flaherty, D.C.; Hoxha, B.; Sun, J.; Gurji, H.; Rodriguez, S.; Bell, G.; Olivencia-Yurvati, A.H.; Mallet, R.T. Pyruvate-fortified cardioplegia evokes myocardial erythropoietin signaling in swine undergoing cardiopulmonary bypass. *Am. J. Physiol. Circ. Physiol.* **2009**, *297*, H1914–H1922. [CrossRef]

34. Ko, Y.; Morita, K.; Nagahori, R.; Kinouchi, K.; Shinohara, G.; Kagawa, H.; Hashimoto, K. Myocardial cyclic AMP augmentation with high-dose PDEIII inhibitor in terminal warm blood cardioplegia. *Ann. Thorac. Cardiovas.c Surg.* **2009**, *15*, 311–317.

35. Bechtel, J.F.; Eichler, W.; Toerber, K.; Weidtmann, B.; Hernandez, M.; Klotz, K.F.; Sievers, H.H.; Bartels, C. The Na+/H+ exchange inhibitor cariporide is washed out of the myocardium by crystalloid cardioplegia. *Thorac. Cardiovasc. Surg.* **2006**, *54*, 317–323. [CrossRef] [PubMed]

36. Klass, O.; Fischer, U.M.; Perez, E.; Easo, J.; Bosse, M.; Fischer, J.H.; Tossios, P.; Mehlhorn, U. Effect of the Na+/H+ exchange inhibitor eniporide on cardiac performance and myocardial high energy phosphates in pigs subjected to cardioplegic arrest. *Ann. Thorac. Surg.* **2004**, *77*, 658–663. [CrossRef]

37. Khan, T.A.; Bianchi, C.; Araujo, E.; Voisine, P.; Xu, S.H.; Feng, J.; Li, J.; Sellke, F.W. Aprotinin preserves cellular junctions and reduces myocardial edema after regional ischemia and cardioplegic arrest. *Circulation* **2005**, *112*, 196–201. [CrossRef]

38. Khan, T.A.; Bianchi, C.; Voisine, P.; Feng, J.; Baker, J.; Hart, M.; Takahashi, M.; Stahl, G.; Sellke, F.W. Reduction of myocardial reperfusion injury by aprotinin after regional ischemia and cardioplegic arrest. *J. Thorac. Cardiovasc. Surg.* **2004**, *128*, 602–608. [CrossRef] [PubMed]

39. Steensrud, T.; Nordhaug, D.; Elvenes, O.; Korvald, C.; Sørlie, D. Superior myocardial protection with nicorandil cardioplegia. *Eur. J. Cardio-Thorac. Surg.* **2003**, *23*, 670–677. [CrossRef]

40. Osipov, R.M.; Robich, M.P.; Feng, J.; Chan, V.; Clements, R.T.; Deyo, R.J.; Szabo, C.; Sellke, F.W. Effect of hydrogen sulfide on myocardial protection in the setting of cardioplegia and cardiopulmonary bypass. *Interact. Cardiovasc. Thorac. Surg.* **2010**, *10*, 506–512. [CrossRef] [PubMed]

41. Powell, S.R.; Nelson, R.L.; Finnerty, J.; Alexander, D.; Pottanat, G.; Kooker, K.; Schiff, R.J.; Moyse, J.; Teichberg, S.; Tortolani, A.J. Zinc-bis-histidinate preserves cardiac function in a porcine model of cardioplegic arrest. *Ann. Thorac. Surg.* **1997**, *64*, 73–80. [CrossRef]

42. Demeekul, K.; Sukumolanan, P.; Bootcha, R.; Panprom, C.; Petchdee, S. A Cardiac Protection of Germinated Brown Rice During Cardiopulmonary Bypass Surgery and Simulated Myocardial Ischemia. *J. Inflamm. Res.* **2021**, *14*, 3307–3319. [CrossRef]

43. Suarez-Pierre, A.; Lui, C.; Zhou, X.; Kearney, S.; Jones, M.; Wang, J.; Thomas, R.P.; Gaughan, N.; Metkus, T.S.; Brady, M.B.; et al. Diazoxide preserves myocardial function in a swine model of hypothermic cardioplegic arrest and prolonged global ischemia. *J. Thorac. Cardiovasc. Surg.* **2020**. [CrossRef]

44. Steensrud, T.; Nordhaug, D.; Husnes, K.V.; Aghajani, E.; Sørlie, D.G. Replacing potassium with nicorandil in cold St. Thomas' Hospital cardioplegia improves preservation of energetics and function in pig hearts. *Ann. Thorac. Surg.* **2004**, *77*, 1391–1397. [CrossRef] [PubMed]

45. Tofukuji, M.; Stamler, A.; Li, J.; Franklin, A.; Wang, S.Y.; Hariawala, M.D.; Sellke, F.W. Effects of magnesium cardioplegia on regulation of the porcine coronary circulation. *J. Surg. Res.* **1997**, *69*, 233–239. [CrossRef] [PubMed]

46. Vähäsilta, T.; Saraste, A.; Kytö, V.; Malmberg, M.; Kiss, J.; Kentala, E.; Kallajoki, M.; Savunen, T. Cardiomyocyte Apoptosis After Antegrade and Retrograde Cardioplegia. *Ann. Thorac. Surg.* **2005**, *80*, 2229–2234. [CrossRef]

47. Uotila, P.; Saraste, A.; Vähäsilta, T.; Kentala, E.; Savunen, T. Stimulated expression of cyclooxygenase-2 in porcine heart after bypass circulation and cardioplegic arrest. *Eur. J. Cardio-Thorac. Surg.* **2001**, *20*, 992–995. [CrossRef]

48. Curro, D.; Bombardieri, G.; Barilaro, C.; Di Francesco, P.; Varano, C.; Possati, G.; Pragliola, C. Time dependence of endothelium-mediated vasodilation by intermittent antegrade warm blood cardioplegia. *Ann. Thorac. Surg.* **1997**, *64*, 1354–1359. [CrossRef]

49. Münch, F.; Purbojo, A.; Kellermann, S.; Janssen, C.; Cesnjevar, R.A.; Rüffer, A.; Czerny, M.; Reser, D.; Eggebrecht, H.; Janata, K.; et al. Improved contractility with tepid modified full blood cardioplegia compared with cold crystalloid cardioplegia in a piglet model. *Eur. J. Cardio-Thorac. Surg.* **2014**, *48*, 236–243. [CrossRef]

50. Chen, Y.; Liu, J.; Li, S.; Yan, F.; Xue, Q.; Wang, H.; Sun, P.; Long, C. Histidine-Tryptophan-Ketoglutarate Solution with Added Ebselen Augments Myocardial Protection in Neonatal Porcine Hearts Undergoing Ischemia/Reperfusion. *Artif. Organs* **2015**, *39*, 126–133. [CrossRef]

51. Kinouchi, K.; Morita, K.; Ko, Y.; Nagahori, R.; Shinohara, G.; Abe, T.; Hashimoto, K. Reversal of oxidant-mediated biochemical injury and prompt functional recovery after prolonged single-dose crystalloid cardioplegic arrest in the infantile piglet heart by terminal warm-blood cardioplegia supplemented with phosphodiesterase III inhibitor. *Gen. Thorac. Cardiovasc. Surg.* **2012**, *60*, 73–81. [CrossRef] [PubMed]

52. Wang, L.; Kinnear, C.; Hammel, J.M.; Zhu, W.; Hua, Z.; Mi, W.; Caldarone, C.A. Preservation of mitochondrial structure and function after cardioplegic arrest in the neonate using a selective mitochondrial KATP channel opener. *Ann. Thorac. Surg.* **2006**, *81*, 1817–1823. [CrossRef] [PubMed]

53. Ando, M.; Murai, T.; Takahashi, Y. The effect of sivelestat sodium on post-cardiopulmonary bypass acute lung injury in a neonatal piglet model. *Interact. Cardiovasc. Thorac. Surg.* **2008**, *7*, 785–788. [CrossRef]

54. Liuba, P.; Johansson, S.; Pesonen, E.; Odermarsky, M.; Kornerup-Hansen, A.; Forslid, A.; Aburawi, E.H.; Higgins, T.; Birck, M.; Perez-de-Sa, V. Coronary flow and reactivity, but not arrhythmia vulnerability, are affected by cardioplegia during cardiopulmonary bypass in piglets. *J. Cardiothorac. Surg.* **2013**, *8*, 157. [CrossRef]

55. Jones, J.; Wilson, K.; Koch, W.; Milano, C. Adenoviral gene transfer to the heart during cardiopulmonary bypass: Effect of myocardial protection technique on transgene expression. *Eur. J. Cardio-Thorac. Surg.* **2002**, *21*, 847–852. [CrossRef]

56. Sayk, F.; Krüger, S.; Bechtel, J.F.; Feller, A.C.; Sievers, H.H.; Bartels, C. Significant damage of the conduction system during cardioplegic arrest is due to necrosis not apoptosis. *Eur J. Cardio-Thorac. Surg.* **2004**, *25*, 801–806. [CrossRef]

57. Portilla-de Buen, E.; Leal, C.; Garcia-Martinez, D.; Cornejo, A.; Zepeda, A.; Aburto, E. Pig heart preservation with antegrade intracellular crystalloid versus antegrade/retrograde miniplegia. *J. Extra-Corpor. Technol.* **2011**, *43*, 130–136.

58. Janssen, C.; Kellermann, S.; Münch, F.; Purbojo, A.; Cesnjevar, R.A.; Rüffer, A. Myocardial Protection During Aortic Arch Repair in a Piglet Model: Beating Heart Technique Compared With Crystalloid Cardioplegia. *Ann. Thorac. Surg.* **2015**, *100*, 1758–1766. [CrossRef]

59. Runge, M.; Hughes, P.; Gøtze, J.P.; Petersen, R.H.; Steinbrüchel, D.A. Evaluation of myocardial metabolism with microdialysis after protection with cold blood- or cold crystalloid cardioplegia. A porcine model. *Scand. Cardiovasc. J.* **2006**, *40*, 186–193. [CrossRef]

60. Nakao, M.; Morita, K.; Shinohara, G.; Kunihara, T. Modified Del Nido Cardioplegia and Its Evaluation in a Piglet Model. *Semin. Thorac. Cardiovasc. Surg.* **2021**, *33*, 84–92. [CrossRef] [PubMed]

61. Nakao, M.; Morita, K.; Shinohara, G.; Kunihara, T. Excellent Restoration of Left Ventricular Compliance After Prolonged Del Nido Single-Dose Cardioplegia in an In Vivo Piglet Model. *Semin. Thorac. Cardiovasc. Surg.* **2020**, *32*, 475–483. [CrossRef]

62. Abe, T.; Morita, K.; Shinohara, G.; Hashimoto, K.; Nishikawa, M. Synergistic effects of remote perconditioning with terminal blood cardioplegia in an in vivo piglet model. *Eur. J. Cardio-Thorac. Surg.* **2017**, *52*, 479–484. [CrossRef] [PubMed]

63. Nakao, M.; Morita, K.; Shinohara, G.; Saito, S.; Kunihara, T. Superior restoration of left ventricular performance after prolonged single-dose del Nido cardioplegia in conjunction with terminal warm blood cardioplegic reperfusion. *J. Thorac. Cardiovasc. Surg.* **2020**. [CrossRef]

64. Dahle, G.O.; Salminen, P.-R.; Moen, C.A.; Eliassen, F.; Jonassen, A.K.; Haaverstad, R.; Matre, K.; Grong, K. Esmolol Added in Repeated, Cold, Oxygenated Blood Cardioplegia Improves Myocardial Function After Cardiopulmonary Bypass. *J. Cardiothorac. Vasc. Anesthesia* **2015**, *29*, 684–693. [CrossRef] [PubMed]

65. Elvenes, O.P.; Korvald, C.; Myklebust, R.; Sørlie, D. Warm retrograde blood cardioplegia saves more ischemic myocardium but may cause a functional impairment compared to cold crystalloid. *Eur. J. Cardio-Thorac. Surg.* **2002**, *22*, 402–409. [CrossRef]

66. Pathi, V.L.; McPhaden, A.R.; Morrison, J.; Belcher, P.R.; Fenner, J.W.; Martin, W.; McQuiston, A.M.; Wheatley, D.J. The effects of cardioplegic arrest and reperfusion on the microvasculature of the heart. *Eur J. Cardio-Thorac. Surg.* **1997**, *11*, 350–357. [CrossRef]

67. Fannelop, T.; Dahle, G.O.; Salminen, P.-R.; Moen, C.A.; Matre, K.; Mongstad, A.; Eliassen, F.; Segadal, L.; Grong, K. Multidose Cold Oxygenated Blood Is Superior to a Single Dose of Bretschneider HTK-Cardioplegia in the Pig. *Ann. Thorac. Surg.* **2009**, *87*, 1205–1213. [CrossRef] [PubMed]

68. Kajimoto, M.; Ledee, D.R.; Olson, A.K.; Isern, N.G.; Robillard-Frayne, I.; Des Rosiers, C.; Portman, M.A. Selective cerebral perfusion prevents abnormalities in glutamate cycling and neuronal apoptosis in a model of infant deep hypothermic circulatory arrest and reperfusion. *J. Cereb. Blood Flow Metab.* **2016**, *36*, 1992–2004. [CrossRef] [PubMed]

69. Chen, Y.; Liu, J.; Li, S.; Li, W.; Yan, F.; Sun, P.; Wang, H.; Long, C. Which is the better option during neonatal cardiopulmonary bypass: HTK solution or cold blood cardioplegia? *ASAIO J.* **2013**, *59*, 69–74. [CrossRef] [PubMed]

70. Gürtler, H. Mittelwerte und Streuungsbereiche diagnostisch nutzbarer Parameter. In *Schweinekrankheiten*, 3rd ed.; Neundorf, R., Seidel, H., Eds.; Ferdinand Enke Verlag: Stuttgart, Germany, 1987; pp. 84–132.

71. Seutter, U. *Einfluß von Rasse, Haltung, Fütterung, Management, Alter und Reproduktionsstadium auf hämatologische und Klinisch-chemische Parameter Beim Schwein*; Ludwig-Maximilians-Universität: München, Germany, 1995.

Systematic Review

A Systematic Review of Therapeutic Approaches Used in Experimental Models of Interstitial Cystitis/Bladder Pain Syndrome

Tadeja Kuret [†], Dominika Peskar [†], Andreja Erman and Peter Veranič *

Institute of Cell Biology, Faculty of Medicine, University of Ljubljana, 1000 Ljubljana, Slovenia;
tadeja.kuret@mf.uni-lj.si (T.K.); dominika.peskar@mf.uni-lj.si (D.P.); andreja.erman@mf.uni-lj.si (A.E.)
* Correspondence: peter.veranic@mf.uni-lj.si
† These two authors contributed equally and share first authorship.

Abstract: Interstitial cystitis/bladder pain syndrome (IC/BPS) is a multifactorial, chronic bladder disorder with limited therapeutic options currently available. The present review provides an extensive overview of therapeutic approaches used in in vitro, ex vivo, and in vivo experimental models of IC/BPS. Publications were identified by electronic search of three online databases. Data were extracted for study design, type of treatment, main findings, and outcome, as well as for methodological quality and the reporting of measures to avoid bias. A total of 100 full-text articles were included. The majority of identified articles evaluated therapeutic agents currently recommended to treat IC/BPS by the American Urological Association guidelines (21%) and therapeutic agents currently approved to treat other diseases (11%). More recently published articles assessed therapeutic approaches using stem cells (11%) and plant-derived agents (10%), while novel potential drug targets identified were proteinase-activated (6%) and purinergic (4%) receptors, transient receptor potential channels (3%), microRNAs (2%), and activation of the cannabinoid system (7%). Our results show that the reported methodological quality of animal studies could be substantially improved, and measures to avoid bias should be more consistently reported in order to increase the value of preclinical research in IC/BPS for potential translation to a clinical setting.

Keywords: interstitial cystitis; bladder pain syndrome; therapeutic approaches; experimental models; in vitro; ex vivo; in vivo

Citation: Kuret, T.; Peskar, D.; Erman, A.; Veranič, P. A Systematic Review of Therapeutic Approaches Used in Experimental Models of Interstitial Cystitis/Bladder Pain Syndrome. *Biomedicines* **2021**, *9*, 865. https://doi.org/10.3390/biomedicines9080865

Academic Editor: Martina Perše

Received: 13 July 2021
Accepted: 20 July 2021
Published: 22 July 2021

Publisher's Note: MDPI stays neutral with regard to jurisdictional claims in published maps and institutional affiliations.

1. Introduction

Interstitial cystitis/bladder pain syndrome (IC/BPS) is a multifactorial, chronic bladder disorder of unknown etiology, generally characterized by discomfort or pain in the bladder and the surrounding pelvic region, associated with increased urinary frequency, urgency, and nocturia [1]. IC/BPS is more frequent in women compared to men with an estimated prevalence of 45–300 per 100,000 women and 8–30 per 100,000 men [2–4]. However, the occurrence of IC/BPS is likely to be underreported due to the complexity of the disease, a variety of different and nonspecific clinical symptoms and signs, and a lack of standardized diagnostic criteria [5,6]. To date, there is no effective therapeutic option available for patients with IC/BPS, and the disease represents an enormous financial burden for the individuals and the economy as a whole [7].

In general, IC/BPS can be categorized into two major subtypes, mainly based on the bladder histological findings [8]. The first type or "classical" IC/BPS with Hunner's lesions (i.e., mucosal lesions accompanied by abnormal capillary structures) is characterized by more severe bladder-centric symptoms, reduced bladder capacity, histological signs of epithelial denudation, inflammatory infiltrates, and edema, while IC/BPS without Hunner's lesions has no obvious bladder etiology, features minimal histological changes, and is frequently accompanied by common systemic comorbidities ("bladder-beyond" pain) [9,10].

Regardless of the IC/BPS subtype, the overall etiology and pathophysiology remain elusive with many different hypotheses proposed over the years, including injury of the bladder epithelium and increased barrier permeability, neurogenic inflammation with mast cell infiltration, and possible autoimmune involvement [11,12]. One of the most common characteristics found in bladder biopsies from IC/PBS patients is denudation or thinning of the bladder urothelium, a specialized type of epithelial tissue that lines the wall in the majority of the urinary tract and plays an important role as a permeability barrier against toxic substances from the urine [13]. In IC/BPS patients, the barrier function is compromised due to various reasons, including the reduction of the glycocalyx layer, consisting of glycoproteins and proteoglycans [14], the disassembly of tight junctions with deregulated expression of certain tight junction proteins (zonula occludens-1 (ZO-1), occludin, and claudins 1, 4, and 8), and reduced expression of specific transmembrane proteins uroplakins [15–17]. The compromised urothelial barrier results in the leakage of urine solutes, such as potassium and urea into the lamina propria, leading to the activation of inflammatory response with increased urothelial release of signaling molecules (e.g., acetylcholine (ACh), adenosine triphosphate (ATP), nitric oxide (NO)) and proinflammatory mediators, such as interleukins (IL)1, IL6, and IL8, tumor necrosis factor alpha (TNFα), and nerve growth factor (NGF), as well as increased nerve fiber density and inflammatory (mast cell) infiltrates, which ultimately contribute to urgency and pain [15,18,19]. Proinflammatory mediators sensitize afferent nerve terminals by activating transient receptor potential (TRP) channels resulting in the release of neuropeptides (e.g., calcitonin gene-related peptide (CGRP) and substance P) that induce mast cell degranulation and further stimulate the release of proinflammatory mediators, leading to a perpetual cycle of inflammation and pain [20–23].

Although various therapeutic options exist for patients with IC/BPS, all of them aim to relieve the symptoms and there is no treatment nor combination of treatments currently available that would be consistently successful in alleviating clinical symptoms and ensuring long-term efficacy. The American Urological Association (AUA) guidelines [24] recommend a stepwise therapeutic approach, in which the first-line therapy includes patient education with daily behavior modification and lifestyle changes. Physical therapy, oral administration of pentosan polysulfate (PPS) or antihistamines, and intravesical application of heparin, lidocaine, or dimethyl sulfoxide (DMSO) constitute second-line therapy. Third-line therapy requires cystoscopy and hydrodistension, while neuromodulation and intravesical injection of botulinum toxin A (BTX-A) are considered as a fourth-line therapy. If a patient does not respond to any of the therapeutic agents, surgical intervention (cystectomy) is needed [24].

The currently available and recommended therapy options for IC/BPS patients are based mostly on empirical studies and suffer from low efficacy. Hence, research on IC/BPS is focusing on the development and evaluation of novel therapeutic options. Since the pathophysiology of IC/BPS is not yet well understood, the development of definitive therapeutic modalities is significantly compromised, and, despite promising preclinical results of various therapeutic agents, only a low percentage reached clinical trials. This might be related to the lack of suitable and validated experimental models that would be able to replicate all aspects of IC/BPS complexity, as well as the inadequate methodological quality of experimental in vivo models and incomplete reporting of relevant information according to published guidelines of animal research [25].

The review aims to give insight into the commonly used experimental in vitro, ex vivo, and in vivo models for IC/BPS, as well as to summarize and discuss the therapeutic approaches used in these models, explain their mechanism of action, and estimate their translational potential. We also aimed to report on the methodological quality of included studies and evaluate whether sufficient measures to avoid the risk of bias were undertaken. The therapeutic approaches identified were categorized into five groups: (i) therapeutic agents currently recommended by AUA guidelines to treat IC/BPS, (ii) therapeutic agents currently approved to treat other diseases, (iii) other intravesical therapy and improved drug delivery systems, (iv) novel emerging therapeutic options and targets, which include stem cell and extracorporeal shock wave therapy (ECSWT), plant-derived agents, or novel potential targets, such as protease-activated receptors (PAR), purinergic receptors, TRP channels, microRNAs, and activation of the cannabinoid system, and (v) other therapeutic agents and targets (Figure 1).

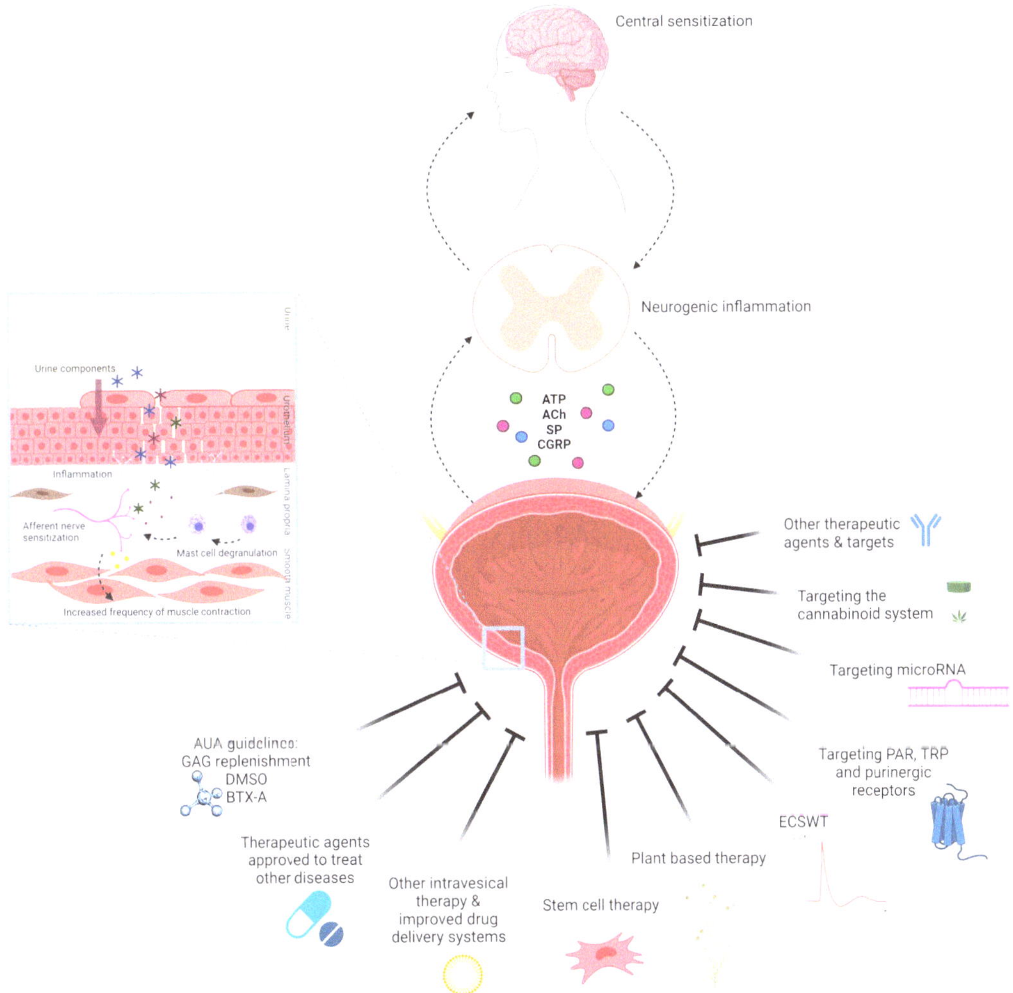

Figure 1. Schematic illustration of IC/BPS pathology and therapeutic approaches evaluated in experimental in vitro, ex vivo, and in vivo models of IC/BPS. Legend: ACh, acetylcholine; ATP, adenosine triphosphate; AUA, the American Urological Association; BTX-A, botulinum toxin A; CGRP, calcitonin gene-related peptide; DMSO, dimethyl sulfoxide; ECSWT, extracorporeal shock wave therapy; GAG, glycosaminoglycan; IC/BPS, interstitial cystitis/bladder pain syndrome; PAR, protease-activated receptors; SP, substance P; TRP, transient receptor potential channels; solid lines indicate the therapeutic approaches for treatment of IC/BPS; dashed arrows indicate proposed sequence of events in IC/BPS pathophysiology. The Figure 1 was created using Biorender.com.

2. Methods

2.1. Search Strategy

A comprehensive literature review was conducted using PubMed, Scopus, and Web of Science databases to identify articles exploring therapeutic options in in vitro, ex vivo, and in vivo experimental models of IC/BPS. We used the following search terms: (("interstitial cystitis" OR "bladder pain syndrome" OR "IC/BPS") AND ("in vitro" OR "ex vivo" OR "in vivo" OR "animal" OR "models") AND ("therapy" OR "treatment")) in different combinations. Only full-text articles in English published from 1 January 2000 until 31 May 2021 were included. As the relationship between IC/PBS and other dysfunctional bladder syndromes in human patients (including the overactive bladder) is less well confirmed, this review is limited to those studies based only on experimental models of IC/BPS.

2.2. Inclusion and Exclusion Criteria and Data Extraction

Articles were reviewed in a two-stage process. The first stage included screening the titles and abstracts of all identified articles. Reviews, editorials, case reports, conference proceedings, notes, and articles not written in English were excluded. Additional exclusion criteria were irrelevant articles describing other diseases and not IC/BPS, articles not including therapeutic agents, and articles not describing an experimental model of IC/BPS. During the second stage, full texts of the remaining studies were evaluated. The reference list of the most relevant studies was also screened to identify any other potentially eligible studies. Two reviewers (T.K. and D.P.) independently assessed the full-text papers to determine if they met the inclusion criteria and selected the final articles to be included in this study. For in vitro and ex vivo studies, we extracted information regarding the experimental design of the study (type of cells used, type, concentration, and time of stimulation and therapy, major findings, and outcome). For in vivo studies, information was extracted for aspects of methodological quality (see below) and experimental design (animal number, species and strain, type, concentration, time and route of administration of IC induction and treatment agent, main findings, and outcome).

2.3. Methodological Quality and Risk of Bias

To determine the methodological quality of published in vivo studies, we defined a 12-point checklist based on published ARRIVE guidelines describing the minimum information that all scientific publications reporting research using animals should include [26]. We specifically focused on the study design (number of animals and experimental groups), experimental animals (species and strain, sex, age, and weight), detailed description of housing and husbandry, and detailed description of the experimental procedure, as well as reporting on measures to avoid the risk of bias (e.g., randomization, sample size calculations, blinding of investigator/caretaker, and blinding of outcome assessment).

3. Results and Discussion

The electronic database (PubMed = 627; Scopus = 230; Web of Science = 481) and reference list search (n = 11) resulted in 1349 articles, of which 159 remained after the removal of duplicates and title/abstract screening. Finally, after assessing the full-text articles for eligibility, a total of 100 full-text articles were included in the present review. A flow diagram of the search and selection process is shown in Figure 2. Seven of the 100 included studies (7%) reported on in vitro models, five (5%) studies used ex vivo models, and 77 (77%) studies included in vivo models. Eleven (11%) studies included in vivo models in combination with in vitro (n = 10) or ex vivo (n = 1) models.

3.1. Experimental Models of IC/BPS

3.1.1. In Vitro and Ex Vivo Models

Since the most consistently described findings in the bladders of IC/PBS patients include abnormalities in the urothelium [27], the majority of identified in vitro models (15/18; 83%) studied either primary urothelial cells, isolated/explanted from human or animal bladders or different urothelial cell lines (i.e., HTB2, HTB4). Most commonly, protamine sulfate (PS), TNFα, lipopolysaccharide (LPS), or H_2O_2 was used in in vitro models to induce urothelial dysfunction and mimic the proinflammatory environment observed in the bladders of IC/BPS patients. Ex vivo models included whole-bladder preparations (5/6; 83%) or bladder detrusor muscle strips (1/6; 17%), isolated from experimental animals. Whole bladders or muscle strips, mounted in organ baths, were stimulated chemically with carbachol, ACh, ATP, capsaicin (TRPV1 receptor agonist) or KCl, or electrically, similar to triggering bladder contractions in vivo. These models were used to evaluate changes in bladder contraction activity induced by pathologic conditions (e.g., acute injury with HCl, H_2O_2, or acrolein), and to explore the nature of neurotransmission and sensitization of afferent pathways [28]. A summary table with the characteristics of each article describing in vitro and ex vivo models is provided (Table S1, Supplementary Materials).

Identification

Records identified through database searching: 1338
PUBMED: 627
Scopus: 230
Web of Science: 481

Additional records identified through other sources: 11

Screening

Records after duplicates removed: 832

Records screened (title and abstract reading): 832

Records excluded: 673
Reviews: 250
Letters/Editorials/Abstract meetings/Case reports: 15
Not in English: 23
Irrelevant (other diseases/conditions): 171
No therapy/treatment described: 216

Eligibility

Full text articles assessed for eligibility: 159

Records excluded: 59
No therapy/treatment described: 15
No experimental model: 44

Included

Studies included: 100

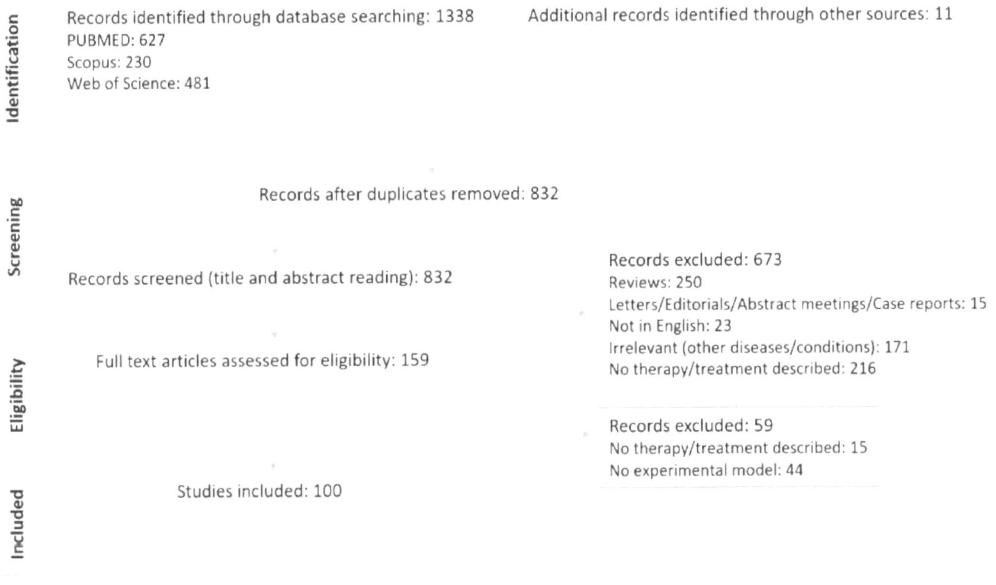

Figure 2. Flow diagram of study search and selection.

3.1.2. In Vivo (Animal) Models

In the present review, all of the in vivo studies ($n = 88$) were conducted on either mice or rats. According to Birder and Andersson, animal models of IC can be categorized into three subtypes, i.e., bladder-centric models, models with complex mechanisms, and stress-induced/natural models [29]. Most of the identified in vivo studies used bladder-centric models (76/88; 86%) with cyclophosphamide (CYP) being the predominant toxic substance for IC induction (29/88; 33%), followed by HCl (9/88; 10%), PS (6/88; 7%), LPS (5/88; 6%), or a combination of different toxins (11/88; 12%). Only a small number of reviewed studies incorporated more complex IC models, such as autoimmune models using immunization of wild-type or transgenic animals for IC induction (7/88; 8%), and stress-induced IC models (5/88; 6%). The majority of in vivo experiments included acute IC (55/88; 63%), while models of chronic IC, characterized by the treatment with bladder-toxic substances for more than 3 days or with more complex mechanisms of induction were described in 34% (30/88) of the reviewed studies (Table S2, Supplementary Materials). Three studies (3%) included both acute and chronic IC models. The most commonly evaluated outcomes of IC induction were nociceptive behavior and mechanical allodynia of the animals (e.g., with the application of von Frey monofilaments), urodynamic parameters with cystometry or void spot assay, and the extent of inflammation in bladder tissues (e.g., histology, immunohistochemistry, qPCR) or urine (levels of secreted proinflammatory mediators). Most of the studies exploited female rodents (77/88; 87%), while male animals were included in 9% of the studies (8/88).

3.2. Types of Treatment Evaluated in Experimental Models

The majority of identified articles included in the present review reported on experimental models, evaluating therapeutic agents currently recommended to treat IC/BPS by AUA guidelines (21/100; 21%), followed by therapeutic agents currently approved to treat other, most commonly chronic inflammatory diseases (11/100; 11%) and improved systems for intravesical drug delivery (6/100; 6%). More recently published articles evaluated therapeutic approaches using stem cells (11/100; 11%), plant-derived agents (10/100; 10%), and ECSWT (3/100, 3%). Novel potential drug targets for IC/BPS identified were PAR (6/100; 6%), purinergic receptors (4/100; 4%), TRP channels (3/100; 3%), microRNAs (2/100; 2%), and activation of the cannabinoid system (7/100; 7%), while other agents and targets (15/100; 15%) included hydroxyfasudil, vitamin D3, growth factors, and adhesion molecules (Figure 3).

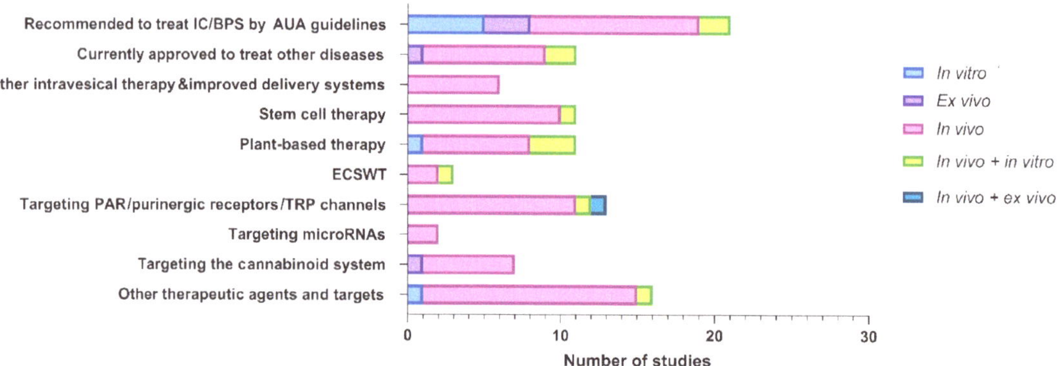

Figure 3. A summary of included experimental models evaluating different types of treatment for IC/BPS. Legend: AUA, the American Urological Association; ECSWT, extracorporeal shock wave therapy; IC/BPS, interstitial cystitis/bladder pain syndrome; PAR, protease-activated receptors; TRP, transient receptor potential.

3.2.1. Therapeutic Agents Recommended by AUA Guidelines for Treatment of IC/BPS
Glycosaminoglycan Replenishment Therapy

According to the hypothesis that damage to the glycosaminoglycan (GAG) layer is among the main causes of IC/BPS symptoms, infusions of exogenous GAG biopolymers (e.g., hyaluronic acid (HA), chondroitin sulfate (CS), and heparin), and PPS intravesically into the bladder have been used in clinical practice for over two decades [30]. Several in vitro mechanistic studies that evaluated GAG replenishment treatment have been published recently, showing the ability of GAGs to decrease urothelial permeability and restore the barrier function; however, confounding results exist regarding their anti-inflammatory effects (Table S1, Supplementary Materials). To evaluate the effect of CS on the barrier function after induction of urothelial damage, Rozzenberg et al. used terminally differentiated porcine urothelial cells, which are morphologically and functionally comparable with the same types of cells in a normal human urothelium. Treatment with CS significantly accelerated the recovery of the barrier function 7 h after acute damage with PS [31]. Rooney et al. showed that high-molecular-weight HA significantly decreased TNFα- and PS-induced IL8 and IL6 production, increased sulfated GAG production, and decreased trans-epithelial permeability without altering tight junction protein expression in the HTB4 urothelial cell line [32]. This was later confirmed by Stellavato et al., showing that HA and CS, alone or in combination, were able to decrease IL6 and IL8 expression, as well as re-establish the expression of ZO-1 in TNFα-treated urothelial cell lines [33]. In contrast, the follow-up study in 2020 revealed that none of the commercially available GAG formulations containing HA or HA with CS were able to attenuate the TNFα-induced production of IL8 and IL6, the expression of GAG synthesis enzymes, or markers of tissue remodeling and pain [34]. Later on, Rooney et al. also reported on a newly developed biphasic system combining cross-linked and native HA in a 1:1 ratio that was able to reduce permeability, while at the same time did not alter the production of proinflammatory cytokines in HTB2 cells [35]. The significant recovery in various cystometric parameters following HA treatment was shown in vivo in H_2O_2-induced IC in female Wistar rats. HA recovered inter-contraction interval, maximal voiding pressure, and the number of pelvic afferent and efferent nerve activities to near-normal levels by directly scavenging H_2O_2 or OH^- activity and decreasing bladder ATP and ACh levels [36]. The immediate effect of intravesical CS on the restoration of bladder permeability and reduced recruitment of inflammatory cells to the suburothelial space was shown in HCl-induced IC in BALB/c mice and Sprague-Dawley (SD) rats [37,38]. Additionally, male SD rats, given a premix of PPS and low-molecular-weight toxic factor, derived from the urine of healthy individuals, showed significantly lower numbers of non-voiding contractions compared to the untreated group [39]. Since the linear GAGs, commonly used in IC/BP therapy, are not able to mimic the normal urothelial hydrophilic surface consisting of a thick glycocalyx layer with large numbers of bound water molecules [40], novel GAG-replenishment strategies are being developed. Greenwood-Van Meerveld et al. reported on restored bladder function and reduced bladder permeability by intravesical instillation of recombinant human proteoglycan 4 (lubricin, rhPRG4), a highly hydrophilic glycoprotein with anti-inflammatory properties in PS-induced IC in female SD rats [41]. The same research group also tested a novel high-molecular-weight GAG biopolymer ("SuperGAG") that was more effective in restoring bladder function and relieving pain compared to CS [42]. Another emerging class of therapeutic GAGs involves semi-synthetic GAG-ethers (SAGE) offering both mucosal restoration and potent analgesic and anti-inflammatory effects. For example, SAGE GM-0111 was tested by several groups demonstrating attenuation of inflammation [43–45]. These novel GAGs offer improved protection of the damaged urothelium, but still encounter many limitations, such as poor urothelial binding and consequent fast clearance with micturition. The synthetic polymer drug delivery systems offer a better accumulation of GAGs, but can potentially weaken normal bladder function by reducing bladder capacity or causing bladder outflow obstruction (BOO) [43].

Dimethyl Sulfoxide (DMSO)

In addition to PPS, a 50% *w*/*w* aqueous solution of DMSO (both recommended as a second-line therapy) is the only drug approved by the FDA for treating IC/BPS [24]. The mechanism of action of DMSO in IC/BPS is not entirely known; however, it is thought to be a combination of anti-inflammatory effects, nerve blockade, and smooth muscle relaxation [46]. Melchior et al. reported that DMSO at concentrations greater than 35% completely inhibits ex vivo bladder contractions, stimulated by the electrical field, ACh, or membrane depolarization [47]. The anti-inflammatory effect of 50% DMSO was shown in URO-OVA mice with activated OT-1 splenocyte-induced acute autoimmune inflammation and URO-OVA/OT-1 transgenic mice with spontaneously developed chronic IC. Three consecutive intravesical DMSO treatments reversed edema and hyperemia, as well as decreased the number of infiltrating $CD8^+$ T cells. A significant downregulation in mRNA levels of proinflammatory mediators (MCP1, IL6, IFNγ, NGF, and TNFα) in acute IC was also observed [48]. Moreover, intravesical instillation of 50% DMSO in adult female Wistar rats with PS-induced acute IC significantly reduced edema, vascular congestion, and polymorphonuclear (PMN) count that persisted for 7 days after treatment. However, mild inflammation with PMN infiltrate and transient edema was provoked in DMSO-instilled normal bladders [49]. These findings might aid in the explanation of the occurrence of urethral irritation/pain, which is the most frequently reported side-effect (48% of patients) of DMSO instillation [50].

Botulinum Toxin A

Botulinum toxin A (BTX-A) is a potent neurotoxin produced by the bacterium *Clostridium botulinum* [51], currently approved by the FDA for the treatment of neurogenic detrusor muscle overactivity and refractory overactive bladder [52,53]. Due to the ability of BTX-A to inhibit ACh release from nerve fibers, resulting in muscle contractions, as well as prevent sensory nerves sensitization and inflammation, its use has been extended in urology also to treat IC/BPS, and it is currently recommended as a fourth-line therapy by AUA guidelines [24,54]. BTX-A application significantly decreased ATP- and capsaicin-induced neuronal activity in an ex vivo model of isolated rat bladders, as determined by decreased release of the sensory neuropeptide calcitonin gene-related peptide (CGRP) [55]. The ability of BTX-A to inhibit the neuropeptide release (CGRPH and substance P) was subsequently confirmed in bladders from normal adult male rats with acute or chronic IC [56]. BTX-A pretreatment of male rats with CYP-induced IC also reduced ATP release from the urothelial side of bladder preparations, as well as suppressed bladder hyperactivity, non-voiding contraction frequency, and COX-2 and EP4 expression [57,58]. Concurrently, these shreds of evidence suggest that the effects of BTX-A on bladder sensory actions might result from a combined inhibition of sensory neurotransmitter release and through modulation of purinergic pathways [57].

3.2.2. Therapeutic Agents Currently Approved to Treat Other Diseases

Several therapeutics approved to treat different chronic pain, inflammatory, and allergic diseases have been evaluated in experimental models of IC/BPS. For example, antihistamines cetirizine and ranitidine significantly reduced chronic pelvic pain allodynia in experimental models of autoimmune IC in BALB/cJ mice [59]. Recently, Grundy et al. discovered that histamine induces mechanical hypersensitivity ex vivo by interacting with histamine H1 receptor and TRPV1, which was blocked in the presence of pyrilamine [60]. Montelukast, a leukotriene D4 receptor antagonist, used to prevent and treat asthma, reestablished uroplakin distribution and tight junction protein expression and decreased inflammatory cell infiltration in PS-induced IC in Wistar albino rats [61]. In a mouse model of IC induced by CYP, administration of carbenoxolone, clinically prescribed to treat digestive ulcers and inflammation, prevented bladder inflammatory changes and urothelial injury, decreased micturition frequency, and increased micturition volume. Further in vitro analysis showed that carbenoxolone reduced CYP metabolite acrolein-induced injury of urothelial cells, isolated from normal mice bladders by decreasing the expression of TRPV4 channels and reducing TRPV4-mediated oxidative stress [62]. Another drug used to treat gastritis, rebamipide, decreased inflammatory cell infiltration, reduced levels of TNFα, IL1β, and IL6, recovered protein expression of uroplakin 3A, accelerated the repair of the damaged urothelium, and suppressed bladder overactivity and nociception in HCl-induced IC in SD rats [63]. Anti-inflammatory hydroxychloroquine (a TLR7/9 antagonist) decreased voiding frequency and volume in a mice model of loxoribine (a selective TLR-7 agonist)-induced IC, suggesting that TLR7 might represent a promising therapeutic target for IC/BPS [64]. Pretreatment with intravesically applied nanocrystalline silver, which is available as an impregnated wound dressing for treatment of burns, significantly decreased infiltration of mast cells, urine levels of histamine, and bladder explant TNFα release in PS/LPS-induced SD rat model of IC [65]. The use of pregabalin and gabapentin, neuromodulators that selectively bind to the alpha-2-delta ($\alpha2\delta$) subunits of voltage-gated Ca^{2+} channels, showed promising preclinical results in experimental models of IC/BPS. Pregabalin treatment decreased hyperalgesia and reduced inflammation by reducing proinflammatory cytokine production and inhibiting NF-κB activation [66], while systemic administration of gabapentin reduced cystitis-related pain and frequency of voiding [67]. Ceftriaxone, a β-lactam antibiotic, diminished visceral hypersensitivity in stress-induced IC rats [68], while neurokinin-1 receptor antagonist aprepitant, used to treat nausea, relieved pelvic pain, urinary symptoms, and bladder inflammation in mice with experimental autoimmune cystitis [69]. Given the substantial costs and time of new drug discovery and development, drug repurposing could represent an attractive option to treat IC/BPS patients, according to the promising results of preclinical research.

3.2.3. Other Intravesical Therapies and Improved Drug Delivery Systems

Intravesically delivered therapeutic agents reduce systemic side-effects and improve treatment effects by maintaining local drug concentration [12]. Due to the significant disadvantages of intravesical drug delivery, such as low permeability of the urothelium and periodical voiding, which results in fast clearance of active substances with urine and subsequent need for repetitive catheterization, novel approaches, such as liposomes and hydrogels, are being developed [12]. For example, intravesical liposome instillation resulted in partially reversed shortening in inter-contraction interval in rat IC model [70]. Presumably, liposomes were able to form a protective film over damaged urothelium and prevent urinary irritants from acting on the afferent branch of the micturition reflex [70]. The superior effects of liposomes on reducing bladder hyperactivity in comparison to PPS and DMSO were later demonstrated by Tyagi et al. [71,72]. Lin et al. demonstrated that intravesical administration of heparin-loaded floating hydrogel extends the residence time of heparin and increases drug efficiency compared to direct intravesical administration of the drug in a rabbit model [73]. Additionally, a pilot study by Rappaport et al. later reported on the safety and efficacy of intravesical instillation of TC-3 hydrogel in combination with BTX-A. IC/BPS patients included in the study reported mild and temporary adverse effects with improvement in pain and bladder function persisting for 12 weeks [74]. Interestingly, Lee et al. developed an intravesical device for sustained drug delivery that can be implanted into and retrieved from the bladder non-surgically through a cystoscope. The device, combining a Nitinol wireframe and drug-loaded silicone tube, provided a sustained and localized lidocaine delivery while moving freely inside the bladder and preventing local irritation [75]. The method was recently upgraded by Xu et al. who used stereolithography (SLA) 3D printing for the fabrication of an intravesical drug delivery device. The SLA method enables the production of solid objects by polymerization of liquid resins under light irradiation, while the drugs can be incorporated into resin before printing. For the in vitro drug release study, lidocaine hydrochloride was added to elastic resin before printing, which provided a linear release of the drug from the solidified device across a 14-day period [76]. Another emerging drug delivery system includes mucoadhesive polymers, which enable greater bioavailability and solubility of poorly soluble drugs. Chitosan is a promising excipient for the development of such systems due to its positive charge and high mucoadhesive properties in acidic urine. Its favorable adhesion and prolonged drug residence time are being extensively researched, especially for the improvement of bladder cancer treatment options [77–80].

3.2.4. Novel Emerging Therapeutic Options and Targets
Stem-Cell Therapy

Stem cells (SCs), including adult stem cells and pluripotent stem cells, such as embryonic stem cells and induced pluripotent stem cells, possess the abilities of self-renewal, proliferation, and differentiation into various cell types. The therapeutic effects of SCs have been extensively researched and preclinically trialed in many diseases, including IC/BPS [81]. To date, SCs of different origins have been tested in animal models of IC/BPS, including human umbilical cord blood, dental pulp, and adipose tissue. SCs injected intravesically or directly into bladder wall ameliorated bladder voiding dysfunction and inflammation, reduced nociceptive behavior, and decreased urothelial damage in rat IC models [82–84]. In addition, combination therapy of adipose tissue-derived SCs and oral PPS showed synergistic effects in normalizing bladder function and reducing inflammatory reaction [85]. Interestingly, urine-derived SCs (USCs) also possess the capacity for multipotent differentiation and can form multilayered tissue-like structures consisting of urothelium and smooth muscles in vivo [86]. Li et al. showed that USC treatment significantly ameliorated urodynamic, inflammatory, oxidative, and apoptotic changes in PS/LPS-induced IC in female SD rats, compared to untreated controls [87]. Recently, Chung et al. compared the therapeutic potency of mesenchymal SCs derived from different sources—urine, bone marrow, adipose tissue, and amniotic fluid, which showed no significant differences in the regeneration of urothelium and smooth muscle. However, urine-derived stem cells showed superior anti-inflammatory properties, compared to SCs derived from other sources [88].

Due to various limitations of using pluripotent embryonic SCs (ESCs) as a treatment option, such as ethical considerations, teratoma development, long-term possibility of carcinogenesis, and potential immunological rejection of transplanted SC [89], novel strategies of ESCs use are being intensively explored. In 2017, Kim et al. reported on the potential use of human ESC (hESC)-derived multipotent mesenchymal SCs (M-MSC) in IC/BPS, with evidence for long-term safety (6 months after transplantation). M-MSCs therapy significantly ameliorated defects in bladder voiding function, reduced visceral hypersensitivity, and expressed superior therapeutic potency compared to adult bone marrow-derived mesenchymal SCs given in the same doses [90]. The results were later confirmed by Lee et al. in the ketamine-induced chronic IC rat model [91]. Intravital imaging of transplanted hESC-derived M-MSCs in PS/LPS-induced rat IC model demonstrated migration of GFP-transfected M-MSCs from the injection site (serosa and muscle layer) to the damaged urothelium and lamina propria, followed by differentiation into various cell types, which correlated with improvement of IC/BPS symptoms [92].

More recently, Inoue et al. successfully generated differentiated urothelial cells (dUCs) from adult human dermal fibroblasts (aHDFs) using direct conversion technology, which enables conversion of differentiated somatic cell line without passing through a pluripotent state. This was achieved by transduction of a set of genes encoding transcriptional factors (e.g., FOXA1, TP63, MYCL, and KLF4) with crucial roles in the development of target cell lineage [93]. Following transduction, dUCs formed epithelial colonies and expressed urothelial specific proteins UP1b, UP2, CDH1, and Krt8/18 with an in vitro conversion rate of 25%. Moreover, transduced aHDFs, transplanted into a murine IC model, were able to convert into dUCs in vivo in the injured bladder urothelium and participate in tissue regeneration, after integrating into the inner bladder surface [94]. The preclinical studies mark SCs as favorable in IC/BPS treatment; however, no ongoing clinical trials on SC therapy of IC/BPS or overactive bladder are currently registered. Due to the several limitations of SC application, the results of animal studies should be carefully interpreted and critically evaluated before designing clinical trials.

Plant-Based Therapy

Plant-derived agents have recently gained a significant amount of interest in treating different inflammatory and chronic diseases due to their low level of toxicity, cost-effectiveness, and easy availability. In IC/BPS, the use of several medicinal plants has been evaluated in experimental models. For example, *Aster tataricus* extract and its main active component Shionone were shown to reduce bladder inflammation in IC rats and decrease pyroptosis in the SV-HUC1 urothelial cell line, by inhibiting the NLRP3–GSDMD pathway [95,96], while intravesical treatment with *Bletilla striata* extract solution was shown to attenuate visceral hypersensitivity and bladder overactivity in zymosan-induced-IC in SD rats [97]. Treatment with *Olea europaea* or *Juniperus procera* leaf extracts in SD rats with stress (water deprivation)-induced IC reduced bladder mast cell infiltration and levels of stress hormones [98], and *Houttuynia cordata* extract inhibited mast cell proliferation and activation, decreased the levels of proinflammatory cytokines IL6, IL8, and TNFα, and reduced inter-contraction intervals in SD rats with CYP-induced IC [99]. Epigallocatechin-3-gallate (EGCG), a major catechin found in green tea, showed an anti-inflammatory effect in stress-induced IC rats [100], as well as urothelial cells, isolated from bladders of IC/BPS patients via inhibition of phosphorylated NF-κB. EGCG also decreased the expression of purinergic receptors and showed antioxidative properties [101]. Adelmidrol, a diethanolamide derivative of natural azelaic acid, ameliorated CYP-induced bladder inflammation and pain by inhibiting the NF-κB pathway and inflammatory mediator levels, as well as reducing mechanical allodynia and NGF levels [102]. Administration of chlorogenic acid, a phenolic compound widely found in fruits and vegetables, significantly attenuated inflammation, by inhibiting the MAPK/NF-κB pathway and decreasing proinflammatory mediators IL6, IL1β, and TNFα in SD rats with CYP-induced IC [103]. More recently, Shih et al. discovered that curcumin administration mitigated bladder injury and reduced the levels of proinflammatory mediators in the combined PS/LPS mice IC model by downregulating the NLRP3 inflammasome/IL-1β-related TGF-β/Smad pathway [104].

Extracorporeal Shock Wave Therapy

Shock waves are sonic pulses that carry energy from an area of positive pressure to the area of negative pressure through a fluid medium (water or gel). Shock waves can be generated using different sources (electrohydraulic, piezoelectric, or electromagnetic type of generators), and they are characterized by the rapid rise (<10 ns) and high peak pressure (up to 100 MPa) of short duration (<10 μs) and a broad range of frequency spectrum (16 to 20 MHz) [105]. Shock wave therapy was originally used for the disintegration of nephroliths and uroliths, and it later provided immense value in the therapy of musculoskeletal disorders by promoting angiogenesis and tissue regeneration [106]. Chen et al. were the first to describe the beneficial effects of extracorporeal shock wave therapy (ECSWT) in a rat IC model. ECSWT, applied to the skin surface above the bladder or directly to exposed bladder dome, successfully attenuated bladder inflammation and oxidative stress, as well as contributed to the preservation of urothelial integrity in CYP-treated rats [107,108] and UPK3A-immunized mice [109]. The combination of ECSW and BTX-A has also produced advantageous results in overactive bladder and IC/BPS animal models. Transient urothelial permeability, following ECSWT, enables more efficient penetration of BTX-A from urine into the submucosa, without the need for an additional injection of the drug into the bladder wall [110,111]. ECSWT has already provided encouraging results in several randomized clinical trials for the treatment of chronic prostatitis-related pain in men [112,113] and also holds promise for translation to clinical use in IC/BPS patients [114].

Targeting Protease-Activated Receptors

One of the mechanisms, involved in inflammation and pain development in IC/BPS is an activation of protease-activated receptors (PAR), present in urothelial cells, nerve fibers, and bladder detrusor muscles. Many serine proteases, which can contribute to PAR activation, can be found in increased concentrations in the urine of IC/BPS patients, including tryptase and thrombin. In in vivo models, activation of PAR1 and PAR4 with specific agonists resulted in prominent bladder edema and PMN cell infiltration [115], while PAR1 receptor blockade improved urodynamic parameters in CYP-induced IC in rats [116]. Moreover, intravesical stimulation of PAR1 with thrombin can cause the urothelial release of macrophage migration inhibitory factor (MIF), a cytokine that acts as a pivotal mediator of acute and chronic inflammation [117]. Kouzoukas et al. discovered that increased abdominal mechanical hypersensitivity to von Frey stimulation, secondary to intravesical instillation of PAR1- and PAR4-activating peptides, was completely abrogated by MIF-antagonist ISO-1 pretreatment. However, PAR activation did not elicit any bladder inflammation, implying MIF involvement predominantly in the development of pain [118]. The pivotal role of MIF in CYP-induced bladder pain was later confirmed by Ma et al., using MIF knockout mice, in which the mechanical hypersensitivity could not be induced, in contrast to wild-type mice [119]. Along with MIF, activation of PAR4 also promotes high-mobility group box 1 protein (HMGB1) release in vitro and in vivo. HMGB1, when released from necrotic, damaged, or immune cells, such as macrophages, acts as one of the damage-associated molecular patterns (DAMPs) that can activate several receptors (i.e., receptor for advanced glycation end-products (RAGE) and Toll-like receptor 4 (TLR4)) and contribute to inflammatory and neuropathic pain [120]. The detrimental effects of HMGB1 in animal IC models were prevented by pretreatment with anti-HMGB1 neutralizing antibody [121,122] or HMGB1 antagonist glycyrrhizin [123], indicating that it could serve as a potential therapeutic target in IC/BPS.

Targeting Purinergic Receptors

Aggravated purinergic signaling has been shown to be one of the main factors in the development of IC/BPS-related bladder hypersensitivity; therefore, it is not surprising that purinergic P1 (activated by adenosine) and P2 (activated by ATP) receptors are rapidly emerging as potential drug targets. In a study by Hiramoto et al., the authors hypothesized that rapid urothelial ATP release, a common observation in IC/BPS patients, can be induced with CYP treatment in mice, followed by the activation of P2X4 and P2X7 receptors and subsequent HMBG1 release, resulting in bladder pain [124]. Administration of a specific P2X7 antagonist A-438079 was shown to markedly reduce CYP-induced nociceptive behavior and inflammation in a mouse model of hemorrhagic cystitis [125]. On the other hand, a study by Aronsson et al. later failed to reproduce the anti-inflammatory effects of P2 receptor blockade using suramin, a nonselective P2 antagonist in CYP-treated SD rats, supposedly due to species differences [126]. The activation of purinergic receptors can also be prevented by lowering the levels of ATP, released through pannexin channels. This was elegantly shown by Beckel et al. in LPS-treated SD rats. Concurrent administration of BB-FCF, a pannexin 1 channel inhibitor, decreased urothelial ATP release in basal and stretched conditions and completely reversed the LPS-induced decrease in inter-contraction interval [127].

In addition to P2 receptors, the blockade of adenosine receptor P1A1 with its antagonist DPCPX was shown to decrease mast cell infiltration in the detrusor of CYP-treated animals [126]. However, confounding results exist regarding the influence of activation or suppression of P1 receptors, specifically adenosine receptor A2a, on inflammatory response and bladder overactivity. Yang et al. reported that inhibition of adenosine A2a receptors with ZM241385, a selective A2a receptor antagonist, significantly alleviated bladder over-activity and hyperalgesia in CYP-treated animals by reducing the sensitivity of TRPV1 in DRG neurons [128], while Ko et al. showed that activation of adenosine A2a receptor with polydeoxyribonucleotide (PDRN), an A2a receptor agonist, improved voiding dysfunction and reduced inflammation and apoptosis [129]. Interestingly, anti-inflammatory effects of PDRN have previously been demonstrated in animal models of various diseases, such as ischemic colitis, gastric ulcers, and Achilles tendon injury [130–132].

Targeting Transient Receptor Potential Channels

Transient receptor potential (TRP) channels are nonselective ion channels, capable of responding to a variety of stimuli, such as mechanical pressure, changes in pH, heat, and different chemical compounds. Since a variety of chemical and physical stimuli can regulate their activation, TRP channels participate in numerous sensory or homeostatic processes, making them promising pharmacological targets. Their role in the development of lower urinary tract dysfunctions has been extensively investigated [133,134], and some of them have also been trialed as potential drug targets in animal IC models. The transcriptional and translational plasticity of TRPA1, TRPV1, and TRPV4 channels in CYP-induced acute and chronic IC models has been demonstrated by Merrill et al. [135], while selective blockade of TRPV4 channels resulted in improved bladder function [136,137]. Moreover, treatment with artemin-neutralizing antibody reversed CYP-induced hyperalgesia in female C57BL/6 mice by regulation of TRPA1 expression. Artemin, a neurotrophic factor released in inflammatory conditions, can sensitize afferent nerve endings in part through upregulation of expression or augmented function of TRPA1 and TRPV1 channels, contributing to bladder pain [138].

Targeting microRNA

MicroRNAs are short, approximately 22 nucleotide long noncoding RNA molecules that can post-transcriptionally regulate gene expression. Various microRNAs can serve as biomarkers of different lower urinary tract diseases, such as bladder cancer [139], bladder outlet obstruction [140], overactive bladder [141], and IC/BPS [142,143], and they have also been recognized as emerging therapeutic targets. In the present review, we identified two studies trialing microRNAs as potential drug targets in animal models of IC/BPS. Song et al. showed that inhibition of miR-132 reduced bladder inflammation and detrusor fibrosis, as well as improved urodynamic parameters in a PS/LPS-induced rat model of IC via regulation of JAK/STAT signaling pathway [144]. Similarly, the application of miR-495-mimic in a rat model of acute IC resulted in JAK3 downregulation and subsequent inhibition of inflammatory response and bladder fibrosis [145].

Activation of the Cannabinoid System

Synthetic and semisynthetic cannabinoids that lack psychotropic effects have gained much interest recently regarding the treatment of chronic inflammatory disorders and pain, including IC/BPS, due to their antiproliferative, anti-inflammatory, and analgesic effects [146]. The effects of cannabinoids are mediated primarily through cannabinoid receptors CB1 and CB2 that have been found to be present in human and rodent bladder urothelium and detrusor smooth muscle cells [147]. In line with this, Hayn et al. observed that ajulemic acid, a mixed CB1/CB2 receptor agonist, inhibited CGRP release ex vivo in capsaicin- and ATP-stimulated whole rat bladders [148]. Tambaro et al. later showed that administration of JWH015, a selective CB2 agonist, significantly reduced leukocyte infiltration and proinflammatory cytokines in bladder interstitium of CD1 mice [149]. Treatment with another selective CB2 agonist GP1a also decreased severity of edema in acrolein-induced cystitis, inhibited mechanical sensitivity, and decreased bladder urinary frequency that may be mediated by inhibition of ERK1/2 pathway [150,151]. Recently, Liu et al. showed that JWH-133, a selective CB2 agonist completely inhibited mechanical hyperalgesia in CYP-induced mice IC, as well as ameliorated bladder inflammation and oxidative stress. The protective effects of CB2 activation against CYP-induced IC could be mediated by autophagy activation since CB2-induced AMPK activation inhibited mTOR signaling, which subsequently activated autophagy [152]. Berger et al. administered beta-caryophyllene (BCP), which is present in cannabis and activates CB2, to mice with LPS-induced IC. BCP reduced bladder inflammation and improved bladder capillary perfusion with comparable effects to the selective CB2 agonist, HU308 [153]. Palmitoylethanolamide, an endogenous lipid chemically related to the endocannabinoid anandamide, which, in addition to PARα, activates CB1 and CB, was found to be able to attenuate pain behavior, voids, and bladder gross damage in a CYP-induced IC mice model [154]. Pharmacological evidence suggests that, in addition to targeting the canonical cannabinoid receptors (e.g., CB1 and CB2), the cannabinoids can also modulate some TRP channels (TRPV1–4, TRPA1, and TRPM8) [155], thus providing a promising multitarget approach for the treatment of IC/BPS. Despite their great potential, however, there are currently no ongoing clinical trials testing their use in patients with IC/BPS.

3.3. Methodological Quality and Risk of Bias

To evaluate the methodological quality of studies included in the present review, we defined a 12-point checklist based on published ARRIVE guidelines describing minimal information required in scientific publications including animal models [26]. The median number of quality items scored was six out of a possible 12 (q25–q75: 2–7) (Table S3, Supplementary Materials). All publications using in vivo experimental models (n = 88) reported the species and strain of the animals included in their study, as well as detailed description of experimental procedures, such as drug formulation, dose, site, and route of administration. Surprisingly, 86/88 (99%) and only 41/87 (47%), 53/87 (61%), and 8/87 (9%) studies included information regarding the sex, exact age, weight, and detailed description of housing and husbandry of the animals, respectively. The total number of animals used was indicated in 70 (80%), while the exact number of experimental (treated and control) groups was reported in 63 (74%) studies. Measures to avoid bias were infrequently reported, with 18 (21%) publications reporting on random allocation of the animals to treatment groups, and none of the studies describing the method of randomization. Blinded assessment of outcome was included in 18 studies (21%), blinding of the investigator/caretaker was reported in four (5%) of the included studies, and one study (1%) described the method used to calculate the sample size (e.g., number of animals per group), a determination required to avoid false outcomes. Our review suggests that the prevalence of measures introduced to reduce the risk of bias and detailed experimental reporting should be substantially increased to improve the reproducibility and interpretability of the studies, as well as to avoid potential false-positive results and overestimates of treatment effects.

4. Conclusions

Currently, the number of in vitro studies on IC/BPS is very limited, and most of them use urothelial cell lines that are transformed and do not form tight monolayers similar to normal urothelium. Although the in vitro experimental design does not reflect the complexity of the in vivo condition, these studies can lead to a better understanding of the pathology of IC/BPS at the cellular and molecular level. Unraveling the exact relationship between altered urothelial, neuronal, smooth muscle, and/or immune signaling, and the clinical symptoms/signs in IC/PBS will be of outmost importance for understanding the disease process and may help to identify promising targets for future treatment.

Preclinical studies involving animal models remain imperative in studies of etiology and pathophysiology of IC/BPS, as well as novel drug target discovery, and they can help to inform the design of clinical trials. However, adequate experimental design and study quality are important factors for successful implementation into a clinical setting. Our results show that the methodological quality of animal studies could be considerably improved and measures to avoid bias should be implemented and adequately reported. Future studies should incorporate a more standardized and rigorous approach for animal modeling in order to increase the value of preclinical research and the translational potential of experimentally evaluated therapies and therapeutic targets for IC/BPS. Despite several limitations of preclinical experimental models, novel conclusions are drawn almost daily, increasing the insight into the complex mechanisms of IC/BPS development, and several novel treatment options are emerging.

Currently, there is no definite therapeutic modality available and recommended that would be consistently successful in all IC/BPS patients. Most patients are treated on the basis of a "trial and error" approach and need to undergo a series of different combinations of therapies, facing potential severe adverse events. Intravesical application of GAG replenishment therapy, DMSO, and BTX-A ensures maximum delivery of active drug ingredients into the bladder; however, repeated catheterizations are required, causing frequent urinary tract infections. A combination of intravesically delivered therapeutic agents with recently improved drug delivery systems, such as liposomes, hydrogels, and biodegradable polymers and/or a simplified approach of 3D printing to manufacture novel indwelling bladder devices will likely contribute to sustained therapeutic effect without the need for repeated instillation. This will diminish systemic side-effects and enable drug delivery over an extended period of time, ensuring a long-lasting therapeutic effect.

As IC/BPS is a multifactorial disease with several proposed mechanisms of pathobiology, it is anticipated that a multitarget therapeutic approach will be required to achieve long-term efficacy. Stem cells, ECSWT, and activation of the cannabinoid system, which modulate several aspects of the diseases, including the inflammatory processes, central sensitization, pain, and tissue repair, have been proven to be effective in several preclinical studies and have a great translational potential. However, current clinical studies are limited to case reports, and large, multicenter, long-term, randomized clinical trials are warranted to elucidate their efficiency and safety in patients with IC/BPS.

Furthermore, phenotyping and stratifying patients into subgroups as a function of clinical signs and bladder histological findings will be particularly important for selection of patients most suitable for a specific therapeutic option. Due to the extremely complex and heterogeneous pathological backgrounds of IC/BPS patients, the currently used "one-size-fits-all" medicine should be replaced with a more personalized approach, also takinginto account the variability in genes, environment, and lifestyle of a particular patient.

Supplementary Materials: The following are available online at https://www.mdpi.com/article/10.3390/biomedicines9080865/s1, Table S1. In vitro and ex vivo experimental models of IC/BPS evaluating different therapeutic options; Table S2. In vivo experimental models of IC/BPS evaluating different therapeutic options; Table S3. Methodological quality and reported measures undertaken to avoid bias.

Author Contributions: Conceptualization, T.K., D.P., A.E. and P.V.; methodology, T.K. and D.P.; validation, T.K. and D.P.; formal analysis, T.K. and D.P.; investigation, T.K. and D.P.; resources, P.V.; data curation, T.K. and D.P.; writing—original draft preparation, T.K. and D.P.; writing—review and editing, T.K., D.P., A.E. and P.V.; visualization, T.K., D.P. and A.E.; supervision, A.E. and P.V.; project administration, A.E. and P.V.; funding acquisition, P.V. All authors read and agreed to the published version of the manuscript.

Funding: This research was funded by the Slovenian Research Agency (ARRS), grant numbers #P3-0108, #J3-2521, and #I0-0022.

Institutional Review Board Statement: Not applicable.

Informed Consent Statement: Not applicable.

Data Availability Statement: No new data were created or analyzed in this study. Data sharing is not applicable to this article.

Conflicts of Interest: The authors declare no conflict of interest.

Abbreviations

A2	adenosine receptor
Ab	antibody
ACh	acetylcholine
ACTH	adrenocorticotropic hormone
AJA	ajulemic acid
APF	antiproliferative factor
ASC	apoptosis-associated speck-like protein
A7R5	smooth muscle cell line
AYPGKF-NH2	PAR4 agonist
BCP	beta-caryophyllene
BTX-A	botulinum toxin A
CAT	catalase
CB	cannabinoid receptor
CBX	carbenexolone
CD	cluster of differentiation
CGRP	calcitonin gene-related peptide
CRH	corticotropin-releasing hormone

CS	chondroitin sulfate
CYP	cyclophosphamide
DMSO	dimethyl sulfoxide
DRG	dorsal root ganglia
ECSWT	extracorporeal shock wave therapy
EGCG	epigallocatechin gallate
eNOS	endothelial nitric oxide synthase
EP	prostaglandin E2 receptor subtype
ERK	extracellular signal-regulated kinase
FcεRIα	high-affinity IgE receptor
FTLK	transcriptional factors FOXA1, TP63, MYCL, and KLF4
GAG	glycosaminoglycan
GM-0111	modified GAG
GSDMD	gasdermin D
GSH	glutathione
HA	hyaluronic acid
H-BLAK	primary human bladder cell line
h-ESC	human embryonic stem cells
HMGB1	high mobility group box 1
HO-1	heme oxygenase-1
H_2O_2	hydrogen peroxide
HTB4, HRB2	human urothelial cells
ICAM	intercellular adhesion molecule
IC/BPS	interstitial cystitis/bladder pain syndrome
ICI	intercontraction interval
IFN-γ	interferon gamma
IκBα	inhibitor of NF-κB
IL	interleukin
iNOS	inducible nitric oxide synthase
JAK	janus kinase
KC	keratinocytes-derived chemokine
LDH	lactate dehydrogenase
LL37	antimicrobial peptide
LPS	lipopolysaccharide
MAPK	mitogen-activated protein kinase
MCP-1	monocyte chemoattractant protein-1
MDA	malondialdehyde
MIF	macrophage migration inhibitory factor
miR	microRNA
MMP9	matrix metalloproteinase 9
M-MSC	multipotent mesenchymal stem cells
MPO	myeloperoxidase
MSC	mesenchymal stem cells
NF-κB	nuclear factor kappa B
NGF	nerve growth factor
NK1R	neurokinin 1 receptor
NLRP3	NLR family pyrin domain-containing 3
NMDAR	N-methyl-D-aspartate receptor
NOX	NAPDH oxidase
NQO-1	NADPH quinine oxidoreductase
NRK-52E	renal tubular epithelial cell line
NVC	non-voiding contraction
OVX	ovariectomized
P1, P2	purinoceptors
P2X	purinergic receptors
p38	mitogen-activated protein kinase
p65	NF-κB subunit

p-AKT	protein kinase B, phosphorylated
PAR	proteinase-activated receptor
PGE2	prostaglandin 2
PMN	polymorphonuclear cells
p-mTOR	mechanistic target of rapamycin, phosphorylated
POMC	pro-opiomelanocortin
PPAR	peroxisome proliferator-activated receptor
PPS	pentosan polysulfate sodium
PS	protamine sulfate
PTX3	pentraxin 3
RANTES	chemokine ligand 5
RhoA	Ras homolog gene family, member A
rhsTM	recombinant human soluble thrombomodulin
ROCK	Rho-associated protein kinase; R
OS	reactive oxygen species;
RT112	3D human bladder epithelium preparation
SAGE	semi-synthetic glycosaminoglycan ethers
SC	stem cells
SD rats	Sprague-Dawley rats
sGC	soluble guanylyl cyclase
MAPK	mitogen-activated protein kinase
MCP-1	monocyte chemoattractant protein-1
MDA	malondialdehyde
SMAD	proteins for signal transduction of the transforming growth factor beta superfamily
TRPV	transient receptor potential channel, vanilloid subgroup
UPK	uroplakin
VEGF	vascular endothelial growth factor
ZO-1	tight junction protein 1

References

1. Homma, Y.; Ueda, T.; Tomoe, H.; Lin, A.T.; Kuo, H.C.; Lee, M.H.; Oh, S.J.; Kim, J.C.; Lee, K.S. Clinical guidelines for interstitial cystitis and hypersensitive bladder updated in 2015. *Int. J. Urol.* **2016**, *23*, 542–549. [CrossRef]
2. Patnaik, S.S.; Laganà, A.S.; Vitale, S.G.; Buttice, S.; Noventa, M.; Gizzo, S.; Valenti, G.; Rapisarda, A.M.C.; La Rosa, V.L.; Magno, C.; et al. Etiology, pathophysiology and biomarkers of interstitial cystitis/painful bladder syndrome. *Arch. Gynecol. Obstet.* **2017**, *295*, 1341–1359. [CrossRef] [PubMed]
3. Hanno, P.; Lin, A.; Nordling, J.; Nyberg, L.; van Ophoven, A.; Ueda, T.; Wein, A. Bladder Pain Syndrome Committee of the International Consultation on Incontinence. *Neurourol. Urodyn.* **2010**, *29*, 191–198. [CrossRef] [PubMed]
4. Clemens, J.Q.; Mullins, C.; Ackerman, A.L.; Bavendam, T.; van Bokhoven, A.; Ellingson, B.M.; Harte, S.E.; Kutch, J.J.; Lai, H.H.; Martucci, K.T.; et al. Urologic chronic pelvic pain syndrome: Insights from the MAPP Research Network. *Nat. Rev. Urol.* **2019**, *16*, 187–200. [CrossRef] [PubMed]
5. Berry, S.H.; Elliott, M.N.; Suttorp, M.; Bogart, L.M.; Stoto, M.A.; Eggers, P.; Nyberg, L.; Clemens, J.Q. Prevalence of symptoms of bladder pain syndrome/interstitial cystitis among adult females in the United States. *J. Urol.* **2011**, *186*, 540–544. [CrossRef] [PubMed]
6. Konkle, K.S.; Berry, S.H.; Elliott, M.N.; Hilton, L.; Suttorp, M.J.; Clauw, D.J.; Clemens, J.Q. Comparison of an interstitial cystitis/bladder pain syndrome clinical cohort with symptomatic community women from the RAND Interstitial Cystitis Epidemiology study. *J. Urol.* **2012**, *187*, 508–512. [CrossRef]
7. Anger, J.T.; Zabihi, N.; Clemens, J.Q.; Payne, C.K.; Saigal, C.S.; Rodriguez, L.V. Treatment choice, duration, and cost in patients with interstitial cystitis and painful bladder syndrome. *Int. Urogynecol. J.* **2011**, *22*, 395–400. [CrossRef] [PubMed]
8. Akiyama, Y.; Hanno, P. Phenotyping of interstitial cystitis/bladder pain syndrome. *Int. J. Urol.* **2019**, *26* (Suppl. 1), 17–19. [CrossRef] [PubMed]
9. Warren, J.W.; Howard, F.M.; Cross, R.K.; Good, J.L.; Weissman, M.M.; Wesselmann, U.; Langenberg, P.; Greenberg, P.; Clauw, D.J. Antecedent nonbladder syndromes in case-control study of interstitial cystitis/painful bladder syndrome. *Urology* **2009**, *73*, 52–57. [CrossRef]
10. Doiron, R.C.; Tolls, V.; Irvine-Bird, K.; Kelly, K.L.; Nickel, J.C. Clinical Phenotyping Does Not Differentiate Hunner Lesion Subtype of Interstitial Cystitis/Bladder Pain Syndrome: A Relook at the Role of Cystoscopy. *J. Urol.* **2016**, *196*, 1136–1140. [CrossRef] [PubMed]
11. Birder, L.A. Pathophysiology of interstitial cystitis. *Int. J. Urol.* **2019**, *26* (Suppl. 1), 12–15. [CrossRef]

12. Lin, Z.; Hu, H.; Liu, B.; Chen, Y.; Tao, Y.; Zhou, X.; Li, M. Biomaterial-assisted drug delivery for interstitial cystitis/bladder pain syndrome treatment. *J. Mater. Chem. B.* **2021**, *9*, 23–34. [CrossRef] [PubMed]

13. Grundy, L.; Caldwell, A.; Brierley, S.M. Mechanisms Underlying Overactive Bladder and Interstitial Cystitis/Painful Bladder Syndrome. *Front. Neurosci.* **2018**, *12*, 931. [CrossRef]

14. Hurst, R.E. Structure, function, and pathology of proteoglycans and glycosaminoglycans in the urinary tract. *World J. Urol.* **1994**, *12*, 3–10. [CrossRef] [PubMed]

15. Keay, S. Cell signaling in interstitial cystitis/painful bladder syndrome. *Cell Signal.* **2008**, *20*, 2174–2179. [CrossRef] [PubMed]

16. Slobodov, G.; Feloney, M.; Gran, C.; Kyker, K.D.; Hurst, R.E.; Culkin, D.J. Abnormal expression of molecular markers for bladder impermeability and differentiation in the urothelium of patients with interstitial cystitis. *J. Urol.* **2004**, *171*, 1554–1558. [CrossRef]

17. Liu, H.T.; Shie, J.H.; Chen, S.H.; Wang, Y.S.; Kuo, H.C. Differences in mast cell infiltration, E-cadherin, and zonula occludens-1 expression between patients with overactive bladder and interstitial cystitis/bladder pain syndrome. *Urology* **2012**, *80*, 225.e13–225.e18. [CrossRef] [PubMed]

18. Parsons, C.L. The role of a leaky epithelium and potassium in the generation of bladder symptoms in interstitial cystitis/overactive bladder, urethral syndrome, prostatitis and gynaecological chronic pelvic pain. *BJU Int.* **2011**, *107*, 370–375. [CrossRef]

19. Winder, M.; Tobin, G.; Zupančič, D.; Romih, R. Signalling molecules in the urothelium. *Biomed. Res. Int.* **2014**, *2014*, 297295. [CrossRef]

20. Liu, H.T.; Tyagi, P.; Chancellor, M.B.; Kuo, H.C. Urinary nerve growth factor level is increased in patients with interstitial cystitis/bladder pain syndrome and decreased in responders to treatment. *BJU Int.* **2009**, *104*, 1476–1481. [CrossRef] [PubMed]

21. Yang, W.; Searl, T.J.; Yaggie, R.; Schaeffer, A.J.; Klumpp, D.J. A MAPP Network study: Overexpression of tumor necrosis factor-α in mouse urothelium mimics interstitial cystitis. *Am. J. Physiol. Renal. Physiol.* **2018**, *315*, F36–F44. [CrossRef] [PubMed]

22. Jiang, Y.H.; Jhang, J.F.; Hsu, Y.H.; Ho, H.C.; Wu, Y.H.; Kuo, H.C. Urine cytokines as biomarkers for diagnosing interstitial cystitis/bladder pain syndrome and mapping its clinical characteristics. *Am. J. Physiol. Renal. Physiol.* **2020**, *318*, F1391–F1399. [CrossRef]

23. Jiang, Y.H.; Jhang, J.F.; Hsu, Y.H.; Ho, H.C.; Wu, Y.H.; Kuo, H.C. Urine biomarkers in ESSIC type 2 interstitial cystitis/bladder pain syndrome and overactive bladder with developing a novel diagnostic algorithm. *Sci. Rep.* **2021**, *11*, 914. [CrossRef] [PubMed]

24. Hanno, P.M.; Erickson, D.; Moldwin, R.; Faraday, M.M. Diagnosis and treatment of interstitial cystitis/bladder pain syndrome: AUA guideline amendment. *J. Urol.* **2015**, *193*, 1545–1553. [CrossRef]

25. Hooijmans, C.R.; Ritskes-Hoitinga, M. Progress in using systematic reviews of animal studies to improve translational research. *PLoS. Med.* **2013**, *10*, e1001482. [CrossRef] [PubMed]

26. Kilkenny, C.; Browne, W.J.; Cuthill, I.C.; Emerson, M.; Altman, D.G. Improving bioscience research reporting: The ARRIVE guidelines for reporting animal research. *PLoS. Biol.* **2010**, *8*, e1000412. [CrossRef]

27. Hauser, P.J.; Dozmorov, M.G.; Bane, B.L.; Slobodov, G.; Culkin, D.J.; Hurst, R.E. Abnormal expression of differentiation related proteins and proteoglycan core proteins in the urothelium of patients with interstitial cystitis. *J. Urol.* **2008**, *179*, 764–769. [CrossRef] [PubMed]

28. Kullmann, F.A.; Daugherty, S.L.; de Groat, W.C.; Birder, L.A. Bladder smooth muscle strip contractility as a method to evaluate lower urinary tract pharmacology. *J. Vis. Exp.* **2014**, e51807. [CrossRef]

29. Birder, L.; Andersson, K.E. Animal Modelling of Interstitial Cystitis/Bladder Pain Syndrome. *Int. Neurourol. J.* **2018**, *22*, S3-9. [CrossRef]

30. Wyndaele, J.J.J.; Riedl, C.; Taneja, R.; Lovasz, S.; Ueda, T.; Cervigni, M. GAG replenishment therapy for bladder pain syndrome/interstitial cystitis. *Neurourol. Urodyn.* **2019**, *38*, 535–544. [CrossRef] [PubMed]

31. Rozenberg, B.B.; Janssen, D.A.W.; Jansen, C.F.J.; Schalken, J.A.; Heesakkers, J. Improving the barrier function of damaged cultured urothelium using chondroitin sulfate. *Neurourol. Urodyn.* **2020**, *39*, 558–564. [CrossRef]

32. Rooney, P.; Srivastava, A.; Watson, L.; Quinlan, L.R.; Pandit, A. Hyaluronic acid decreases IL-6 and IL-8 secretion and permeability in an inflammatory model of interstitial cystitis. *Acta. Biomater.* **2015**, *19*, 66–75. [CrossRef] [PubMed]

33. Stellavato, A.; Pirozzi, A.V.A.; Diana, P.; Reale, S.; Vassallo, V.; Fusco, A.; Donnarumma, G.; De Rosa, M.; Schiraldi, C. Hyaluronic acid and chondroitin sulfate, alone or in combination, efficiently counteract induced bladder cell damage and inflammation. *PLoS ONE* **2019**, *14*, e0218475. [CrossRef] [PubMed]

34. Rooney, P.; Ryan, C.; McDermott, B.J.; Dev, K.; Pandit, A.; Quinlan, L.R. Effect of Glycosaminoglycan Replacement on Markers of Interstitial Cystitis In Vitro. *Front. Pharmacol.* **2020**, *11*, 575043. [CrossRef]

35. Rooney, P.R.; Kannala, V.K.; Kotla, N.G.; Benito, A.; Dupin, D.; Loinaz, I.; Quinlan, L.R.; Rochev, Y.; Pandit, A. A high molecular weight hyaluronic acid biphasic dispersion as potential therapeutics for interstitial cystitis. *J. Biomed. Mater. Res. B. Appl. Biomater.* **2021**, *109*, 864–876. [CrossRef]

36. Yeh, C.H.; Chiang, H.S.; Chien, C.T. Hyaluronic acid ameliorates bladder hyperactivity via the inhibition of H2O2-enhanced purinergic and muscarinic signaling in the rat. *Neurourol. Urodyn.* **2010**, *29*, 765–770. [CrossRef] [PubMed]

37. Hauser, P.J.; Buethe, D.A.; Califano, J.; Sofinowski, T.M.; Culkin, D.J.; Hurst, R.E. Restoring barrier function to acid damaged bladder by intravesical chondroitin sulfate. *J. Urol.* **2009**, *182*, 2477–2482. [CrossRef] [PubMed]

38. Engles, C.D.; Hauser, P.J.; Abdullah, S.N.; Culkin, D.J.; Hurst, R.E. Intravesical chondroitin sulfate inhibits recruitment of inflammatory cells in an acute acid damage "leaky bladder" model of cystitis. *Urology* **2012**, *79*, 483.e13–483.e17. [CrossRef]

39. Rajasekaran, M.; Stein, P.; Parsons, C.L. Toxic factors in human urine that injure urothelium. *Int. J. Urol.* **2006**, *13*, 409–414. [CrossRef]

40. Hurst, R.E.; Zebrowski, R. Identification of proteoglycans present at high density on bovine and human bladder luminal surface. *J. Urol.* **1994**, *152*, 1641–1645. [CrossRef]

41. Greenwood-Van Meerveld, B.; Mohammadi, E.; Latorre, R.; Truitt, E.R., 3rd; Jay, G.D.; Sullivan, B.D.; Schmidt, T.A.; Smith, N.; Saunders, D.; Ziegler, J.; et al. Preclinical Animal Studies of Intravesical Recombinant Human Proteoglycan 4 as a Novel Potential Therapy for Diseases Resulting From Increased Bladder Permeability. *Urology* **2018**, *116*, 230.e231–230.e237. [CrossRef] [PubMed]

42. Towner, R.A.; Greenwood-Van Meerveld, B.; Mohammadi, E.; Saunders, D.; Smith, N.; Sant, G.R.; Shain, H.C.; Jozefiak, T.H.; Hurst, R.E. SuperGAG biopolymers for treatment of excessive bladder permeability. *Pharmacol. Res. Perspect.* **2021**, *9*, e00709. [CrossRef] [PubMed]

43. Jensen, M.M.; Jia, W.; Schults, A.J.; Isaacson, K.J.; Steinhauff, D.; Green, B.; Zachary, B.; Cappello, J.; Ghandehari, H.; Oottamasathien, S. Temperature-responsive silk-elastinlike protein polymer enhancement of intravesical drug delivery of a therapeutic glycosaminoglycan for treatment of interstitial cystitis/painful bladder syndrome. *Biomaterials* **2019**, *217*, 119293. [CrossRef]

44. Oottamasathien, S.; Jia, W.J.; McCoard, L.; Slack, S.; Zhang, J.X.; Skardal, A.; Job, K.; Kennedy, T.P.; Dull, R.O.; Prestwich, G.D. A Murine Model of Inflammatory Bladder Disease: Cathelicidin Peptide Induced Bladder Inflammation and Treatment With Sulfated Polysaccharides. *J. Urol.* **2011**, *186*, 1684–1692. [CrossRef] [PubMed]

45. Lee, W.Y.; Savage, J.R.; Zhang, J.X.; Jia, W.J.; Oottamasathien, S.; Prestwich, G.D. Prevention of Anti-microbial Peptide LL-37-Induced Apoptosis and ATP Release in the Urinary Bladder by a Modified Glycosaminoglycan. *PLoS ONE* **2013**, *8*, e77854. [CrossRef]

46. Rawls, W.F.; Cox, L.; Rovner, E.S. Dimethyl sulfoxide (DMSO) as intravesical therapy for interstitial cystitis/bladder pain syndrome: A review. *Neurourol. Urodyn.* **2017**, *36*, 1677–1684. [CrossRef]

47. Melchior, D.; Packer, C.S.; Johnson, T.C.; Kaefer, M. Dimethyl sulfoxide: Does it change the functional properties of the bladder wall? *J. Urol.* **2003**, *170*, 253–258. [CrossRef]

48. Kim, R.; Liu, W.; Chen, X.; Kreder, K.J.; Luo, Y. Intravesical dimethyl sulfoxide inhibits acute and chronic bladder inflammation in transgenic experimental autoimmune cystitis models. *J. Biomed. Biotechnol.* **2011**, *2011*, 937061. [CrossRef]

49. Soler, R.; Bruschini, H.; Truzzi, J.C.; Martins, J.R.; Camara, N.O.; Alves, M.T.; Leite, K.R.; Nader, H.B.; Srougi, M.; Ortiz, V. Urinary glycosaminoglycans excretion and the effect of dimethyl sulfoxide in an experimental model of non-bacterial cystitis. *Int. Braz. J. Urol.* **2008**, *34*, 503–511; discussion 511. [CrossRef]

50. Rössberger, J.; Fall, M.; Peeker, R. Critical appraisal of dimethyl sulfoxide treatment for interstitial cystitis: Discomfort, side-effects and treatment outcome. *Scand. J. Urol. Nephrol.* **2005**, *39*, 73–77. [CrossRef]

51. Erbguth, F.J. Historical notes on botulism, Clostridium botulinum, botulinum toxin, and the idea of the therapeutic use of the toxin. *Mov. Disord.* **2004**, *19* (Suppl. 8), S2–S6. [CrossRef] [PubMed]

52. Reitz, A.; Stohrer, M.; Kramer, G.; Del Popolo, G.; Chartier-Kastler, E.; Pannek, J.; Burgdorfer, H.; Gocking, K.; Madersbacher, H.; Schumacher, S.; et al. European experience of 200 cases treated with botulinum-A toxin injections into the detrusor muscle for urinary incontinence due to neurogenic detrusor overactivity. *Eur. Urol.* **2004**, *45*, 510–515. [CrossRef] [PubMed]

53. Chen, J.L.; Kuo, H.C. Clinical application of intravesical botulinum toxin type A for overactive bladder and interstitial cystitis. *Investig Clin. Urol.* **2020**, *61*, S33–S42. [CrossRef]

54. Dan Spinu, A.; Gabriel Bratu, O.; Cristina Diaconu, C.; Maria Alexandra Stanescu, A.; Bungau, S.; Fratila, O.; Bohiltea, R.; Liviu Dorel Mischianu, D. Botulinum toxin in low urinary tract disorders—Over 30 years of practice. *Exp. Ther. Med.* **2020**, *20*, 117–120. [CrossRef] [PubMed]

55. Rapp, D.E.; Turk, K.W.; Bales, G.T.; Cook, S.P. Botulinum toxin type a inhibits calcitonin gene-related peptide release from isolated rat bladder. *J. Urol.* **2006**, *175*, 1138–1142. [CrossRef]

56. Lucioni, A.; Bales, G.T.; Lotan, T.L.; McGehee, D.S.; Cook, S.P.; Rapp, D.E. Botulinum toxin type A inhibits sensory neuropeptide release in rat bladder models of acute injury and chronic inflammation. *BJU Int.* **2008**, *101*, 366–370. [CrossRef]

57. Smith, C.P.; Vemulakonda, V.M.; Kiss, S.; Boone, T.B.; Somogyi, G.T. Enhanced ATP release from rat bladder urothelium during chronic bladder inflammation: Effect of botulinum toxin A. *Neurochem. Int.* **2005**, *47*, 291–297. [CrossRef]

58. Chuang, Y.C.; Yoshimura, N.; Huang, C.C.; Wu, M.; Chiang, P.H.; Chancellor, M.B. Intravesical botulinum toxin A administration inhibits COX-2 and EP4 expression and suppresses bladder hyperactivity in cyclophosphamide-induced cystitis in rats. *Eur. Urol.* **2009**, *56*, 159–166. [CrossRef]

59. Bicer, F.; Altuntas, C.Z.; Izgi, K.; Ozer, A.; Kavran, M.; Tuohy, V.K.; Daneshgari, F. Chronic pelvic allodynia is mediated by CCL2 through mast cells in an experimental autoimmune cystitis model. *Am. J. Physiol. Renal Physiol.* **2015**, *308*, F103–F113. [CrossRef]

60. Grundy, L.; Caldwell, A.; Garcia Caraballo, S.; Erickson, A.; Schober, G.; Castro, J.; Harrington, A.M.; Brierley, S.M. Histamine induces peripheral and central hypersensitivity to bladder distension via the histamine H(1) receptor and TRPV1. *Am. J. Physiol. Renal Physiol.* **2020**, *318*, F298–F314. [CrossRef]

61. Çetinel, S.; Çanıllıoğlu, Y.E.; Çikler, E.; Sener, G.; Ercan, F. Leukotriene D4 receptor antagonist montelukast alleviates protamine sulphate-induced changes in rat urinary bladder. *BJU Int.* **2011**, *107*, 1320–1325. [CrossRef]

62. Zhang, X.L.; Gao, S.; Tanaka, M.; Zhang, Z.; Huang, Y.R.; Mitsui, T.; Kamiyama, M.; Koizumi, S.; Fan, J.L.; Takeda, M.; et al. Carbenoxolone inhibits TRPV4 channel-initiated oxidative urothelial injury and ameliorates cyclophosphamide-induced bladder dysfunction. *J. Cell. Mol. Med.* **2017**, *21*, 1791–1802. [CrossRef] [PubMed]

63. Funahashi, Y.; Yoshida, M.; Yamamoto, T.; Majima, T.; Takai, S.; Gotoh, M. Intravesical application of rebamipide promotes urothelial healing in a rat cystitis model. *J. Urol.* **2014**, *192*, 1864–1870. [CrossRef] [PubMed]
64. Ichihara, K.; Aizawa, N.; Akiyama, Y.; Kamei, J.; Masumori, N.; Andersson, K.E.; Homma, Y.; Igawa, Y. Toll-like receptor 7 is overexpressed in the bladder of Hunner-type interstitial cystitis, and its activation in the mouse bladder can induce cystitis and bladder pain. *Pain* **2017**, *158*, 1538–1545. [CrossRef] [PubMed]
65. Boucher, W.; Stern, J.M.; Kotsinyan, V.; Kempuraj, D.; Papaliodis, D.; Cohen, M.S.; Theoharides, T.C. Intravesical nanocrystalline silver decreases experimental bladder inflammation. *J. Urol.* **2008**, *179*, 1598–1602. [CrossRef]
66. Boudieu, L.; Mountadem, S.; Lashermes, A.; Meleine, M.; Ulmann, L.; Rassendren, F.; Aissouni, Y.; Sion, B.; Carvalho, F.A.; Ardid, D. Blocking $\alpha(2)\delta$-1 Subunit Reduces Bladder Hypersensitivity and Inflammation in a Cystitis Mouse Model by Decreasing NF-kB Pathway Activation. *Front. Pharmacol.* **2019**, *10*, 133. [CrossRef]
67. Yoshizumi, M.; Watanabe, C.; Mizoguchi, H. Gabapentin reduces painful bladder hypersensitivity in rats with lipopolysaccharide-induced chronic cystitis. *Pharmacol. Res. Perspect.* **2021**, *9*, e00697. [CrossRef]
68. Holschneider, D.P.; Wang, Z.; Chang, H.Y.; Zhang, R.; Gao, Y.L.; Guo, Y.M.; Mao, J.; Rodriguez, L.V. Ceftriaxone inhibits stress-induced bladder hyperalgesia and alters cerebral micturition and nociceptive circuits in the rat: A multidisciplinary approach to the study of urologic chronic pelvic pain syndrome research network study. *Neurourol. Urodyn.* **2020**, *39*, 1628–1643. [CrossRef]
69. Liu, B.K.; Jin, X.W.; Lu, H.Z.; Zhang, X.; Zhao, Z.H.; Shao, Y. The Effects of Neurokinin-1 Receptor Antagonist in an Experimental Autoimmune Cystitis Model Resembling Bladder Pain Syndrome/Interstitial Cystitis. *Inflammation* **2019**, *42*, 246–254. [CrossRef]
70. Fraser, M.O.; Chuang, Y.C.; Tyagi, P.; Yokoyama, T.; Yoshimura, N.; Huang, L.; De Groat, W.C.; Chancellor, M.B. Intravesical liposome administration–a novel treatment for hyperactive bladder in the rat. *Urology* **2003**, *61*, 656–663. [CrossRef]
71. Tyagi, P.; Chancellor, M.; Yoshimura, N.; Huang, L. Activity of different phospholipids in attenuating hyperactivity in bladder irritation. *BJU Int.* **2008**, *101*, 627–632. [CrossRef] [PubMed]
72. Tyagi, P.; Hsieh, V.C.; Yoshimura, N.; Kaufman, J.; Chancellor, M.B. Instillation of liposomes vs dimethyl sulphoxide or pentosan polysulphate for reducing bladder hyperactivity. *BJU Int.* **2009**, *104*, 1689–1692. [CrossRef]
73. Lin, T.S.; Zhao, X.Z.; Zhang, Y.F.; Lian, H.B.; Zhuang, J.L.; Zhang, Q.; Chen, W.; Wang, W.; Liu, G.X.; Guo, S.H.; et al. Floating Hydrogel with Self-Generating Micro-Bubbles for Intravesical Instillation. *Materials* **2016**, *9*, 5. [CrossRef] [PubMed]
74. Rappaport, Y.H.; Zisman, A.; Jeshurun-Gutshtat, M.; Gerassi, T.; Hakim, G.; Vinshtok, Y.; Stav, K. Safety and Feasibility of Intravesical Instillation of Botulinum Toxin-A in Hydrogel-based Slow-release Delivery System in Patients With Interstitial Cystitis-Bladder Pain Syndrome: A Pilot Study. *Urology* **2018**, *114*, 60–65. [CrossRef]
75. Lee, H.; Cima, M.J. An intravesical device for the sustained delivery of lidocaine to the bladder. *J. Control. Release* **2011**, *149*, 133–139. [CrossRef]
76. Xu, X.; Goyanes, A.; Trenfield, S.J.; Diaz-Gomez, L.; Alvarez-Lorenzo, C.; Gaisford, S.; Basit, A.W. Stereolithography (SLA) 3D printing of a bladder device for intravesical drug delivery. *Mater. Sci. Eng. C. Mater. Biol. Appl.* **2021**, *120*, 111773. [CrossRef] [PubMed]
77. Lupo, N.; Jalil, A.; Nazir, I.; Gust, R.; Bernkop-Schnürch, A. In vitro evaluation of intravesical mucoadhesive self-emulsifying drug delivery systems. *Int. J Pharm.* **2019**, *564*, 180–187. [CrossRef] [PubMed]
78. Kolawole, O.M.; Lau, W.M.; Khutoryanskiy, V.V. Methacrylated chitosan as a polymer with enhanced mucoadhesive properties for transmucosal drug delivery. *Int. J. Pharm.* **2018**, *550*, 123–129. [CrossRef]
79. Ali, M.S.; Metwally, A.A.; Fahmy, R.H.; Osman, R. Chitosan-coated nanodiamonds: Mucoadhesive platform for intravesical delivery of doxorubicin. *Carbohydr. Polym.* **2020**, *245*, 116528. [CrossRef]
80. Li, G.; He, S.; Schätzlein, A.G.; Weiss, R.M.; Martin, D.T.; Uchegbu, I.F. Achieving highly efficient gene transfer to the bladder by increasing the molecular weight of polymer-based nanoparticles. *J. Control. Release* **2021**, *332*, 210–224. [CrossRef]
81. Dayem, A.A.; Kim, K.; Lee, S.B.; Kim, A.; Cho, S.G. Application of Adult and Pluripotent Stem Cells in Interstitial Cystitis/Bladder Pain Syndrome Therapy: Methods and Perspectives. *J. Clin. Med.* **2020**, *9*, 766. [CrossRef] [PubMed]
82. Song, M.; Lim, J.; Yu, H.Y.; Park, J.; Chun, J.Y.; Jeong, J.; Heo, J.; Kang, H.; Kim, Y.; Cho, Y.M.; et al. Mesenchymal Stem Cell Therapy Alleviates Interstitial Cystitis by Activating Wnt Signaling Pathway. *Stem. Cells* **2015**, *24*, 1648–1657. [CrossRef]
83. Hirose, Y.; Yamamoto, T.; Nakashima, M.; Funahashi, Y.; Matsukawa, Y.; Yamaguchi, M.; Kawabata, S.; Gotoh, M. Injection of Dental Pulp Stem Cells Promotes Healing of Damaged Bladder Tissue in a Rat Model of Chemically Induced Cystitis. *Cell Transplant.* **2016**, *25*, 425–436. [CrossRef]
84. Furuta, A.; Yamamoto, T.; Igarashi, T.; Suzuki, Y.; Egawa, S.; Yoshimura, N. Bladder wall injection of mesenchymal stem cells ameliorates bladder inflammation, overactivity, and nociception in a chemically induced interstitial cystitis-like rat model. *Int. Urogynecol. J.* **2018**, *29*, 1615–1622. [CrossRef] [PubMed]
85. Kim, B.S.; Chun, S.Y.; Lee, E.H.; Chung, J.W.; Lee, J.N.; Ha, Y.S.; Choi, J.Y.; Song, P.H.; Kwon, T.G.; Han, M.H.; et al. Efficacy of combination therapy with pentosan polysulfate sodium and adipose tissue-derived stem cells for the management of interstitial cystitis in a rat model. *Stem. Cell. Res.* **2020**, *45*, 101801. [CrossRef] [PubMed]
86. Bharadwaj, S.; Liu, G.; Shi, Y.; Wu, R.; Yang, B.; He, T.; Fan, Y.; Lu, X.; Zhou, X.; Liu, H.; et al. Multipotential differentiation of human urine-derived stem cells: Potential for therapeutic applications in urology. *Stem. Cells.* **2013**, *31*, 1840–1856. [CrossRef] [PubMed]
87. Li, J.; Luo, H.; Dong, X.; Liu, Q.; Wu, C.; Zhang, T.; Hu, X.; Zhang, Y.; Song, B.; Li, L. Therapeutic effect of urine-derived stem cells for protamine/lipopolysaccharide-induced interstitial cystitis in a rat model. *Stem. Cell. Res. Ther.* **2017**, *8*, 107. [CrossRef]

88. Chung, J.W.; Chun, S.Y.; Lee, E.H.; Ha, Y.S.; Lee, J.N.; Song, P.H.; Yoo, E.S.; Kwon, T.G.; Chung, S.K.; Kim, B.S. Verification of mesenchymal stem cell injection therapy for interstitial cystitis in a rat model. *PLoS ONE* **2019**, *14*, e0226390. [CrossRef]

89. Rusu, E.; Necula, L.; Neagu, A.; Alecu, M.; Stan, C.; Albulescu, R.; Tanase, C. Current status of stem cell therapy: Opportunities and limitations. *Turk. J. Biol.* **2016**, *40*, 955–967. [CrossRef]

90. Kim, A.; Yu, H.Y.; Lim, J.; Ryu, C.M.; Kim, Y.H.; Heo, J.; Han, J.Y.; Lee, S.; Bae, Y.S.; Kim, J.Y.; et al. Improved efficacy and in vivo cellular properties of human embryonic stem cell derivative in a preclinical model of bladder pain syndrome. *Sci. Rep.* **2017**, *7*, 8872. [CrossRef]

91. Lee, S.W.; Ryu, C.M.; Shin, J.H.; Choi, D.; Kim, A.; Yu, H.Y.; Han, J.Y.; Lee, H.Y.; Lim, J.; Kim, Y.H.; et al. The Therapeutic Effect of Human Embryonic Stem Cell-Derived Multipotent Mesenchymal Stem Cells on Chemical-Induced Cystitis in Rats. *Int. Neurourol. J.* **2018**, *22*, S34–S45. [CrossRef]

92. Ryu, C.M.; Yu, H.Y.; Lee, H.Y.; Shin, J.H.; Lee, S.; Ju, H.; Paulson, B.; Lee, S.; Kim, S.; Lim, J.; et al. Longitudinal intravital imaging of transplanted mesenchymal stem cells elucidates their functional integration and therapeutic potency in an animal model of interstitial cystitis/bladder pain syndrome. *Theranostics* **2018**, *8*, 5610–5624. [CrossRef]

93. Huang, P.; He, Z.; Ji, S.; Sun, H.; Xiang, D.; Liu, C.; Hu, Y.; Wang, X.; Hui, L. Induction of functional hepatocyte-like cells from mouse fibroblasts by defined factors. *Nature* **2011**, *475*, 386–389. [CrossRef]

94. Inoue, Y.; Kishida, T.; Kotani, S.I.; Akiyoshi, M.; Taga, H.; Seki, M.; Ukimura, O.; Mazda, O. Direct conversion of fibroblasts into urothelial cells that may be recruited to regenerating mucosa of injured urinary bladder. *Sci. Rep.* **2019**, *9*, 13850. [CrossRef]

95. Wang, X.; Fan, L.; Yin, H.; Zhou, Y.; Tang, X.; Fei, X.; Tang, H.; Peng, J.; Ren, X.; Xue, Y.; et al. Protective effect of Aster tataricus extract on NLRP3-mediated pyroptosis of bladder urothelial cells . *J. Cell Mol. Med.* **2020**, *24*, 13336–13345. [CrossRef] [PubMed]

96. Wang, X.; Yin, H.; Fan, L.; Zhou, Y.Q.; Tang, X.L.; Fei, X.J.; Tang, H.L.; Peng, J.; Zhang, J.J.; Xue, Y.; et al. Shionone alleviates NLRP3 inflammasome mediated pyroptosis in interstitial cystitis injury. *Int. Immunopharmacol.* **2021**, *90*, 107132. [CrossRef] [PubMed]

97. Liu, Y.C.; Lee, W.T.; Liang, C.C.; Lo, T.S.; Hsieh, W.C.; Lin, Y.H. Beneficial effect of Bletilla striata extract solution on zymosan-induced interstitial cystitis in rat. *Neurourol. Urodyn.* **2021**, *40*, 763–770. [CrossRef]

98. Nassir, A.M.; Ibrahim, I.A.A.; Afify, M.A.; ElSawy, N.A.; Imam, M.T.; Shaheen, M.H.; Basyuni, M.A.; Bader, A.; Azhar, R.A.; Shahzad, N. Olea europaea subsp. Cuspidata and Juniperus procera Hydroalcoholic Leaves' Extracts Modulate Stress Hormones in Stress-Induced Cystitis in Rats. *Urol. Sci.* **2019**, *30*, 151–156. [CrossRef]

99. Li, W.; Yang, F.; Zhan, H.; Liu, B.; Cai, J.; Luo, Y.; Zhou, X. Houttuynia cordata Extract Ameliorates Bladder Damage and Improves Bladder Symptoms via Anti-Inflammatory Effect in Rats with Interstitial Cystitis. *Evid. Based. Complement. Alternat. Med.* **2020**, *2020*, 9026901. [CrossRef] [PubMed]

100. Bazi, T.; Hajj-Hussein, I.A.; Awwad, J.; Shams, A.; Hijaz, M.; Jurjus, A. A modulating effect of epigallocatechin gallate (EGCG), a tea catechin, on the bladder of rats exposed to water avoidance stress. *Neurourol. Urodyn.* **2013**, *32*, 287–292. [CrossRef] [PubMed]

101. Liu, M.; Xu, Y.F.; Feng, Y.; Yang, F.Q.; Luo, J.; Zhai, W.; Che, J.P.; Wang, G.C.; Zheng, J.H. Epigallocatechin gallate attenuates interstitial cystitis in human bladder urothelium cells by modulating purinergic receptors. *J. Surg. Res.* **2013**, *183*, 397–404. [CrossRef]

102. Ostardo, E.; Impellizzeri, D.; Cervigni, M.; Porru, D.; Sommariva, M.; Cordaro, M.; Siracusa, R.; Fusco, R.; Gugliandolo, E.; Crupi, R.; et al. Adelmidrol plus sodium hyaluronate in IC/BPS or conditions associated to chronic urothelial inflammation. A translational study. *Pharmacol. Res.* **2018**, *134*, 16–30. [CrossRef]

103. Luo, J.; Yang, C.; Luo, X., Yang, Y.; Li, J.; Song, B.; Zhao, J.; Li, L. Chlorogenic acid attenuates cyclophosphamide-induced rat interstitial cystitis. *Life Sci.* **2020**, *254*, 117590. [CrossRef] [PubMed]

104. Shih, H.J.; Chang, C.Y.; Lai, C.H.; Huang, C.J. Therapeutic effect of modulating the NLRP3-regulated transforming growth factor-β signaling pathway on interstitial cystitis/bladder pain syndrome. *Biomed. Pharmacother.* **2021**, *138*, 111522. [CrossRef]

105. d'Agostino, M.C.; Craig, K.; Tibalt, E.; Respizzi, S. Shock wave as biological therapeutic tool: From mechanical stimulation to recovery and healing, through mechanotransduction. *Int. J. Surg.* **2015**, *24*, 147–153. [CrossRef] [PubMed]

106. Wang, C.J. An overview of shock wave therapy in musculoskeletal disorders. *Chang. Gung. Med. J.* **2003**, *26*, 220–232. [PubMed]

107. Chen, Y.T.; Yang, C.C.; Sun, C.K.; Chiang, H.J.; Chen, Y.L.; Sung, P.H.; Zhen, Y.Y.; Huang, T.H.; Chang, C.L.; Chen, H.H.; et al. Extracorporeal shock wave therapy ameliorates cyclophosphamide-induced rat acute interstitial cystitis though inhibiting infammation and oxidative stress-in vitro and in vivo experiment studies. *Am. J. Transl. Res.* **2014**, *6*, 631–648.

108. Wang, H.J.; Lee, W.C.; Tyagi, P.; Huang, C.C.; Chuang, Y.C. Effects of low energy shock wave therapy on inflammatory moleculars, bladder pain, and bladder function in a rat cystitis model. *Neurourol. Urodyn.* **2017**, *36*, 1440–1447. [CrossRef]

109. Li, H.; Zhang, Z.; Peng, J.; Xin, Z.; Li, M.; Yang, B.; Fang, D.; Tang, Y.; Guo, Y. Treatment with low-energy shock wave alleviates pain in an animal model of uroplakin 3A-induced autoimmune interstitial cystitis/painful bladder syndrome. *Investig. Clin. Urol.* **2019**, *60*, 359–366. [CrossRef]

110. Chuang, Y.C.; Huang, T.L.; Tyagi, P.; Huang, C.C. Urodynamic and Immunohistochemical Evaluation of Intravesical Botulinum Toxin A Delivery Using Low Energy Shock Waves. *J. Urol.* **2016**, *196*, 599–608. [CrossRef] [PubMed]

111. Nageib, M.; Zahran, M.H.; El-Hefnawy, A.S.; Barakat, N.; Awadalla, A.; Aamer, H.G.; Khater, S.; Shokeir, A.A. Low energy shock wave-delivered intravesical botulinum neurotoxin-A potentiates antioxidant genes and inhibits proinflammatory cytokines in rat model of overactive bladder. *Neurourol. Urodyn.* **2020**, *39*, 2447–2454. [CrossRef] [PubMed]

112. Rayegani, S.M.; Razzaghi, M.R.; Raeissadat, S.A.; Allameh, F.; Eliaspour, D.; Abedi, A.R.; Javadi, A.; Rahavian, A.H. Extracorporeal Shockwave Therapy Combined with Drug Therapy in Chronic Pelvic Pain Syndrome: A Randomized Clinical Trial. *Urol. J.* **2020**, *17*, 185–191. [CrossRef] [PubMed]

113. Skaudickas, D.; Telksnys, T.; Veikutis, V.; Aniulis, P.; Jievaltas, M. Extracorporeal shock wave therapy for the treatment of chronic pelvic pain syndrome. *Open Med.* **2020**, *15*, 580–585. [CrossRef] [PubMed]

114. Chuang, Y.C.; Meng, E.; Chancellor, M.; Kuo, H.C. Pain reduction realized with extracorporeal shock wave therapy for the treatment of symptoms associated with interstitial cystitis/bladder pain syndrome-A prospective, multicenter, randomized, double-blind, placebo-controlled study. *Neurourol. Urodyn.* **2020**, *39*, 1505–1514. [CrossRef] [PubMed]

115. Saban, R.; D'Andrea, M.R.; Andrade-Gordon, P.; Derian, C.K.; Dozmorov, I.; Ihnat, M.A.; Hurst, R.E.; Davis, C.A.; Simpson, C.; Saban, M.R. Mandatory role of proteinase-activated receptor 1 in experimental bladder inflammation. *BMC. Physiol.* **2007**, *7*, 4. [CrossRef]

116. Monjotin, N.; Gillespie, J.; Farrié, M.; Le Grand, B.; Junquero, D.; Vergnolle, N. F16357, a novel protease-activated receptor 1 antagonist, improves urodynamic parameters in a rat model of interstitial cystitis. *Br. J. Pharmacol.* **2016**, 2224–2236. [CrossRef] [PubMed]

117. Vera, P.L.; Wolfe, T.E.; Braley, A.E.; Meyer-Siegler, K.L. Thrombin Induces Macrophage Migration Inhibitory Factor Release and Upregulation in Urothelium: A Possible Contribution to Bladder Inflammation. *PLoS ONE* **2010**, *5*, e15904. [CrossRef]

118. Kouzoukas, D.E.; Meyer-Siegler, K.L.; Ma, F.; Westlund, K.N.; Hunt, D.E.; Vera, P.L. Macrophage Migration Inhibitory Factor Mediates PAR-Induced Bladder Pain. *PLoS ONE* **2015**, *10*, e0127628. [CrossRef] [PubMed]

119. Ma, F.; Kouzoukas, D.E.; Meyer-Siegler, K.L.; Hunt, D.E.; Leng, L.; Bucala, R.; Vera, P.L. MIF mediates bladder pain, not inflammation, in cyclophosphamide cystitis. *Cytokine X.* **2019**, *1*, 100003. [CrossRef]

120. Sims, G.P.; Rowe, D.C.; Rietdijk, S.T.; Herbst, R.; Coyle, A.J. HMGB1 and RAGE in inflammation and cancer. *Annu. Rev. Immunol.* **2010**, *28*, 367–388. [CrossRef] [PubMed]

121. Tanaka, J.; Yamaguchi, K.; Ishikura, H.; Tsubota, M.; Sekiguchi, F.; Seki, Y.; Tsujiuchi, T.; Murai, A.; Umemura, T.; Kawabata, A. Bladder pain relief by HMGB1 neutralization and soluble thrombomodulin in mice with cyclophosphamide-induced cystitis. *Neuropharmacology* **2014**, *79*, 112–118. [CrossRef]

122. Irie, Y.; Tsubota, M.; Maeda, M.; Hiramoto, S.; Sekiguchi, F.; Ishikura, H.; Wake, H.; Nishibori, M.; Kawabata, A. HMGB1 and its membrane receptors as therapeutic targets in an intravesical substance P-induced bladder pain syndrome mouse model. *J. Pharmacol. Sci.* **2020**, *143*, 112–116. [CrossRef]

123. Kouzoukas, D.E.; Ma, F.; Meyer-Siegler, K.L.; Westlund, K.N.; Hunt, D.E.; Vera, P.L. Protease-Activated Receptor 4 Induces Bladder Pain through High Mobility Group Box-1. *PLoS ONE* **2016**, *11*, e0152055. [CrossRef]

124. Hiramoto, S.; Tsubota, M.; Yamaguchi, K.; Okazaki, K.; Sakaegi, A.; Toriyama, Y.; Tanaka, J.; Sekiguchi, F.; Ishikura, H.; Wake, H.; et al. Cystitis-Related Bladder Pain Involves ATP-Dependent HMGB1 Release from Macrophages and Its Downstream $H_2S/Ca_v3.2$ Signaling in Mice. *Cells* **2020**, *9*, 1748. [CrossRef]

125. Martins, J.P.; Silva, R.B.; Coutinho-Silva, R.; Takiya, C.M.; Battastini, A.M.; Morrone, F.B.; Campos, M.M. The role of P2X7 purinergic receptors in inflammatory and nociceptive changes accompanying cyclophosphamide-induced haemorrhagic cystitis in mice. *Br. J. Pharmacol.* **2012**, *165*, 183–196. [CrossRef] [PubMed]

126. Aronsson, P.; Johnsson, M.; Vesela, R.; Winder, M.; Tobin, G. Adenosine receptor antagonism suppresses functional and histological inflammatory changes in the rat urinary bladder. *Auton. Neurosci.* **2012**, *171*, 49–57. [CrossRef] [PubMed]

127. Beckel, J.M.; Daugherty, S.L.; Tyagi, P.; Wolf-Johnston, A.S.; Birder, L.A.; Mitchell, C.H.; de Groat, W.C. Pannexin 1 channels mediate the release of ATP into the lumen of the rat urinary bladder. *J. Physiol.* **2015**, *593*, 1857–1871. [CrossRef] [PubMed]

128. Yang, Y.; Zhang, H.; Lu, Q.; Liu, X.; Fan, Y.; Zhu, J.; Sun, B.; Zhao, J.; Dong, X.; Li, L. Suppression of adenosine A(2a) receptors alleviates bladder overactivity and hyperalgesia in cyclophosphamide-induced cystitis by inhibiting TRPV1. *Biochem. Pharmacol.* **2021**, *183*, 114340. [CrossRef] [PubMed]

129. Ko, I.G.; Jin, J.J.; Hwang, L.; Kim, S.H.; Kim, C.J.; Won, K.Y.; Na, Y.G.; Kim, K.H.; Kim, S.J. Adenosine A(2A) Receptor Agonist Polydeoxyribonucleotide Alleviates Interstitial Cystitis-Induced Voiding Dysfunction by Suppressing Inflammation and Apoptosis in Rats. *J. Inflamm. Res.* **2021**, *14*, 367–378. [CrossRef]

130. Kim, S.-E.; Ko, I.-G.; Jin, J.-J.; Hwang, L.; Kim, C.-J.; Kim, S.-H.; Han, J.-H.; Jeon, J.W. Polydeoxyribonucleotide Exerts Therapeutic Effect by Increasing VEGF and Inhibiting Inflammatory Cytokines in Ischemic Colitis Rats. *BioMed. Res. Int.* **2020**, *2020*, 2169083. [CrossRef]

131. Ko, I.-G.; Kim, S.-E.; Jin, J.-J.; Hwang, L.; Ji, E.-S.; Kim, C.-J.; Han, J.-H.; Hong, I.T.; Kwak, M.S.; Yoon, J.Y.; et al. Combination therapy with polydeoxyribonucleotide and proton pump inhibitor enhances therapeutic effectiveness for gastric ulcer in rats. *Life Sci.* **2018**, *203*, 12–19. [CrossRef]

132. Kho, J.H.; Ko, I.G.; Jin, J.J.; Hwang, L.; Kim, S.H.; Chung, J.Y.; Hwang, T.J.; Han, J.H. Polydeoxyribonucleotide Ameliorates Inflammation and Apoptosis in Achilles Tendon-Injury Rats. *Int. Neurourol. J.* **2020**, *24*, 79–87. [CrossRef] [PubMed]

133. Franken, J.; Uvin, P.; De Ridder, D.; Voets, T. TRP channels in lower urinary tract dysfunction. *Br. J. Pharmacol.* **2014**, *171*, 2537–2551. [CrossRef] [PubMed]

134. Deruyver, Y.; Voets, T.; De Ridder, D.; Everaerts, W. Transient receptor potential channel modulators as pharmacological treatments for lower urinary tract symptoms (LUTS): Myth or reality? *BJU Int.* **2015**, *115*, 686–697. [CrossRef]

135. Merrill, L.; Girard, B.M.; May, V.; Vizzard, M.A. Transcriptional and Translational Plasticity in Rodent Urinary Bladder TRP Channels with Urinary Bladder Inflammation, Bladder Dysfunction, or Postnatal Maturation. *J. Mol. Neurosci.* **2012**, *48*, 744–756. [CrossRef] [PubMed]
136. Merrill, L.; Vizzard, M.A. Intravesical TRPV4 blockade reduces repeated variate stress-induced bladder dysfunction by increasing bladder capacity and decreasing voiding frequency in male rats. *Am. J. Physiol. Regul. Integr. Comp. Physiol.* **2014**, *307*, R471–R480. [CrossRef]
137. Kawasaki, S.; Soga, M.; Sakurai, Y.; Nanchi, I.; Yamamoto, M.; Imai, S.; Takahashi, T.; Tsuno, N.; Asaki, T.; Morioka, Y.; et al. Selective blockade of transient receptor potential vanilloid 4 reduces cyclophosphamide-induced bladder pain in mice. *Eur. J. Pharmacol.* **2021**, 174040. [CrossRef] [PubMed]
138. DeBerry, J.J.; Saloman, J.L.; Dragoo, B.K.; Albers, K.M.; Davis, B.M. Artemin Immunotherapy Is Effective in Preventing and Reversing Cystitis-Induced Bladder Hyperalgesia via TRPA1 Regulation. *J. Pain* **2015**, *16*, 628–636. [CrossRef]
139. Braicu, C.; Cojocneanu-Petric, R.; Chira, S.; Truta, A.; Floares, A.; Petrut, B.; Achimas-Cadariu, P.; Berindan-Neagoe, I. Clinical and pathological implications of miRNA in bladder cancer. *Int. J. Nanomed.* **2015**, *10*, 791–800. [CrossRef] [PubMed]
140. Gheinani, A.H.; Kiss, B.; Moltzahn, F.; Keller, I.; Bruggmann, R.; Rehrauer, H.; Fournier, C.A.; Burkhard, F.C.; Monastyrskaya, K. Characterization of miRNA-regulated networks, hubs of signaling, and biomarkers in obstruction-induced bladder dysfunction. *JCI Insight* **2017**, *2*, e89560. [CrossRef]
141. Fırat, E.; Aybek, Z.; Akgün, Ş.; Küçüker, K.; Akça, H.; Aybek, H. Exploring biomarkers in the overactive bladder: Alterations in miRNA levels of a panel of genes in patients with OAB. *Neurourol. Urodyn.* **2019**, *38*, 1571–1578. [CrossRef]
142. Gheinani, A.H.; Burkhard, F.C.; Monastyrskaya, K. Deciphering microRNA code in pain and inflammation: Lessons from bladder pain syndrome. *Cell Mol. Life Sci.* **2013**, *70*, 3773–3789. [CrossRef] [PubMed]
143. Arai, T.; Fuse, M.; Goto, Y.; Kaga, K.; Kurozumi, A.; Yamada, Y.; Sugawara, S.; Okato, A.; Ichikawa, T.; Yamanishi, T.; et al. Molecular pathogenesis of interstitial cystitis based on microRNA expression signature: miR-320 family-regulated molecular pathways and targets. *J. Hum. Genet.* **2018**, *63*, 543–554. [CrossRef] [PubMed]
144. Song, Y.J.; Cao, J.Y.; Jin, Z.; Hu, W.G.; Wu, R.H.; Tian, L.H.; Yang, B.; Wang, J.; Xiao, Y.; Huang, C.B. Inhibition of microRNA-132 attenuates inflammatory response and detrusor fibrosis in rats with interstitial cystitis via the JAK-STAT signaling pathway. *J. Cell Biochem.* **2019**, *120*, 9147–9158. [CrossRef] [PubMed]
145. Hou, Y.; Li, H.; Huo, W. MicroRNA-495 alleviates ulcerative interstitial cystitis via inactivating the JAK-STAT signaling pathway by inhibiting JAK3. *Int. Urogynecol. J.* **2021**, *32*, 1253–1263. [CrossRef]
146. Hedlund, P.; Gratzke, C. The endocannabinoid system—A target for the treatment of LUTS? *Nat. Rev. Urol.* **2016**, *13*, 463–470. [CrossRef] [PubMed]
147. Tyagi, V.; Philips, B.J.; Su, R.; Smaldone, M.C.; Erickson, V.L.; Chancellor, M.B.; Yoshimura, N.; Tyagi, P. Differential expression of functional cannabinoid receptors in human bladder detrusor and urothelium. *J. Urol.* **2009**, *181*, 1932–1938. [CrossRef]
148. Hayn, M.H.; Ballesteros, I.; de Miguel, F.; Coyle, C.H.; Tyagi, S.; Yoshimura, N.; Chancellor, M.B.; Tyagi, P. Functional and immunohistochemical characterization of CB1 and CB2 receptors in rat bladder. *Urology* **2008**, *72*, 1174–1178. [CrossRef]
149. Tambaro, S.; Casu, M.A.; Mastinu, A.; Lazzari, P. Evaluation of selective cannabinoid CB(1) and CB(2) receptor agonists in a mouse model of lipopolysaccharide-induced interstitial cystitis. *Eur. J. Pharmacol.* **2014**, *729*, 67–74. [CrossRef]
150. Wang, Z.Y.; Wang, P.; Bjorling, D.E. Activation of cannabinoid receptor 2 inhibits experimental cystitis. *Am. J. Physiol. Regul. Integr. Comp. Physiol.* **2013**, *304*, R846–R853. [CrossRef] [PubMed]
151. Wang, Z.Y.; Wang, P.; Bjorling, D.E. Treatment with a cannabinoid receptor 2 agonist decreases severity of established cystitis. *J. Urol.* **2014**, *191*, 1153–1158. [CrossRef] [PubMed]
152. Liu, Q.; Wu, Z.; Liu, Y.; Chen, L.; Zhao, H.; Guo, H.; Zhu, K.; Wang, W.; Chen, S.; Zhou, N.; et al. Cannabinoid receptor 2 activation decreases severity of cyclophosphamide-induced cystitis via regulating autophagy. *Neurourol. Urodyn.* **2020**, *39*, 158–169. [CrossRef] [PubMed]
153. Berger, G.; Arora, N.; Burkovskiy, I.; Xia, Y.F.; Chinnadurai, A.; Westhofen, R.; Hagn, G.; Cox, A.; Kelly, M.; Zhou, J.; et al. Experimental Cannabinoid 2 Receptor Activation by Phyto-Derived and Synthetic Cannabinoid Ligands in LPS-Induced Interstitial Cystitis in Mice. *Molecules* **2019**, *24*, 4239. [CrossRef]
154. Pessina, F.; Capasso, R.; Borrelli, F.; Aveta, T.; Buono, L.; Valacchi, G.; Fiorenzani, P.; Di Marzo, V.; Orlando, P.; Izzo, A.A. Protective effect of palmitoylethanolamide in a rat model of cystitis. *J. Urol.* **2015**, *193*, 1401–1408. [CrossRef] [PubMed]
155. Muller, C.; Morales, P.; Reggio, P.H. Cannabinoid Ligands Targeting TRP Channels. *Front. Mol. Neurosci.* **2018**, *11*, 487. [CrossRef] [PubMed]

 *biomedicines* 

Article

Carollia perspicillata: A Small Bat with Tremendous Translational Potential for Studies of Brain Aging and Neurodegeneration

Mark Stewart [1,2], Timothy Morello [1], Richard Kollmar [3,4] and Rena Orman [1,*]

[1] Departments of Physiology & Pharmacology, SUNY Downstate Health Sciences University, Brooklyn, NY 11203, USA; mark.stewart@downstate.edu (M.S.); timothy.morello@downstate.edu (T.M.)
[2] Departments of Neurology, SUNY Downstate Health Sciences University, Brooklyn, NY 11203, USA
[3] Departments of Cell Biology, SUNY Downstate Health Sciences University, Brooklyn, NY 11203, USA; richard.kollmar@downstate.edu
[4] Departments of Otolaryngology, SUNY Downstate Health Sciences University, Brooklyn, NY 11203, USA
[*] Correspondence: rena.orman@downstate.edu

Abstract: As the average human lifespan lengthens, the impact of neurodegenerative disease increases, both on the individual suffering neurodegeneration and on the community that supports those individuals. Studies aimed at understanding the mechanisms of neurodegeneration have relied heavily on observational studies of humans and experimental studies in animals, such as mice, in which aspects of brain structure and function can be manipulated to target mechanistic steps. An animal model whose brain is structurally closer to the human brain, that lives much longer than rodents, and whose husbandry is practical may be valuable for mechanistic studies that cannot readily be conducted in rodents. To demonstrate that the long-lived Seba's short-tailed fruit bat, *Carollia perspicillata*, may fit this role, we used immunohistochemical labeling for NeuN and three calcium-binding proteins, calretinin, parvalbumin, and calbindin, to define hippocampal formation anatomy. Our findings demonstrate patterns of principal neuron organization that resemble primate and human hippocampal formation and patterns of calcium-binding protein distribution that help to define subregional boundaries. Importantly, we present evidence for a clear prosubiculum in the bat brain that resembles primate prosubiculum. Based on the similarities between bat and human hippocampal formation anatomy, we suggest that *Carollia* has unique advantages for the study of brain aging and neurodegeneration. A captive colony of *Carollia* allows age tracking, diet and environment control, pharmacological manipulation, and access to behavioral, physiological, anatomical, and molecular evaluation.

Keywords: prosubiculum; calbindin; calretinin; parvalbumin; fruit bat

Citation: Stewart, M.; Morello, T.; Kollmar, R.; Orman, R. *Carollia perspicillata*: A Small Bat with Tremendous Translational Potential for Studies of Brain Aging and Neurodegeneration. *Biomedicines* **2021**, *9*, 1454. https://doi.org/10.3390/biomedicines9101454

Academic Editor: Martina Perše

Received: 15 September 2021
Accepted: 10 October 2021
Published: 13 October 2021

Publisher's Note: MDPI stays neutral with regard to jurisdictional claims in published maps and institutional affiliations.

1. Introduction

As the world's population ages, the public health impact of neurodegenerative disease increases. Understanding the reasons for age-related neurodegeneration and discovering mechanisms by which neurons are protected are important in addressing this growing crisis. Past studies of brain aging can be broadly grouped into two categories: (1) observational studies of anatomy, physiology, molecular biology, and behavior at different ages or stages of disease in a large variety of species, and (2) experimental studies, predominantly in mice, that evaluate the same variables but with direct manipulation of genes, proteins or pharmacology to simulate, accelerate or reverse neurodegeneration. Whereas this work has led to considerable progress [1,2], the numerous differences between mouse brain and human brain have limited the translation of experimental findings in mice for diagnostic or therapeutic tools for humans [3,4].

The hippocampal formation in humans and some rodent models is severely impacted by neurodegenerative disease (e.g., [5–12]). Age-related neuronal loss has been shown to impact inhibitory neuronal in multiple brain regions [13–17]. The calcium-binding proteins, parvalbumin, calbindin, and calretinin, which label inhibitory neurons in many brain structures [18–20], are valuable markers for age-related changes in inhibitory circuits.

There have been studies of aspects of hippocampal formation anatomy and physiology in bats, including animals from *Myotis*, *Rousettus*, and *Epomophorus* (e.g., [21–23]), but the hippocampal formation of *Carollia* has not been described in detail [24,25]. Here we present anatomical details of *Carollia* hippocampal formation using immunohistochemistry for NeuN, calretinin, parvalbumin, and calbindin, including evidence of a clear prosubiculum, a subregion that is otherwise best seen in the primate brain [26–29].

We propose the bat as a novel model for studies of the neurobiology of aging and neurodegenerative disease. Bats demonstrate extreme longevity [30–34], making them an excellent animal model of aging. The short-tailed fruit bat, *Carollia perspicillata*, lives up to 13 years in captivity [35–37], remains reproductively active throughout life (females) [38,39], and its neuroanatomy resembles that of primates more closely than the neuroanatomy of rodents does (e.g., [24,25,28,40–43]). Importantly, *Carollia* are small in body size and their husbandry has been refined sufficiently that they are practical to maintain in captivity [35–37].

2. Methods

ANIMALS AND HUSBANDRY: As described previously [24,40,44], our use of animals conformed to the standards set forth in the NIH Guide for the Care and Use of Laboratory Animals [45] and was approved by the SUNY Downstate Institutional Animal Care and Use Committee.

Our colony, which has been in existence >30 years, consists of approximately 150 members with active breeding that can be controlled and timed. Our basic husbandry of *Carollia* consists of a caging system with open feeding and closed roosting sections of 20 ft^3 and 16.7 ft^3, respectively [35–37]. We keep animals in three configurations: male animals only, female animals only, and mixed gender. Ten to twenty animals are housed in each cage with continuous access to water. The room is kept on a 12:12 light:dark cycle. Animals receive a blended fruit mix that contains apricot nectar, peaches, monkey chow, supplemental vitamins and minerals, and emulsified corn oil [46]. The diet is periodically enriched with pieces of whole fruits. The daily per diem housing cost for bats at our institution is identical to the cost for mice. The food costs are somewhat higher because of the cost differential between our blended fruit mix ingredients and standard rodent chow, but the costs of our food resemble the cost of specialized rodent diets (approximately \$50–60/animal/year).

EXPERIMENTAL DESIGN: We collected 4 sagittal and 2 coronal series of brain sections using NeuN as a general neuronal marker from 6 adult bats (4 female and 2 male). Animals ranged in age from 4–12 months of age and had a body mass range of 15–25 g. On serially adjacent or near-adjacent (within 75 μm) sections in 2 brains, NeuN labeling was paired with labeling for 1 of 3 calcium-binding proteins—calretinin, parvalbumin, or calbindin.

IMMUNOHISTOCHEMISTRY: Each animal was anesthetized with urethane (0.02–0.04 mL 20% wt/vol solution in water per animal given subcutaneously) and perfused with cold phosphate-buffered saline (PBS, 0.1 M, pH 7.4: NaCl 1.37 × 10^{-1} M, KCl 2.68 × 10^{-3} M, Na$_2$HPO$_4$ 1.014 × 10^{-2} M, KH$_2$PO$_4$ 1.76 × 10^{-3} M), followed by cold 4% wt/vol paraformaldehyde in PBS. After a variable post-fixation period depending upon the planned staining, brains were washed multiple times and cryoprotected with 30% sucrose. Sagittal or coronal sections were cut at 35 μm thickness with a freezing microtome (Thermo Fisher Scientific Microm HM 430, Waltham, MA, USA). Sections were collected into wells containing PBS (pH 7.2, NaCl 1.54 × 10−1 M, Na2HPO4 7.68 × 10−3 M, NaH2PO4 9.08 × 10−3 M), and stored at 4 °C until processed. Sections were incubated with primary antibodies overnight (16–20 h) at 4 °C. The primary antibodies included mono- and polyclonal antibodies against NeuN, parvalbumin, calbindin, and calretinin (see Table 1 of antibodies for sources and RRID numbers of primary and secondary antibodies). After washing with PBS, secondary antibodies, diluted in blocking buffer, were added and incubated in the dark for 2 h.

After washing, sections were counterstained with 4′,6-diamidino-2-phenylindole (DAPI) (Invitrogen, D357) for 20 min and mounted with ProLong Diamond Antifade Mountant (Invitrogen, P36961).

Table 1. Table of Antibodies.

Target	Species	Immunogen	Clonality	Isotype	Working Dilution	Source (Cat. #)	RRID
		PRIMARY ANTIBODIES					
Calbindin	Rabbit	recombinant rat calbindin D-28k	Poly	(antiserum)	1/3000	Swant (CB 38)	AB_10000340
Calretinin	Rabbit	recombinant human calretinin containing a 6-his tag at the N-terminal	Poly	(antiserum)	1/3000	Swant (CR7697)	AB_2619710
NeuN	Mouse	purified cell nuclei from mouse brain	Mono	IgG1	1/2000	Millipore (MAB377)	AB_2298772
Parvalbumin	Rabbit	recombinant rat parvalbumin	Poly	(antiserum)	1/3000	Swant (PV 27)	AB_2631173
		SECONDARY ANTIBODIES					
Mouse IgG	Goat		Poly	IgG	1/500	Jackson (115-545-003)	AB_2338840
Rabbit IgG	Goat		Poly	IgG	1/500	Jackson (111-295-003)	AB_2338022

There was no labeling in the absence of primary antibody. Our controls for primary specificity include (a) matches between the sequenced bat genomes [47] and the sequences of antigens used for antibody production (published by supplier), (b) detection of the correct molecular weight band on Western blots of brain tissue, and (c) similar labeling patterns in bats as have been published in other species.

IMAGING: Samples were imaged with an Axio Observer 7/LSM 800 inverted compound microscope and Zen Blue version 2.3 software (Carl Zeiss Microscopy, Thornwood, NY, USA). Sections were imaged with 5×/0.16 and 10×/0.45 plan-apochromatic objectives in widefield mode. Details were imaged with a 63×/1.4 plan-apochromatic objective in confocal mode. Tiled 63× overview images were taken with tile sizes at 512 × 512, pixel dwell at 1 µs, 2 times line averaging, and the pinhole set to 1 Airy unit. Laser intensity and master gain were adjusted for optimal contrast per antibody.

3. Results

The brain of *Carollia* resembles the mouse brain in its overall size. A large cerebellum is located behind the forebrain [25], and the rostro-caudal extent of the adult forebrain is about 10 mm and its largest dimension from side to side is about 9 mm.

Previously, we published serial images of *Carollia* brains that were sectioned in coronal [25] and sagittal [24] planes, and gave basic identification of hippocampal formation regions in each instance. An important feature that can be extracted from these series is that the main axis of the dorsal hippocampal formation in *Carollia* forms an angle that is approximately 30 degrees off the coronal plane (running from rostro-medial to caudo-lateral) and approximately 60 degrees off the sagittal plane. Neither plane of section, therefore, cuts an ideal transverse section, but the sagittal plane is closer to a transverse plane of section for dorsal hippocampal formation in *Carollia* brain.

Two other advantages of the sagittal plane of section warrant comment. The first is that the greater roundness of the *Carollia* brain [24,25] compared with rodent brain [41,42] can make replicating coronal, horizontal, or other oblique (e.g., perfectly transverse for the main axis of the dorsal hippocampal formation) planes of section challenging, whereas there is never a question about reproducing the sagittal plane of section. Second, as described below, the modest tilt of the sagittal plane of section off what would be an ideal transverse plane results in a favorable elongation of the hippocampal cross-section that facilitates subregional definition.

Figure 1 illustrates NeuN labeling of sections of dorsal hippocampal formation. In each, NeuN was paired with one of three calcium binding proteins—calretinin, parvalbumin, or calbindin. The basic hippocampal subregional identifications can be made and are illustrated using the changes in density of NeuN labeled neurons, layer thickness, and the change in appearance from 3-layered (dentate gyrus, CA3, CA1, subiculum) to 6-layered cortex (presubiculum, entorhinal cortex, neocortex). The labeling patterns for the three calcium binding proteins demonstrate the different distributions of labeled cells and processes in each subregion, but also reinforce subregional boundary identification.

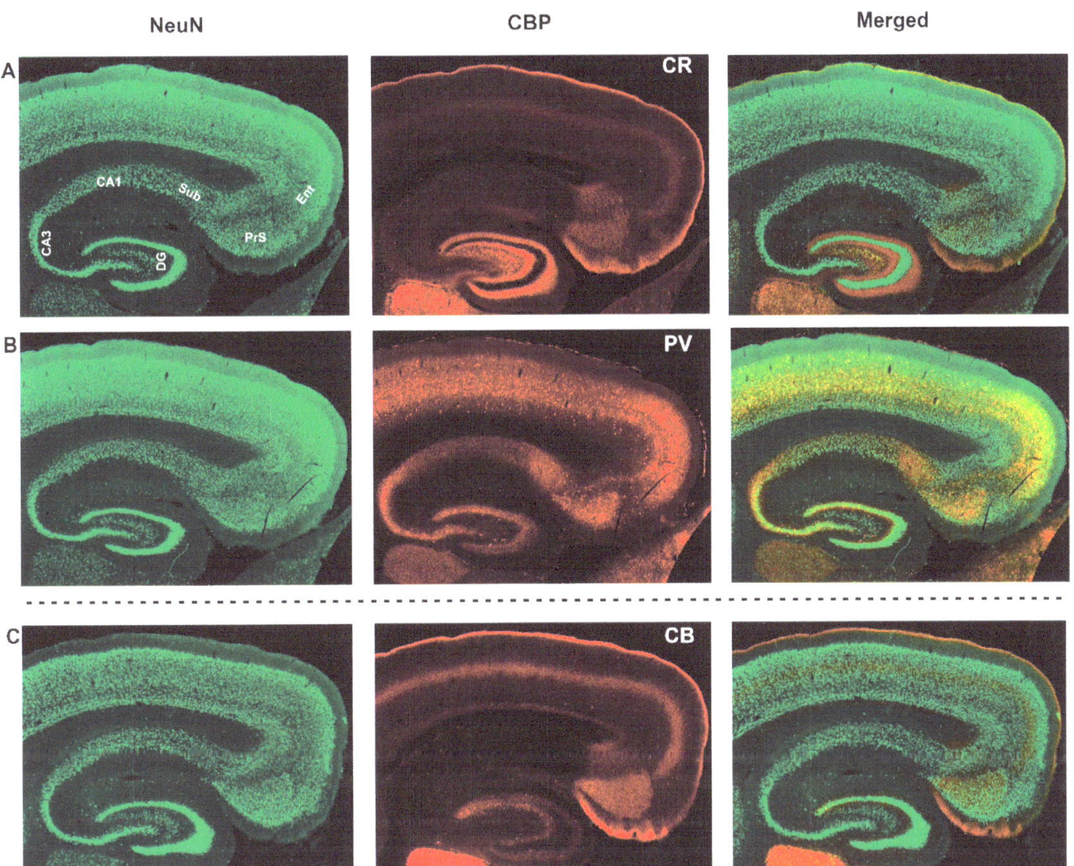

Figure 1. Hippocampal formation in the brain of *Carollia perspicillata* labeled with NeuN, calretinin, parvalbumin, and calbindin. Widefield images of sagittal sections of Carollia brain showing three pairs of labeling (by row) with NeuN and a calcium binding protein (CBP). Top row (**A**): NeuN with calretinin. Subregional labels are indicated on the NeuN only section (top left). Calretinin labeling alone is shown in the top center (CBP/CR) and the merged image is shown at the top right. Middle row (**B**): NeuN and parvalbumin (PV) labeling from the adjacent section. Bottom row (**C**): NeuN and calbindin (CB) labeling from a near-adjacent section from the same brain. Dotted-line separator indicates that the bottom row is not adjacent to the middle row. The distributions of the calcium binding proteins help to delineate regions and to highlight differences in the distributions of inhibitory neurons and processes. Additional abbreviations: DG—dentate gyrus, CA3—part of regio inferior of cornu Ammonis, CA1—regio superior of cornu Ammonis, Sub—subiculum, PrS—presubiculum, Ent—entorhinal cortex.

Calretinin and closely related calbindin clearly label dentate gyrus, hilus, and help to differentiate the upper and lower blades of dentate (especially calbindin). Calretinin and calbindin also strongly label the presubiculum and entorhinal cortex. Both markers label cortical layer I, and calbindin labeling is more pronounced in the deeper layers, which helps to distinguish parasubiculum in between presubiculum and entorhinal cortex.

Parvalbumin is arguably the most valuable label in defining dorsal hippocampal formation subregional boundaries in *Carollia* brain (Figure 2). The density of parvalbumin-immunoreactive fibers is evidently heavier in CA3 compared with CA1 with pyramidal cells (appearing as "holes" in the parvalbumin image) enwrapped by labeled processes. CA2 can be distinguished from its neighbor, CA3, on the basis of increased spread of parvalbumin-labeled processes into the apical dendritic zone, and the change in packing of NeuN-labeled cells. CA2 can be distinguished from CA1 on the basis of the much lower density of parvalbumin-labeled processes and the distinctly looser packing of CA1 neurons.

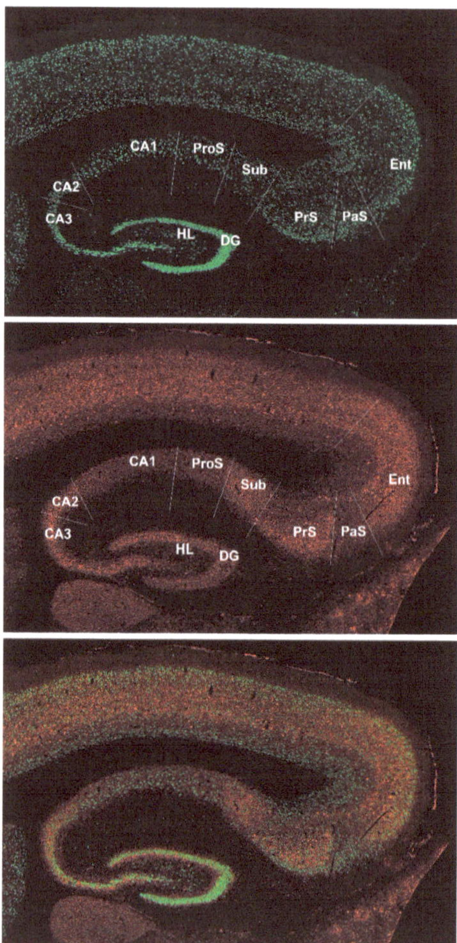

Figure 2. Hippocampal formation in the brain of *Carollia perspicillata* labeled with NeuN and parvalbumin. Tiled confocal 63× sagittal composite section of Carollia brain labeled with NeuN (top panel—green channel) and parvalbumin (middle panel—red channel) to show subregional boundaries in greater detail. A merged image is shown in the bottom panel. The subregional boundaries are described in the Results and are drawn on the figure as white lines. The sagittal plane of this section was determined to be 2.45 mm from the lateral edge and 2.07 mm from the midline. Additional abbreviations: DG—dentate gyrus, HL—dentate hilus, CA3—part of regio inferior of cornu Ammonis, CA2—part of regio inferior of cornu Ammonis, CA1—regio superior of cornu Ammonis, ProS—prosubiculum, Sub—subiculum, PrS—presubiculum, PaS—parasubiculum, Ent—entorhinal cortex.

Coming from the entorhinal cortex in toward Ammon's horn, parvalbumin labeling identifies the parasubiculum as a wedge of cortex in between entorhinal cortex and presubiculum with relatively lower level of parvalbumin labeling that distinguishes parasubiculum. The heavily labeled presubiculum abruptly transitions into the broad 3-layer cortex of subiculum (Figure 2).

Parvalbumin also reveals a prosubiculum in *Carollia* brain. The prosubiculum has a cell-layer breadth and packing density that resembles the subiculum (adjacent to the presubiculum) but contrasts with CA1. The differentiation of prosubiculum from subiculum can be made by parvalbumin labeling: the subiculum is much more heavily labeled than prosubiculum, and the separation of presubiculum from subiculum is obvious in the figure.

4. Discussion

The brain of *Carollia perspicillata* is remarkable for many reasons. Here, we illustrate the basic anatomy of the dorsal hippocampal formation. Key features are (1) the compact cell layer (along the alvear–pial axis) in area CA3 in contrast to the broader cell layer of area CA1 and (2) the presence of a clear prosubiculum, both of which are prominent features of primate brains, but not rodent brains. These features strongly support the argument that *Carollia* brain is closer in neuroanatomical features and organization to human brain than rodent brain is to human brain (see also [43,48]). The relative size, the dorso-ventral extent, and the location of the hippocampal formation in relation to the dorsal surface of the brain give *Carollia* hippocampal formation the experimental access advantages of rodent hippocampus. The long natural life span of *Carollia* (approximately 13 years in captivity) and the ability to maintain these animals in controlled conditions lead us to conclude that *Carollia* is a unique model for studies of many neurological issues including brain aging and neurodegeneration.

4.1. Prosubiculum in Bat Brain

Prosubiculum is not labeled in the majority of reports on rodent hippocampal formation structure, including our own (e.g., [41,42,49–54]). As such, specific functional attributes of prosubicular cells and the network connectivity of prosubiculum are not well understood. Prosubiculum has, however, been distinguished from subiculum in rodent brain using transcriptomics [55] and in situ hybridization data [56], and mouse and human may be more similar than they first appear [28,56].

Our data show a very clear separation of prosubiculum from both CA1 and subiculum and that prosubiculum in sagittal sections of *Carollia* brain is of sufficient size for studies of cell function and connectivity. Our boundaries compare to those reported with a host of other markers in rodent and primate brain [28]. Clear subregional identification is invaluable because subiculum is among the earliest structures to be impacted in Alzheimer's disease [7] and prosubiculum has been identified as a specific target for amyloid plaque deposition [29].

Due to the dorso-ventral orientation of the hippocampus in the bat, as opposed to the more rostro-caudal orientation of the hippocampus in the rodent, there is no plane of section in rat or mouse that can be used as a simple comparison to the bat. However, there are numerous examples in atlases, publications, and websites with sections from mouse and rat brain if one was interested in attempting such comparisons for overall structure or more specific comparisons of specific labels.

4.2. Carollia as a New Animal Model of Brain Aging and Neurodegeneration

The neuroanatomical features of *Carollia* hippocampal formation and its natural long life makes *Carollia* an important alternative model for studies of the neurobiology of aging, neurodegenerative disease, and normal hippocampal function.

It is critically important to answer the question: are bats subject to neurodegenerative diseases? Given that bats naturally live much longer (>10 years) than rats and mice that are commonly used as research models (typically up to 3 years old), there is much more time for cumulative evidence of neurodegenerative disease. This might make bats a valuable model for neuropathology. If, on the other hand, evidence of neurodegenerative disease in such aged animals is absent, bats' resistance to neurodegenerative changes will add to their appeal as a model for longevity.

A critical feature of *Carollia* in favor of its utility as an animal model is that these animals can be easily kept in captivity, which offers a controlled environment with the ability to control stress levels, diet and other community variables that can influence normal aging and the development of neurodegenerative disease. Whether bats represent a better animal model of neurodegenerative disease or the resistance to neurodegenerative disease is a focus of our work, and the answer will help to define the utility of the bat.

Author Contributions: M.S.—designed experiments, collected and analyzed data, prepared manuscript. T.M.—collected data, prepared manuscript. R.K.—analyzed data, prepared manuscript. R.O.—conceived the study, designed experiments, collected and analyzed data, prepared manuscript. All authors have read and agreed to the published version of the manuscript.

Funding: This study was funded in part by support from philanthropic contributions and the University.

Institutional Review Board Statement: The study was conducted according to the guidelines of the Declaration of Helsinki and was approved by the Institutional Animal Care and Use Committee (protocol number: 15-10450 and approval date: 21 November 2020).

Informed Consent Statement: Not applicable.

Data Availability Statement: Data images available upon appropriate request.

Acknowledgments: The authors wish to thank Inna Rozenberg for technical assistance and Olga Papirova of the SUNY Research Foundation for operational support.

Conflicts of Interest: The authors declare no conflict of interest.

References

1. Beach, T.G. A Review of Biomarkers for Neurodegenerative Disease: Will They Swing Us Across the Valley? *Neurol. Ther.* **2017**, *6*, 5–13. [CrossRef]
2. Ramachandran, A.K.; Das, S.; Joseph, A.; Shenoy, G.G.; Alex, A.T.; Mudgal, J. Neurodegenerative Pathways in Alzheimer's Disease: A Review. *Curr. Neuropharmacol.* **2021**, *19*, 679–692. [CrossRef]
3. Driscoll, I.; Sutherland, R.J. The aging hippocampus: Navigating between rat and human experiments. *Rev. Neurosci.* **2005**, *16*, 87–121. [CrossRef] [PubMed]
4. Stewart, M. And when I die … What time should I expect it? *J. Physiol.* **2021**, *599*, 1729–1730. [CrossRef] [PubMed]
5. Trujillo-Estrada, L.; Davila, J.C.; Sanchez-Mejias, E.; Sanchez-Varo, R.; Gomez-Arboledas, A.; Vizuete, M.; Vitorica, J.; Gutierrez, A. Early neuronal loss and axonal/presynaptic damage is associated with accelerated amyloid-beta accumulation in AbetaPP/PS1 Alzheimer's disease mice subiculum. *J. Alzheimers' Dis.* **2014**, *42*, 521–541. [CrossRef] [PubMed]
6. Angulo, S.L.; Orman, R.; Neymotin, S.A.; Liu, L.; Buitrago, L.; Cepeda-Prado, E.; Stefanov, D.; Lytton, W.W.; Stewart, M.; Small, S.A.; et al. Tau and amyloid-related pathologies in the entorhinal cortex have divergent effects in the hippocampal circuit. *Neurobiol. Dis.* **2017**, *108*, 261–276. [CrossRef] [PubMed]
7. Carlesimo, G.A.; Piras, F.; Orfei, M.D.; Iorio, M.; Caltagirone, C.; Spalletta, G. Atrophy of presubiculum and subiculum is the earliest hippocampal anatomical marker of Alzheimer's disease. *Alzheimers' Dement.* **2015**, *1*, 24–32. [CrossRef] [PubMed]
8. Ma, C.; Wang, G.Z.; Braak, H. Pathological changes of the retrosplenial cortex in senile dementia of Alzheimer type. *Chin. Med. J.* **1994**, *107*, 119–123.
9. Pengas, G.; Williams, G.B.; Acosta-Cabronero, J.; Ash, T.W.; Hong, Y.T.; Izquierdo-Garcia, D.; Fryer, T.D.; Hodges, J.R.; Nestor, P.J. The relationship of topographical memory performance to regional neurodegeneration in Alzheimer's disease. *Front. Aging Neurosci.* **2012**, *4*, 17. [CrossRef]
10. Nestor, P.J.; Fryer, T.D.; Ikeda, M.; Hodges, J.R. Retrosplenial cortex (BA 29/30) hypometabolism in mild cognitive impairment (prodromal Alzheimer's disease). *Eur. J. Neurosci.* **2003**, *18*, 2663–2667. [CrossRef]
11. Robertson, R.T.; Baratta, J.; Yu, J.; LaFerla, F.M. Amyloid-beta expression in retrosplenial cortex of triple transgenic mice: Relationship to cholinergic axonal afferents from medial septum. *Neuroscience* **2009**, *164*, 1334–1346. [CrossRef] [PubMed]
12. Fisher, E.M.C.; Bannerman, D.M. Mouse models of neurodegeneration: Know your question, know your mouse. *Sci. Transl. Med.* **2019**, *11*. [CrossRef] [PubMed]
13. Fuhrer, T.E.; Palpagama, T.H.; Waldvogel, H.J.; Synek, B.J.L.; Turner, C.; Faull, R.L.; Kwakowsky, A. Impaired expression of GABA transporters in the human Alzheimer's disease hippocampus, subiculum, entorhinal cortex and superior temporal gyrus. *Neuroscience* **2017**, *351*, 108–118. [CrossRef] [PubMed]

14. Kwakowsky, A.; Calvo-Flores Guzman, B.; Pandya, M.; Turner, C.; Waldvogel, H.J.; Faull, R.L. GABAA receptor subunit expression changes in the human Alzheimer's disease hippocampus, subiculum, entorhinal cortex and superior temporal gyrus. *J. Neurochem.* **2018**, *145*, 374–392. [CrossRef] [PubMed]

15. Mikkonen, M.; Alafuzoff, I.; Tapiola, T.; Soininen, H.; Miettinen, R. Subfield- and layer-specific changes in parvalbumin, calretinin and calbindin-D28K immunoreactivity in the entorhinal cortex in Alzheimer's disease. *Neuroscience* **1999**, *92*, 515–532. [CrossRef]

16. Ahn, J.H.; Hong, S.; Park, J.H.; Kim, I.H.; Cho, J.H.; Lee, T.K.; Lee, J.C.; Chen, B.H.; Shin, B.N.; Bae, E.J.; et al. Immunoreactivities of calbindinD28k, calretinin and parvalbumin in the somatosensory cortex of rodents during normal aging. *Mol. Med. Rep.* **2017**, *16*, 7191–7198. [CrossRef]

17. Freund, T.F.; Buzsaki, G. Interneurons of the hippocampus. *Hippocampus* **1996**, *6*, 347–470. [CrossRef]

18. Petilla Interneuron Nomenclature, G.; Ascoli, G.A.; Alonso-Nanclares, L.; Anderson, S.A.; Barrionuevo, G.; Benavides-Piccione, R.; Burkhalter, A.; Buzsaki, G.; Cauli, B.; Defelipe, J.; et al. Petilla terminology: Nomenclature of features of GABAergic interneurons of the cerebral cortex. *Nat. Rev. Neurosci.* **2008**, *9*, 557–568. [CrossRef]

19. DeFelipe, J.; Lopez-Cruz, P.L.; Benavides-Piccione, R.; Bielza, C.; Larranaga, P.; Anderson, S.; Burkhalter, A.; Cauli, B.; Fairen, A.; Feldmeyer, D.; et al. New insights into the classification and nomenclature of cortical GABAergic interneurons. *Nat. Rev. Neurosci.* **2013**, *14*, 202–216. [CrossRef]

20. Baimbridge, K.G.; Celio, M.R.; Rogers, J.H. Calcium-binding proteins in the nervous system. *Trends Neurosci.* **1992**, *15*, 303–308. [CrossRef]

21. Eliav, T.; Geva-Sagiv, M.; Yartsev, M.M.; Finkelstein, A.; Rubin, A.; Las, L.; Ulanovsky, N. Nonoscillatory Phase Coding and Synchronization in the Bat Hippocampal Formation. *Cell* **2018**, *175*, 1119–1130.e15. [CrossRef]

22. Gatome, C.W.; Mwangi, D.K.; Lipp, H.P.; Amrein, I. Hippocampal neurogenesis and cortical cellular plasticity in Wahlberg's epauletted fruit bat: A qualitative and quantitative study. *Brain Behav. Evol.* **2010**, *76*, 116–127. [CrossRef]

23. Cotter, J.R.; Laemle, L.K. Cholecystokinin (CCK)-like immunoreactivity in the brain of the little brown bat (*Myotis lucifugus*). *J. Hirnforsch* **1990**, *31*, 87–97.

24. Orman, R.; Kollmar, R.; Stewart, M. Claustrum of the short-tailed fruit bat, *Carollia perspicillata*: Alignment of cellular orientation and functional connectivity. *J. Comp. Neurol.* **2017**, *525*, 1459–1474. [CrossRef]

25. Scalia, F.; Rasweiler, J.J.; Scalia, J.; Orman, R.; Stewart, M. *Forebrain Atlas of the Short-Tailed Fruit Bat, Carollia perpicillata*; Springer: New York, NY, USA, 2013; p. 1.

26. Blatt, G.J.; Rosene, D.L. Organization of direct hippocampal efferent projections to the cerebral cortex of the rhesus monkey: Projections from CA1, prosubiculum, and subiculum to the temporal lobe. *J. Comp. Neurol.* **1998**, *392*, 92–114. [CrossRef]

27. Braak, H.; Del Tredici, K. From the Entorhinal Region via the Prosubiculum to the Dentate Fascia: Alzheimer Disease-Related Neurofibrillary Changes in the Temporal Allocortex. *J. Neuropathol. Exp. Neurol.* **2020**, *79*, 163–175. [CrossRef]

28. Ding, S.L. Comparative anatomy of the prosubiculum, subiculum, presubiculum, postsubiculum, and parasubiculum in human, monkey, and rodent. *J. Comp. Neurol.* **2013**, *521*, 4145–4162. [CrossRef] [PubMed]

29. Marshall, G.A.; Kaufer, D.I.; Lopez, O.L.; Rao, G.R.; Hamilton, R.L.; DeKosky, S.T. Right prosubiculum amyloid plaque density correlates with anosognosia in Alzheimer's disease. *J. Neurol. Neurosurg. Psychiatry* **2004**, *75*, 1396–1400. [CrossRef] [PubMed]

30. Podlutsky, A.J.; Khritankov, A.M.; Ovodov, N.D.; Austad, S.N. A new field record for bat longevity. *J. Gerontol. A Biol. Sci. Med. Sci.* **2005**, *60*, 1366–1368. [CrossRef] [PubMed]

31. Ball, H.C.; Levari-Shariati, S.; Cooper, L.N.; Aliani, M. Comparative metabolomics of aging in a long-lived bat: Insights into the physiology of extreme longevity. *PLoS ONE* **2018**, *13*, e0196154. [CrossRef]

32. Brunet-Rossinni, A.K. Reduced free-radical production and extreme longevity in the little brown bat (Myotis lucifugus) versus two non-flying mammals. *Mech. Ageing Dev.* **2004**, *125*, 11–20. [CrossRef]

33. Huang, Z.; Jebb, D.; Teeling, E.C. Blood miRNomes and transcriptomes reveal novel longevity mechanisms in the long-lived bat, Myotis myotis. *BMC Genom.* **2016**, *17*, 906. [CrossRef]

34. Seim, I.; Fang, X.; Xiong, Z.; Lobanov, A.V.; Huang, Z.; Ma, S.; Feng, Y.; Turanov, A.A.; Zhu, Y.; Lenz, T.L.; et al. Genome analysis reveals insights into physiology and longevity of the Brandt's bat Myotis brandtii. *Nat. Commun.* **2013**, *4*, 2212. [CrossRef]

35. Rasweiler, J.J.t.; Badwaik, N.K. Improved procedures for maintaining and breeding the short-tailed fruit bat (*Carollia perspicillata*) in a laboratory setting. *Lab. Anim.* **1996**, *30*, 171–181. [CrossRef] [PubMed]

36. Rasweiler Iv, J.J.; Badwaik, N.K. The laboratory environment for maintaining and breeding some bats in the Family Phyllostomidae. In *Bats in Captivity*, 1st ed.; Barnard, S.M., Ed.; Logos Press: Washington, DC, USA, 2009; pp. 345–356.

37. Skrinyer, A.J.; Faure, P.A.; Dannemiller, S.; Ball, H.C.; Delaney, K.H.; Orman, R.; Stewart, M.; Cooper, L.N. Care and husbandry of bats, the world's only flying mammals. *Lab. Anim. Sci. Pro.* **2017**, *5*, 24–27.

38. Rasweiler, J.J.t.; Badwaik, N.K.; Mechineni, K.V. Ovulation, fertilization, and early embryonic development in the menstruating fruit bat, *Carollia perspicillata*. *Anat. Rec.* **2011**, *294*, 506–519. [CrossRef]

39. Rasweiler, J.J.t.; Cretekos, C.J.; Behringer, R.R. The short-tailed fruit bat *Carollia perspicillata*: A model for studies in reproduction and development. *Cold Spring Harb. Protoc.* **2009**, *2009*, pdb emo118. [CrossRef]

40. Smith, J.B.; Alloway, K.D.; Hof, P.R.; Orman, R.; Reser, D.H.; Watakabe, A.; Watson, G.D.R. The relationship between the claustrum and endopiriform nucleus: A perspective towards consensus on cross-species homology. *J. Comp. Neurol.* **2019**, *527*, 476–499. [CrossRef] [PubMed]

41. Paxinos, G.; Franklin, K.B.J. *The Mouse Brain in Stereotaxic Coordinates*, 2nd ed.; Elsevier Academic Press: Amsterdam, The Netherlands; Boston, MA, USA, 2004.
42. Paxinos, G.; Watson, C. *The Rat Brain in Stereotaxic Coordinates*, 6th ed.; Academic Press/Elsevier: Amsterdam, The Netherlands; Boston, MA, USA, 2007.
43. Ding, S.L.; Van Hoesen, G.W. Organization and Detailed Parcellation of Human Hippocampal Head and Body Regions Based on a Combined Analysis of Cyto- and Chemoarchitecture. *J. Comp. Neurol.* **2015**, *523*, 2233–2253. [CrossRef] [PubMed]
44. Orman, R. Claustrum: A case for directional, excitatory, intrinsic connectivity in the rat. *J. Physiol. Sci.* **2015**, *65*, 533–544. [CrossRef] [PubMed]
45. Committee for the Update of the Guide for the Care and Use of Laboratory Animals. *Guide for the Care and Use of Laboratory Animals*, 8th ed.; The National Academies Press: Washington, DC, USA, 2011.
46. Rasweiler, J.J.t.; Cretekos, C.J.; Behringer, R.R. Feeding short-tailed fruit bats (*Carollia perspicillata*). *Cold Spring Harb. Protoc.* **2009**, *2009*, pdb-prot5159. [CrossRef]
47. Jebb, D.; Huang, Z.; Pippel, M.; Hughes, G.M.; Lavrichenko, K.; Devanna, P.; Winkler, S.; Jermiin, L.S.; Skirmuntt, E.C.; Katzourakis, A.; et al. Six reference-quality genomes reveal evolution of bat adaptations. *Nature* **2020**, *583*, 578–584. [CrossRef] [PubMed]
48. Zeidman, P.; Maguire, E.A. Anterior hippocampus: The anatomy of perception, imagination and episodic memory. *Nat. Rev. Neurosci.* **2016**, *17*, 173–182. [CrossRef]
49. Taube, J.S. Electrophysiological properties of neurons in the rat subiculum in vitro. *Exp. Brain Res.* **1993**, *96*, 304–318. [CrossRef] [PubMed]
50. Witter, M.P.; Groenewegen, H.J. The subiculum: Cytoarchitectonically a simple structure, but hodologically complex. *Prog. Brain Res.* **1990**, *83*, 47–58. [CrossRef] [PubMed]
51. Harris, E.; Witter, M.P.; Weinstein, G.; Stewart, M. Intrinsic connectivity of the rat subiculum: I. Dendritic morphology and patterns of axonal arborization by pyramidal neurons. *J. Comp. Neurol.* **2001**, *435*, 490–505. [CrossRef] [PubMed]
52. Kunitake, A.; Kunitake, T.; Stewart, M. Differential modulation by carbachol of four separate excitatory afferent systems to the rat subiculum in vitro. *Hippocampus* **2004**, *14*, 986–999. [CrossRef] [PubMed]
53. Orman, R.; Von Gizycki, H.; Lytton, W.W.; Stewart, M. Local axon collaterals of area CA1 support spread of epileptiform discharges within CA1, but propagation is unidirectional. *Hippocampus* **2008**, *18*, 1021–1033. [CrossRef]
54. Naggar, I.; Stewart, M.; Orman, R. High Frequency Oscillations in Rat Hippocampal Slices: Origin, Frequency Characteristics, and Spread. *Front. Neurol.* **2020**, *11*, 326. [CrossRef]
55. Ding, S.L.; Yao, Z.; Hirokawa, K.E.; Nguyen, T.N.; Graybuck, L.T.; Fong, O.; Bohn, P.; Ngo, K.; Smith, K.A.; Koch, C.; et al. Distinct Transcriptomic Cell Types and Neural Circuits of the Subiculum and Prosubiculum along the Dorsal-Ventral Axis. *Cell Rep.* **2020**, *31*, 107648. [CrossRef]
56. Bienkowski, M.S.; Sepehrband, F.; Kurniawan, N.D.; Stanis, J.; Korobkova, L.; Khanjani, N.; Clark, K.; Hintiryan, H.; Miller, C.A.; Dong, H.W. Homologous laminar organization of the mouse and human subiculum. *Sci. Rep.* **2021**, *11*, 3729. [CrossRef] [PubMed]

Article

Novel Endometrial Cancer Models Using Sensitive Metastasis Tracing for CXCR4-Targeted Therapy in Advanced Disease

Esperanza Medina-Gutiérrez [1,2], María Virtudes Céspedes [1], Alberto Gallardo [3], Elisa Rioja-Blanco [1,2], Miquel Àngel Pavón [4,5,6], Laura Asensio-Puig [7], Lourdes Farré [4,7], Lorena Alba-Castellón [1,2], Ugutz Unzueta [1,2,8,9], Antonio Villaverde [8,9,10], Esther Vázquez [8,9,10,*], Isolda Casanova [1,2,8,*] and Ramon Mangues [1,2,8,*]

[1] Institut d'Investigació Biomèdica Sant Pau (IIB-Sant Pau), 08041 Barcelona, Spain; emedina@santpau.cat (E.M.-G.); mcespedes@santpau.cat (M.V.C.); erioja@santpau.cat (E.R.-B.); lalba@santpau.cat (L.A.-C.); uunzueta@santpau.cat (U.U.)
[2] Institut de Recerca contra la Leucèmia Josep Carreras, 08025 Barcelona, Spain
[3] Department of Pathology, Hospital de la Santa Creu i Sant Pau, 08041 Barcelona, Spain; agallardoa@santpau.cat
[4] Institut Català d'Oncologia (ICO), 08908 L'Hospitalet de Llobregat, Spain; mpavon@iconcologia.net (M.À.P.); lfarre@iconcologia.net (L.F.)
[5] CIBER en Epidemiología y Salud Pública (CIBERESP), 28029 Madrid, Spain
[6] Institut d'Investigació Biomèdica de Bellvitge (IDIBELL), 08908 L'Hospitalet de Llobregat, Spain
[7] Program Against Cancer Therapeutic Resistance (ProCURE), Catalan Institute of Oncology (ICO), Oncobell Program, Bellvitge Biomedical Research Institute (IDIBELL), 08908 L'Hospitalet del Llobregat, Spain; lasensio@idibell.cat
[8] CIBER de Bioingeniería, Biomateriales y Nanomedicina (CIBER-BBN), 28029 Madrid, Spain; antoni.villaverde@uab.cat
[9] Departament de Genètica i de Microbiologia, Universitat Autònoma de Barcelona, 08193 Bellaterra, Spain
[10] Institut de Biotecnologia i de Biomedicina, Universitat Autònoma de Barcelona, 08193 Bellaterra, Spain
* Correspondence: esther.vazquez@uab.cat (E.V.); icasanova@santpau.cat (I.C.); rmangues@santpau.cat (R.M.)

Citation: Medina-Gutiérrez, E.; Céspedes, M.V.; Gallardo, A.; Rioja-Blanco, E.; Pavón, M.À.; Asensio-Puig, L.; Farré, L.; Alba-Castellón, L.; Unzueta, U.; Villaverde, A.; et al. Novel Endometrial Cancer Models Using Sensitive Metastasis Tracing for CXCR4-Targeted Therapy in Advanced Disease. *Biomedicines* **2022**, *10*, 1680. https://doi.org/10.3390/biomedicines10071680

Academic Editor: Martina Perše

Received: 24 May 2022
Accepted: 10 July 2022
Published: 12 July 2022

Publisher's Note: MDPI stays neutral with regard to jurisdictional claims in published maps and institutional affiliations.

Abstract: Advanced endometrial cancer (EC) lacks therapy, thus, there is a need for novel treatment targets. CXCR4 overexpression is associated with a poor prognosis in several cancers, whereas its inhibition prevents metastases. We assessed CXCR4 expression in EC in women by using IHC. Orthotopic models were generated with transendometrial implantation of CXCR4-transduced EC cells. After in vitro evaluation of the CXCR4-targeted T22-GFP-H6 nanocarrier, subcutaneous EC models were used to study its uptake in tumor and normal organs. Of the women, 91% overexpressed CXCR4, making them candidates for CXCR4-targeted therapies. Thus, we developed CXCR4[+] EC mouse models to improve metastagenesis compared to current models and to use them to develop novel CXCR4-targeted therapies for unresponsive EC. It showed enhanced dissemination, especially in the lungs and liver, and displayed 100% metastasis penetrance at all clinically relevant sites with anti-hVimentin IHC, improving detection sensitivity. Regarding the CXCR4-targeted nanocarrier, 60% accumulated in the SC tumor; therefore, selectively targeting CXCR4[+] cancer cells, without toxicity in non-tumor organs. Our CXCR4[+] EC models will allow testing of novel CXCR4-targeted drugs and development of nanomedicines derived from T22-GFP-H6 to deliver drugs to CXCR4[+] cells in advanced EC. This novel approach provides a therapeutic option for women with metastatic, high risk or recurrent EC that have a dismal prognosis and lack effective therapies.

Keywords: advanced endometrial cancer; orthotopic model; metastasis; CXCR4-targeted nanoparticles; animal model

1. Introduction

Endometrial cancer (EC) is the most common cancer of the female genital tract, and the sixth most diagnosed cancer in women [1,2]. Approximately 3–14% of EC patients are younger than 40 and want to spare their fertility [3]. The standard treatment for low grade EC is surgery, with prior fertility-sparing treatments when required [4,5]. Still, advanced stages or high-risk EC patients currently lack effective therapies. Even though EC has a five year survival rate of 80% [6], it decreases dramatically to a 56% or 20% rate when locoregional or distal metastasis occurs [7,8]. This dismal prognosis is due to the inability of cisplatin/paclitaxel chemotherapy or radiotherapy to effectively block metastatic dissemination [9].

In this regard, there is a need to develop animal models that mimic the metastatic progression and pathological features found in EC in humans, which could improve the knowledge of the molecular drivers of EC metastases in different organs and allow the adequate testing of drugs that may block dissemination during their preclinical development [10]. Nevertheless, among the available orthotopic preclinical models only a few replicate the advanced stages of EC, and when they do, they still show important limitations for testing novel drugs. Thus, some models do not reach a 100% engraftment to ensure the generation of primary tumor in all implanted mice and/or display high interindividual variability in metastatic foci development at clinically relevant sites [11–15]. However, most reported metastatic EC models are heterothopic since the location of EC cells implantation is other than the endometrium. Therefore, they are unable to mimic the primary tumor microenvironment and are scarcely relevant in the investigation of the possible molecular or cellular pathways that drive EC metastases in humans [16–22].

An additional and highly relevant issue among the published disseminated EC models, is the low sensitivity of most of the techniques used to detect metastatic foci at the different sites. Most researchers detect metastases identifying foci by using optical microscopy in tissue sections after hematoxylin-eosin (H&E) staining, whereas others generate cell-line-derived bioluminescent models that allow in vivo and ex vivo EC tracking. Nevertheless, neither method can reliably detect single cells or small cell clusters that infiltrate the organs where metastases are expected. This low sensitivity leads to the loss of relevant information on tumor cell arrival and colonization of distal organs. Therefore, there is a need to improve the metastatic cell detection threshold to unequivocally spot tumor cells that infiltrate distant organs at the single cell level.

The CXCR4 chemokine receptor is overexpressed at mRNA and protein levels in EC, compared to hyperplasia and a normal endometrium, being proposed to play a role in tumor progression in EC [23]. In fact, the CXCR4/CXCL12 pathway has been associated with EC progression [24,25]. This is consistent with reports demonstrating that overexpression of this receptor is associated with poor prognosis and/or enhanced metastases in at least 20 cancers, both solid and hematological [26,27]. In EC, there are inconclusive works regarding CXCR4 expression as a possible prognostic factor [28,29], similar to findings in breast cancer, in which CXCR4 expression shows a different prognosis depending on the tumor subtype [30,31]. Therefore, the role of CXCR4 in EC is still controversial regarding its possible effect on tumor growth or metastatic dissemination [27,32,33]. Interestingly, in subcutaneous EC models, CXCR4 expression increases EC cell engraftment and tumor growth rate [24,34–36]. However, to date, the study of the effect of CXCR4 overexpression on metastatic dissemination in EC mouse models has not been addressed yet, whereas in a majority of evaluated cancers, CXCR4 overexpression is associated with a worse prognosis [26].

Here, we generated a new CXCR4-overexpressing (CXCR4$^+$) orthotopic EC mouse model, using a novel surgical procedure, and set up a highly sensitive immunohistochemical tumor cell marker to detect small metastatic foci and single cancer cells in clinically relevant organs. Using this methodology, we have demonstrated that CXCR4 overexpression enhances metastatic dissemination in EC. Next, we developed a CXCR4$^+$ subcutaneous EC model to be used to demonstrate a highly specific accumulation of the CXCR4-targeted nanocarrier T22-GFP-H6 in EC. We had previously produced this nanocarrier [37] and also reported high tumor uptake in cancer models other than EC [38–40]. Thus, these novel EC models could be a new resource for developing CXCR4-targeted therapies that provide a non-surgical therapeutic option for women who lack effective therapies, such as those in advanced stages or high-risk EC patients.

2. Materials and Methods

2.1. Endometrial Cancer Patient Samples

Patient samples were collected under the *Hospital de la Santa Creu i de Sant Pau* Ethics Committee, with informed consent. It included a retrospective collection of 102 archived paraffined tissue samples, including normal and tumor tissue, from 79 women with endometrial cancer. Cylindrical tissue cores of 3 mm diameter were extracted from different paraffin blocks and re-embedded into single recipients (microarrays), to perform CXCR4 immunohistochemistry.

2.2. Cell Culture

2.2.1. Cell Line Constructs and Stable Cell Line Generation

AN3CA cell line (ATCC, Manassas, VA, USA) was cultured in DMEM and F12 medium 1:1; HEC1A cell line (ATCC, USA) and ARK-2 (kindly provided by Dr. Matias-Guiu) were cultured in high glucose DMEM medium. All of them were maintained in medium supplemented with 10% fetal bovine serum and penicillin/streptomycin, at 37 °C and 5% CO_2. Cells were tested for *Mycoplasma* sp. contamination using LookOut Mycoplasma PCR Detection Kit (Sigma-Aldrich, Taufkirchen, Germany) once every three months.

2.2.2. Lentiviral Transduction

For CXCR4 and/or luciferase expression, AN3CA was transduced with either pLentiIII-UbC-CXCR4-Luciferase or pLentiIII-UbC-Luciferase plasmids using lentivirus. After 48 h, cells were selected with puromycin (1 µg/mL for AN3CA, 3 µg/mL for HEC1A and 5 µg/mL for ARK-2).

2.2.3. Flow Cytometry

Membrane CXCR4 Assessment

CXCR4 membrane expression was determined using anti-CXCR4 antibody (PE-Cy5 mouse anti-human CD184, BD Biosciences, Franklin Lakes, NJ, USA) and its control (PE-Cy5 mouse IgG2a, K isotype control, BD), by flow cytometry in FACScalibur (BD Biosciences).

To obtain homogeneous cell populations, with high intensity of CXCR4 in the membrane in a large percent of the population, cells were sorted using FACSAria (BD Biosciences).

Internalization Assay

1 mL of a suspension of 500,000 $CXCR4^+$ $Luciferase^+$ AN3CA cells/mL was seeded in 6 well plates for 24 h. Then, they were exposed to T22-GFP-H6, a protein-based nanocarrier that targets CXCR4, at different concentrations (1, 10, 50, 100. 400 nM) for 1, 6 or 24 h. The competition assay for CXCR4-dependent internalization was performed by preincubating the cells with 1 µM of the CXCR4 antagonist AMD3100 for 1 h before exposure to T22-GFP-H6.

Quantification of nanocarrier internalization in cells (GFP^+ cells) was performed using flow cytometry. After cell detachment with trypsin, cells were washed with PBS, and treated with trypsin-EDTA (1 mg/mL, Life Technologies, Carlsbad, CA, USA) to remove nonspecific binding of T22-GFP-H6 binding to cell membrane.

Results were analyzed with software CellQuest Pro and expressed as percentage of fluorescent cells or mean fluorescence intensity.

2.2.4. Cell Viability Assay

100 µL 250,000 $CXCR4^+$ AN3CA cells/mL per well were seeded in 96 well plates. Then, 24 h later, they were incubated with T22-GF-H6 in concentrations ranging from 10 nM

to 400 nM. The competition assay was performed by preincubating the cells with 1 μM CXCR4 antagonist AMD3100 for 1 h. Cell viability was assessed using Cell Proliferation Kit II (XTT, Roche, Basel, Switzerland). Absorbance at 490 nm was read 48 h later (FLUOstar OPTIMA, BMG Labtech, Ortenberg, Germany). Data were shown as percentage of viable cells as compared to the viability of buffer exposed cells.

2.2.5. Cell Blocks

Cells were seeded in 150 cm^2 cell culture flasks. Once they reached a 70–80% confluence, cells were washed, trypsinized, collected and washed in PBS. After 5 min centrifugation at 400× g, cell blocks were obtained mixing gently 2 drops of human plasma and 2 drops of human thrombin. Cell pellets were fixed in paraformaldehyde 4% and embedded in paraffin to further CXCR4 analysis by immunocytochemistry.

2.3. In Vivo Experiments

Five-week-old female Swiss nude (Crl:NU(Ico)-*Foxn1^{nu}*) or NSG (NOD.Cg-*Prkdc^{scid} Il2rg^{tm1Wjl}/SzJ*) mice (Charles River, France) were used to develop in vivo subcutaneous and orthotopic models, respectively. After an acclimatization period of 7 days, they were housed in groups of five, in individually ventilated cage units (15.40 × 7.83 × 6.30 inch, Sealsafe Plus GM500, Techniplast, West Chester, PA, USA), with individual poplar chips bedding and its cellulose bag as environmental enrichment (Sodispan). Mice were maintained under specific pathogen-free conditions and a light/dark cycle of 12 h. They were fed with irradiated food (14% protein, 4% fat, Teklad Global diet, ENVIGO, Indianapolis, IN, USA) and reverse osmosis autoclaved water ad libitum. During surgery, eye hydration was maintained with a saline drip, and hypothermia was avoided using a heating blanket under a drape to keep body temperature at 37 °C. All the animal procedures were performed using sterile material in a type II biosafety cabinet (BIO-II-A/P, Telstar, Barcelona, Spain).

Randomizations of animals in experimental groups were performed using dice. The investigator was not blinded to group allocation during in vivo manipulation of animals; nevertheless, they were blinded for the histological analysis.

All mice were euthanized by cervical dislocation once they arrived at the experimental endpoint or reached endpoint criteria (10–20% body weight loss, signs of pain or distress such as abnormal postures, ulcers, alopecia, ruffled fur, abnormal breathing, abnormal activity, coma, ataxia, tremors). All the in vivo procedures were approved by the *Hospital de la Santa Creu i de Sant Pau* Animal Ethics Committee and Generalitat de Catalunya (FUE-2018-00819447, 24 April 2019), and performed according to European Council directives (CEA-OH/9721/2).

2.3.1. Orthotopic EC Models

NSG mice were anesthetized with 100 mg/kg ketamine (Ketolar, Pfizer, New york, NY, USA) and 10 mg/kg xylazine (Rompun, Bayer, Leverkusen, Germany). The lower abdomen was shaved and swabbed with povidone. A medial laparotomy was performed to expose the right uterine horn. Its irrigation system was separated, and a ligature was performed in the proximal section of the horn using 7/0 Optilene (BBraun, Melsungen, Germany).

After randomization (n = 4/group), either 10^6 CXCR4$^+$ AN3CA or CXCR4$^-$ AN3CA cells resuspended in 25 ul of culture medium were inoculated through the myometrium into the endometrial cavity using a 29G Hamilton syringe (Microliter Serie 800, Hamilton, Reno, NV, USA).

Overall animal health conditions were monitored three times a week. All mice were left to survival, and they were sacrificed as soon as they reached human endpoint criteria (55 ± 17 days; mean ± SD). In addition to the previously stated endpoint criteria, for this experiment, high primary tumor or peritoneal carcinomatosis growth (determined by palpation or a maximum whole body bioluminiscence of 2×10^{10} p/s/cm^2/sr) as well as abdominal distension, were observed.

2.3.2. Subcutaneous CXCR4$^+$ EC Models

Swiss nude mice were anesthetized with 2% isoflurane, and their backs were swabbed with povidone-iodine. In both flanks, 10^7 CXCR4$^+$ AN3CA cells were inoculated subcutaneously (SC). Overall animal health and tumor growth were monitored twice a week, using a caliper and applying the formula V = width^2 × length/2. For the setup experiment ($n = 4$), endpoint was set at 600 mm^3, which was reached 15–30 days after inoculation.

2.3.3. In Vivo T22-GFP-H6 Biodistribution

Nanocarrier biodistribution was assessed in the SC EC mouse model following the experimental design shown in Figure S1. A total of 19 animals with subcutaneous tumor development were randomized in control ($n = 3$) or treated groups ($n = 4$) for different time points (2, 5, 24 and 48 h) once tumors reached 150–200 mm^3. All of them received intravenously 200 μL of buffer or 200 μg T22-GFP-H6, respectively. Fluorescence intensity emitted by the GFP domain of T22-GFP-H6 was measured ex vivo, after euthanasia at 2, 5 or 24 h in SC tumors and non-tumor organs (liver, kidney, lung, spleen). Emitted fluorescence intensity was expressed as the average radiant efficiency $[(p/s/cm^2/str)/(mW/cm^2)]$ since this reflects the accumulation of T22-GFP-H6 in each tissue, once the autofluorescence of control mouse tissues is subtracted.

2.3.4. Necropsy and Histological Examination

After necropsy and ex vivo assessment of biodistribution or bioluminescence emission, tumor and non-tumor organs were collected and fixed in 4% formaldehyde for histopathological or immunohistochemical evaluation, during which the investigator was blinded for the animal group.

All organs were stained with hematoxylin-eosin (H&E) to evaluate tumor histologic architecture (stromal, vascular development or possible necrosis) and possible toxicity or nanocarrier accumulation.

2.4. Bioluminescence Intensity Assessment

Luciferase expression in vitro was assessed by seeding 500,000 cells per well on 6 well plates. Bioluminescence emission was expressed as radiance photons $(p/s/cm^2/sr)$ and registered using IVIS Spectrum 200 equipment, adding 200 μL D-luciferin firefly potassium salt 1.0 g (15 mg/mL, Perkin Elmer, Waltham, MA, USA). Images were analyzed using Living Image v.4.7.3. software.

To assess the correct implantation of cells in vivo, mice were injected intraperitoneally with firefly D-luciferin (2.25 mg/mouse, Perkin Elmer), which is a substrate for luciferase which is expressed in CXCR4$^+$ EC cells. Mice were anesthetized with 3% isoflurane in oxygen to capture bioluminescence intensity 5 min after injection. Images were analyzed using Living Image v.4.7.3. software.

2.5. Immunocytochemistry and Immunohistochemistry

Immunocytochemical (ICC) and immunohistochemical (IHC) staining were performed in a DAKO Autostainer Link48 (Agilent, Santa Clara, CA, USA) following the manufacturer's instructions, using 4 μm paraffin sections of cell blocks or organs.

The human CXCR4 staining pattern in cells was assessed using anti-CXCR4 (1:200; Abcam, ab124824) in all cell blocks, patient tissue samples and tumor and non-tumor organs. After assessing human vimentin (V9, Dako, IGA63061-2) expression in subcutaneous tumors, it was used to determine the presence or absence of tumor cells in non-tumor tissues. Ki67 (Dako GA62661-2) was used to determine the proliferation profile of subcutaneous tumors.

All slides were examined by a blind observer under an optic microscope and representative pictures were taken using an Olympus DP73 camera. Liver sections were processed with cellSens software (Olympus, RRID:SCR_014551, Tokyo, Japan), while lung sections were analyzed with QuPath [41], both at 200× magnification.

2.6. Statistical Analysis

In vitro experiments (internalization, XTT) were performed in biological triplicates. Differences between groups were analyzed using the Mann–Whitney test, prior determination of normality by the Shapiro–Wilk test using SPSS software v.23. Differences were considered statistically significant at $p \leq 0.05$. In vivo experiments (mice models setup, nanocarrier biodistribution) were performed in quadruplicates, defined by previous experiments regarding the interindividual variability among mice in terms of tumor growth and metastatic load. No animal exclusion was done. Randomization was performed using dice; odd numbers were assigned to group 1, while even numbers were assigned to group 2. Cohen's delta was performed to assess the size of differences in lung and liver metastatic load between animal groups, using R software v.3.4.4. The effect size was considered large when $d > 0.8$ and medium when $d > 0.5$ [42,43]. All results were expressed as the mean $\pm$ standard error (s.e.m.).

3. Results

3.1. Immunohistochemical Evaluation of CXCR4 Expression in EC Patient Samples

To assess CXCR4 expression levels and the localization pattern in EC tumors, we performed an IHC staining of this protein using tissue samples from 79 patients diagnosed with EC at *Hospital de la Santa Creu i Sant Pau*. CXCR4 was highly overexpressed in 91.6% of tumor tissue, as compared to the level displayed by healthy cells in endometrial tissue. Moreover, a majority of EC patients (64.4%) overexpressed CXCR4 in their membrane (Figure 1). Patients with membrane CXCR4 overexpression become candidates for targeting of CXCR4[+] EC cells, including selective drug delivery, through receptor-mediated internalization, as a therapeutic strategy.

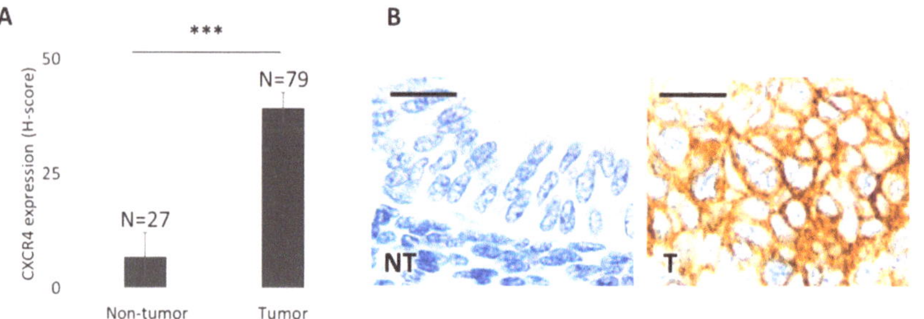

Figure 1. CXCR4 expression in tissue microarrays obtained from endometrial cancer patients. (**A**) CXCR4 is overexpressed in tumor tissue ($n = 79$), as compared to adjacent healthy endometrial tissue ($n = 27$), that shows almost undetectable CXCR4 expression levels by immunohistochemistry (IHC) (Mann–Whitney test, *** $p = 0.000$; mean $\pm$ s.e.m). (**B**) CXCR4 protein expression after IHC is negligible in healthy endometrial tissue, whereas it is highly overexpressed and mostly located in the cell membrane of tumor endometrial tissue. Bar: 20 µm. NT: non-tumor, healthy tissue; T: tumor tissue.

3.2. Generation of CXCR4[+] Luciferase[+] Human EC Cell Lines

We evaluated CXCR4 expression levels in two endometrioid EC human cell lines, AN3CA and HEC1A, and one serous EC human cell line, ARK2. All three showed negligible levels of CXCR4 as measured both by flow cytometry and immunocytochemistry (ICC) (Figure S2). Therefore, they were transduced with CXCR4-Luciferase lentiviral vectors.

All transduced EC cell lines, including AN3CA, ARK2 and HEC1, became CXCR4[+] Luciferase[+] and therefore displayed high bioluminescent emission (Figure S3A). However, the highest expression of membrane CXCR4 was achieved by the bioluminescent AN3CA cell line (Figure S2B). In this sense, while the transduced, luminescent AN3CA cell line maintained high CXCR4 expression in the membrane, as seen by flow cytometry and ICC, transduced HEC1A cells showed a high CXCR4 expression. However, according to CXCR4 ICC, its expression pattern was mostly cytoplasmatic. Similarly, ARK-2 showed an increase in CXCR4 levels when transduced, but these levels did not remain intense or steady over passages. Thus, the CXCR4$^+$ Luciferase$^+$ AN3CA cell line was the only one that maintained membrane CXCR4 overexpression over time, therefore being selected for further in vitro and in vivo work.

3.3. Development of a Subcutaneous Tumor Model Bearing Human EC Cells in Swiss Nude and Follow-Up Markers of Cancer Cell Growth

We developed a CXCR4$^+$ subcutaneous model, implanting 10^7 CXCR4$^+$ Luciferase$^+$ AN3CA cells in the flank of Swiss nude mice to validate tumor cells engraftment and growth (Figure S3B). We obtained a 100% engraftment rate, and luciferase emission was assessed over time. The bioluminescence emitted by the SC tumor correlated with tumor size, even though there was a certain variability in tumor growth rate among mice (Figure S4A).

After necropsy, tumor histology was evaluated by H&E, showing mostly EC epithelial cells jointly with stromal and vascular tissue areas with no signs of necrosis neither macroscopically nor microscopically (Figure S4B). Tumors also showed high Ki67 expression that indicated high proliferative activity and also a high membrane CXCR4 levels in cancer epithelial cells as we had already observed in vitro. Moreover, the human mesodermal marker vimentin was expressed, showing a highly intense staining in in EC epithelial cells, and a lack of staining in all mouse cells, including stroma and endothelium (Figure S4B).

3.4. Bioluminescent Follow-Up of Primary Tumor and Metastatic Dissemination in a Novel Orthotopic Model of Advanced EC in NSG Mice

We developed a unique procedure to generate a novel orthotopic mouse model of endometrial cancer by ligation of the horn and transmyometrial injection of CXCR4$^+$ Luciferase$^+$ or CXCR4$^-$ AN3CA cells (Figure 2). The ligature avoided vaginal leakage of the injected cell suspension (Figure 2A). Bioluminiscence allowed us to know whether local or distant tissues were colonized by EC cells. Nevertheless, a histological analysis or an IHC evaluation yielded a more precise identification and quantitation of the disseminated EC cells, as reported in Figure 2. Thus, bioluminescence proved to be useful in identifying the presence of EC cells in distant organs (Figure 2B), despite being only a qualitative marker, showing that our model replicates the sites of metastases observed in advanced EC patients. This fact entails an improvement over previously reported models, because of its take rate of 100% of the inoculated mice in all clinically relevant organs, including the uterus (primary tumor) and ovaries, the peritoneal cavity, lymph nodes, and the liver and lungs (Figure 2C).

3.5. Marker-Guided Comparison of Metastatic Yield in the EC Intrauterine Orthotopic Models Generated from CXCR4$^-$ or CXCR4$^+$ EC Cells in NSG Mice

Anti-human CXCR4 or anti-human vimentin antibodies were shown to be specific and highly sensitive markers in tumor cells in our CXCR4$^+$ model. Nevertheless, the mesodermal marker for human vimentin allowed the most sensitive and precise staining for the detection of microscopic metastases, when compared with CXCR4 IHC staining or H&E staining (Figure 3). Anti-human vimentin provides an easier and highly reliable quantification of foci number and their area in paraffined tissue slides, that can be standardized using image analysis software such as Image-J, cellSens or QuPath.

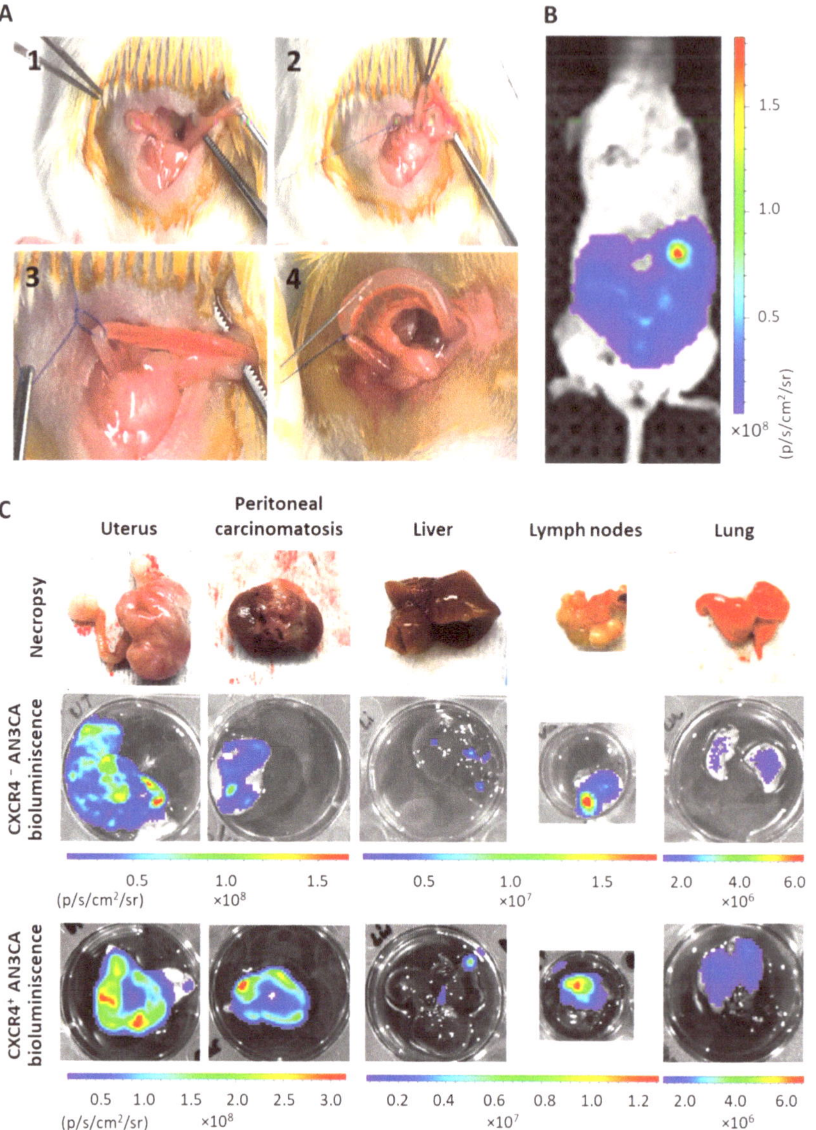

Figure 2. Procedure to orthotopically inoculate human cell lines in NSG mice to obtain an aggressive endometrial cancer model that metastasizes in all clinically relevant organs. (**A**) Representative photographs of the taken procedure, starting by showing uterus exposure (1), followed by right horn ligature without clamping the arterial irrigation system (2 and 3) and intraluminal transmyometrial injection of 10^6 CXCR4$^+$ Luciferase$^+$ AN3CA or CXCR4$^-$ Luciferase$^+$ AN3CA cells suspended on culture medium using a Hamilton syringe. (**B**) Representative image of Luciferase bioluminiscence emission in a mouse implanted with Luciferase$^+$ AN3CA cells, indicating the correct engraftment of tumor cells inside the uterus. (**C**) Representative images of relevant organs bearing metastases after necropsy and their ex vivo emission of bioluminiscence, registered by IVIS Spectrum.

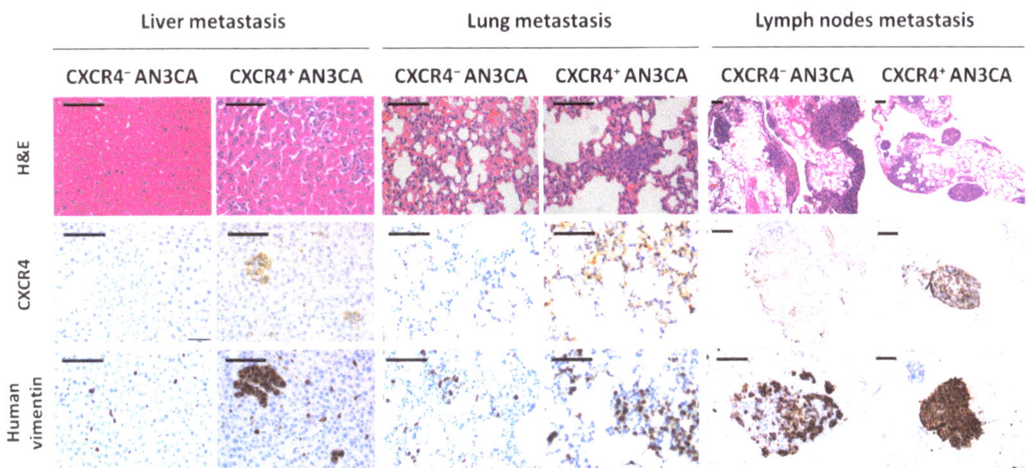

Figure 3. Dissemination pattern of the xenograft orthotopic model of advanced endometrial cancer generated in NSG mice and comparison of dissemination between models derived from CXCR4$^+$ and CXCR4$^-$ AN3CA cells. Histology of liver, lungs and lymph nodes with possible metastatic colonization after hematoxilin-eosin (H&E) staining of tissue sections or using immunohistochemical staining for anti-human CXCR4 or anti-human vimentin. Note the dramatically higher sensitivity of the anti-human vimentin IHC staining, to spot single or clustered cancer cells in the mouse tissues as compared to anti-CXCR4 IHC or H&E-stained tissues. Bar: 100 μm.

The high sensitivity of human vimentin staining allows the comparison of the metastatic dissemination pattern and the metastatic load, as single cell foci invading the tissue or cell clusters, generated in the orthotopic model derived from CXCR4$^+$ AN3CA cells with those generated by the orthotopic model derived from CXCR4$^-$ AN3CA cells (Figure 3). In fact, when comparing the metastases in clinically relevant sites such as the liver, lungs and lymph nodes, after IHC staining with anti-human vimentin, we observed that the total liver foci number was higher ($d = 0.92$) in the CXCR4$^+$ AN3CA model than in the CXCR4$^-$ AN3CA model, particularly in clustered cells foci ($d = 2.3$). Similarly, the total area occupied by the metastatic foci in the lung was higher in the CXCR4$^+$ AN3CA model than in the CXCR4$^-$ AN3CA model ($d = 1.26$) (Table 1). Therefore, CXCR4 overexpression was substantively associated with a more metastatic and aggressive phenotype in the ORT model than the one achieved with CXCR4$^-$ AN3CA cells.

Table 1. Comparison of metastatic dissemination in advanced endometrial cancer orthotopic models derived from CXCR4$^+$ or CXCR4$^-$ AN3CA cells.

Inoculated Cell Line	Liver Mets					Lung Mets
	Total Foci		Single Cell Foci	Clustered Cells Foci		Invaded Tissue Area (%)
	Number	Area (μm^2)	Number	Number	Area (μm^2)	
CXCR4$^-$ AN3CA	10.2 ± 6.7 [a]	615.5 ± 429.5 [b]	9.4 ± 6.3	0.9 ± 0.5 [c]	4486.4 ± 2728.0	11.5 ± 4.2 [d]
CXCR4$^+$ AN3CA	24.2 ± 8.3 [a]	2536.8 ± 1746.8 [b]	15.6 ± 6.5	9.7 ± 2.7 [c]	5305.0 ± 3517.9	26.1 ± 7.0 [d]

The xenograft orthotopic models of advanced endometrial cancer were generated in NSG mice by implantation of CXCR4$^+$ or CXCR4$^-$ AN3CA cells ($n = 4$/group). Results are reported as mean ± s.e.m. of number or area of metastatic foci per mouse counting medium power microscope fields (200×, 10 fields) in liver or lung sections, after IHC staining using anti-human vimentin. Assessment of the effect size between groups was performed using Cohen's delta (d) test. [a] $d = 0.92$; [b] $d = 0.75$; [c] $d = 2.3$; [d] $d = 1.26$.

3.6. In Vitro Uptake of the Fluorescent T22-GFP-H6 Nanocarrier and Its Cytotoxicity in Human EC Cell Lines

Once we determined that CXCR4-overexpression increases the metastatic load in the ORT EC model derived from the AN3CA EC cell line, we wanted to know if the T22-GFP-H6 protein-based nanocarrier could selectively internalize in CXCR4$^+$ EC cells. This nanocarrier was previously developed in our group [37], and it targets the CXCR4 receptor, and therefore the CXCR4-overexpressing cancer cells (Figure S5). We observed that the T22-GFP-H6 showed a time and concentration-dependent internalization in the transduced EC cell line CXCR4$^+$ Luciferase$^+$ AN3CA (Figure 4A). It achieved a high internalization efficiency, as 80% of cells were GFP$^+$ after 6 h of exposure to 10 nM T22-GFP-H6. It also showed CXCR4-dependent internalization, as parental CXCR4$^-$ cells showed no nanocarrier internalization. Similarly, when CXCR4$^+$ cells were preincubated with the CXCR4 antagonist AMD3100 1 h prior to nanocarrier exposure (100 nM T22-GFP-H6), internalization was significantly reduced, while completely blocked at lower concentrations (10 nM T22-GFP-H6) (Figure 4B). Internalization of the nanocarrier in the CXCR4$^+$ Luciferase$^+$ AN3CA cell line measured as mean fluorescent intensity (MFI) (Figure S6) showed similar levels of T22-GFP-H6 uptake and also demonstrated competition with AMD3100 as described above.

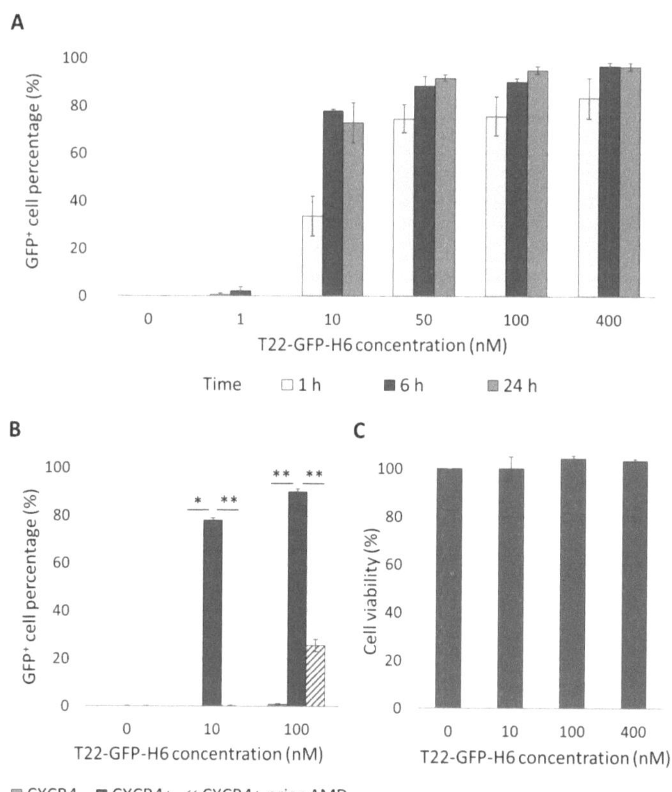

Figure 4. In vitro CXCR4-dependent internalization and lack of cytotoxicity of T22-GFP-H6 nanocarrier in human endometrial cancer cell line AN3CA. (**A**) T22-GFP-H6 internalization in CXCR4$^+$ AN3CA cells at 1, 6 and 24 h and different concentrations, expressed as percentage of GFP$^+$ cells. (**B**) Blockage of T22-GFP-H6 internalization at 6 h, measured by flow cytometry, in CXCR4$^-$ AN3CA cells, CXCR4$^+$ AN3CA cells and CXCR4$^+$ AN3CA cells after 1h pretreatment with 1 μM of the CXCR4 antagonist AMD3100 (Mann-Whitney test; * $p < 0.05$, ** $p < 0.01$; $n = 3$; mean ± s.e.m). (**C**) T22-GFP-H6 cytotoxicity on CXCR4$^+$ AN3CA cells as measured by XTT viability test at 48 h. ($n = 3$; mean ± s.e.m).

Furthermore, T22-GFP-H6 showed no cytotoxic effect on the CXCR4[+] AN3CA cell line, as measured by the XTT viability test (Figure 4C). Thus, AN3CA is a good candidate to develop CXCR4[+] mice models to test the in vivo effect of CXCR4-targeted nanoparticles.

3.7. Biodistribution of CXCR4-Targeted Nanocarrier T22-GFP-H6 in a CXCR4[+] Subcutaneous EC Model

We evaluated the biodistribution of the T22-GFP-H6 nanocarrier in the CXCR4[+] AN3CA subcutaneous model, measuring its uptake in tumor tissue and in non-tumor organs. For this purpose, we measured the GFP fluorescence emitted along time in the different organs after a single bolus of 200 µg T22-GFP-H6, administered intravenously and compared to the control, as previously described [39].

The nanocarrier was uptaken by the subcutaneous tumor starting 2 h after injection, showing a fluorescence peak at 5 h, which decreased by 24 h (Figure 5A). The total amount of T22-GFP-H6 accumulated in tumor tissue, as measured by the area under the curve of fluorescence intensity over time, was over 60% of the administered dose (Figure 5B). In contrast, H&E staining showed no signs of cell death or histological alteration in tumor tissue at neither of those time points (2, 5 and 24 h) nor at 48 h (Figure 5C).

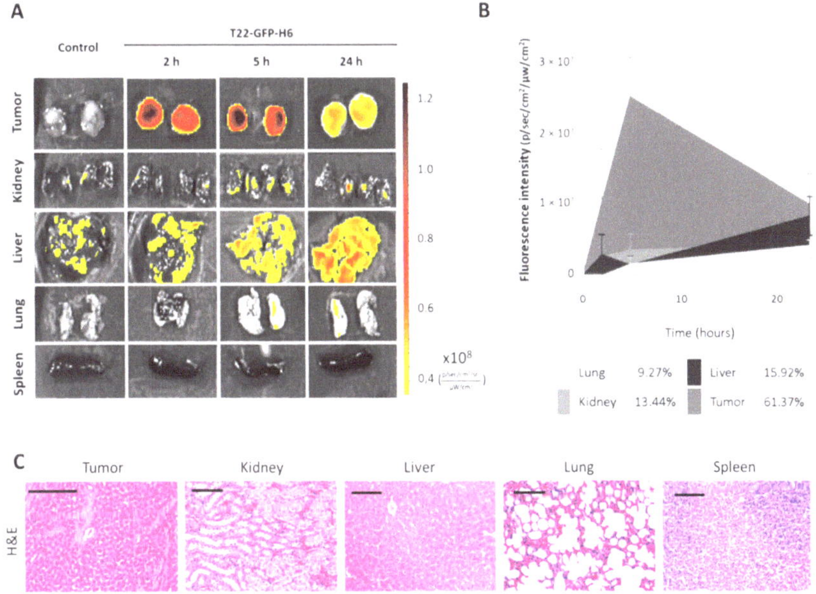

Figure 5. In vivo biodistribution and toxicity assessment of nanocarrier T22-GFP-H6 in a subcutaneous mouse model derived from CXCR4+ AN3CA cells. (**A**) Representative images of fluorescence emitted by the nanocarrier after 2, 5 and 24 h after intravenous injection of 200 µg of T22-GFP-H6. (**B**) Area under the curve representation of fluorescence emitted over time by tumor and non-tumor tissues, and their respective percentage of nanocarrier uptake out of the total fluorescence emitted by T22-GFP-H6 in all tissues ($n = 4$/group; mean ± s.e.m). (**C**) Hematoxylin-eosin staining of tumor and non-tumor organs 48 h after administration of the nanocarrier. Bar: 100 µm.

Non-target organs such as the kidneys, liver, lungs and spleen, also showed normal tissue architecture (Figure 5C) and low (9–16% range) nanocarrier accumulation, as measured by fluorescence emission (Figure 5A,B).

Thus, our nanoparticle has been proven to selectively biodistribute selectively to tumors, in absence of toxicity in tumor or in normal organs.

4. Discussion

Aiming to offer a therapeutic option for the treatment of endometrial cancer (EC) patients currently lacking an effective therapy [9], we demonstrated that 64.4% of EC patients overexpress high membrane CXCR4 expression, therefore making them candidates for CXCR4-targeted therapies. Thus, we generated a highly metastatic model to demonstrate that CXCR4 enhances EC metastases, becoming a useful resource for developing CXCR4 targeted therapies against high risk or advanced CXCR4$^+$ EC. This novel orthotopic (ORT) CXCR4$^+$ EC model, derived from the CXCR4$^+$ Luciferase$^+$ EC AN3CA cell line, was set up using a new surgical procedure that yields full metastatic penetrance. In addition, we have used an anti-human vimentin antibody to detect metastatic foci, proving this method to be highly sensitive, enabling us to detect even single metastatic cells. Using this model, we have demonstrated that CXCR4-overexpression enhances EC metastatic dissemination to the lungs and liver. Moreover, we have also developed a CXCR4$^+$ subcutaneous (SC) model which we next used to demonstrate that the protein-based T22-GFP-H6 nanocarrier that targets CXCR4$^+$ cells, displays a high selectivity in its accumulation in CXCR4$^+$ EC tumor tissues. The generated EC ORT and SC models can be used for the preclinical testing of novel drug delivery approaches that target CXCR4$^+$ EC cells for the treatment of high risk or advanced CXCR4$^+$ EC, as well as to study the mechanisms of tumor growth and metastatic dissemination that are dependent on CXCR4 overexpression signaling. Following, we are describing the improvement introduced by our findings, compared to previously published EC models.

4.1. CXCR4 Expression Pattern in EC Patients

First, the IHC evaluation of a series of 79 EC patients showed a high CXCR4 receptor expression in 91.6% of primary tumors, as compared to the low level detected in non-tumor endometrium. Remarkably, predominant membrane expression of CXCR4 was detected in 64.4% of patients. Nevertheless, no metastatic orthotopic endometrial cancer model has been described to overexpress it yet, to our knowledge. This reveals a gap in treatment research for EC patients, since a large proportion of them could be candidates for treatments targeted to CXCR4, given its high membrane expression. We have taken advantage of this relevant clinical feature to develop novel CXCR4$^+$ EC models, validating its usefulness using a targeted drug delivery approach based in nanoparticles.

4.2. Development of an Aggressive CXCR4$^+$ Advanced EC Metastatic Model

The reason for the selection of the AN3CA cell line to generate the advanced metastatic EC model was mainly its high membrane CXCR4 overexpression when transduced, which more closely replicates its expression pattern in patients, as compared to HEC1A or ARK-2 cell lines. Membrane expression of this receptor drives CXCR4-dependent metastatic progression in other tumor types [26], while its role was still unknown in EC. All tested cell lines also expressed luciferase to allow in vivo and ex vivo assessment of primary tumor engraftment and metastatic dissemination follow-up.

The generation for this ORT EC model in immunosuppressed NSG mice applies a novel and easy procedure that consists of the ligation of the horn of the uterus before the transmyometrial injection of the human EC cells. We have produced this model that mimics the highly aggressive behavior observed in advanced EC in humans. Moreover, when CXCR4$^+$ AN3CA cells were inoculated, all primary tumor and metastasis maintained the high CXCR4 membrane expression observed in culture. This can be related to the fact that the AN3CA cell line derives from a lymph node EC metastasis, displays high microsatellite instability, *TP53* mutations and *PTEN* deletions and shows resistance to cisplatin and paclitaxel [36,44], all features that are associated with EC aggressiveness [45,46] and relapse [10]. Consistently, the assessment of primary tumors histology showed high grade undifferentiated tumors, with loss of uterine tissue architecture, that clearly resembles cancer progression in patients.

In addition, this novel CXCR4$^+$ ORT EC model develops metastases in all clinically relevant sites with 100% penetrance, involving ovaries, abdominal lymph nodes, the peritoneum, liver and lungs, as identified by luminescence emission. Our results clearly improve the models previously described as advanced or metastatic, because they did not accurately replicate the dissemination pattern observed in humans since they were generated through EC cell implantation at heterotopic sites (non-ORT implantation). This kind of cell implantation does not mimic the EC tumor microenvironment [10], yielding, therefore, low rates of tumor engraftment or metastases development, which in turn, limits or precludes the study of the mechanisms of EC metastatic dissemination.

4.3. CXCR4 Overexpression Is Associated with Enhanced Metastatic Dissemination in EC

We have also shown that a high level of CXCR4 overexpression in the EC cell membrane (CXCR4$^+$) is substantively associated with an increase in metastasis development in the lungs and liver, by comparing the metastatic load between the AN3CA derived CXCR4$^+$ and CXCR4$^-$ ORT models. This is the first demonstration of the capacity of the overexpression of this receptor to enhance the spread of metastases in EC, which is consistent with previous reports describing the role of CXCR4 on migration in vitro [24,35] and with CXCR4 receptor being overexpressed at mRNA and protein levels in EC, compared to hyperplasia and normal endometrium (Buchynska et al., 2021; Liu et al., 2016; Sun et al., 2017). It is also consistent with the proposed role of CXCR4 in EC progression (Buchynska et al., 2021; Liu et al., 2016) [23,25,36].

4.4. Use of Highly Sensitive Human-Vimentin as EC Tumor Cells Marker to Detect Metastatic Foci

Recording the luminescence emitted by cancer cells can spot metastatic foci at the different sites, only when they are large enough and close to the surface of the measured tissue section. Thus, this method is rather qualitative as a tracker of EC foci because of the low penetration of light in tissues.

In contrast, using the immunohistochemical (IHC) detection of human vimentin in tissue sections allowed the easy, reliable and highly sensitive quantification of EC cells in all clinically relevant metastatic organs. Vimentin detection has been previously used as a marker to identify epithelial cancer cells in primary EC tumors [12,15,22], but, to our knowledge, this is the first time that human vimentin is used to detect and quantify the number and area of EC metastatic foci. In this regard, it represents a huge increase in sensitivity as compared to IHC detection of metastases using anti-human CXCR4 or H&E staining. Thus, human vimentin reliably identifies all human tumor cells in the studied section of affected organs, including single EC cells infiltrating the tissue (which anti-CXCR4 or H&E cannot) and small or large size metastatic foci.

The large advantage for human vimentin in scoring metastatic foci could be related to the high affinity of the antibody used for its reaction with the human vimentin epitope. This avoids background staining since it shows no cross reaction with mouse vimentin in tumor stroma or blood vessels, and also because vimentin is expressed on all transformed EC epithelial cells since the endometrium derives from the mesoderm layer in the embryo [47]. This approach dramatically improves the methods used to identify metastatic foci in previously reported metastatic EC models because of the low sensitivity of the used marker or staining (H&E) [12,13,15,17,18,20,22].

This novel application of human vimentin IHC staining could be used, jointly with the detection of molecular drivers of EC dissemination, to investigate the effect of drugs on metastatic colonization, regarding single or cluster cell arrival to the tissue, organ colonization or foci growth at different metastatic sites. Image software for quantitating single or multiple foci in a region of interest or in the whole tissue section will make it even easier to reach this goal.

4.5. Development of a CXCR4 Subcutaneous Tumor Model and Its Use to Evaluate Targeting of Protein-Based Nanocarriers to CXCR4⁺ EC Cells

With the final aim of developing an effective antitumor and antimetastatic drug against this cancer, we started to explore the selectivity of the biodistribution to the tumor tissue and the possible toxicity of the protein-based nanocarrier T22-GFP-H6, which was previously developed by our group and targets CXCR4 [37]. To that purpose, it was first necessary to demonstrate in vitro that T22-GFP-H6 internalization in CXCR4⁺ EC cells was exclusively dependent on CXCR4. Secondly, we developed a CXCR4⁺ SC EC model to evaluate the percent of the administered dose that reaches the tumor tissue, since the subcutaneous tumor model is more suited than the orthotopic model for this specific purpose.

Thus, we performed in vitro experiments that showed that T22-GFP-H6 reached a highly selective, concentration-dependent and CXCR4-mediated internalization in CXCR4⁺ cells. Consistently, this nanocarrier does not internalize in CXCR4⁻ cells; moreover, pre-treating CXCR4⁺ cells with the CXCR4 antagonist AMD3100 showed an effective blockage of T22-GFP-H6 internalization. In addition, the nanocarrier displayed neither antitumor activity in EC nor cytotoxicity in normal cells in vitro.

To validate these results in vivo, we used the CXCR4⁺ bioluminescent AN3CA EC cell line to develop a SC EC model with high CXCR4 membrane expression. It produced a 100% rate of tumor engraftment and induced aggressive tumor growth since most of the transformed EC epithelial cells were cycling as demonstrated by Ki67 expression in most cells. It also showed high membrane CXCR4 expression, which has been reported to increase proliferation in SC EC models derived from other cell lines [24,34–36]. Moreover, we found a strong correlation between tumor luminescent emission and tumor growth rate.

We used this model to demonstrate that a single intravenous bolus of T22-GFP-H6 leads to its selective accumulation in CXCR4⁺ EC tumors. Thus, 60% of the total emitted fluorescence, out of the total administered dose, is registered in tumor tissue, indicating its CXCR4-dependent internalization. In contrast, its fluorescence level registered in normal organs/tissues was much lower despite showing detectable fluorescence above background (liver, kidney, lung or spleen), while no histological alterations were detected in these or other normal organs.

On this basis, we believe that T22-GFP-H6 nanocarrier is suitable for the development of nanomedicine-based targeted drug delivery approaches for EC therapy since it is not cytotoxic and is highly selective towards CXCR4⁺ EC cells, both in vitro and in vivo. We are now planning to load the nanocarrier with one or a combination of drugs, expecting to achieve a highly selective and sustained delivery of the drug/drugs specifically to tumor tissue with low or negligible drug delivery in normal tissues. This is supported by the high uptake of the nanocarrier by tumor tissue, which starts at 2 h; it lasts until 24 h, and has a peak at 5 h, reaching its accumulation in tumors at much higher levels than those achieved in non-tumor organs. Additional support for its development as a nanomedicine for CXCR4⁺ EC comes from our previous work reporting a similar biodistribution of T22-GFP-H6 in other solid tumors or hematological cancers [38–40,48].

4.6. Future Contribution of the Novel Models for the Development of Targeted Therapies in Advanced EC

The developed CXCR4⁺ SC EC tumor model could be used to evaluate the antitumor activity of novel CXCR4-targeted drugs, since SC tumor models are widely used in preclinical drug development for EC therapy [10]. However, they do not mimic the microenvironment where EC develops, nor can they be used to assess the effect of drugs on metastatic dissemination. In contrast, the novel ORT CXCR4⁺ EC model developed here ensures a correct implantation of cancer cells in the endometrium, avoiding their leakage, while maintaining the tumor microenvironment where EC develops. It also overexpresses the CXCR4 receptor in their membrane, being able to help in the preclinical development of antimetastatic drugs that target the CXCR4 receptor. Although CXCR4 inhibitors are in clinical trials for other diseases, they have not been tested in EC yet [49]. Our model, being able to assess the antimetastatic effect as well as to possibly predict drug responses to drugs that target this receptor, could have a clinical translation for the treatment of advanced CXCR4$^+$ EC in patients. It could also permit us to study the impact of the tested drug on the different processes activated by CXCR4 overexpression during metastasis development (e.g., myometrial or lymphovascular invasion, or colonization of different organs), as well as the evaluation of their underlying mechanisms, measuring their effect on the signaling pathways driving them, since they may closely mimic the clinicopathological features, and the molecular and cellular mechanisms observed in humans with CXCR4-overexpressing advanced EC.

In addition to its orthotopic implantation, the metastatic model above improves previous metastatic models reported by other authors since it achieves 100% metastasis penetrance in all relevant organs. Moreover, to our knowledge, we report for the first time the quantification of the number and area of metastatic foci, or the total area occupied by these foci in different organs using a highly sensitive cancer cell marker in an ORT EC model, as compared to the mere description of the presence or absence of metastases. Thus, our approach allows the application of statistical tests, with enough power, to determine if there are, or are not, significant differences in metastatic load among the compared groups. Therefore, we believe that both a high metastatic penetrance and a thorough characterization of the number and size of the developed metastatic foci are essential to perform antimetastatic drug evaluation, an aspect seldomly found in published metastatic EC models. A penetrance lower than 100% cannot reliably ensure that the antimetastatic effect achieved is due to the effect of the tested drug or to the lack of EC engraftment in primary tumor or metastatic sites. In addition, a lack of description regarding the number and size of metastases reduces the ability to identify the possible antimetastatic effect of the tested drug.

Importantly, the development of the SC CXCR4$^+$ EC model allowed us to prove that our previously developed protein nanocarrier shows high accumulation in CXCR4$^+$ EC tumors because of its capacity to selectively target the CXCR4 receptor. Based on these results, further development of a nanomedicine loading T22-GFP-H6 with drugs of choice for EC would be fully feasible. Then, our next goal will be to generate antimetastatic nanomedicines that achieve targeted drug delivery, to test them in the novel CXCR4$^+$ SC and advanced ORT EC models. We expect these nanomedicines to selectively deliver to EC tissues a payload drug of choice (e. g., genotoxic, microtubule inhibitor, toxin . . .) with high cytotoxic activity in EC in vitro, expecting first to inhibit tumor growth in the CXCR4$^+$ SC EC model, and then, to achieve the control of metastatic dissemination in the CXCR4$^+$ ORT EC model. A successful preclinical development will, then, initiate an effort to reach clinical translation for patients in most need of new therapies—medically inoperable patients or those who bear highly metastatic, high risk or recurrent EC—selecting only as candidates for treatment, patients bearing CXCR4$^+$ EC, since they are more likely to respond.

5. Conclusions

In conclusion, we have developed a CXCR4-overexpresing (CXCR4$^+$) advanced EC model that improves previously reported models, regarding its CXCR4 overexpression resembling human tumors, and its metastatic penetrance, which was 100% in all clinically relevant sites. Using highly sensitive IHC to detect human vimentin, we reliably identified single EC cells and EC metastatic clusters invading tissues, improving the precision of the metastatic foci detection in EC, in terms of both foci number and size. In addition, we have used this model and this IHC method to demonstrate for the first time that CXCR4-overexpression enhances metastatic dissemination in EC. This model can be used as a resource for the development of therapeutic approaches that target CXCR4, which is overexpressed in EC, especially aimed at patients who currently lack an effective therapy. This way, CXCR4$^+$ EC patients will benefit from CXCR4-targeted therapies, including CXCR4 inhibitors or CXCR4-targeted drug delivery approaches.

Supplementary Materials: The following supporting information can be downloaded at: https://www.mdpi.com/article/10.3390/biomedicines10071680/s1: Figure S1. Experimental design of the in vivo evaluation of T22-GFP-H6 biodistribution; Figure S2. Comparison of the percentage of CXCR4+ cells with membrane expression after transduction with CXCR4-Luciferase constructs in human endometrial cancer cell lines; Figure S3. Luciferase activity in transduced human endometrial cancer cell lines AN3CA, ARK2 and HEC1A; Figure S4. Tumor growth and phenotypic characterization of the human CXCR4+ AN3CA cell line-derived subcutaneous cancer mouse model; Figure S5. Schematic design and functional versatility of CXCR4-targeted nanocarrier T22-GFP-H6; Figure S6. In vitro CXCR4-dependent internalization of T22-GFP-H6 in CXCR4+ EC AN3CA cells.

Author Contributions: Conceptualization, E.M.-G., M.V.C., A.V., E.V., I.C. and R.M.; Data curation, R.M.; Formal analysis, E.M.-G., M.À.P., L.A.-P. and R.M.; Funding acquisition, U.U., A.V., E.V. and R.M.; Investigation, E.M.-G., A.G. and E.R.-B.; Methodology, E.M.-G., M.V.C., E.R.-B. and L.F.; Project administration, U.U., A.V., E.V., I.C. and R.M.; Resources, U.U., A.V., E.V. and R.M.; Supervision, L.A.-C., I.C. and R.M.; Validation, E.M.-G., A.G., E.R.-B., L.A.-C. and U.U.; Visualization, E.M.-G., M.À.P., I.C. and R.M.; Writing—original draft, E.M.-G. and R.M.; Writing—review and editing, M.V.C., M.À.P., U.U., A.V., E.V. and R.M. All authors have read and agreed to the published version of the manuscript.

Funding: This work was supported by Instituto de Salud Carlos III (ISCIII, Spain; Co-funding from FEDER, European Union) (PI21/00159, PIPI18/00650 and EU COST Action CA 17140 to Ramon Mangues and PI20/00400 to Ugutz Unzueta); Agencia Estatal de Investigación (AEI, Spain) and Fondo Europeo de Desarrollo Regional (FEDER, European Union) (grant BIO2016-76063-R, AEI/FEDER, UE to Antonio Villaverde and grant PID2019-105416RBI00/AEI/10.13039/501100011033 to Esther Vázquez); CIBER-BBN (Spain) (CB06/01/1031 and 4NanoMets to Ramon Mangues, VENOM4CANCER to Antonio Villaverde, NANOREMOTE to Esther Vázquez, and NANOSCAPE to Ugutz Unzueta]; AGAUR (Spain) 2017-SGR-865 to Ramon Mangues, and 2017SGR-229 to Antonio Villaverde; Josep Carreras Leukemia Research Institute (Spain) (P/AG to Ramon Mangues). Elisa Rioja-Blanco was supported by a predoctoral fellowship from AGAUR (Spain) (2020FI_B2 00168 and 2018FI_B2_00051) co-funded by European Social Fund (ESF investing in your future, European Union). Lorena Alba-Castellón was supported by a postdoctoral fellowship from AECC (Spanish Association of Cancer Research, Spain). Antonio Villaverde received an Icrea Academia Award (Spain). Ugutz Unzueta and Mª Virtudes Céspedes were also supported by Miguel Servet fellowships (CP19/00028 and CPII20/00007, respectively) from Instituto de Salud Carlos III (Spain) co-funded by European Social Fund (ESF investing in your future, European Union). The bioluminescent follow-up of cancer cells and toxicity studies have been performed in the ICTS-141007 Nanbiosis Platform, using its CIBER-BBN Nanotoxicology Unit (http://www.nanbiosis.es/portfolio/u18-nanotoxicology-unit/, accessed on 15 May 2021). Protein production has been partially performed with the ICTS "NANBIOSIS", more specifically with the Protein Production Platform of CIBER-BBN/IBB (http://www.nanbiosis.es/unit/u1-protein-production-platform-ppp/, accessed on 15 May 2021).

Institutional Review Board Statement: All the animal experiments were approved by the Institutional Animal Ethics Committee of Hospital de Sant Pau and approved by the local government (Generalitat de Catalunya) (protocol number 10234; date of approval: 24 February 2019).

Informed Consent Statement: Informed consent was obtained from all subjects involved in the study.

Data Availability Statement: The datasets generated during and/or analyzed during the current study, as well as additional information and data, are available from the corresponding author upon reasonable request.

Conflicts of Interest: Antonio Villaverde, Esther Vázquez, Ugutz Unzueta, Ramon Mangues, and Isolda Casanova are cited as inventors in PCT/EP2012/050513 covering Targeted Delivery of Therapeutic Molecules to CXCR4 Cells, and in PCT/EP2018/061732, covering Therapeutic Nanostructured Proteins. All other authors report no conflict of interest in this work.

References

1. World Cancer Research Fund. Endometrial Cancer Statistics 2018. Available online: https://www.wcrf.org/dietandcancer/cancer-trends/endometrial-cancer-statistics (accessed on 4 April 2020).
2. Sung, H.; Ferlay, J.; Siegel, R.L.; Laversanne, M.; Soerjomataram, I.; Jemal, A.; Bray, F. Global Cancer Statistics 2020: GLOBOCAN Estimates of Incidence and Mortality Worldwide for 36 Cancers in 185 Countries. *CA Cancer J. Clin.* **2021**, *71*, 209–249. [CrossRef]
3. Yu, M.; Wang, Y.; Yuan, Z.; Zong, X.; Huo, X.; Cao, D.-Y.; Yang, J.-X.; Shen, K. Fertility-Sparing Treatment in Young Patients with Grade 2 Presumed Stage IA Endometrioid Endometrial Adenocarcinoma. *Front. Oncol.* **2020**, *10*, 1437. [CrossRef]
4. Cavaliere, A.F.; Perelli, F.; Zaami, S.; D'Indinosante, M.; Turrini, I.; Giusti, M.; Gullo, G.; Vizzielli, G.; Mattei, A.; Scambia, G.; et al. Fertility Sparing Treatments in Endometrial Cancer Patients: The Potential Role of the New Molecular Classification. *Int. J. Mol. Sci.* **2021**, *22*, 12248. [CrossRef] [PubMed]
5. Gullo, G.; Etrusco, A.; Cucinella, G.; Perino, A.; Chiantera, V.; Laganà, A.S.; Tomaiuolo, R.; Vitagliano, A.; Giampaolino, P.; Noventa, M.; et al. Fertility-Sparing Approach in Women Affected by Stage I and Low-Grade Endometrial Carcinoma: An Updated Overview. *Int. J. Mol. Sci.* **2021**, *22*, 11825. [CrossRef] [PubMed]
6. Jemal, A.; Bray, F.; Center, M.M.; Ferlay, J.; Ward, E.; Forman, D. Global cancer statistics. *CA Cancer J. Clin.* **2011**, *61*, 69–90. [CrossRef] [PubMed]
7. Braun, M.M.; Overbeek-Wager, E.A.; Grumbo, R.J. Diagnosis and Management of Endometrial Cancer. *Am. Fam. Physician* **2016**, *93*, 468–474. [PubMed]
8. Santaballa, A.; Matías-Guiu, X.; Redondo, A.; Carballo, N.; Gil, M.; Gómez, C.; Gorostidi, M.; Gutierrez, M.; Gónzalez-Martín, A. SEOM clinical guidelines for endometrial cancer (2017). *Clin. Transl. Oncol.* **2018**, *20*, 29–37. [CrossRef] [PubMed]
9. Campos, S.M.; Cohn, D.E. Treatment of Metastatic Endometrial Cancer. *UpToDate* **2021**, *26*, 15–23. Available online: https://www.uptodate.com/contents/treatment-of-metastatic-endometrial-cancer (accessed on 20 December 2021).
10. Van Nyen, T.; Moiola, C.P.; Colas, E.; Annibali, D.; Amant, F. Modeling Endometrial Cancer: Past, Present, and Future. *Int. J. Mol. Sci.* **2018**, *19*, 2348. [CrossRef]
11. Cabrera, S.; Llauradó, M.; Castellví, J.; Fernandez, Y.; Alameda, F.; Colás, E.; Ruiz, A.; Doll, A.; Schwartz, S., Jr.; Carreras, R.; et al. Generation and characterization of orthotopic murine models for endometrial cancer. *Clin. Exp. Metastasis* **2012**, *29*, 217–227. [CrossRef]
12. Fedorko, A.M.; Kim, T.H.; Broaddus, R.; Schmandt, R.; Chandramouli, G.V.; Kim, H.I.; Jeong, J.-W.; Risinger, J.I. An immune competent orthotopic model of endometrial cancer with metastasis. *Heliyon* **2020**, *6*, e04075. [CrossRef]
13. Konings, G.F.; Saarinen, N.; Delvoux, B.; Kooreman, L.; Koskimies, P.; Krakstad, C.; Fasmer, K.E.; Haldorsen, I.S.; Zaffagnini, A.; Häkkinen, M.R.; et al. Development of an Image-Guided Orthotopic Xenograft Mouse Model of Endometrial Cancer with Controllable Estrogen Exposure. *Int. J. Mol. Sci.* **2018**, *19*, 2547. [CrossRef]
14. Pillozzi, S.; Fortunato, A.; De Lorenzo, E.; Borrani, E.; Giachi, M.; Scarselli, G.; Arcangeli, A.; Noci, I. Over-Expression of the LH Receptor Increases Distant Metastases in an Endometrial Cancer Mouse Model. *Front. Oncol.* **2013**, *3*, 285. [CrossRef]
15. Winship, A.L.; Van Sinderen, M.; Donoghue, J.; Rainczuk, K.; Dimitriadis, E. Targeting Interleukin-11 Receptor-α Impairs Human Endometrial Cancer Cell Proliferation and Invasion In Vitro and Reduces Tumor Growth and Metastasis In Vivo. *Mol. Cancer Ther.* **2016**, *15*, 720–730. [CrossRef]
16. Chen, H.Y.; Chiang, Y.-F.; Huang, J.-S.; Huang, T.-C.; Shih, Y.-H.; Wang, K.-L.; Ali, M.; Hong, Y.-H.; Shieh, T.-M.; Hsia, S.-M. Isoliquiritigenin Reverses Epithelial-Mesenchymal Transition Through Modulation of the TGF-β/Smad Signaling Pathway in Endometrial Cancer. *Cancers* **2021**, *13*, 1236. [CrossRef]
17. Doll, A.; Gonzalez, M.; Abal, M.; Llaurado, M.; Rigau, M.; Colas, E.; Monge, M.; Xercavins, J.; Capella, G.; Diaz, B.; et al. An orthotopic endometrial cancer mouse model demonstrates a role for RUNX1 in distant metastasis. *Int. J. Cancer* **2009**, *125*, 257–263. [CrossRef]
18. Haldorsen, I.S.; Popa, M.; Fonnes, T.; Brekke, N.; Kopperud, R.; Visser, N.C.; Rygh, C.B.; Pavlin, T.; Salvesen, H.B.; Mc Cormack, E.; et al. Multimodal Imaging of Orthotopic Mouse Model of Endometrial Carcinoma. *PLoS ONE* **2015**, *10*, e0135220. [CrossRef]
19. Hanekamp, E.E.; Gielen, S.C.; van Oosterhoud, S.A.; Burger, C.W.; Grootegoed, J.; Huikeshoven, F.J.; Blok, L.J. Progesterone receptors in endometrial cancer invasion and metastasis: Development of a mouse model. *Steroids* **2003**, *68*, 795–800. [CrossRef]
20. Kato, M.; Onoyama, I.; Yoshida, S.; Cui, L.; Kawamura, K.; Kodama, K.; Hori, E.; Matsumura, Y.; Yagi, H.; Asanoma, K.; et al. Dual-specificity phosphatase 6 plays a critical role in the maintenance of a cancer stem-like cell phenotype in human endometrial cancer. *Int. J. Cancer* **2020**, *147*, 1987–1999. [CrossRef]
21. Popli, P.; Richters, M.M.; Chadchan, S.B.; Kim, T.H.; Tycksen, E.; Griffith, O.; Thaker, P.H.; Griffith, M.; Kommagani, R. Splicing factor SF3B1 promotes endometrial cancer progression via regulating KSR2 RNA maturation. *Cell Death Dis.* **2020**, *11*, 842. [CrossRef]
22. Unno, K.; Ono, M.; Winder, A.D.; Maniar, K.P.; Paintal, A.S.; Yu, Y.; Wei, J.-J.; Lurain, J.R.; Kim, J.J. Establishment of Human Patient-Derived Endometrial Cancer Xenografts in NOD scid Gamma Mice for the Study of Invasion and Metastasis. *PLoS ONE* **2014**, *9*, e116064. [CrossRef]
23. Liu, P.; Long, P.; Huang, Y.; Sun, F.; Wang, Z. CXCL12/CXCR4 axis induces proliferation and invasion in human endometrial cancer. *Am. J. Transl. Res.* **2016**, *11*, 1719.
24. Teng, F.; Tian, W.-Y.; Wang, Y.-M.; Zhang, Y.-F.; Guo, F.; Zhao, J.; Gao, C.; Xue, F.-X. Cancer-associated fibroblasts promote the progression of endometrial cancer via the SDF-1/CXCR4 axis. *J. Hematol. Oncol.* **2016**, *9*, 8. [CrossRef]

25. Buchynska, L.G.; Movchan, O.M.; Iurchenko, N.P. Expression of chemokine receptor CXCR4 in tumor cells and content of CXCL12+-fibroblasts in endometrioid carcinoma of endometrium. *Exp. Oncol.* **2021**, *43*, 135–141. [PubMed]
26. Kircher, M.; Herhaus, P.; Schottelius, M.; Buck, A.K.; Werner, R.A.; Wester, H.-J.; Keller, U.; Lapa, C. CXCR4-directed theranostics in oncology and inflammation. *Ann. Nucl. Med.* **2018**, *32*, 503–511. [CrossRef] [PubMed]
27. Felix, A.S.; Edwards, R.; Bowser, R.; Linkov, F. Chemokines and Cancer Progression: A Qualitative Review on the Role of Stromal Cell-derived Factor 1-alpha and CXCR4 in Endometrial Cancer. *Cancer Microenviron.* **2010**, *3*, 49–56. [CrossRef] [PubMed]
28. Mizokami, Y.; Kajiyama, H.; Shibata, K.; Ino, K.; Kikkawa, F.; Mizutani, S. Stromal cell-derived factor-1?-induced cell proliferation and its possible regulation by CD26/dipeptidyl peptidase IV in endometrial adenocarcinoma. *Int. J. Cancer* **2004**, *110*, 652–659. [CrossRef] [PubMed]
29. Gelmini, S.; Mangoni, M.; Castiglione, F.; Beltrami, C.; Pieralli, A.; Andersson, K.L.; Fambrini, M.; Taddei, G.L.; Serio, M.; Orlando, C. The CXCR4/CXCL12 axis in endometrial cancer. *Clin. Exp. Metastasis* **2009**, *26*, 261–268. [CrossRef]
30. Lefort, S.; Thuleau, A.; Kieffer, Y.; Sirven, P.; Bieche, I.; Marangoni, E.; Vincent-Salomon, A.; Mechta-Grigoriou, F. CXCR4 inhibitors could benefit to HER2 but not to triple-negative breast cancer patients. *Oncogene* **2016**, *36*, 1211–1222. [CrossRef] [PubMed]
31. Lyu, L.; Zheng, Y.; Hong, Y.; Wang, M.; Deng, Y.; Wu, Y.; Xu, P.; Yang, S.; Wang, S.; Yao, J.; et al. Comprehensive analysis of the prognostic value and immune function of chemokine-CXC receptor family members in breast cancer. *Int. Immunopharmacol.* **2020**, *87*, 106797. [CrossRef] [PubMed]
32. Krikun, G. The CXL12/CXCR4/CXCR7 axis in female reproductive tract disease: Review. *Am. J. Reprod. Immunol.* **2018**, *80*, e13028. [CrossRef]
33. Walentowicz-Sadlecka, M.; Sadlecki, P.; Bodnar, M.; Marszalek, A.; Walentowicz, P.; Sokup, A.; Wilińska-Jankowska, A.; Grabiec, M. Stromal Derived Factor-1 (SDF-1) and Its Receptors CXCR4 and CXCR7 in Endometrial Cancer Patients. *PLoS ONE* **2014**, *9*, e84629. [CrossRef]
34. Huang, Y.; Ye, Y.; Long, P.; Zhao, S.; Zhang, L. Silencing of CXCR4 and CXCR7 expression by RNA interference suppresses human endometrial carcinoma growth in vivo. *Am. J. Transl. Res.* **2017**, *9*, 1896–1904.
35. Sirohi, V.K.; Popli, P.; Sankhwar, P.; Kaushal, J.B.; Gupta, K.; Manohar, M.; Dwivedi, A. Curcumin exhibits anti-tumor effect and attenuates cellular migration via Slit-2 mediated down-regulation of SDF-1 and CXCR4 in endometrial adenocarcinoma cells. *J. Nutr. Biochem.* **2017**, *44*, 60–70. [CrossRef]
36. Sun, Y.; Yoshida, T.; Okabe, M.; Zhou, K.; Wang, F.; Soko, C.; Saito, S.; Nikaido, T. Isolation of Stem-Like Cancer Cells in Primary Endometrial Cancer Using Cell Surface Markers CD133 and CXCR4. *Transl. Oncol.* **2017**, *10*, 976–987. [CrossRef]
37. Villaverde, A.; Unzueta, U.; Céspedes, M.V.; Ferrer-Miralles, N.; Casanova, I.; Cedano, J.; Corchero, J.L.; Domingo-Espín, J.; Mangues, R.; Vázquez, E. Intracellular CXCR4+ cell targeting with T22-empowered protein-only nanoparticles. *Int. J. Nanomed.* **2012**, *7*, 4533–4544. [CrossRef]
38. Céspedes, M.V.; Unzueta, U.; Álamo, P.; Gallardo, A.; Sala, R.; Casanova, I.; Pavon, M.A.; Mangues, M.A.; Trías, M.; Pousa, A.L.; et al. Cancer-specific uptake of a liganded protein nanocarrier targeting aggressive CXCR4 + colorectal cancer models. *Nanomed. Nanotechnol. Biol. Med.* **2016**, *12*, 1987–1996. [CrossRef]
39. Falgàs, A.; Pallarès, V.; Unzueta, U.; Céspedes, M.V.; Arroyo-Solera, I.; Moreno, M.J.; Sierra, J.; Gallardo, A.; Mangues, M.A.; Vázquez, E.; et al. A CXCR4-targeted nanocarrier achieves highly selective tumor uptake in diffuse large B-cell lymphoma mouse models. *Haematologica* **2019**, *105*, 741–753. [CrossRef]
40. Rioja-Blanco, E.; Arroyo-Solera, I.; Álamo, P.; Casanova, I.; Gallardo, A.; Unzueta, U.; Serna, N.; Sánchez-García, L.; Quer, M.; Villaverde, A.; et al. Self-assembling protein nanocarrier for selective delivery of cytotoxic polypeptides to CXCR4+ head and neck squamous cell carcinoma tumors. *Acta Pharm. Sin. B* **2021**, *12*, 2578–2591. [CrossRef]
41. Bankhead, P.; Loughrey, M.B.; Fernández, J.A.; Dombrowski, Y.; McArt, D.G.; Dunne, P.D.; McQuaid, S.; Gray, R.T.; Murray, L.J.; Coleman, H.G.; et al. QuPath: Open source software for digital pathology image analysis. *Sci. Rep.* **2017**, *7*, 16878. [CrossRef]
42. Cohen, J. *Statistical Power Analysis for the Behavioral Sciences*, 2nd ed.; L. Erlbaum Associates: Hillsdale, NJ, USA, 1988.
43. Ialongo, C. Understanding the effect size and its measures. *Biochem. Med.* **2016**, *26*, 150–163. [CrossRef]
44. Friel, A.; Sergent, P.; Patnaude, C.; Szotek, P.P.; Oliva, E.; Scadden, D.T.; Seiden, M.V.; Foster, R.; Rueda, B.R. Functional analyses of the cancer stem cell-like properties of human endometrial tumor initiating cells. *Cell Cycle* **2008**, *7*, 242–249. [CrossRef]
45. The Cancer Genome Atlas Research Network; Kandoth, C.; Schultz, N.; Cherniack, A.D.; Akbani, R.; Liu, Y.; Shen, H.; Robertson, A.-G.; Pashtan, I.; Shen, R.; et al. Integrated genomic characterization of endometrial carcinoma. *Nature* **2013**, *497*, 67–73. [CrossRef]
46. Piulats, J.M.; Guerra, E.; Gil-Martín, M.; Roman-Canal, B.; Gatius, S.; Sanz-Pamplona, R.; Velasco, A.; Vidal, A.; Matias-Guiu, X. Molecular approaches for classifying endometrial carcinoma. *Gynecol. Oncol.* **2016**, *145*, 200–207. [CrossRef]
47. Tomiyama, L.; Kamino, H.; Fukamachi, H.; Urano, T. Precise epitope determination of the anti-vimentin monoclonal antibody V9. *Mol. Med. Rep.* **2017**, *16*, 3917–3921. [CrossRef]
48. Pallarès, V.; Unzueta, U.; Falgàs, A.; Sánchez-García, L.; Serna, N.; Gallardo, A.; Morris, G.A.; Alba-Castellón, L.; Álamo, P.; Sierra, J.; et al. An Auristatin nanoconjugate targeting CXCR4+ leukemic cells blocks acute myeloid leukemia dissemination. *J. Hematol. Oncol.* **2020**, *13*, 36. [CrossRef]
49. Walenkamp, A.M.E.; Lapa, C.; Herrmann, K.; Wester, H.-J. CXCR4 Ligands: The Next Big Hit? *J. Nucl. Med.* **2017**, *58*, 77S–82S. [CrossRef]

 *biomedicines*

Article

A Novel Murine Multi-Hit Model of Perinatal Acute Diffuse White Matter Injury Recapitulates Major Features of Human Disease

Patricia Renz [1,2], Andreina Schoeberlein [1,2], Valérie Haesler [1,2], Theoni Maragkou [3], Daniel Surbek [1,2] and Amanda Brosius Lutz [1,2,*]

1   Department for BioMedical Research, University of Bern and Switzerland, 3010 Bern, Switzerland
2   Department of Obstetrics and Gynecology, Division of Feto-Maternal Medicine University Hospital, University of Bern, 3010 Bern, Switzerland
3   Institute of Pathology, University of Bern, 3010 Bern, Switzerland
*   Correspondence: amanda.brosiuslutz@insel.ch

Abstract: The selection of an appropriate animal model is key to the production of results with optimal relevance to human disease. Particularly in the case of perinatal brain injury, a dearth of affected human neonatal tissue available for research purposes increases the reliance on animal models for insight into disease mechanisms. Improvements in obstetric and neonatal care in the past 20 years have caused the pathologic hallmarks of perinatal white matter injury (WMI) to evolve away from cystic necrotic lesions and toward diffuse regions of reactive gliosis and persistent myelin disruption. Therefore, updated animal models are needed that recapitulate the key features of contemporary disease. Here, we report a murine model of acute diffuse perinatal WMI induced through a two-hit inflammatory–hypoxic injury paradigm. Consistent with diffuse human perinatal white matter injury (dWMI), our model did not show the formation of cystic lesions. Corresponding to cellular outcomes of dWMI, our injury protocol produced reactive microgliosis and astrogliosis, disrupted oligodendrocyte maturation, and disrupted myelination.. Functionally, we observed sensorimotor and cognitive deficits in affected mice. In conclusion, we report a novel murine model of dWMI that induces a pattern of brain injury mirroring multiple key aspects of the contemporary human clinical disease scenario.

Keywords: perinatal brain injury; diffuse injury; white matter injury; mouse model; gliosis; myelination failure; two-hit model

Citation: Renz, P.; Schoeberlein, A.; Haesler, V.; Maragkou, T.; Surbek, D.; Brosius Lutz, A. A Novel Murine Multi-Hit Model of Perinatal Acute Diffuse White Matter Injury Recapitulates Major Features of Human Disease. *Biomedicines* **2022**, *10*, 2810. https://doi.org/10.3390/biomedicines10112810

Academic Editor: Juan Sahuquillo

Received: 15 September 2022
Accepted: 27 October 2022
Published: 4 November 2022

Publisher's Note: MDPI stays neutral with regard to jurisdictional claims in published maps and institutional affiliations.

1. Introduction

The WHO estimates that one in ten live births occurs preterm (prior to 37 weeks gestation) worldwide, with major regional differences [1]. While ongoing progress in neonatal care has improved the survival of preterm infants, increased survival in this population is accompanied by a climbing disease burden due to the chronic disabilities associated with prematurity [1]. Approximately 10% of preterm survivors suffer permanent motor impairment, while up to 25–50% of preterm infants exhibit other neurodevelopmental deficits, including cognitive, learning, and social-behavioral disabilities [1]. White matter injury is the predominant form of brain injury in survivors of preterm birth, resulting from preterm-birth-associated perinatal inflammatory and hypoxic–ischemic insults due to clinical scenarios, such as chorioamnionitis, neonatal sepsis, pulmonary immaturity, and perinatal hemodynamic instability [2]. Preclinical and clinical studies emphasize the importance of numerous perinatal hits in the etiology of WMI, with initial insults sensitizing the developing brain to subsequent injury [2,3].

Whereas large foci of necrosis were commonly identified in preterm white matter injury tissue prior to 2000, thanks to advancements in obstetric and neonatal care, diffuse non-necrotic lesions have become the predominant disease pathology in recent decades [4]. These diffuse lesions are best characterized histopathologically by the presence of reactive astrocytes and microglia, disrupted maturation of pre-oligodendrocytes, and consequent myelination failure [1]. While subsequent dysmaturational events lead to long-term abnormalities in brain development, including gray matter abnormalities, in the acute phase, cortical gray matter, as well as axons, are generally spared [1].

The key to understanding the pathophysiology underlying these diffuse white matter lesions is the maturation-dependent vulnerability of the oligodendrocyte lineage [5,6]. Susceptibility to preterm white matter injury peaks between 23 and 32 weeks postconception, a period during which the pre-oligodendrocyte (pre-OL), a pre-myelinating oligodendrocyte progenitor, predominates in human cerebral white matter. Compared with earlier and later developmental stages of the oligodendrocyte lineage, the pre-OL is highly vulnerable to oxidative stress [2,6]. The temporal appearance and spatial distribution of this cell type correlate with the magnitude and location of WMI in humans and in experimental studies [5]. Taken together, the evidence thus far generates a working hypothesis of disease pathogenesis in which multiple perinatal insults during the vulnerable phase of oligodendrocyte cell lineage development generate reactive glia and impair oligodendrocyte lineage maturation. The result is long-term myelination failure and dysmaturational events, culminating in a spectrum of motor, cognitive, and behavioral disabilities [1].

Despite years of study, the molecular underpinnings of white matter injury remain incompletely understood, therapeutic windows are poorly defined, and therapeutic options are extremely limited [1,3,7]. Given the evident paucity of human pathologic specimens from affected infant brains available for research, reliance on experimental models that reproduce key features of human disease is even greater than in other areas of study. Numerous mammalian models (mouse, rat, sheep, and non-human primate) have been developed to study white matter injury disease mechanisms and potential therapeutic interventions [8–14]. Many of these models have been developed to model the focal cystic necrosis produced by earlier WMI. Murine models that reproduce the two-hit etiology of WMI and produce lesions mimicking diffuse white matter injury are needed to ensure relevance to contemporary disease patterns and to allow the incorporation of molecular genetic tools established in mice. Here, we present a mouse model of acute diffuse perinatal white matter injury based on a two-hit hypoxic–inflammatory insult during the period of pre-OL predominance in murine white matter corresponding to 23–32 weeks postconception in humans [15]. Our model generates reactive changes in astrocytes and microglia, pre-OL maturation failure, defects in myelination, and behavioral deficits in the absence of cystic white matter lesions or extensive primary gray matter injury, thus recapitulating multiple key features of human acute diffuse WMI.

2. Materials and Methods

2.1. Animals

All animal procedures were approved by the Veterinary Department of the Canton of Bern, Switzerland (Protocol reference number: BE19/85), and the animals were maintained under standard housing conditions.

2.2. Animal Model

C57BL/6 mouse pups on postnatal day (P) 2 weighing 1.5–1.8 g were randomly divided into injured and control group. Lipopolysaccharide (LPS; Escherichia coli strain O55:B5, Sigma Aldrich, St. Louis, MO, USA) was diluted in sterile saline to a working concentration of 0.2 mg/mL, and injected at a dose of 2 mg/kg subcutaneously (s.c.) between the scapulae. Healthy control pups received an equivalent weight-adjusted saline injection. Following injection, pups were returned to their home cages. Hypoxia (8% O_2/92% N_2, 3 L/min) was performed 6 h after LPS injection in a temperature-controlled isolette (34 °C) for 25 min under continuous O_2 monitoring. Healthy control pups were removed from dams and exposed to room air during this time (Figure 1). LPS dose and hypoxia duration were titrated to optimize for significant myelination defects and minimal animal mortality.

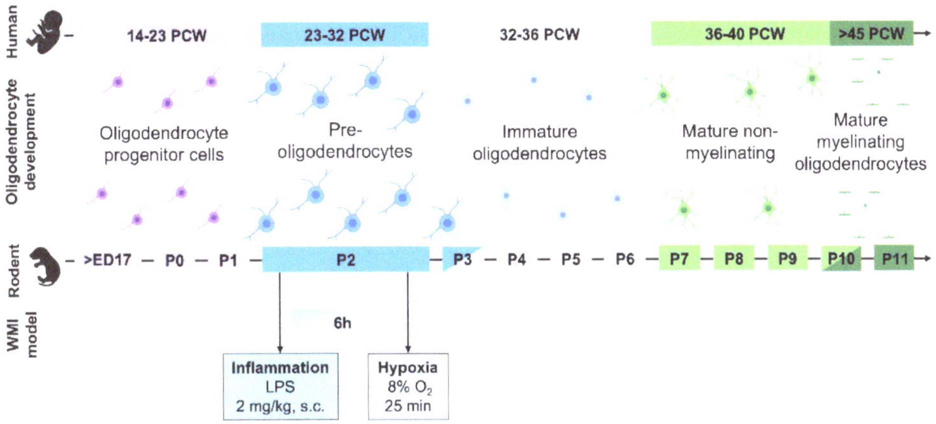

Figure 1. Murine 2-hit inflammatory–hypoxic model for acute diffuse perinatal white matter injury. WMI was induced in postnatal day 2 (P2) mice pups by s.c. injection of LPS, followed by exposure to hypoxia for 25 min 6 h later. PCW, postconception week; ED, embryonic day; P, postnatal day. Created with BioRender.com with data from [8].

2.3. Tissue Preparation

Mice pups at P3 (1 day post injury, 1 dpi) and P4 (2 days post injury, 2 dpi) were euthanized using rapid decapitation. Brains were removed and directly fixed in 4% paraformaldehyde (PFA) for 20 h at 4 °C. Mice pups at P11 (9 days post injury, 9 dpi) were euthanized with a terminal dose of sodium pentobarbital (150 mg/kg body weight, i.p.; Esconarkon, Streuli Tiergesundheit AG, Switzerland) and transcardially perfused with phosphate-buffered saline (PBS). Brains were fixed in 4% PFA for 22 h at 4 °C. Following fixation, brains were transferred to PBS. Following removal of the cerebelli and olfactory bulbs, brains were dehydrated through sequential emersion in ethanol and xylene and embedded in paraffin. Brains were sectioned into 6 μM slices using an HM 340E rotary microtome (Thermo Fisher Scientific, Waltham, MA, USA). Brains were sectioned in the coronal plane at the level of the hippocampus corresponding with provided illustrations. This plane was chosen because it allows analysis of the corpus callosum, as well as the internal capsule, regions typically affected in human WMI. Sections outside of this plane or that exhibited lost tissue integrity were excluded from our analysis (Supplementary Table S1).

2.4. Hematoxylin and Eosin (H and E) Staining

Paraffin-embedded tissue sections were deparaffinized and rehydrated. Tissue sections were then immersed in Mayer's hematoxylin (Sigma Aldrich) for 12 min, washed with tap water for 2 min, immersed in eosin (Sigma Aldrich) for 1 min, and washed quickly in 95% ethanol. Tissue sections were then dehydrated by immersion in serial baths of increasing ethanol concentrations, followed by clearing in xylene twice and mounting using Eukitt.

2.5. Histologic Analysis

Paraffin-embedded H-and-E-stained control and injured brains were evaluated by a blinded neuropathologist for the presence of macro- and microscopic cystic lesions (microcysts, previously defined as focal lesions with a diameter of approximately 1 mm) [16,17]. At least 2 brain sections per animal were examined.

2.6. Immunohistochemistry

Paraffin-embedded tissue sections were deparaffinized and rehydrated. Sections were then heated in a pressure cooker in 0.1 M sodium citrate buffer for 12 min for antigen retrieval. Sections were washed with Tris-buffered saline (TBS) Tween® 20 (Sigma Aldrich) and blocked with 10% goat serum and 1% bovine serum albumin (BSA) (Sigma Aldrich) in TBS for 1 h at room temperature. Subsequently, sections were incubated overnight at 4 °C with primary antibodies against the following proteins: Olig2 (1:200, ab109186, Abcam, Cambridge, UK), Ki67 (1:100, 550609, BD, Pharmingen, BD Biosciences, Franklin Lakes, NJ, USA), CNPase (1:200, c5922, Sigma Aldrich), NG2 (ab5320, 1:100, Milipore, Merck KGaA, Darmstadt, Germany), MBP (1:200, ab40390, Abcam), Iba1 (1:2000, ab178846, Abcam), GFAP (1:500, mab360, Millipore), and NF200 (1:1000, smi31r100, Novus Biologicals, Centennial, CO, USA). After washing, the sections were incubated with the appropriate secondary antibody, either Alexa fluor® 488-conjugated or Alexa fluor® 594-conjugated (Thermo Fisher Scientific), for 1 h at room temperature in the dark. The tissue was counterstained with 4′,6-diamidino-2-phenylindole (DAPI) (Sigma Aldrich). NF200 stains were incubated with peroxidase-labeled secondary antibody (1:100; DAKO, Glostrup, Denmark), followed by diaminobenzamidine and EnVision+ System HRP (DAKO) for visualization. The tissue was counterstained with hematoxylin. Cell death was quantified using a terminal deoxynucleotidyl transferase dUTP nick-end labeling (TUNEL) in situ cell death detection kit (TMR rot, Sigma Aldrich). Blocked sections were stained according to the manufacturer's protocol (Chapter 3.3.4, Version 17, Sigma-Aldrich). After TUNEL staining, sections were stained with primary antibodies against the oligodendrocyte marker Olig2 (diluted in 10% goat serum and 1% BSA in PBS; incubated overnight at 4 °C). Sections were then incubated with Alexa fluor® 488-conjugated secondary antibody (Thermo Fisher Scientific) for 1 h at room temperature and then counterstained with DAPI.

2.7. Image Analysis

All quantifications were carried out blinded to the experimental group. Images were either acquired with a DM6000 B microscope (Leica Microsystems, Wetzlar, Germany) or scanned using a Panoramic 250 Flash II slide scanner (3DHISTECH, Budapest, Hungary). The regions of the corpus callosum, globus pallidus and internal capsule, and cerebral cortex were outlined. For quantification of Ki67, TUNEL, CNPase, and NG2, cells were counted manually within the region of the corpus callosum. For MBP and GFAP, staining was quantified using ImageJ Software v1.47 (Rasband, W.S., National Institutes of Health, Bethesda, MD, USA, http://imagej.nih.gov/ij (accessed on 4 February 2022). Briefly, using a custom macro, the signal area within the brain region of interest and above a defined signal threshold was quantified as the percentage of the defined area. For Iba1 staining, cells were counted manually in each region of interest. The reactivity index was quantified in a multistep process according to [18]. Briefly, using a custom Image J macro, we first determined a signal area at a signal threshold of 40 (approximation for the whole cell area). Subsequently, a second signal area (approximation for the cell body area) was determined at a signal threshold of 90, a minimum size of 150 pixels, and a circularity index of 0.1–1. Cell body fraction was defined as the cell body area divided by the whole cell area. For IHC experiments, two brain sections were quantified and the results averaged to generate each datapoint. In rare cases, we were only able to quantify IHC results for one brain section because of insufficient tissue preservation.

2.8. Hindlimb Foot Angle

Foot angle was measured in video recordings of walking pups (13 days post injury, 13 dpi) by drawing a line from the center of the heel through the middle (longest) toe. The measurement was only performed when a pup took a full step in a straight line, and both hindfeet were flat on the ground. Three to five foot angles were measured, and the average angle was calculated for each pup [19].

2.9. Rotarod

Motor coordination and balance were determined in a rotarod test for injured and healthy control mice (28 days post injury, 28 dpi). Mice were trained on one day to remain on a rotating rod (five rotations per minute (rpm)) over a five-minute period. The following day, the rotational speed was increased continuously from 15 to 33 rpm, alternating between forward and reverse rotation modes. Trials were terminated when the mice fell off the rotarod, clung to the bar for two full rotations, or remained on the bar for five minutes. The procedure was repeated for a total of three trials, each separated by 15 min. The mean latency to fall was analyzed.

2.10. Novel Object Recognition

A novel object recognition task was used to assess recognition memory at 28 dpi [20]. Mice were placed in an open field where two identical objects were placed. Four hours later, one object was replaced by a new one at the same position, and the mice explored the open arena in the presence of both the new and a familiar object. The total time spent exploring the objects was limited to 20 s, with a maximum time of 10 min. Object types were placed on a randomly assigned side (left or right) of the open field to eliminate any possible side or object preference. Novel object recognition was demonstrated by discrimination ratios (novel object interaction/total interaction with both objects) above 0.5.

2.11. Statistical Analysis

Given the sample size < 30 for all of our analyses and, therefore, the inability to reliably test for normal distribution of data, we applied the Mann–Whitney test for the pairwise comparison of continuous variables throughout the manuscript. Significance was determined at $p < 0.05$.

3. Results

3.1. Perinatal Inflammation–Hypoxia Does Not Lead to the Formation of Macroscopic or Microscopic Cystic Lesions

Contemporary cohorts of human postmortem tissue affected by perinatal white matter injury reveal a strong decline in the burden of necrotic lesions in periventricular white matter tracts in comparison with older cohorts [1]. We, therefore, began with a histologic evaluation of H-and-E-stained paraffin sections from our mouse model. This analysis, performed blinded at 1 dpi, 2 dpi, and 9 dpi, did not detect the presence of matrix changes corresponding to cystic lesions in the gray or white matter of any control or injured brains at the timepoints examined (Figure 2A). In contrast to earlier disease cohorts, studies of modern human perinatal white matter injury demonstrate only minimal neuronal loss in the cortical and deep gray matter regions [17]. When we quantified cell death in the cerebral cortex using TUNEL staining, we observed a transient significant increase in cortical cell death at 1 dpi. We did not observe this increase at 2 and 9 dpi (Figure 2B).

3.2. Perinatal Inflammation–Hypoxia Insult Induces White Matter Hypomyelination

A defining feature of perinatal white matter injury is white matter hypomyelination [1,21]. We evaluated myelination using immunohistochemistry for myelin basic protein (MBP) at 9 dpi in three brain regions: the corpus callosum, the cerebral cortex, and the region of the internal capsule and globus pallidus (Figure 3A). These experiments revealed a significant decrease in MBP immunoreactivity in the corpus callosum region and overlying cortex (Figure 3B,D). Myelination remained unchanged in the internal capsule and globus pallidus region (Figure 3D). Quantification of axon density in the corpus callosum at 9 dpi using immunohistochemistry for neurofilament (NF200) did not reveal any significant differences between injured and control tissue (Supplementary Figure S1).

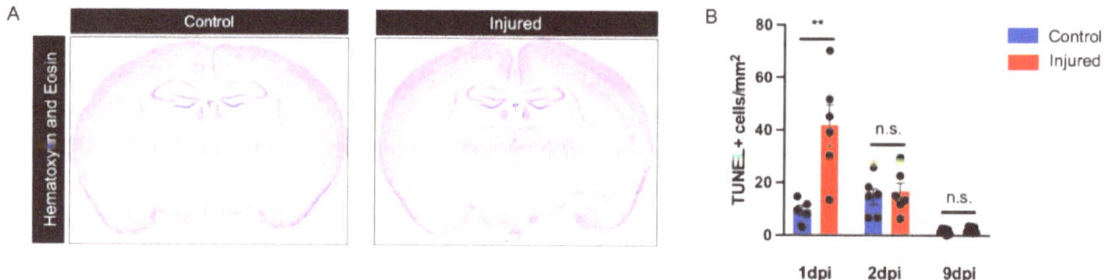

Figure 2. Evaluation of cystic lesion formation and cortical cell death. (**A**) Representative images of hematoxylin and eosin staining of control and injured brains at 9 dpi. (**B**) Quantification of cell death in mouse cortex using TUNEL staining in control and injured mice at 1 dpi (U = 1, p = 0.0087), 2 dpi (U = 18, p >0.999), and 9 dpi (U = 18, p = 0.4557). Sample amounts: 1 dpi: control n = 5, injured n = 6; 2 dpi: control n = 6, injured n = 6; 9 dpi: control n = 11, injured n = 9. Data are presented as mean ± SEM; n.s., not significant; ** p < 0.01.

3.3. Perinatal Inflammation–Hypoxia Leads to Impaired Oligodendrocyte Lineage Differentiation

Studies of human postmortem tissue affected by WMI have revealed that impaired oligodendrocyte maturation rather than a loss of mature oligodendrocytes, underlies white matter hypomyelination in this disease [2,4,6,16]. We, therefore, sought to understand the cellular dynamics underlying decreased myelination in the corpus callosum 9 days after injury in our disease model. We began by quantifying cell proliferation and cell death within the oligodendrocyte lineage in this region using Olig2 (oligodendrocyte lineage marker) immunohistochemistry combined with Ki67 (cell proliferation marker) immunostaining and TUNEL (cell death marker) labeling. These experiments indicated a significant decrease in Olig2+ cell proliferation at 1 and 2 dpi and a significant increase in Olig2+ cell death at 1 and 9 dpi (Figure 4A–D). Examination of the actual numbers of cells dying revealed an approximate 10–15% decrease in the cell proliferation of Olig2+ cells in the injured versus control corpus callosum samples and a rate of Olig2+ cell death in this region of approximately 3–5% in comparison with approximately 2% in healthy control mice. Despite these differences, the total number of NG2+ oligodendrocyte precursor cells, as well as the total number of Olig2+ cells, remained unchanged at these timepoints according to our analysis (Figure 4E–H). When we examined oligodendrocyte lineage differentiation using immunohistochemical labeling for CNPase, a marker of oligodendrocyte maturation beyond the pre-OL stage, we observed a significant decrease in the percentage of Olig2+ cells in the corpus callosum expressing CNPase in injured mice versus control mice, suggesting impaired oligodendrocyte lineage maturation (Figure 4G–I).

3.4. Perinatal Inflammation–Hypoxia Induces Reactive Changes in Microglia and Astrocytes

Another hallmark of human perinatal white matter injury is reactive changes in microglia and astrocytes [4,17,21]. We examined this feature in our disease model, for the corpus callosum, the cerebral cortex, and the region of the internal capsule and globus pallidus. Using immunohistochemistry for Iba1, we quantified the density of Iba1+ cells in each region of interest, as well as the fraction of the total microglia cell area (Iba1+ area) occupied by the cell body (cell body fraction) as a proxy for microglial reactivity. This analysis revealed an initial significant increase in the number of microglia occupying the corpus callosum and cerebral cortex regions of injured mice occurring within 24 h of injury (Figure 5B,E). This increase in cell number was no longer significant in comparison with control mice one day later in the corpus callosum but persisted in the cerebral cortex at 2 dpi (Figure 5E). The globus pallidus and internal capsule region showed no significant increase in microglia number in injured mice in comparison with control mice at the timepoints examined (Figure 5E).

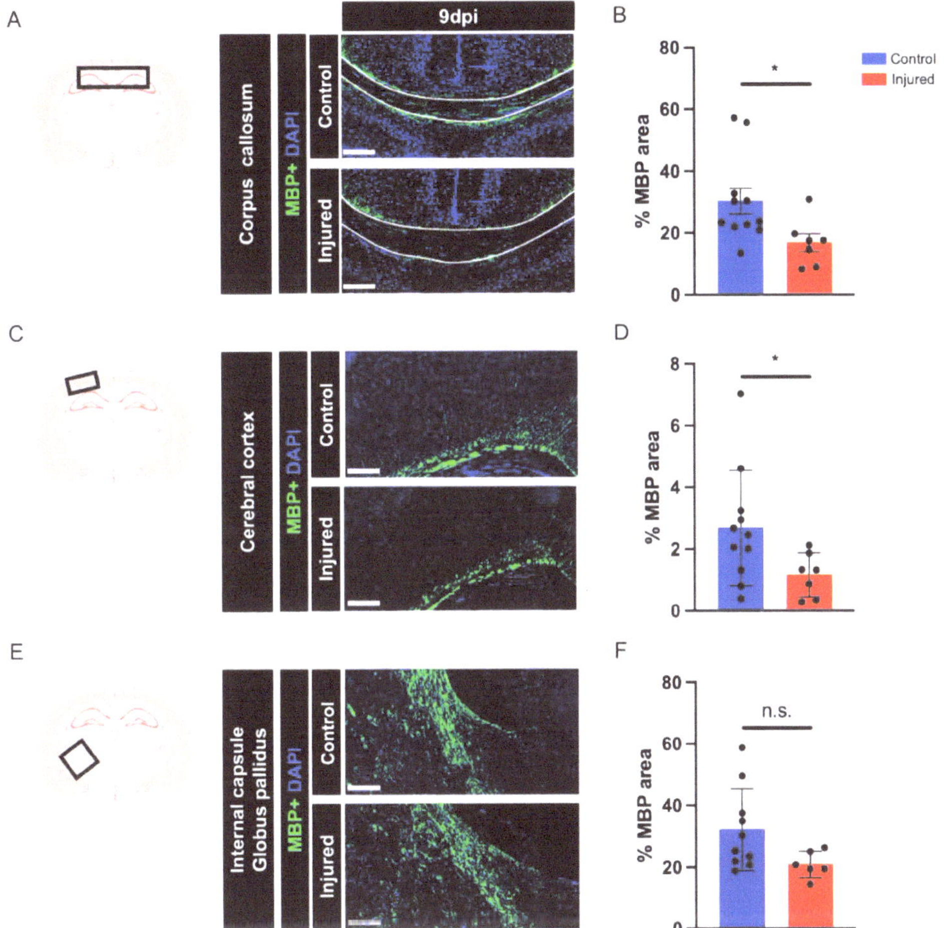

Figure 3. Quantification of myelination defects. Coronal brain sections and representative images indicating the brain regions used for quantification: corpus callosum (**A**), cerebral cortex (**C**), and internal capsule and globus pallidus (**E**). Quantification of MBP immunohistochemistry of injured and control brains at 9 dpi in the corpus callosum (U = 2, *p* = 0.0347) (**B**), cerebral cortex (U = 7, *p* = 0.0513) (**D**), and internal capsule and globus pallidus (U = 1, *p* = 0.0159) (**F**). Scale bar: 200 μm. Sample amounts: 9 dpi: control *n* = 11, injured *n* = 7. Data are presented as mean ± SEM; n.s., not significant; * *p* < 0.05.

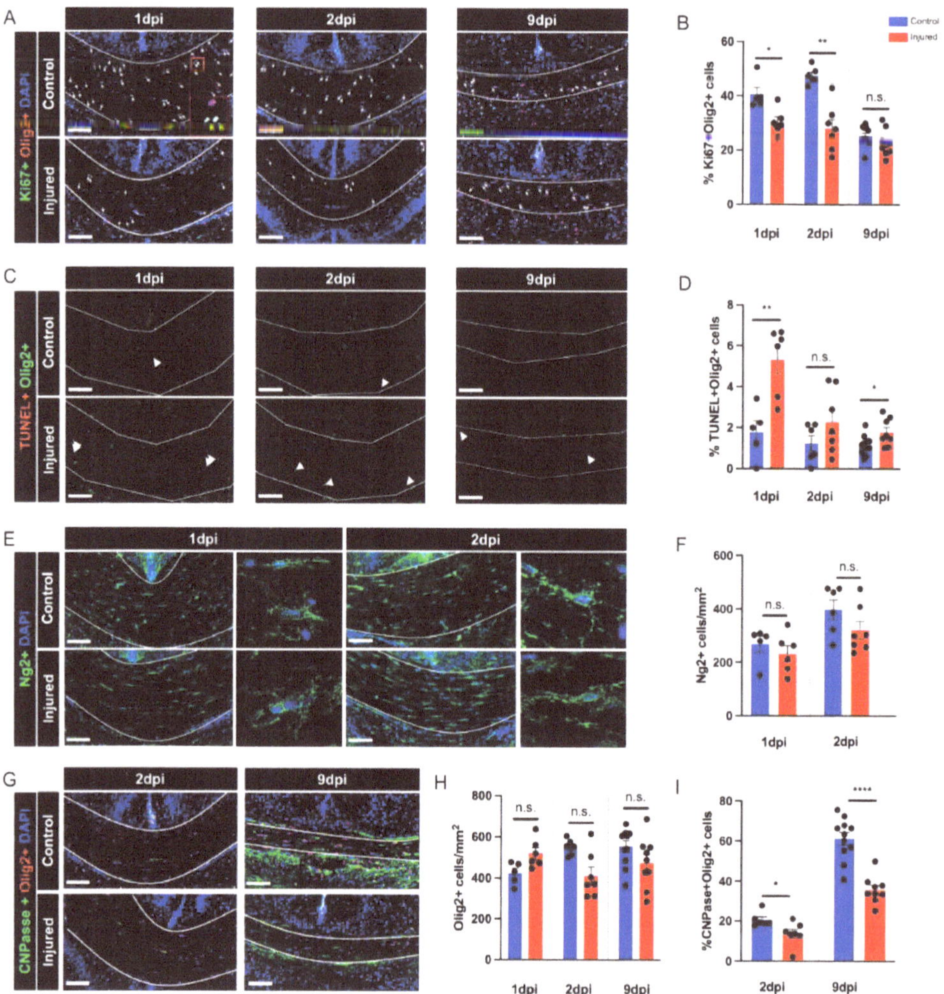

Figure 4. Oligodendrocyte cell lineage dynamics underlying myelination failure. (**A**,**B**) Representative images and quantification of oligodendrocyte cell lineage proliferation (Olig2+Ki67+) in the corpus callosum at 1 dpi (U = 2, p = 0.0173), 2 dpi (U = 0, p = 0.0012), and 9 dpi (U = 31, p = 0.1754). Arrows indicate double staining for oligodendrocyte marker Olig2 (red) and proliferation marker Ki67 (green). (**C**,**D**) Representative images and quantification of oligodendrocyte cell lineage death (Olig2+TUNEL+) at 1 dpi (U = 1, p = 0.0087), 2 dpi (U = 14, p = 0.3450), and 9 dpi (U = 17, p = 0.0259). Arrows indicate double staining for Olig2 (green) and TUNEL staining for DNA fragmentation (red). (**E**,**F**) Representative images and quantification of oligodendrocyte precursor (OPC) cell numbers (NG2+) at 1 dpi (U = 9, p = 0.3290) and 2 dpi (U = 10, p = 0.1305). (**G–I**) Representative images and quantification of oligodendrocyte maturation (Olig2+ and Olig2+CNPase+) at 2 dpi (U = 2, p <0.0001) and 9 dpi (U = 6, p = 0.0350). Scale bar: 100 μm. Sample amounts: 1 dpi: control n = 5, injured n = 6; 2 dpi: control n = 6, injured n = 7; 9 dpi: control n = 11, injured n = 9. Data are presented as mean ± SEM; n.s., not significant; * p < 0.05, ** p < 0.01, and **** p < 0.0001.

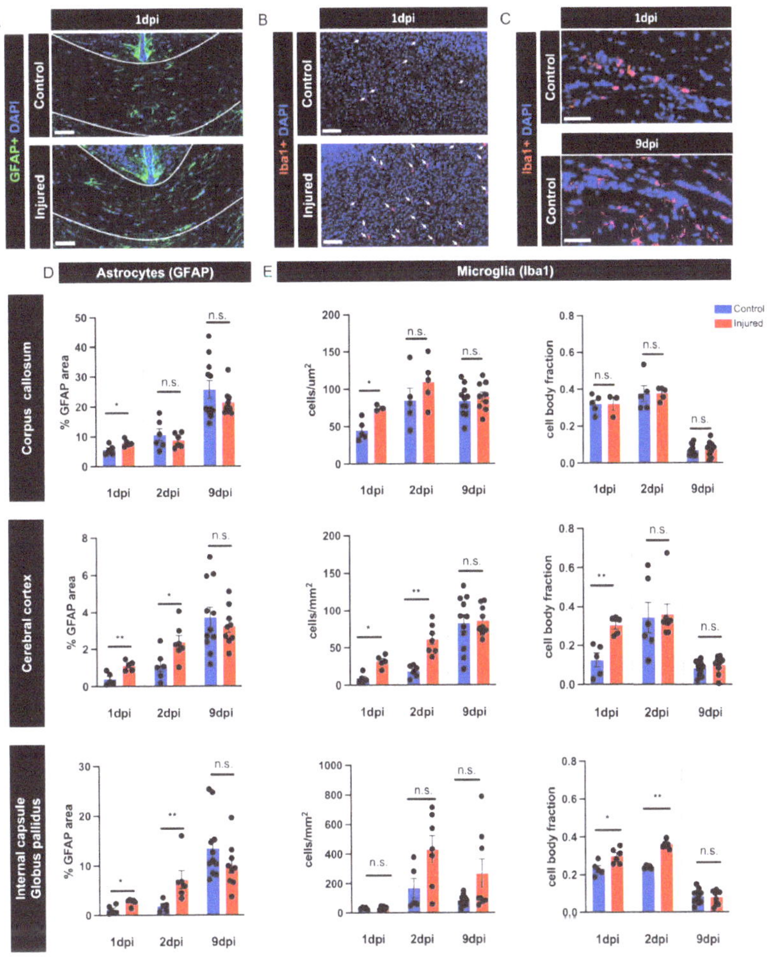

Figure 5. Quantifying reactive changes in astrocytes and microglia. (**A,B**) Representative images of control and injured brains at 1 dpi labeled with GFAP in the corpus callosum (**A**) and Iba1 in the cortex (**B**). (**C**) Representative images of the corpus callosum from control mice at 1 dpi and 9 dpi. (**D**) Quantification of GFAP immunohistochemistry in mouse corpus callosum at 1 dpi (U = 3, p = 0.0347), 2 dpi (U = 15, p = 0.6991), and 9 dpi (U = 44, p = 0.7103); in cerebral cortex at 1 dpi (U = 1, p = 0.0159), 2 dpi (U = 7, p = 0.0513), and 9 dpi (U = 46, p = 0.8238); and in internal capsule and globus pallidus at 1 dpi (U = 2, p = 0.0173), 2 dpi (U = 1, p = 0.0087), and 9 dpi (U = 31, p = 0.1754). (**E**) Quantification of Iba1 immunohistochemistry for cells/mm^2 and cell body fraction (cbf) in mouse corpus callosum at 1 dpi (cells/mm^2: U = 0, p = 0.0357; cbf: U = 7, p > 0.999), 2 dpi (cells/mm^2: U = 6, p = 0.222; cbf: U = 9, p = 0.5476), and 9 dpi (cells/mm^2: U = 42, p = 0.6027; cbf: U = 45, p = 0.7664); in cerebral cortex at 1 dpi (cells/mm^2: U = 1, p = 0.0159; cbf: U =0, p = 0.0079), 2 dpi (cells/mm^2: U = 0, p = 0.0012; cbf: U = 14, p = 0.3660), and 9 dpi (cells/mm^2: U = 47, p = 0.8820; cbf: U = 34, p = 0.2610); and in internal capsule and globus pallidus at 1 dpi (cells/mm^2: U = 13, p = 0.7922; cbf: U = 3, p = 0.0303), 2 dpi (cells/mm^2: U = 5, p = 0.0480; cbf: U = 0, p = 0.0025); and 9 dpi (cells/mm^2: U = 31, p = 0.3100; cbf: U = 49, p > 0.999). Sample amounts: 1 dpi: control n = 5, injured n = 5; 2 dpi: control n = 3–6, injured n = 7; 9 dpi: control n = 9, injured n = 9. Scale bar: 100 μm. Data are presented as mean ± SEM; n.s., not significant; * p < 0.05, ** p < 0.01.

When we examined the cell body fraction of Iba1-immunoreactive cells in injured mice, we noted a rapid increase in cell body fractions in the cerebral cortex, as well as the globus pallidus and internal capsule region, that were significant in comparison with control mice (Figure 5E). The cell body fraction of microglia in the corpus callosum did not exhibit a significant increase upon injury. Instead, microglia from both control and injured brains in this region displayed increased cell body fractions. The magnitude of the cell body fraction for microglia in the control corpus callosum was comparable to that observed for the microglia of the injured cerebral cortex and globus pallidus and internal capsule at 1 dpi (Figure 5E). When we examined cell morphology, the microglia from the corpus callosum of control brains appeared amoeboid, with few and short cellular processes (Figure 5C). By 9 dpi, microglia exhibited a mature ramified resting morphology and, therefore, a low cell body fraction across all the brain regions examined in both control and injured mice (Figure 5E).

We quantified the percentage area occupied by GFAP immunoreactivity as a measure of astrocyte reactivity. This analysis revealed a significant increase in astrocyte reactivity in injured compared with control mice in all three brain regions examined at 1 dpi (Figure 5A,D). This difference was no longer significant at 2 dpi in the corpus callosum but persisted at 2 dpi in the cerebral cortex, as well as the globus pallidus and internal capsule region (Figure 5D).

3.5. Perinatal Inflammation–Hypoxia Results in Sensorimotor and Cognitive Deficits

Perinatal white matter manifests as a broad spectrum of motor, cognitive, and behavioral deficits in affected individuals [1]. We aimed to test whether our disease model also yielded motor and cognitive deficits using a neonatal locomotion assay, a rotarod test, and a novel object recognition test [19,20,22]. The neonatal locomotion assay analysis revealed a significant increase in ambulation angle in pups at 13 dpi subjected to our disease model in comparison with control pups, indicating impaired sensorimotor development (Figure 6A). The rotarod test analysis demonstrated that the motor performances of the injured mice were significantly impaired compared with the performances of control mice at 28 dpi (Figure 6B). In addition, novel object testing revealed impaired recognition memory in injured mice (Figure 6C).

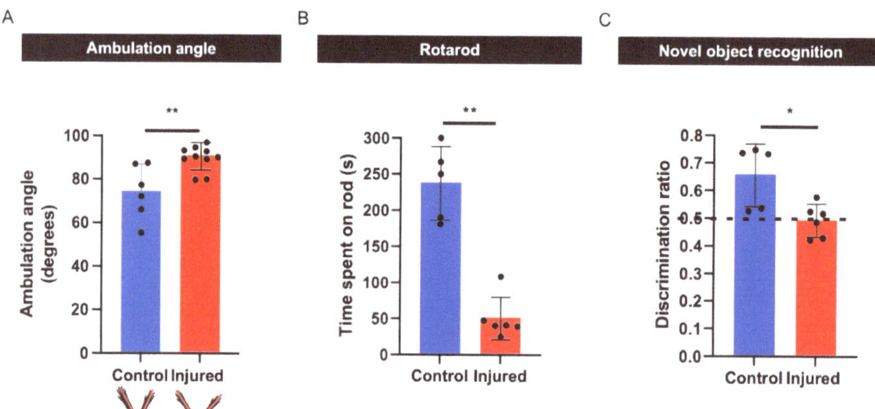

Figure 6. Evaluation of behavior. (**A**) Quantification of hindlimb ambulation angle at 13 dpi, with control $n = 6$ and injured $n = 11$ (U = 4, $p = 0.0019$). (**B**) Quantification of rotarod test, with control $n = 5$ and injured $n = 6$. (**C**) Quantification of novel object recognition task, with $n = 5$ and $n = 6$ (U = 2, $p = 0.0173$). Data are presented as mean ± SEM; * $p < 0.05$ and ** $p < 0.01$.

4. Discussion

4.1. Perinatal Inflammation and Hypoxia Produce Diffuse Perinatal White Matter Injury

Two main etiologies of acute perinatal white matter injury are recognized: hypoxia-ischemia and inflammation [1]. White matter injury results when these insults occur in regions where the oligodendrocyte lineage is predominated by pre-OL [2]. Clinically, hypoxic insults may occur in the context of placental insufficiency, birth asphyxia, impaired cerebral autoregulation, and neonatal pulmonary insufficiency. Inflammatory insults may be caused by perinatal clinical scenarios, including chorioamnionitis and neonatal sepsis. Animal models of perinatal white matter injury have been developed across various species. While large-animal models more closely model human physiology and offer the potential for instrumentation, rodent models continue to offer key advantages in terms of time, cost, and ease of manipulation. Models driven by neuroinflammation achieve myelination defects following single or repeated administrations of inflammatory stimuli, including pro-inflammatory cytokines, bacterial-mimicking LPS, viral-mimicking Poly I:C, or live infectious organisms, during the window of white matter vulnerability [23–25]. The most well-known hypoxia–ischemia-driven model of perinatal white matter injury is the classical Rice–Vanucci model, which achieves brain injury through a hypoxic–ischemic insult in postnatal day 7 neonatal rat pups by combining unilateral carotid artery ligation with an episode of systemic hypoxia [26]. In this model, the combination of hypoxia and ischemia is required to produce brain injury. Rats are able to survive periods of hypoxia without brain damage, and the unilateral induction of ischemia through carotid artery ligation can be compensated for by contralateral circulation thanks to the rats' complete Circle of Willis [27]. This model has been modified in recent years to optimize injury timing relative to the time course of oligodendrocyte lineage development in rodent white matter [27]. Steps to adapt this model to mice have revealed significant strain-specific differences in sensitivity to hypoxic–ischemic injury [28]. The injuries produced in these hypoxia–ischemia models often result in cystic white matter lesions that are not representative of modern WMI pathology [23]. Whether a common mechanism leads to myelination defects in models driven by both disease etiologies (inflammation v. hypoxia–ischemia) remains unclear.

Evidence from human and animal studies has demonstrated that inflammation potentiates the effects of subsequent brain insults and that WMI often results following multiple perinatal insults [24,26,29–32]. This notion of increased susceptibility to brain injury following a sensitizing perinatal insult has obvious clinical implications. Increased ability to recognize the occurrence of a first hit may represent an opportunity for intervention to protect the brain from or prevent the occurrence of a second insult [3]. Recapitulating a two-hit disease etiology in animal models employed for the study of disease mechanisms and therapeutic interventions is important to increase the likelihood of successful translation.

The current model is a multi-hit injury paradigm involving two successive systemic acute insults, one inflammatory and one hypoxic. Despite involving two distinct insults, the severity of the resulting injury was confined to diffuse changes in glial reactivity and myelination outcomes rather than the production of cystic lesions.

4.2. Impaired Oligodendrocyte Differentiation Appears to Underlie Perinatal Hypoxia–Inflammation-Induced Myelination Defects

Hypomyelination in cerebral white matter is a requisite finding for the diagnosis of perinatal white matter injury [1]. In humans, the corpus callosum and internal capsule regions are particularly affected [17]. We observed significant reductions in myelination in the corpus callosum and cortex at 9 dpi in our disease model. In contrast, we found no significant decrease in myelination, as evaluated by MBP immunoreactivity, in the internal capsule and globus pallidus region, although this region displayed a robust induction of microglia and astrocyte reactivity. Reductions in cortical myelin in human white matter injury have not been well-characterized. The lack of a myelination phenotype in the internal capsule region of our injured brains could be due to region-specific differences in susceptibility to injury in humans versus rodents [33]. Alternatively, there may be defects in myelination in this region that require other methods (e.g., electron microscopy) of detection.

Years of research have led to the consensus that the myelination defects seen in perinatal white matter injury stem from impaired maturation of the oligodendrocyte lineage rather than the death of mature oligodendrocytes [5,6,16]. In line with these findings, we found evidence for impaired maturation of the oligodendrocyte lineage in the corpus callosum at 9 dpi. The acute response of the oligodendrocyte lineage to injury has also been studied in human tissue affected by WMI, revealing initial rapid degeneration of the pre-OL population followed by replenishment of this cell population through a phase of increased oligodendrocyte precursor proliferation [6,16,34,35]. The blocked differentiation of this regenerated pre-OL pool underlies the ensuing persistent myelination defects [16]. We examined the death and proliferation of Olig2+ cells at 24 and 48 h after injury. The increase in Olig2+ cell death that we observed (2–5%) is very modest in comparison to the magnitude of cell death observed at early timepoints in human tissue [6,35]. Instead of increased Olig2+ cell proliferation, we observed a significant decrease in Olig2+ cell proliferation at 1 and 2 dpi. Interestingly, these cellular dynamics were not pronounced enough to significantly affect the total number of OPCs or Olig2+ cells at the timepoints examined, suggesting that they were relatively small and transient phenomena. One possibility is that an initial wave of cell death and compensatory proliferation already occurred by 24 h after injury or that OPC proliferation occurred between the 2 and 9 dpi timepoints. Indeed, experimental models have suggested a quick rise in pre-OL degeneration peaking within the first 12 h after injury and a phase of OPC proliferation several days after injury [6,31,36]. Alternatively, it could be that pre-OL death is more pronounced in the severe form of perinatal white matter injury involving tissue necrosis, cystic lesion formation, and significant neuronal cell death and that our failure to observe this phenomenon reflected the milder nature of our insult. At 9 dpi, NF200 immunohistochemistry did not reveal axonal loss that could underlie the observed myelination defects. To fully investigate a role for secondary demyelination, axon studies should be performed at later timepoints and include specific markers of axon pathology.

4.3. Observed Reactive Changes in Microglia and Astrocytes after Perinatal Inflammation–Hypoxia Appear to Be Transient

In human tissue, the duration of reactive gliosis and blocked oligodendrocyte maturation remains somewhat unresolved due to a dearth of neuropathological data from human infants. Nonetheless, the scant evidence that has been gathered suggests that these features persist for at least months after term-equivalent age [4,17].

In line with studies of human postmortem brain tissue affected by WMI, we observed evidence of microglia and astrocyte reactivity in our disease model. These changes, determined by an increased GFAP+ signal area and an increased Iba1+ microglia cell body fraction relative to control animals, were no longer detectable by 9 dpi, suggesting that reactive gliosis was transient in our model. The specific gene expression changes characteristic of the reactive astrocytes and microglia that form in WMI are incompletely understood [37–39]. Evaluation of these changes was beyond the scope of our study. Unveiling the molecular identity and functional properties of these cells is a fascinating topic for future study. Our results may also reflect the more robust regenerative capacity of mice in comparison with humans [40]. It is also possible that additional postnatal insults might contribute to sustaining gliosis in human infants with white matter injury.

4.4. Glial Reactivity in Developing Versus Mature Brains

Glial reactivity has, to date, mostly been studied in mature brains. We are still learning about differences in the capacity of microglia and astrocytes to respond to injury and the nature of these responses in perinatal versus adult brains [37]. Whether transient physiologic reactive changes in microglia and astrocytes are a part of normal development is not fully understood. Proliferative microglia of the developing white matter are known to exhibit an amoeboid reactive-like morphology under physiological conditions [37]. Recent single-cell-RNA-sequencing studies have identified molecular markers of this population, the so-called proliferation-associated microglia, or PAMs, and have highlighted similarities in gene expression with disease-associated microglia (DAMs) in the adult brain [41–43]. We observed an increased cell body fraction in microglia in control corpus callosum samples in our disease model at 1 and 2 dpi, potentially reflecting the presence of this PAM population. The effect of this microglia subtype on astrocyte reactivity in developing white matter has not yet been investigated. Furthermore, the relevance of these cells to human WMI and how this microglial population influences the response of white matter to perinatal insults is an area of active investigation.

5. Conclusions

The identification of animal models that faithfully reproduce key features of human disease is essential to the advancement of research on these diseases, especially those diseases for which human tissue is difficult to procure. Even with improved animal models, no single model recapitulates all aspects of human pathophysiology. Models must be carefully chosen for the research question at hand and validated, when possible, with postmortem tissue. The use of human cellular models of white matter injury offers an exciting model system complementary to the use of animal models for furthering our understanding of this disease and confirming the relevance of findings in these animal models to human physiology [44]. We propose a mouse model of acute diffuse perinatal white matter injury that produces myelination defects, impaired oligodendrocyte maturation, reactive gliosis, and motor and cognitive deficits, allcharacteristics of contemporary human white matter injury. Limitations of our study include a histologic analysis up to only 9 days after injury, limited power of statistical analyses due to the low sample number, and missing molecular characterizations of astrocytes and microglia. Our model did not result in the formation of cystic white matter lesions, making it amenable to future studies assessing the pathophysiology of and novel therapeutic approaches for modern perinatal white matter injury.

Supplementary Materials: The following supporting information can be downloaded at: https://www.mdpi.com/article/10.3390/biomedicines10112810/s1, Figure S1: Evaluation of neurofilament density; Table S1: Summary of experimental cohort.

Author Contributions: Conceptualization, P.R., A.S. and A.B.L.; methodology, P.R., T.M., D.S. and A.B.L.; validation, P.R. and A.B.L.; formal analysis, P.R., T.M., D.S. and A.B.L.; investigation, P.R. and V.H.; writing—original draft preparation, P.R. and A.B.L.; writing—review and editing, A.S., D.S. and A.B.L.; visualization, P.R. and A.B.L.; supervision, A.S., D.S. and A.B.L.; project administration, A.B.L.; funding acquisition, D.S. and A.B.L. All authors have read and agreed to the published version of the manuscript.

Funding: This work was supported by the SGGG/Bayer Research Grant 2019 (A.B.L.), the UniBE Initiator Award 2020 (A.B.L.), and an intramural grant from the departmental research fund 2018–2022.

Institutional Review Board Statement: The animal study protocol was approved by the Ethics Committee of the Canton of Bern (reference number: BE19/85).

Data Availability Statement: The data presented in this study are available on request from the corresponding author.

Acknowledgments: We express our appreciation to Vera Tscherrig and Marianne Jörger-Messerli for helpful scientific input and experimental support. In addition, we are grateful to Smita Saxena and her lab members for guidance with the rotarod test.

References

1. Volpe, J.J. Dysmaturation of Premature Brain: Importance, Cellular Mechanisms, and Potential Interventions. *Pediatr. Neurol.* **2019**, *95*, 42–66. [CrossRef] [PubMed]
2. Back, S.A. White matter injury in the preterm infant: Pathology and mechanisms. *Acta Neuropathol.* **2017**, *134*, 331–349. [CrossRef] [PubMed]
3. Kaindl, A.M.; Favrais, G.; Gressens, P. Molecular mechanisms involved in injury to the preterm brain. *J. Child Neurol.* **2009**, *24*, 1112–1118. [CrossRef] [PubMed]
4. Buser, J.R.; Maire, J.; Riddle, A.; Gong, X.; Nguyen, T.; Nelson, K.; Luo, N.L.; Ren, J.; Struve, J.; Sherman, L.S.; et al. Arrested preoligodendrocyte maturation contributes to myelination failure in premature infants. *Ann. Neurol.* **2012**, *71*, 93–109. [CrossRef] [PubMed]
5. Volpe, J.J.; Kinney, H.C.; Jensen, F.E.; Rosenberg, P.A. The developing oligodendrocyte: Key cellular target in brain injury in the premature infant. *Int. J. Dev. Neurosci.* **2011**, *29*, 423–440. [CrossRef]
6. Back, S.A.; Han, B.H.; Luo, N.L.; Chricton, C.A.; Xanthoudakis, S.; Tam, J.; Arvin, K.L.; Holtzman, D.M. Selective vulnerability of late oligodendrocyte progenitors to hypoxia-ischemia. *J. Neurosci.* **2002**, *22*, 455–463. [CrossRef]
7. Penn, A.A.; Gressens, P.; Fleiss, B.; Back, S.A.; Gallo, V. Controversies in preterm brain injury. *Neurobiol. Dis.* **2016**, *92*, 90–101. [CrossRef]
8. Rumajogee, P.; Bregman, T.; Miller, S.P.; Yager, J.Y.; Fehlings, M.G. Rodent Hypoxia-Ischemia Models for Cerebral Palsy Research: A Systematic Review. *Front. Neurol.* **2016**, *7*, 57. [CrossRef]
9. Silbereis, J.C.; Huang, E.J.; Back, S.A.; Rowitch, D.H. Towards improved animal models of neonatal white matter injury associated with cerebral palsy. *Dis. Model. Mech.* **2010**, *3*, 678–688. [CrossRef]
10. Thomi, G.; Joerger-Messerli, M.; Haesler, V.; Muri, L.; Surbek, D.; Schoeberlein, A. Intranasally Administered Exosomes from Umbilical Cord Stem Cells Have Preventive Neuroprotective Effects and Contribute to Functional Recovery after Perinatal Brain Injury. *Cells* **2019**, *8*, 855. [CrossRef]
11. Joerger-Messerli, M.S.; Thomi, G.; Haesler, V.; Keller, I.; Renz, P.; Surbek, D.V.; Schoeberlein, A. Human Wharton's Jelly Mesenchymal Stromal Cell-Derived Small Extracellular Vesicles Drive Oligodendroglial Maturation by Restraining MAPK/ERK and Notch Signaling Pathways. *Front. Cell Dev. Biol.* **2021**, *9*, 622539. [CrossRef] [PubMed]
12. Thomi, G.; Surbek, D.; Haesler, V.; Joerger-Messerli, M.; Schoeberlein, A. Exosomes derived from umbilical cord mesenchymal stem cells reduce microglia-mediated neuroinflammation in perinatal brain injury. *Stem. Cell Res. Ther.* **2019**, *10*, 105. [CrossRef] [PubMed]
13. Li, J.; Yawno, T.; Sutherland, A.; Loose, J.; Nitsos, I.; Bischof, R.; Castillo-Melendez, M.; McDonald, C.A.; Wong, F.Y.; Jenkin, G.; et al. Preterm white matter brain injury is prevented by early administration of umbilical cord blood cells. *Exp. Neurol.* **2016**, *283*, 179–187. [CrossRef] [PubMed]
14. Mike, J.K.; Wu, K.Y.; White, Y.; Pathipati, P.; Ndjamen, B.; Hutchings, R.S.; Losser, C.; Vento, C.; Arellano, K.; Vanhatalo, O.; et al. Defining Longer-Term Outcomes in an Ovine Model of Moderate Perinatal Hypoxia-Ischemia. *Dev. Neurosci.* **2022**, *44*, 277–294. [CrossRef] [PubMed]
15. Semple, B.D.; Blomgren, K.; Gimlin, K.; Ferriero, D.M.; Noble-Haeusslein, L.J. Brain development in rodents and humans: Identifying benchmarks of maturation and vulnerability to injury across species. *Prog. Neurobiol.* **2013**, *106-107*, 1–16. [CrossRef]
16. Billiards, S.S.; Haynes, R.L.; Folkerth, R.D.; Borenstein, N.S.; Trachtenberg, F.L.; Rowitch, D.H.; Ligon, K.L.; Volpe, J.J.; Kinney, H.C. Myelin abnormalities without oligodendrocyte loss in periventricular leukomalacia. *Brain Pathol.* **2008**, *18*, 153–163. [CrossRef]
17. Pierson, C.R.; Folkerth, R.D.; Billiards, S.S.; Trachtenberg, F.L.; Drinkwater, M.E.; Volpe, J.J.; Kinney, H.C. Gray matter injury associated with periventricular leukomalacia in the premature infant. *Acta Neuropathol.* **2007**, *114*, 619–631. [CrossRef]
18. Hovens, I.B.; Nyakas, C.; Schoemaker, R.G. A novel method for evaluating microglial activation using ionized calcium-binding adaptor protein-1 staining: Cell body to cell size ratio. *Neuroimmunol. Neuroinflammation* **2014**, *1*, 82–88. [CrossRef]
19. Feather-Schussler, D.N.; Ferguson, T.S. A Battery of Motor Tests in a Neonatal Mouse Model of Cerebral Palsy. *J. Vis. Exp.* **2016**. [CrossRef]
20. Bevins, R.A.; Besheer, J. Object recognition in rats and mice: A one-trial non-matching-to-sample learning task to study 'recognition memory'. *Nat. Protoc.* **2006**, *1*, 1306–1311. [CrossRef]
21. Banker, B.Q.; Larroche, J.C. Periventricular leukomalacia of infancy. A form of neonatal anoxic encephalopathy. *Arch Neurol.* **1962**, *7*, 386–410. [CrossRef] [PubMed]
22. Bellantuono, I.; de Cabo, R.; Ehninger, D.; Di Germanio, C.; Lawrie, A.; Miller, J.; Mitchell, S.J.; Navas-Enamorado, I.; Potter, P.K.; Tchkonia, T.; et al. A toolbox for the longitudinal assessment of healthspan in aging mice. *Nat. Protoc.* **2020**, *15*, 540–574. [CrossRef] [PubMed]
23. Favrais, G.; van de Looij, Y.; Fleiss, B.; Ramanantsoa, N.; Bonnin, P.; Stoltenburg-Didinger, G.; Lacaud, A.; Saliba, E.; Dammann, O.; Gallego, J.; et al. Systemic inflammation disrupts the developmental program of white matter. *Ann. Neurol.* **2011**, *70*, 550–565. [CrossRef] [PubMed]
24. Van Steenwinckel, J.; Schang, A.L.; Sigaut, S.; Chhor, V.; Degos, V.; Hagberg, H.; Baud, O.; Fleiss, B.; Gressens, P. Brain damage of the preterm infant: New insights into the role of inflammation. *Biochem. Soc. Trans.* **2014**, *42*, 557–563. [CrossRef] [PubMed]
25. Hagberg, H.; Peebles, D.; Mallard, C. Models of white matter injury: Comparison of infectious, hypoxic-ischemic, and excitotoxic insults. *Ment. Retard. Dev. Disabil. Res. Rev.* **2002**, *8*, 30–38. [CrossRef] [PubMed]

26. Rice, J.E., 3rd; Vannucci, R.C.; Brierley, J.B. The influence of immaturity on hypoxic-ischemic brain damage in the rat. *Ann. Neurol.* **1981**, *9*, 131–141. [CrossRef]
27. Vannucci, S.J.; Back, S.A. The Vannucci Model of Hypoxic-Ischemic Injury in the Neonatal Rodent: 40 years later. *Dev. Neurosci.* **2022**. [CrossRef]
28. Sheldon, R.A.; Sedik, C.; Ferriero, D.M. Strain-related brain injury in neonatal mice subjected to hypoxia-ischemia. *Brain Res.* **1998**, *810*, 114–122. [CrossRef]
29. Korzeniewski, S.J.; Romero, R.; Cortez, J.; Pappas, A.; Schwartz, A.G.; Kim, C.J.; Kim, J.S.; Kim, Y.M.; Yoon, B.H.; Chaiworapongsa, T.; et al. A "multi-hit" model of neonatal white matter injury: Cumulative contributions of chronic placental inflammation, acute fetal inflammation and postnatal inflammatory events. *J. Perinat. Med.* **2014**, *42*, 731–743. [CrossRef]
30. Leviton, A.; Fichorova, R.N.; O'Shea, T.M.; Kuban, K.; Paneth, N.; Dammann, O.; Allred, E.N. Two-hit model of brain damage in the very preterm newborn: Small for gestational age and postnatal systemic inflammation. *Pediatr. Res.* **2013**, *73*, 362–370. [CrossRef]
31. van Tilborg, E.; Achterberg, E.J.M.; van Kammen, C.M.; van der Toorn, A.; Groenendaal, F.; Dijkhuizen, R.M.; Heijnen, C.J.; Vanderschuren, L.; Benders, M.; Nijboer, C.H.A. Combined fetal inflammation and postnatal hypoxia causes myelin deficits and autism-like behavior in a rat model of diffuse white matter injury. *Glia* **2018**, *66*, 78–93. [CrossRef] [PubMed]
32. Adén, U.; Favrais, G.; Plaisant, F.; Winerdal, M.; Felderhoff-Mueser, U.; Lampa, J.; Lelièvre, V.; Gressens, P. Systemic inflammation sensitizes the neonatal brain to excitotoxicity through a pro-/anti-inflammatory imbalance: Key role of TNFalpha pathway and protection by etanercept. *Brain Behav. Immun.* **2010**, *24*, 747–758. [CrossRef] [PubMed]
33. Back, S.A.; Riddle, A.; McClure, M.M. Maturation-dependent vulnerability of perinatal white matter in premature birth. *Stroke* **2007**, *38*, 724–730. [CrossRef] [PubMed]
34. Folkerth, R.D.; Trachtenberg, F.L.; Haynes, R.L. Oxidative injury in the cerebral cortex and subplate neurons in periventricular leukomalacia. *J. Neuropathol. Exp. Neurol.* **2008**, *67*, 677–686. [CrossRef] [PubMed]
35. Haynes, R.L.; Folkerth, R.D.; Keefe, R.J.; Sung, I.; Swzeda, L.I.; Rosenberg, P.A.; Volpe, J.J.; Kinney, H.C. Nitrosative and oxidative injury to premyelinating oligodendrocytes in periventricular leukomalacia. *J. Neuropathol. Exp. Neurol.* **2003**, *62*, 441–450. [CrossRef]
36. Ness, J.K.; Romanko, M.J.; Rothstein, R.P.; Wood, T.L.; Levison, S.W. Perinatal hypoxia-ischemia induces apoptotic and excitotoxic death of periventricular white matter oligodendrocyte progenitors. *Dev. Neurosci.* **2001**, *23*, 203–208. [CrossRef]
37. Delahaye-Duriez, A.; Dufour, A.; Bokobza, C.; Gressens, P.; Van Steenwinckel, J. Targeting Microglial Disturbances to Protect the Brain From Neurodevelopmental Disorders Associated With Prematurity. *J. Neuropathol. Exp. Neurol.* **2021**, *80*, 634–648. [CrossRef]
38. Nobuta, H.; Ghiani, C.A.; Paez, P.M.; Spreuer, V.; Dong, H.; Korsak, R.A.; Manukyan, A.; Li, J.; Vinters, H.V.; Huang, E.J.; et al. STAT3-mediated astrogliosis protects myelin development in neonatal brain injury. *Ann. Neurol.* **2012**, *72*, 750–765. [CrossRef]
39. Shiow, L.R.; Favrais, G.; Schirmer, L.; Schang, A.L.; Cipriani, S.; Andres, C.; Wright, J.N.; Nobuta, H.; Fleiss, B.; Gressens, P.; et al. Reactive astrocyte COX2-PGE2 production inhibits oligodendrocyte maturation in neonatal white matter injury. *Glia* **2017**, *65*, 2024–2037. [CrossRef]
40. Li, J.; Pan, L.; Pembroke, W.G.; Rexach, J.E.; Godoy, M.I.; Condro, M.C.; Alvarado, A.G.; Harteni, M.; Chen, Y.W.; Stiles, L.; et al. Conservation and divergence of vulnerability and responses to stressors between human and mouse astrocytes. *Nat. Commun.* **2021**, *12*, 3958. [CrossRef]
41. Hammond, T.R.; Dufort, C.; Dissing-Olesen, L.; Giera, S.; Young, A.; Wysoker, A.; Walker, A.J.; Gergits, F.; Segel, M.; Nemesh, J.; et al. Single-Cell RNA Sequencing of Microglia throughout the Mouse Lifespan and in the Injured Brain Reveals Complex Cell-State Changes. *Immunity* **2019**, *50*, 253–271. [CrossRef] [PubMed]
42. Li, Q.; Cheng, Z.; Zhou, L.; Darmanis, S.; Neff, N.F.; Okamoto, J.; Gulati, G.; Bennett, M.L.; Sun, L.O.; Clarke, L.E.; et al. Developmental Heterogeneity of Microglia and Brain Myeloid Cells Revealed by Deep Single-Cell RNA Sequencing. *Neuron* **2019**, *101*, 207–223.e210. [CrossRef] [PubMed]
43. Volpe, J.J. Microglia: Newly discovered complexity could lead to targeted therapy for neonatal white matter injury and dysmaturation. *J. Neonatal. Perinatal. Med.* **2019**, *12*, 239–242. [CrossRef] [PubMed]
44. Pașca, A.M.; Park, J.Y.; Shin, H.W.; Qi, Q.; Revah, O.; Krasnoff, R.; O'Hara, R.; Willsey, A.J.; Palmer, T.D.; Pașca, S.P. Human 3D cellular model of hypoxic brain injury of prematurity. *Nat. Med.* **2019**, *25*, 784–791. [CrossRef]

Article

Development and Characterization of a Subcutaneous Implant-Related Infection Model in Mice to Test Novel Antimicrobial Treatment Strategies

Charlotte Wittmann [1], Niels Vanvelk [1], Anton E. Fürst [2], T. Fintan Moriarty [1] and Stephan Zeiter [1,*]

[1] AO Research Institute Davos, 7270 Davos, Switzerland
[2] Equine Department–Vetsuisse Faculty, University of Zurich, 8057 Zurich, Switzerland
*   Correspondence: stephan.zeiter@aofoundation.org; Tel.: +41-81-414-2311

Abstract: Orthopedic-device-related infection is one of the most severe complications in orthopedic surgery. To reduce the associated morbidity and healthcare costs, new prevention and treatment modalities are continuously under development. Preclinical in vivo models serve as a control point prior to clinical implementation. This study presents a mouse model of subcutaneously implanted titanium discs, infected with *Staphylococcus aureus*, to fill a gap in the early-stage testing of antimicrobial biomaterials. Firstly, three different inocula were administered either pre-adhered to the implant or pipetted on top of it following implantation to test their ability to reliably create an infection. Secondly, the efficacy of low-dose (25 mg/kg) and high-dose (250 mg/kg) cefazolin administered systemically in infection prevention was assessed. Lastly, titanium implants were replaced by antibiotic-loaded bone cement (ALBC) discs to investigate the efficacy of local antibiotics in infection prevention. The efficacy in infection prevention of the low-dose perioperative antibiotic prophylaxis (PAP) depended on both the inoculum and inoculation method. Bacterial counts were significantly lower in animals receiving the high dose of PAP. ALBC discs with or without the additional PAP proved highly effective in infection prevention and provide a suitable positive control to test other prevention strategies.

Keywords: implant-related infection; mouse model; biomaterial testing

Citation: Wittmann, C.; Vanvelk, N.; Fürst, A.E.; Moriarty, T.F.; Zeiter, S. Development and Characterization of a Subcutaneous Implant-Related Infection Model in Mice to Test Novel Antimicrobial Treatment Strategies. *Biomedicines* **2023**, *11*, 40. https://doi.org/10.3390/biomedicines11010040

Academic Editor: Martina Perše

Received: 21 November 2022
Revised: 19 December 2022
Accepted: 21 December 2022
Published: 24 December 2022

1. Introduction

Orthopedic-device-related infection (ODRI) represents one of the most severe complications in orthopedic surgery. Infection results in increased postoperative pain, delayed healing, loss of function and the potential amputation of the affected limb [1]. Furthermore, it represents a significant socioeconomic burden causing a substantial increase in cost and healthcare resource utilization when compared to non-infected cases [2,3]. Infection rates vary between 0.5 and 40% depending on various factors including host physiology (e.g., diabetes mellitus) and injury characteristics (e.g., Gustilo–Anderson type) [4–6]. Among isolated bacteria from ODRI, *Staphylococcus aureus* (*S. aureus*) are most prevalent and almost exclusively used in preclinical studies of ODRI [7,8]. The presence of foreign material, such as a fracture fixation device, has been described as an independent risk factor for infection [4]. Elek et al., showed that the presence of sutures in human volunteers decreased the minimal infecting dose with a *Staphylococcus* strain from 5×10^6 to 3×10^2 [9]. In a guinea pig model, Zimmerli et al., found the minimal infecting dose of *Staphylococcus aureus* (*S. aureus*) in the presence of tissue cages to be 1×10^2 colony-forming Units (CFU), while in the absence of any foreign material 1×10^8 CFU did not reliably cause an infection [10]. These studies thus demonstrate the important contribution of a foreign body in infection development.

In orthopedic surgery, the use of perioperative antibiotic prophylaxis (PAP) is a common practice to decrease the incidence of postoperative infection [11]. However, impaired vascularization resulting from the initial trauma and biofilm formation on the foreign body could prevent a sufficiently high antimicrobial concentration at the surgical site. Increasing antibiotic doses raises the risk of systemic toxicity and side effects. Local application of antibiotics directly to the surgical site can support prophylaxis and can reach a higher concentration with minimal systemic side effects [12]. Polymethylmethacrylate (PMMA) and collagen sponges are commercially available carriers for local antibiotic delivery and are currently used routinely in orthopedic surgery to prevent ODRI [13]. Disadvantages of these devices are the exothermic process of PMMA polymerization which poses the risk of thermal necrosis of the bone and prohibits the incorporation of heat sensitive antibiotics [14]. Additionally, PMMA is not biodegradable and has to be removed at a later surgical revision.

Preclinical in vivo models serve as a critical control point prior to the translation of any new procedure or intervention. These models can be grouped according to their complexity [15], with one option being to make animal models as clinically relevant and translational as possible, whereby the clinical scenario ought to be replicated as closely as possible [8]. Fully mimicking clinical conditions, however, requires a high level of expertise, appropriate equipment, and may also increase burden upon the animal and may not be required for all studies. For these reasons, models of high complexity are not justifiable for every stage of medical device development and less complex animal models can be utilized for early screening or proof of concept studies. For example, placing an implant subcutaneously in a non-functional position is a simple procedure and renders the implant as a foreign body, without any effort to function as an orthopedic device. Placing an implant subcutaneously does, however, include the host's response to novel biomaterials that may be sufficient at an early stage of product development or proof of concept studies.

Previous research by our own group, evaluating an antimicrobial coating by placing the coated implant in a subcutaneous pocket on the back of mice, revealed the lack of a fully characterized and standardized murine model. The impact of a single injection of antibiotics prior to surgery was not evaluated in that study (unpublished data).

This study presents the design and characterization of a subcutaneous implant-associated infection model in mice to fill that gap in the early-stage testing of antimicrobial biomaterials. The primary objective was to determine the optimal bacterial dose and inoculation method to establish a reliable *S. aureus* infection. The secondary objective was to investigate the efficacy of systemic and local antibiotics in infection prevention using clinical standard prophylactic antibiotics and commercially available antibiotic releasing biomaterials.

2. Materials and Methods

2.1. Study Overview

The study was approved by the Ethical Committee of the Canton of Grisons, Switzerland (TVB 20_2019 and 02E_2020). All procedures were performed in an Association of Assessment and Accreditation of Laboratory Animals Care International (AAALAC)-approved facility and according to the Swiss animal protection law and regulation.

Sixty-five female C57BL/6N mice (Charles River Laboratories, Germany) were included across three study phases (Table 1 summarizes groups and number of animals per group).

The first phase (dose-finding phase, group 1) aimed to determine the optimal route of administration and bacterial inoculum required to reliably create an infection. One titanium implant was inserted subcutaneously on either side of the spine. Bacteria were introduced either pre-adhered to the implant surface or an equal number were pipetted in suspension on top of the implant once placed in the tissue pocket. Bacteria were given in one of three bacterial doses (High, 1×10^6 CFU/mL; Medium, 1×10^4 CFU/mL; Low, 1×10^3 CFU/mL) (Figure 1). The surgeon was blinded to the route of administration until pipetting was performed or the pre-inoculated implant was presented. Afterwards, the animals were observed for three days before scheduled euthanasia.

Table 1. Overview of study design.

Group	Prophylaxis	Inoculation Dose [1]	Inoculation Method	Implant(s)/Animal	Group Size
		Dose-finding phase			
1	None	High * Medium Low	Pre-adhered/ pipetted	2 titanium discs	7 6 7
		Systemic antibiotic prophylaxis phase			
2 3	Cefazolin 25 mg/kg Cefazolin 250 mg/kg	Medium Low Low	Pre-adhered/ Pipetted Pre-adhered/ pipetted	2 titanium discs 2 titanium discs	6 6 8
		Local antibiotic prophylaxis phase			
4	None	Low	Pipetted	1 ALBC	10
5	Cefazolin 250 mg/kg	Low	Pipetted	1 ALBC	11
6	None	Low	Pipetted	1 titanium disc	4
Total					65

[1],* High, 1×10^6 CFU/mL; Medium, 1×10^4 CFU/mL; Low, 1×10^3 CFU/mL, ALBC, Antibiotic Loaded Bone Cement.

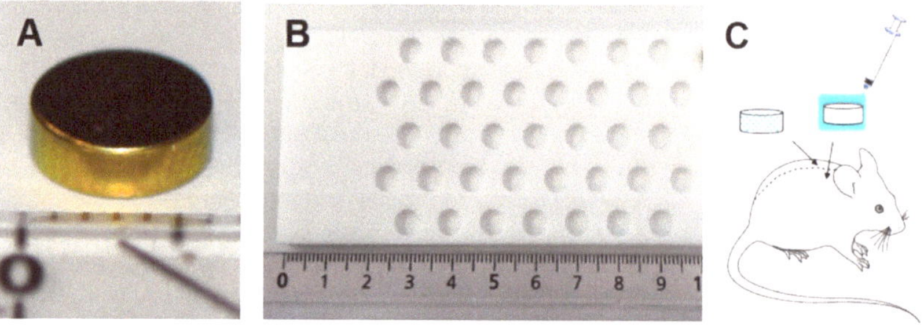

Figure 1. Implant design and implantation. (**A**) Titanium implant used in the study. (**B**) By using a custom-made Teflon mold, PMMA discs of the same shape and size as the titanium discs were obtained for the third phase. (**C**) Illustration of the two implants placed bilateral of the spine during the first and second phase. The implant on the left (dotted) represents the pre-inoculated implant. The box around the sterile implant demonstrates the inoculum pipetted (syringe) after implantation. In the third phase PMMA discs or titanium implants were inserted unilaterally, and the inoculum was pipetted.

In the second phase (systemic antibiotic prophylaxis phase, group 2 and 3), the efficacy of two different doses of systemically administered cefazolin (Labatec-Pharma SA) in infection prevention was evaluated to mimic the clinical situation where PAP is routinely started prior to surgery. The first dose (25 mg/kg) was extrapolated from clinical practice, where administration of 2 to 3 g (25 mg/kg to 37.5 mg/kg) is recommended. The second dose (250 mg/kg) was based on research from Stavrakis et al., showing increased antimicrobial efficacy in an implant-associated infection model in mice with high dose antibiotics (200 mg/kg) [16]. A single injection of antibiotics was injected subcutaneously approximately 15 min prior to surgical incision. Afterwards, animals were infected as in the dose-finding phase (either pre-adhered to the implant surface or an equal number were pipetted).

In the local antibiotic prophylaxis phase (groups 4 to 6), antibiotic-loaded bone cement (ALBC) discs were implanted instead of titanium implants unilaterally on either the left or right side of the spine to study the efficacy of local antibiotics in infection prevention. The side of the implant was randomly assigned. Immediately after implantation, the lowest inoculum to reliably cause an infection in the dose-finding phase was pipetted onto the implant. A proportion of the animals received additional systemic antibiotics prior to implantation of the ALBC disc. Four additional animals received one titanium implant with the same inoculum and functioned as a control group.

The surgeon as well as the person collecting and analyzing data (CFU and body weight) were blinded. Animals were operated in phases based on the study design mentioned above on seven different surgery days (6 to 12 animals/day). Animals within the same cage were randomly allocated to the predefined systematically rotated group assignment.

2.2. Animal Welfare and Ethical Approvals

Female C57BL/6N mice (Charles River Laboratories, Germany) of 13–19 weeks of age were included in the study. The mean weight at inclusion was 233 ± 1.3 g. The animals were housed in groups of three to five animals in individually ventilated cages (Techniplast and Allentown, 530 cm^2 ground floor) with a 12 h light/dark cycle (7 a.m. to 7 p.m.). They were allowed ad libitum access to autoclaved tap water and food (Kliba Nafag, Provimini Kliba SA, 3436 EXS12 S/R Entretien Extrude, Alleinfuttermittel für Mäuse und Ratten). Postoperative monitoring consisted of scoring certain parameters twice a day by a veterinarian. These included behavior, external appearance, breathing, excretions, wound healing, and bodyweight (Supplementary Material Figure S1).

2.3. Implants

The metal implants used in this study were composed of medical grade titanium alloy niobium (TAN Grade NB, L. Klein SA, Switzerland), and were 2 mm thick with a diameter of 6 mm. Further processing involved cleaning, degreasing and rotofinishing with a particle size of 3 × 2.5 × 3 mm. Finally, the surface was anodized (TioCol™ KKS Ultraschall AG, Frauholzring 29, 6422 Steinen, Switzerland) to a gold-colored finish.

ALBC was prepared to the same dimensions as the titanium discus using a custom-made mold (Figure 1). The bone cement was made of PMMA (Palacos® R+G, Heraeus Medical, Philipp-Reis-Straße 8-13, 61273 Wehrheim, Germany), loaded with gentamicin (12.25 mg/g bone cement) and prepared according to the manufacturer's guidelines.

2.4. Surgical Intervention and Euthanasia

The surgical procedure was performed under general anesthesia and analgesia. Induction of anesthesia was achieved by exposing the animals to 8% sevoflurane in 100% oxygen after transferring them to an induction box. Sevoflurane was reduced to approximately 2.5% in 100% oxygen to maintain anesthesia. Oxygen flow was set to 700 mL/minute during the surgical procedure. Preoperative analgesia included subcutaneous injections of buprenorphine (0.03 mg/kg, Bupaq® ad us. vet., Injektionslösung; Streuli Pharma AG, Bahnhofstrasse 7, 8730 Uznach, Switzerland) and carprofen (5 mg/kg, Rimadyl® ad us. vet., Injektionslösung; Zoetis Schweiz GmbH, Rue de la Jeunesse 2, 2800 Delémont, Switzerland). One ml of prewarmed Lactated Ringer's solution (Ringer Spüllösung Ecobag® 250 mL; B. Braun Medical AG, Seesatz 17, 6204 Sempach, Switzerland) was injected subcutaneously between the shoulder blades for temperature management and to replace perioperative fluid loss. The surgical field was clipped from the tail-base to the neck and aseptically prepared with Hibiscrub® (chlorhexidindigluconat 40 mg per 1 mL; CPS Cito Pharma Service, Gschwaderstrasse 35/D, 8610 Uster, Switzerland) and 96% ethanolum (Softasept N; B. Braun Medical AG, Seesatz 17, 6204 Sempach, Switzerland). The animal was then positioned on the surgery table and draped with Dermadrape® (Tiaset). On each side, an incision of 0.4 cm was made 1 cm lateral to the spine, avoiding bridging between sites. Subsequently, the implants were positioned subcutaneously with minimal tissue damage, taking care not to contaminate the implant during placement. Depending on the group, bacteria were pre-adhered onto the implant or pipetted onto their surface directly after implantation. The subcutis was closed with simple interrupted sutures using 6-0 Monocryl (Monocryl®, Ethicon, Johnson & Johnson AG, Gubelstrasse 34, 6300 Zug,

Switzerland). The skin was closed with an intradermal suture using 5-0 Vicryl (Vicryl®, Ethicon, Johnson & Johnson AG, Gubelstrasse 34, 6300 Zug, Switzerland). Postoperative analgesia included paracetamol (acetaminophen) of 1.9 mg/mL added to the drinking water for 3 days. Body temperature was maintained during preparation and surgery using heating pads, and all animals received eye ointment.

2.5. Bacteria and Inoculum Preparation

A clinical *S. aureus* strain (JAR060131), isolated from a patient with an infected hip prosthesis, was used [17]. The strain is broadly antibiotic susceptible, except for resistance to penicillin. It is available at the Swiss Culture Collection, with accession number CCOS 890. The bacterial inocula were individually prepared in phosphate-buffered saline solution (PBS, Merck KGaA, Frankfurterstrasse 250, 64293 Darmstadt, Germany) for each surgery. Three different bacterial inocula were used in this study: high, 1×10^6 CFU/mL; medium, 1×10^4 CFU/mL; and low, 1×10^3 CFU/mL. A quantitative culture of each inoculum was performed immediately after preparation to check the accuracy of the prepared inoculum. The inoculum was pre-adhered by dipping the implant in the inoculum or 20 µL of inoculum was pipetted on top of it after implantation.

2.6. Quantitative Bacteriology at Euthanasia

All animals were kept at 4 °C after euthanasia and dissected within 4 h. At dissection, the surgical wound was first inspected followed by dissection of the implant and surrounding soft tissue. Post-mortem quantitative bacterial cultures were performed in all animals for the soft tissue adjacent to the implant and the implants itself. Soft tissue samples were homogenized in PBS using a homogenizer (Omni TH, tissue homogenizer TH-02/TH21649) until the tissue was a fine suspension and implants were sonicated in a Bandelin Ultrasonic water bath (Model RK 510 H) for 3 min. Serial tenfold dilutions of both solutions were plated on Tryptic soy agar (TSA) plates (Liofilchem srl, Via Scozia, 64026 Roseto degli Abruzzi TE, Italy). Bacterial growth was checked to determine if it was *S. aureus* using the latex agglutination test (Staphaurex™, Thermo Fisher Scientific Inc, Neuhofstrasse 11, 4153 Reinach TechCenter, 4153 Basel, Switzerland). The TSA plates were kept at room temperature for an additional 24 h to check for any slow-growing contaminants.

2.7. Statistical Analysis

Statistical analysis was performed using the Kruskal–Wallis test for comparisons between multiple groups. In the case of a statistically significant difference, multiple comparisons were performed using Dunn's multiple comparisons test. The Mann–Whitney test was used for comparing two groups.

3. Results

3.1. Animal Welfare

All animals recovered uneventfully from general anesthesia (average duration: 41 min) following surgery (average duration: 21 min). For the first 24 h, the animals had a score of up to three which declined thereafter. No difference between groups was observed. One animal of the dose-finding phase had to be excluded due to wound dehiscence with the subsequent exposure of the implant, but otherwise, all animals reached the scheduled euthanasia timepoint and showed no clinical signs of systemic infection.

In the dose-finding phase, animals infected with the low bacterial inoculum gained 0.9 ± 1.2 g, while animals infected with the medium and high inoculum lost 0.8 ± 1.1 g and 0.3 ± 0.9 g, respectively.

Animals included in the systemic antibiotic prophylaxis phase and treated with a low dose of systemic antibiotics lost 0.5 ± 1.0 g when infected with the low inoculum and 0.3 ± 0.9 g when infected with the medium inoculum. In the group receiving the high dose of systemic antibiotics, mice lost 1.3 ± 2.3 g. Animals being treated only with ALBC lost 1.2 ± 0.4 g, while the animals treated with a combination of local and systemic antibiotics lost 0.8 ± 0.7 g. Mice in the control group lost 1.2 ± 0.2 g. There was no significant difference in weight change between the groups within the different study phases.

3.2. Bacteriology

3.2.1. Dose Finding

The goal of this phase was to determine the optimal route of administration and bacterial inoculum required to reliably create an infection. None of the 19 mice included in the dose-finding phase were able to eradicate the infection (Figure 2). For both inoculation methods, CFU counts of soft tissue samples were significantly higher in animals receiving the high bacterial inoculum when compared to animals receiving the low inoculum ($p = 0.021$ for pipette inoculation, $p = 0.004$ for disc inoculation). A similar trend was observed in the CFU counts recovered from the surface of the implants, but this was only significant in the disc inoculation method ($p = 0.050$ for medium versus low inoculum).

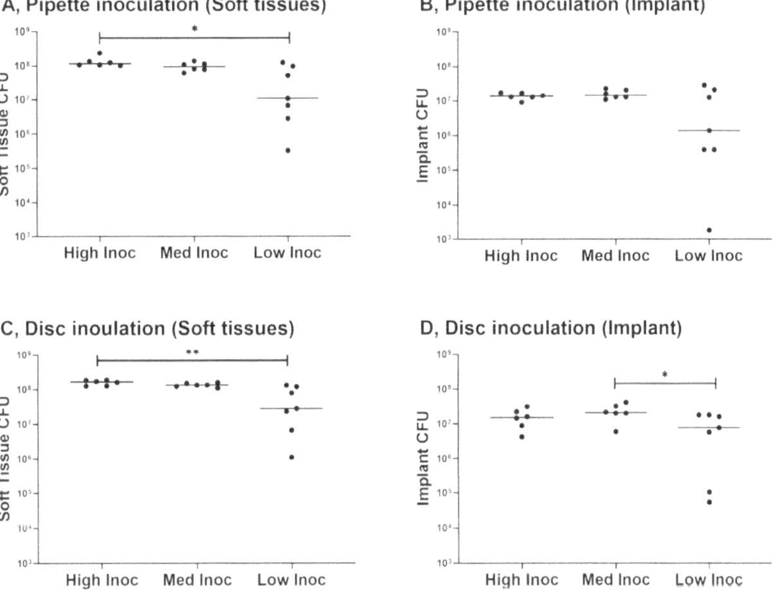

Figure 2. Results of dose-finding phase. Results are grouped according to the sample type and inoculation method for the high (n = 7), medium (n = 6) and low (n = 7) inoculum; (**A**) Pipette inoculation, soft tissue sample, * statistical significance between the high and low inoculum ($p = 0.02$); (**B**) Pipette inoculation, implant sample; (**C**) Disc inoculation, soft tissue sample, ** statistical significance between the high and low inoculum ($p = 0.004$); (**D**) Disc inoculation, implant sample, * statistical significance between the medium and low inoculum ($p = 0.05$). The detected CFU count is displayed on the y-axis; the three utilized bacterial inocula are represented on the x-axis.

CFU counts were generally comparable between the different inoculation methods. However, for the medium inoculum CFU, counts in soft-tissue samples were higher after disc inoculation ($p = 0.0368$). In this first phase, all three inocula and two inoculation methods reliably created an infection, with lower bacterial counts in the low inoculum groups.

3.2.2. Systemic Antibiotic Prophylaxis

This phase aimed to investigate the efficacy of systemic antibiotics in the prevention of infection. CFU counts of both the soft tissue and implant samples were comparable between the animals receiving the low and medium inoculum and being treated with 25 mg/kg of cefazolin. A significant difference in CFU counts between animals infected with the low inoculum and treated with either 25 mg/kg or 250 mg/kg of cefazolin was found in both sample types after disc inoculation (p = 0.020 for implant sample, p = 0.010 for soft tissue sample). For this phase of the trial, no significant difference in CFU counts was found between pipette inoculation and disc inoculation. In one out of eight animals receiving 250 mg/kg of cefazolin subcutaneously prior to surgery, no bacterial growth was detectable. The other mice were not able to completely clear the infection. Both the implant and soft tissue samples remained negative in one of the mice after pipette inoculation, while another mouse was able to clear the infection of both samples after disc inoculation (Figure 3).

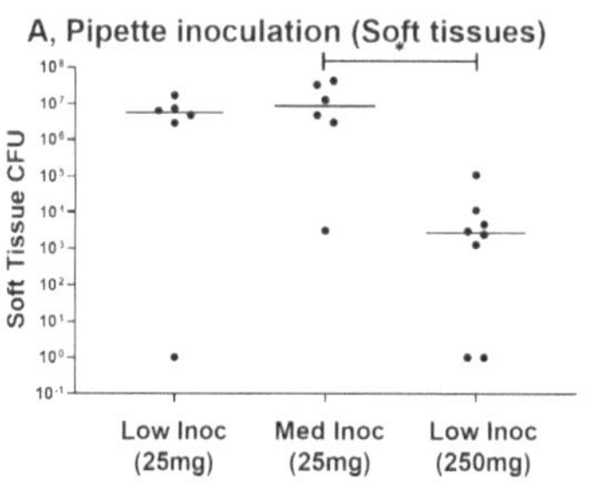

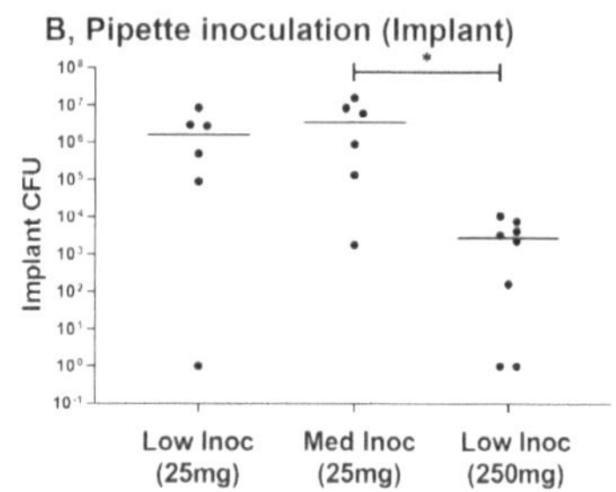

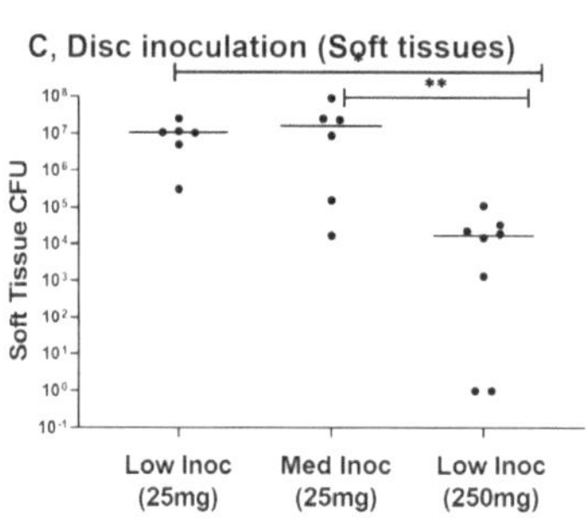

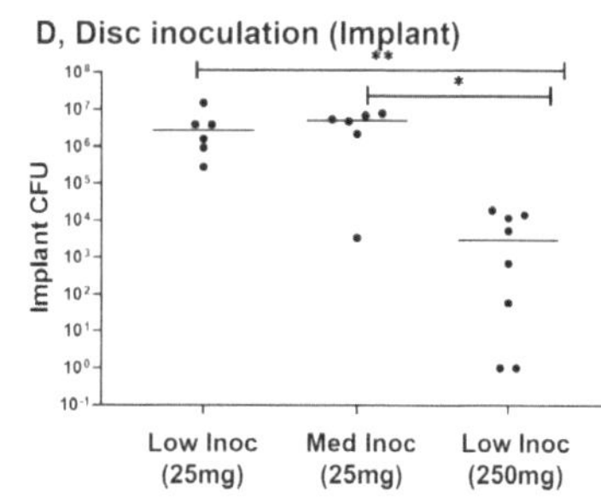

Figure 3. Results of systemic antibiotic prophylaxis phase. Results are grouped according to inoculation method and sample type for the low inoculum, treated with 25 mg (n = 6) or 250 mg (n = 8) of cefazolin and the medium inoculum, treated with 25 mg of cefazolin (n = 6). (**A**) Pipette inoculation, soft tissue sample, * statistical significance between the low inoculum treated with 250 mg of cefazolin and the medium inoculum treated with 25 mg of cefazolin (p = 0.01); (**B**) Pipette inoculation, implant sample, * statistical significance between the low inoculum treated with 250 mg of cefazolin and the medium inoculum treated with 25 mg of cefazolin (p = 0.03); (**C**) Disc inoculation, soft tissue sample, * statistical significance between the low inoculum treated with 250 mg of cefazolin and the low inoculum treated with 25 mg of cefazolin (p = 0.01), ** statistical significance between the low inoculum treated with 250 mg of cefazolin and the medium inoculum treated with 25 mg of cefazolin (p = 0.02); (**D**) Disc inoculation, implant sample, * statistical significance between the low inoculum treated with 250 mg of cefazolin and the medium inoculum treated with 25 mg of cefazolin (p = 0.01), ** statistical significance between the low inoculum treated with 250 mg of cefazolin and the low inoculum treated with 25 mg of cefazolin (p = 0.02). The CFU count is reflected on the y-axis; the x-axis displays inoculum and antibiotic concentration being used.

3.2.3. Local Antibiotic Prophylaxis

Control animals receiving the titanium implant without any systemic antibiotics presented with a high CFU count on all samples. In the group receiving no systemic antibiotic prophylaxis, bacterial growth was observed in three soft tissue samples. ALBC discs were not infected. In the group of mice receiving additional systemic antibiotic prophylaxis, one animal was not able to eradicate the infection in both the soft tissue and implant samples. All other animals cleared the infection. There was no statistically significant difference in CFU count between the groups with ALBC alone and ALBC combined with subcutaneous cefazolin (Figure 4).

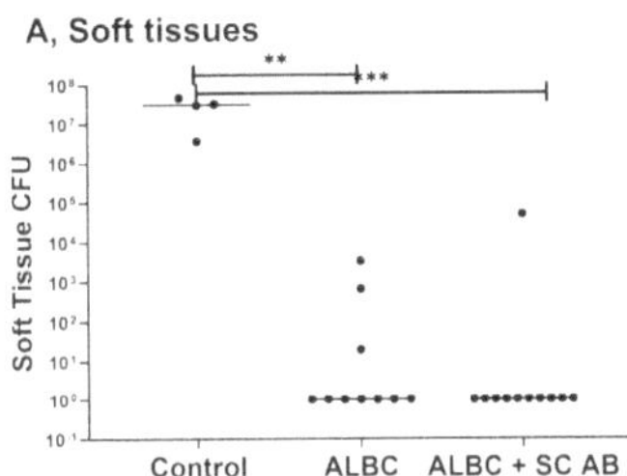

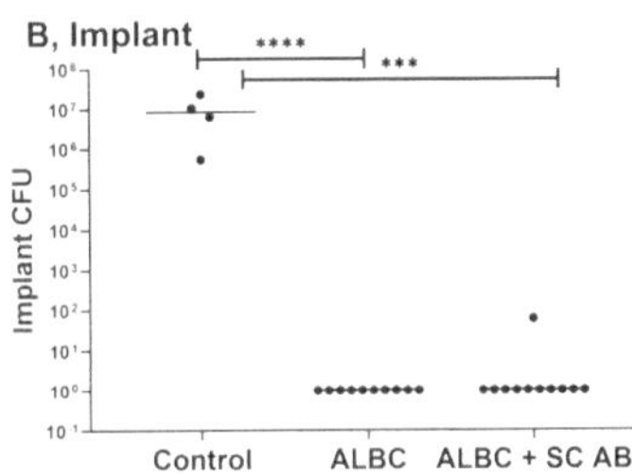

Figure 4. Results of local antibiotic prophylaxis phase. Results are grouped according to sample type for the control group (n = 4) and groups with ALBC with (n = 10) or without (n = 11) additional antibiotics. (**A**) Soft tissue sample, ** statistical significance between the control group and group with ALBC without additional antibiotics (p = 0.004), *** statistical significance between the control group and group with additional antibiotics (p = 0.0005); (**B**) Implant sample, *** statistical significance between the control group and group with additional antibiotics (p = 0.0002), **** statistical significance between the control group and group with ALBC without additional antibiotics (p < 0.0001). The CFU count is presented on the y-axis. The x-axis is reflecting the implant type, as well as additionally administered systemic antibiotic prophylaxis

4. Discussion

This study presents the characterization of a low complexity subcutaneous implant-associated infection model in mice. This model includes the host's response to novel biomaterials and can be applied in future research on future strategies in the prevention and treatment of ODRI.

The dose-finding phase of this study revealed that an *S. aureus* bacterial inoculum of 1×10^3 CFU/mL was sufficient to create a reproducible infection. In in vivo testing of antimicrobial coatings, it is critical that an appropriate bacterial burden is introduced to the surgical site. Bacterial inocula used in dose-finding studies typically vary between 1×10^2 and 1×10^8 CFU/site [18]. If the bacterial burden is too high, it does not adequately reflect the clinical situation of sterile surgery, where bacterial numbers are in the low hundreds and mostly airborne [19,20]. Furthermore, high inocula may mask the ability of a novel coating to prevent infection due to the overwhelming bacterial numbers typically not present in any clinical scenario. In contrast, the 1×10^3 CFU/mL dose established in the present study produced a sufficient but variable infection in all animals without leading to sepsis, making this a safe and suitable bacterial dose for future biomaterial testing.

This study also investigated different methods of bacterial administration, representing the different routes through which surgical implants might become infected during surgery. Pre-adhered bacteria reflect implants having been contaminated on the instrument tray by means of circulating airborne bacteria [20], whereas the pipetted bacterial inoculum represents bacteria being transferred during surgery or entering the wound from the skin soon afterwards. Generally, both methods produced comparable CFU counts, and a difference was only found with the medium inoculum in the dose-finding phase, where inoculation with pre-adhered bacteria resulted in higher CFU counts. The validity of both methods allows the researcher to adapt the protocol according to their needs. Pipetting the inoculum provides the advantages of negating implant preparation and the ability to test coated materials; however, containing the bacterial suspension within the surgical site during closure can be challenging. Pre-inoculated implants provide the unique advantage of reducing surgical time and staff but requires special care while handling to not disturb the bacterial layer and it cannot be performed on implants with antimicrobial coatings. Both approaches create a reliable infection for use in a wide range of research applications.

In the second phase of this study, the efficacy of PAP in infection prevention was investigated in this model to mimic the common practice of PAP in orthopedic surgery to decrease the incidence of postoperative infection. In a relevant model of ODRI, PAP alone would decrease the bacterial burden, but not avoid/clear the infection in all cases. In human medicine, the recommended dose for prophylactic cefazolin administration varies from 2 to 3 g (25 mg/kg to 37.5 mg/kg in obese patients) [11]. In this model, monotherapy with a 25 mg/kg dose of cefazolin was not effective in clearing the infection; however, significantly lower CFU counts were found after this treatment for infections created with the medium inoculum. In this way, the model reflects the clinical situation where infection often develops despite systemic antibiotic prophylaxis. Since the dose extrapolated from clinical practice was not effective in eliminating bacterial burden, the antimicrobial dose was increased to 250 mg/kg. However, this was still ineffective in completely clearing the infection and a significant difference in CFU counts when compared to treatment with the 25 mg/kg dose of cefazolin after infection development with the low inoculum was found only after disc inoculation. Species differences could account for the discrepancy between the required antimicrobial dose. Firstly, the serum half-life of cefazolin in mice is much lower than in humans (~2 h in humans vs. ~23 min in mice) [11,21]. Secondly, serum protein binding is higher in humans, leaving mice with a higher component of unbound and active drug [16]. These factors could explain the limited effect on infection prevention of systemic antibiotics monotherapy in this study. This shows that for future research in mice intravenous administration of systemic antibiotics might be preferable over subcutaneous administration.

For the successful translation of new interventions, a clinically relevant preclinical model needs to include a positive control reflecting the current state of the art to judge the effectiveness of the new treatment modality in comparison. Therefore, in the third phase of this preclinical study, locally applied gentamicin impregnated PMMA was included, as it is currently used routinely in orthopedic surgery to prevent ODRI. In soft tissue samples, the gentamicin-loaded bone cement completely cleared the infection in seven animals and markedly reduced the bacterial burden of the remaining three animals. The implants of all ten animals were culture negative. All samples were tested for gentamicin resistance and proven negative. Therefore, the local application of gentamicin using a PMMA carrier provides a significant advantage over the sole use of systemic PAP and provides a benchmark for the future testing of new materials.

Following previously conducted research, where an additional systemic antimicrobial injection in combination with gentamicin-loaded bone cement resulted in the lowest revision rate of total hip arthroplasty [22], an additional cefazolin injection (250 mg/kg) was given prior to surgery. The combination of systemic antibiotic prophylaxis as well as ALBC implants was highly efficacious in reducing the bacterial burden. However, one animal was highly infected in both obtained samples. No bacteria could be cultured on the implants of the remaining animals. Samples were checked for resistance against gentamicin with negative results. A possible explanation for this is an ineffective eradication of bacteria with a subsequent recolonization or contamination of the implant at the time of harvesting. Nonetheless, this model proved to be suitable for the testing of local antimicrobial strategies.

5. Limitations

This study has several limitations. Firstly, this study is valid for early-stage testing of antimicrobial biomaterials targeting infection prophylaxis. Therefore, implant and tissues were harvested on day 3 post infection. However, additional models are needed to evaluate the novel antimicrobial biomaterials for infection therapy at later time points. Secondly, cefazolin and gentamicin serum concentrations were not measured, which could have revealed insufficient levels and explained persistently infected animals. Only one type of implant was used with a uniform site over all included animals. The type of material, shape and size of the implant might have an impact on infection development and antibiotic efficacy. In future studies, other clinically relevant bacteria, especially Gram-negative, should also be investigated to enhance clinical relevance of this model. Lastly, as with every animal model, the pharmacokinetics and pharmacodynamics must be considered when extrapolating these results to humans.

6. Conclusions

This study presents a reproducible mouse infection model, which can be adapted for future research on novel strategies in the prevention and treatment of ODRI. Gentamicin-loaded PMMA discs with or without the additional systemic administration of cefazolin at a dose of 250 mg/kg proved highly effective in infection prevention and provide a suitable positive control to test other prevention strategies.

Supplementary Materials: The following supporting information can be downloaded at: https://www.mdpi.com/article/10.3390/biomedicines11010040/s1, Figure S1: Score sheet.

Author Contributions: Conceptualization, C.W., T.F.M. and S.Z.; methodology, C.W., T.F.M. and S.Z.; analysis, C.W. and N.V.; writing—original draft preparation, C.W. and N.V.; writing—review and editing, C.W., N.V., A.E.F., T.F.M. and S.Z.; supervision, A.E.F., T.F.M. and S.Z. All authors have read and agreed to the published version of the manuscript.

Funding: This study was funded by the Arbeitsgemeinschaft für Osteosynthesefragen (AO) Foundation (Grant number: AR2020_08).

Institutional Review Board Statement: The animal study protocol was approved by the Ethical Committee of the Canton of Grisons, Switzerland (TVB 20_2019 and 02E_2020).

Informed Consent Statement: Not applicable.

Data Availability Statement: Not applicable.

Acknowledgments: The authors would like to thank Iris Keller for technical assistance throughout the study.

Conflicts of Interest: The authors declare no conflict of interest.

References

1. Metsemakers, W.J.; Kuehl, R.; Moriarty, T.F.; Richards, R.G.; Verhofstad, M.H.J.; Borens, O.; Kates, S.; Morgenstern, M. Infection after fracture fixation: Current surgical and microbiological concepts. *Injury* **2018**, *49*, 511–522. [CrossRef] [PubMed]
2. Metsemakers, W.J.; Smeets, B.; Nijs, S.; Hoekstra, H. Infection after fracture fixation of the tibia: Analysis of healthcare utilization and related costs. *Injury* **2017**, *48*, 1204–1210. [CrossRef] [PubMed]
3. Chu, V.H.; Crosslin, D.R.; Friedman, J.Y.; Reed, S.D.; Cabell, C.H.; Griffiths, R.I.; Masselink, L.E.; Kaye, K.S.; Corey, G.R.; Reller, L.B.; et al. Staphylococcus aureus bacteremia in patients with prosthetic devices: Costs and outcomes. *Am. J. Med.* **2005**, *118*, 1416. [CrossRef] [PubMed]
4. Trampuz, A.; Widmer, A.F. Infections associated with orthopedic implants. *Curr. Opin. Infect Dis.* **2006**, *19*, 349–356. [CrossRef]
5. Gustilo, R.B.; Anderson, J.T. Prevention of infection in the treatment of one thousand and twenty-five open fractures of long bones: Retrospective and prospective analyses. *J. Bone Jt. Surg. Am.* **1976**, *58*, 453–458. [CrossRef]
6. Kortram, K.; Bezstarosti, H.; Metsemakers, W.J.; Raschke, M.J.; Van Lieshout, E.M.M.; Verhofstad, M.H.J. Risk factors for infectious complications after open fractures; a systematic review and meta-analysis. *Int. Orthop.* **2017**, *41*, 1965–1982. [CrossRef]
7. Torbert, J.T.; Joshi, M.; Moraff, A.; Matuszewski, P.E.; Holmes, A.; Pollak, A.N.; O'Toole, R.V. Current bacterial speciation and antibiotic resistance in deep infections after operative fixation of fractures. *J. Orthop. Trauma* **2015**, *29*, 7–17. [CrossRef]

8.   Vanvelk, N.; Morgenstern, M.; Moriarty, T.F.; Richards, R.G.; Nijs, S.; Metsemakers, W.J. Preclinical in vivo models of fracture-related infection: A systematic review and critical appraisal. *Eur. Cell Mater.* **2018**, *36*, 184–199. [CrossRef]

9.   Elek, S.D.; Conen, P.E. The virulence of Staphylococcus pyogenes for man; a study of the problems of wound infection. *Br. J. Exp. Pathol.* **1957**, *38*, 573–586.

10.   Zimmerli, W.; Waldvogel, F.A.; Vaudaux, P.; Nydegger, U.E. Pathogenesis of foreign body infection: Description and characteristics of an animal model. *J. Infect Dis.* **1982**, *146*, 487–497. [CrossRef]

11.   Bratzler, D.W.; Dellinger, E.P.; Olsen, K.M.; Perl, T.M.; Auwaerter, P.G.; Bolon, M.K.; Fish, D.N.; Napolitano, L.M.; Sawyer, R.G.; Slain, D.; et al. Clinical practice guidelines for antimicrobial prophylaxis in surgery. *Am. J. Health Syst. Pharm.* **2013**, *70*, 195–283. [CrossRef] [PubMed]

12.   ter Boo, G.J.; Grijpma, D.W.; Moriarty, T.F.; Richards, R.G.; Eglin, D. Antimicrobial delivery systems for local infection prophylaxis in orthopedic- and trauma surgery. *Biomaterials* **2015**, *52*, 113–125. [CrossRef] [PubMed]

13.   Metsemakers, W.J.; Fragomen, A.T.; Moriarty, T.F.; Morgenstern, M.; Egol, K.A.; Zalavras, C.; Obremskey, W.T.; Raschke, M.; McNally, M.A.; Fracture-Related Infection consensus, g. Evidence-Based Recommendations for Local Antimicrobial Strategies and Dead Space Management in Fracture-Related Infection. *J. Orthop. Trauma* **2020**, *34*, 18–29. [CrossRef] [PubMed]

14.   Samara, E.; Moriarty, T.F.; Decosterd, L.A.; Richards, R.G.; Gautier, E.; Wahl, P. Antibiotic stability over six weeks in aqueous solution at body temperature with and without heat treatment that mimics the curing of bone cement. *Bone Jt. Res.* **2017**, *6*, 296–306. [CrossRef] [PubMed]

15.   Moriarty, T.F.; Harris, L.G.; Mooney, R.A.; Wenke, J.C.; Riool, M.; Zaat, S.A.J.; Moter, A.; Schaer, T.P.; Khanna, N.; Kuehl, R.; et al. Recommendations for design and conduct of preclinical in vivo studies of orthopedic device-related infection. *J. Orthop. Res.* **2019**, *37*, 271–287. [CrossRef]

16.   Stavrakis, A.I.; Niska, J.A.; Shahbazian, J.H.; Loftin, A.H.; Ramos, R.I.; Billi, F.; Francis, K.P.; Otto, M.; Bernthal, N.M.; Uslan, D.Z.; et al. Combination prophylactic therapy with rifampin increases efficacy against an experimental Staphylococcus epidermidis subcutaneous implant-related infection. *Antimicrob. Agents Chemother.* **2014**, *58*, 2377–2386. [CrossRef]

17.   Campoccia, D.; Montanaro, L.; Moriarty, T.F.; Richards, R.G.; Ravaioli, S.; Arciola, C.R. The selection of appropriate bacterial strains in preclinical evaluation of infection-resistant biomaterials. *Int. J. Artif Organs.* **2008**, *31*, 841–847. [CrossRef]

18.   Busscher, H.J.; Woudstra, W.; van Kooten, T.G.; Jutte, P.; Shi, L.; Liu, J.; Hinrichs, W.L.J.; Frijlink, H.W.; Shi, R.; Liu, J.; et al. Accepting higher morbidity in exchange for sacrificing fewer animals in studies developing novel infection-control strategies. *Biomaterials* **2020**, *232*, 119737. [CrossRef]

19.   Lidwell, O.M.; Lowbury, E.J.; Whyte, W.; Blowers, R.; Stanley, S.J.; Lowe, D. Airborne contamination of wounds in joint replacement operations: The relationship to sepsis rates. *J. Hosp Infect* **1983**, *4*, 111–131. [CrossRef]

20.   Diab-Elschahawi, M.; Berger, J.; Blacky, A.; Kimberger, O.; Oguz, R.; Kuelpmann, R.; Kramer, A.; Assadian, O. Impact of different-sized laminar air flow versus no laminar air flow on bacterial counts in the operating room during orthopedic surgery. *Am. J. Infect Control* **2011**, *39*, e25–e29. [CrossRef]

21.   Lee, F.H.; Pfeffer, M.; Van Harken, D.R.; Smyth, R.D.; Hottendorf, G.H. Comparative pharmacokinetics of ceforanide (BL-S786R) and cefazolin in laboratory animals and humans. *Antimicrob. Agents Chemother.* **1980**, *17*, 188–192. [CrossRef] [PubMed]

22.   Engesaeter, L.B.; Lie, S.A.; Espehaug, B.; Furnes, O.; Vollset, S.E.; Havelin, L.I. Antibiotic prophylaxis in total hip arthroplasty: Effects of antibiotic prophylaxis systemically and in bone cement on the revision rate of 22,170 primary hip replacements followed 0–14 years in the Norwegian Arthroplasty Register. *Acta Orthop Scand.* **2003**, *74*, 644–651. [CrossRef] [PubMed]

Article

Local Experimental Intracerebral Hemorrhage in Rats

Ekaterina Vasilevskaya *, Aleksandr Makarenko, Galina Tolmacheva, Irina Chernukha, Anastasiya Kibitkina and Liliya Fedulova

V.M. Gorbatov Federal Research Center for Food Systems, Russian Academy of Sciences, 109316 Moscow, Russia; vivarium@vniimp.ru (A.M.); tgs2991@yandex.ru (G.T.); imcher@inbox.ru (I.C.); anastasjya@list.ru (A.K.); fedulova@vniimp.ru (L.F.)
* Correspondence: e.vasilevskaya@fncps.ru; Tel.: +7-495-676-9211

Abstract: (1) Background: Hemorrhagic stroke is a lethal disease, accounting for 15% of all stroke cases. However, there are very few models of stroke with a hemorrhagic etiology. Research work is devoted to studying the development of cerebrovascular disorders in rats with an intracerebral hematoma model. The aim of this study was to conduct a comprehensive short-term study, including neurological tests, biochemical blood tests, and histomorphological studies of brain structures. (2) Methods: The model was reproduced surgically by traumatizing the brain in the capsula interna area and then injecting autologous blood. Neurological deficit was assessed according to the Mc-Grow stroke-index scale, motor activity, orientation–exploratory behavior, emotionality, and motor functions. On Day 15, after the operation, hematological and biochemical blood tests as well as histological studies of the brain were performed. (3) Results: The overall lethality of the model was 43.7%. Acute intracerebral hematoma in rats causes marked disorders of motor activity and functional impairment, as well as inflammatory processes in the nervous tissue, which persist for at least 14 days. (4) Conclusions: This model reflects the situation observed in the clinic and reproduces the main diagnostic criteria for acute disorders of cerebral circulation.

Keywords: biomodel; stroke; hemorrhage; neurology; rats; behavior

Citation: Vasilevskaya, E.; Makarenko, A.; Tolmacheva, G.; Chernukha, I.; Kibitkina, A.; Fedulova, L. Local Experimental Intracerebral Hemorrhage in Rats. *Biomedicines* **2021**, *9*, 585. https://doi.org/10.3390/biomedicines9060585

Academic Editor: Martina Perše

Received: 15 April 2021
Accepted: 19 May 2021
Published: 21 May 2021

Publisher's Note: MDPI stays neutral with regard to jurisdictional claims in published maps and institutional affiliations.

1. Introduction

Acute and chronic disorders of cerebral circulation are problems that affect many people. To develop adequate methods for the prevention of various stroke subtypes and rehabilitation after the disease, it is necessary to conduct comprehensive research, including using animal biomodels [1,2]. The success of such studies depends on the choice of an experimental model since an inadequate model can lead to limitations that compromise the results and analysis. In addition, extrapolating the results of animal models to humans can be unreliable. Rodents are mainly used for stroke modelling due to cost considerations, ethical considerations, the availability of standardized neurobehavioral assessments, and the simplicity of the physiological monitoring [3–5].

Modern publications mention two main types of models for higher nervous activity disorders: the use of chemical agents for post-stroke syndrome development in animals and surgical models of spontaneous stroke [6]. Both mice and rats are used as research objects. However, due to the larger size of the latter (in adulthood, rats weigh about 8–10 times more than mice), experiments with rats guarantee several practical advantages, especially concerning surgical procedures. Models of the brain and spinal cord injury using rats have a great translational value [7]. In addition, rats and mice showed significant differences in the plasticity of their hippocampal and cortical neurons [8]. It was found that the rate of neurogenesis in the hippocampus of sexually mature rats is much higher than in mice and, importantly, in rats, new cells mature about 2 weeks earlier than in mice; in addition, the probability of their activation is ten times higher [9].

There also are rat strains susceptible to stroke (spontaneously hypertensive SHR and stroke-prone spontaneously hypertensive SHRSP rats) that can be used as an experimental model, characterized by a high frequency of spontaneous strokes as well as increased sensitivity to experimentally induced focal cerebral ischemia [10]. The advantages of using SHR rats in stroke research are concomitant hypertension presence and the development of reproducible infarction of adequate size after distal middle cerebral artery occlusion. At the same time, this model has a significant disadvantage—SHR and SHRSP rats are expensive, and a high mortality of the animals is observed at a certain age; in addition, these rats are resistant to therapy [11]. Furthermore, SHRSP rats usually have ischemic, or less frequently, hemorrhagic stroke in the cortex, rather than in the brainstem, cerebellum, or basal ganglia, as in patients with hypertension [12].

It is worth noting the relatively small number of stroke models with a hemorrhagic etiology, this while hemorrhagic stroke accounts for approximately 15% of all stroke cases and leads to high mortality. The main models were developed in the twentieth century and are based on direct injection of autologous blood or bacterial collagenase into various brain regions (autologous blood injection model and bacterial collagenase injection model) [3,13–16]. However, they are most often used for large animals (primates, pigs, and rabbits) and less often for rats and mice [17–19]. One of the disadvantages of the autologous blood injection model is the use of an anticoagulant agent, which leads to additional internal injuries and retrograde penetration of heparinized blood through the reverse channel, as well as into non-target areas of the brain [18]. The method developed earlier [20] for reproducing experimental acute intracerebral hemorrhage allows us to recreate limited damage to brain structures localized in the capsula interna area, which is the most reliable and close to the human model, but not all aspects of this model have been considered.

Acute stroke period assessment is a critical stage used to understand stroke severity, treatment options, and prognosis, and it serves as the main tool for identification biomarkers, but the recovery period and its characteristics are also of great importance [21]. To better analyze the pathophysiology of the acute and recovery periods in the hemorrhagic stroke model according to [20], for adequate in vivo studies, a comprehensive short-term study was conducted, including neurological tests, biochemical blood analysis, and histomorphological studies of brain structures.

2. Materials and Methods

2.1. Animals

The study was carried out on male Wistar rats ($n = 30$), 10 weeks old, and weighing 240 ± 25 g, obtained from the Federal State Budgetary Institution of Science "Scientific Center for Biomedical Technologies of the Federal Medical And Biological Agency", Andreevka Branch (Moscow Region, Solnechnogorsk District, Andreevka Settlement).

Animals were acclimatized to handling and the new facilities in harmonious groups for seven days prior to the experimental work. Visual inspection of the animals showed a normal health status over this time. A normal health status was confirmed by visual inspection of the animals and daily weighting during acclimatization. Animals were provided with cage substrate (Lignocel BK 8–15/LIGNOCEL, J. Rettenmaier & Sohne GMBH+CO KG, Sulzbach-Rosenberg, Germany) and a plastic shelter (Tecniplast, Milan, Italy). In the 3 days before surgery, animals were provided with nesting material for nest assembly. The cages were changed once every 6–7 days (but not less than three days before behavioural testing).

No supplemental diet was used in this experiment; after surgery, the rats' diet (ad libitum) consisted of a complete feed compound (Laboratorkorm, Moscow, Russia).

Rats were kept in T-IV polycarbonate cages (Tecniplast, Milan, Italy), in groups of 3–4 individuals. The animals were kept under controlled environmental conditions: air temperature $21 \pm 2\ ^\circ$C and relative humidity 50–60%. Lighting in the experimental rooms was artificial (12 h/day, 12 h/night), and daylight was inverted (light period from 6 p.m. to 6 a.m.); behavioral tests were performed in the dark period under dim red-lighting conditions, because rats are more active during the dark period [22].

2.2. Animal Model

To study the cerebrovascular disorders' development, the rats were divided into the following groups:

1.   Intact—intact animals kept under equal conditions ($n = 6$).
2.   Sham—sham-operated animals ($n = 8$) underwent anesthesia and surgery with craniotomy without brain traumatization.
3.   Model—operated on animals with reproduced intracerebral hematoma ($n = 16$)

Surgical operations were accompanied by the appropriate veterinary and medical procedures for preoperative preparation of the animals and postoperative care [23]. All non-sterile materials potentially in contact with the operating field (instruments, suture material, bandages, napkins, cotton wool, etc.) were sterilized with hot steam (121 °C, 20 min). The equipment and materials used in the operation (surgical table, stereotaxis, etc.) were treated with a disinfectant solution (0.5% solution of chlorhexidine biglucanate in 70% ethyl alcohol), followed by rinsing with sterile water.

Before the operation, the animals were subjected to food deprivation for 6 h, without restricting access to water. Fasting prior to surgery was used as preconditions against a variety of complications in preclinical models [24].

Immediately before the operation, the animal was anesthetized for 40 min with a mixture of Xila (Interchemie werken, De Adelaar, B.V, Castenray, the Netherlands) in a 10 mg/kg dose and Zoletil 100 (Virbac, Carros, France) in a 20 mg/kg dose intramuscularly [25].

Then, the animals were placed in an empty cage with a paper towel pad and closely monitored until anesthesia occurs. Anesthesia was verified by the disappearance of the reaction to pain stimuli (prick of the paw) and inhibition of the corneal reflex.

During general anesthesia and in the immediate post-operative period, the animal's body temperature was maintained by insulation with rectal temperature control (VET1-R, Parthner-Agro SPB, Sankt-Peterburg, Russia). To warm the animals during and after the operation, we used a heating device for animals (Zoomed HM-1, Zoomed, Zelenograd, Russia) with a circulating-warm-water blanket (40 $\times$ 60 cm), with a feedback heating system that cuts out when a normal body temperature is reached; the heating temperature was 38.5 °C.

Pulse oximetry (Utech UT100, Utech, Shanghai, China) was used to ensure physiological stability under anesthesia, registering the tissue oxygen saturation and heart rate with a sensor placed on the tail.

Each animal was fixed with its back upon a surgical table and tied with loop-like knots behind its paws. To prevent the cornea from drying out, an eye gel Oftagel (Santen OY, Osaka, Finland) was dripped into the eyes. In the places of planned incisions, fur was removed, and the surgical field was treated from the center to the periphery with sterile swabs moistened with disinfectant. After treatment, the operating field was isolated with sterile wipes.

During the operation, a sagittal dissection of the integumentary tissues in the frontal, parietal, and occipital regions was performed; the periosteum was removed from the bones of the cranial vault. Modelling of unilateral stroke was carried out according to [20,26]. Under the technique, holes were drilled in the projection of the inner capsule using the stereotaxis device RWD 68025 and microdrill RWD 78001 (RWD Life Science Co., LTD, Shenzhen, Guangdong, China), with coordinates H = 4.0 mm, L = 3.0 mm, and AP = 1.5 mm from bregma according to [27] in the area of the right hemispheres; the diameter of the trepanation hole was 2 mm. Then the dura mater was pierced using a sharpened cannula needle (diameter 0.8 mm) with a rubber retainer and immersed to the required depth (4 mm). Brain traumatization in the inner capsule was carried out with a mandrel, using clockwise rotational movements (5–6 times), resulting in "cone-shaped" undercutting of the brain tissue and damage to the vessels in the area of the capsula interna (Figure 1).

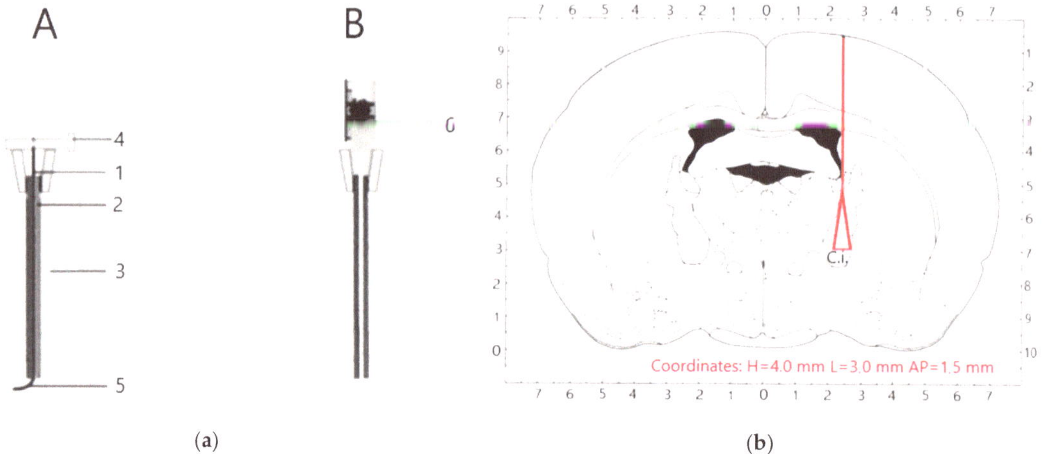

(a)　　　　　　　　　　　　　　　　(b)

Figure 1. Scheme of the procedure of acute hemorrhagic stroke in the capsula interna: (**a**) Mandrel device: A—device is in position with extended mandrel; B—device after injection in the structure of the brain: 1—tungsten stylet-knife; 2—guiding needle cannula; 3—holder of the stereotaxic device; 4—latch top portion mandrel-knife; 5—cutting the bottom end mandrel-knife [20]. (**b**) Stereoscopic model [27] for simulating acute hemorrhagic stroke in the internal capsule. The red line and the red triangular area correspond to the post-surgery location of the damage. C.i.—Capsula Interna.

An additional 100 µL of autologous blood was injected into the indicated area 2–3 min after trauma with a syringe (the blood was taken from the tail vein). Undercutting the vessels with a mandrel and introducing autologous blood into the site of injury is necessary to standardize the morphology, size, and location of the blood deposits [27]. The next incision was sutured with an intermittent suture with needles triangular in cross-section, and the suture was treated with an alcoholic solution of brilliant green.

In sham-operated animals, 5 mm-long skin incisions were made, holes were drilled in the skull, and stitches were applied as in group 3 rats.

After surgery, each animal was placed in a postoperative box until it was completely out of anesthesia.

Inter-animal housing, feeding, and handling practices before and after stroke was ensured. Animals were returned to the same group of animals they were with before surgery within 24 h after stroke. Each cage contained 2 sham-operated and 2 stroke animals, because social housing of rats with a healthy (non-stroke) partner immediately after stroke reduces mortality compared to housing with a stroke partner. Post-surgery care regarding bedding included cage substrate (Lignocel BK 8–15/LIGNOCEL, J. Rettenmaier & Sohne GMBH + CO KG, Sulzbach-Rosenberg, Germany) and nests.

Animals were monitored at least 6 times a day during the first 72 h post-stroke; the sensorimotor deficit and motor skills, water and food consumption, weight, and presence/absence of urine/feces were monitored, so too the surgical wound, breathing, and handling behavior were assessed [28].

To avoid dehydration, 1–2 mL/100 g of warm (35–37 °C) sterile Ringer's solution was administered subcutaneously after the operation. Loose pellets were supplied for seven days post-stroke on the cage floor.

Further postoperative care included daily monitoring of the animal conditions: post-surgery pain in rats was assessed observing the following: decreased activity, piloerection, an ungroomed appearance, self-mutilation, abnormal stance or a hunched posture, respiration rapid and shallow with grunting or chattering on expiration, dilated pupils, porphyrin secretion ("red tears") around the eyes and nose, vocalization and unusually aggressive when handled, changes in feeding activity, and group behavior or grooming [29–31].

If signs of pain and distress were detected [26], Xylazine in a 5–12 mg/kg dose was administered subcutaneously every 2–4 h [32] for the1st and 2nd days post-surgery for all animals, and special attention was paid to the condition of the sutures.

Animal body weights were measured daily using an electronic laboratory balance (Adventurer Pro AV 2101, OHAUS, Parsippany, NJ, USA).

Humane endpoint criteria were selected [28,30,33]:

(a)   Animal not moving, unresponsive to stimulation, or in a lateral recumbent position.
(b)   Weight loss exceeding 20% beyond 48 h post-stroke despite all efforts to supplement fluid and diet.
(c)   Respiratory distress persisting beyond the first 48 h.
(d)   Intermittent abnormal motor activity, suggestive of seizure, tonic clonic seizures, persisting beyond first 72 h, presence of barrel rolling.
(e)   No recovery of weight towards the pre-stroke level 7 days post-stroke.

Animals were euthanized using carbon dioxide in a VetTech installation (VetTech Solutions Ltd., Cheshire, UK), following Directive 2010/63/EU of the European Parliament and the European Union Council for Protection of Animals used for scientific purposes. The rats were kept and exposed to all manipulations in compliance with Directive 2010/63/EU of the European Parliament and of the Council. The research was approved by the bioethical commission of the V.M. Gorbatov Federal Research Centre for Food Systems of the Russian Academy of Sciences (protocol #04/2017, dated 22 September 2017).

2.3. Neurological Deficit Assessment

To study the rats' neurological status on Days 0 (prior to surgery), 2, 4, 7, 10, and 13 of the experiment, the McGrow stroke-index scale was used with the modifications of I. V. Gannushkina [34,35] and A. E. Kulchikov [36]. Neurological symptoms according to Gannushkina [34] were recorded in points: animals with mild symptoms were scored up to 2.5 points (lethargy, slowness of movement, weakness of the limbs, tremor, unilateral or bilateral ptosis, and Manege movements), with severe manifestations of neurological disorders scoring from 3 to 10 points (paresis and/or paralysis of the lower extremities, lateral position, and comatose state). Following Kulchikov [36], neurological testing was performed by evaluating the postural reactions, flexion reflex, paw placement reaction, and testing in an open field and pulling up on a crossbar; when testing, involuntary innate behavioral reactions were assigned a score: for values from 1 to 5 points, a mild degree of brain damage was diagnosed, from 6 to 9, it was diagnosed as a medium, and from 10 to 20, it was diagnosed as severe (Table 1). The group average score was calculated from the summary points of each rat.

Table 1. Neurological deficit in the animals after assessment, using the McGrow stroke-index scale with A. E. Kulchikov's modification [36].

Manipulations	Neurological Status Assessment in Points
The animal was brought to the table, the back of the front legs should touch the edge of the table.	Normally, the animal puts both paws on the table, if the animal does not place the paralyzed limb on the table—1 point; does not put both limbs—2 points.
The animal was fixed by the tail above the table and gradually lowered so that the vibrissa touched the table.	Normally, the animal tries to grab the edge of the table with its front paws, if there is no reaction, 2 points are assigned.
The animal was fixed by the tail above the table and gradually lowered so that the vibrissa did not touch the edge of the table.	Normally, the animal tries to grab onto the edge of the table with its front paws, if there is no reaction, 2 points are assigned.
The animal was tested under standard "Open field" conditions.	If the animal makes circular movements in the direction of the paralyzed limb around itself, without going beyond the square—1 point; if circular movements in the direction opposite to the paralyzed limb—2 points
	If the animal makes a circular motion within the central squares, then it is assigned 2 points.
Examination of the eyeballs.	If, after modeling the pathology, ptosis of one eye occurs—1 point; if ptosis of both eyes—2 points; if exophthalmos develops—2 points.
The animal was fixed by the tail and gradually raised.	If the hind legs are crossed—2 points.
Holding on a horizontal rod.	Normally, the animal pulls itself up on the crossbar, if the animal does not pull up and falls—2 points.
Prick in the paw.	If the animal withdraws its paw with a defensive reaction—0 points; if there is no defensive reaction and there is a withdrawal—1 point; if both reactions are absent—2 points.

Motor activity, exploratory behavior, and emotionality were evaluated in the "Open field" test. The studies were carried out on Days 0 (prior to surgery), 2, 3, 6, 7, 10, 11, and 13 of the experiment. The "Open field" installation is a square box (100×100 cm) with sides of 45 cm high. The installation arena is divided into 25 squares (20×20 cm), necessary for visual registration of the movements of the tested animals. The arena has 16 holes with a diameter of 20 mm at the intersection of the lines of sectors (mink). According to the standard method, the animal was placed in the center of the arena and the behavioral acts were recorded for three minutes: horizontal motor activity—the number of crossed squares; vertical motor activity—the number of racks; the number of examined minks ("mink reflex"); grooming; and the number of acts of defecation (boluses). The arena was cleaned with disinfectant after testing each animal.

Motor functions were evaluated in the test of holding the animals on a horizontal rod on Days 0 (prior to surgery), 2, 6, 10, and 13 of the experiment. The installation consisted of a horizontal crossbar (2 mm in diameter) placed at a height of 60 cm from the table. The time of holding the animals on the crossbar was recorded, and the animal's inability to pull its hind legs up to the crossbar was recorded as limb weakness and a manifestation of neurological deficit. Rats' activity was recorded on a Nikon DT5600 Kit (Nikon, Tokyo, Japan) and automatically processed using RealTimer software (OpenScience, Moscow, Russia).

The experiment was completed on the 15th day, which was associated with a decrease in the severity of the clinical disease in rats on Days 9–14, according to [37]. At the end of the experiment, the rats were euthanized in a euthanasia chamber (VetTech Solutions Ltd., Cheshire, UK) following EU Directive 2010/63/EU. Blood was collected from the stunned animals from the right atrium in EDTA tubes for hematological studies and standard glass tubes for subsequent serum sampling after centrifugation ($1300 \times g$ for 5 min).

The hematological analysis was performed on an Abacus Junior Vet 2.7 automatic analyzer (Diatron Messtechnik GmbH, Wiener Neudorf, Austria) using Diatron reagent kits. Biochemical analysis of the blood serum [38] was performed on a BioChem SA analyzer (HTI, North Attleboro, MA, USA) using reagent kits (HTI, North Attleboro, MA, USA).

To study the morphological parameters of cells, the brain was immediately removed from the cranium and fixed in 10% buffered formalin (HistoSafe, BioVitrum, Sankt-Peterburg, Russia) for 2 hours. The fixed brain was cut coronally through the needle entry site (defined on the surface of the brain) as well as 2 mm in the front and 2 mm behind this plane. Sections of the frontal cerebral cortex 10 microns thick were prepared on a Microm HM 525 cryotome (Carl Zeiss, Oberkochen, Germany). Every 10th section from the rostral to the caudal part of the residual hematoma cavity was stained with hematoxylin–eosin and according to the Nissl method. For the Nissl staining method, sections were defated in xylene (BioVitrum, Sankt-Peterburg, Russia) 2 × 3 min and rehydrated in descending alcohols (100%/95%/70% isopropanol, BioVitrum, Sankt-Peterburg, Russia) for 3 min each. Next, sections were rinsed in distilled water and stained in 0.1% cresyl violet solution (Sigma, Ronkonkoma, NY, USA) for 5 min (warmed up in 37 °C), and then rinsed quickly in distilled water. Then, sections were differentiated in ethyl alcohol (70%/100%/100%, BioVitrum, Sankt-Peterburg, Russia) for 5 min, dehydrated in isopropanol (100%, BioVitrum, Sankt-Peterburg, Russia) for 3 min, cleared in xylene (BioVitrum, Sankt-Peterburg, Russia) 2 × 3 min. Sections were mounted in Bio Mount HM (Bio-Optica, Milan, Italy).

Morphometric studies were performed using a BX 51 microscope (Olympus, Tokyo, Japan). The functional state of the surviving nerve cells was judged based on changes in the area and perimeter of the nuclei and perikaryon of the neurons, the death of neurons and glial cells, and the neuroglial index. Morphometric studies were carried out in a test zone with a size of $270 \times 270 \ \mu m^2$. Cells with signs of cytolysis and karyolysis, and cells with homogeneously stained acidophilic nuclei, devoid of nucleoli, were considered degenerative [39].

2.4. Statistics

Statistical processing of the results was performed in the program STATISTICA version 10.0 (Statsoft, Tulsa, OK, USA). Intergroup comparison was performed using a one-way ANOVA median test and the ANOVA method using the Kruskal–Wallis criterion, with pairwise comparison of groups using the nonparametric Mann–Whitney test. The results are presented as the median (Me) and interquartile range ((P25–P75)). Morphological data are expressed as the mean $\pm$ SEM and were analyzed by a two-tailed Student's t-test for unpaired data. A p value of 0.05 or less was considered statistically significant.

3. Results
3.1. Mortality

Within two weeks after the start of the experiment, no deaths of intact animals and sham-operated rats were observed. In the model group of animals, the animals' mortality amounted to 43.7% for the entire period of the experiment (Table 2); during the 1st day after the operation, 12.5% of the rats with the hemorrhagic stroke model died; further deaths were noted on Days 5, 10, 12, 13, and 14—up to 10% of the animals.

Table 2. Animal mortality during the experiment.

Animal Group	Day 1	Day 5	Day 10	Day 12	Day 13	Day 14	Total Mortality
1—Intact	0/6	0/6	0/6	0/6	0/6	0/6	0/6
2—Sham	0/8	0/8	0/8	0/8	0/8	0/8	0/8
3—Model	2/16	1/14	1/13	1/12	1/11	1/10	7/16

In all cases, mortality was caused by massive hemorrhage into the internal capsule of the brain: the brain was hyperemic, and a subdural hematoma was found on the surface of the right hemisphere of the brain under the dura mater, occupying the parietal and temporal regions (Figure 2) of all animals. A sagittal section revealed massive hemorrhage under the dura mater on the right and left from the lateral and basal regions of both hemispheres. In 12.5% of the dead animals, a thrombus was found under the dura mater (Figure 2c), spreading from the modelling area in the caudal direction to the occipital lobe. Histological studies (Figure 2e,f) revealed the polymorphic nature of the structural and pathomorphological disorders. A significant number of neurons were in a state of acute edema, with swelling of the bodies and their nuclei: neurons with deformed nuclei and signs of their destruction were present. There were dead neurons with signs of autolysis, many diffusely located dead glial cells, pycnosis of the nuclei of some gliocytes, and hyperchromatosis of the vast majority of glial cells.

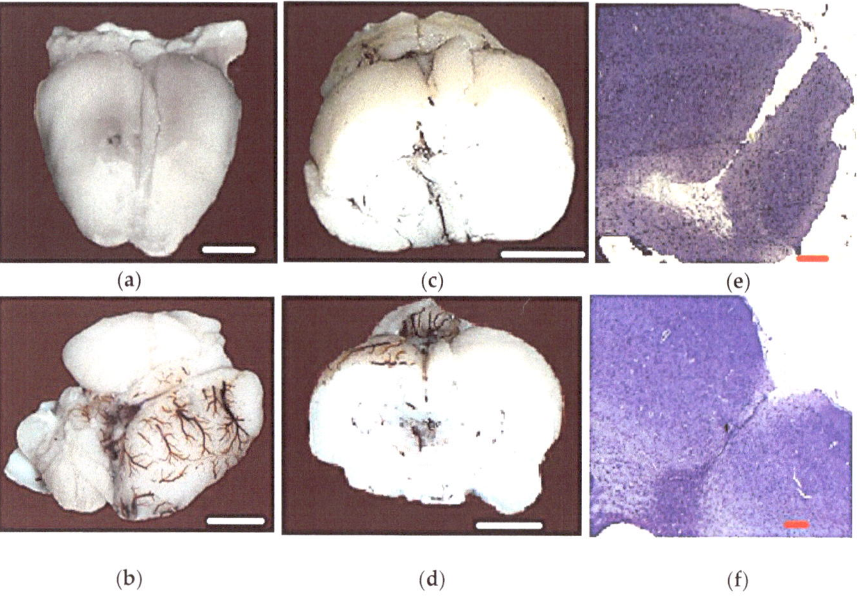

Figure 2. Macro-and microscopic structure of the dead animal brains. (**a**–**d**) Hemorrhages in the structure of the brain; (**e**,**f**) hemorrhage in the capsula interna area: hematoxylin–eosin staining (20×). The lines in the figures represent the scale: white lines—35 mm; red lines—100 microns.

3.2. Weight

When forming groups (Day 0 of the experiment), the Intact animals' average weight was 247.0 (227.8–261) g. In the Sham and Model rats, the average weights were 218.0 (209.5–220.8) g and 239.0 (229.3–255.3) g and were reduced relative to the values of Group 1 by 11.7% and 3.2%, respectively ($p < 0.01$).

The mass dynamics in Intact rats throughout the experiment (Figure 3) were favorable; the mass increased by an average of 2.5 ± 0.9 g per day, and the total weight gain by the end of the experiment was 13.5%—285.5 (285.0–304.8).

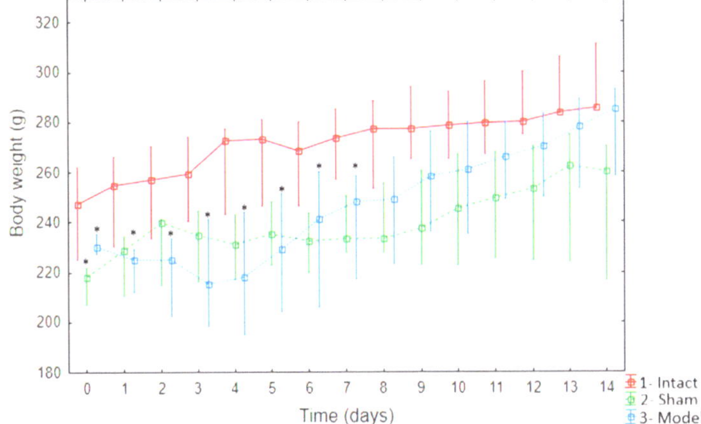

Figure 3. Mass dynamics during the experiment (* $p < 0.05$ in comparison to Group 1—Intact animals).

Sham animals showed a decreased body mass on Day 1 by 10.2% compared to Intact animals; the weight of the animals was 228.5 (211.3–235.5) g ($p = 0.058$). On Day 9, the decrease was up to 14.3% and amounted to 237.5 (225.3–260.3) g ($p = 0.083$); on Day 13, the decrease was up to 7.4% and amounted to 262.5 (228.0–274.5) g ($p = 0.054$). Model animals showed a decrease in weight relative to the Intact animals in the period from 2 to 10 days of the experiment (Figure 3). On Days 2 and 3 of the experiment, the weight decreased by 10.7% and 13.9% ($p < 0.02$), and the weights were 229.5 (209.3–236.0) g and 223.0 (201.0–236.8) g, respectively. Weight loss on Day 4 reached 20.5% ($p = 0.006$), and on Days 5 and 6, weight loss was 16.9% and 15.5% ($p < 0.02$), and the values were 216.5 (196.8–240.3) g, 227.0 (199.0–249.0) g and 220.0 (202.0–256.0) g, respectively. On Day 7, the weight loss reached 20.7%, amounting to 217.0 (204.0–256.0) g ($p = 0.008$). Observations on Days 8–10 revealed a slowdown in weight loss. On Day 8, the weight decreased by 17.0% ($p = 0.022$); on Day 9, the weight decreased 14.8% ($p = 0.032$); and on Day 10, the weight decreased 9.5% ($p = 0.085$)—the values were 230.0 (208.0–259.0) g, 236.0 (199.0–263.0) g, and 252 (208.8–274.0) g, respectively.

On Day 13, the average weight of the Model animals was 285.0 (258.0–293.0) g and did not differ significantly from the values of the Intact (285.5 (285.0–304.7) g) and Sham (260.0 (224.7–269.0) g) animals.

3.3. Neurological Assessment

Prior to surgery, there were no absence of neurological deficit symptoms in Intact, Sham, and Model animals.

Results of the animals' neurological status analysis showed the absence of neurological deficit symptoms in Intact animals during the experiment.

In Sham animals, there were no neurological disorders both on Day 2 after surgery and during follow-up (Table 3).

Table 3. Results of the assessment of neurological deficit according to [26]. In the table, colored cells give the percentage of animals with recorded symptoms. Group average score presented as Me, (P25–P75).

Symptoms	Points	Day 2		Day 4		Day 7		Day 10		Day 13	
		Sham	Model	Sham	Model	Sham	Model	Sham	Model	Sham	Model
Lethargy, slowness of movement	0.5	62	71	50	64	37	31	25	33	12	40
Tremor	1.0	12	28	12	14	12	15	12	17	12	10
Unilateral partial ptosis	1.0	0	50	0	50	0	38	0	33	0	20
Bilateral partial ptosis	1.5	12	0	12	0	12	8	12	8	12	0
Inability to pull back a limb while holding it	1.5	0	71	0	71	0	69	0	33	0	20
Unilateral ptosis	1.5	0	21	0	36	0	23	0	17	0	10
Bilateral ptosis	1.5	0	28	0	14	0	15	0	8	0	10
Manage movements	2.0	0	57	0	50	0	38	0	17	0	10
1 limb paresis	2.0	0	0	0	0	0	0	0	0	0	0
2 limb paresis	3.0	0	43	0	57	0	46	0	17	0	10
1 limb paralysis	3.0	0	28	0	14	0	15	0	8	0	10
2 limb paralysis	4.0	0	0	0	0	0	0	0	0	0	0
Coma	7.0	0	28	0	14	0	15	0	17	0	10
Fatal outcome	10.0	0	12	0	0	0	7	0	8	0	17
Group average score	Me	0.5	8.5	0.2	6.0	0.0	5.0	0.0	1.0	0.0	0.7
	P25	0.0	4.7	0.0	1.0	0.0	1.0	0.0	0.0	0.0	0.0
	P75	0.5	11.6	0.5	8.5	0.1	8.7	0.0	9.5	0.0	7.7

Symptoms such as lethargy, slow motion, and tremors were reported in the Sham animals. The total score, when assessed on the McGrow scale as modified by I.V. Ganushkina, in the group on Day 2 was 0.5 (0.0–0.5), and in A.E. Kulchikov's modification, it was 2.0 (2.0–2.0) (Figure 4). The neurological deficit in this group was characterized as mild and tended to regress on Day 4. On day 13, 88% of the animals did not have any symptoms of neurological deficit.

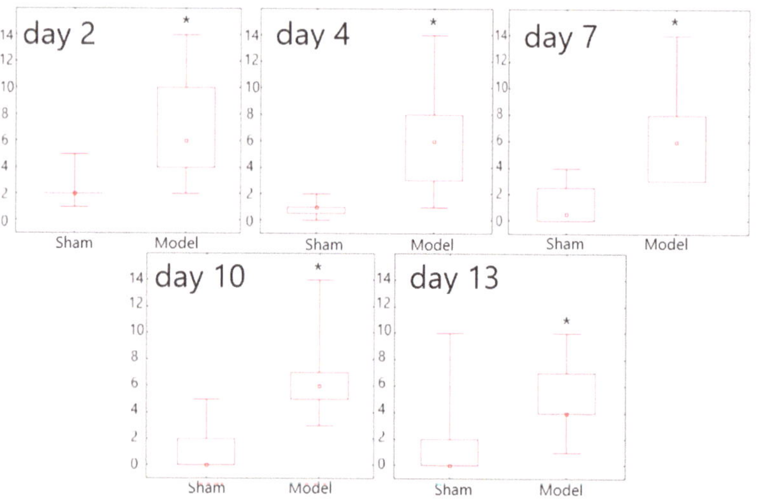

Figure 4. Neurological deficit analysis according to [35]. The data are presented in the form of a swing plot with P25–P75 boundaries and marks of the minimum and maximum values of the dataset. * $p < 0.05$ in comparison to Group 2—Sham animals.

In Model animals, from Day 2 after the operation, various neurological symptoms corresponding to a severe deficit were observed, with regression by Day 13 (Table 3). The total scores on Days 2 and 13 on the McGraw scale in I.V. Ganushkina's modification in the group were 8.5 (4.7–11.6) and 0.7 (0.0–7.7), and according to the scale modified by A.E. Kulchikov, the scores were 6.0 (4.2–9.5) and 5.0 (4.0–6.5) (Figure 4), respectively.

The study of the exploratory behavior in the "Open field" test (Table 4) in Intact rats showed that from Day 7 of the experiment, there was a gradual decrease in the horizontal (from 24.0 (16.8–32.0) to 15.5 (12.8–19.8)) and vertical activity (from 12.0 (8.3–13.5) to 5.0 (2.5–6.8)), exploratory behavior (from 11.0 (6.8–14.5) to 4.0 (1.8–5.5)), and grooming (6.0 (5.3–6.8) to 3.0 (3.0–3.8)), which is associated with the habituation of animals to the experimental setup and their transition to the behavior of 'patrolling' familiar territory. However, there were no statistically significant differences relative to Day 0 for the analyzed parameters in animals of this group.

Table 4. Motor activity and orientation–exploratory behavior of rats in the "Open field" test. The results are presented as Me, (P$_{25}$–P$_{75}$). In the table, colored cells allow to focus on the most important parameters.

Groups	Day 0	Day 2	Day 3	Day 6	Day 7	Day 10	Day 11	Day 13
Horizontal Activity, Number of Crossed Squares								
1—Intact	45.5 (32.8–55.3)	35.0 (19.3–54.5)	44.0 (32.8–54.5)	20.5 (9.8–50.8)	24.0 (16.8–32.0)	19.0 (8.5–36.3)	12.5 (5.8–17.8)	15.5 (12.8–19.8)
2—Sham	36.0 (31.3–39.5)	4.0 (2.0–9.3)	5.0 * (2.8–7.0)	9.5 (4.8–23.5)	15.0 (10.3–29.5)	16.5 (12.0–19.5)	16.5 (9.3–26.5)	37.0 (11.3–43.3)
3—Model	50.5 (32.3–53.5)	0.5 * (0.0–4.0)	5.5 * (0.0–23.0)	17.0 (6.3 –40.8)	17.0 (6.0–33.0)	24.0 (12.0–54.0)	40.0 * (26.0–43.0)	22.0 (10.3–45.3)
Vertical Activity, Number of Racks.								
1—Intact	11.0 (5.8–12.5)	12.0 (7.0–17.0)	15.0 (10.8–17.8)	12.5 (6.0–21.3)	12.0 (8.3–13.5)	7.5 (5.3–11.3)	6.5 (4.5–7.0)	5.0 (2.5–6.8)
2—Sham	5.0 (2.8–9.3)	1.0 (0.8–1.0)	1.0 * (0.8–1.3)	2.5 (1.0–7.0)	2.5 * (1.8–4.0)	3.0 (1.5–6.0)	3.0 (1.0–4.3)	1.0 (0.0–3.3)
3—Model	12.5 (6.3–16.5)	0.0 * (0.0–0.0)	1.0 * (0.0–1.0)	4.0 (1.0–6.0)	4.0 * (1.0–7.0)	3.0 (1.0–10.0)	5.0 (3.0–11.0)	3.5 (0.8–6.5)
Mink Reflex, Number of Examined Minks								
1—Intact	11.0 (8.8–12.5)	13.5 (12.3–17.0)	14.5 (13.3–15.8)	7.0 (5.0–12.8)	11.0 (6.8–14.5)	6.5 (4.0–11.3)	5.0 (4.3–5.0)	4.0 (1.8–5.5)
2—Sham	9.5 (7.5–11.3)	3.5 (1.0–5.5)	1.5 * (1.0–2.3)	3.0 (1.8–6.0)	2.5 (1.0–6.8)	3.0 (0.8–4.3)	2.5 (0.0–4.3)	4.5 (0.8–9.0)
3—Model	10.0 (8.5–18.0)	0.0 * (0.0–2.0)	0.0 * (0.0–2.0)	4.0 (1.3–6.8)	3.0 * (1.0–7.0)	4.0 (1.0 7.0)	3.0 (1.0 6.0)	3.0 (0.8 6.0)
Grooming, Number of Acts								
1—Intact	5.0 (3.3–6.0)	11.0 (6.5–12.5)	6.5 (6.0–7.0)	5.0 (2.5–6.8)	6.0 (5.3–6.8)	6.5 (4.5–7.8)	3.5 (3.0–4.0)	3.0 (3.0–3.8)
2—Sham	4.0 (3.0–5.0)	0.0 * (0.0–1.0)	0.5 * (0.0–1.0)	4.0 (2.8–4.3)	4.0 (3.0–4.3)	3.5 (1.8–5.5)	3.5 (1.8–5.3)	2.0 (0.0–4.0)
3—Model	4.0 (2.0–4.8)	0.5 * (0.0–4.8)	1.0 * (0.0–4.0)	4.0 (0.0 –5.8)	4.0 (2.0–5.0)	1.0 (0.0–8.0)	1.0 (1.0–9.0)	3.0 (0.8–4.3)

$* p < 0.05$ in comparison to Group 1—Intact animals.

In Sham animals, compared with Intact animals, there was a significant decrease in horizontal activity on Days 2 and 3 of the experiment to 90% ($p = 0.064$; $p = 0.023$, respectively); relative to day 0, this indicator decreased by 88.9% and 86.1%, respectively ($p < 0.001$). Further analysis showed that on Day 7, the horizontal motor activity of the Sham rats was correlated with the intact group and the results of Day 0 ($p = 1,000$). Vertical locomotor activity relative to intact rats decreased on Days 2 and 3 by 12.0 times ($p = 0.063$) and 15.0 times ($p < 0.001$), and on Day 7 by 4.8 times ($p = 0.014$). It was noted that relative to Day 0, the vertical activity of the Sham rats decreased by 5 times ($p = 0.032$) on Day 2 after surgery. The number of hole examinations decreased on Days 2 and 3 by 3.9 ($p = 0.076$) and 9.7 times ($p = 0.021$), respectively. Concerning Day 0, the incidence of hole examination decreased by 83.3% on Day 3 ($p < 0.05$). Grooming in 62.5% of Sham animals was absent on Day 2, and on Day 3, it was absent in 50% of the animals and in general decreased by 13.0 times relative to the Intact animals ($p = 0.021$), by 87.5% relative to Day 0 ($p = 0.025$). The number of grooming acts reached the values of the Intact group on Day 6.

The horizontal activity of the Model animals relative to the Intact ones significantly decreased on Days 2 and 3 by 98.6% ($p = 0.002$) and 87.5% ($p = 0.019$), relative to Day 0, and the activity decreased up to 10 times ($p < 0.001$) on Day 3. From Days 6 to 10, the horizontal motor activity of the Model rats was correlated with the Intact rats ($p = 1.000$). On Day 11, a sharp increase in the activity of the animals was registered, by 3.0 times ($p = 0.022$) compared to the Intact ones. There were no vertical stands on Day 2 of the experiment ($p < 0.001$ relative to the Intact animals and 0 days before the stroke). With further observation on Days 3, 6, and 7, the vertical activity of the animals was still significantly reduced relative to the Intact ones (12.0 times ($p < 0.001$); 4.2 times ($p = 0.072$); 3.0 times ($p = 0.022$), respectively). The number of minks on Days 2 and 3 was significantly less than in the Intact animals ($p < 0.001$); the number of minks was still reduced relative to the Intact ones on Day 6, by 1.75 times ($p = 0.170$), and on Day 7 by 3.7 times ($p = 0.053$). By Day 14, the exploratory activity was comparable to the Intact ones. Grooming was absent in 50.0% of the animals on Day 2; 21.4% had super-intensive grooming. In total, the number of grooming acts decreased 22.0 times ($p < 0.001$) relative to the Intact group. On Day 3, grooming was reduced by 6.5 times ($p < 0.001$); 21.4% of the animals retained super-intensive grooming; and on Days 10 and 11, super-intensive grooming was noted in 38.0%. There were no statistically significant differences relative to Day 0 for grooming in animals of this group (Table 4).

There were no significant differences in the indicators of horizontal motor activity between Sham and Model rats. However, in Model animals relative to Sham rats on Day 2, horizontal motor activity decreased up to 8 times, then on Days 6 and 11 increased by 1.8 and 2.4 times. In the complete absence of vertical activity and mink reflex on Days 2 and 3 in Model animals, horizontal activity in Sham rats was reduced, but not absent. The number of acts of defecation in animals of all groups did not differ significantly.

In the test of holding on a horizontal rod, the retention time on Day 0 of the experiment was 29.5 (14.5–40.0) s for Intact animals, 30.5 (25.2–39.2) s for Sham rats, and 31.5 (13.2–47.7) s for Model rats, and did not differ significantly.

On Day 2, in Sham animals, the retention time was 15.0 (10.0–20.0) s and was reduced relative to Day 0 by 2.0 times ($p = 0.455$), compared to 20.5 (10.5–35.7), and in Intact animals, it was reduced by 26.8% ($p = 1.000$). On Day 6, the retention time of Sham rats was 18.5 (14.0–30.5) s, and did not differ from the intact values of 19.5 (13.7–29.0) s. On days 10 and 13, the Sham rats were held on the bar for 31.0 (14.2–43.7) s and 26.0 (21.0–36.2) s, and in Intact rats, it was 31.5 (16.5–39.7) s and 15.0 (9.7–27.0) s.

On Day 2, 50% of Model animals showed impaired motor functions; the retention time was 2.0 (0.0–6.7) s and was reduced relative to Day 0 by 15.7 times ($p < 0.001$). The retention time on Day 2 for model rats was also reduced in comparison to Intact rats (Day 2) by 10.2 times ($p = 0.011$); in comparison to the Sham rats (Day 2), by 7.5 times ($p = 0.027$). On Day 6, 35.7% of Model animals showed limb weakness; retention time increased 6.7 times relative to 2 days ($p = 0.023$) and amounted to 13.5 (4.7–50.5) s, which did not differ significantly from Intact and Sham rats. On Day 10, motor dysfunctions were found in 33.3% of rats; the rats hung on the bar for 8.0 (6.0–30.0) s, which was 3.9 times less than the values of Intact and Sham animals ($p = 0.406$ and $p = 0.198$, respectively). On Day 13, motor disorders were observed in 30.0% of the Model animals; the retention time was 8.5 (5.0–15.5) s, which was 3.1 times less than the values of Sham rats ($p = 0.013$).

3.4. Results of the Hematological and Biochemistry Analysis

The hematological parameters of the Intact animals were within the physiological norm and corresponded to healthy Wistar rats of the given sex and age (Table 5).

Table 5. Results of the animals' blood hematological analysis at the end of the experiment. The results are presented as Me, (P_{25}–P_{75}). In the table, colored cells allow to focus on the most important parameters.

Parameters	Group of Animals		
	1—Intact	2—Sham	3—Model
LEU, 10^9/L	7.45 (6.16–9.70)	6.04 (5.22–7.50)	7.90 (5.72–8.25)
LYM, 10^9/L	7.00 (5.54–8.22)	5.01 (4.47–6.03)	6.02 (5.22–6.44)
Relative LYM, %	91.25 (87.95–92.85)	84.10 (77.95–87.78)	78.80 (74.40–86.35)
MID, 10^9/L	0.09 (0.08–0.13)	0.04 (0.03–0.05)	0.17 [#] (0.15–0.31)
Relative MID, %	1.60 (0.90–2.00)	0.60 (0.60–0.70)	2.25 [#] (1.93–3.98)
GRAN, 10^9/L	0.57 (0.33–1.03)	0.95 (0.76–1.17)	1.19 (0.56–1.86)
Relative GRAN, %	6.55 (5.55–10.55)	15.30 (10.30–21.90)	17.00 (8.53–23.83)
RBC, 10^{12}/L	7.69 (7.65–8.17)	7.82 (7.67–8.00)	7.43 (7.25–8.00)
HGB, g/L	130.50 (125.50–136.50)	129.50 (127.25–144.25)	131.00 (127.25–137.75)
HCT, %	39.79 (38.16–41.15)	39.16 (38.11–39.61)	39.08 (38.32–39.95)
MCV, mkm^3	50.00 (49.00–51.00)	49.00 (48.75–50.25)	52.00 [#] (51.00–52.00)
PLT, 10^9/L	733.50 (709.00–809.50)	751.50 (660.75–826.00)	759.00 (719.00–829.25)
PCT, %	0.51 (0.47–0.56)	0.48 (0.43–0.54)	0.50 (0.48–0.55)

[#]—$p < 0.05$ in comparison to 2—Sham animals; Abbreviations: LEU—leukocytes; LYM—lymphocytes; GRAN—granulocytes; MID—a mixture of monocytes, eosinophils, basophils and immature cells; RBC—red blood cells; HGB—hemoglobin; HCT—hematocrit; PLT—platelets; PCT—plateletcrit; MCV—average erythrocyte volume.

In Sham animals relative to the Intact ones, a decrease in leukocytes content by 18.9% ($p = 0.793$) was revealed, including a decrease in the lymphocytes' relative content by 7.8% ($p = 0.362$), and a decreased amount of monocytes, eosinophils, basophils, and immature cells by 62.5% ($p = 0.386$) against the background of an increase in the granulocytes' relative content by 2.3 times ($p = 0.226$).

In Model rats, relative to Intact ones, an increase in the content of monocytes, eosinophils, basophils, and immature cells was noted, by 1.89 times ($p = 0.002$), and the relative content of a mixture of monocytes, eosinophils, basophils, and immature cells increased by 1.4 times ($p = 0.340$). Granulocytes increased by 2.6 times ($p = 0.126$), with a decrease in the relative content of lymphocytes by 13.6% ($p = 0.089$) in Model rats. In addition, an increase in the average volume of erythrocytes by 4.0% ($p = 0.191$) was revealed. Relative to the Sham animals, the Model rats showed an increase in the relative content of a mixture of monocytes, eosinophils, basophils, and immature cells by 3.77 times ($p < 0.001$), granulocytes by 11.0% ($p = 0.226$), as well as an increase in the average volume of erythrocytes of 6.1 % ($p = 0.011$).

The biochemical parameters of the Intact animals' serum (Table 6) were within the physiological norm and corresponded to the parameters of healthy Wistar rats of the given sex and age.

Table 6. Results of animal blood serum biochemical analysis at the end of the experiment. The results are presented as Me, (P$_{25}$–P$_{75}$). In the table, colored cells allow to focus on the most important parameters.

Parameters	Group of Animals		
	1—Intact	2—Sham	3—Model
Total protein, g/L	65.30 (64.65–66.90)	63.15 (62.20–65.25)	68.10 (66.30–68.75)
Albumin, g/L	45.20 (43.70–45.60)	43.25 (42.38–43.95)	41.60 * (41.40–41.85)
Glucose, mmol/L	10.80 (9.10–13.58)	16.55 * (16.00–17.60)	17.97 [#] (14.05–19.3)
Bilirubin indirect, µmol/L	3.20 (2.90–3.40)	3.28 (3.15–4.25)	3.50 (3.10–3.80)
Bilirubin direct, µmol/L	1.80 (1.65–1.90)	1.75 (1.68–1.95)	1.90 (1.85–2.15)
Creatinine, µmol/L	54.50 (51.75–56.50)	42.38 * (42.00–47.33)	46.00 (44.75–48.25)
Urea, mmol/L	6.15 (5.62–6.57)	6.60 (5.53–7.56)	6.07 (5.91–6.64)
AST, E/L	132.65 (130.17–133.97)	128.34 (117.39–149.97)	136.20 (115.24–143.12)
ALT, E/L	31.00 (26.50–34.00)	31.50 (28.00–33.18)	17.30 *, [#] (16.00–25.00)
ALP, E/L	175.50 (158.40–199.50)	146.05 (131.05–173.55)	166.90 (154.85–177.40)
GGT, E/L	2.65 (2.52–2.84)	2.03 (1.87–2.26)	2.43 (2.23–2.70)
LDH, E/L	404.56 (383.39–461.42)	521.34 (456.84–624.32)	556.00 (328.13–595.01)
Cholesterol, mmol/L	1.57 (1.47–1.67)	1.79 (1.71–1.92)	1.85 * (1.70–2.21)
Triglycerides, mmol/L	1.20 (0.85–1.55)	1.25 (1.08–1.41)	1.40 (1.34–1.55)

* $p < 0.05$ in comparison to 1—Intact animals; [#] $p < 0.05$ in comparison to 2—Sham animals. Abbreviations: GGT—gamma-glutamyl transpeptidase; ALT—alanine aminotransferase; LDH—lactate dehydrogenase; AST—aspartate aminotransferase; ALP—alkaline phosphatase.

In Sham animals relative to the Intact ones, there was an increase in glucose content by 34.7% ($p = 0.027$), creatinine by 22.2% ($p = 0.027$), and decrease in GGT activity by 23.4% ($p = 0.114$).

In Model animals relative to the Intact ones, the albumin content decreased by 8% ($p = 0.005$) and creatinine by 15.6% ($p = 0.159$), with an increase in glucose by 66.4% ($p = 0.029$) and cholesterol by 17, 8% ($p = 0.044$); there was decreased ALT activity by 44.2% ($p = 0.014$). Relative to Sham rats, the ALT activity decreased by 45.1% ($p = 0.007$).

3.5. Results of the Histopathological Analysis

When studying the sensorimotor cortex cerebral hemispheres of the Intact animals, the neocortex structure did not have pronounced signs of pathological process (Figure 5(1a)). Pyramidal neurons were characterized by tigroid background in the cytoplasm, located evenly around the nucleus. Some neurons with a tigroid background were found in dispersed clump form. The neurons' nuclei occupied a central position in the cells' cytoplasm. The neurons' groups with a tigroid background and weak staining of the nuclei were encountered in the neocortex. The shape of the pyramidal neurons was not changed, and the nuclei were large. There were single hyperchromic neurons with shrunken perikarya. Indistinct pericellular edema was observed around such cells. Most often, one nucleolus was present in the nuclei of the pyramidal neurons. No pathological changes in the glial cells and blood vessels were revealed. Signs of perivascular edema were observed around individual blood vessels (Figure 5(1a)).

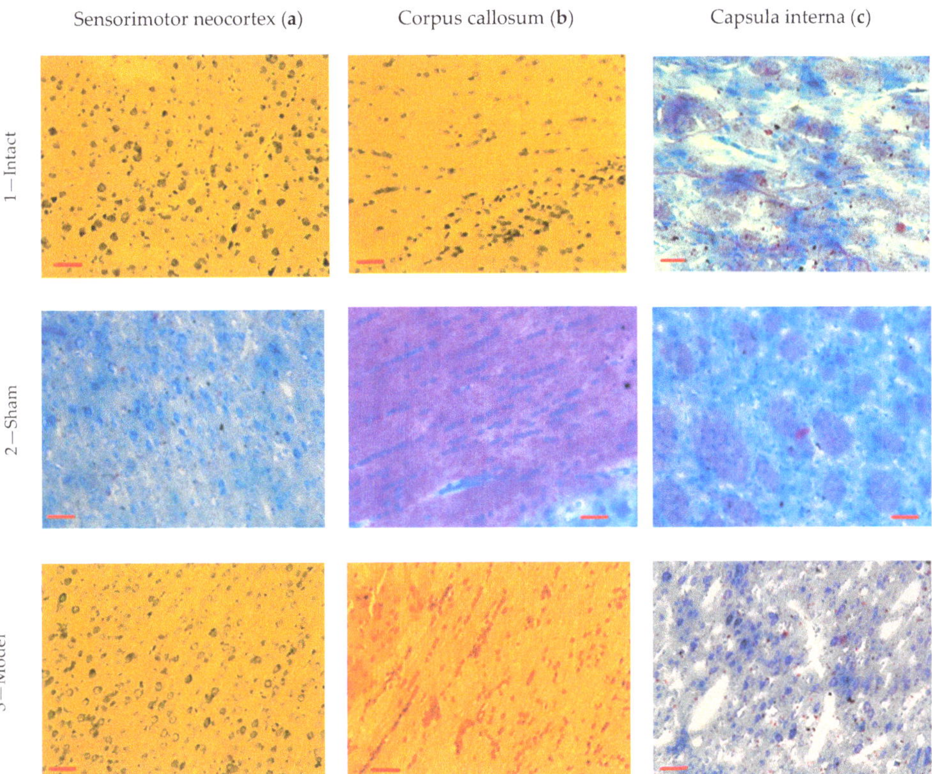

Sensorimotor neocortex (a)   Corpus callosum (b)   Capsula interna (c)

Figure 5. Histological brains structure (200×). Nissl/Hematoxylin–eosin staining. Scale bar 50 μm.

In the corpus callosum, glial cells were arranged in groups along the main direction of the nerve fibers in a given brain structure. The white matter in this interhemispheric zone of the animals' brains was basophilic and was not characterized by acidophilic or oxyphilic staining. Glial cell nuclei had a normal shape without signs of pathological changes and processes (Figure 5(1b)). The nuclei of these cells contained single nucleoli. Mild manifestations of perivascular edema were observed around individual blood vessels.

In the capsula interna, diffusely located glial cells were noted (Figure 5(1c)) among the bundles of nerve fibers; however, proliferation phenomena and glial cell hypertrophy were not detected. Some nerve fibers had acidophilic staining against the background of dominant basophilia. There were local perivascular edema and the development of individual perivascular edema in the area of the inner capsule.

In the neocortex of Sham rats, it was found that significant structural changes did not develop in the cerebral vessels; only a slight increase in the perivascular space was noted.

Individual vessels spasm was noted. There were no pathological changes in the vessels and glia; the nuclei of the gliocytes had a normal shape. At the same time, pericellular edema was observed around the glial cells. Pyramidal neurons were characterized by a tigroid background in the cytoplasm located in significant amounts around the nucleus. The nuclei were centrally located in the bodies of the neurons (Figure 5(2a)).

In the corpus callosum, glial cells were arranged in strictly oriented sequences along with groups of nerve fibers. The white matter in this zone had acidophilic staining. The shape of the glial cell's nuclei was normal, without pathological changes (Figure 5(2b)). In the capsula interna, glial cells were located diffusely (Figure 5(2c)); the proliferation and hypertrophy of glial cells were not established.

In Model animals with a model of left-sided hemorrhagic stroke, in the neocortex, the polymorphic nature of the structural changes was observed. In the stroke hemisphere, most neurons were in a state of acute edema and swelling of the bodies and their nuclei (Figure 5(3a)). Neurons with deformed nuclei and signs of destruction were present. There were dead neurons groups with signs of autolysis. Acute pericellular edema was noted, especially in the area of the apical dendrites. Perivascular edema was also present. Glial cells were also characterized by structural changes. Pycnosis of some cell nuclei was noted. Many dead cells were revealed, which were diffusely located in the cortex. At the same time, hyperchromatosis of the remaining cells was noted (Figure 5(3b)). Pericellular edema around individual cells was clearly visible.

In the corpus callosum, glial cells were located chaotically. There were no signs of active proliferation. Often these cells were kept in small groups. The white matter in this zone had a staining level similar to the previous groups. Glial cell nuclei were small in size with hyperchromic staining (Figure 5(3b)).

In the capsula interna, glial cells were oriented in small groups; around them, there was pericellular edema, and perivascular edema was also noted (Figure 5(3c)).

When analyzing the neurons' morphometric parameters in the V layer of the sensorimotor neocortex in the cerebral cortex (ipsilateral and contralateral) of Intact rats, it was found that the pyramidal neurons' area averaged 347.39 ± 6.94 μm, and the area of the nuclei was 159.13 ± 2.65 μm (Table 7). The number of dead neurons in the neocortex sensorimotor area did not exceed 3.0%; the number of dead gliocytes was 2.6%. The neuroglial index averaged 0.82 ± 0.05 units.

Table 7. The neurons' morphometric parameters in the V layer of the sensorimotor neocortex.

Parameters			Group of Animals			
		1—Intact	2—Sham		3—Model	
			Ipsi	Contr	Ipsi	Contr
Area of the pyramidal neurons, μm	Mean	347.39	351.54	337.45	447.46 *	388.85
	SEM	6.94	5.52	5.21	11.45	9.72
Area of the nuclei, μm	Mean	159.13	139.37	121.53	237.49 *	171.95
	SEM	2.65	2.11	3.41	5.50	3.59
Neuroglial index	Mean	0.82	0.85	0.83	0.92	0.96
	SEM	0.05	0.02	0.02	0.01	0.02

* $p < 0.05$ in comparison to Group 1—Intact animals.

In Sham animals, slight changes in the cells and nuclei areas of the giant pyramidal neurons in the V layer of the sensorimotor neocortex were revealed, both in the ipsi and in the contralateral hemisphere. In this case, the pyramidal neurons' areas were 351.54 ± 5.52 μm^2 and 337.45 ± 5.21 μm^2, and the areas of the giant pyramidal neurons' nuclei were 139.37 ± 2.11 μm^2 and 121.53 ± 3.41 μm^2, respectively. The number of dead neurons did not exceed 25.7%; the number of dead gliocytes was 14.2%. The neuroglial index averaged 0.84 ± 0.02 units.

At the same time, in Model animals, the neurons' area in the V layer of the sensorimotor neocortex in the ipsilateral hemisphere was 447.46 ± 11.45 μm^2, in the contralateral hemisphere—388.85 ± 9.72 μm^2; i.e., it increased sharply relative to the intact values by 28.8% and 10.7% ($p = 0.089$), respectively. The areas of the pyramidal neurons' nuclei increased, respectively, up to 237.49 ± 5.50 μm^2 and 171.96 ± 3.59 μm^2 (by 33.0% and 7.5%). The percentages of dead neurons and gliocytes were 44.9% and 13.4% for the ipsilateral hemisphere and 40.6% and 16.3% for the contralateral hemisphere, respectively. The neuroglial index was 0.92 ± 0.01 units for the ipsilateral hemisphere and 0.96 ± 0.02 units for the contralateral hemisphere.

4. Discussion

As a rule, most animal models do not seek to model the entire pathological process, focusing on their individual aspects under the strict control of all other circumstances. At the same time, the goal is to better understand the mechanisms of pathogenesis, as well as to test the effectiveness of various approaches to therapy and prevention. Since none of the models can recreate all aspects of hemorrhagic stroke, it is advisable to use different models when testing the means and methods of therapy. The model with autologous blood injection stimulates the development of stroke by the hematoma mechanism. The advantage of these models is the ability to control the volume of blood injected, which contributes to good reproducibility. In addition, in such models, the volume of blood injected correlates with the degree of brain tissue damage and the severity of the functional disorders [6].

However, when analyzing the data and interpreting the results obtained in these models, it is necessary to take into account that they reproduce a pathology with a constant hematoma volume, while in the clinic, a secondary increase in the hematoma is possible. In the present study, we characterized the development of disorders in hemorrhagic damage to the tissues of the inner capsule of the left hemisphere of the rat brain when administered autologous blood by the method [20].

In the model we studied, the death of animals was observed during the 14-day experiment; the total mortality of the model was 43.7%. The critical days were Days 1, 5, and 12–14; it is expected that during this period, the probability of death increases when this model is replicated. In people with hemorrhagic stroke, the mortality rate during the first month is more than 30% [6,40].

In deceased animals in the capsula interna region of the ipsilateral hemisphere, signs of acute pericellular edema were detected, especially in the apical dendrites; also, a significant number of dead neurons with signs of autolysis; acute edema of neurons with swelling of their nuclei; a lot of diffusely located dead glial cells with hyperchromatosis of the vast majority of glial cells; and, in some glyocytes, pycnosis of the nuclei. So, the model causes neuronal death and is accompanied by pronounced gliosis and glial cell proliferation due to the development of a local inflammatory process, which was also noted [41,42] and can be used for verification methods of treating brain edema and secondary inflammation after intracerebral hemorrhage [37,43]. Thus, an inflammatory response occurs early in patients with cerebral hemorrhage and influences and predicts the disease course [44,45].

Various neurological disorders were observed in the surviving animals throughout the entire period of the experiment. On Day 2 after the operation, a comatose state was registered in 28% of the animals (they subsequently died); the remaining animals were found to have paralysis and paresis of the contralateral limbs, weakness and insensitivity of the limbs, clockwise Manege movements, unilateral and bilateral ptosis, and exophthalmos. In animals, severe motor disorders were registered, manifested in the inability to hold on to a horizontal rod, which persisted up to 13 days in 30% of rats; at the same time, the retention time on the rod was significantly reduced. Neurological deficit led to inhibition of the periods of adaptation and development of research behavior. It is important to note that the increase in horizontal activity from Day 6 to day 11 was due to the Manege movements of rats and is imaginary. During observations, there were manifestations of both over-intensive grooming (on Days 2 and 11 in 21% and 38% of rats, respectively) and its absence (in 50% and 30% of animals, respectively). Violations persisted until the end of the study; 10% of the rats remained in a serious condition, 50% of the animals had an average degree of neurological deficit, and 40% of the animals had mild symptoms of neurological manifestations. The observed neurological disorders correspond to the characteristic symptoms of stroke patients, and their safety correlates quite well with 70–75% of disability in survivors of hemorrhagic stroke [46].

Brain section histomorphometric studies of the animals in the stroke model revealed a significant number for both neurons (up to 45%) and glia (up to 14%); in the ipsilateral hemisphere, most neurons were in a state of acute edema, with swelling of bodies and their nuclei; significant numbers of neurons with deformed nuclei and signs of destruction were detected. There was acute recellular edema in the apical dendrites, glial cells with structural changes, and pycnosis of the nuclei, also described in [47]. Such inflammatory processes led to a shift in the leukocyte formula in the direction of increasing the relative content of granulocytes and a mixture of monocytes, eosinophils, basophils, and immature cells, with a decrease in the relative content of lymphocytes. In addition, there was an increase in the average volume of red blood cells, which may be a delayed response to acute hematoma in stroke modelling. The increase in glucose and cholesterols is associated with the mechanism of compensation for the increased metabolic needs of the body as a whole, and especially the brain. A decrease in serum albumin, creatinine, as well as ALT activity, are predictors of functional outcomes acute stroke [48,49] and confirms the validity of the model.

Post-stroke weight loss of rats increased up to 10 days after the simulation, and by the end of the experiment, it corresponded to the weight of the Intact animals. Weight loss is associated with the development of acute cerebral circulatory disorders and brain edema, which affects the feeding behavior of animals due to compression of the anterior hypothalamus. Thus, body weight is an indirect indicator of hemorrhagic brain damage, which is also confirmed by other studies [50].

For comparison, a group of Sham-operated animals was introduced into the experiment, which underwent an identical craniotomy without damage to the brain structures. No deaths of rats were recorded. In Sham-operated animals, the weight loss on Day 1 after surgery was caused by the consequences of the anesthesia and craniotomy, which also led to the appearance of symptoms of mild neurological deficit. On Days 2 and 3 after trepanation, vertical and horizontal activity, research behavior, and grooming of rats, as well as retention time on the horizontal bar, were significantly reduced. By the end of the experiment, 12.5% of the rats showed lethargy of movement, tremor, weakness of the limbs, and unilateral partial ptosis, as well as weight loss. In other animals, all the analyzed parameters corresponded to the values of the Intact animals. Histological studies of the brain of Sham-operated rats showed no pathological changes in blood vessels, neurons, and glia. However, blood analyses revealed minor changes in the leucocyte count (decrease in the relative content of lymphocytes of a mixture of monocytes, eosinophils, basophils, and immature cells and an increase in the relative content of granulocytes), minor hyperglycemia, and an increase in the creatinine level and GGT activity. It can point to kidney injury after anesthesia [51] or inflammatory processes at the site of injury since the animals were not kept in sterile conditions.

It is worth noting that if it is necessary to conduct molecular studies of the brain, blood sampling from the heart should be carried out in the animal under anesthesia (for example, Zoletil/Xyla) after the rib cage dissection, after taking blood from the right atrium, with the animal immediately transcardially perfused with heparinized saline (0.9%) followed by a periodate lysine paraformaldehyde solution [52], to dissect and isolate the needed brain structures.

5. Conclusions

Reproduction of an acute intracerebral hematoma in the left hemisphere of the rat brain according to the method in [20] causes an acute functional deficit, which is manifested by a pronounced violation of motor functions, paralysis, paresis, weakness and insensitivity of the limbs, Manege movements, ptosis, and exophthalmos. These symptoms persist for at least 14 days. In parallel, there was marked edema, neuronal death, and gliosis. The data obtained prove that this model fully reflects the observed clinical situation and reproduces the main diagnostic criteria for acute cerebral circulatory disorders.

It is known that up to 85% of hemorrhagic strokes develop by the mechanism of hematoma and only about 15% by the mechanism of diapedesis [45]. In this regard, the model with the introduction of autologous blood, which is the method of [20], is in many aspects more relevant. The advantages of these models are good reproducibility of the size and position of the hematoma, preservation of near-hematomic brain tissue, and the ability to change the severity of the structural damage and neurological consequences [26].

A comprehensive study of the local intracerebral hematoma modelling performed in this paper [18] will allow not only to better understand the pathophysiology of the acute period in this model but also to improve the quality of in vivo studies aimed at correcting acute cerebral circulatory disorders.

Author Contributions: Conceptualization, E.V. and L.F.; methodology, E.V. and G.T.; software, L.F.; validation, A.M. and I.C.; formal analysis, A.M.; investigation, G.T., L.F., E.V., A.K., I.C. and A.M.; resources and data curation, L.F.; writing—original draft preparation, E.V. and A.M.; writing—review and editing, A.K. and G.T.; visualization, I.C.; supervision, A.K.; project administration, L.F. All authors have read and agreed to the published version of the manuscript.

Funding: This research was funded by Russian Science Foundation, grant number 19-76-10034.

Institutional Review Board Statement: The study was conducted according to the guidelines of the Declaration of Helsinki, and approved by the Ethics Committee of V.M. Gorbatov Federal Research Centre for Food Systems of the Russian Academy of Sciences (protocol code 04/2017, date of approval 22 September 2017).

Informed Consent Statement: Not applicable.

Data Availability Statement: Not applicable.

Conflicts of Interest: The authors declare no conflict of interest.

References

1. Imai, T.; Iwata, S.; Miyo, D.; Nakamura, S.; Shimazawa, M.; Hara, H. A novel free radical scavenger, NSP-116, ameliorated the brain injury in both ischemic and hemorrhagic stroke models. *J. Pharmacol. Sci.* **2019**, *141*, 119–126. [CrossRef] [PubMed]
2. Tsymbalyuk, I.; Manuilov, A.; Popov, K.; Basov, A. Metabolic Correction of the Ischemia Reperfusive Injury with Sodium Dichloroacetate in Vascular Isolation of the Liver in Experiment. *Novosti Khirurgii* **2017**, *25*, 447–453. [CrossRef]
3. Liu, H.; Sun, X.; Zou, W.; Leng, M.; Zhang, B.; Kang, X.; He, T.; Wang, H. Scalp acupuncture attenuates neurological deficits in a rat model of hemorrhagic stroke. *Complement. Ther. Med.* **2017**, *32*, 85–90. [CrossRef]
4. Vahidinia, Z.; Tameh, A.A.; Nejati, M.; Beyer, C.; Talaei, S.A.; Moghadam, S.E.; Atlasi, M.A. The protective effect of bone marrow mesenchymal stem cells in a rat model of ischemic stroke via reducing the C-Jun N-terminal kinase expression. *Pathol. Res. Pr.* **2019**, *215*, 152519. [CrossRef]
5. Basov, A.A.; Kozin, S.V.; Bikov, I.M.; Popov, K.A.; Moiseev, A.V.; Elkina, A.A.; Dzhimak, S.S. Changes in Prooxidant-Antioxidant System Indices in the Blood and Brain of Rats with Modelled Acute Hypoxia Which Consumed a Deuterium Depleted Drinking Diet. *Biol. Bull.* **2019**, *46*, 531–535. [CrossRef]
6. Krafft, P.R.; Bailey, E.L.; Lekic, T.; Rolland, W.B.; Altay, O.; Tang, J.; Wardlaw, J.M.; Zhang, J.H.; Sudlow, C.L.M. Etiology of Stroke and Choice of Models. *Int. J. Stroke* **2012**, *7*, 398–406. [CrossRef] [PubMed]
7. Kjell, J.; Olson, L. Rat models of spinal cord injury: From pathology to potential therapies. *Dis. Model. Mech.* **2016**, *9*, 1125–1137. [CrossRef]
8. Snyder, J.S.; Choe, J.S.; Clifford, M.A.; Jeurling, S.I.; Hurley, P.; Brown, A.; Kamhi, J.F.; Cameron, H.A. Adult-Born Hippocampal Neurons Are More Numerous, Faster Maturing, and More Involved in Behavior in Rats than in Mice. *J. Neurosci.* **2009**, *29*, 14484–14495. [CrossRef] [PubMed]
9. Lazarov, O.; Hollands, C. Hippocampal neurogenesis: Learning to remember. *Prog. Neurobiol.* **2016**, *138–140*, 1–18. [CrossRef]
10. Rubattu, S.; Stanzione, R.; Bianchi, F.; Cotugno, M.; Forte, M.; Della Ragione, F.; Fioriniello, S.; D'Esposito, M.; Marchitti, S.; Madonna, M.; et al. Reduced brain UCP2 expression mediated by microRNA-503 contributes to increased stroke susceptibility in the high-salt fed stroke-prone spontaneously hypertensive rat. *Cell Death Dis.* **2017**, *8*, e2891. [CrossRef]
11. Yao, H.; Nabika, T. Standards and pitfalls of focal ischemia models in spontaneously hypertensive rats: With a systematic review of recent articles. *J. Transl. Med.* **2012**, *10*, 139. [CrossRef] [PubMed]
12. Iida, S.; Baumbach, G.L.; Lavoie, J.L.; Faraci, F.M.; Sigmund, C.D.; Heistad, D.D. Spontaneous Stroke in a Genetic Model of Hypertension in Mice. *Stroke* **2005**, *36*, 1253–1258. [CrossRef]
13. Ren, C.; Sy, C.; Gao, J.; Ding, Y.; Ji, X. Animal Stroke Model: Ischemia–Reperfusion and Intracerebral Hemorrhage. *Methods Mol. Biol.* **2016**, *1462*, 373–390. [CrossRef]
14. James, M.L.; Warner, D.S.; Laskowitz, D.T. Preclinical Models of Intracerebral Hemorrhage: A Translational Perspective. *Neurocritical Care* **2007**, *9*, 139–152. [CrossRef] [PubMed]
15. Durukan, A.; Tatlisumak, T. Rodent Models of Hemorrhagic Stroke. *Curr. Pharm. Des.* **2008**, *14*, 352–358. [CrossRef]
16. MacLellan, C.L.; Silasi, G.; Auriat, A.M.; Colbourne, F. Rodent Models of Intracerebral Hemorrhage. *Stroke* **2010**, *41*, S95–S98. [CrossRef]

17. Sansing, L.H.; Kasner, S.E.; McCullough, L.; Agarwal, P.; Welsh, F.A.; Kariko, K. Autologous Blood Injection to Model Spontaneous Intracerebral Hemorrhage in Mice. *J. Vis. Exp.* **2011**, *54*, e2618. [CrossRef]
18. O'Lynnger, T.; Mao, S.; Hua, Y.; Xi, G. Blood Injection Intracerebral Hemorrhage Rat Model. In *Cerebral Ischemic Reperfusion Injuries (CIRI)*; Springer: Cham, Switzerland, 2019; pp. 257–262.
19. Marinkovic, I.; Strbian, D.; Mattila, O.; Abo-Ramadan, U.; Tatlisumak, T. A novel combined model of intracerebral and intraventricular hemorrhage using autologous blood-injection in rats. *Neuroscience* **2014**, *272*, 286–294. [CrossRef]
20. Makarenko, A.N.; Kositsyn, N.S.; Pasikova, N.V.; Svinov, M.M. Simulation of local cerebral hemorrhage in different brain structures of experimental animals. *Zhurnal Vysshei Nervnoi Deiatelnosti Imeni I P Pavlova* **2003**, *52*, 765–768.
21. Kaiser, E.E.; Waters, E.S.; Fagan, M.M.; Scheulin, K.M.; Platt, S.R.; Jeon, J.H.; Fang, X.; Kinder, H.A.; Shin, S.K.; Duberstein, K.J.; et al. Characterization of tissue and functional deficits in a clinically translational pig model of acute ischemic stroke. *Brain Res.* **2020**, *1736*, 146778. [CrossRef]
22. Hawkins, P.; Golledge, H.D.R. The 9 to 5 Rodent—Time for Change? Scientific and animal welfare implications of circadian and light effects on laboratory mice and rats. *J. Neurosci. Methods* **2018**, *300*, 20–25. [CrossRef]
23. Fish, R.E.; Brown, M.; Danneman, P.J.; Karas, A.Z. *Anesthesia and Analgesia in Laboratory Animals*; Academic Press: Cambridge, MA, USA, 2008; p. 672.
24. Longchamp, A.; Harputlugil, E.; Corpataux, J.-M.; Ozaki, C.K.; Mitchell, J.R. Is Overnight Fasting before Surgery Too Much or Not Enough? How Basic Aging Research Can Guide Preoperative Nutritional Recommendations to Improve Surgical Outcomes: A Mini-Review. *Gerontology* **2017**, *63*, 228–237. [CrossRef] [PubMed]
25. Lester, P.A.; Moore, R.M.; Shuster, K.A.; Myers, D.D. Anesthesia and Analgesia. In *The Laboratory Rabbit, Guinea Pig, Hamster, and other Rodents*; Academic Press: Cambridge, MA, USA, 2012; Chapter 2; pp. 33–56.
26. Barth, A.; Guzman, R.; Andres, R.H.; Mordasini, P.; Barth, L.; Widmer, H.R. Experimental intracerebral hematoma in the rat. *Restor. Neurol. Neurosci.* **2007**, *25*, 1–7. [PubMed]
27. Paxinos, G.; Watson, C. *The Rat Brain in Stereotaxic Coordinates*; Academic Press: Cambridge, MA, USA, 1998; p. 242.
28. Du Sert, N.P.; Alfieri, A.; Allan, S.M.; Carswell, H.V.; Deuchar, G.A.; Farr, T.D.; Flecknell, P.; Gallagher, L.; Gibson, C.L.; Haley, M.J.; et al. The IMPROVE Guidelines (Ischaemia Models: Procedural Refinements of in vivo Experiments). *Br. J. Pharmacol.* **2017**, *37*, 3488–3517. [CrossRef]
29. Recognition and Alleviation of Distress in Laboratory Animals, Committee on Recognition and Alleviation of Distress in Laboratory Animals, National Research Council. 2008; p. 132. Available online: https://grants.nih.gov/grants/olaw/NAS_distress_report.pdf (accessed on 8 November 2020).
30. National Research Council (US) Committee on Pain and Distress in Laboratory Animals. *Recognition and Alleviation of Pain and Distress in Laboratory Animals. Recognition and Assessment of Pain, Stress, and Distress*; National Academies Press: Washington, DC, USA, 1992; Chapter 4; pp. 32–52.
31. Philips, B.H.; Weisshaar, C.L.; Winkelstein, B.A. Use of the Rat Grimace Scale to Evaluate Neuropathic Pain in a Model of Cervical Radiculopathy. *Comp. Med.* **2017**, *67*, 34–42.
32. Greene, S.A. Pros and cons of using α-2 agonists in small animal anesthesia practice. *Clin. Tech. Small Anim. Pr.* **1999**, *14*, 10–14. [CrossRef]
33. Helgers, S.O.A.; Talbot, S.R.; Riedesel, A.-K.; Wassermann, L.; Wu, Z.; Krauss, J.K.; Häger, C.; Bleich, A.; Schwabe, K. Body weight algorithm predicts humane endpoint in an intracranial rat glioma model. *Sci. Rep.* **2020**, *10*, 9020. [CrossRef]
34. Gannushkina, I.V.; Shafranova, V.P.; Ryasina, T.V. *Functional Brain Angioarchitectonics*; Meditsina: Moscow, Russia, 1977; p. 40.
35. Leonidov, N.B.; Jakovlev, R.Y.; Rakita, D.R.; Lisichkin, G.V. Agent with Anti-Stroke Action, and Method for Preparing it. Russian Federation Patent RU 2521404 C1, 20 June 2006. Available online: https://patents.s3.yandex.net/RU2521404C1_20140627.pdf (accessed on 8 November 2020).
36. Kulchikov, A.E.; Makarenko, A.N.; Novikova, Y.L.; Dobychina, E.E. Method of Neurologic Detection for Small Laboratory Animals Suffering From Cerebral Affections. Russian Federation Patent RU 2327227 C2, 20 June 2008. Available online: https://patents.s3.yandex.net/RU2327227C2_20080620.pdf (accessed on 8 November 2020).
37. Zheng, H.; Chen, C.; Zhang, J.; Chunli, C. Mechanism and Therapy of Brain Edema after Intracerebral Hemorrhage. *Cerebrovasc. Dis.* **2016**, *42*, 155–169. [CrossRef] [PubMed]
38. Chernukha, I.M.; Fedulova, L.V.; Kotenkova, E.A.; Takeda, S.; Sakata, R. Hypolipidemic and anti-inflammatory effects of aorta and heart tissues of cattle and pigs in the atherosclerosis rat model. *Anim. Sci. J.* **2018**, *89*, 784–793. [CrossRef]
39. Belenichev, I.F.; Mazur, I.A.; Kolesnik, Y.M.; Abramov, A.V.; Bukhtiyarova, N.V.; Sidorova, I.V. Thiotriazonin influence on histomorphological changes of cortical and hippocampus neurons in the post-stroke period. *News Med. Pharm.* **2007**, *5*, 14–25.
40. González-Pérez, A.; Gaist, D.; Wallander, M.-A.; McFeat, G.; García-Rodríguez, L.A. Mortality after hemorrhagic stroke: Data from general practice (The Health Improvement Network). *Neurology* **2013**, *81*, 559–565. [CrossRef]
41. Altumbabic, M.; Peeling, J.; Del Bigio, M.R. Intracerebral Hemorrhage in the Rat: Effects of Hematoma Aspiration. *Stroke* **1998**, *29*, 1917–1923. [CrossRef] [PubMed]
42. Derugin, N.; Wendland, M.; Muramatsu, K.; Roberts, T.P.L.; Gregory, G.; Ferriero, D.M.; Vexler, Z.S. Evolution of Brain Injury After Transient Middle Cerebral Artery Occlusion in Neonatal Rats. *Stroke* **2000**, *31*, 1752–1761. [CrossRef] [PubMed]
43. Di Napoli, M.; Slevin, M.; Popa-Wagner, A.; Singh, P.; Lattanzi, S.; Divani, A.A. Monomeric C-Reactive Protein and Cerebral Hemorrhage: From Bench to Bedside. *Front. Immunol.* **2018**, *9*, 1921. [CrossRef]

44. Lattanzi, S.; Brigo, F.; Trinka, E.; Cagnetti, C.; Di Napoli, M.; Silvestrini, M. Neutrophil-to-Lymphocyte Ratio in Acute Cerebral Hemorrhage: A System Review. *Transl. Stroke Res.* **2019**, *10*, 137–145. [CrossRef] [PubMed]
45. Lattanzi, S.; Di Napoli, M.; Ricci, S.; Divani, A.A. Matrix Metalloproteinases in Acute Intracerebral Hemorrhage. *Neurotherapeutics* **2020**, *17*, 484–496. [CrossRef] [PubMed]
46. Intercollegiate Stroke Working Party. *National Clinical Guideline for Stroke*, 4th ed.; Royal College of Physicians: London, UK, 2016; Available online: http://citeseerx.ist.psu.edu/viewdoc/download?doi=10.1.1.476.6093&rep=rep1&type=pdf (accessed on 8 November 2020).
47. Li, L.; Lou, X.; Zhang, K.; Yu, F.; Zhao, Y.; Jiang, P. Hydrochloride fasudil attenuates brain injury in ICH rats. *Transl. Neurosci.* **2020**, *11*, 75–86. [CrossRef]
48. Nair, R.; Radhakrishnan, K.; Chatterjee, A.; Gorthi, S.P.; Prabhu, V.A. Serum Albumin as a Predictor of Functional Outcomes Following Acute Ischemic Stroke. *J. Vasc. Interv. Neurol.* **2018**, *10*, 65–68.
49. Gao, F.; Chen, C.; Lu, J.; Zheng, J.; Ma, X.-C.; Yuan, X.-Y.; Huo, K.; Han, J.-F. De Ritis ratio (AST/ALT) as an independent predictor of poor outcome in patients with acute ischemic stroke. *Neuropsychiatr. Dis. Treat.* **2017**, *13*, 1551–1557. [CrossRef]
50. Rogers, D.C.; Campbell, C.A.; Stretton, J.L.; Mackay, K.B. Correlation Between Motor Impairment and Infarct Volume After Permanent and Transient Middle Cerebral Artery Occlusion in the Rat. *Stroke* **1997**, *28*, 2060–2066. [CrossRef]
51. Moore, E.; Bellomo, R.; Nichol, A. Biomarkers of acute kidney injury in anesthesia, intensive care and major surgery: From the bench to clinical research to clinical practice. *Minerva Anestesiol.* **2010**, *76*, 425–440. [PubMed]
52. Gage, G.J.; Kipke, D.R.; Shain, W. Whole Animal Perfusion Fixation for Rodents. *J. Vis. Exp.* **2012**, *65*, e3564. [CrossRef] [PubMed]

 *biomedicines*

Article

Model of Acute Liver Failure in an Isolated Perfused Porcine Liver—Challenges and Lessons Learned

Joshua Hefler [1], Sanaz Hatami [2,3], Aducio Thiesen [4], Carly Olafson [4,5], Kiarra Durand [4,5], Jason Acker [4,5], Constantine J. Karvellas [6,7], David L. Bigam [1], Darren H. Freed [3,8] and Andrew Mark James Shapiro [1,3,9,*]

[1] Division of General Surgery, Department of Surgery, Faculty of Medicine & Dentistry, University of Alberta, Edmonton, AB T6G 2R3, Canada
[2] Department of Medicine, Faculty of Medicine & Dentistry, University of Alberta, Edmonton, AB T6G 2R3, Canada
[3] Canadian Donation & Transplantation Research Program, Edmonton, AB T6G 2R3, Canada
[4] Department of Laboratory Medicine & Pathology, Faculty of Medicine & Dentistry, University of Alberta, Edmonton, AB T6G 2R3, Canada
[5] Canadian Blood Services, Edmonton, AB T6G 2R3, Canada
[6] Division of Gastroenterology, Department of Medicine, Faculty of Medicine & Dentistry, University of Alberta, Edmonton, AB T6G 2R3, Canada
[7] Department of Critical Care Medicine, Faculty of Medicine & Dentistry, University of Alberta, Edmonton, AB T6G 2R3, Canada
[8] Division of Cardiac Surgery, Department of Surgery, Faculty of Medicine & Dentistry, University of Alberta, Edmonton, AB T6G 2R3, Canada
[9] Clinical Islet Transplant Program, University of Alberta, Edmonton, AB T6G 2R3, Canada
* Correspondence: jshapiro@ualberta.ca

Citation: Hefler, J.; Hatami, S.; Thiesen, A.; Olafson, C.; Durand, K.; Acker, J.; Karvellas, C.J.; Bigam, D.L.; Freed, D.H.; Shapiro, A.M.J. Model of Acute Liver Failure in an Isolated Perfused Porcine Liver—Challenges and Lessons Learned. *Biomedicines* **2022**, *10*, 2496. https://doi.org/10.3390/biomedicines10102496

Academic Editor: Martina Perše

Received: 26 July 2022
Accepted: 2 October 2022
Published: 6 October 2022

Publisher's Note: MDPI stays neutral with regard to jurisdictional claims in published maps and institutional affiliations.

Abstract: Acute liver failure (ALF) is a rare but devastating disease associated with substantial morbidity and a mortality rate of almost 45%. Medical treatments, apart from supportive care, are limited and liver transplantation may be the only rescue option. Large animal models, which most closely represent human disease, can be logistically and technically cumbersome, expensive and pose ethical challenges. The development of isolated organ perfusion technologies, originally intended for preservation before transplantation, offers a new platform for experimental models of liver disease, such as ALF. In this study, female domestic swine underwent hepatectomy, followed by perfusion of the isolated liver on a normothermic machine perfusion device. Five control livers were perfused for 24 h at 37 °C, while receiving supplemental oxygen and nutrition. Six livers received toxic doses of acetaminophen given over 12 h, titrated to methemoglobin levels. Perfusate was sampled every 4 h for measurement of biochemical markers of injury (e.g., aspartate aminotransferase [AST], alanine aminotransferase [ALT]). Liver biopsies were taken at the beginning, middle, and end of perfusion for histological assessment. Acetaminophen-treated livers received a median dose of 8.93 g (8.21–9.75 g) of acetaminophen, achieving a peak acetaminophen level of 3780 μmol/L (3189–3913 μmol/L). Peak values of ALT (76 vs. 105 U/L; $p = 0.429$) and AST (3576 vs. 4712 U/L; $p = 0.429$) were not significantly different between groups. However, by the end of perfusion, histology scores were significantly worse in the acetaminophen treated group ($p = 0.016$). All acetaminophen treated livers developed significant methemoglobinemia, with a peak methemoglobin level of 19.3%, compared to 2.0% for control livers ($p = 0.004$). The development of a model of ALF in the ex vivo setting was confounded by the development of toxic methemoglobinemia. Further attempts using alternative agents or dosing strategies may be warranted to explore this setting as a model of liver disease.

Keywords: acute liver failure; porcine model; ex situ machine perfusion; acetaminophen; carbon tetrachloride

1. Introduction

Acetaminophen toxicity is the most common cause of acute liver failure (ALF) in North America [1]. Despite its low occurrence (10 people per million annually), it is associated with high morbidity and a devastating mortality rate of almost 45% [2]. Early presentations (within 24 h) may respond to treatment with N-acetylcysteine (NAC), which replenishes the liver's glutathione reserve and facilitates the conversion of acetaminophen into non-toxic metabolites [3,4]. NAC can delay progression, but, otherwise, management is largely supportive, allowing the liver time to recover. Liver transplantation may be considered based on the severity of injury, as judged by the King's College criteria, and the patient's pre-morbid status [5]. While survival rate is high after transplantation (75% at 5 years), transplantation itself is associated with its own morbidity and requires lifelong immunosuppression, which is not to mention the limited supply of suitable organs [6,7].

Acetaminophen is a common over-the-counter analgesic. It is primarily metabolized in the liver, where it undergoes either glucuronidation or sulfonation by cytochrome P450 enzymes to generate water soluble metabolites that can be readily excreted in the urine [8]. A small proportion is oxidized into N-acetyl-p-benzoquinone (NAPQI). NAPQI is a strong oxidizing compound that generates oxidative stress and causes cellular dysfunction by forming acetaminophen-protein adducts [9]. NAPQI can be further metabolized into non-toxic compound by glutathione dependent pathways [10]. However, when the main pathways of acetaminophen metabolism become saturated and glutathione stores are depleted, accumulation of NAPQI can lead to hepatocyte death and severe liver injury [9].

Pre-clinical models used to study acetaminophen toxicity in the liver vary from hepatocytes cultured in vitro, to small (e.g., rodents) and large animal (e.g., dogs, primates, pigs) models. More advanced models, including use of bio-artificial liver and 3D cell culture, have also been employed [10–12]. Porcine models are typically preferred as large animal models, due to their similarity to humans in terms of liver anatomy, histology and metabolism, as well as logistic considerations (e.g., size, handling, availability) [13,14]. Previous methods to induce ALF with acetaminophen in pigs have required the animals to be maintained under anesthesia for the duration of the experiment (typically > 24 h), which are expensive, technically challenging, and confounded by ethical considerations [15,16].

Ex vivo normothermic machine perfusion (NMP) may provide an alternative approach and more ethically acceptable platform for modeling ALF by isolating the injury to the target organ rather than the entire animal. NMP is a technology that aims to maintain normal liver physiology and metabolism outside of the body [17]. It was developed with the intent of preserving and assessing grafts before transplantation, such as marginal grafts donated after cardiovascular death. Its clinical utility was demonstrated in a randomized controlled trial published in 2018 by Nasralla and colleagues, where minimal liver injury was observed in grafts subjected to NMP, despite lower discard rate and longer preservation time, when compared to traditional cold static storage [18]. Related preclinical research has focused on graft treatment while on the ex vivo circuit, including 'de-fatting' steatotic livers, gene therapy, and eradication of hepatitis C [19–22] Several other exciting potential applications of NMP outside of transplantation, such as use in toxicology studies, cancer research, and liver support remain to be explored [23].

Development of reproducible, translatable ALF models are needed to facilitate the search for novel treatments. An ALF model in an isolated organ has several potential advantages over in vivo models, including improving animal welfare and extending treatment duration. However, the impact that the lack of other, potentially compensatory mechanisms, provided by other organ systems will have is unknown, as are the effects of interaction with artificial circuit components. Herein, we describe our approach to creating a large animal ex vivo model of ALF using acetaminophen, as well as another common agent, carbon tetrachloride, and describe the limitations we encountered.

2. Materials and Methods

This study used domestic swine, 3–4 months of age, weighing 40–50 kg, at which point their livers are similar in size to an adult human (~1200 g), while still being easily manageable in our surgical facility. Only female animals were used as male animals are routinely castrated to prevent aggression and boar taint (if used for meat), which can lead to visceral fat accumulation, among other physiological changes. Animals were supplied by the Swine Research and Technology Centre at the University of Alberta. Animals were obtained from different litters, randomly housed, and allocated to treatment groups. This study was conducted in accordance with the Canadian Council on Animal Care Guidelines and Policies with approval from the Animal Care and Use Committee (Health Sciences) for the University of Alberta (AUP00001036; approved 04/07/2014), which ensure the ethical treatment of research animals.

2.1. Liver Procurement

Hepatectomy was performed in fasted, anesthetized pigs, as in our previously established protocol [24,25]. In brief, animals were premedicated with 20 mg/kg of ketamine (Ketaset; Zoetis Canada Inc., Kirkland, QC, Canada) and 0.05 mg/kg of atropine (Atro-SA; Rafter 8 Products Inc., Calgary, AB, Canada), followed by endotracheal intubation using direct laryngoscopy. During the procedure, animals were maintained on inhalational anesthesia (3–4% isoflurane; Frensenius Kabi Canada, Toronto, ON, Canada). An orogastric tube was also inserted to decompress the stomach.

Animals were prepped with a 7.5% povidone-iodine solution and draped appropriately. Surgery was conducted under aseptic technique, including the use of sterilized instruments. An 8 Fr venous cannula was inserted into the right internal jugular vein for fluid administration. A midline laparotomy was performed to expose the intra-abdominal contents. The bowel was displaced from the abdomen to expose to the infra-renal aorta, which was then dissected and encircled with 0-silk ties (Perma Hand; Ethicon Inc., Bridgewater, NJ, USA) for ease of future cannulation. The bowel was returned to abdomen and rolled laparotomy sponges were place beneath the liver to elevate it into the surgical field. The fundus of the gallbladder was grasped with an Allis clamp and retracted superiorly to facilitate dissection from the gallbladder fossa. The cystic duct and artery were then ligated and the gallbladder was disposed of. The porta hepatis was exposed and the common bile duct was identified and ligated just prior to it entering the pancreas. The main portal vein was then dissected free of its peritoneal covering and tagged with a silk tie. A portion of the upper intra-abdominal aorta was also dissected and tagged with a silk tie for subsequent clamp placement during abdominal aortic flush.

Blood to be used in the perfusion was collected via the right atrium. To facilitate this, a midline sternotomy was performed and the heart was exposed and freed from the pericardium and associated tissues. The superior portion of the right atrium was encircled with a 4-0 polypropylene suture (Prolene Blu; Ethicon Inc., Bridgewater, NJ, USA). An incision was made in the atrium above the suture and a 28 French two-stage venous cannula was inserted and secured with a vascular torniquet. Blood was collected into a sterile container to the point of complete exsanguination (1.5–2 L).

Upon completion of exsanguination, an aortic cross-clamp was placed on the upper portion of the abdominal aorta previously identified to prevent flow of the preservation solution into the chest. The infra-renal aorta was then cannulated and 2 L of cold (4 °C) histidine-tryptophan-ketoglutarate solution (Custodiol® HTK; Essential Pharmaceuticals LLC., Durham, NC, USA) was used to flush the intra-abdominal organs. Ice was placed into the abdomen to cool the organs and the inferior vena cava above the diaphragm was incised to vent the flush solution. Following completion of the intra-abdominal flush, the liver was resected and weighed on the back table.

2.2. Normothermic Machine Perfusion (NMP)

Livers were perfused according to our previously established protocol [24,25]. A schematic of our NMP circuit is provided in Supplementary Figure S1. Our custom designed circuit consisted of an EOS ECMO oxygenator (Sorin Group Canada Inc., Burnaby, BC, Canada), a MYOtherm XP heat exchanger (Medtronic of Canada Ltd., Brampton, ON, Canada), and two BPX-80 Bi-Medicus centrifugal pumps (Medtronic of Canada Ltd., Brampton, ON, Canada). Perfusate draining from the liver was collected in a filtered CARD EVO cardiotomy reservoir (Sorin Group Canada Inc., Burnaby, BC, Canada) before recirculation through the pumps and was heated using a CW-05G water bath (Lab Companion, Jeio Tech Inc., Billerica, MA, USA). Hepatic artery and portal venous flows were measured using Transonic Clamp-on Tubing Flowsensors (6PXL and 7PXL; Transonic Systems Inc., Ithaca, NY, USA) and pressures were measured using TruWave pressure transducers (Edwards Life Sciences Canada Inc., Mississauga, ON, Canada). The centrifugal pumps were computer-controlled to maintain desired hepatic artery pressure and portal venous flow. Inflows of oxygen and carbon dioxide were titrated to maintain a partial pressure of arterial oxygen between 100 and 120 mmHg and a partial pressure of carbon dioxide between 35 and 45 mmHg.

The circuit was primed with a 1:1 ratio of whole blood to modified Krebs-Henseleit solution (glucose [5 mM], sodium chloride [85 mM], potassium chloride [5 mM], calcium chloride [1 mM], magnesium chloride [1 mM], sodium bicarbonate [25 mM], sodium phosphate monobasic [1 mM], sodium pyruvate [5 mM], and 8% bovine serum albumin). Piperacillin/tazobactam (3.375 g; Sandoz Canada, Boucherville, QC, Canada), methylprednisolone (500 mg; Solu-Medrol, Pfizer Canada Inc., Kirkland, QC, Canada), and sodium heparin (5000 U; Fresenius Kabi Canada, Toronto, ON, Canada) were added prior to attaching the organ. During perfusion, the organ received infusions of regular insulin (2 U/h; Humulin R; Eli Lilly & Company, Toronto, ON, Canada) and sodium heparin (1000 U/h). Piperacillin/tazobactam and methylprednisolone were re-dosed after 12 h. Perfusate pH and glucose was maintained within physiological range (7.35–7.45 and 6–10 mmol/L).

Prior to attaching the liver to the circuit, it was flushed with 2 L of normal saline. The hepatic artery and portal vein were canulated with an 8 Fr arterial cannula (Medtronic of Canada Ltd., Brampton, ON, Canada) and 1/4' polycarbonate tubing connector, respectively. The common bile duct was cannulated with tubing that drained to a reservoir outside of the organ chamber. The perfusate was heated to maintain the organ at 37 °C. Arterial and portal venous flows were gradually increased to physiological values, as the organ warmed.

2.3. Experimental Protocol

Control livers were run for 24 h ($n = 5$). Acetaminophen-treated livers ($n = 6$) received doses of acetaminophen (Sigma-Aldrich Canada, Oakville, ON, Canada) dissolved directly into the perfusate, starting 1 h after beginning perfusion. Acetaminophen was dosed at 30-min intervals up to 12 h, with the dose titrated to avoid excessive methemoglobinemia. Acetaminophen-treated livers were intended to run for 24 h, but were stopped early as necessitated by perfusion parameters and circuit volume. These livers similarly received additives necessary to maintain physiological blood gas parameters. Since we were unable to achieve sufficient toxicity without causing methemoglobinemia, further perfusions were discontinued, which lead to unequal numbers between groups. An additional experimental group treated with carbon tetrachloride (CCl_4; Sigma-Aldrich Canada, Oakville, ON, Canada) as the hepatotoxic agent was also attempted, but discontinued after two perfusions, due to circuit and perfusate incompatibility.

2.4. Perfusate Biochemistry

Perfusate was sampled from the hepatic artery and portal venous inflows and vena cava outflow every four hours and as needed to measure hemoglobin, methemoglobin, pH, electrolytes, lactate, glucose, and partial pressures of oxygen and carbon dioxide using a point-of-care blood gas machine (ABL Flex Analyzer; Radiometer Canada, Mississauga, ON, Canada). Perfusate sampled from the vena cava outflow was also measured for alanine aminotransferase (ALT), aspartate aminotransferase (AST), lactate dehydrogenase (LDH), and total and conjugated bilirubin using a Beckman Coulter Unicel Dxc800 Syncron (Beckman Coulter Canada LP, Mississauga, ON, Canada).

2.5. Histology

Incisional liver biopsies were taken at 2, 12 and 24 h into perfusion from the right lateral lobe, a portion of which was fixed in 10% formalin. These samples were then embedded in paraffin, stained with hematoxylin and eosin, and examined in a blinded fashion by an expert pathologist (AT). Samples were rated on the presence of necrosis, vacuolization, and congestion, as previously reported [26].

2.6. Chromogenic and Enzyme-Linked Immunosorbent Assays

Both malondialdehyde (MDA) and glutathione were measured in tissue using colorimetric assay kits (ab118970, Abcam PLC, Cambridge, UK; Invitrogen EIAGSHC, Fisher Scientific Co., Edmonton, AB, Canada), run according to manufacturer's instructions. Samples and standards were run in duplicate to a 96-well microplate and absorbance was read on a spectrophotometer (Multiskan SkyHigh; Fisher Scientific Co., Edmonton, AB, Canada) measuring optical density at 532 and 405 nm, respectively. Oxidized low-density lipoprotein (oxLDL) and haptoglobin were measured in perfusate samples using porcine-specific enzyme-linked immunosorbent assays (ELISA) (MBS2508909, MyBioSource Inc., San Diego, CA, USA; ab205091; Abcam PLC, Cambridge, UK), run according to manufacturer's instructions.

For determination of free hemoglobin (an indicator of hemolysis), perfusate samples were centrifuged (2200× g, 10 min, 4 °C) and the supernatant in was diluted in Drabkin's reagent (0.61 mmol/L potassium ferricyanide, 0.77 mmol/L potassium cyanide, 1.03 mmol/L potassium dihydrogen phosphate, and 0.1% [*vol/vol*] TritonX-100 [Sigma-Aldrich Canada, Oakville, ON, Canada]). Trilevel Hb controls (StanBio Laboratory, Boerne, TX, USA) were used for quality control. Absorbances were read on a spectrophotometer (SpectraMax 384 Plus, Molecular Devices Corp., Sunnyvale, CA, USA) at 540 nm and free hemoglobin concentration was determined using the calculation previously described in the literature [27].

2.7. In Vitro Testing of Acetaminophen Metabolites and CCl$_4$

The effect of acetaminophen metabolites on methemoglobinemia was tested in vitro by adding them to a sample of perfusate (1:1 whole blood with modified Krebs-Henseleit). P-aminophenol (Sigma Aldrich Canada, Oakville, ON, Canada) was added in a concentration of 100 µg/mL and let sit at room temperature. Methemoglobin was measured, as described above, at 30 and 90 min. N-acetyl-p-benzoquinone imine (NAPQI) (Sigma Aldrich Canada, Oakville, ON, Canada) was added to a separate sample of perfusate at a concentration of 125 µg/mL and methemoglobin was again measured after 90 and 180 min at room temperature. For comparison, acetaminophen (Sigma Aldrich Canada, Oakville, ON, Canada) was added to perfusate at a concentration of 10 mg/mL and methemoglobin was measured after 60 min. The effect of CCl$_4$ on hemolysis was tested in vitro by adding CCl$_4$ (Sigma Aldrich Canada, Oakville, ON, Canada) to whole blood in a concentration of 10 µL/mL and measuring plasma free hemoglobin after 30 min at room temperature, as detailed above. Free hemoglobin was measured in porcine whole blood after 30 min at room temperature without the addition of CCl$_4$ for control comparison.

2.8. Statistical Analysis

Data are reported as medians (interquartile range). The Mann U Whitney test was used to compare single measures between the two groups, as well as repeated measures at different time points. A *p*-value of <0.05 was considered significant. All analyses were performed using GraphPad Prism v9 (GraphPad Software Inc., San Diego, CA, USA).

3. Results

3.1. Perfusion Parameters

All control livers were perfused for 24 h (*n* = 5), while acetaminophen-treated livers (*n* = 6) we perfused for a median time of 892 (566–1316) minutes. Peak hepatic artery flow and pressure were 480.1 mL/min and 59.9 mmHg in the untreated group and 565.3 mL/min and 50.1 mmHg in the acetaminophen treated group (*p* = 0.931 and 0.046, respectively) (Supplementary Figure S2). Peak portal venous flow and pressure were 902.1 mL/min and 7.4 mmHg in the untreated group and 759.2 mL/min and 8.2 mmHg in the acetaminophen treated group (*p* = 0.017 and 0.662, respectively). Median hourly bile production was no different between groups (7.7 mL/h vs. 7.0 mL/h; *p* = 0.610).

3.2. Perfusate Biochemistry

Acetaminophen-treated livers received a median dose of 8.93 g (8.21–9.75 g) of acetaminophen, achieving a median peak acetaminophen level of 3780 µmol/L (3189–3913 µmol/L). The acetaminophen concentration remained above the clinical threshold for toxicity (1000 µmol/L) up to 12 h of perfusion (Figure 1). Peak values of ALT (76 vs. 105 U/L; *p* = 0.429), AST (3576 vs. 4712 U/L; *p* = 0.429), and LDH (2389 vs. 2496 U/L; *p* = 0.537) were not significantly different between groups (Figure 2). There was a trend toward peak bilirubin being higher in the acetaminophen-treated group (45 vs. 57 µmol/L; *p* = 0.056). End lactate was significantly higher (1.0 vs. 4.4 mmol/L; *p* = 0.011) and end pH was significantly lower (7.42 vs. 7.31; *p* = 0.030) in the acetaminophen-treated group, despite using more THAM for pH correction (33.0 vs. 55.5 mL; *p* = 0.004) (Figure 3). There was also a significant difference in perfusate INR, which, over the course of the perfusion, fell by 1.65 points (to 1.65 [1.53–2.45]) in the control group and rose by 0.45 points (to 4.0 [3.93–4.45]) in the acetaminophen-treated group (*p* = 0.029).

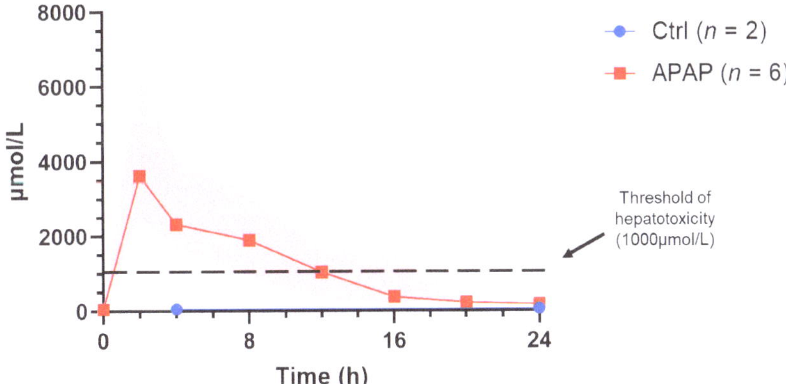

Figure 1. Acetaminophen (n-acetyl-para-aminophenol, APAP) concentration in perfusate over the duration of liver perfusion compared to control (Ctrl) livers without addition of APAP. The highlighted area represents the IQR.

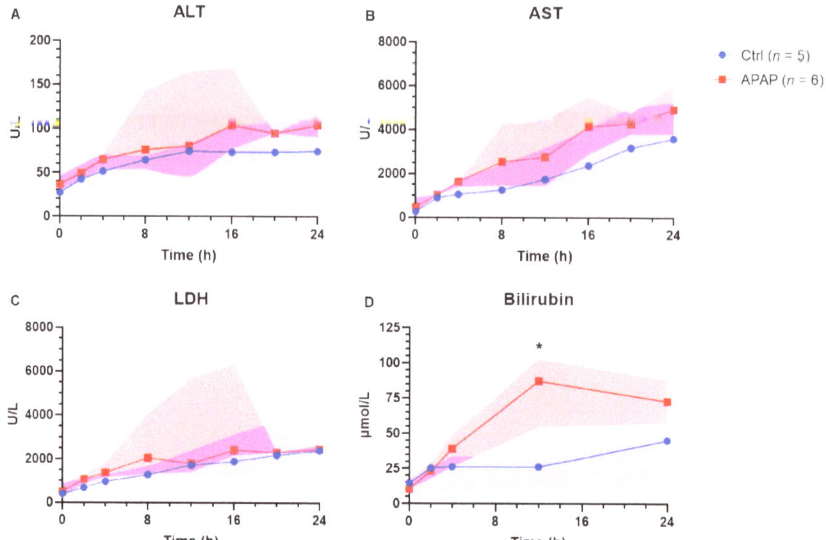

Figure 2. Perfusate concentrations of (**A**) alanine transaminase (ALT), (**B**) aspartate transaminase (AST), (**C**) lactate dehydrogenase (LDH), and (**D**) total bilirubin over the duration of perfusion in livers treated with acetaminophen (n-acetyl-para-aminophenol, APAP) compared to those without (Ctrl). The highlighted area represents IQR. * = $p < 0.05$.

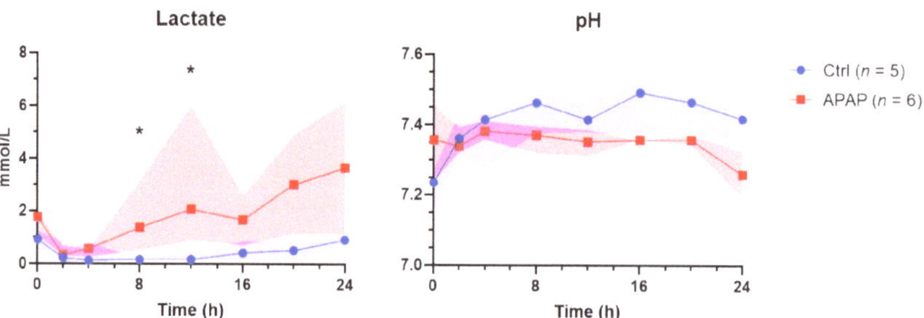

Figure 3. Perfusate (**A**) lactate and (**B**) pH over the duration of perfusion in livers treated with acetaminophen (n-acetyl-para-aminophenol, APAP) compared to those without (Ctrl). The highlighted area represents IQR. * = $p < 0.05$.

3.3. Histology

Histological scoring was similar between groups at the start of perfusion ($p = 0.900$). However, by the end was significantly worse in the acetaminophen treated group ($p = 0.016$) (Figure 4). Specifically, sub-scores of necrosis and vacuolization were significantly higher in the treated group by the end of perfusion ($p = 0.016$ and 0.001, respectively).

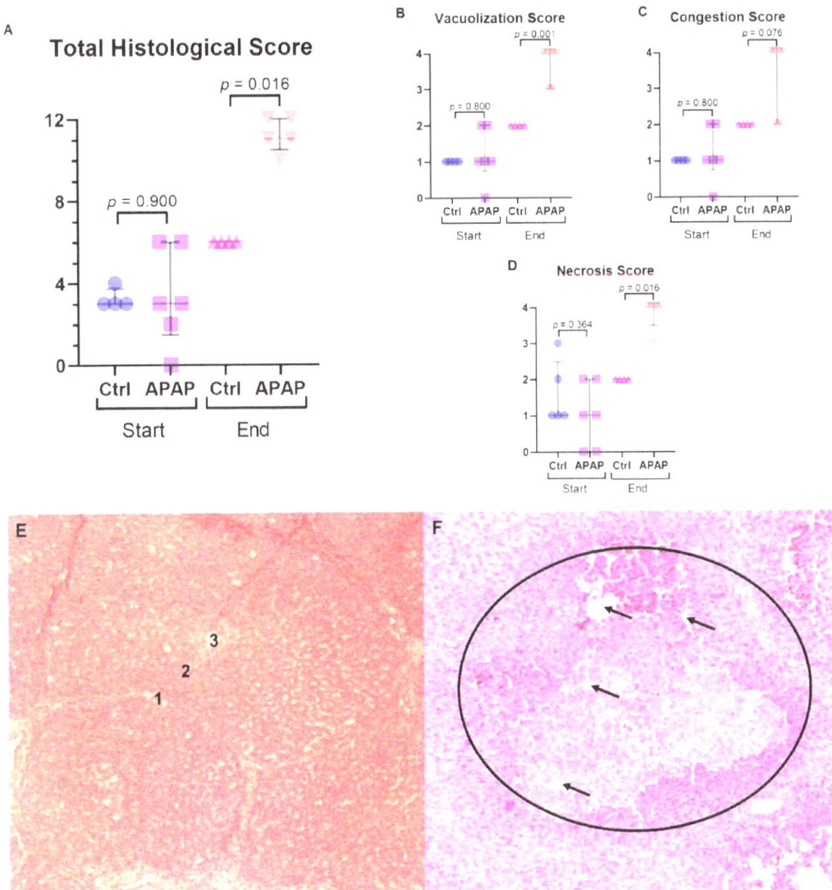

Figure 4. Histological scoring for control (Ctrl) and acetaminophen-treated (n acetyl para aminophenol, APAP) livers. Comparisons made at the start and end of perfusion based on (**A**) total histological score and sub-scores of (**B**) vacuolization, (**C**) congestion, and (**D**) necrosis. The data is presented as median and IQR. A representative image showing preserved hepatic architecture is shown (**E**), with zones 1–3 readily identifiable. This is contrasted with an image from an acetaminophen treated liver (**F**) showing pan-lobular hepatocyte necrosis and congestion (circle) and diffuse sinusoidal dilation (arrows).

3.4. Measures of Oxidative Stress

Tissue MDA as a marker of oxidative stress was significantly increased in the acetaminophen-treated group at the end of perfusion compared to controls (21.9 [21.7–22.5] vs. 30.4 [29.1–31.6] nmol; $p = 0.036$) (Figure 5A). Perfusate oxLDL, a circulating marker of oxidative stress, was also found to be increased in the acetaminophen-treated group at the end compared to the start (1.81 [1.43–2.27] vs. 8.84 [6.95–11.72] nmol/mL; $p = 0.004$), though was not significantly different from end concentration of the controls (4.36 [2.27–7.52] nmol/mL; $p = 0.191$) (Figure 5B). As well, tissue glutathione content, which buffers against oxidative stress, was correspondingly decreased in acetaminophen-treated livers at the end of perfusion (31.0 [24.8–33.1] vs. 7.51 [6.4–12.0] μM/mg tissue; $p = 0.004$) (Figure 5C).

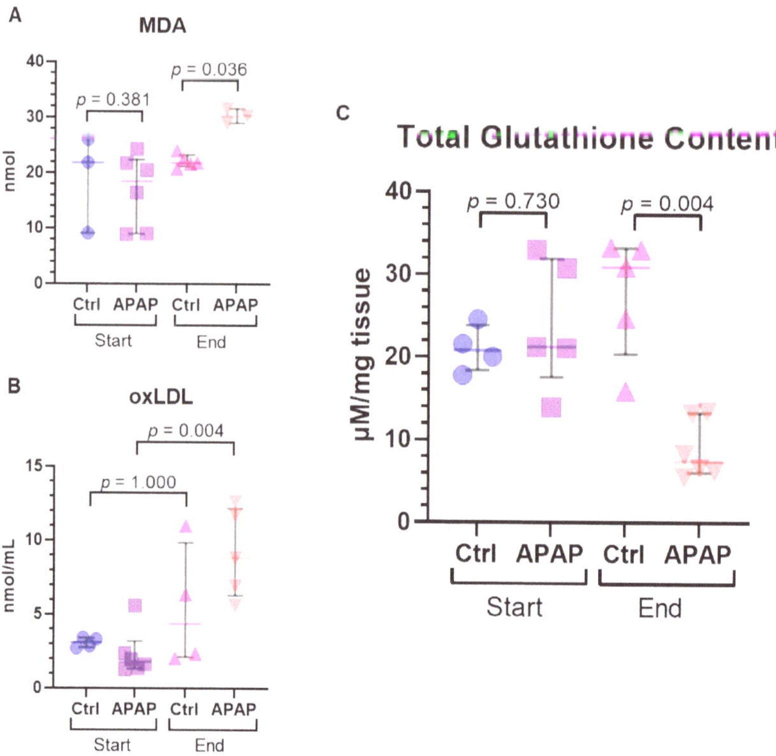

Figure 5. Markers of oxidative stress during liver perfusion in livers treated with acetaminophen (n-acetyl-para-aminophenol, APAP) compared to those without (Ctrl): (**A**) tissue malondialdehyde (MDA), (**B**) perfusate oxidized low-density lipoprotein (oxLDL), and (**C**) tissue glutathione. The data is presented as median and IQR.

3.5. Methemoglobinemia and Hemolysis

Methemoglobinemia, a well described complication of acetaminophen toxicity in in vivo animal models, was also observed in our ex vivo model (Figure 6) [28]. The median peak methemoglobulin level was 19.3% for acetaminophen-treated livers, compared to 2.0% for control livers (p = 0.004). Free hemoglobin, a measure of hemolysis, was significantly lower in controls at the midpoint of perfusion (12 h) compared to similar time points in acetaminophen-treated livers (0.96 [0.72–1.43] vs. 2.37 [2.10–7.65] g/L; p = 0.017), though was not significantly different at the end of perfusion (1.41 [1.08-3.02] vs. 2.15 [1.74–7.60] g/L; p = 0.247) (Figure 7A). Hemolysis was further evident in the median drop in hemoglobin over the course of the perfusion, which was 19.0 g/L for the acetaminophen-treated group and only 1.0 g/L for the control livers (p = 0.008). Median perfusate haptoglobin at the endpoint was significantly lower in the acetaminophen-treated group (27.56 [22.73–30.80] vs. 4.67 [2.90–6.09] µg/mL; p = 0.024) (Figure 7B).

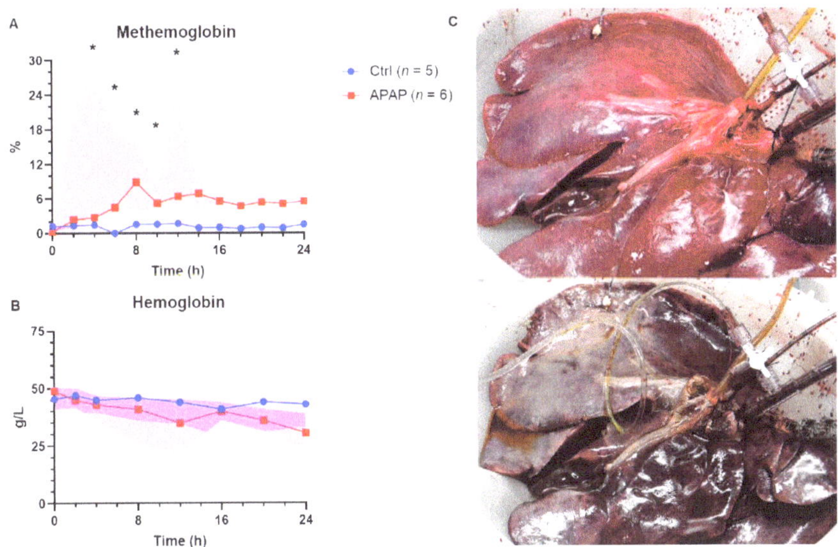

Figure 6. Effect of acetaminophen of perfusate (**A**) methemoglobin and (**B**) hemoglobin concentrations in livers treated with acetaminophen (n-acetyl-para-aminophenol, APAP) compared to those without (Ctrl). The highlighted area represents IQR; * = $p < 0.05$. Perfused liver before (top) and after (bottom) development of methemoglobinemia (**C**).

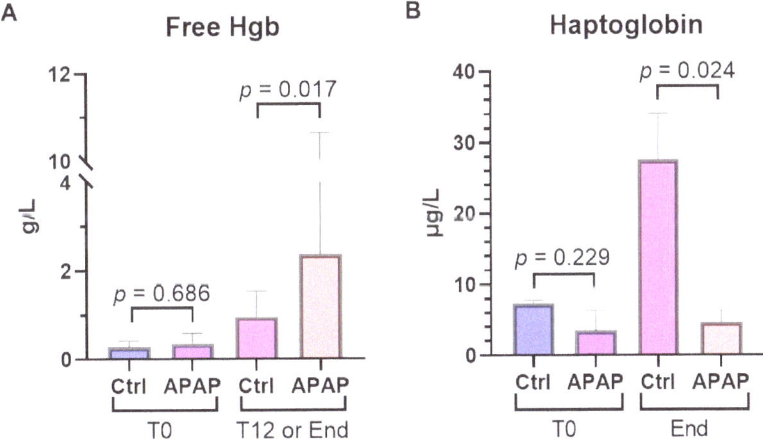

Figure 7. Perfusate (**A**) free hemoglobin (Hgb) and (**B**) haptoglobin over the course of liver normothermic perfusion in livers treated with acetaminophen (n-acetyl-para-aminophenol, APAP) compared to those without (Ctrl). The data is presented as median and IQR.

It had been suggested that the metabolite of acetaminophen responsible for methemoglobinemia was PAP [29]. The addition 100 µg/mL of PAP in vitro to a sample of perfusate resulted in significant methemoglobin compared to the control after 30 (1.4% [0.8–1.6%] vs. 6.6% [5.2–9.6%]; $p = 0.008$) and 90 min (1.5% [1.5–1.8%] vs. 17.4% [14.9–20.2%]; $p = 0.008$) at room temperature (Figure 8A, Supplementary Figure S3). Adding acetaminophen (10 mg/mL) and NAPQI (125 µg/mL) to perfusate in vitro did not result in the development of significant methemoglobinemia (Figure 8B,C).

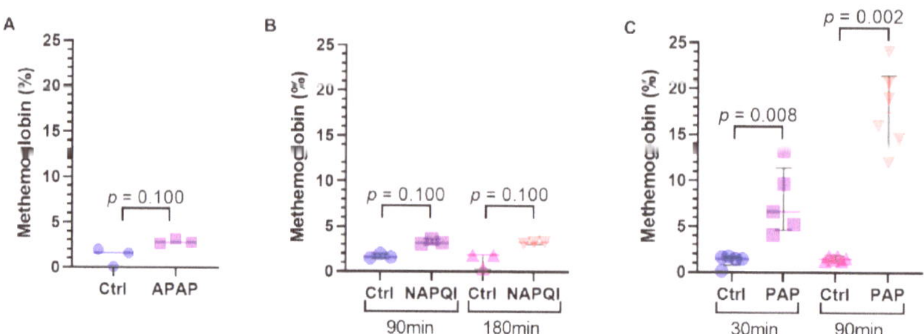

Figure 8. Effects of (**A**) acetaminophen and its metabolites, (**B**) n-acetyl-p-benzoquinone imine (NAPQI), and (**C**) p-aminophenol (PAP) on the development of methemoglobinemia in vitro. The data is presented as median and IQR.

3.6. Carbon Tetrachloride (CCl₄)

CCl_4 was trialed as an alternative to acetaminophen in two perfused livers, but similar challenges were encountered. The first liver was treated with 10 mL of CCl_4 and a second liver with 5 mL. The first was perfused for 1160 min and the second for 661 min. Both had higher peak ALT (518 and 298 U/L), AST (12,929 and 15,271 U/L), and LDH (8,912 and 8,486 U/L) compared to controls. End pH (6.99 and 7.13) and lactate (10.5 and 9.9 mmol/L) were similarly abnormal. Comparisons with controls were not statistically significant owing to the small sample size.

Hemolysis was particularly problematic in the CCl_4-treated livers. Free hemoglobin was measured at 8.5 and 5.0 g/L by the end of perfusion. Hemoglobin concentration also decreased by 25 g/L in both perfusions. CCl_4 as the causative agent of hemolysis was again demonstrated in vitro. CCl_4 added to whole blood at a concentration of 10 μL/mL resulted in free hemoglobin of 13.67 g/L (5.41–15.48 g/L) after 30 min, compared to 0.37 g/L (0.30–0.38 g/L) in controls ($p = 0.008$) (Supplementary Figure S4A,B). In addition to the hemolysis, incompatibilities with circuit components (mainly the polycarbonate stopcocks and tubing connectors) made further attempts untenable (Supplementary Figure S4C).

4. Discussion

Our study set out to develop a large animal model of ALF in an isolated, perfused liver to abrogate ethical challenges, cost and complexity of toxicity studies in the intact pig. We were able to show that the addition of acetaminophen in the ex vivo setting resulted in massive liver injury. However, we found that the degree of injury was not well represented by perfusate biochemistry and was confounded by the presence of substantial methemoglobinemia and an overall increase in hemolysis. Additionally, we were able to demonstrate in vitro that the cause of the methemoglobinemia was due to PAP, as opposed to acetaminophen itself or its main hepatotoxic metabolite, NAPQI. We also performed precursory studies with CCl_4 as an alternative hepatotoxic agent and identified major limitations with hemolysis for its use in the ex vivo setting.

Methemoglobinemia is a well described complication of acetaminophen toxicity in large animal models, including swine and canines [30,31]. However, it is only rarely seen in clinical cases of acetaminophen overdose [32]. This has made it challenging to find relevant preclinical models, as below a certain dose, animals tend to recover without developing lasting liver injury, while above it, they succumb to the effects of methemoglobinemia. Accumulation of methemoglobin differs from the mechanism by which acetaminophen is known to cause hepatocyte injury and fulminant liver failure, which results from an accumulation of NAPQI by cytochrome P450 metabolism of acetaminophen [33]. In contrast, PAP has been proposed as the metabolite responsible for methemoglobinemia and occurs as a result of deacetylation of acetaminophen [29]. Our in vitro studies confirmed that PAP readily induces methemoglobin formation in blood-base perfusate, whereas exposure to acetaminophen or NAPQI does not. The only swine models that have been able to overcome this challenge with methemoglobinemia have done so using titrated doses of acetaminophen over several hours [15,16].

Despite titrating acetaminophen dosing to methemoglobin level in the ex vivo setting, we were not able to avoid onset of experiment-limiting methemoglobinemia as the organ continued to be perfused. Methemoglobin develops as a result of the oxidation of the iron molecule in hemoglobin from the ferrous (Fe^{2+}) to the ferric state (Fe^{3+}) [34]. Whether porcine blood is more prone to methemoglobin formation, the additional cellular stress arising from exposure to the artificial components of the ex vivo circuit likely further contributed to its development [35]. Supporting this is the observation of methemoglobinemia developing after extended perfusion of discarded human livers [36].

Though the use of CCl_4 in models of both acute and chronic liver injury has been well described, we found it unsuitable as an agent for hepatotoxicity in the ex vivo setting [37]. Doses were chosen based on a porcine model of ALF, in which CCl_4 was injected directly into the portal vein [38]. However, this resulted in significant hemolysis, which both confounded the injury pattern and hampered perfusion of the liver in our model. CCl_4 is typically dosed intraperitoneally or per os, which may mitigate its hemolytic effects in reported models [37]. We were unable to find mention of hemolysis amongst studies using CCl_4 to induce liver failure. In addition, CCl_4 proved to be quite a powerful solvent such that its mere infusion through polycarbonate stopcocks resulted in their degradation (Supplementary Figure S4). A corollary of this observation is that CCl_4 may be useful for models of hemolytic anemia.

There are several alternative agents that could be considered to induce ALF. D-galactosamine has commonly been used, often in combination with lipopolysaccharide, to induce ALF in animal models [39]. As an amino sugar it is likely to be compatible with ex vivo perfusion components. However, it is not clear the degree to which immune-mediated damage is necessary to induce fulminant liver failure and whether that could be achieved in an isolated organ. In addition, it has been reported to have some batch-to-batch variability in potency, owing to is isolation from a biological source, which may affect model reproducibility depending on the dose required. Thioacetamide and azoxymethane are agents that have been used in vivo in rodent models [40,41]. However, there is limited experience with large animal models. As small molecule, biologically active chemicals, both would first warrant in vitro testing to ensure compatibility with porcine blood and circuit components. A-amanitin, the cyclic peptide produced by the *Amanita* genus of mushrooms, may be a more suitable candidate, as its structure is unlikely to cause unfavourable interactions with blood or artificial materials and, with its mechanism of RNA polymerase inhibition, it would be directly hepatotoxic [42]. There is even some experience with in vivo models in swine and non-human primates [43,44]. However, the cost of the molecule ($250–350 USD/mg) may be somewhat of a deterrent. Alternative measurements of liver injury that are more sensitive may be of use in developing consistency amongst such models, but ultimately they will need to show benefit by some clinically accepted parameter (such as transaminases or lactate clearance) to justify clinical translation.

Among important considerations for future use of isolated perfused organs as models of disease with the aim of achieving reproducibility are the development of standardized protocols for organ procurement, including the use of appropriate cold preservation solutions and cooling techniques if the organ is to be kept cold or minimizing warm ischemia if the organ is to be perfused immediately. As well, appropriate surgical skills are necessary to ensure organ procurement occurs efficiently, and the risks of undue injury are minimized. Similarly, protocols for organ perfusion itself, such perfusion pressures or flows and drug additives, should be consistent across the experiment and researchers should be comfortable with the technical aspects of their setup to allow for basic troubleshooting. If circuit components are to be reused (which is typical practice in the experimental setting), they should be cleaned thoroughly to avoid contamination between experiments and replaced entirely after a certain period of time or number of uses. Otherwise, routine experimental practices, such as maintaining standards for animal care and welfare, randomizing and blinding when possible, and using the same reagents between experiments, should be followed

Efficient, reproducible large animal models of ALF are needed to develop effective treatments to screen for new interventional therapies that improve patient outcomes and reduce the need for liver transplantation. Ex vivo liver NMP offers a new platform on which to implement such models, with the additional benefit of reducing animal suffering and safely extending the duration of experiments. However, our early experience identified several challenging factors that hindered development of a reproducible model using two well-known hepatotoxic agents in this setting. Our finding from the present study provides useful insight to other investigators in the field, and description of this approach will provide additional information for researchers in the field of liver injury to improve their understanding of models of organ injury and aid the development of efficient animal models. Future studies altering the mode of delivery or trialing novel agents may be able to overcome these barriers to make use of NMP as a model for ALF.

Supplementary Materials: The following supporting information can be downloaded at: https://www.mdpi.com/article/10.3390/biomedicines10102496/s1, Supplementary Figure S1: Normothermic Machine Perfusion Schematic Diagram; Supplementary Figure S2: Perfusion Parameters; Supplementary Figure S3: Methemoglobinemia in vitro; Supplementary Figure S4: Carbon Tetrachloride in vitro.

Author Contributions: Conceptualization, D.L.B. and A.M.J.S.; methodology, J.H., S.H., C.J.K., D.H.F. and D.L.B.; analysis, J.H., S.H., A.T., C.O., K.D. and J.A.; resources, J.A., D.H.F. and A.M.J.S.; writing—original draft preparation, J.H.; writing—review and editing, J.H., S.H., A.T., C.O., K.D., J.A., C.J.K., D.H.F., D.L.B. and A.M.J.S.; supervision, C.J.K., D.H.F., D.L.B. and A.M.J.S.; funding acquisition, A.M.J.S. All authors have read and agreed to the published version of the manuscript.

Funding: This research was funded by the Canadian Donation and Transplantation Research Program, the Canadian Liver Foundation, the University Hospital Foundation and the Edmonton Civic Employees Charitable Assistance Fund. The APC was paid by general research funds from A.M.J.S.

Institutional Review Board Statement: This study was conducted in accordance with the Canadian Council on Animal Care Guidelines and Policies with approval from the Animal Care and Use Committee (Health Sciences) for the University of Alberta (AUP00001036; approved 04/07/2014).

Data Availability Statement: The data presented in this study are available in the article and supplementary material.

Acknowledgments: Thank you to the technicians at the Ray Rajotte Surgical Medical Research Institute for their making this research possible. Thank you to Rena Pawlick for assistance with editing and revising the manuscript.

Conflicts of Interest: The authors declare no conflict of interest. The funders had no role in the design of the study; in the collection, analyses, or interpretation of data; in the writing of the manuscript; or in the decision to publish the results.

References

1. Lee, W.M. Etiologies of Acute Liver Failure. *Semin. Liver Dis.* **2008**, *28*, 142–152. [CrossRef] [PubMed]
2. Khashab, M.; Tector, A.J.; Kwo, P.Y. Epidemiology of acute liver failure. *Curr. Gastroenterol. Rep.* **2007**, *9*, 66–73. [CrossRef] [PubMed]
3. Darweesh, S.K.; Ibrahim, M.F.; El-Tahawy, M.A. Effect of N-Acetylcysteine on Mortality and Liver Transplantation Rate in Non-Acetaminophen-Induced Acute Liver Failure: A Multicenter Study. *Clin. Drug Investig.* **2017**, *37*, 473–482. [CrossRef] [PubMed]
4. Lee, W.M.; Hynan, L.S.; Rossaro, L.; Fontana, R.J.; Stravitz, R.T.; Larson, A.M.; Davern, T.J.; Murray, N.G.; McCashland, T.; Reisch, J.S.; et al. Intravenous N-Acetylcysteine Improves Transplant-Free Survival in Early Stage Non-Acetaminophen Acute Liver Failure. *Gastroenterology* **2009**, *137*, 856–864. [CrossRef]
5. O'Grady, J.G.; Alexander, G.J.; Hayllar, K.M.; Williams, R. Early indicators of prognosis in fulminant hepatic failure. *Gastroenterology* **1989**, *97*, 439–445. [CrossRef]
6. Lidofsky, S.D. Liver transplantation for fulminant hepatic failure. *Gastroenterol. Clin. N. Am.* **1993**, *22*, 257–269. [CrossRef]

7. Ostapowicz, G.; Fontana, R.J.; Schiødt, F.V.; Larson, A.; Davern, T.J.; Han, S.H.; McCashland, T.M.; Shakil, A.O.; Hay, J.E.; Hynan, L.; et al. Results of a Prospective Study of Acute Liver Failure at 17 Tertiary Care Centers in the USA. *Ann. Intern. Med.* **2002**, *137*, 947–954. [CrossRef]
8. McGill, M.R.; Jaeschke, H. Metabolism and Disposition of Acetaminophen: Recent Advances in Relation to Hepatotoxicity and Diagnosis. *Pharm. Res.* **2013**, *30*, 2174–2187. [CrossRef]
9. Mazaleuskaya, L.L.; Sangkuhl, K.; Thorn, C.F.; FitzGerald, G.A.; Altman, R.B.; Klein, T.E. PharmGKB summary: Pathways of acetaminophen metabolism at the therapeutic versus toxic doses. *Pharm. Genom.* **2015**, *25*, 416–426. [CrossRef]
10. Moyer, A.M.; Fridley, B.L.; Jenkins, G.D.; Batzler, A.J.; Pelleymounter, L.L.; Kalari, K.R.; Ji, Y.; Chai, Y.; Nordgren, K.K.; Weinshilboum, R.M. Acetaminophen-NAPQI Hepatotoxicity: A Cell Line Model System Genome-Wide Association Study. *Toxicol. Sci.* **2011**, *120*, 33–41. [CrossRef]
11. Xue, Y.-L.; Zhao, S.-F.; Zhang, Z.-Y.; Wang, Y.-F.; Li, X.-J.; Huang, X.-Q.; Luo, Y.; Huang, Y.-C.; Liu, C.-G. Effects of a bioartificial liver support system on acetaminophen induced acute liver failure canines. *World J. Gastroenterol.* **1999**, *5*, 308–311. [CrossRef] [PubMed]
12. Lauschke, V.M.; Hendriks, D.F.; Bell, C.C.; Andersson, T.B.; Ingelman-Sundberg, M. Novel 3D Culture Systems for Studies of Human Liver Function and Assessments of the Hepatotoxicity of Drugs and Drug Candidates. *Chem. Res. Toxicol.* **2016**, *29*, 1936–1955. [CrossRef] [PubMed]
13. Eberlova, L.; Maleckova, A.; Mik, P.; Tonar, Z.; Jirik, M.; Mirka, H.; Palek, R.; Leupen, S.; Liska, V. Porcine Liver Anatomy Applied to Biomedicine. *J. Surg. Res.* **2020**, *250*, 70–79. [CrossRef] [PubMed]
14. Wenzel, N.; Blasczyk, R.; Figueiredo, C. Animal Models in Allogenic Solid Organ Transplantation. *Transplantology* **2021**, *2*, 412–424. [CrossRef]
15. Thiel, C.; Thiel, K.; Etspueler, A.; Morgalla, M.; Rubitschek, S.; Schmid, S.; Steurer, W.; Königsrainer, A.; Schenk, M. A Reproducible Porcine Model of Acute Liver Failure Induced by Intrajejunal Acetaminophen Administration. *Eur. Surg. Res.* **2011**, *46*, 118–126. [CrossRef]
16. Newsome, P.N.; Henderson, N.C.; Nelson, L.J.; Dabos, C.; Filippi, C.; Bellamy, C.; Howie, F.; Clutton, R.E.; King, T.; Lee, A.; et al. Development of an invasively monitored porcine model of acetaminophen-induced acute liver failure. *BMC Gastroenterol.* **2010**, *10*, 34. [CrossRef]
17. van Beekum, C.J.; Vilz, T.O.; Glowka, T.R.; von Websky, M.W.; Kalff, J.C.; Manekeller, S. Normothermic Machine Perfusion (NMP) of the Liver—Current Status and Future Perspectives. *Ann. Transplant.* **2021**, *24*, e931664. [CrossRef]
18. Nasralla, D.; for the Consortium for Organ Preservation in Europe; Coussios, C.C.; Mergental, H.; Akhtar, M.Z.; Butler, A.J.; Ceresa, C.D.L.; Chiocchia, V.; Dutton, S.J.; García-Valdecasas, J.C.; et al. A randomized trial of normothermic preservation in liver transplantation. *Nature* **2018**, *557*, 50–56. [CrossRef]
19. Nagrath, D.; Xu, H.; Tanimura, Y.; Zuo, R.; Berthiaume, F.; Avila, M.; Yarmush, R.; Yarmush, M.L. Metabolic preconditioning of donor organs: Defatting fatty livers by normothermic perfusion ex vivo. *Metab. Eng.* **2009**, *11*, 274–283. [CrossRef]
20. Bishawi, M.; Roan, J.-N.; Milano, C.A.; Daneshmand, M.A.; Schroder, J.N.; Chiang, Y.; Lee, F.H.; Brown, Z.D.; Nevo, A.; Watson, M.; et al. A normothermic ex vivo organ perfusion delivery method for cardiac transplantation gene therapy. *Sci. Rep.* **2019**, *9*, 8029. [CrossRef]
21. Galasso, M.; Feld, J.J.; Watanabe, Y.; Pipkin, M.; Summers, C.; Ali, A.; Qaqish, R.; Chen, M.; Ribeiro, R.V.P.; Ramadan, K.; et al. Inactivating hepatitis C virus in donor lungs using light therapies during normothermic ex vivo lung perfusion. *Nat. Commun.* **2019**, *10*, 1–12. [CrossRef] [PubMed]
22. Goldaracena, N.; Spetzler, V.N.; Echeverri, J.; Kaths, J.M.; Cherepanov, V.; Persson, R.; Hodges, M.R.; Janssen, H.L.A.; Selzner, N.; Grant, D.R.; et al. Inducing Hepatitis C Virus Resistance After Pig Liver Transplantation-A Proof of Concept of Liver Graft Modification Using Warm Ex Vivo Perfusion. *Am. J. Transplant.* **2017**, *17*, 970–978. [CrossRef] [PubMed]
23. Hefler, J.; Marfil-Garza, B.A.; Dadheech, N.; Shapiro, A.J. Machine Perfusion of the Liver: Applications Beyond Transplantation. *Transplantation* **2020**, *104*, 1804–1812. [CrossRef] [PubMed]
24. Bral, M.; Gala-Lopez, B.; Thiesen, A.; Hatami, S.; Bigam, D.L.; Freed, D.M.; Shapiro, A.J. Determination of Minimal Hemoglobin Level Necessary for Normothermic Porcine Ex Situ Liver Perfusion. *Transplantation* **2018**, *102*, 1284–1292. [CrossRef]
25. Nostedt, J.J.; Churchill, T.; Ghosh, S.; Thiesen, A.; Hopkins, J.; Lees, M.C.; Adam, B.; Freed, D.H.; Shapiro, A.M.J.; Bigam, D.L. Avoiding initial hypothermia does not improve liver graft quality in a porcine donation after circulatory death (DCD) model of normothermic perfusion. *PLoS ONE* **2019**, *14*, e0220786. [CrossRef]
26. Xie, F.; Li, Z.-P.; Wang, H.-W.; Fei, X.; Jiao, Z.-Y.; Tang, W.-B.; Tang, J.; Luo, Y.-K. Evaluation of Liver Ischemia-Reperfusion Injury in Rabbits Using a Nanoscale Ultrasound Contrast Agent Targeting ICAM-1. *PLoS ONE* **2016**, *11*, e0153805. [CrossRef]
27. Davis, B.H.; Jungerius, B.; On behalf of International Council for the Standardization of Haematology (ICSH). International Council for Standardization in Haematology technical report 1-2009: New reference material for haemiglobincyanide for use in standardization of blood haemoglobin measurements. *Int. J. Lab. Hematol.* **2010**, *32*, 139–141. [CrossRef]
28. Dargue, R.; Zia, R.; Lau, C.; Nicholls, A.W.; Dare, T.O.; Lee, K.; Jalan, R.; Coen, M.; Wilson, I.D. Metabolism and Effects on Endogenous Metabolism of Paracetamol (Acetaminophen) in a Porcine Model of Liver Failure. *Toxicol. Sci. Off. J. Soc. Toxicol.* **2020**, *175*, 87–97. [CrossRef]
29. McConkey, S.E.; Grant, D.M.; Cribb, A.E. The role of para-aminophenol in acetaminophen-induced methemoglobinemia in dogs and cats. *J. Veter- Pharmacol. Ther.* **2009**, *32*, 585–595. [CrossRef]

30. Henne-Bruns, D.; Artwohl, J.; Broelsch, C.; Kremer, B. Acetaminophen-induced acute hepatic failure in pigs: Controversical results to other animal models. *Res. Exp. Med.* **1988**, *188*, 463–472. [CrossRef]
31. Gazzard, B.G.; Hughes, R.D.; Mellon, P.J.; Portmann, B.; Williams, R. A dog model of fulminant hepatic failure produced by paracetamol administration. *Br. J. Exp. Pathol.* **1975**, *56*, 408–411. [PubMed]
32. Rianprakaisang, T.; Blumenberg, A.; Hendrickson, R.G. Methemoglobinemia associated with massive acetaminophen ingestion: A case series. *Clin. Toxicol.* **2020**, *58*, 495–497. [CrossRef] [PubMed]
33. Ben-Shachar, R.; Chen, Y.; Luo, S.; Hartman, C.; Reed, M.; Nijhout, H.F. The biochemistry of acetaminophen hepatotoxicity and rescue: A mathematical model. *Theor. Biol. Med. Model.* **2012**, *9*, 55. [CrossRef]
34. Warren, O.U.; Blackwood, B. Acquired Methemoglobinemia. *N. Engl. J. Med.* **2019**, *381*, 1158. [CrossRef]
35. Al-Fares, A.; Pettenuzzo, T.; Del Sorbo, L. Extracorporeal life support and systemic inflammation. *Intensiv. Care Med. Exp.* **2019**, *25*, 1–14. [CrossRef]
36. Tingle, S.J.; Ibrahim, I.; Thompson, E.R.; Bates, L.; Sivaharan, A.; Bury, Y.; Figuereido, R.; Wilson, C. Methaemoglobinaemia Can Complicate Normothermic Machine Perfusion of Human Livers. *Front. Surg.* **2021**, *8*, 634777. [CrossRef] [PubMed]
37. Rahman, T.M.; Hodgson, H.J.F. Animal models of acute hepatic failure. *Int. J. Exp. Pathol.* **2000**, *81*, 145–157. [CrossRef] [PubMed]
38. Yuasa, T.; Yamamoto, T.; Rivas-Carrillo, J.D.; Chen, Y.; Navarro-Alvarez, N.; Soto-Guiterrez, A.; Noguchi, H.; Matsumoto, S.; Tanaka, N.; Kobayashi, N. Laparoscopy-Assisted Creation of a Liver Failure Model in Pigs. *Cell Transplant.* **2008**, *17*, 187–193. [CrossRef]
39. Apte, U. Galactosamine. In *Encyclopedia of Toxicology*, 3rd ed.; Wexler, P., Ed.; Oxford University Press: Oxford, UK, 2014; pp. 689–690.
40. Wallace, M.C.; Hamesch, K.; Lunova, M.; Kim, Y.; Weiskirchen, R.; Strnad, P.; Friedman, S.L. Standard Operating Procedures in Experimental Liver Research: Thioacetamide model in mice and rats. *Lab. Anim.* **2015**, *49*, 21–29. [CrossRef]
41. Matkowskyj, K.A.; Marrero, J.A.; Carroll, R.E.; Danilkovich, A.V.; Green, R.M.; Benya, R.V. Azoxymethane-induced fulminant hepatic failure in C57BL/6J mice: Characterization of a new animal model. *Am. J. Physiol.-Gastrointest. Liver Physiol.* **1999**, *277*, G455–G462. [CrossRef]
42. Ferriero, R.; Nusco, E.; De Cegli, R.; Carissimo, A.; Manco, G.; Brunetti-Pierri, N. Pyruvate dehydrogenase complex and lactate dehydrogenase are targets for therapy of acute liver failure. *J. Hepatol.* **2018**, *69*, 325–335. [CrossRef] [PubMed]
43. Takada, Y.; Ishiguro, S.; Fukunaga, K.; Gu, M.; Taniguchi, H.; Seino, K.I.; Yuzawa, K.; Otsuka, M.; Todoroki, T.; Fukao, K. Increased intracranial pressure in a porcine model of fulminant hepatic failure using amatoxin and endotoxin. *J. Hepatol.* **2001**, *34*, 825–831. [CrossRef]
44. Zhou, P.; Xia, J.; Guo, G.; Huang, Z.X.; Lu, Q.; Li, L.; Li, H.X.; Shi, Y.J.; Bu, H. A Macaca mulatta model of fulminant hepatic failure. *World J. Gastroenterol.* **2012**, *18*, 435–444. [CrossRef] [PubMed]

biomedicines

MDPI

Article

Potential Mucosal Irritation Discrimination of Surface Disinfectants Employed against SARS-CoV-2 by *Limacus flavus* Slug Mucosal Irritation Assay

Marco Alfio Cutuli [1], Antonio Guarnieri [1], Laura Pietrangelo [1], Irene Magnifico [1], Noemi Venditti [1], Laura Recchia [1], Katia Mangano [2], Ferdinando Nicoletti [2], Roberto Di Marco [1,*] and Giulio Petronio Petronio [1]

[1] Department of Medicine and Health Science "V. Tiberio", Università degli Studi del Molise, 8600 Campobasso, Italy; m.cutuli@studenti.unimol.it (M.A.C.); a.guarnieri@studenti.unimol.it (A.G.); laura.pietrangelo@unimol.it (L.P.); i.magnifico@studenti.unimol.it (I.M.); n.venditti@studenti.unimol.it (N.V.); laura.recchia@unimol.it (L.R.); giulio.petroniopetronio@unimol.it (G.P.P.)

[2] Department of Biomedical and Biotechnological Sciences, University of Catania, 95123 Catania, Italy; kmangano@unict.it (K.M.); ferdinic@unict.it (F.N.)

* Correspondence: roberto.dimarco@unimol.it; Tel.: +39-087-440-4712

Citation: Cutuli, M.A.; Guarnieri, A.; Pietrangelo, L.; Magnifico, I.; Venditti, N.; Recchia, L.; Mangano, K.; Nicoletti, F.; Di Marco, R.; Petronio Petronio, G. Potential Mucosal Irritation Discrimination of Surface Disinfectants Employed against SARS-CoV-2 by *Limacus flavus* Slug Mucosal Irritation Assay. *Biomedicines* 2021, 9, 424. https://doi.org/10.3390/biomedicines9040424

Academic Editor: Martina Perše

Received: 31 March 2021
Accepted: 12 April 2021
Published: 14 April 2021

Publisher's Note: MDPI stays neutral with regard to jurisdictional claims in published maps and institutional affiliations.

Abstract: Preventive measures have proven to be the most effective strategy to counteract the spread of the SARS-CoV-2 virus. Among these, disinfection is strongly suggested by international health organizations' official guidelines. As a consequence, the increase of disinfectants handling is going to expose people to the risk of eyes, mouth, nose, and mucous membranes accidental irritation. To assess mucosal irritation, previous studies employed the snail *Arion lusitanicus* as the mucosal model in Slug Mucosal Irritation (SMI) assay. The obtained results confirmed snails as a suitable experimental model for their anatomical characteristics superimposable to the human mucosae and the different easily observed readouts. Another terrestrial gastropod, *Limacus flavus*, also known as " Yellow slug ", due to its larger size and greater longevity, has already been proposed as an SMI assay alternative model. In this study, for the first time, in addition to the standard parameters recorded in the SMI test, the production of yellow pigment in response to irritants, unique to the snail *L. flavus*, was evaluated. Our results showed that this species would be a promising model for mucosal irritation studies. The study conducted testing among all those chemical solutions most commonly recommended against the SARS-CoV-2 virus.

Keywords: animal models; biomarkers; SARS-CoV-2; slug mucosal irritation assay; *Limacus flavus*; yellow pigment; linear discriminant analysis

1. Introduction

The disinfection practice involves the actuation of physical procedures and/or chemical or biological products to eliminate pathogenic microorganisms. Therefore, it appears the first preventive practice against the development and spreading of infectious diseases caused by various pathogens, among them the SARS-CoV-2 virus [1–3].

The SARS-CoV-2 pandemic was immediately declared a global health emergency by the World Health Organization (WHO). A mortality rate between 2 and 2.5%, 122 million cases worldwide, and more than 2.7 million deaths (WHO update 21 March 2021) confirm the infection's severity [4]. The virus transmission generally occurs from human to human through airborne, Flügge droplets, and contact with contaminated surfaces [5,6].

The SARS-CoV-2, being an enveloped virus, is poorly resistant to acids, detergents, disinfectants, drying, and heat, thus it is quite susceptible to disinfection [7–9]

Among disinfectants, chemicals are the most widely used, due to the wide availability of products, the cost–benefit ratio, the broad spectrum of action, and the possible employing on many surfaces and objects. In this context, several institutions, both national [10] and international [11,12], have drawn lists of products for surfaces and not on people counteract the coronavirus SARS-CoV-2 when used according to label directions.

The most commonly recommended substances are alcohols, chlorine compounds, hydrogen peroxide, phenols, iodine-based substances, and Quaternary Ammonium Compounds (QACs), each with a different mechanism of action:

- Alcohols, cross-linking, coagulation, and clumping, acting mainly as an aqueous emulsion on membrane proteins;
- Chlorine compounds, oxidation of proteins, carbohydrates, and lipids even at low concentrations;
- Hydrogen peroxide, oxidation of cell membrane components through the formation of peroxide radicals;
- Phenols, cross-linking, coagulating, clumping destroying the cell wall, and blocking of enzymatic activity;
- Iodine-based substances interfere at the level of the respiratory chain by blocking the transport of electrons, thus decreasing the oxygen supply of aerobic microorganisms;
- QACs act as ionic compounds that irreversibly bind membrane phospholipids, altering their permeability [13].

The effectiveness of an antiseptic product should be completed within thirty seconds to one minute. Indeed, the longer is the time required for disinfection to be effective, the greater is the risk that the user will not follow the correct application procedure required.

Since the pandemic earliest moments, the effectiveness of preventive and control measures reported by national and international guidelines in significantly reducing nosocomial infections has been demonstrated [2]. Among these, chemical disinfection played a crucial role in increasing the number of people who frequently handled disinfectants and a higher risk of accidental eyes, mouth, nose, and mucous membranes irritation.

Localized at the interface with the external environment, the mucosa plays a fundamental role as a protective barrier towards microorganisms and substances of a different nature. For this reason, it is not uncommon that oral, nasal, gastrointestinal, vaginal, and rectal mucous membranes can be intentionally or accidentally exposed to xenobiotics that irritate, with microlesions formation, compromising the protective barrier function, especially towards pathogenic microorganisms [14–16].

For toxicological evaluation on mucous membranes, vertebrates are mainly used in laboratory routines [17,18]. Although animal experimentation guarantees reliable data, it raises ethical, legal, and scientific high impact implications. For this reason, directives limiting the use of mammals and vertebrates have been issued [19]. Due to these limitations, researchers have begun to use invertebrates for in vivo model studies [20].

To cope with this scenario, a new in vivo assay, "Slug Mucosal Irritation" (SMI), has been proposed, and the snail *Arion lusitanicus* (*A. lusitanicus*) was the first invertebrate used as an alternative to mammals for mucosal toxicology studies [21].

The usage of snails as an alternative model finds its fundamental in mucosal tissue's anatomy and physiology, which possesses numerous characteristics overlapping with its human counterpart. Moreover, the snail body conformation can be easily analyzed outside instead of inside the organism. From a structural standpoint, the mucosal tissue is non-keratinized with a monolayer outer epithelium composed of non-ciliated cells with microvilli and mucus-secreting glandular cells [22].

In addition to its locomotor, lubricating, and antidehydration functions, mucus is an essential component in external insults or damage protection [23]. These unique properties depend primarily on mucins, molecules that can hydrate and swell up to 100 times in a few fractions of a second [24]. Mucus secretory mechanism depends on chemical or mechanical stimulation of smooth muscle located near the mucus-secreting cells that produce apocrine secretion granules. Upon reaching the extracellular environment, the passage of ions and water allows rapid hydration of mucins [25].

Several comparative test studies Based on *A. lusitanicus* mucus production, showing a high predictive degree of chemicals irritant potential, have been developed.

Among these, the more relevant were focused on nasal discomfort [26], the tolerability of certain excipients on the children's skin [27], chemicals that cause severe eye damage and irritation [28,29], and also cosmetic formulations biocompatibility [30–34].

Furthermore, the snail model overcomes some of in vitro cell culture assays limitations; due to the lack of a protective barrier formed by mucus, in vitro cell lines are more sensitive to chemicals than the in vivo mucosa. Another significant advantage is assessing tissue damage, which can be evaluated by microscopic examination and the release of several markers, including proteins and enzymes [35–37].

The high predictivity of the SMI test was also demonstrated in compared studies with the rabbit mucosal model since overlapping results were observed between the two models [38,39].

Limacus flavus (*L. flavus*) (phylum: Mollusca, class: Gastropoda, order: Stylommatophora, family: Limacidae), also known as "Yellow slug" is a pulmonate, hermaphroditic, synanthropic terrestrial gastropod that is autochthonous to the Mediterranean region and widespread worldwide. Colouration varies from yellowish to pinkish-orange, with grey-green spots. It has light blue-grey tubercles, a short crest, and a yellow-white belly (or feet). The mucus produced by the body is yellow, while the belly one is transparent [40,41].

Studies by Cook et al. on the anatomy and histochemistry of the mucus-producing glands of *Limax pseudoflavus*, a snail species closely related to *L. flavus*, demonstrated diffuse yellow granular cells on the dorsal surface. These glands are responsible for the yellow colouration of the dorsal mucus in *L. pseudoflavus* [42]. Moreover, other studies conducted by Chang et al. on the epithelial and mucus-producing cells have shown notable similarities between *L. flavus* and *A. lusitanicus* in mucus production [43].

The use of *L. flavus* in the SMI test has already been reported in some previous studies about polymers biocompatibility [34,44].

Regarding the chemicals' irritation potential, Dhondt et al. in 2006 compared the use of *L. flavus* versus *A. lusitanicus* by the SMI test to evaluate 28 eye-referring chemicals. In this work, the authors justify this snail's choice based on several criteria, such as larger size and greater longevity than *A. lusitanicus*. The results showed more outstanding mucus production for *L. flavus* than *A. lusitanicus* [45].

In this study, for the first time, in addition to the standard parameters recorded in the SMI test (i.e., mucus production, weight change, protein content, and LDH production), the secretion of the characteristic yellow mucus pigment of *L. flavus* was evaluated as a readout of the irritation response.

Thus, the study aimed to evaluate whether the characteristics of *L. flavus* could represent reliable readouts for the SMI test. To this end, the assay was implemented and adapted to *L. flavus* features to discriminate the main biocide substances used for surfaces and objects disinfection. The disinfectant solutions most commonly used and recommended for surfaces use only were investigated in this study, among the plethora of chemicals listed in the official guidelines produced by national and international health organizations to counter the spread of SARS CoV-2 [10–12].

2. Materials and Methods

2.1. Chemicals

Phosphate Buffered Saline (PBS) pH 7.4 (cat no. P3813 Sigma-Aldrich), benzalkonium chloride (BAC) 1% (w/v) (cat no.12060 Sigma-Aldrich, St. Louis, MO, USA), ethanol (EtOH) 70% (v/v) (cat no. 1.07017 Merck-Supelco, Darmstadt, Germany), isopropanol 95% (v/v) (cat. no. 34863 Sigma-Aldrich, St. Louis, MO, USA), isopropanol 70% (v/v) (cat. no. 34863 Sigma-Aldrich, St. Louis, MO, USA), sodium hypochlorite 0.1% (v/v) (Honeywell Fluka^{TM} cat. no. 71696 Thermo Fisher Scientific, Waltham, MA, USA), hydrogen peroxide 3% (v/v) (cat. no H1009 Sigma-Aldrich, St. Louis, MO, USA), chlorhexidine 1% (w/v) (cat. no 282227 Sigma-Aldrich, St. Louis, MO, USA), and iodopovidone 10% (w/v) (cat. no PVP1 Sigma-Aldrich, St. Louis, MO, USA). All substances were diluted to the final concentration in sterile deionized water.

2.2. Collection, Housing, and Identification of Limacus flavus Specimens

Snails were collected in the proximity of Trivento municipality (Campobasso, Molise Italy) and then housed at a temperature between 16 and 18 °C with a photoperiod of 12-12. Each specimen was carefully inspected for macroscopic lesions or damage to the tubercles to exclude ineligible samples.

A preliminary snail identification in the field was carried out by macroscopic evaluation to confirm the body color (green, brown and greyish), the texture of the mantle (a network of yellow spots of ovoid shape), and the tentacles (bluish color). Additionally, both specimens size and position of the respiratory pore (pneumostome) located behind the mantle's midline were evaluated [46].

2.3. SMI Assay

The assay was conducted following Adriaens et al. [28] with some modifications designed to optimize the unique characteristics of the *L. flavus* species. Snails with a weight between 3 and 5 g were selected. Three days before testing, they were housed in a ventilated plastic terrarium lined with PBS (pH 7.4) soaked paper towels and fitted with a wire mesh to drain mucus. Snails before testing were weighed and placed each in a Petri dish. The degree of irritability of the tested biocides was assessed by exposing the snails to 100 µL of each test substance (Figure 1a).

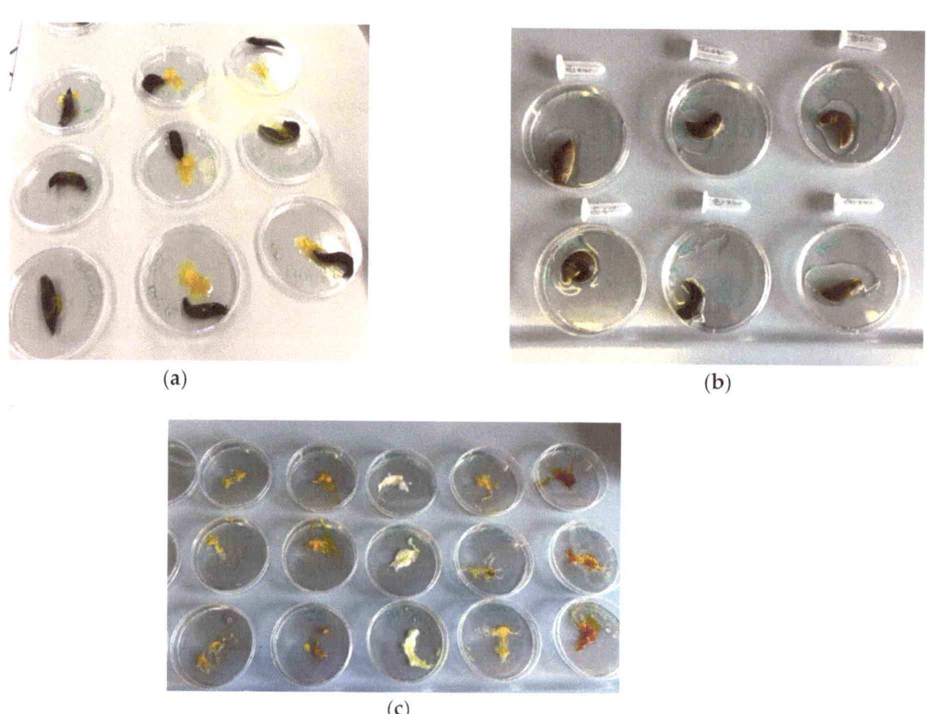

(a)

(b)

(c)

Figure 1. (**a**) Snails during exposure with 100 µL of the test substances; (**b**) snails in contact with PBS after treatment; and (**c**) Petri dishes containing the mucus produced by snails during treatment.

Unlike the SMI protocol proposed by Adriens et al. [28], test substances were applied to the back and not the snails' underside. Additionally, as already pointed out by Dhondt et al. [45], given the high mucus production in response to chemicals stimulation, the contact period with the tested substance was reduced from 60 to 15 min. After 15 min, snails were weighed to assess weight loss and transferred to another Petri dish in contact with 1 mL of PBS for 45 min (Figure 1b). Subsequently, samples were collected and frozen at $-80\ ^\circ$C for further analysis.

Treatment-induced weight changes were expressed as percentage weight/weight ($\% \times w/w$) according to the formula:

Bodyweight variation = snail weight after treatment/snail weight before treatment $\times$ 100.

Petri dishes in which snails were placed in contact with the test substances were weighed before and after the test to determine the mucus produced by the irritant stimulus (Figure 1c).

The amount of mucus produced after the 15-min contact period ($\% \ w/w$) was calculated as follows:

Mucus production = mucus produced after treatment/snail weight before treatment $\times$ 100.

Three snails were used for each substance tested, and the experiment was repeated three times independently (three biological and technical replicates).

2.4. Mucus Analysis

The mucus produced from each snail after treatment and collected in PBS, henceforth referred to as "sample", was evaluated for protein quantification, lactate dehydrogenase (LDH) activity, and UV-visible absorbance (λ_{max} 420 nm).

2.4.1. Protein Quantification

Quantification was performed using the Pierce BCA Protein Assay Kit (cat. No. 23227 Thermo-Scientific, Waltham, MA, USA) following the manufacturer's instructions. Serial dilutions of BSA from 2000 to 20 µg/mL were performed to construct the calibration curve. Of each sample 25 µL were added to 200 µL of the BCA working solution in a microplate. After an incubation period at 37 $^\circ$C for 30 min, measurements were made with VICTOR3 model1420 Multilabel Counter (PerkinElmer, Waltham, MA, USA) at a wavelength of 562 nm. Results were expressed as µg/mL and normalized for the initial weight of each snail.

2.4.2. LDH

LDH enzyme activity was measured according to Adriaens et al. [35] by the LDH activity assay kit (cat. No. MAK066 Sigma-Aldrich, St. Louis, MO, USA) according to the manufacturer's instruction. LDH activity is reported as nmol/min/mL – milliunits/mL. One LDH activity unit was defined as the amount of enzyme that catalyzed the conversion of lactate to pyruvate to generate 1.0 mole of NADH per minute at 37 $^\circ$C.

2.4.3. UV–Visible Spectra

Due to the yellow color of the mucus produced by *L. flavus*, test samples were diluted 1:5 (V/V) in deionized water and subjected first to spectrophotometric reading from 400 to 600 nm with 1 nm intervals (Lambda25 UV/Vis PerkinElmer); generating UV–VIS spectra for each substance tested.

Curves analysis displayed a maximum absorbance peak for all substances tested at λ_{max} 420 nm, so subsequently, all samples were tested for λ_{max} 420 nm.

2.5. Statistical Analysis

Data were expressed as mean $\pm$ standard deviation (S.D.) of three biological replicates from three independent experiments. Prism Graph Pad 6 software was used for the one-way ANOVA test, followed by multiple Bonferroni correction tests and linear regression analysis.

Linear Discriminant Analysis

As previously reported by Adriens et al. [17], linear discriminant analysis (LDA) was used as a classification prediction model.

In this study, this PM was applied for discrimination according to mucosal irritation potential of nine surface and object disinfectants prescribed to counteract the spread of SARS-CoV-2. Nine observations (three technical and three biological replicates) were considered for each variable (weight variations, mucus production, protein quantification, LDH activity, and λ_{max} 420 nm to set up the dataset. Two separate LDA models were then developed, one based on the linear combination of the first four variables without λ_{max} 420 nm (LDA w/o λ_{max} 420 nm), the other included the λ_{max} 420 nm spectrophotometric variable (LDA λ_{max} 420 nm). Comparing each model's confusion matrices (LDA w/o λ_{max} 420 nm and LDA λ_{max} 420 nm) allowed assessment of agreement between observed and predicted categories.

Finally, PBS (control) squared distances (D2) of the LDA λ_{max} 420 nm analysis were used for classification of the tested substances by a numerical score (0–7) and a grade (none to exceptionally high). The significance level was set at $p = 0.05$. XLSTAT software v.2021.1 (Addinsoft Paris, France) was used for LDA analysis.

3. Results

3.1. SMI Assay

3.1.1. Bodyweight Variation

The mean change in weight of snails exposed to the tested disinfectants and controls (PBS pH 7.4, sodium hypochlorite 0.1%, hydrogen peroxide 3%, chlorhexidine 1%, povidone iodine 10%(w/v), EtOH 70% (v/v), BAC 1% (w/v), isopropanol 70%, and isopropanol 95% ranged from 100% to 78.78% (Figure 2).

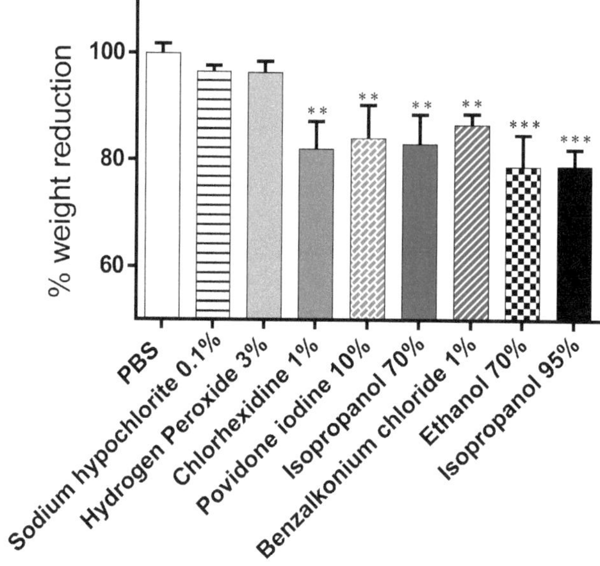

Figure 2. Change in body weight induced by the substances tested. Data are presented as mean ± S.D. and expressed as a percentage of the initial body weight. *** $p < 0.001$, ** $p < 0.01$, significance from control (PBS).

No change in body weight was observed in snails treated with PBS buffer used as control (100 ± 1.84). In the groups exposed to disinfectants, a weight change was observed caused by the biocidal substances' irritant action on the snail mucosa (lower percentage values indicate more significant weight loss), among all chemicals tested, EtOH 70% (*v*/*v*) and isopropanol 95% (*v*/*v*) significantly induced higher snail body weight variation by 78.78% ± 5.93% and 78.95% ± 3.10%, respectively. Hydrogen peroxide and sodium hypochlorite had the most negligible influence on body weight change, 96.37% ± 2.05% and 96.53% ± 1.15% (Table 1).

Table 1. SMI test results for all disinfectants tested. Data are expressed as mean ± standard deviation (S.D.), each value with its unit of measure; [a,b] protein quantification and LDH were normalized to the weight of the snails before treatment.

Chemicals	Bodyweight Variation (*w*/*w*%)	Mucus Production (*w*/*w*%)	Protein Quantification (μg/mL)/g [a]	LDH (iu/l)/g [b]	λ_{max} 420 (nm)
PBS	100.00 ± 1.84	0.30 ± 0.41	20.07 ± 8.68	-	0.028 ± 0.006
Sodium Hypochlorite 0.1%	96.53 ± 1.15	3.73 ± 1.27	28.72 ± 13.36	0.57 ± 0.15	0.035 ± 0.005
Hydrogen Peroxide 3%	96.37 ± 2.05	5.03 ± 1.68	56.47 ± 15.21	0.47 ± 0.12	0.048 ± 0.006
Chlorhexidine 1%	82.04 ± 5.19	10.23 ± 2.28	104.90 ± 29.35	2.10 ± 0.73	0.103 ± 0.016
Povidone Iodine 10%	84.09 ± 6.22	14.29 ± 2.07	139.11 ± 51.58	2.28 ± 0.16	0.117 ± 0.016
EtOH 70%	78.78 ± 5.93	14.63 ± 2.85	68.64 ± 28.85	3.62 ± 0.75	0.070 ± 0.012
BAC 1%	86.68 ± 1.98	12.36 ± 3.32	230.02 ± 103.95	3.64 ± 0.90	0.301 ± 0.015
Isopropanol 70%	83.04 ± 5.54	11.58 ± 3.98	275.53 ± 76.64	3.51 ± 0.51	0.330 ± 0.020
Isopropanol 95%	78.95 ± 3.10	19.96 ± 4.08	396.87 ± 73.77	5.73 ± 1.13	0.394 ± 0.019

3.1.2. Mucus Production

Data on the amount of mucus produced given as a percentage of the snails' initial weight are shown in Figure 3. Treatment with PBS poorly stimulated mucus production 0.3% ± 0.41%. All substances tested, except hydrogen peroxide (5.03% ± 1.68%) and sodium hypochlorite (3.73% ± 1.27%), induced a significant increase in mucus secretion: EtOH 70% (14.63% ± 2.85%); isopropanol 70% (11.58% ± 3.98%); chlorhexidine 1% (10.23% ± 2.28%); and iodopovidone (14.28% ± 2.07%). Isopropanol 95% was the substance that more stimulated mucus production 19.96% ± 4.08% (Table 1).

3.1.3. Protein Quantification

Data from protein quantification in samples taken after a 15 min exposure period with the different treatments and subsequent contact with PBS for 1 h were normalized to the snails' initial weight (Figure 4).

Moderate protein release into the mucus (20.07 ± 8.68 μg/mL) was recorded in the PBS control group. Treatment with 95% isopropanol, 1% BAC, and 70% isopropanol induced significant protein release compared with the control. The lowest protein concentrations among the biocidal substances tested were determined in the samples from the groups exposed to hydrogen peroxide and hypochlorite (Table 1).

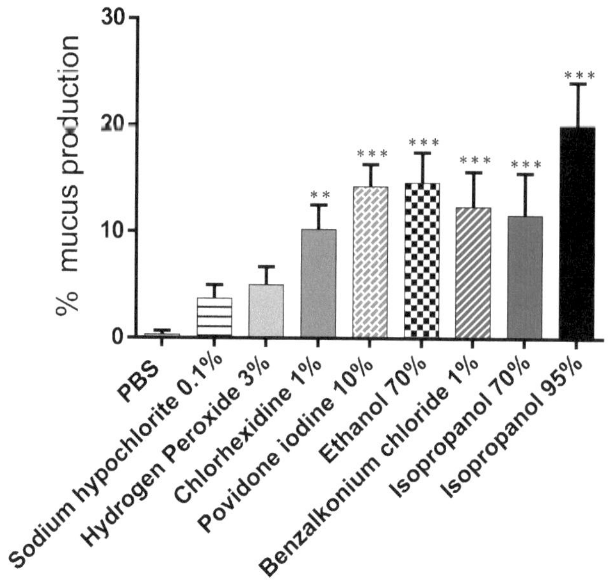

Figure 3. Amount of mucus produced by snails after a 15-min contact period with test substances (percentage relative to initial snail weight). Data are presented as mean ± S.D. *** $p < 0.001$, ** $p < 0.01$ significance from negative control (PBS).

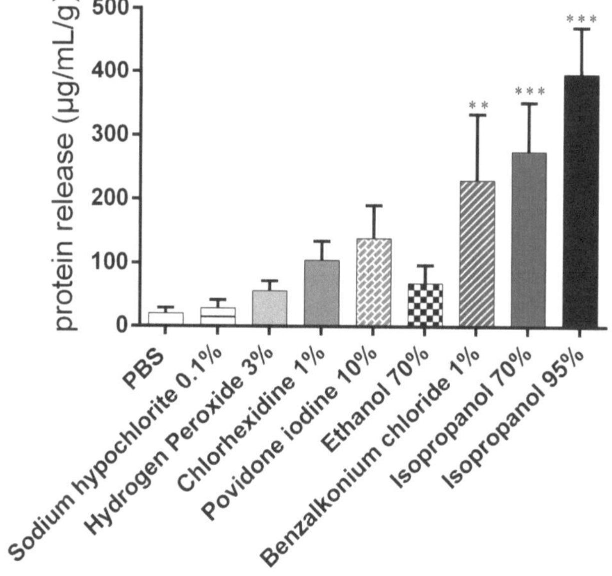

Figure 4. Proteins released from snail mucosa after a 15 min exposure period with the different treatments and subsequent contact with PBS for 45 minutes. Data are presented as mean values ± S.D. and expressed in µg/mL. *** $p < 0.001$, ** $p < 0.01$ significance from negative control (PBS).

3.1.4. LDH

The LDH activity in mucus correlates with cellular damage caused by snail exposure to test substances.

For PBS-treated snails, no LDH activity was recorded (Figure 5). On the other hand, 95% isopropanol exposure showed significant activity (5.73 ± 1.13 IU/L.G./g), hydrogen peroxide, and sodium hypochlorite induced less LDH activity 0.47 ± 0.12 and 0.57 ± 0.15IU/L.G./g, respectively.

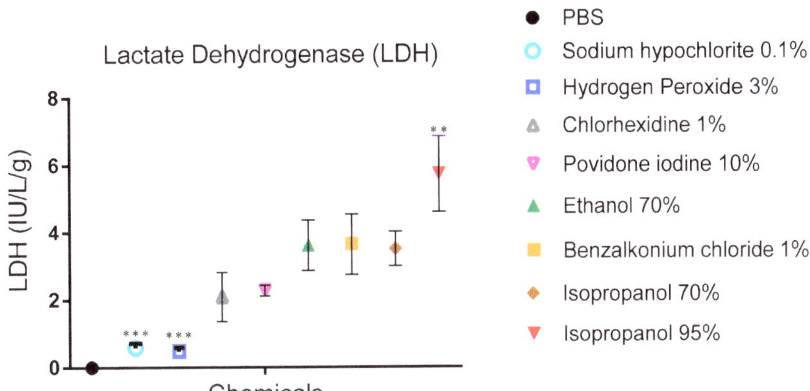

Figure 5. LDH activity in PBS samples after a period of contact with the different biocides for 15 min. Data are presented as mean values and expressed in units/mL PBS per gram body weight. Two-way ANOVA comparing tested substances versus BAC 1% (mean LDH 3.6 IU/LG/g) as already reported by Adriaens et al. [17]. *** $p < 0.001$, ** $p < 0.01$ significance from control (BAC 1%).

3.1.5. UV–VIS Spectra

Table 1 shows the λ_{max} 420 nm values. Lower absorbance values were found for the PBS (0.028 0.006), sodium hypochlorite 0.1% (0.035 ± 0.005), and hydrogen peroxide 3% (0.048±0.006) groups while the highest values were found for the isopropanol 95% (0.394 ± 0.019) isopropanol 70% (0.330 ± 0.020) and BAC 1% (0.0301 ± 0.015) groups.

Additionally, a linear correlation between λ_{max} 420 nm and mucus protein was observed (Figure 6).

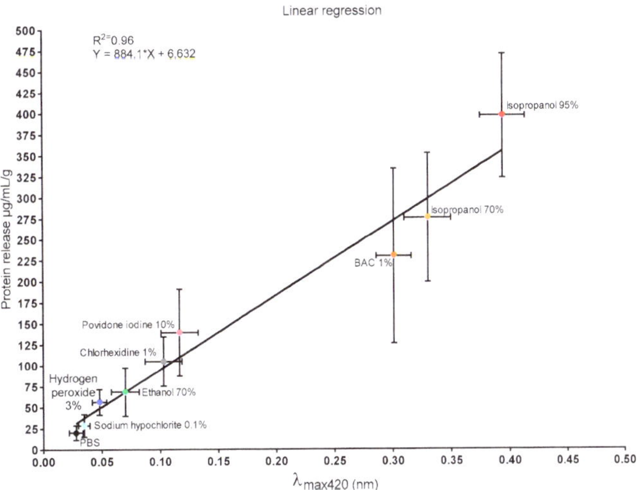

Figure 6. Linear regression analysis between λ_{max} 420 nm values and mucus protein concentration ($R^2 = 0.96$; Y = 884.1 * X + 6.632). Data are expressed as mean ± standard deviation (vertical bars protein concentration, horizontal bars λ_{max} 420 nm).

3.2. Linear Discriminant Analysis

The centroid plot of LDA without λ_{max} 420 nm is shown in Figure 7. On the axes were the variables linear combinations with the highest discriminated percentage extracted from the analysis. On the *x*-axis was factor one (F1) with 89.44% discrimination and on the *y*-axis was factor two (F2) with 8.72%. The sum of the two factors (F1 and F2) reached a total bias of 98.16%.

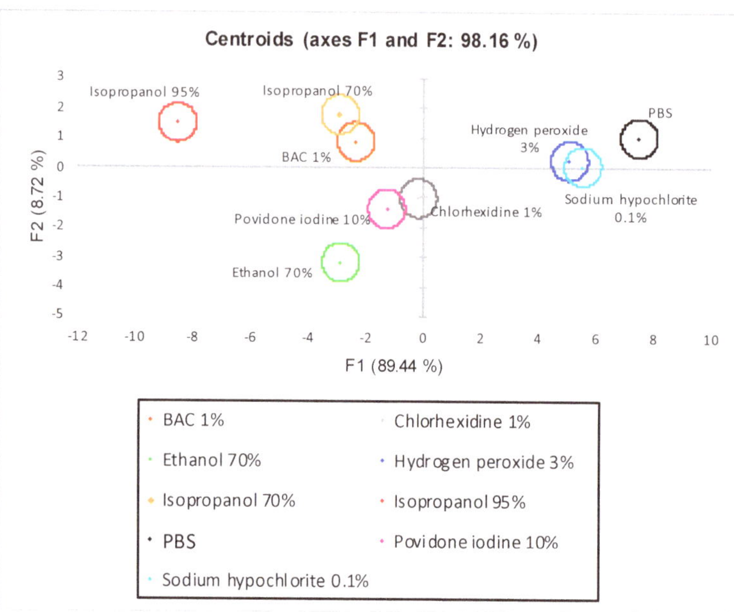

Figure 7. Centroids graph for linear discriminant analysis (LDA) w/o λ_{max} 420 nm. F1 (*x*-axis) and F2 (*y*-axis) are the factor axes extracted from the original variables. For each factor, the percentage of discrimination, both individual and cumulative, is reported in brackets.

Of the nine disinfectants tested, three were discriminated entirely (PBS, EtOH 70%, and isopropanol 95%), while the remaining were grouped into three distinct groups (hydrogen peroxide 3%-sodium hypochlorite 0.1, chlorhexidine 1%, povidone-iodine 10%, and BAC 1%-isopropanol 70%).

The centroid plot of the LDA with λ_{max} 420 nm is shown in Figure 8. The axes are shown the linear combinations of the variables under investigation with the highest discriminated percentage extracted from the analysis. On the *x*-axis is factor one (F1) with 95.62% discrimination and on the *y*-axis is factor two (F2) with 4.09%. The sum of the two factors (F1 and F2) achieved total percentage discrimination of 99.71%. All nine disinfectants tested were completely discriminated.

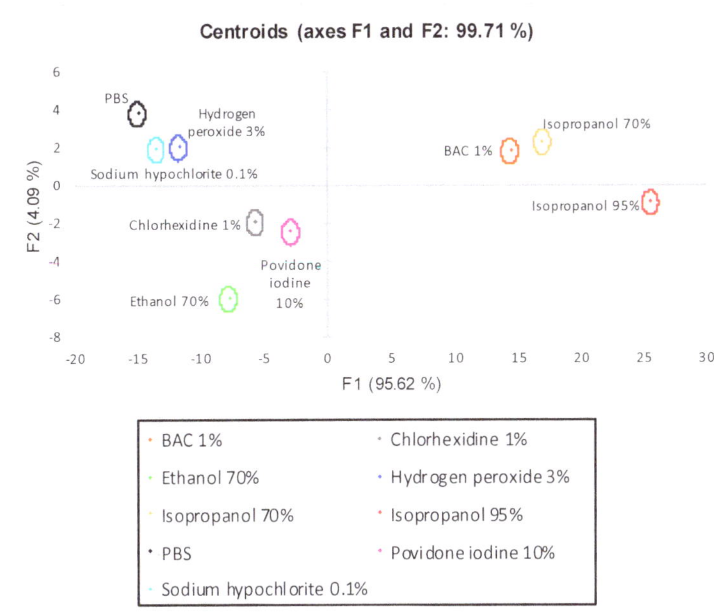

Figure 8. Centroids graph for LDA λ_{max} 420 nm. F1 (*x*-axis) and F2 (*y*-axis) are the factor axes extracted from the original variables. For each factor, the percentage of discrimination, both individual and cumulative, is reported in brackets.

Table 2 shows the percentage of correct predictions broken down for each class (disinfectants tested) for the two models analyzed (λ_{max} 420 and w/o λ_{max} 420) together with the percentage differences.

Table 2. Training sample percentage of correct predictions for the LDA analysis performed. LDA λ_{max} 420: analysis with λ_{max} 420 variable; LDA w/o λ_{max} 420: analysis without λ_{max} 420; % correct LDA λ_{max} 420—% correct LDA w/o λ_{max} 420: the difference between the two analysis.

Chemicals	% Correct LDA λ_{max} 420	% Correct LDA w/o λ_{max} 420	Δ% Correct (% Correct LDA λ_{max} 420– % Correct LDA w/o λ_{max} 420)
PBS	100	100	–
Sodium Hypochlorite 0.1%	88.89	77.78	11.11
Hydrogen Peroxide 3%	77.77	66.67	11.11
Chlorhexidine 1%	88.89	88.89	–
Povidone Iodine 10%	100	66.67	33.33
EtOH 70%	100	88.89	11.11
BAC 1%	77.78	22.22	55.55
Isopropanol 70%	100	77.78	22.22
Isopropanol 95%	100	100	–
Total	**92.59%**	**76.54%**	**16.05**

The introduction of the. Variable λ_{max} 420 resulted in a percentage increase in total correct predictions of 16.05%. The classes that showed the most significant increase in prediction accuracy were BAC 1% (55.55% increase), povidone-iodine 10% (33.33%), and isopropanol 70% (22.22%). Sodium hypochlorite 0.1%, hydrogen peroxide 3%, and EtOH 70% had a modest increase (11.11%), while PBS, chlorhexidine 1%, and isopropanol 95% had no increase.

4. Discussion

The main SARS-CoV-2 routes of transmission are interhuman airborne and contact with contaminated surfaces. Thus, healthcare and community prevention measures include regular use of surface disinfectants and hand sanitizers [47–49].

The persistence of coronavirus on surfaces has caused the intensive use of chemical disinfectants by healthcare professions and the general public [50,51].

According to Van Doremalen et al. [52], SARS-CoV-2 could remain viable in aerosols for three hours with a low infectious load reduction. The viable virus was shown to be more stable on plastic and stainless steel and could be detected up to 72 h after deposition, although a substantial decrease in viral load was observed. A rough estimate of the median half-life was 5.6 h on steel and 6.8 h on plastic. No viable virus was detected after 4 h on copper or after 24 h on cardboard.

Due to the vast resonance of the pandemic and the high number of confirmed cases worldwide [4], people easily take inappropriate actions caused by fear. Numerous studies have reported improper use of chemical surface disinfectants prescribed against the spread of SARS-CoV-2 by not properly following the manufacturers' instructions [53], with an increased poisoning risk rate during the year 2020 [54].

Besides, a prepandemic study conducted by Casey et al. [55] had already established a correlation between routine use of chemical disinfectants and mucosal irritant effects together with respiratory health. This study had concluded that the risks of mucosal irritation and asthma are higher in healthcare workers. Thus, repeated exposure to disinfectants had to be evaluated as an occupational disease risk factor when drafting healthcare disinfection protocols [56].

In this scenario, the study aimed at the discrimination according to mucosal irritation potential of nine disinfectants used for objects and surfaces against the spread of SARS-CoV-2 [10–12].

To this end, the land snail *L. flavus* was evaluated as a test organism in place of *A. lusitanicus*. In addition to the characteristics common to *A. lusitanicus* (i.e., mucosal epithelium, easily observable irritant effect, vulnerability to mechanical, or chemical damage), this species possesses unique features that promote its adoption as an "ideal" candidate for the evaluation of irritant potential [43–45].

As in the *A. lusitanicus* model, mucus production, weight changes, protein content, and LDH activity were evaluated as the main readouts [57].

For the first time in this study, mucus staining was used as a readout to assess mucosal irritant potential. Although the pigment's chemical nature is unknown, this phenomenon has already been described in other land snail species. Some snails of the genus *Ariolimax* have bright yellow colouration [58]. *Arion fasciatus* (Nilsson) also possesses yellow-orange mantle pigments lost when reared on carrots, lettuce, or paper [59]. Moreover, besides the studies conducted by Cook et al. [42] on *L. pseudoflavus*, yellow pigmentations were evaluated by Seki et al. for the taxonomic distinction of two sibling snails species *Bradybaena pellucida* and *B. similaris* [60].

UV–Vis spectra of mucus samples retrieved from the disinfectant-treated snail (data not shown) revealed a maximum absorbance peak at 420 nm (λ_{max}). These absorbance values were correlated with the respective data obtained from mucus protein quantification (Figure 6). Linear correlation analysis demonstrated a direct proportionality between the two variables ($R^2 = 0.96$). These findings support the hypothesis that, similar to proteins [61], the pigment is released under stressful conditions.

Evaluation of mucosal irritant potential based on the comparison of individual readouts provided a difficult results interpretation (Table 1). However, a correlation between yellow pigmentation and protein release was demonstrated (Figure 6, the same was not true for % weight variations (Figure 2), % mucus production (Figure 3), and LDH activity (Figure 5).

For this reason, the mucosal irritation potential discrimination of disinfectants (classes) under investigation was assessed by a classification prediction model (LDA) that accounted for both the individual readouts (observations) and a statistically significant combination of them $p < 0.05$.

Two different LDA analyses were performed; in the first one (LDA w/o λ_{max} 420), four observations common to both snail species (% weight variations, % mucus production, LDH, and protein quantification) were taken into account. This model's total discrimination rate, defined as the sum of observations linear combinations (axes F1 and F2), was 98.16%. Nevertheless, only three of the nine observations were fully discriminated (Figure 7).

In the second model (LDA λ_{max} 420), yellow mucus pigmentation (λ_{max} 420) unique to *L. flavus* was included. The introduction of absorbance data into the model has increased both its discriminatory power (axes F1 and F2 99.71%) and predictive ability. Indeed, all nine observations were discriminated (Figure 8).

The predictive ability of both models has been reported in Table 2. The correct predictions (% correct) are summarized in percentages and divided both for each class (disinfectants tested) and total prediction. Comparison of the processed models ($\Delta\%$ correct) showed an increase in predictive ability for both total classes (16.05%) and six of the nine chemicals tested.

These findings demonstrate the contribution of the data obtained from the spectrophotometric analysis of *L. flavus* pigment in class discrimination according to their mucosal irritation potential.

Due to the better discrimination and predictive ability of the LDA λ_{max} 420 model, its D^2 data (Mahalanobis distance) from PBS (D^2 PBS LDA λ_{max} 420) were used to classify the tested disinfectants according to mucosal irritation potential (Table 3).

Table 3. Disinfectant discrimination according to the mucosal irritation potential.

Chemicals	D^2 PBS LDA λ_{max} 420	Range	Score	Grade
PBS	5	0 to 5	0	none
Sodium Hypochlorite 0.1%	11	11 to 20	1	low
Hydrogen Peroxide 3%	20			
Chlorhexidine 1%	132	20 to 132	2	mild
Povidone Iodine 10%	155	132 to 155	3	average
EtOH 70%	196	155 to 196	4	average-high
BAC 1%	878	196 to 878	5	high
Isopropanol 70%	1031	878 to 1031	6	very high
Isopropanol 95%	1676	1031 to 1676	7	extremely high

This parameter was chosen for its statistical meaning. When the D^2 value between an observation and the group's centre (calculated as mean) was the smallest, that observation could be classified into that group. The LDA provided D^2 values for each group, called the linear discriminant function. For each observation, the group with the smallest D^2 had the largest linear discriminant function, and the observation was classified into that group [62].

Since PBS was used as a control (no irritation potential), its Mahalanobis distance values were selected for classification.

By the D^2 range values, a classification among the tested disinfectants was set up (Table 3) according to a numerical score (from 0 to 7) and an effects grade (from none to exceptionally high).

According to Dhont et al., *L. flavus* could be used to assess the irritant potential of chemicals without affecting the *A. lusitanicus* SMI concordance and specificity [45].

Nevertheless, in the present study, several chemical disinfectants at different concentrations prescribed by international disinfection protocols have been tested [10–12].

For this reason, the *L. flavus* SMI protocol was optimized to introduce a new readout (λ_{max} 420).

Among the substances tested, only EtOH, chlorhexidine, and isopropanol have been previously analyzed by SMI studies (except PBS and BAC employed as control substances) [17,28,45,57,63]. For these substances, given the different concentrations used, a direct comparison of mucosa irritant potentials was not possible.

Nevertheless, although the concentration of isopropanol tested in previous studies was much lower than the current one (10% vs. 70 and 95%), it was already tagged as an irritant confirming snail mucosa high toxicity [28].

Regarding BAC 1%, both the classification proposed in this work (*L. flavus* SMI) and that proposed by previous works (*A. lusitanicus* SMI) were in agreement (very irritating to mucous membranes) [17].

5. Conclusions

In conclusion, our results confirmed that this native Mediterranean species might be a viable alternative in the SMI assay [43–45]. Furthermore, for the first time in this study, *L. flavus* mucus "yellow pigment" was evaluated as a new readout. Although further studies on the chemical composition and release patterns of this pigment are needed to elucidate the mechanisms of response to chemical insults, it demonstrated a strong correlation with protein secretion under stress conditions and a remarkable capacity in discriminating the mucosal irritant potential of nine surface disinfectants used to counteract the spread of SARS-CoV-2.

Author Contributions: Conceptualization, R.D.M., G.P.P. and M.A.C.; methodology, M.A.C., G.P.P., L.P., N.V., L.R. and F.N.; validation, R.D.M. and F.N.; formal analysis, L.R. and K.M.; investigation, G.P.P., M.A.C., L.P. and A.G.; resources, R.D.M.; data curation, I.M. and L.R.; writing—original draft preparation, M.A.C., G.P.P. and A.G.; writing—review and editing, R.D.M., G.P.P. and M.A.C.; supervision, K.M., G.P.P. and F.N.; funding acquisition, R.D.M. All authors have read and agreed to the published version of the manuscript.

Funding: This research received no external funding.

Institutional Review Board Statement: Ethical review and approval were waived for this study, due to the use of invertebrate animals.

Informed Consent Statement: Not applicable.

Data Availability Statement: Not applicable.

Conflicts of Interest: The authors declare no conflict of interest.

References

1. Drexler, M. *What You Need to Know about Infectious Disease*; National Academies Press (US): Washington, DC, USA, 2014.
2. Lai, X.; Wang, X.; Yang, Q.; Xu, X.; Tang, Y.; Liu, C.; Tan, L.; Lai, R.; Wang, H.; Zhang, X. Will healthcare workers improve infection prevention and control behaviors as COVID-19 risk emerges and increases, in China? *Antimicrob. Resist. Infect. Control* **2020**, *9*, 1–9. [CrossRef] [PubMed]
3. Kim, S.W.; Kang, S.I.; Shin, D.H.; Oh, S.Y.; Lee, C.W.; Yang, Y.; Son, Y.K.; Yang, H.-S.; Lee, B.-H.; An, H.-J. Potential of Cell-Free Supernatant from *Lactobacillus plantarum* NIBR97, Including Novel Bacteriocins, as a Natural Alternative to Chemical Disinfectants. *Pharmaceuticals* **2020**, *13*, 266. [CrossRef] [PubMed]
4. WHO. Coronavirus (COVID-19) Dashboard. 2021. Available online: https://covid19.who.int/ (accessed on 22 March 2021).
5. Malik, Y.A. Properties of coronavirus and SARS-CoV-2. *Malays. J. Pathol.* **2020**, *42*, 3–11. [PubMed]
6. Petronio, G.P.; di Marco, R.; Costagliola, C. Do Ocular fluids represent a transmission route of SARS-CoV-2 infection? *Front. Med.* **2020**, *7*, 1–4.

7. Carraturo, F.; del Giudice, C.; Morelli, M.; Cerullo, V.; Libralato, G.; Galdiero, E.; Guida, M. Persistence of SARS-CoV-2 in the environment and COVID-19 transmission risk from environmental matrices and surfaces. *Environ. Pollut.* **2020**, 115010. [CrossRef]
8. García-Ávila, F.; Valdiviezo-Gonzales, L.; Cadme-Galabay, M.; Gutiérrez-Ortega, H.; Altamirano-Cárdenas, L.; Zhindón-Arévalo, C.; del Pino, L.F. Considerations on water quality and the use of chlorine in times of SARS-CoV-2 (COVID-19) pandemic in the community. *Case Stud. Chem. Environ. Eng.* **2020**, 100049. [CrossRef]
9. Hussain, S.; Cheema, M.J.M.; Motahhir, S.; Iqbal, M.M.; Arshad, A.; Waqas, M.S.; Usman Khalid, M.; Malik, S. Proposed Design of Walk-Through Gate (WTG): Mitigating the Effect of COVID-19. *Appl. Syst. Innov.* **2020**, *3*, 41. [CrossRef]
10. Gruppo di Lavoro ISS Prevenzione e Controllo delle Infezioni. Indicazioni ad Interim per la Sanificazione Degli Ambienti Interni nel Contesto Sanitario e Assistenziale per Prevenire la Trasmissione di SARS-CoV 2. 2020. Available online: https://www.iss.it/documents/20126/0/Rapporto+ISS+COVID-19+n.+20_2020+REV+2.pdf/afbbb63a-1f0f-9a11-6229-a130 09cacd96?t=1594641502569 (accessed on 22 March 2021).
11. WHO. Infection Prevention and Control Guidance for Long-Term Care Facilities in the Context of COVID-19: Interim Guidance. 2021. Available online: https://apps.who.int/iris/bitstream/handle/10665/338481/WHO-2019-nCoV-IPC_long_term_care-2021.1-eng.pdf?sequence=1&isAllowed=y (accessed on 22 March 2021).
12. *Disinfectants for Use Against SARS-CoV-2 (COVID-19)*; U.S. Environmental Protection Agency: Washington, DC, USA, 2020.
13. Maris, P. Modes of action of disinfectants. *Rev. Sci. Et Tech. (Int. Off. Epizoot.)* **1995**, *14*, 47–55. [CrossRef] [PubMed]
14. Chesné, J.; Cardoso, V.; Veiga-Fernandes, H. Neuro-immune regulation of mucosal physiology. *Mucosal Immunol.* **2019**, *12*, 10–20. [CrossRef]
15. Marini, A.; Hengge, U. Important viral and bacterial infections of the skin and mucous membrane. *Der Internist* **2009**, *50*, 160–170. [CrossRef]
16. Magnifico, I.; Petronio, G.; Venditti, N.; Cutuli, M.A.; Pietrangelo, L.; Vergalito, F.; Mangano, K.; Zella, D.; di Marco, R. Atopic dermatitis as a multifactorial skin disorder. Can the analysis of pathophysiological targets represent the winning therapeutic strategy? *Pharmaceuticals* **2020**, *13*, 411. [CrossRef] [PubMed]
17. Adriaens, E.; Remon, J.P. Evaluation of an alternative mucosal irritation test using slugs. *Toxicol. Appl. Pharmacol.* **2002**, *182*, 169–175. [CrossRef] [PubMed]
18. Mangano, K.; Vergalito, F.; Mammana, S.; Mariano, A.; de Pasquale, R.; Meloscia, A.; Bartollino, S.; Guerra, G.; Nicoletti, F.; di Marco, R. Evaluation of hyaluronic acid-P40 conjugated cream in a mouse model of dermatitis induced by oxazolone. *Exp. Ther. Med.* **2017**, *14*, 2439–2444. [CrossRef] [PubMed]
19. Olsson, I.A.S.; Silva, S.P.d.; Townend, D.; Sandøe, P. Protecting animals and enabling research in the European Union: An overview of development and implementation of directive 2010/63/EU. *ILAR J.* **2017**, *57*, 347–357. [CrossRef]
20. Cutuli, M.A.; Petronio Petronio, G.; Vergalito, F.; Magnifico, I.; Pietrangelo, L.; Venditti, N.; di Marco, R. Galleria mellonella as a consolidated in vivo model hosts: New developments in antibacterial strategies and novel drug testing. *Virulence* **2019**, *10*, 527–541. [CrossRef]
21. Van Houte, D.E. *De Ontwikkeling van een Alternatieve Mucosale Irritatietest*; Optimization and Validation of An Alternative Mucosal Irritation Test. 2000; Pharmaceutical Care Ghent University: Ghent, Belgium, 2005.
22. Jensen, K.; Engelke, S.; Simpson, S.J.; Mayntz, D.; Hunt, J. Balancing of specific nutrients and subsequent growth and body composition in the slug *Arion lusitanicus*. *Physiol. Behav.* **2013**, *122*, 84–92. [CrossRef]
23. South, A. *Terrestrial slugs: Biology, Ecology and Control*; Chapman & Hall: London, UK, 1992.
24. Verdugo, P.; Deyrup-Olsen, I.; Aitken, M.; Villalon, M.; Johnson, D. Molecular mechanism of mucin secretion: I. The role of intragranular charge shielding. *J. Dent. Res.* **1987**, *66*, 506–508. [CrossRef]
25. Deyrup-Olsen, I.; Martin, A.W. Surface exudation in terrestrial slugs. *Comp. Biochem. Physiol. Part C Comp. Pharmacol.* **1982**, *72*, 45–51. [CrossRef]
26. Lenoir, J.; Bachert, C.; Remon, J.-P.; Adriaens, E. The slug mucosal irritation (SMI) assay: A tool for the evaluation of nasal discomfort. *Toxicol. Vitr.* **2013**, *27*, 1954–1961. [CrossRef] [PubMed]
27. Kendall, R.; Lenoir, J.; Gerrard, S.; Scheuerle, R.L.; Slater, N.K.; Tuleu, C. Using the Slug Mucosal Irritation assay to investigate the tolerability of tablet excipients on human skin in the context of the use of a nipple shield delivery system. *Pharm. Res.* **2017**, *34*, 687–695. [CrossRef] [PubMed]
28. Adriaens, E.; Guest, R.; Willoughby Sr, J.; Fochtman, P.; Kandarova, H.; Verstraelen, S.; van Rompay, A. CON4EI: Slug Mucosal Irritation (SMI) test method for hazard identification and labelling of serious eye damaging and eye irritating chemicals. *Toxicol. Vitr.* **2018**, *49*, 77–89. [CrossRef]
29. Ceulemans, J.; Vermeire, A.; Adriaens, E.; Remon, J.P.; Ludwig, A. Evaluation of a mucoadhesive tablet for ocular use. *J. Control. Release* **2001**, *77*, 333–344. [CrossRef]
30. Lenoir, J.; Adriaens, E.; Remon, J.P. New aspects of the Slug Mucosal Irritation assay: Predicting nasal stinging, itching and burning sensations. *J. Appl. Toxicol.* **2011**, *31*, 640–648. [CrossRef] [PubMed]
31. Petit, J.-Y.; Doré, V.; Marignac, G.; Perrot, S. Assessment of ocular discomfort caused by 5 shampoos using the Slug Mucosal Irritation test. *Toxicol. Vitr.* **2017**, *40*, 243–247. [CrossRef]
32. Adriaens, E.; Ameye, D.; Dhondt, M.; Foreman, P.; Remon, J.P. Evaluation of the mucosal irritation potency of co-spray dried Amioca®/poly (acrylic acid) and Amioca®/Carbopol® 974P mixtures. *J. Control. Release* **2003**, *88*, 393–399. [CrossRef]

33. De Cock, L.J.; Lenoir, J.; de Koker, S.; Vermeersch, V.; Skirtach, A.G.; Dubruel, P.; Adriaens, E.; Vervaet, C.; Remon, J.P.; de Geest, B.G. Mucosal irritation potential of polyelectrolyte multilayer capsules. *Biomaterials* **2011**, *32*, 1967–1977. [CrossRef] [PubMed]

34. Porfiryeva, N.N.; Nasibullin, S.F.; Abdullina, S.G.; Tukhbatullina, I.K.; Moustafine, R.I.; Khutoryanskiy, V.V. Acrylated Eudragit® E PO as a novel polymeric excipient with enhanced mucoadhesive properties for application in nasal drug delivery. *Int. J. Pharm.* **2019**, *562*, 241–248. [CrossRef] [PubMed]

35. Adriaens, E.; Dierckens, K.; Bauters, T.G.; Nelis, H.J.; van Goethem, F.; Vanparys, P.; Remon, J.P. The mucosal toxicity of different benzalkonium chloride analogues evaluated with an alternative test using slugs. *Pharm. Res.* **2001**, *18*, 937–942. [CrossRef]

36. Adriaens, E.; Remon, J.P. Mucosal irritation potential of personal lubricants relates to product osmolality as detected by the slug mucosal irritation assay. *Sex. Transm. Dis.* **2008**, *35*, 512–516. [CrossRef]

37. Vandamme, K.; Melkebeek, V.; Cox, E.; Deforce, D.; Lenoir, J.; Adriaens, E.; Vervaet, C.; Remon, J.P. Influence of reaction medium during synthesis of Gantrez® AN 119 nanoparticles for oral vaccination. *Eur. J. Pharm. Biopharm.* **2010**, *74*, 202–208. [CrossRef]

38. Callens, C.; Adriaens, E.; Dierckens, K.; Remon, J.P. Toxicological evaluation of a bioadhesive nasal powder containing a starch and Carbopol® 974 P on rabbit nasal mucosa and slug mucosa. *J. Control. Release* **2001**, *76*, 81–91. [CrossRef]

39. Dhondt, M.M.; Adriaens, E.; Van Roey, J.; Remon, J.P. The evaluation of the local tolerance of vaginal formulations containing dapivirine using the Slug Mucosal Irritation test and the rabbit vaginal irritation test. *Eur. J. Pharm. Biopharm.* **2005**, *60*, 419–425. [CrossRef] [PubMed]

40. Forcart, L. Limacus maculatus (Kaleniczenko) und *Limacus flavus* (Linnaeus). *Mitt. Dtsch. Malakozool. Ges.* **1986**, *38*, 21–23.

41. Chelazzi, G.; le Voci, G.; Parpagnoli, D. Relative importance of airborne odours and trails in the group homing of *Limacus flavus* (Linnaeus) (Gastropoda, Pulmonata). *J. Molluscan Stud.* **1988**, *54*, 173–180. [CrossRef]

42. Cook, A.; Shirbhate, R. The mucus producing glands and the distribution of the cilia of the pulmonate slug *Limax pseudoflavus*. *J. Zool.* **1983**, *201*, 97–116. [CrossRef]

43. Chang, N.-S. Ultrastructural and Histochemical Studies on the Epithelial Cells and Mucus-producing Cells of Korean Slug (*Limax flavus* L.). *Appl. Microsc.* **1988**, *18*, 1–20.

44. Khutoryanskaya, O.V.; Mayeva, Z.A.; Mun, G.A.; Khutoryanskiy, V.V. Designing temperature-responsive biocompatible copolymers and hydrogels based on 2-hydroxyethyl (meth) acrylates. *Biomacromolecules* **2008**, *9*, 3353–3361. [CrossRef] [PubMed]

45. Dhondt, M.M.; Adriaens, E.; Pinceel, J.; Jordaens, K.; Backeljau, T.; Remon, J.P. Slug species-and population-specific effects on the end points of the Slug Mucosal Irritation test. *Toxicol. Vitr.* **2006**, *20*, 448–457. [CrossRef] [PubMed]

46. Castillo, V.M. *Limacus flavus* (Linnaeus, 1758 Linnaeus, 1758 Linnaeus, 1758): Antecedentes de la especie Antecedentes de la especie. *Amici Molluscarum* **2020**, *27*, 21–26.

47. WHO. *Cleaning and Disinfection of Environmental Surfaces in the Context of COVID-19: Interim Guidance, 15 May 2020*; World Health Organization: Geneva, Switzaerland, 2020; Available online: https://apps.who.int/iris/handle/10665/332096 (accessed on 22 March 2021).

48. Lauritano, D.; Moreo, G.; Limongelli, L.; Nardone, M.; Carinci, F. Environmental Disinfection Strategies to Prevent Indirect Transmission of SARS-CoV2 in Healthcare Settings. *Appl. Sci.* **2020**, *10*, 6291. [CrossRef]

49. Al-Gheethi, A.; Al-Sahari, M.; Abdul Malek, M.; Noman, E.; Al-Maqtari, Q.; Mohamed, R.; Talip, B.A.; Alkhadher, S.; Hossain, M. Disinfection Methods and Survival of SARS-CoV-2 in the Environment and Contaminated Materials: A Bibliometric Analysis. *Sustainability* **2020**, *12*, 7378. [CrossRef]

50. Ijaz, M.; Brunner, A.; Sattar, S.; Nair, R.C.; Johnson-Lussenburg, C. Survival characteristics of airborne human coronavirus 229E. *J. Gen. Virol.* **1985**, *66*, 2743–2748. [CrossRef] [PubMed]

51. Safari, Z.; Fouladi-Fard, R.; Vahidmoghadam, R.; Hosseini, M.R.; Mohammadbeigi, A.; Omidi Oskouei, A.; Rezaali, M.; Ferrante, M.; Fiore, M. Awareness and Performance towards Proper Use of Disinfectants to Prevent COVID-19: The Case of Iran. *Int. J. Environ. Res. Public Health* **2021**, *18*, 2099. [CrossRef] [PubMed]

52. Van Doremalen, N.; Bushmaker, T.; Morris, D.H.; Holbrook, M.G.; Gamble, A.; Williamson, B.N.; Tamin, A.; Harcourt, J.L.; Thornburg, N.J.; Gerber, S.I. Aerosol and surface stability of SARS-CoV-2 as compared with SARS-CoV-1. *N. Engl. J. Med.* **2020**, *382*, 1564–1567. [CrossRef] [PubMed]

53. Chang, A.; Schnall, A.H.; Law, R.; Bronstein, A.C.; Marraffa, J.M.; Spiller, H.A.; Hays, H.L.; Funk, A.R.; Mercurio-Zappala, M.; Calello, D.P. Cleaning and disinfectant chemical exposures and temporal associations with COVID-19—national poison data system, United States, 1 January–31 March 2020. *Morb. Mortal. Wkly. Rep.* **2020**, *69*, 496. [CrossRef] [PubMed]

54. Babić, Ž.; Turk, R.; Macan, J. Toxicological aspects of increased use of surface and hand disinfectants in Croatia during the COVID-19 pandemic: A preliminary report. *Arch. Ind. Hyg. Toxicol.* **2020**, *71*, 261–264. [CrossRef]

55. Casey, M.L.; Hawley, B.; Edwards, N.; Cox-Ganser, J.M.; Cummings, K.J. Health problems and disinfectant product exposure among staff at a large multispecialty hospital. *Am. J. Infect. Control* **2017**, *45*, 1133–1138. [CrossRef]

56. Kim, J.-H.; Hwang, M.-Y.; Kim, Y.-j. A Potential Health Risk to Occupational User from Exposure to Biocidal Active Chemicals. *Int. J. Environ. Res. Public Health* **2020**, *17*, 8770. [CrossRef]

57. Adriaens, E.; Bytheway, H.; de Wever, B.; Eschrich, D.; Guest, R.; Hansen, E.; Vanparys, P.; Schoeters, G.; Warren, N.; Weltens, R. Successful prevalidation of the slug mucosal irritation test to assess the eye irritation potency of chemicals. *Toxicol. Vitr.* **2008**, *22*, 1285–1296. [CrossRef]

58. Gordon, D.G. *Field Guide to the Slug*; Sasquatch Books: Seattle, WA, USA, 1994.

59. Jordaens, K.; Riel, P.V.; Geenen, S.; Verhagen, R.; Backeljau, T. Food-induced body pigmentation questions the taxonomic value of colour in the self-fertilizing slug *Carinarion* spp. *J. Molluscan Stud.* **2001**, *67*, 161–167. [CrossRef]
60. Seki, K.; Wiwegweaw, A.; Asami, T. Fluorescent pigment distinguishes between sibling snail species. *Zool. Sci.* **2008**, *25*, 1212–1219. [CrossRef] [PubMed]
61. Adriaens, E.; Remon, J.P. Gastropods as an evaluation tool for screening the irritating potency of absorption enhancers and drugs. *Pharm. Res.* **1999**, *16*, 1240–1244. [CrossRef] [PubMed]
62. Izenman, A.J. Linear discriminant analysis. In *Modern Multivariate Statistical Techniques*; Springer: New York, NY, USA, 2013; pp. 237–280.
63. Adriaens, E.; Dhondt, M.; Remon, J.P. Refinement of the Slug Mucosal Irritation test as an alternative screening test for eye irritation. *Toxicol. Vitr.* **2005**, *19*, 79–89. [CrossRef] [PubMed]

biomedicines

Article

Molecular Abnormalities in BTBR Mice and Their Relevance to Schizophrenia and Autism Spectrum Disorders: An Overview of Transcriptomic and Proteomic Studies

Polina Kisaretova [1,2], **Anton Tsybko** [1], **Natalia Bondar** [1] and **Vasiliy Reshetnikov** [1,3,*]

1. Institute of Cytology and Genetics, Siberian Branch of Russian Academy of Sciences, Prospekt Akad. Lavrentyeva 10, Novosibirsk 630090, Russia
2. Department of Natural Sciences, Novosibirsk State University, Pirogova Street 2, Novosibirsk 630090, Russia
3. Department of Biotechnology, Sirius University of Science and Technology, 1 Olympic Avenue, Sochi 354340, Russia
* Correspondence: vasiliyreshetnikov@bionet.nsc.ru

Abstract: Animal models of psychopathologies are of exceptional interest for neurobiologists because these models allow us to clarify molecular mechanisms underlying the pathologies. One such model is the inbred BTBR strain of mice, which is characterized by behavioral, neuroanatomical, and physiological hallmarks of schizophrenia (SCZ) and autism spectrum disorders (ASDs). Despite the active use of BTBR mice as a model object, the understanding of the molecular features of this strain that cause the observed behavioral phenotype remains insufficient. Here, we analyzed recently published data from independent transcriptomic and proteomic studies on hippocampal and corticostriatal samples from BTBR mice to search for the most consistent aberrations in gene or protein expression. Next, we compared reproducible molecular signatures of BTBR mice with data on postmortem samples from ASD and SCZ patients. Taken together, these data helped us to elucidate brain-region-specific molecular abnormalities in BTBR mice as well as their relevance to the anomalies seen in ASDs or SCZ in humans.

Keywords: BTBR; transcriptome; proteome; cortex; hippocampus; ASD; schizophrenia

Citation: Kisaretova, P.; Tsybko, A.; Bondar, N.; Reshetnikov, V. Molecular Abnormalities in BTBR Mice and Their Relevance to Schizophrenia and Autism Spectrum Disorders: An Overview of Transcriptomic and Proteomic Studies. *Biomedicines* **2023**, *11*, 289. https://doi.org/10.3390/biomedicines11020289

Academic Editor: Martina Perše

Received: 20 December 2022
Revised: 17 January 2023
Accepted: 18 January 2023
Published: 20 January 2023

1. Introduction

Autism spectrum disorders (ASDs) are a group of lifelong neurodevelopmental conditions characterized by dysfunction of communication and of social interaction as well as restricted repetitive behavior and interests [1]. Despite high variation of ASD prevalence across the world, the general tendency has been a dramatic increase of this prevalence in recent decades [2]. The global burden of ASDs is substantial and has continued to grow over the past three decades as well [3].

Schizophrenia (SCZ) is associated with such symptoms (present for at least 6 months) as delusions, hallucinations, disorganized speech and behavior, and dysfunctions of social and occupational domains [4]. Although ASDs and SCZ are recognized as separate disorders with divergent clinical profiles, growing evidence suggests that SCZ is a neurodevelopmental disorder as well [5–9]. Moreover, the convergence of ASDs and SCZ at certain levels supports the idea that their phenotypes may overlap [10,11]. Specific genetic risk alleles shared among intellectual disability, ASD, SCZ, attention deficit/hyperactivity disorder, and bipolar disorder have led investigators to propose the model of neurodevelopment continuum, in which these diseases represent a range of outcomes that flow from aberrant brain development [12]. Despite some studies showing that the burden of DNA copy number variants (CNVs) positively correlates with the severity of childhood neurodevelopmental disorders [13–15] (in line with the continuum hypothesis), this model may be an oversimplification of the diagnostic conundrum. At the same time, a close connection between ASDs and SCZ is likely. SCZ alone is three-to-six times more common in people with ASDs than in controls, as demonstrated in two meta-analyses [16,17]. A recent systematic review showed that ASDs and SCZ significantly overlap in behavior, perception, cognition, some biomarkers, and genetic risk factors [18]. These findings indicate that SCZ and autism are on the same continuum.

Given that estimated heritability is 50%, genetic factors are the main contributors to ASD etiology [19]. Nonetheless, autism has substantial phenotypic heterogeneity, which arises from multiple genetic sources and many of them overlap [20]. From the perspective of the complex genetic architecture of autism, Iakoucheva et al. have proposed the omnigenic model of ASDs, which posits inseparability of effects of genes which have an impact on the trait directly and gene modifiers which act indirectly through gene regulation [21]. On the other hand, on the basis of 28 meta-analyses, Qiu et al. recently identified 12 significant single-nucleotide polymorphisms (SNPs) in nine candidate genes [22], most of which fit the definition of "core" genes. Thus, the probability of identification of at least several key genes for the ASD phenotype is still not negligible. In this context, transcriptomic and proteomic approaches provide insights into molecular characteristics to help bridge the gap between genes and functions. There are several transcriptomic articles about postmortem brain samples from ASD subjects and much fewer proteomic ones [23]. The number of comparative transcriptomic studies on brain samples from both ASD and SCZ patients is relatively small [24–27], and there is only a single systematic multi-omics study on ASD and SCZ [28].

Both ASD and SCZ are polygenic disorders, which means that a great variety of genes engaged in different processes are involved in phenotype development. In ASD alterations in expression of glutamate decarboxylases GAD65/GAD67 and GABAA and GABAB receptor subunits are considered the most reproducible. For SCZ, the most consistent genes are trophic factor *NRG1* and post-synaptic tyrosine kinase receptor *ERBB4*. Genetic studies identified genes affected in both disorders (*DISC1*, *NRXN1*, *NLGN3-4*, *SHANK3*, and *CNTNAP2*) [29].

One of the most widely used animal models of idiopathic ASDs is the BTBR T$^+$ Itpr3tf/J (BTBR) inbred mouse strain. The phenotype of BTBR mice is relevant to major diagnostic symptoms of ASDs, including reduced social interactions and stereotyped behavior [30–32]. BTBR mice share neuroanatomical features with a subgroup of ASD patients and have a complex molecular phenotype [33], but the underlying genetic abnormalities are unclear and still being investigated. To date, only one study has matched the transcriptome of BTBR mice with transcriptomic data from ASD patients [34]. So far, only some features of SCZ have been modeled in the BTBR strain. Social withdrawal as a consequence of disrupted sociability is the main early symptom of both ASDs and SCZ. In this context, BTBR mice have been employed for research into the social dysfunction relevant to SCZ [35]. Reduced GABA and increased glutamate concentrations in the prefrontal cortex (PFC)—as well as weakened maturation of GABA circuits accompanied by impairments of multisensory integration—reflect the features of an SCZ-like phenotype in BTBR mice [35,36]. Moreover, one of the major genetic risk factors of SCZ, a spontaneous deletion of the *Disc1* gene, is also present in BTBR mice [37]. Because BTBR mice are often used to explore various therapeutic strategies aimed at alleviating autistic-like symptoms, it is important to understand the degree of genetic convergence between this animal model and both ASD and SCZ patients.

Here, we collate recently published data from independent transcriptomic and proteomic studies on brain samples from ASD and SCZ patients as well as the popular autistic-like animal model: BTBR mice. The main purpose of this review is to combine these datasets in one bioinformatic analysis. First, we examined these data to identify common differentially expressed genes (DEGs). Next, we explored proteomic datasets to identify differentially expressed proteins. Finally, these genes and proteins were characterized with respect to enrichment in gene–gene and protein–protein interaction networks to investigate the link with canonical pathways and biological processes implicated in the etiology of both ASDs and SCZ and shared with BTBR mice.

2. Materials and Methods

2.1. Analysis of Publicly Available Transcriptomic and Proteomic Data

To find reproducible alterations in the transcriptome of BTBR mice, we conducted literature searches on Google Scholar and PubMed (NCBI). We utilized the following search query: (BTBR), AND (cortex), OR (striatum), OR (hippocampus); AND (RNA-seq) or (microarray). This search was completed on 26 June 2022 and produced 1642 hits on Google Scholar and 661 hits on PubMed. After screening the papers, seven studies were found to meet the following criteria: (1) the experiment was performed on mice of BTBR and C57BL/6 strains; (2) tissue samples of the cortex, striatum, or hippocampus were examined; (3) a differential gene expression analysis of BTBR mice compared to C57BL/6 mice was performed; (4) proteomic or transcriptomic analysis (RNA-seq or microarray) was used (Table 1); and (5) the paper was written in the English language.

Table 1. The transcriptomic and proteomic studies on BTBR mice included in the analysis.

Strains	Age	Brain Region	Method	Threshold	Raw Data	Reference
BTBR T + Itpr3tf/J (BTBR)/C57BL/6J	12 w	Hippocampus	RNA-seq	p value < 0.05 *	PRJNA533538	[38]
BTBR T+ Itpr3tf/J (BTBR)/C57BL/6J	3–5 m	Hippocampus	Microarray	Rank Product (RP) non-parametric method was used	GSE81501	[39]
BTBR T + Itprtf/J mice/C57BL/6J	4 m	Hippocampus and cortex	Microarray	Z-ratio value of ± 1.50 and/or a Z-test value $p < 0.05$	N/A	[40]
BTBR/B6	8 w	Cortex	RNA-seq	$|\log_2 FC| \geq 1$ and $p_{adj} \leq 0.05$	N/A	[41]
BTBR T + tf/J/C57BL/6J	8–10 w	Striatum (bregma −0.58–1.53).	RNA-seq	p value < 0.05 *	GSE138539	[42]
BTBRTF/ArtRbrc mice Compared to C57BL/6J Mice	7 w	Striatum + Cortex	Microarray	$p_{adj} \leq 0.05$	GSE156646	[34]
BTBR T + Itprtf/J mice/C57BL/6J	4 m	Hippocampus	iTRAQ LC–MS/MS	fold change > 0.2	N/A	[40]
BTBR T + Itprtf/J mice/C57BL/6J	8 w	Cortex	iTRAQ LC–MS/MS	fold change > 0.5, $p < 0.05$	N/A	[43]

Threshold column shows the significance cut off used for DEG assessment. In the column Raw Data can be found GEO project IDs if available or N/A (not available) if not. * RNA-seq studies with available raw data were reanalyzed for a p value cutoff of <0.05; in other papers, the cutoff of the original study was accepted; w: weeks, m: months.

2.2. Functional Annotation of Reproducible Genes in BTBR Mice

With our inclusion criteria, we found three transcriptomic studies on the hippocampus and four on the cortex and/or striatum and two proteomic studies on the cortex and hippocampus (Table 1). Between the hippocampal datasets, we compared DEG sets, and changes in expression that were unidirectional between at least two datasets were designated as reproducible in BTBR mice. Because the striatum and frontal cortex are closely related structures and have many functional connections with each other and because some transcriptomic studies were performed on a tissue common between these structures, we decided to combine the data obtained from the striatum and cortex. In total, DEG sets from the cortex and/or striatum were compared between four datasets, and the criteria for reproducible changes in expression were the same as those for the hippocampus (changes in expression that were unidirectional between at least two datasets).

Gene ontology (GO) enrichment analysis of the reproducible DEGs was conducted using the enrichGO function from the Cluster-Profiler (v4.0.5) R package (v4.1.0). Basic workflow is shown in Figure 1.

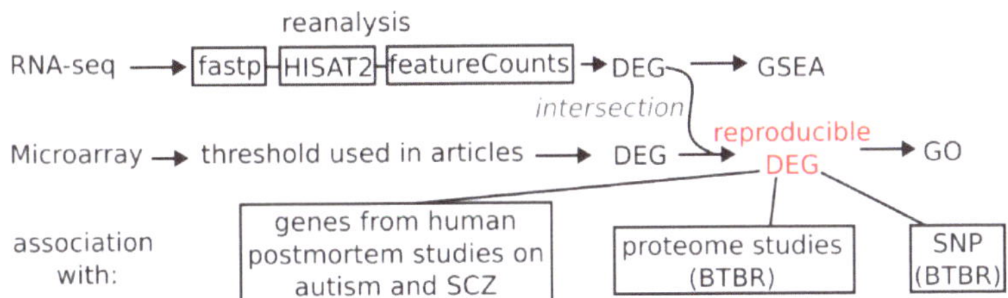

Figure 1. Basic workflow scheme. These steps were applied separately for hippocampus and cortex + striatum datasets resulting in two reproducible differentially expressed genes lists. For reanalysis of available RNA-seq data first adapters and low-quality reads were trimmed using fastp program, then aligned to mm10 genome with HISAT2 (2.2.1) software and summarized to gene counts with featureCounts (v2.0) function using genecode vM22 annotation.

2.3. Association of a Genetic Background and Gene Expression

We tested how genetic variants of BTBR mice affect gene expression. For this purpose, we employed BTBR mice's sequenced genome data available in the Trust Sanger Institute mouse genome project (https://www.sanger.ac.uk/data/mouse-genomes-project/ accessed on 1 November 2022). The final list of mutations (SNPs, insertions, deletions, CNVs, and structural variants) included 785,695 rs. We analyzed the potential significance of these rs using the Ensembl Variant Effect Predictor Web interface with default parameters [44]. Next, we determined which of the genes affected by these mutations are expressed (at least 10 counts in each sample) in the mouse cortex and hippocampus. To this end, we compared genes having "high prediction score" mutations with genes from previously published datasets on C57BL/6 adult mice [45,46], which included 14,989 and 14,901 genes expressed in the hippocampus and PFC, respectively. The list of expressed genes carrying high-prediction-score mutations was analyzed for enrichment with GO terms (by means of the enrichGO function from the Cluster-Profiler (v4.0.5) R package v4.1.0) and compared with the set of reproducible DEGs.

2.4. Reanalysis of the Published RNA-Seq Data from the Cortex and Hippocampus of BTBR Mice

Raw data from two studies that involved RNA-seq to evaluate transcriptomic changes were reanalyzed via gene set enrichment analysis (GSEA) and via examination of alternative splicing changes. The sequencing data were preprocessed in fastp v0.20.1 [47] to remove adapters and low-quality sequences. The preprocessed data were mapped to the *Mus musculus* GRCm38 reference genome assembly in the HISAT2 aligner software, v2.2.1 [48]. The HISAT2 alignments were quantified by means of featureCounts v2.0 [49]. The quality of the sequencing data was assessed using FastQC (v0.11.9) and Picard Collec-tRnaSeqMetrics (v2.18.7) software. The aligned data having fragments per kilobase of transcript per million fragments mapped (FPKM) > 0.1 were then converted into per-gene count tables with the help of GENCODE vM22 gene annotation data. Genes were then subjected to an analysis of differential gene expression via the DESeq2 R v4.1.0 package [50]. Genes with a p value < 0.05 were designated as statistically significant DEGs.

GSEA was conducted using the gseGO function of the ClusterProfiler (v4.0.5) R package (v4.1.0). Genes were ranked by log2 (fold change) from DESeq2 results. In the results, a normalized enrichment score indicated whether the genes were mostly up- or downregulated in a given gene set.

2.5. A Comparison between BTBR-Related Genes and Genes Associated with Human Autism or SCZ

We compared data from our study with results of two meta-analyses of RNA-seq data obtained from five independent cortical datasets (postmortem tissue samples) of ASD patients and two meta-analyses of microarray data obtained from eight independent cortical datasets of SCZ patients (Table 2). Gene orthologs were identified by means of BioMart (https://www.ensembl.org/biomart/martview/ accessed on 1 November 2022). Genes that were differentially expressed ($p < 0.05$) unidirectionally between two meta-analyses were regarded as associated with either ASD or SCZ.

Table 2. Publicly available meta-analyses of postmortem cortical transcriptomes of SCZ and ASD patients.

Disease	Type of Review	Data ID	Method	Threshold	Reference
ASD	Meta-analysis	GSE28475 GSE28521 GSE36192	Microarray/RNA-seq	$p < 0.05$	[51]
ASD	Meta-analysis	GSE64018 GSE30573	RNA-seq	$p < 0.05$	[52]
SCZ	Meta-analysis	GSE17612 GSE21935 GSE25673 GSE12649 GSE21338	Microarray	$p < 0.05$	[53]
SCZ	Meta-analysis	GSE17612 GSE21138 +and four others	Microarray	q-value < 0.05	[54]

Threshold column shows the significance cut off used for DEG assessment. DATA ID shows GEO accession numbers of experiments used in the study. ASD—autism spectrum disorder; SCZ—schizophrenia.

3. Results

3.1. Transcriptome Aberrations in the Hippocampus of BTBR Mice

The comparison of sets of up- and downregulated genes from various authors revealed 56 upregulated genes and 135 downregulated genes whose expression changed in at least two of the three studies included in the analysis (Table S1). In this gene set, we detected one enriched GO term at a false discovery rate (FDR) of <0.05: "oxidoreductase activity" (related to genes *Adi1*, *Ptgs2*, *Alox8*, and *Alox12b*). Expression of 4 upregulated genes (*Blvrb*, *Scg5*, *Serpina3n*, and *Anxa5*) and 14 downregulated genes (*C1qb*, *Spink8*, *Entpd4*, *Pop4*, *Alg1*, *Rpl29*, *Ccnd1*, *Mt3*, *Zfp131*, *6330403K07Rik*, *Nudt19*, *C1ql2*, *Evc2*, and *Cetn4*) changed unidirectionally among all three datasets (Figure 2a).

Among these genes, there are genes encoding complement components (*C1qb* and *C1ql2*), inhibitors of serine peptidases (*Serpina3n* and *Spink8*), and a neuron-specific gene (*Scg5* encoding neuroendocrine protein 7B2 involved in the regulation of the corticotropin secretory pathway). Consistent changes in the expression of these genes in a number of studies indicated that they constitute a molecular signature of the hippocampal transcriptome of BTBR mice. Nonetheless, functions of the protein products of these genes are diverse because they are not components of a specific pathway(s).

A comparison of the transcriptome data with the proteome data showed that only two genes undergo unidirectional expression changes at both mRNA and protein levels. One of them is the Wfs1 gene—whose protein product participates in protein biosynthesis, stabilization, folding, maturation, and secretion [55]—and the other gene is Clcn6, which codes for a chloride/proton exchanger playing an important role in autophagic-lysosomal function [56]. Of note, mitochondrial phosphate carrier (Slc25a3) mRNA was underexpressed while the protein was overexpressed.

Next, to identify the functional pathways or signatures in the set of all expressed genes, we performed GSEA and found that in the hippocampus of BTBR mice, there was overexpression of genes associated with GO terms (biological processes) "sensory perception of smell" and "olfactory receptor activity" and underexpression of genes related to regulation of "corticotropin secretion" (Figure 2e, Table S3).

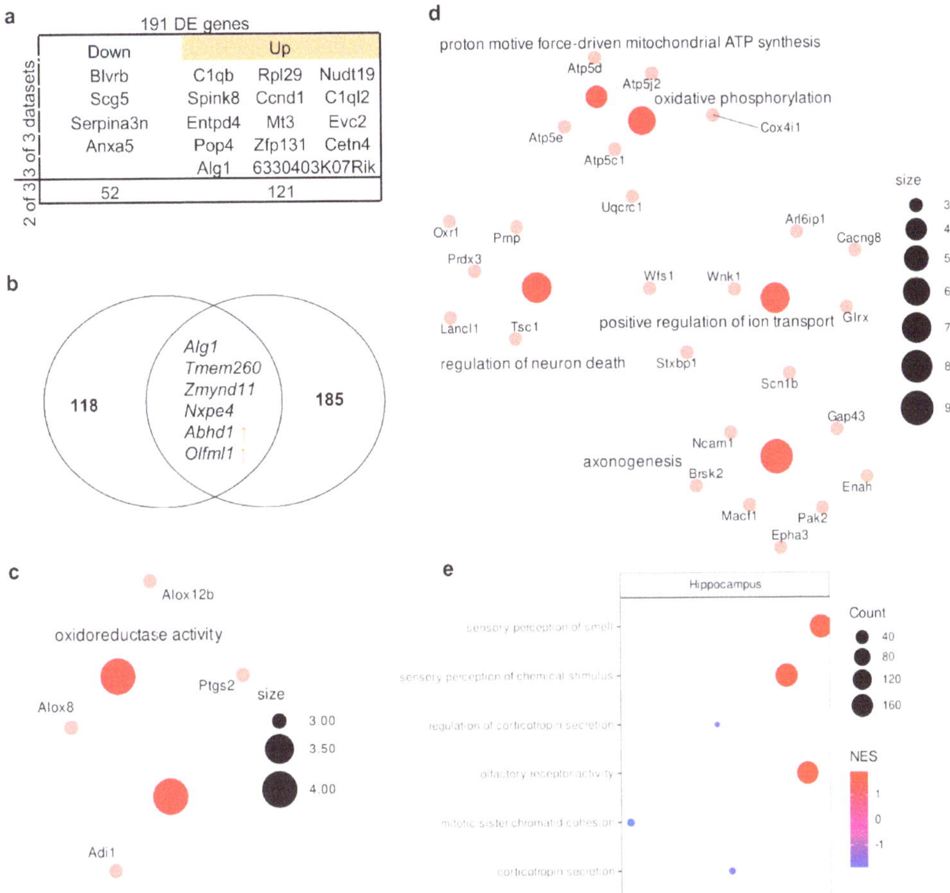

Figure 2. Molecular signatures in the hippocampus of BTBR mice. (**a**) Reproducible up- and downregulated genes in the hippocampus. Out of a total of 191 DE genes, 4 were downregulated in all 3 datasets, and 52 were downregulated in 2 out of 3; 14 were upregulated in all 3 datasets, and 121 were upregulated in 2 out of 3. (**b**) The overlap between the set of genes carrying a "high prediction score" mutation and the set of DEGs. Out of 124 "high prediction score" genes only 6 overlapped with reproducible DE genes set (191 gene). (**c**) The GO category enriched in the set of DEGs. Size of the node represents number of genes in the category. (**d**) The GO categories enriched in the set of differentially expressed proteins according to data from ref. [40]. Size of the node represents number of genes in the category. (**e**) Top three up- and downregulated GSEA categories in the set of all expressed genes according to reanalysis of PRJNA533538 data. NES: normalized enrichment score. Size of the circle (count) represents number of genes in the category, color of the circle represents normalized enrichment score (*NES*).

3.2. Transcriptome Alterations in the Corticostriatal Area of BTBR Mice

We identified 79 upregulated genes and 113 downregulated genes whose expressions were found to change in at least two of the four studies included in the analysis (Table S2). In this gene set, there were two enriched GO terms with FDR < 0.05: "actin cytoskeleton" (related to such genes as *Crocc*, *Myh6*, *Myh7*, and *Myo7a*) and "humoral immune response" (related to genes *C1qb*, *C1qc*, *B2m*, *Masp2*, *Cci17*, *Slpi*, *Ptprc*, and *Pglyrp1*; Figure 3c). The expression of six genes changed unidirectionally among all datasets (Figure 3a); among these genes, an increase in expression (*Scg5* and *Serpina3n*) and a decrease in expression (*C1qb*, *Pop4*, and *Nudt19*) were also found in all hippocampal datasets, suggesting that these changes are consistent across different experimental designs and not specific to one brain structure. The expression of *Anxa3*, which encodes a calcium-dependent phospholipid-binding protein involved in signal transduction pathways, was low only in the corticostriatal region

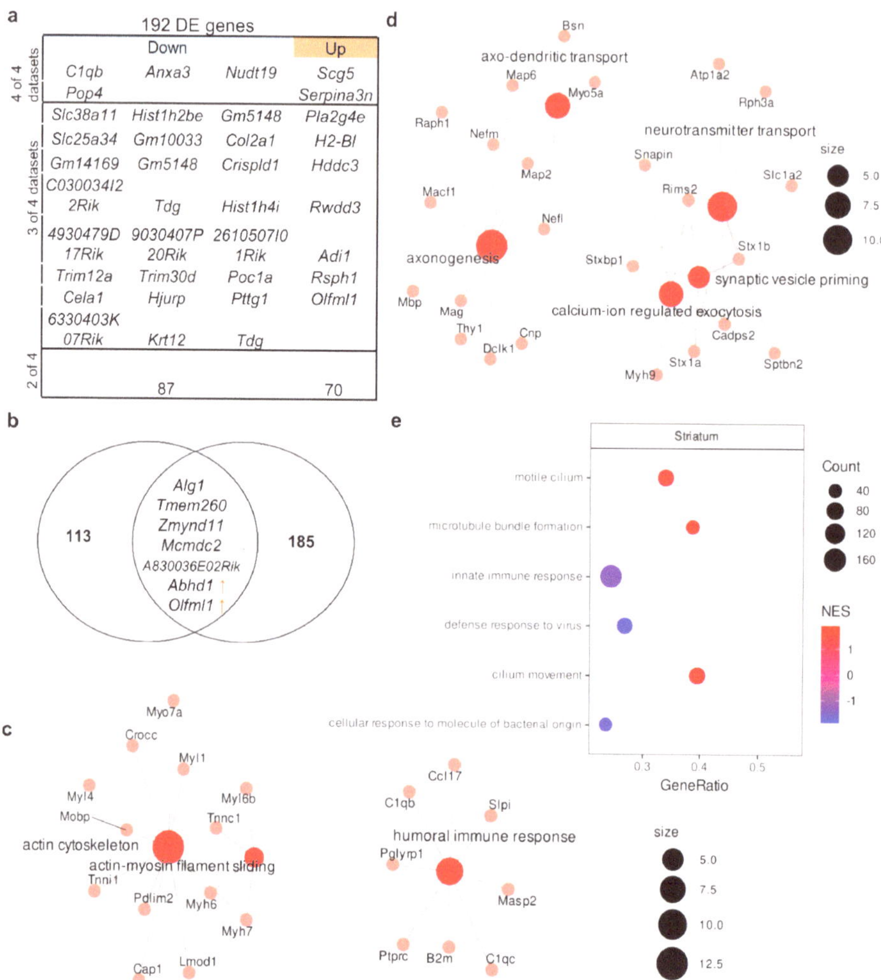

Figure 3. Molecular signatures in the corticostriatal area of BTBR mice. Out of a total of 192 DE genes, 4 were downregulated in all 4 datasets, and 24 were downregulated in 3 out of 4 and 87 in 2 out of 3. In all datasets, 2 genes were upregulated, 7 genes were upregulated in 3 out of 4 datasets, and 70 were upregulated in 2 out of 4. (**a**) The comparison of DEGs from the corticostriatal area. (**b**) The overlap between the set of genes carrying a high-prediction-score mutation and the set of DEGs. Out of 120 "high prediction score" genes, only 7 overlapped with reproducible DE genes set (192 gene). (**c**) The GO category enriched in the set of DEGs. Size of the node represents number of genes in the category. (**d**) GO categories enriched in the set of differentially expressed proteins on the basis of data from Ref. [43]. Size of the node represents number of genes in the category. (**e**) Top three up- and downregulated GSEA categories in the set of all expressed genes according to the reanalysis of GSE138539 data. NES: normalized enrichment score. Size of the circle (count) represents number of genes in the category, color of the circle represents normalized enrichment score (NES).

To understand the similarity of the molecular signatures between the hippocampal transcriptome and corticostriatal transcriptome, we compared consistent changes in the expression of genes (a change in expression unidirectional between at least two datasets) and noticed that the expression of 55 genes (13 upregulated and 42 downregulated) changed in the same way between these brain structures (Figure 4). At the same time, a comparison of proteomic changes between the cortex and hippocampus suggested that protein expression of only four genes (*Macf1*, *Bsn*, *Psd3*, and *Chchd3*) changed in the same direction between these brain regions. Of note, a comparison of the transcriptome data with the proteome data did not uncover expression alterations that were unidirectional between mRNA and protein levels in either the cortex or striatum.

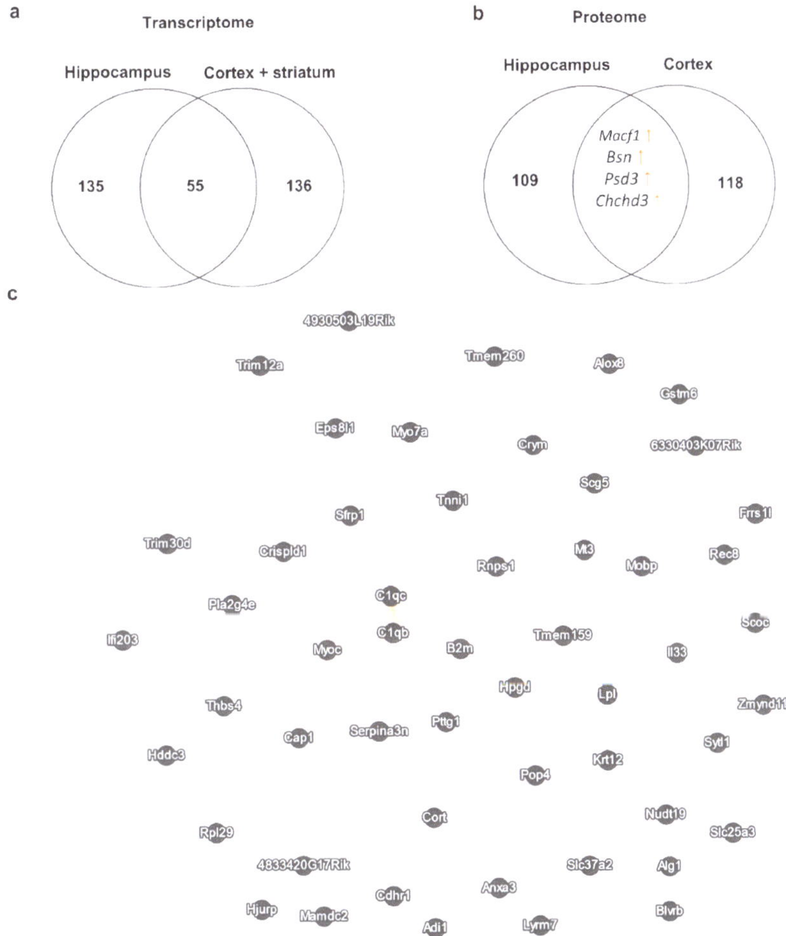

Figure 4. DEGs in the corticostriatal area and hippocampus. (**a**) This Venn diagram depicts the overlap between sets of reproducible corticostriatal and hippocampal genes changing their expression unidirectionally between these brain regions. Numbers on the diagram represent how many genes are unique for the dataset and how many overlap. (**b**) This Venn diagram illustrates the overlap between sets of cortical and hippocampal proteins changing their expression unidirectionally between these brain regions. Numbers on the diagram represent how many genes are unique for the dataset and how many overlap. (**c**) Gene mania network representation of the 55 reproducible corticostriatal and hippocampal genes changing their expression unidirectionally between these brain regions; each gene manifested expression changes in at least two datasets from the hippocampus and corticostriatal area.

Next, to identify functional pathways or signatures in the set of all expressed genes, we performed GSEA and noticed that in the striatum of BTBR mice, there was upregulation of genes associated with the GO terms (biological processes) "motile cilium" and "microtube bundle formation" and downregulation of genes related to "innate immune response" (Figure 3e, Table S4)

3.3. Genetic Characteristics of BTBR Mice

We found 378 mutations that had a high prediction score (Table S3). This list included 261 mutations that lead to a frameshift, 108 mutations that affect splice donor sites, and four mutations located in the 3′ untranslated region and resulting in the loss of a stop codon. Among the genes expressed in the hippocampus and cortex, only 124 and 120 genes had at least one mutation with a high-prediction-score predictor impact. These sets of genes turned out to be not enriched with any GO terms. Unexpectedly, we found only six DEGs in the hippocampus and seven DEGs in the corticostriatal area that were also in the set of expressed genes containing high-prediction-score mutations (Figures 1b and 2b). Among these genes, the expression of *Alg1*, *Tmem260*, and *Zmynd11* was low, while the expression of *Abhd1* and *Olfml1* was high in both brain structures. In these genes, the presence of high-prediction-score mutations—which can lead to a frameshift or affect a splice site—can result in a defective protein product. Thus, regardless of the level of expression of these genes, the functionality of the encoded protein products may be impaired.

3.4. The Comparison between DEGs from Postmortem ASD or SCZ Human Tissue Samples and DEGs from BTBR Mice

After comparing the results of meta-analyses of postmortem samples from patients with SCZ or autism, we identified 226 ASD-related genes and 155 SCZ-related genes in the frontal cortex (Tables S6 and S7). The comparison of these gene sets with the reproducible DEGs found in the corticostriatal area of BTBR mice yielded only a small overlap (Figure 5). Only the expression of *Lpl* proved to be upregulated in postmortem samples from SCZ patients and in BTBR mice. *Lpl* codes for an enzyme called lipoprotein lipase, which plays a key part in the brain energy balance.

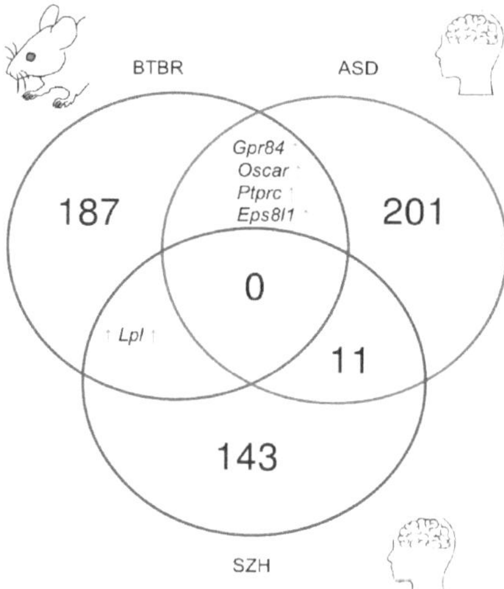

Figure 5. The comparison of reproducible DEGs between the corticostriatal area of BTBR mice and postmortem samples from the frontal cortex of SCZ and ASD patients. Numbers on the diagram represent how many genes are unique for the dataset and how many overlap.

The set of DEGs from BTBR mice shared four genes with the set of ASD-related genes (*Gpr84*, *Oscar*, *Ptprc*, and *Eps8l1*), but their expression alterations were not unidirectional (upregulated in ASD and downregulated in BTBR mice). *Gpr84* and *Ptprc* are microglia-specific genes [57] and encode receptors involved in the regulation of neuroinflammation. *Ptprc* encodes common lymphocyte antigen CD45 playing an important role in T cell activation [58]. *Gpr84* participates in the modulation of the inflammatory response, and upregulation of its expression has been documented during a response to inflammatory conditions and stimuli [59].

4. Discussion

Our review includes transcriptomic and proteomic data from the hippocampus and corticostriatal area. Functional and neuroanatomical changes in these brain structures are most often reported in patients with an ASD. Much evidence points to atypical cortical morphology, volume [60,61], and grey and white matter thickness [62–65] as well as microstructure abnormalities [66] in individuals with an ASD. A number of studies indicate abnormal enlargement [67–74], shape asymmetry [75], and altered functional connectivity [76] in the hippocampus of ASD subjects. Similar alterations of shapes [77,78], volumes [79–83], microstructure [84], and connectivity [85,86] have been found in basal ganglia, especially in striatal subregions.

Neuroanatomical and functional changes of the cortex [87–90], hippocampus [91,92], and striatum [93–95] are also present in animal models of ASDs but can vary from one model to another [96]. Neuroanatomic features of BTBR mice as a model of idiopathic ASDs are relatively well characterized. Apart from their most striking feature—the absence of the corpus callosum [97]—BTBR mice also have a deficient dorsal hippocampal commissure, smaller hippocampal volume [98,99], and dramatically reduced thickness and volume of the cortex, particularly in the PFC [99]. Smaller gray matter volume in various cortical and subcortical areas has been reported in MRI studies [98,100,101]. It is worth noting structural [100] and functional [102,103] deviations in the striatum of BTBR mice. Marked changes in cortical and subcortical connectivity in BTBR mice have been registered as well [89,102,104]. Overall, the aforementioned impairments not only are consistent with the behavioral phenotype of BTBR mice but also recapitulate neuroimaging hallmarks of autism. Considerable changes in neuroplasticity inevitably underlie these neuroanatomical abnormalities.

We found that the transcription of genes *Scg5* and *Serpina3n* is overactive and that the transcription of *C1qb*, *Pop4*, and *Nudt19* is low in all datasets from the hippocampus, cortex, and striatum of BTBR mice. This finding indicates that *Scg5*, *Serpina3n*, *C1qb*, *Pop4*, and *Nudt19* are universal transcriptional markers of BTBR mice.

The 7B2 protein, the product of the *Scg5* gene, is known as a secreted chaperone whose expression is restricted to neurons and neuroendocrine and endocrine cells [105,106]. 7B2 not only is crucial for peptide hormone storage [107,108] but also has antiaggregant properties and is capable of reducing the fibrillation of aggregating proteins [109,110]. Some contradictory pieces of evidence are suggestive of aberrant 7B2 levels in patients with Alzheimer's disease [110–112]. Furthermore, Helwig et al. [110] have documented colocalization of 7B2 with α-synuclein deposits in brain samples from patients with Parkinson's disease. Currently, there is no proof of the involvement of 7B2 in the pathogeneses of neurodevelopmental disorders. In one study on autistic subjects in the Japanese population, researchers did not find a correlation between a deletion in chromosomal region 15q11–q13 (containing *SCG5*) and autism [113].

In the present work, we noticed unidirectional changes between *Wfs1* transcription and protein levels. WFS1 is a membrane protein that is vital for the transfer of vesicular cargo proteins from the endoplasmic reticulum to the Golgi apparatus and is associated with diabetes [114]. There is evidence that *Wfs1* regulates proper folding of 7B2 [115]. Although the link between *Wfs1* and autism is still unproven, it has been demonstrated that a mutation in *WFS1* causes Wolfram syndrome, which is associated with bipolar disorder and SCZ [116–118]. Wolfram syndrome itself is characterized by various neurological problems, including ataxia, seizures, hypersomnolence, brain stem atrophy, peripheral and autonomic neuropathy, and an inability or decreased ability to sense taste and odors [119]. Recently, it was shown that WFS1 directly interacts with SCG5 vesicular cargo protein in pancreatic β-cells. In the brain, they potentially could be involved in the process of sorting neuropeptide cargo proteins into the COPII vesicles. However, this is yet to be proved [114]. Collectively, these data imply that *Scg5* and *Wfs1* are relevant to proteostatic processes, and it is possible that in the neuronal system of BTBR mice, proteostasis is compromised.

Serpina3n encodes the SerpinA3N protein belonging to the serpin superfamily of protease inhibitors. Upregulation of transcription of *Serpina3n* is a strong marker of reactive astrogliosis [120]. Some research articles show anti-inflammatory and neuroprotective effects of SerpinA3N [121–123], whereas in one study, the opposite was demonstrated [124]. SERPINA3 upregulation has been detected in postmortem samples of the cortex from SCZ patients [125]. It is not clear whether high levels of SerpinA3N can also accompany other neurodevelopmental disorders, but we can hypothesize that elevated transcription of *Serpina3n* in different brain structures of BTBR mice is an indicator of neuroinflammatory processes.

Our findings suggest that the complement and coagulation cascade signaling pathway are also affected in the brain of BTBR mice because the transcription of *C1qb* proved to be low in all the analyzed brain structures. Together with other components of the classical complement cascade (C1qa, C1qc, C2, and C4), C1qb is expressed by uninjured peripheral nerves and is known to play a crucial part in myelin clearance after peripheral nerve injury [126]. Reactivation of compliment expression can induce or propagate inflammation and may be detrimental to peripheral nerves [127]. *C1qb* is also implicated in the proinflammatory response in the central nervous system [128–131]. It is of note that in DBA/2J mice, which exhibit a deficit of social interaction and are known as a model of SCZ-related behavior, *C1qb* transcription is lower in the cortex, hippocampus, and hypothalamus as compared to C57BL/6N mice [132]. Consequently, changes in the transcription of *C1qb* in BTBR mice also support the involvement of inflammation in the pathogenesis of their autistic-like phenotype. These results are also consistent with previous studies in which BTBR mice have been shown to have an impaired immune response [133,134].

Pop4 is a subunit of a ribonucleoprotein enzyme called ribonuclease P (RNase P): an essential enzyme that catalyzes the removal of the 5′ leader sequence from precursor tRNAs [135]. There is lack of information on the role of Pop4 in neuronal functions and neuropathology. In general, the evidence of RNase P participation in neuronal function is scarce [136–139]. In a paper by Cai et al. [137], it is demonstrated that a knockdown of *Rpph1* diminishes dendritic spine density in primary culture of hippocampal pyramidal neurons. One report suggests that in the PFC of autism patients, transcription of *RPP25* is low [140]. The weak *Pop4* transcription in the brain of BTBR mice also suggests that RNase P may be implicated in the development of the autistic-like phenotype in these animals.

Nudt19 is a member of the Nudix hydrolase family with an RNA-decapping activity [141,142]. Additionally, Nudt19 activity controls coenzyme A degradation [143,144] and fatty acid oxidation in hepatic cells [145]. In the hepatocellular carcinoma cell line, Nudt19 activated the mTORC1–P70S6K signaling pathway [146]. To date, there have been only two studies on Nudt19 in the context of neuronal functions. On the basis of whole-genome microarray analyses, Arisi et al. [147] proposed that Nudt19 is a potential biomarker of the early stage of Alzheimer's-disease-like neurodegeneration in mice. Recently, an analysis of the striatal proteome of depression-susceptible and anxiety-susceptible and -insusceptible rat cohorts detected Nudt19 among abnormally expressed proteins [148]. Although the exact functions of Nudt19 inside neuronal cells are still unclear, the regulation of cellular bioenergetics and of mTOR signaling by Nudt19 is an intriguing topic and implies a major role in neuronal functions and neuropathology as well.

Aside from *Wfs1*, our comparison of transcriptomic and proteomic data revealed unidirectional expression changes for the *Clcn6* gene in the hippocampus. Mutations in *Clcn6* in mice lead to mild lysosomal storage abnormalities, whereas in humans, a *CLCN6* mutation causes a severe developmental delay with pronounced neurological and neurodegenerative problems [149]. Proper functioning of Cl^-/H^+ exchangers is important for adequate activities of endosomes and lysosomes. Even in the presence of mild lysosomal abnormalities, in $Clcn6^{-/-}$ mice, abnormal storage of some substances causes pathological enlargement of proximal axons [150]. The marked astrocytosis in the cortex and lowered corpus callosum volume are also seen in $Clcn6^{-/-}$ mice [151].

Our proteomic data indicate that changes in the protein expression of *Macf1*, *Bsn*, *Psd3*, and *Chchd3* are unidirectional between the cortex and hippocampus. The *Chchd3* gene encodes a mitochondrial protein located in the intermembrane space and essential for maintaining crista integrity and mitochondrial function [152]. One article points to an association between CNVs in *CHCHD3* and an ASD [153].

Protein products of genes *Macf1*, *Bsn*, and *Psd3* are associated with cellular morphology and synaptic function. Microtubule–actin crosslinking factor 1 (MACF1) is a cytoskeleton-crosslinking protein that interacts with microtubules and F-actin to modulate the polarization of cells and coordination of cellular movements [154]. Furthermore, MACF1 participates in the Wnt–β-catenin signaling pathway. It is not surprising that MACF1 is important for cell migration, which requires continuous reconstruction of the cytoskeleton [155,156]. A number of reports indicate a contribution of MACF1 mutations to different neurological disorders including SCZ [154]. In the context of an ASD as a neurodevelopmental disorder involving aberrant neuronal migration [157], it is highly likely that MACF1 deregulation may also participate in the pathological changes. Pleckstrin and sec7 domain-containing 3 (PSD3), also known as EFA6R, regulates localization of small GTPase ARF6, thereby promoting a cytoskeletal rearrangement [158]. At least in humans, PSD3 localization is restricted to the PFC [159], and *PSD3* has been identified among candidate genes related to Alzheimer's disease pathophysiology [160]. PSD3 was identified as one candidate of ASD-associated genes in a duplicated locus of chromosomal region 8p22-21.3 in ASD patients [161].

Bassoon (Bsn) is the presynaptically localized scaffolding protein that is a negative regulator of presynaptic autophagy and of the ubiquitin–proteasome system in the presynapse [162]. Research on bassoon knockout mice and cultured neurons indicates that this protein is a key regulator of synaptic vesicle proteostasis [163,164]. A conditional knockout of *Bsn* exclusively in forebrain excitatory neurons enhances hippocampal excitability and neurogenesis [165], which may be TrkB-dependent [166]. Participation of bassoon in a number of neuropathologies has been demonstrated. Inflammation-induced accumulation of bassoon in the central nervous systems of mice and humans boosts neurotoxic processes in multiple sclerosis [167]. A rare mutation in *BSN* correlates with a familial type of SCZ [168].

The abovementioned reproducible genes identified in transcriptomic and proteomic datasets fit well the functional pathways revealed by GSEA. For example, "sensory perception of smell" and "olfactory receptor activity" are consistent with the involvement of *Wfs1* in olfaction [169–171]. This finding is in good agreement with neuroanatomical anomalies in olfactory bulbs [98,100] and low capacity for odor discrimination in BTBR mice [172–174]. Likewise, the underexpression of genes associated with the regulation of corticotropin secretion is in agreement with corticosteroid dysregulation [175,176] and excessive stress hormone responses in BTBR mice [177,178]. The protein products of genes *Wfs1* and *Scg5* may directly take part in the above-mentioned pathways. Similarly, changes in the expression of *Macf1*, *Bsn*, and *Psd3* are consistent with "motile cilium" and "microtube bundle formation" pathways, and the *C1qb* transcription change is concordant with "innate immune response" uncovered by GSEA. Again, in BTBR mice, we can see relevant phenotypic changes linked with neuronal morphology and migration [97,179,180] and with the immune system [181–183].

Our comparison of the results of meta-analyses of DEGs in postmortem samples from SCZ or autism patients with reproducible DEGs in the corticostriatal area of BTBR mice yielded only a small overlap. Despite the substantial overlap between ASD and SCZ genetic risk factors in humans [18,28], such convergence is absent in mice. Nonetheless, it should be taken into account that the comparison of results on gene and protein expression between BTBR mice and postmortem samples from individuals with SCZ or autism has a number of limitations in terms of data interpretation. The first set of limitations has to do with the heterogeneity of experimental data both in humans (age, severity of the disease, genetics, and comorbidities) and in mice (age, coordinates of brain structures, sample preparation, and various data analysis algorithms). Postmortem data also suffer from small sample sizes and lack of ethnic diversity because they are composed primarily of subjects with European or North American genetic backgrounds [184]. Differences in the design of experiments mean that we can analyze only the most pronounced aberrations. The second limitation is brain morphofunctional differences between rodents and humans. These dissimilarities are observed both in adulthood and in the neonatal period, when the stage of mouse brain development corresponds to the stage of development in the third trimester of pregnancy in humans [185]. Presently, it is not possible to unambiguously match different regions of the cerebral cortex between rodents and humans. On the other hand, the "multi-omics" approach employed here to characterize genetic architecture of BTBR mice uncovered some convergence between genes and pathways implicated in ASDs and in the autistic-like behavior of BTBR mice. Mahony and O'Ryan [23], in their review of proteomic, transcriptomic, and epigenomic data from postmortem brain samples from ASD patients, identified four canonical pathways enriched within seven or more independent datasets. mTOR signaling, oxidative phosphorylation, adipogenesis, and complement response pathways, which were revealed by Mahony and O'Ryan, are relevant to some genetic hallmarks of BTBR mice. For example, the observed alterations of transcription of *Serpina3n*, *C1qb*, and *Nudt19* and of protein expression of *Chchd3* all point to the involvement of mTOR signaling, of the complement response, and of mitochondrial function in the pathogenesis. These expression changes in genes are likely different from those in ASD patients but lead to similar phenotypic manifestations in BTBR mice. It can be hypothesized that we have an example of convergent development of an ASD-like phenotype and an ASD phenotype.

5. Conclusions

In this review, for the first time, we identified reproducible alterations of gene and protein expression in different regions of the brains of BTBR mice. Our findings indicate that these molecular signatures do not reproduce the expression changes observed in postmortem samples from ASD and SCZ patients. In addition, these changes in expression do not correlate well with the genetic background of these animals. To expand our knowledge about the molecular signatures of BTBR mice, transcriptomic and epigenomic studies on individual cell populations in different brain regions are required. These data will allow us to assess the contribution of epigenetic features to the magnitude of gene expression as well as cell-specific anomalies.

Supplementary Materials: The following supporting information can be downloaded at: https://www.mdpi.com/article/10.3390/biomedicines11020289/s1. Table S1: The list of most reproducible DE genes in the hippocampus of BTBR mice; Table S2: The list of most reproducible DE genes in the cortex and striatum of BTBR mice; Table S3: Gene set enrichment analysis of all expressed genes in the hippocampus; Table S4: Gene set enrichment analysis of all expressed genes in the striatum; Table S5: The list of "high-prediction-score" mutations; Table S6: The list of SZH-related genes; Table S7: The list of ASD-related genes.

Author Contributions: Conceptualization, methodology, analysis, writing, V.R.; analysis, P.K.; writing—original draft preparation, A.T.; review and editing, N.B.; All authors have read and agreed to the published version of the manuscript.

Funding: The study was supported by a publicly funded project (FWNR-2022-0002, Russia).

Institutional Review Board Statement: Not applicable.

Informed Consent Statement: Not applicable.

Data Availability Statement: Data supporting the findings of this study are available within the article and Supplementary Materials here.

Conflicts of Interest: The authors declare no conflict of interest.

References

1. Lai, M.-C.; Lombardo, M.V.; Baron-Cohen, S. Autism. *Lancet* **2014**, *383*, 896–910. [CrossRef] [PubMed]
2. Chiarotti, F.; Venerosi, A. Epidemiology of Autism Spectrum Disorders: A Review of Worldwide Prevalence Estimates Since 2014. *Brain Sci.* **2020**, *10*, 274. [CrossRef] [PubMed]
3. Li, Z.; Yang, L.; Chen, H.; Fang, Y.; Zhang, T.; Yin, X.; Man, J.; Yang, X.; Lu, M. Global, Regional and National Burden of Autism Spectrum Disorder from 1990 to 2019: Results from the Global Burden of Disease Study 2019. *Epidemiol. Psychiatr. Sci.* **2022**, *31*, e33. [CrossRef]
4. Owen, M.J.; Sawa, A.; Mortensen, P.B. Schizophrenia. *Lancet* **2016**, *388*, 86–97. [CrossRef] [PubMed]
5. Marenco, S.; Weinberger, D.R. The Neurodevelopmental Hypothesis of Schizophrenia: Following a Trail of Evidence from Cradle to Grave. *Dev. Psychopathol.* **2000**, *12*, 501–527. [CrossRef]
6. Fatemi, S.H.; Folsom, T.D. The Neurodevelopmental Hypothesis of Schizophrenia, Revisited. *Schizophr. Bull.* **2009**, *35*, 528–548. [CrossRef]
7. Insel, T.R. Rethinking Schizophrenia. *Nature* **2010**, *468*, 187–193. [CrossRef]
8. Murray, R.M.; Bhavsar, V.; Tripoli, G.; Howes, O. 30 Years on: How the Neurodevelopmental Hypothesis of Schizophrenia Morphed into the Developmental Risk Factor Model of Psychosis. *Schizophr. Bull.* **2017**, *43*, 1190–1196. [CrossRef]
9. Egerton, A.; Grace, A.A.; Stone, J.; Bossong, M.G.; Sand, M.; McGuire, P. Glutamate in Schizophrenia: Neurodevelopmental Perspectives and Drug Development. *Schizophr. Res.* **2020**, *223*, 59–70. [CrossRef]
10. King, B.H.; Lord, C. Is Schizophrenia on the Autism Spectrum? *Brain Res.* **2011**, *1380*, 34–41. [CrossRef]
11. de Lacy, N.; King, B.H. Revisiting the Relationship between Autism and Schizophrenia: Toward an Integrated Neurobiology. *Annu. Rev. Clin. Psychol.* **2013**, *9*, 555–587. [CrossRef] [PubMed]
12. Owen, M.J.; O'Donovan, M.C. Schizophrenia and the Neurodevelopmental Continuum:Evidence from Genomics. *World Psychiatry* **2017**, *16*, 227–235. [CrossRef] [PubMed]
13. Cooper, G.M.; Coe, B.P.; Girirajan, S.; Rosenfeld, J.A.; Vu, T.H.; Baker, C.; Williams, C.; Stalker, H.; Hamid, R.; Hannig, V.; et al. A Copy Number Variation Morbidity Map of Developmental Delay. *Nat. Genet.* **2011**, *43*, 838–846. [CrossRef] [PubMed]
14. Girirajan, S.; Brkanac, Z.; Coe, B.P.; Baker, C.; Vives, L.; Vu, T.H.; Shafer, N.; Bernier, R.; Ferrero, G.B.; Silengo, M.; et al. Relative Burden of Large CNVs on a Range of Neurodevelopmental Phenotypes. *PLoS Genet.* **2011**, *7*, e1002334. [CrossRef]
15. Kirov, G.; Rees, E.; Walters, J.T.R.; Escott-Price, V.; Georgieva, L.; Richards, A.L.; Chambert, K.D.; Davies, G.; Legge, S.E.; Moran, J.L.; et al. The Penetrance of Copy Number Variations for Schizophrenia and Developmental Delay. *Biol. Psychiatry* **2014**, *75*, 378–385. [CrossRef]
16. Zheng, Z.; Zheng, P.; Zou, X. Association between Schizophrenia and Autism Spectrum Disorder: A Systematic Review and Meta-Analysis. *Autism Res. Off. J. Int. Soc. Autism Res.* **2018**, *11*, 1110–1119. [CrossRef]
17. Lai, M.-C.; Kassee, C.; Besney, R.; Bonato, S.; Hull, L.; Mandy, W.; Szatmari, P.; Ameis, S.H. Prevalence of Co-Occurring Mental Health Diagnoses in the Autism Population: A Systematic Review and Meta-Analysis. *Lancet. Psychiatry* **2019**, *6*, 819–829. [CrossRef]
18. Jutla, A.; Foss-Feig, J.; Veenstra-VanderWeele, J. Autism Spectrum Disorder and Schizophrenia: An Updated Conceptual Review. *Autism Res. Off. J. Int. Soc. Autism Res.* **2022**, *15*, 384–412. [CrossRef]
19. Sandin, S.; Lichtenstein, P.; Kuja-Halkola, R.; Larsson, H.; Hultman, C.M.; Reichenberg, A. The Familial Risk of Autism. *JAMA* **2014**, *311*, 1770–1777. [CrossRef] [PubMed]
20. Warrier, V.; Zhang, X.; Reed, P.; Havdahl, A.; Moore, T.M.; Cliquet, F.; Leblond, C.S.; Rolland, T.; Rosengren, A.; Rowitch, D.H.; et al. Genetic Correlates of Phenotypic Heterogeneity in Autism. *Nat. Genet.* **2022**, *54*, 1293–1304. [CrossRef]
21. Iakoucheva, L.M.; Muotri, A.R.; Sebat, J. Getting to the Cores of Autism. *Cell* **2019**, *178*, 1287–1298. [CrossRef] [PubMed]
22. Qiu, S.; Qiu, Y.; Li, Y.; Cong, X. Genetics of Autism Spectrum Disorder: An Umbrella Review of Systematic Reviews and Meta-Analyses. *Transl. Psychiatry* **2022**, *12*, 249. [CrossRef] [PubMed]
23. Mahony, C.; O'Ryan, C. Convergent Canonical Pathways in Autism Spectrum Disorder from Proteomic, Transcriptomic and DNA Methylation Data. *Int. J. Mol. Sci.* **2021**, *22*, 10757. [CrossRef] [PubMed]
24. Ellis, S.E.; Panitch, R.; West, A.B.; Arking, D.E. Transcriptome Analysis of Cortical Tissue Reveals Shared Sets of Downregulated Genes in Autism and Schizophrenia. *Transl. Psychiatry* **2016**, *6*, e817. [CrossRef]
25. Gandal, M.J.; Zhang, P.; Hadjimichael, E.; Walker, R.L.; Chen, C.; Liu, S.; Won, H.; van Bakel, H.; Varghese, M.; Wang, Y.; et al. Transcriptome-Wide Isoform-Level Dysregulation in ASD, Schizophrenia, and Bipolar Disorder. *Science* **2018**, *362*, eaat8127. [CrossRef]
26. Gandal, M.J.; Haney, J.R.; Parikshak, N.N.; Leppa, V.; Ramaswami, G.; Hartl, C.; Schork, A.J.; Appadurai, V.; Buil, A.; Werge, T.M.; et al. Shared Molecular Neuropathology across Major Psychiatric Disorders Parallels Polygenic Overlap. *Science* **2018**, *359*, 693–697. [CrossRef]

27. Guan, J.; Cai, J.J.; Ji, G.; Sham, P.C. Commonality in Dysregulated Expression of Gene Sets in Cortical Brains of Individuals with Autism, Schizophrenia, and Bipolar Disorder. *Transl. Psychiatry* **2019**, *9*, 152. [CrossRef]

28. Nomura, J.; Mardo, M.; Takumi, T. Molecular Signatures from Multi-Omics of Autism Spectrum Disorders and Schizophrenia. *J. Neurochem.* **2021**, *159*, 647–659. [CrossRef]

29. Gao, R.; Penzes, P. Common Mechanisms of Excitatory and Inhibitory Imbalance in Schizophrenia and Autism Spectrum Disorders. *Curr. Mol. Med.* **2015**, *15*, 146–167. [CrossRef]

30. Scattoni, M.L.; Martire, A.; Cartocci, G.; Ferrante, A.; Ricceri, L. Reduced Social Interaction, Behavioural Flexibility and BDNF Signalling in the BTBR T+tf/J Strain, a Mouse Model of Autism. *Behav. Brain Res.* **2013**, *251*, 35–40. [CrossRef]

31. Ellegood, J.; Crawley, J.N. Behavioral and Neuroanatomical Phenotypes in Mouse Models of Autism. *Neurother. J. Am. Soc. Exp. Neurother.* **2015**, *12*, 521–533. [CrossRef] [PubMed]

32. Reshetnikov, V.V.; Ayriyants, K.A.; Ryabushkina, Y.A.; Sozonov, N.G.; Bondar, N.P. Sex-Specific Behavioral and Structural Alterations Caused by Early-Life Stress in C57BL/6 and BTBR Mice. *Behav. Brain Res.* **2021**, *414*, 113489. [CrossRef] [PubMed]

33. Meyza, K.Z.; Blanchard, D.C. The BTBR Mouse Model of Idiopathic Autism—Current View on Mechanisms. *Neurosci. Biobehav. Rev.* **2017**, *76*, 99–110. [CrossRef] [PubMed]

34. Mizuno, S.; Hirota, J.-N.; Ishii, C.; Iwasaki, H.; Sano, Y.; Furuichi, T. Comprehensive Profiling of Gene Expression in the Cerebral Cortex and Striatum of BTBRTF/ArtRbrc Mice Compared to C57BL/6J Mice. *Front. Cell. Neurosci.* **2020**, *14*, 595607. [CrossRef]

35. Bove, M.; Ike, K.; Eldering, A.; Buwalda, B.; de Boer, S.F.; Morgese, M.G.; Schiavone, S.; Cuomo, V.; Trabace, L.; Kas, M.J.H. The Visible Burrow System: A Behavioral Paradigm to Assess Sociability and Social Withdrawal in BTBR and C57BL/6J Mice Strains. *Behav. Brain Res.* **2018**, *344*, 9–19. [CrossRef]

36. Gogolla, N.; Takesian, A.E.; Feng, G.; Fagiolini, M.; Hensch, T.K. Sensory Integration in Mouse Insular Cortex Reflects GABA Circuit Maturation. *Neuron* **2014**, *83*, 894–905. [CrossRef]

37. Clapcote, S.J.; Roder, J.C. Deletion Polymorphism of Disc1 Is Common to All 129 Mouse Substrains: Implications for Gene-Targeting Studies of Brain Function. *Genetics* **2006**, *173*, 2407–2410. [CrossRef]

38. Gasparini, S.; Del Vecchio, G.; Gioiosa, S.; Flati, T.; Castrignano, T.; Legnini, I.; Licursi, V.; Ricceri, L.; Scattoni, M.L.; Rinaldi, A.; et al. Differential Expression of Hippocampal Circular RNAs in the BTBR Mouse Model for Autism Spectrum Disorder. *Mol. Neurobiol.* **2020**, *57*, 2301–2313. [CrossRef]

39. Provenzano, G.; Corradi, Z.; Monsorno, K.; Fedrizzi, T.; Ricceri, L.; Scattoni, M.L.; Bozzi, Y. Comparative Gene Expression Analysis of Two Mouse Models of Autism: Transcriptome Profiling of the BTBR and En2$^{-/-}$ Hippocampus. *Front. Neurosci.* **2016**, *10*, 396. [CrossRef]

40. Daimon, C.M.; Jasien, J.M.; Wood, W.H.; Zhang, Y.; Becker, K.G.; Silverman, J.L.; Crawley, J.N.; Martin, B.; Maudsley, S. Hippocampal Transcriptomic and Proteomic Alterations in the BTBR Mouse Model of Autism Spectrum Disorder. *Front. Physiol.* **2015**, *6*, 324. [CrossRef]

41. Wang, M.; He, J.; Zhou, Y.; Lv, N.; Zhao, M.; Wei, H.; Li, R. Integrated Analysis of MiRNA and MRNA Expression Profiles in the Brains of BTBR Mice. *Int. J. Dev. Neurosci. Off. J. Int. Soc. Dev. Neurosci.* **2020**, *80*, 221–233. [CrossRef] [PubMed]

42. Oron, O.; Getselter, D.; Shohat, S.; Reuveni, E.; Lukic, I.; Shifman, S.; Elliott, E. Gene Network Analysis Reveals a Role for Striatal Glutamatergic Receptors in Dysregulated Risk-Assessment Behavior of Autism Mouse Models. *Transl. Psychiatry* **2019**, *9*, 257. [CrossRef] [PubMed]

43. Wei, H.; Ma, Y.; Liu, J.; Ding, C.; Hu, F.; Yu, L. Proteomic Analysis of Cortical Brain Tissue from the BTBR Mouse Model of Autism: Evidence for Changes in STOP and Myelin-Related Proteins. *Neuroscience* **2016**, *312*, 26–34. [CrossRef] [PubMed]

44. McLaren, W.; Gil, L.; Hunt, S.E.; Riat, H.S.; Ritchie, G.R.S.; Thormann, A.; Flicek, P.; Cunningham, F. The Ensembl Variant Effect Predictor. *Genome Biol.* **2016**, *17*, 122. [CrossRef] [PubMed]

45. Reshetnikov, V.V.; Kisaretova, P.E.; Ershov, N.I.; Shulyupova, A.S.; Oshchepkov, D.Y.; Klimova, N.V.; Ivanchihina, A.V.; Merkulova, T.I.; Bondar, N.P. Genes Associated with Cognitive Performance in the Morris Water Maze: An RNA-Seq Study. *Sci. Rep.* **2020**, *10*, 22078. [CrossRef]

46. Reshetnikov, V.V.; Kisaretova, P.E.; Ershov, N.I.; Merkulova, T.I.; Bondar, N.P. Social Defeat Stress in Adult Mice Causes Alterations in Gene Expression, Alternative Splicing, and the Epigenetic Landscape of H3K4me3 in the Prefrontal Cortex: An Impact of Early-Life Stress. *Prog. Neuropsychopharmacol. Biol. Psychiatry* **2021**, *106*, 110068. [CrossRef]

47. Bolger, A.M.; Lohse, M.; Usadel, B. Trimmomatic: A Flexible Trimmer for Illumina Sequence Data. *Bioinformatics* **2014**, *30*, 2114–2120. [CrossRef]

48. Kim, D.; Langmead, B.; Salzberg, S.L. HISAT: A Fast Spliced Aligner with Low Memory Requirements. *Nat. Methods* **2015**, *12*, 357–360. [CrossRef]

49. Liao, Y.; Smyth, G.K.; Shi, W. FeatureCounts: An Efficient General Purpose Program for Assigning Sequence Reads to Genomic Features. *Bioinformatics* **2014**, *30*, 923–930. [CrossRef]

50. Love, M.I.; Huber, W.; Anders, S. Moderated Estimation of Fold Change and Dispersion for RNA-Seq Data with DESeq2. *Genome Biol.* **2014**, *15*, 550. [CrossRef]

51. Forés-Martos, J.; Catalá-López, F.; Sánchez-Valle, J.; Ibáñez, K.; Tejero, H.; Palma-Gudiel, H.; Climent, J.; Pancaldi, V.; Fañanás, L.; Arango, C.; et al. Transcriptomic Metaanalyses of Autistic Brains Reveals Shared Gene Expression and Biological Pathway Abnormalities with Cancer. *Mol. Autism* **2019**, *10*, 17. [CrossRef] [PubMed]

52. Rahman, M.R.; Petralia, M.C.; Ciurleo, R.; Bramanti, A.; Fagone, P.; Shahjaman, M.; Wu, L.; Sun, Y.; Turanli, B.; Arga, K.Y.; et al. Comprehensive Analysis of RNA-Seq Gene Expression Profiling of Brain Transcriptomes Reveals Novel Genes, Regulators, and Pathways in Autism Spectrum Disorder. *Brain Sci.* **2020**, *10*, 747. [CrossRef]

53. Manchia, M.; Piras, I.S.; Huentelman, M.J.; Pinna, F.; Zai, C.C.; Kennedy, J.L.; Carpiniello, B. Pattern of Gene Expression in Different Stages of Schizophrenia: Down-Regulation of NPTX2 Gene Revealed by a Meta-Analysis of Microarray Datasets. *Eur. Neuropsychopharmacol. J. Eur. Coll. Neuropsychopharmacol.* **2017**, *27*, 1054–1063. [CrossRef]

54. Qin, W.; Liu, C.; Sodhi, M.; Lu, H. Meta-Analysis of Sex Differences in Gene Expression in Schizophrenia. *BMC Syst. Biol.* **2016**, *10* (Suppl. S1), 9. [CrossRef]

55. Li, L.; Venkataraman, L.; Chen, S.; Fu, H. Function of WFS1 and WFS2 in the Central Nervous System: Implications for Wolfram Syndrome and Alzheimer's Disease. *Neurosci. Biobehav. Rev.* **2020**, *118*, 775–783. [CrossRef] [PubMed]

56. He, H.; Cao, X.; Yin, F.; Wu, T.; Stauber, T.; Peng, J. West Syndrome Caused by a Chloride/Proton Exchange-Uncoupling CLCN6 Mutation Related to Autophagic-Lysosomal Dysfunction. *Mol. Neurobiol.* **2021**, *58*, 2990–2999. [CrossRef] [PubMed]

57. McKenzie, A.T.; Wang, M.; Hauberg, M.E.; Fullard, J.F.; Kozlenkov, A.; Keenan, A.; Hurd, Y.L.; Dracheva, S.; Casaccia, P.; Roussos, P.; et al. Brain Cell Type Specific Gene Expression and Co-Expression Network Architectures. *Sci. Rep.* **2018**, *8*, 8868. [CrossRef]

58. Altin, J.G.; Sloan, E.K. The Role of CD45 and CD45-Associated Molecules in T Cell Activation. *Immunol. Cell Biol.* **1997**, *75*, 430–445. [CrossRef]

59. Marsango, S.; Barki, N.; Jenkins, L.; Tobin, A.B.; Milligan, G. Therapeutic Validation of an Orphan G Protein-Coupled Receptor: The Case of GPR84. *Br. J. Pharmacol.* **2022**, *179*, 3529–3541. [CrossRef]

60. Pappaianni, E.; Siugzdaite, R.; Vettori, S.; Venuti, P.; Job, R.; Grecucci, A. Three Shades of Grey: Detecting Brain Abnormalities in Children with Autism Using Source-, Voxel- and Surface-Based Morphometry. *Eur. J. Neurosci.* **2018**, *47*, 690–700. [CrossRef]

61. Crucitti, J.; Hyde, C.; Enticott, P.G.; Stokes, M.A. A Systematic Review of Frontal Lobe Volume in Autism Spectrum Disorder Revealing Distinct Trajectories. *J. Integr. Neurosci.* **2022**, *21*, 57. [CrossRef] [PubMed]

62. Hardan, A.Y.; Muddasani, S.; Vemulapalli, M.; Keshavan, M.S.; Minshew, N.J. An MRI Study of Increased Cortical Thickness in Autism. *Am. J. Psychiatry* **2006**, *163*, 1290–1292. [CrossRef] [PubMed]

63. Mahajan, R.; Mostofsky, S.H. Neuroimaging Endophenotypes in Autism Spectrum Disorder. *CNS Spectr.* **2015**, *20*, 412–426. [CrossRef] [PubMed]

64. Libero, L.E.; Burge, W.K.; Deshpande, H.D.; Pestilli, F.; Kana, R.K. White Matter Diffusion of Major Fiber Tracts Implicated in Autism Spectrum Disorder. *Brain Connect.* **2016**, *6*, 691–699. [CrossRef]

65. Lucibello, S.; Bertè, G.; Verdolotti, T.; Lucignani, M.; Napolitano, A.; D'Abronzo, R.; Cicala, M.G.; Pede, E.; Chieffo, D.; Mariotti, P.; et al. Cortical Thickness and Clinical Findings in Prescholar Children with Autism Spectrum Disorder. *Front. Neurosci.* **2021**, *15*, 776860. [CrossRef] [PubMed]

66. Donovan, A.P.A.; Basson, M.A. The Neuroanatomy of Autism—A Developmental Perspective. *J. Anat.* **2017**, *230*, 4–15. [CrossRef]

67. Sparks, B.F.; Friedman, S.D.; Shaw, D.W.; Aylward, E.H.; Echelard, D.; Artru, A.A.; Maravilla, K.R.; Giedd, J.N.; Munson, J.; Dawson, G.; et al. Brain Structural Abnormalities in Young Children with Autism Spectrum Disorder. *Neurology* **2002**, *59*, 184–192. [CrossRef]

68. Schumann, C.M.; Hamstra, J.; Goodlin-Jones, B.L.; Lotspeich, L.J.; Kwon, H.; Buonocore, M.H.; Lammers, C.R.; Reiss, A.L.; Amaral, D.G. The Amygdala Is Enlarged in Children but not Adolescents with Autism; the Hippocampus Is Enlarged at All Ages. *J. Neurosci. Off. J. Soc. Neurosci.* **2004**, *24*, 6392–6401. [CrossRef]

69. Rojas, D.C.; Peterson, E.; Winterrowd, E.; Reite, M.L.; Rogers, S.J.; Tregellas, J.R. Regional Gray Matter Volumetric Changes in Autism Associated with Social and Repetitive Behavior Symptoms. *BMC Psychiatry* **2006**, *6*, 56. [CrossRef]

70. Groen, W.; Teluij, M.; Buitelaar, J.; Tendolkar, I. Amygdala and Hippocampus Enlargement during Adolescence in Autism. *J. Am. Acad. Child Adolesc. Psychiatry* **2010**, *49*, 552–560. [CrossRef]

71. Hasan, K.M.; Walimuni, I.S.; Frye, R.E. Global Cerebral and Regional Multimodal Neuroimaging Markers of the Neurobiology of Autism: Development and Cognition. *J. Child Neurol.* **2013**, *28*, 874–885. [CrossRef] [PubMed]

72. Zuo, C.; Wang, D.; Tao, F.; Wang, Y. Changes in the Development of Subcortical Structures in Autism Spectrum Disorder. *Neuroreport* **2019**, *30*, 1062–1067. [CrossRef] [PubMed]

73. Reinhardt, V.P.; Iosif, A.-M.; Libero, L.; Heath, B.; Rogers, S.J.; Ferrer, E.; Nordahl, C.; Ghetti, S.; Amaral, D.; Solomon, M. Understanding Hippocampal Development in Young Children with Autism Spectrum Disorder. *J. Am. Acad. Child Adolesc. Psychiatry* **2020**, *59*, 1069–1079. [CrossRef] [PubMed]

74. Li, G.; Chen, M.-H.; Li, G.; Wu, D.; Lian, C.; Sun, Q.; Rushmore, R.J.; Wang, L. Volumetric Analysis of Amygdala and Hippocampal Subfields for Infants with Autism. *J. Autism Dev. Disord.* **2022**, 1–15. [CrossRef] [PubMed]

75. Richards, R.; Greimel, E.; Kliemann, D.; Koerte, I.K.; Schulte-Körne, G.; Reuter, M.; Wachinger, C. Increased Hippocampal Shape Asymmetry and Volumetric Ventricular Asymmetry in Autism Spectrum Disorder. *NeuroImage. Clin.* **2020**, *26*, 102207. [CrossRef]

76. Qin, B.; Wang, L.; Cai, J.; Li, T.; Zhang, Y. Functional Brain Networks in Preschool Children with Autism Spectrum Disorders. *Front. Psychiatry* **2022**, *13*, 896388. [CrossRef]

77. Qiu, A.; Adler, M.; Crocetti, D.; Miller, M.I.; Mostofsky, S.H. Basal Ganglia Shapes Predict Social, Communication, and Motor Dysfunctions in Boys with Autism Spectrum Disorder. *J. Am. Acad. Child Adolesc. Psychiatry* **2010**, *49*, 539–551. [CrossRef]

78. Schuetze, M.; Park, M.T.M.; Cho, I.Y.; MacMaster, F.P.; Chakravarty, M.M.; Bray, S.L. Morphological Alterations in the Thalamus, Striatum, and Pallidum in Autism Spectrum Disorder. *Neuropsychopharmacol. Off. Publ. Am. Coll. Neuropsychopharmacol.* **2016**, *41*, 2627–2637. [CrossRef]
79. Voelbel, G.T.; Bates, M.E.; Buckman, J.F.; Pandina, G.; Hendren, R.L. Caudate Nucleus Volume and Cognitive Performance: Are They Related in Childhood Psychopathology? *Biol. Psychiatry* **2006**, *60*, 942–950. [CrossRef]
80. Langen, M.; Durston, S.; Staal, W.G.; Palmen, S.J.M.C.; van Engeland, H. Caudate Nucleus Is Enlarged in High-Functioning Medication-Naive Subjects with Autism. *Biol. Psychiatry* **2007**, *62*, 262–266. [CrossRef]
81. Estes, A.; Shaw, D.W.W.; Sparks, B.F.; Friedman, S.; Giedd, J.N.; Dawson, G.; Bryan, M.; Dager, S.R. Basal Ganglia Morphometry and Repetitive Behavior in Young Children with Autism Spectrum Disorder. *Autism Res. Off. J. Int. Soc. Autism Res.* **2011**, *4*, 212–220. [CrossRef] [PubMed]
82. Sussman, D.; Leung, R.C.; Vogan, V.M.; Lee, W.; Trelle, S.; Lin, S.; Cassel, D.B.; Chakravarty, M.M.; Lerch, J.P.; Anagnostou, E.; et al. The Autism Puzzle: Diffuse but Not Pervasive Neuroanatomical Abnormalities in Children with ASD. *NeuroImage Clin.* **2015**, *8*, 170–179. [CrossRef] [PubMed]
83. Van Rooij, D.; Anagnostou, E.; Arango, C.; Auzias, G.; Behrmann, M.; Busatto, G.F.; Calderoni, S.; Daly, E.; Deruelle, C.; Di Martino, A.; et al. Cortical and Subcortical Brain Morphometry Differences between Patients with Autism Spectrum Disorder and Healthy Individuals across the Lifespan: Results from the ENIGMA ASD Working Group. *Am. J. Psychiatry* **2018**, *175*, 359–369. [CrossRef] [PubMed]
84. Kuo, H.-Y.; Liu, F.-C. Pathological Alterations in Striatal Compartments in the Human Brain of Autism Spectrum Disorder. *Mol. Brain* **2020**, *13*, 83. [CrossRef] [PubMed]
85. Di Martino, A.; Kelly, C.; Grzadzinski, R.; Zuo, X.-N.; Mennes, M.; Mairena, M.A.; Lord, C.; Castellanos, F.X.; Milham, M.P. Aberrant Striatal Functional Connectivity in Children with Autism. *Biol. Psychiatry* **2011**, *69*, 847–856. [CrossRef] [PubMed]
86. Janouschek, H.; Chase, H.W.; Sharkey, R.J.; Peterson, Z.J.; Camilleri, J.A.; Abel, T.; Eickhoff, S.B.; Nickl-Jockschat, T. The Functional Neural Architecture of Dysfunctional Reward Processing in Autism. *NeuroImage Clin.* **2021**, *31*, 102700. [CrossRef]
87. Rubenstein, J.L.R. Annual Research Review: Development of the Cerebral Cortex: Implications for Neurodevelopmental Disorders. *J. Child Psychol. Psychiatry* **2011**, *52*, 339–355. [CrossRef]
88. Meechan, D.W.; Maynard, T.M.; Tucker, E.S.; Fernandez, A.; Karpinski, B.A.; Rothblat, L.A.; LaMantia, A.-S. Modeling a Model: Mouse Genetics, 22q11.2 Deletion Syndrome, and Disorders of Cortical Circuit Development. *Prog. Neurobiol.* **2015**, *130*, 1–28. [CrossRef]
89. Fenlon, L.R.; Liu, S.; Gobius, I.; Kurniawan, N.D.; Murphy, S.; Moldrich, R.X.; Richards, L.J. Formation of Functional Areas in the Cerebral Cortex Is Disrupted in a Mouse Model of Autism Spectrum Disorder. *Neural Dev.* **2015**, *10*, 10. [CrossRef]
90. Takumi, T.; Tamada, K.; Hatanaka, F.; Nakai, N.; Bolton, P.F. Behavioral Neuroscience of Autism. *Neurosci. Biobehav. Rev.* **2020**, *110*, 60–76. [CrossRef]
91. Schmeisser, M.J. Translational Neurobiology in Shank Mutant Mice–Model Systems for Neuropsychiatric Disorders. *Ann. Anat. Anat. Anz. Off. Organ Anat. Ges.* **2015**, *200*, 115–117. [CrossRef] [PubMed]
92. Bostrom, C.; Yau, S.-Y.; Majaess, N.; Vetrici, M.; Gil-Mohapel, J.; Christie, B.R. Hippocampal Dysfunction and Cognitive Impairment in Fragile-X Syndrome. *Neurosci. Biobehav. Rev.* **2016**, *68*, 563–574. [CrossRef] [PubMed]
93. Fuccillo, M. V Striatal Circuits as a Common Node for Autism Pathophysiology. *Front. Neurosci.* **2016**, *10*, 27. [CrossRef] [PubMed]
94. Peter, S.; De Zeeuw, C.I.; Boeckers, T.M.; Schmeisser, M.J. Cerebellar and Striatal Pathologies in Mouse Models of Autism Spectrum Disorder. *Adv. Anat. Embryol. Cell Biol.* **2017**, *224*, 103–119. [CrossRef] [PubMed]
95. Thabault, M.; Turpin, V.; Maisterrena, A.; Jaber, M.; Egloff, M.; Galvan, L. Cerebellar and Striatal Implications in Autism Spectrum Disorders: From Clinical Observations to Animal Models. *Int. J. Mol. Sci.* **2022**, *23*, 2294. [CrossRef]
96. Tsurugizawa, T. Translational Magnetic Resonance Imaging in Autism Spectrum Disorder from the Mouse Model to Human. *Front. Neurosci.* **2022**, *16*, 872036. [CrossRef]
97. Meyza, K.Z.; Defensor, E.B.; Jensen, A.L.; Corley, M.J.; Pearson, B.L.; Pobbe, R.L.H.; Bolivar, V.J.; Blanchard, D.C.; Blanchard, R.J. The BTBR T+ Tf/J Mouse Model for Autism Spectrum Disorders-in Search of Biomarkers. *Behav. Brain Res.* **2013**, *251*, 25–34. [CrossRef]
98. Dodero, L.; Damiano, M.; Galbusera, A.; Bifone, A.; Tsaftsaris, S.A.; Scattoni, M.L.; Gozzi, A. Neuroimaging Evidence of Major Morpho-Anatomical and Functional Abnormalities in the BTBR T+TF/J Mouse Model of Autism. *PLoS ONE* **2013**, *8*, e76655. [CrossRef]
99. Faraji, J.; Karimi, M.; Lawrence, C.; Mohajerani, M.H.; Metz, G.A.S. Non-Diagnostic Symptoms in a Mouse Model of Autism in Relation to Neuroanatomy: The BTBR Strain Reinvestigated. *Transl. Psychiatry* **2018**, *8*, 234. [CrossRef]
100. Ellegood, J.; Babineau, B.A.; Henkelman, R.M.; Lerch, J.P.; Crawley, J.N. Neuroanatomical Analysis of the BTBR Mouse Model of Autism Using Magnetic Resonance Imaging and Diffusion Tensor Imaging. *Neuroimage* **2013**, *70*, 288–300. [CrossRef]
101. Pagani, M.; Damiano, M.; Galbusera, A.; Tsaftaris, S.A.; Gozzi, A. Semi-Automated Registration-Based Anatomical Labelling, Voxel Based Morphometry and Cortical Thickness Mapping of the Mouse Brain. *J. Neurosci. Methods* **2016**, *267*, 62–73. [CrossRef]
102. Sforazzini, F.; Bertero, A.; Dodero, L.; David, G.; Galbusera, A.; Scattoni, M.L.; Pasqualetti, M.; Gozzi, A. Altered Functional Connectivity Networks in Acallosal and Socially Impaired BTBR Mice. *Brain Struct. Funct.* **2016**, *221*, 941–954. [CrossRef] [PubMed]

103. Squillace, M.; Dodero, L.; Federici, M.; Migliarini, S.; Errico, F.; Napolitano, F.; Krashia, P.; Di Maio, A.; Galbusera, A.; Bifone, A.; et al. Dysfunctional Dopaminergic Neurotransmission in Asocial BTBR Mice. *Transl. Psychiatry* **2014**, *4*, e427. [CrossRef] [PubMed]
104. Miller, V.M.; Gupta, D.; Neu, N.; Cotroneo, A.; Boulay, C.B.; Seegal, R.F. Novel Inter-Hemispheric White Matter Connectivity in the BTBR Mouse Model of Autism. *Brain Res.* **2013**, *1513*, 26–33. [CrossRef] [PubMed]
105. Mbikay, M.; Seidah, N.G.; Chrétien, M. Neuroendocrine Secretory Protein 7B2: Structure, Expression and Functions. *Biochem. J.* **2001**, *357*, 329–342. [CrossRef] [PubMed]
106. Chaplot, K.; Jarvela, T.S.; Lindberg, I. Secreted Chaperones in Neurodegeneration. *Front. Aging Neurosci.* **2020**, *12*, 268. [CrossRef] [PubMed]
107. Laurent, V.; Kimble, A.; Peng, B.; Zhu, P.; Pintar, J.E.; Steiner, D.F.; Lindberg, I. Mortality in 7B2 Null Mice Can Be Rescued by Adrenalectomy: Involvement of Dopamine in ACTH Hypersecretion. *Proc. Natl. Acad. Sci. USA* **2002**, *99*, 3087–3092. [CrossRef]
108. Peinado, J.R.; Laurent, V.; Lee, S.-N.; Peng, B.W.; Pintar, J.E.; Steiner, D.F.; Lindberg, I. Strain-Dependent Influences on the Hypothalamo-Pituitary-Adrenal Axis Profoundly Affect the 7B2 and PC2 Null Phenotypes. *Endocrinology* **2005**, *146*, 3438–3444. [CrossRef]
109. Lee, S.-N.; Lindberg, I. 7B2 Prevents Unfolding and Aggregation of Prohormone Convertase 2. *Endocrinology* **2008**, *149*, 4116–4127. [CrossRef]
110. Helwig, M.; Hoshino, A.; Berridge, C.; Lee, S.-N.; Lorenzen, N.; Otzen, D.E.; Eriksen, J.L.; Lindberg, I. The Neuroendocrine Protein 7B2 Suppresses the Aggregation of Neurodegenerative Disease-Related Proteins. *J. Biol. Chem.* **2013**, *288*, 1114–1124. [CrossRef]
111. Iguchi, H.; Chan, J.S.; Seidah, N.G.; Chrétien, M. Evidence for a Novel Pituitary Protein (7B2) in Human Brain, Cerebrospinal Fluid and Plasma: Brain Concentrations in Controls and Patients with Alzheimer's Disease. *Peptides* **1987**, *8*, 593–598. [CrossRef] [PubMed]
112. Winsky-Sommerer, R.; Grouselle, D.; Rougeot, C.; Laurent, V.; David, J.-P.; Delacourte, A.; Dournaud, P.; Seidah, N.G.; Lindberg, I.; Trottier, S.; et al. The Proprotein Convertase PC2 Is Involved in the Maturation of Prosomatostatin to Somatostatin-14 but Not in the Somatostatin Deficit in Alzheimer's Disease. *Neuroscience* **2003**, *122*, 437–447. [CrossRef] [PubMed]
113. Kato, C.; Tochigi, M.; Koishi, S.; Kawakubo, Y.; Yamamoto, K.; Matsumoto, H.; Hashimoto, O.; Kim, S.-Y.; Watanabe, K.; Kano, Y.; et al. Association Study of the Commonly Recognized Breakpoints in Chromosome 15q11-Q13 in Japanese Autistic Patients. *Psychiatr. Genet.* **2008**, *18*, 133–136. [CrossRef] [PubMed]
114. Wang, L.; Liu, H.; Zhang, X.; Song, E.; Wang, Y.; Xu, T.; Li, Z. WFS1 Functions in ER Export of Vesicular Cargo Proteins in Pancreatic β-Cells. *Nat. Commun.* **2021**, *12*, 6996. [CrossRef]
115. Tein, K.; Kasvandik, S.; Kõks, S.; Vasar, E.; Terasmaa, A. Prohormone Convertase 2 Activity Is Increased in the Hippocampus of Wfs1 Knockout Mice. *Front. Mol. Neurosci.* **2015**, *8*, 45. [CrossRef]
116. Swift, R.G.; Polymeropoulos, M.H.; Torres, R.; Swift, M. Predisposition of Wolfram Syndrome Heterozygotes to Psychiatric Illness. *Mol. Psychiatry* **1998**, *3*, 86–91. [CrossRef]
117. Swift, M.; Swift, R.G. Wolframin Mutations and Hospitalization for Psychiatric Illness. *Mol. Psychiatry* **2005**, *10*, 799–803. [CrossRef]
118. Munshani, S.; Ibrahim, E.Y.; Domenicano, I.; Ehrlich, B.E. The Impact of Mutations in Wolframin on Psychiatric Disorders. *Front. Pediatr.* **2021**, *9*, 718132. [CrossRef]
119. Urano, F. Wolfram Syndrome: Diagnosis, Management, and Treatment. *Curr. Diab. Rep.* **2016**, *16*, 6. [CrossRef]
120. Zamanian, J.L.; Xu, L.; Foo, L.C.; Nouri, N.; Zhou, L.; Giffard, R.G.; Barres, B.A. Genomic Analysis of Reactive Astrogliosis. *J. Neurosci. Off. J. Soc. Neurosci.* **2012**, *32*, 6391–6410. [CrossRef]
121. Haile, Y.; Carmine-Simmen, K.; Olechowski, C.; Kerr, B.; Bleackley, R.C.; Giuliani, F. Granzyme B-Inhibitor Serpina3n Induces Neuroprotection in Vitro and in Vivo. *J. Neuroinflamm.* **2015**, *12*, 157. [CrossRef] [PubMed]
122. Zheng, P.; Bai, Q.; Feng, J.; Zhao, B.; Duan, J.; Zhao, L.; Liu, N.; Ren, D.; Zou, S.; Chen, W. Neonatal Microglia and Proteinase Inhibitors-Treated Adult Microglia Improve Traumatic Brain Injury in Rats by Resolving the Neuroinflammation. *Bioeng. Transl. Med.* **2022**, *7*, e10249. [CrossRef] [PubMed]
123. Zhang, Y.; Chen, Q.; Chen, D.; Zhao, W.; Wang, H.; Yang, M.; Xiang, Z.; Yuan, H. SerpinA3N Attenuates Ischemic Stroke Injury by Reducing Apoptosis and Neuroinflammation. *CNS Neurosci. Ther.* **2022**, *28*, 566–579. [CrossRef]
124. Xi, Y.; Liu, M.; Xu, S.; Hong, H.; Chen, M.; Tian, L.; Xie, J.; Deng, P.; Zhou, C.; Zhang, L.; et al. Inhibition of SERPINA3N-Dependent Neuroinflammation Is Essential for Melatonin to Ameliorate Trimethyltin Chloride-Induced Neurotoxicity. *J. Pineal Res.* **2019**, *67*, e12596. [CrossRef]
125. Murphy, C.E.; Kondo, Y.; Walker, A.K.; Rothmond, D.A.; Matsumoto, M.; Shannon Weickert, C. Regional, Cellular and Species Difference of Two Key Neuroinflammatory Genes Implicated in Schizophrenia. *Brain. Behav. Immun.* **2020**, *88*, 826–839. [CrossRef] [PubMed]
126. Chrast, R.; Verheijen, M.H.G.; Lemke, G. Complement Factors in Adult Peripheral Nerve: A Potential Role in Energy Metabolism. *Neurochem. Int.* **2004**, *45*, 353–359. [CrossRef]
127. Ramaglia, V.; Daha, M.R.; Baas, F. The Complement System in the Peripheral Nerve: Friend or Foe? *Mol. Immunol.* **2008**, *45*, 3865–3877. [CrossRef]

128. Spielman, L.; Winger, D.; Ho, L.; Aisen, P.S.; Shohami, E.; Pasinetti, G.M. Induction of the Complement Component C1qB in Brain of Transgenic Mice with Neuronal Overexpression of Human Cyclooxygenase-2. *Acta Neuropathol.* **2002**, *103*, 157–162. [CrossRef]
129. Swanberg, M.; Duvefelt, K.; Diez, M.; Hillert, J.; Olsson, T.; Piehl, F.; Lidman, O. Genetically Determined Susceptibility to Neurodegeneration Is Associated with Expression of Inflammatory Genes. *Neurobiol. Dis.* **2006**, *24*, 67–88. [CrossRef]
130. Kraft, A.D.; McPherson, C.A.; Harry, G.J. Association Between Microglia, Inflammatory Factors, and Complement with Loss of Hippocampal Mossy Fiber Synapses Induced by Trimethyltin. *Neurotox. Res.* **2016**, *30*, 53–66. [CrossRef]
131. Wu, C.; Bendriem, R.M.; Freed, W.J.; Lee, C.-T. Transcriptome Analysis of Human Dorsal Striatum Implicates Attenuated Canonical WNT Signaling in Neuroinflammation and in Age-Related Impairment of Striatal Neurogenesis and Synaptic Plasticity. *Restor. Neurol. Neurosci.* **2021**, *39*, 247–266. [CrossRef]
132. Ma, L.; Kulesskaya, N.; Võikar, V.; Tian, L. Differential Expression of Brain Immune Genes and Schizophrenia-Related Behavior in C57BL/6N and DBA/2J Female Mice. *Psychiatry Res.* **2015**, *226*, 211–216. [CrossRef]
133. Careaga, M.; Schwartzer, J.; Ashwood, P. Inflammatory Profiles in the BTBR Mouse: How Relevant Are They to Autism Spectrum Disorders? *Brain. Behav. Immun.* **2015**, *43*, 11–16. [CrossRef]
134. Mutovina, A.; Ayriyants, K.; Mezhlumyan, E.; Ryabushkina, Y.; Litvinova, E.; Bondar, N.; Khantakova, J.; Reshetnikov, V. Unique Features of the Immune Response in BTBR Mice. *Int. J. Mol. Sci.* **2022**, *23*, 15577. [CrossRef] [PubMed]
135. Phan, H.-D.; Lai, L.B.; Zahurancik, W.J.; Gopalan, V. The Many Faces of RNA-Based RNase P, an RNA-World Relic. *Trends Biochem. Sci.* **2021**, *46*, 976–991. [CrossRef]
136. Lipovich, L.; Dachet, F.; Cai, J.; Bagla, S.; Balan, K.; Jia, H.; Loeb, J.A. Activity-Dependent Human Brain Coding/Noncoding Gene Regulatory Networks. *Genetics* **2012**, *192*, 1133–1148. [CrossRef]
137. Cai, Y.; Sun, Z.; Jia, H.; Luo, H.; Ye, X.; Wu, Q.; Xiong, Y.; Zhang, W.; Wan, J. Rpph1 Upregulates CDC42 Expression and Promotes Hippocampal Neuron Dendritic Spine Formation by Competing with MiR-330-5p. *Front. Mol. Neurosci.* **2017**, *10*, 27. [CrossRef] [PubMed]
138. Gu, R.; Wang, L.; Tang, M.; Li, S.-R.; Liu, R.; Hu, X. LncRNA Rpph1 Protects Amyloid-β Induced Neuronal Injury in SK-N-SH Cells via MiR-122/Wnt1 Axis. *Int. J. Neurosci.* **2020**, *130*, 443–453. [CrossRef] [PubMed]
139. Gu, R.; Liu, R.; Wang, L.; Tang, M.; Li, S.-R.; Hu, X. LncRNA RPPH1 Attenuates Aβ(25-35)-Induced Endoplasmic Reticulum Stress and Apoptosis in SH-SY5Y Cells via MiR-326/PKM2. *Int. J. Neurosci.* **2021**, *131*, 425–432. [CrossRef]
140. Huang, H.-S.; Cheung, I.; Akbarian, S. RPP25 Is Developmentally Regulated in Prefrontal Cortex and Expressed at Decreased Levels in Autism Spectrum Disorder. *Autism Res. Off. J. Int. Soc. Autism Res.* **2010**, *3*, 153–161. [CrossRef]
141. Song, M.-G.; Bail, S.; Kiledjian, M. Multiple Nudix Family Proteins Possess MRNA Decapping Activity. *RNA* **2013**, *19*, 390–399. [CrossRef]
142. Sharma, S.; Grudzien-Nogalska, E.; Hamilton, K.; Jiao, X.; Yang, J.; Tong, L.; Kiledjian, M. Mammalian Nudix Proteins Cleave Nucleotide Metabolite Caps on RNAs. *Nucleic Acids Res.* **2020**, *48*, 6788–6798. [CrossRef]
143. Shumar, S.A.; Kerr, E.W.; Geldenhuys, W.J.; Montgomery, G.E.; Fagone, P.; Thirawatananond, P.; Saavedra, H.; Gabelli, S.B.; Leonardi, R. Nudt19 Is a Renal CoA Diphosphohydrolase with Biochemical and Regulatory Properties That Are Distinct from the Hepatic Nudt7 Isoform. *J. Biol. Chem.* **2018**, *293*, 4134–4148. [CrossRef]
144. Kerr, E.W.; Shumar, S.A.; Leonardi, R. Nudt8 Is a Novel CoA Diphosphohydrolase That Resides in the Mitochondria. *FEBS Lett.* **2019**, *593*, 1133–1143. [CrossRef]
145. Görigk, S.; Ouwens, D.M.; Kuhn, T.; Altenhofen, D.; Binsch, C.; Damen, M.; Khuong, J.M.-A.; Kaiser, K.; Knebel, B.; Vogel, H.; et al. Nudix Hydrolase NUDT19 Regulates Mitochondrial Function and ATP Production in Murine Hepatocytes. *Biochim. Biophys. Acta. Mol. Cell Biol. Lipids* **2022**, *1867*, 159153. [CrossRef]
146. Lan, C.; Wang, Y.; Su, X.; Lu, J.; Ma, S. LncRNA LINC00958 Activates MTORC1/P70S6K Signalling Pathway to Promote Epithelial-Mesenchymal Transition Process in the Hepatocellular Carcinoma. *Cancer Investig.* **2021**, *39*, 539–549. [CrossRef]
147. Arisi, I.; D'Onofrio, M.; Brandi, R.; Felsani, A.; Capsoni, S.; Drovandi, G.; Felici, G.; Weitschek, E.; Bertolazzi, P.; Cattaneo, A. Gene Expression Biomarkers in the Brain of a Mouse Model for Alzheimer's Disease: Mining of Microarray Data by Logic Classification and Feature Selection. *J. Alzheimer's Dis.* **2011**, *24*, 721–738. [CrossRef]
148. Cai, X.; Yang, C.; Chen, J.; Gong, W.; Yi, F.; Liao, W.; Huang, R.; Xie, L.; Zhou, J. Proteomic Insights into Susceptibility and Resistance to Chronic-Stress-Induced Depression or Anxiety in the Rat Striatum. *Front. Mol. Biosci.* **2021**, *8*, 730473. [CrossRef]
149. Polovitskaya, M.M.; Barbini, C.; Martinelli, D.; Harms, F.L.; Cole, F.S.; Calligari, P.; Bocchinfuso, G.; Stella, L.; Ciolfi, A.; Niceta, M.; et al. A Recurrent Gain-of-Function Mutation in CLCN6, Encoding the ClC-6 Cl^-/H^+-Exchanger, Causes Early-Onset Neurodegeneration. *Am. J. Hum. Genet.* **2020**, *107*, 1062–1077. [CrossRef]
150. Poët, M.; Kornak, U.; Schweizer, M.; Zdebik, A.A.; Scheel, O.; Hoelter, S.; Wurst, W.; Schmitt, A.; Fuhrmann, J.C.; Planells-Cases, R.; et al. Lysosomal Storage Disease upon Disruption of the Neuronal Chloride Transport Protein ClC-6. *Proc. Natl. Acad. Sci. USA* **2006**, *103*, 13854–13859. [CrossRef]
151. Pressey, S.N.R.; O'Donnell, K.J.; Stauber, T.; Fuhrmann, J.C.; Tyynelä, J.; Jentsch, T.J.; Cooper, J.D. Distinct Neuropathologic Phenotypes after Disrupting the Chloride Transport Proteins ClC-6 or ClC-7/Ostm1. *J. Neuropathol. Exp. Neurol.* **2010**, *69*, 1228–1246. [CrossRef]
152. Darshi, M.; Mendiola, V.L.; Mackey, M.R.; Murphy, A.N.; Koller, A.; Perkins, G.A.; Ellisman, M.H.; Taylor, S.S. ChChd3, an Inner Mitochondrial Membrane Protein, Is Essential for Maintaining Crista Integrity and Mitochondrial Function. *J. Biol. Chem.* **2011**, *286*, 2918–2932. [CrossRef]

153. Lionel, A.C.; Crosbie, J.; Barbosa, N.; Goodale, T.; Thiruvahindrapuram, B.; Rickaby, J.; Gazzellone, M.; Carson, A.R.; Howe, J.L.; Wang, Z.; et al. Rare Copy Number Variation Discovery and Cross-Disorder Comparisons Identify Risk Genes for ADHD. *Sci. Transl. Med.* **2011**, *3*, 95ra75. [CrossRef]

154. Moffat, J.J.; Ka, M.; Jung, E.-M.; Smith, A.L.; Kim, W.-Y. The Role of MACF1 in Nervous System Development and Maintenance. *Semin. Cell Dev. Biol.* **2017**, *69*, 9–17. [CrossRef]

155. Ka, M.; Jung, E.-M.; Mueller, U.; Kim, W.-Y. MACF1 Regulates the Migration of Pyramidal Neurons via Microtubule Dynamics and GSK-3 Signaling. *Dev. Biol.* **2014**, *395*, 4–18. [CrossRef]

156. Dobyns, W.B.; Aldinger, K.A.; Ishak, G.E.; Mirzaa, G.M.; Timms, A.E.; Grout, M.E.; Dremmen, M.H.G.; Schot, R.; Vandervore, L.; van Slegtenhorst, M.A.; et al. MACF1 Mutations Encoding Highly Conserved Zinc-Binding Residues of the GAR Domain Cause Defects in Neuronal Migration and Axon Guidance. *Am. J. Hum. Genet.* **2018**, *103*, 1009–1021. [CrossRef]

157. Courchesne, E.; Gazestani, V.H.; Lewis, N.E. Prenatal Origins of ASD: The When, What, and How of ASD Development. *Trends Neurosci.* **2020**, *43*, 326–342. [CrossRef]

158. Kanamarlapudi, V. Exchange Factor EFA6R Requires C-Terminal Targeting to the Plasma Membrane to Promote Cytoskeletal Rearrangement through the Activation of ADP-Ribosylation Factor 6 (ARF6). *J. Biol. Chem.* **2014**, *289*, 33378–33390. [CrossRef]

159. Gonzalez, D.A.; Jia, T.; Pinzón, J.H.; Acevedo, S.F.; Ojelade, S.A.; Xu, B.; Tay, N.; Desrivières, S.; Hernandez, J.L.; Banaschewski, T.; et al. The Arf6 Activator Efa6/PSD3 Confers Regional Specificity and Modulates Ethanol Consumption in Drosophila and Humans. *Mol. Psychiatry* **2018**, *23*, 621–628. [CrossRef]

160. Quan, X.; Liang, H.; Chen, Y.; Qin, Q.; Wei, Y.; Liang, Z. Related Network and Differential Expression Analyses Identify Nuclear Genes and Pathways in the Hippocampus of Alzheimer Disease. *Med. Sci. Monit. Int. Med. J. Exp. Clin. Res.* **2020**, *26*, e919311. [CrossRef]

161. Dong, P.; Xu, Q.; An, Y.; Zhou, B.-R.; Lu, P.; Liu, R.-C.; Xu, X. A Novel 1.0 Mb Duplication of Chromosome 8p22-21.3 in a Patient with Autism Spectrum Disorder. *Child Neurol. Open* **2015**, *2*, 1–6. [CrossRef]

162. Gundelfinger, E.D.; Karpova, A.; Pielot, R.; Garner, C.C.; Kreutz, M.R. Organization of Presynaptic Autophagy-Related Processes. *Front. Synaptic Neurosci.* **2022**, *14*, 829354. [CrossRef]

163. Hoffmann-Conaway, S.; Brockmann, M.M.; Schneider, K.; Annamneedi, A.; Rahman, K.A.; Bruns, C.; Textoris-Taube, K.; Trimbuch, T.; Smalla, K.-H.; Rosenmund, C.; et al. Parkin Contributes to Synaptic Vesicle Autophagy in Bassoon-Deficient Mice. *Elife* **2020**, *9*, e56590. [CrossRef]

164. Montenegro-Venegas, C.; Guhathakurta, D.; Pina-Fernandez, E.; Andres-Alonso, M.; Plattner, F.; Gundelfinger, E.D.; Fejtova, A. Bassoon Controls Synaptic Vesicle Release via Regulation of Presynaptic Phosphorylation and CAMP. *EMBO Rep.* **2022**, *23*, e53659. [CrossRef]

165. Annamneedi, A.; Caliskan, G.; Müller, S.; Montag, D.; Budinger, E.; Angenstein, F.; Fejtova, A.; Tischmeyer, W.; Gundelfinger, E.D.; Stork, O. Ablation of the Presynaptic Organizer Bassoon in Excitatory Neurons Retards Dentate Gyrus Maturation and Enhances Learning Performance. *Brain Struct. Funct.* **2018**, *223*, 3423–3445. [CrossRef]

166. Annamneedi, A.; Del Angel, M.; Gundelfinger, E.D.; Stork, O.; Çalışkan, G. The Presynaptic Scaffold Protein Bassoon in Forebrain Excitatory Neurons Mediates Hippocampal Circuit Maturation: Potential Involvement of TrkB Signalling. *Int. J. Mol. Sci.* **2021**, *22*, 7944. [CrossRef]

167. Schattling, B.; Engler, J.B.; Volkmann, C.; Rothammer, N.; Woo, M.S.; Petersen, M.; Winkler, I.; Kaufmann, M.; Rosenkranz, S.C.; Fejtova, A.; et al. Bassoon Proteinopathy Drives Neurodegeneration in Multiple Sclerosis. *Nat. Neurosci.* **2019**, *22*, 887–896. [CrossRef]

168. Chen, C.-H.; Huang, Y.-S.; Liao, D.-L.; Huang, C.-Y.; Lin, C.-H.; Fang, T.-H. Identification of Rare Mutations of Two Presynaptic Cytomatrix Genes BSN and PCLO in Schizophrenia and Bipolar Disorder. *J. Pers. Med.* **2021**, *11*, 1057. [CrossRef]

169. Takeda, K.; Inoue, H.; Tanizawa, Y.; Matsuzaki, Y.; Oba, J.; Watanabe, Y.; Shinoda, K.; Oka, Y. WFS1 (Wolfram Syndrome 1) Gene Product: Predominant Subcellular Localization to Endoplasmic Reticulum in Cultured Cells and Neuronal Expression in Rat Brain. *Hum. Mol. Genet.* **2001**, *10*, 477–484. [CrossRef]

170. Marshall, B.A.; Permutt, M.A.; Paciorkowski, A.R.; Hoekel, J.; Karzon, R.; Wasson, J.; Viehover, A.; White, N.H.; Shimony, J.S.; Manwaring, L.; et al. Phenotypic Characteristics of Early Wolfram Syndrome. *Orphanet J. Rare Dis.* **2013**, *8*, 64. [CrossRef]

171. Bischoff, A.N.; Reiersen, A.M.; Buttlaire, A.; Al-Lozi, A.; Doty, T.; Marshall, B.A.; Hershey, T. Selective Cognitive and Psychiatric Manifestations in Wolfram Syndrome. *Orphanet J. Rare Dis.* **2015**, *10*, 66. [CrossRef]

172. Moy, S.S.; Nadler, J.J.; Poe, M.D.; Nonneman, R.J.; Young, N.B.; Koller, B.H.; Crawley, J.N.; Duncan, G.E.; Bodfish, J.W. Development of a Mouse Test for Repetitive, Restricted Behaviors: Relevance to Autism. *Behav. Brain Res.* **2008**, *188*, 178–194. [CrossRef]

173. Yang, M.; Abrams, D.N.; Zhang, J.Y.; Weber, M.D.; Katz, A.M.; Clarke, A.M.; Silverman, J.L.; Crawley, J.N. Low Sociability in BTBR T+tf/J Mice Is Independent of Partner Strain. *Physiol. Behav.* **2012**, *107*, 649–662. [CrossRef]

174. Arakawa, H. Somatosensorimotor and Odor Modification, Along with Serotonergic Processes Underlying the Social Deficits in BTBR T+ Itpr3(Tf)/J and BALB/CJ Mouse Models of Autism. *Neuroscience* **2020**, *445*, 144–162. [CrossRef]

175. Frye, C.A.; Llaneza, D.C. Corticosteroid and Neurosteroid Dysregulation in an Animal Model of Autism, BTBR Mice. *Physiol. Behav.* **2010**, *100*, 264–267. [CrossRef]

176. Silverman, J.L.; Yang, M.; Turner, S.M.; Katz, A.M.; Bell, D.B.; Koenig, J.I.; Crawley, J.N. Low Stress Reactivity and Neuroendocrine Factors in the BTBR T+tf/J Mouse Model of Autism. *Neuroscience* **2010**, *171*, 1197–1208. [CrossRef]

177. Benno, R.; Smirnova, Y.; Vera, S.; Liggett, A.; Schanz, N. Exaggerated Responses to Stress in the BTBR T+tf/J Mouse: An Unusual Behavioral Phenotype. *Behav. Brain Res.* **2009**, *197*, 462–465. [CrossRef]
178. Gould, G.G.; Burke, T.F.; Osorio, M.D.; Smolik, C.M.; Zhang, W.Q.; Onaivi, E.S.; Gu, T.-T.; DeSilva, M.N.; Hensler, J.G. Enhanced Novelty-Induced Corticosterone Spike and Upregulated Serotonin 5-HT1A and Cannabinoid CB1 Receptors in Adolescent BTBR Mice. *Psychoneuroendocrinology* **2014**, *39*, 158–169. [CrossRef]
179. Cheng, N.; Khanbabaei, M.; Murari, K.; Rho, J.M. Disruption of Visual Circuit Formation and Refinement in a Mouse Model of Autism. *Autism Res. Off. J. Int. Soc. Autism Res.* **2017**, *10*, 212–223. [CrossRef]
180. Luo, T.; Ou, J.-N.; Cao, L.-F.; Peng, X.-Q.; Li, Y.-M.; Tian, Y.-Q. The Autism-Related LncRNA MSNP1AS Regulates Moesin Protein to Influence the RhoA, Rac1, and PI3K/Akt Pathways and Regulate the Structure and Survival of Neurons. *Autism Res. Off. J. Int. Soc. Autism Res.* **2020**, *13*, 2073–2082. [CrossRef]
181. Heo, Y.; Zhang, Y.; Gao, D.; Miller, V.M.; Lawrence, D.A. Aberrant Immune Responses in a Mouse with Behavioral Disorders. *PLoS ONE* **2011**, *6*, e20912. [CrossRef]
182. Bakheet, S.A.; Alzahrani, M.Z.; Nadeem, A.; Ansari, M.A.; Zoheir, K.M.A.; Attia, S.M.; Al-Ayadhi, L.Y.; Ahmad, S.F. Resveratrol Treatment Attenuates Chemokine Receptor Expression in the BTBR T+tf/J Mouse Model of Autism. *Mol. Cell. Neurosci.* **2016**, *77*, 1–10. [CrossRef]
183. Uddin, M.N.; Yao, Y.; Mondal, T.; Matala, R.; Manley, K.; Lin, Q.; Lawrence, D.A. Immunity and Autoantibodies of a Mouse Strain with Autistic-like Behavior. *Brain Behav. Immun. - Health* **2020**, *4*, 100069. [CrossRef]
184. Palmer-Aronsten, B.; Sheedy, D.; McCrossin, T.; Kril, J. An International Survey of Brain Banking Operation and Characterization Practices. *Biopreserv. Biobank.* **2016**, *14*, 464–469. [CrossRef]
185. Lupien, S.J.; McEwen, B.S.; Gunnar, M.R.; Heim, C. Effects of Stress throughout the Lifespan on the Brain, Behaviour and Cognition. *Nat. Rev. Neurosci.* **2009**, *10*, 434–445. [CrossRef]

Article

Early-Onset Glaucoma in *egl1* Mice Homozygous for *Pitx2* Mutation

Bindu Kodati [1,†], Shawn A. Merchant [2,†], J. Cameron Millar [1] and Yang Liu [1,*]

[1]   North Texas Eye Research Institute, Department of Pharmacology and Neuroscience, University of North Texas Health Science Center, Fort Worth, TX 76107, USA; bindu.kodati@unthsc.edu (B.K.); cameron.millar@unthsc.edu (J.C.M.)

[2]   Department of Psychology and Neuroscience, Baylor University, Waco, TX 76798, USA; shawn_merchant1@baylor.edu

*   Correspondence: yang.liu@unthsc.edu

†   These authors contributed equally to this work.

Abstract: Mutations in *PITX2* cause Axenfeld–Rieger syndrome, with congenital glaucoma as an ocular feature. The *egl1* mouse strain carries a chemically induced *Pitx2* mutation and develops early-onset glaucoma. In this study, we characterized the glaucomatous features in *egl1* mice. The eyes of *egl1* and C57BL/6J control mice were assessed by slit lamp examination, total aqueous humor outflow facility, intraocular pressure (IOP) measurement, pattern electroretinography (PERG) recording, and histologic and immunohistochemistry assessment beginning at 3 weeks and up to 12 months of age. The *egl1* mice developed elevated IOP as early as 4 weeks old. The IOP elevation was variable and asymmetric within and between the animals. The aqueous humor outflow facility was significantly reduced in 12-month-old animals. PERG detected a decreased response at 2 weeks after the development of IOP elevation. Retinal ganglion cell (RGC) loss was detected after 8 weeks of IOP elevation. Slit lamp and histologic evaluation revealed corneal opacity, iridocorneal adhesions (anterior synechiae), and ciliary body atrophy in *egl1* mice. Immunohistochemistry assessment demonstrated glial cell activation and RGC axonal injury in response to IOP elevation. These results show that the eyes of *egl1* mice exhibit anterior segment dysgenesis and early-onset glaucoma. The *egl1* mouse strain may represent a useful model for the study of congenital glaucoma.

Keywords: glaucoma; *Pitx2*; mouse model

Citation: Kodati, B.; Merchant, S.A.; Millar, J.C.; Liu, Y. Early-Onset Glaucoma in *egl1* Mice Homozygous for *Pitx2* Mutation. *Biomedicines* **2022**, *10*, 516. https://doi.org/10.3390/biomedicines10030516

Academic Editor: Martina Perše

Received: 12 January 2022
Accepted: 19 February 2022
Published: 22 February 2022

Publisher's Note: MDPI stays neutral with regard to jurisdictional claims in published maps and institutional affiliations.

1. Introduction

Glaucoma is a multifactorial optic neuropathy characterized by retinal ganglion cell (RGC) loss, which accompanies optic nerve axonal injury. The disease leads to irreversible blindness if left untreated. Elevated intraocular pressure (IOP) is a major risk factor for glaucoma development and progression [1]. Congenital glaucoma, a major cause of childhood blindness, is a severe form of glaucoma that occurs in children from birth to the age of 3 years (infantile glaucoma) or after 3 years of age (juvenile glaucoma). According to the American Academy of Ophthalmology, congenital glaucoma is diagnosed only in 25% of babies who manifest the condition at birth and may occur in as many as 1 in 10,000 live births [2,3]. The increase in IOP in congenital glaucoma is due to developmental abnormalities in the anterior chamber angle, including trabeculodysgenesis, which leads to increased aqueous humor outflow resistance and thus elevated IOP [4,5]. Despite extensive research on congenital glaucoma, the molecular and cellular pathological events that occur are still unclear.

Genetic studies have been conducted to identify genes associated with congenital glaucoma. Research has shown that mutations in certain genes, such as *PITX2*, *FOXC1*, *PAX6*, and *CYP1B1*, are associated with congenital glaucoma [6–11]. Among these genes, extensive work has been carried out to identify the functions of *PITX2*. *PITX2* is involved in the Nodal/Sonic Hedgehog pathway, which determines the polarity and asymmetrical expression during the early development of the mesoderm derived organs as well as the eye, tooth, and umbilicus [12–14]. *PITX2* produces three isoforms (*PITX2a*, *PITX2b*, and *PITX2c*), which consist of similar bicoid-like homeodomain 2 (HD) transcription factors that play an important role in fetal and embryonic development. Missense mutations, or heterozygous defects of the *PITX2* gene in the HD region, cause DNA binding impairment, thereby altering the transactivation of transcription factors, which can lead to ocular deficits by gain or loss of function [15–17]. The expression of *PITX2* is higher in the anterior segment than in either the retina or the sclera during the first 9.5 weeks of intrauterine development. *PITX2* expression is further increased in the iridocorneal complex during week 18 [18]. Intragenic deletions and decreased mRNA expression of *PITX2* have been observed in patients with Axenfeld–Rieger syndrome affected by advanced glaucoma [7,19,20]. Using transcription activator-like effector nuclease (TALEN)-mediated genome editing, studies on the zebrafish for the generation of $pitx2^{-/-}$ lines have also revealed the importance of *pitx2* in ocular development. Transcriptome studies have identified the molecular changes associated with disease pathology [21]. Similarly, other studies on zebrafish and mouse models have also shown that, as in Axenfeld–Rieger syndrome, *Pitx2* mutations can lead to abnormalities in the development of the anterior segment of the eye as well as protuberant umbilicus, dental hypoplasia, and facial dysmorphism [22–24].

Certain mouse strains develop spontaneous glaucoma at a young age. These strains are valuable tools for studying the mechanisms of early-onset glaucoma in humans. A recent study described that mice heterozygous for the *Pitx2* deletion ($Pitx2^{+/-}$) modeled the major ocular features of Axenfeld–Rieger syndrome and associated glaucoma [25]. In addition, a mouse strain homozygous for the *Pitx2* mutation named the *egl1* strain has been established in an N-ethyl-N-nitrosourea chemical mutagenesis screening program [26,27]. On the basis of the whole-exome sequencing results of the *egl1* strain, *Pitx2* p.R115L knock-in mice were further generated using CRISPR/Cas9 technology and characterized for a glaucomatous phenotype [26]. The establishment of these models provides a powerful tool to explore mechanisms of glaucoma pathogenesis. Each mouse strain is unique in its development, genotype, and genetic background. The *egl1* mouse strain is commercially available as an early-onset glaucoma mouse model. However, the characterization of this strain has yet to be completed.

In this study, we further characterized the glaucomatous phenotype in the *egl1* mouse strain. We assessed the anterior segment morphology, aqueous humor outflow facility, IOP elevation, and RGC and optic nerve head degeneration. Our study provides insightful information for using the *egl1* mouse strain as an early-onset glaucoma model.

2. Materials and Methods

2.1. Animals

C57BL/6J and C57BL/6J-$Pitx2^{egl1}$/Boc (*egl1* mutant) mice were obtained from the Jackson Laboratory (Bar Harbor, ME, USA). All animal experiments were performed in accordance with the National Institutes of Health guide for the care and use of laboratory animals and were approved by the Institutional Animal Care and Use Committee of the University of North Texas Health Science Center (IACUC 2020-0023). Mice (both sexes) aged 3 weeks to 12 months were examined. Thirteen C57BL/6J mice (7 females and 6 males) and 60 *egl1* mice (33 females and 27 males) were randomized into the study, and data were obtained from 26 and 116 eyes from C57BL/6J and *egl1* mutant mice, respectively. Exclusion criteria included fluid leaking during aqueous humor outflow facility measurement and central corneal opacity affecting pattern electroretinography recording. Each individual eye was considered the experimental unit in this study. Each experiment includes similar numbers of female and male mice.

All animals were housed in individually ventilated cages (IVCs, polysulfone material with 500 cm^2 floor space) (Allentown, Allentown, NJ, USA) at a temperature of 21 to 24 °C and humidity of 40–45%. Lights were turned on at 0630 h, and a 12 h light/12 h dark cycle was maintained. Same-sex littermates were housed together at the maximum density of 5 mice per cage. As bedding, a 1/8-inch corn cob (The Andersons, Maumee, OH, USA) was provided. Shredded paper and enrichment rectangles were used as nesting and enrichment (The Andersons). Mice were fed an irradiated mouse diet (5LG4, LabDiet, St. Louis, MO, USA) and provided reverse osmosis filtered drinking water *ad libitum*. All materials, including IVCs, lids, feeders, bedding, nesting, and enrichment, were autoclaved before use. Sentinel mice were negative for at least all Federation of Laboratory Animal Science Associations (FELASA)-relevant murine infectious agents, as monitored by the Department of Laboratory Animal Medicine on campus.

2.2. Intraocular Pressure Measurement

Intraocular pressure (IOP) was measured non-invasively using the TonoLab impact tonometer (Colonial Medical Supply, Franconia, NH, USA) as described previously [28,29]. Briefly, mice were placed in a soft plastic cone (Braintree Scientific, Inc., Braintree, MA, USA) and gently restrained in a plastic mouse restrainer (Colonial Medical Supply, Londonderry, NH, USA). IOP was measured after mice were acclimated. All measurements were performed during the same 3 h time window (1–4 pm) during the lights-on phase of the day; the average of 4–6 measurements was used as the IOP value. IOP was measured twice a week until pressure elevation was detected, and once a week thereafter. Total IOP exposure for each individual eye was determined by the determination of the area under the IOP–time curve (AUC). Six C57BL/6J and twenty-two *egl1* mice were included in this experiment, and IOP was monitored at age 3 through 8 weeks old.

2.3. Slit Lamp Examination

Anterior segments of mouse eyes were examined with a slit lamp (SL-D7; Topcon, Tokyo, Japan), and images were taken with a digital camera (D100; Nikon, Tokyo, Japan). Slit lamp examinations were performed on conscious animals. Six C57BL/6J and twenty-two *egl1* mice were examined longitudinally at age 4 weeks to 12 months.

2.4. Anterior Segment Histologic Examination

For histologic examination, animals were euthanized by exposure to 10% to 30% cage volume/min carbon dioxide. After death was confirmed, eyes were enucleated and fixed in 10% neutral formalin (Electron Microscopy Sciences, Hatfield, PA, USA) overnight. Eyes were then dehydrated with ethanol and xylene and embedded in paraffin. Sagittal sections (5 µm) were prepared, mounted on glass microscope slides, and stained with hematoxylin and eosin (H&E) for structural evaluation. Three C57BL/6J and six *egl1* mice were examined before and 4 weeks after intraocular pressure elevation.

2.5. Immunofluorescent Staining

For immunofluorescent staining, eyes were fixed in 4% paraformaldehyde (Electron Microscopy Sciences) in phosphate-buffered saline (PBS) for 2 h at 4 °C and cryo-preserved after 10%, 20%, and 30% sucrose (Thermo Scientific, Rockford, IL, USA) sequential cryo-protection. Cryosections of mouse eyes were blocked with PBS-based SuperBlock (Thermo Scientific) for 2 h at room temperature (RT) and incubated overnight at 4 °C with primary antibodies against glial fibrillary acidic protein (GFAP, 1:500 dilution; Cell Signaling #3670, Danvers, CA, USA) [30], ionized calcium-binding adaptor 1 (Iba-1, 1:500 dilution, Fujifilm Cellular Dynamics 019-19741, Madison, WI, USA) [31], or neurofilament H (NF-H, 1:1000 dilution, Abcam ab8135, Waltham, MA, USA) [32]. After 3 rinses in PBS, sections were further incubated with Alexa488 or TRITC conjugated secondary antibodies (Life Technologies, Carlsbad, CA, USA) against rabbit (for Iba-1 and NF-H) or mouse (for GFAP) IgGs for 1 h at RT. The sections were rinsed again and mounted in ProLong Gold anti-fade reagent with DAPI (Thermo Scientific). Non-primary control staining was performed using PBS instead of primary antibodies. Images were viewed and captured using a Zeiss LSM 510 META confocal microscope. The fluorescence intensity for Iba-1 and GFAP were analyzed using ImageJ software in a masked manner. Two C57BL/6J and seven *egl1* mice (2–3 mice per time point) were included in this experiment.

2.6. Aqueous Humor (AH) Outflow Facility Measurement

The *egl1* mice at ages 2 weeks and 1 year were used for aqueous humor outflow facility measurement, which was performed using a constant flow infusion method established previously [33–36]. In brief, mice were anesthetized by an intraperitoneal injection of a cocktail of ketamine/xylazine (100/10 mg/kg, respectively; maintenance: $1/2 \times$ to $1/4 \times$ induction dose). One drop of 0.5% proparacaine HCl was applied for corneal anesthesia (Alcaine, Alcon, Fort Worth, TX, USA). The anterior chamber of each eye was cannulated with a 32-gauge needle attached to tubing connected to a pressure transducer (BLPR2; World Precision Instruments (WPI), Sarasota, FL, USA) and a glass microsyringe (Hamilton Company, Reno, NV, USA) filled with sterile PBS and loaded onto a microdialysis infusion pump (SP101i; WPI). The eyes were infused at a flow rate of 0.1 µL/min initially for approximately 30 min to stabilize the pressure registered by the pressure transducer. On pressure stabilization, 3 pressure readings, spaced 5 min apart, were obtained over the following 10 min period. The flow rate was then increased to 0.2 µL/min, and following 5 min for stabilization, 3 pressure readings were obtained in a similar manner. The process was then repeated at flow rates of 0.3, 0.4, and 0.5 µL/min. Mean stabilized pressure flow rate curves were generated for each eye and fit using simple linear regression. The total aqueous humor outflow facility was calculated as the reciprocal of the slope of each respective curve. All measurements were conducted in a single masked manner. Seven *egl1* mice (3 aged 2 months and 4 aged 12 months) were assessed.

2.7. Retinal Ganglion Cell Function Assessment

Pattern electroretinography (PERG) was performed in 6 *egl1* mice using the JORVEC System (Intelligent Hearing Systems, Miami, FL, USA) as described previously [37]. The *egl1* mice with severe corneal opacity that blocked central cornea were excluded from the PERG study. Briefly, mice were anesthetized with intraperitoneal injections of a mixture of ketamine and xylazine (100 and 10 mg/kg, respectively). The PERG responses were recorded from a stainless-steel needle (Grass, West Warwick, RI, USA) placed in the snout subcutaneously. Pattern stimuli consisting of contrast-reversing gratings with a spatial frequency of 0.05 cycles/deg and maximum contrast were displayed on two custom-made tablets. The contrast reversal frequency was 1 Hz. A total of 2232 responses were averaged. The amplitude was measured from the positive peak to the negative trough, and the latency was the time to the peak of the response.

2.8. Retinal Ganglion Cell Quantification

Quantification of retinal ganglion cells (RGCs) was performed using immunostained retinal whole mounts [38]. Briefly, mice (5 C57BL/6J and 21 *egl1* mice) were euthanized by exposure to 100% $CO_{2(g)}$. Following the cessation of breathing and heartbeat, eyes were enucleated and fixed in 4% paraformaldehyde (Electron Microscopy Sciences) for 2 h at 4 °C. Retinas were dissected from fixed eyes and blocked with 0.3% Triton X-100 in PBS containing 2% goat serum for 2 h. Retinas were incubated in rabbit polyclonal RBPMS antibody (1:200, diluted in 0.3% Triton X-100 in PBS, GeneTex GTX118619, Irvine, CA, USA) [39] overnight at 4 °C. Following washes in PBS, retinas were further incubated in AlexaFluor596 goat-anti-rabbit (1:1000, diluted in 0.1% Triton X-100 in PBS) overnight at 4 °C. After washes with PBS, the retinas were cut into four quadrants and mounted on glass slides (Fischer Scientific, Pittsburgh, PA, USA). Eight images were taken from the peripheral and mid-peripheral regions in four quadrants of each retina. The number of cells from each image (0.0867 mm^2 retina area) was counted using Adobe Photoshop software V22.2 (Adobe Systems, Inc., San Jose, CA, USA). The average of all counts from each retina was used as the number of RGCs presented in each eye. Cell counts were performed in a masked manner.

2.9. Statistical Analysis

One-way ANOVA followed by a Tukey post hoc test was performed to analyze intragroup differences. The unpaired Student's t-test was used for the comparison of differences between the two groups. The correlation between RGC counts and IOP exposure was assessed by Pearson's correlation coefficient. Data are presented as means ± SEM, and $p < 0.05$ was considered statistically significant.

3. Results

3.1. Anterior Segment Morphology in egl1 Mice

We performed slit lamp examination on C57/BL/6J (Figure 1A) and *egl1* mice (Figure 1B–I) aged 4 weeks to 12 months. The *egl1* mice developed anterior segment abnormalities of various types and severities (Figure 1C–I). Approximately 60% of eyes showed localized faint opacity in the paracentral zone of the cornea (Figure 1C). The onset of the corneal lesion was as early as 4 weeks old and appeared independent of mechanical stimulation. Some corneas of aged *egl1* mice remained clear, despite repeated IOP measurements (Figure 1B), similar to those of C57BL/6J wildtype animals (Figure 1A). The corneal lesion progressed to dense corneal opacity with neovascularization in some animals as they aged (Figure 1D,E). In a few animals aged 12 months, we observed the diffuse opacity covering the entire cornea (Figure 1F).

The *egl1* mice also developed iris defects including anterior synechiae (Figure 1G–I), atrophy (Figure 1D), and pupil deviation (Figure 1E). We further examined the anterior segment histology of 8-week-old young adult mice. In the *egl1* mice with normal IOP, the iridocorneal angle remained open, and the ciliary body appeared normal (Figure 1K) compared to C57BL/6J wildtype mice (Figure 1J). In eyes with elevated IOP (Figure 1L), there were anterior synechiae and ciliary body atrophy. Despite the ocular abnormalities, *egl1* animals appeared similar in body size and coat color compared to C57BL/6J mice.

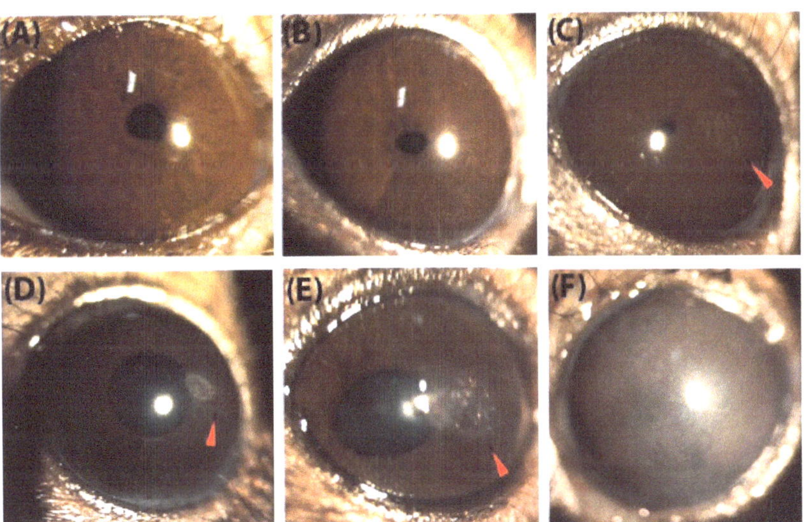

Figure 1. *Cont.*

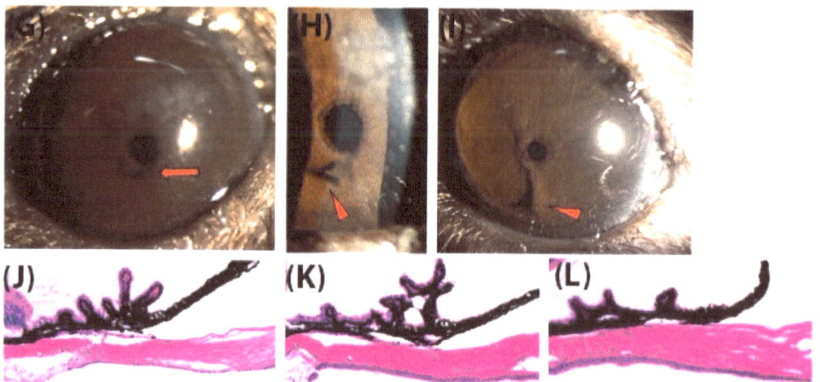

Figure 1. Anterior segment morphology in *egl1* mice. (**A–I**) Representative slit lamp image from C57BL/6J (**A**) and *egl1* mice (**B–I**). (**A**) Representative slit lamp image showing clear and avascular cornea and centrally located pupil in uniformly shaped iris. (**B**) Normal cornea from an *egl1* mouse. (**C–I**) Anterior segment lesions in *egl1* mice. The *egl1* mice developed a variety of anterior segment lesions (indicated by triangles) such as localized faint opacity in paracentral zone of the cornea (**C**), dense corneal opacity with neovascularization (**D,E**), iris thinning (**D,E**) and pupil deviation (**E**), diffuse opacity covering the entire cornea (**F**), and iris synechiae (the iris adhesion to the cornea) (**G–I**). (**H**) Enlarged view of area indicated by arrow in (**G**). (**J–L**) Representative H&E staining images of cross-section of the ciliary body and iridocorneal angle from C57BL/6J (**J**), *egl1* mouse with normal IOP (**K**) and after 4 weeks of IOP elevation (**L**). The iridocorneal angle remained open and ciliary body appeared normal triangular in C57BL/6J and the *egl1* mice with normal IOP. Anterior synechiae (the iris attached in the iridocorneal angle) and ciliary body atrophy were seen in eyes with elevated IOP. Scale bar = 50 μm.

3.2. IOP Elevation in egl1 Mice

The *egl1* mice were maintained on the C57BL/6J background; thus, C57BL/6J mice were used as control animals in this analysis. We monitored the conscious IOP longitudinally in a cohort of 22 *egl1* mice and 6 C57BL/6J mice using the TonoLab tonometer (Figure 2A–C). At 3 weeks of age, *egl1* mice exhibited an IOP of 15.9 ± 0.5 mmHg (mean ± SEM; n = 44) which was similar to age-matched C57BL/6J control mice (14.0 ± 1.3 mmHg; mean ± SEM; n = 12). A significant elevation of IOP was observed in *egl1* mice as early as 4 weeks old and remain elevated at all ages examined. The IOP values were 19.2 ± 0.6, 21.7 ± 0.8, and 21.4 ± 0.8 mmHg (mean ± SEM; n = 44) at 4, 6, and 8 weeks of age, respectively, as shown in Figure 2D. The IOP values remained unchanged in C57BL/6J mice, which were 15.1 ± 0.8, 14.8 ± 1.0, and 15.7 ± 0.9 mmHg (mean ± SEM; n = 12) at 4, 6 and 8 weeks of age, respectively.

The IOP elevation in *egl1* mice was variable and asymmetric. By age 8 weeks, 55% of mice examined exhibited IOPs under 20 mmHg (Figure 2E). The elevated IOP values also demonstrated a wide range in mice of the same age. Some mice had IOPs greater than 35 mmHg at all time points examined. Furthermore, individual animals showed dramatic asymmetry in IOPs, which differed by more than 15 mmHg.

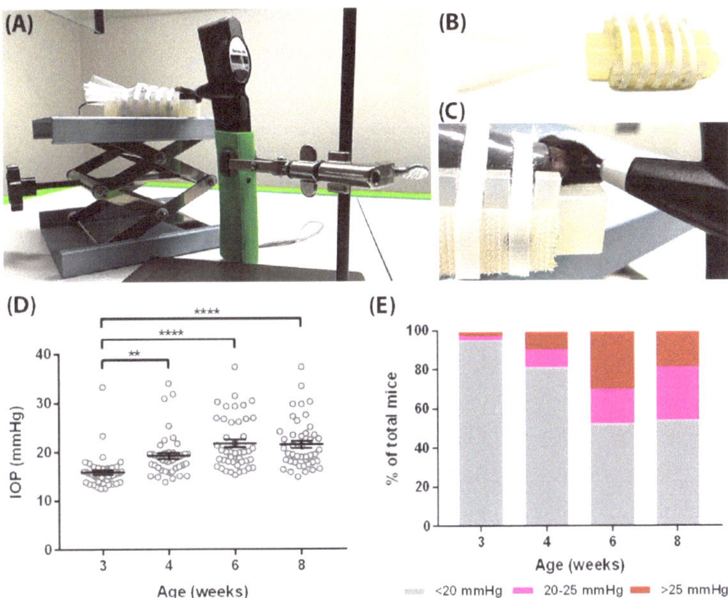

Figure 2. IOP elevation in *egl1* mice. (**A**) Conscious IOP measurement setup. (**B**) Plastic cone and mouse restrainer for IOP measurement. (**C**) Close view of conscious IOP measurement. (**D**) The IOP was monitored in *egl1* mice longitudinally. Significant IOP elevation was detected at 4 weeks of age and remained at all ages examined. Data are presented as means $\pm$ SEM ($n = 44$). **: $p < 0.01$, ****: $p < 0.001$. (**E**) The IOP values from each time point were plotted by three ranges, <20 mmHg, 20–25 mmHg, and >25 mmHg. The IOP values demonstrated a wide range in mice of the same age.

3.3. Aqueous Humor Outflow Facility in egl1 Mice

The AH circulation plays a key role in IOP regulation. We measured total aqueous humor outflow facility in young (2 months) and aged (1 year) *egl1* mice. The outflow facility in young animals was 30.25 ± 1.34 nL/min/mmHg (mean $\pm$ SEM; $n = 5$); The aged animals demonstrated significantly reduced outflow facility (17.07 ± 1.44 nL/min/mmHg, mean $\pm$ SEM; $n = 7$, $p < 0.001$), indicating a significantly higher outflow resistance in the aged animals (Figure 3A). The outflow facility and IOP measurement showed a trend of negative correlation, but this did not achieve statistical significance (Figure 3B).

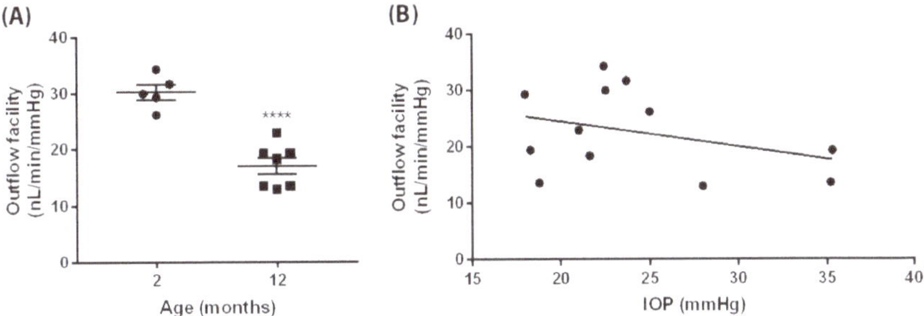

Figure 3. Total aqueous humor outflow facility in young and aged *egl1* animals. (**A**) The aged animals demonstrated significantly lower outflow facility compared to the young animals. Data are presented as means $\pm$ SEM ($n = 5$ and 7 for 2- and 12-months old animals, respectively). ****: $p < 0.001$. (**B**) Individual outflow facility was plotted against IOP measurement. The outflow facility and IOP measurement show a trend of negative correlation, but this did not achieve statistical significance.

3.4. RGC Death in egl1 Mice

The degeneration of RGCs is a key feature of glaucoma. We quantified the RGC loss in response to IOP elevation in *egl1* mice to assess the glaucomatous neurodegeneration. The *egl1* mice with normal IOPs showed similar numbers of RGCs ($263 \pm 6/0.0867$ mm^2; mean $\pm$ SEM; $n = 17$) in the retina, compared to C57BL/6J control mice ($270 \pm 8/0.0867$ mm^2; mean $\pm$ SEM; $n = 10$). As expected, prolonged exposure to high IOP induced RGC loss in *egl1* mice. RGC numbers ($217 \pm 14/0.0867$ mm^2; mean $\pm$ SEM; $n = 14$) were significantly decreased in mice with 8 weeks of IOP elevation (Figure 4A,B).

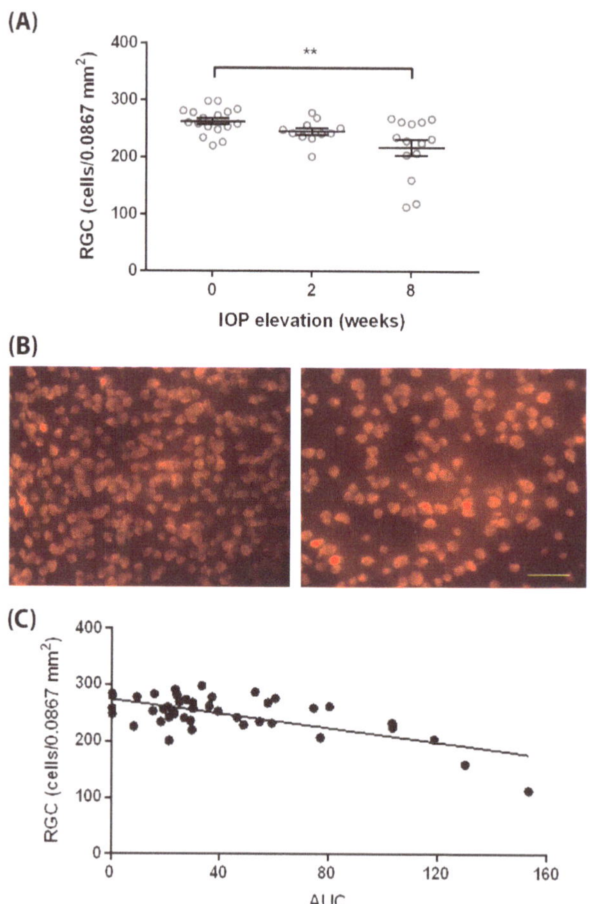

Figure 4. RGC degeneration in *egl1* mice. (**A**) Quantification of RGCs was performed before and at 2 and 8 weeks after IOP elevation. Significant RGC loss was observed after 8 weeks of IOP elevation. Data are presented as means $\pm$ SEM ($n = 11$–17). ******: $p < 0.01$. (**B**) Representative images show RBPMS (red) immunolabeled retinal whole mounts from mice before (**left**) and 8 weeks after (**right**) IOP elevation. The density of RGCs, demonstrated by RBPMS positive cells, decreased after 8 weeks of IOP elevation. Scale bar = 50 μm. (**C**) Individual RGC count was plotted against IOP exposure described as AUC. There was a moderate negative correlation between RGC counts and IOP exposure. R square: 0.43, $p < 0.001$.

To determine the correlation between RGC death and IOP elevation, we further plotted the RGC numbers against IOP exposure calculated as the area under the IOP–time curve.

The RGC counts showed a moderate negative correlation (R square: 0.43, $p < 0.001$) with IOP exposure (Figure 4C).

3.5. RGC Functional Loss in egl1 Mice

To further examine glaucomatous neurodegeneration in *egl1* animals, we performed PERG recording in mice with normal and elevated IOP. The *egl1* animals demonstrated a normal PERG waveform with a positive peak elicited at 89.6 ± 2.4 ms, followed by a negative trough (Figure 5A). The PERG amplitude in the *egl1* mice with normal IOP was 23.0 ± 3.2 µV (mean ± SEM; $n = 5$). After 2 weeks of IOP elevation, there was a significant decrease in PERG amplitude (9.2 ± 1.3 µV) (Figure 5B) as well as an increase in latency (155.7 ± 21.6 ms, mean ± SEM; $n = 6$), which indicates damaged RGC function (Figure 5C).

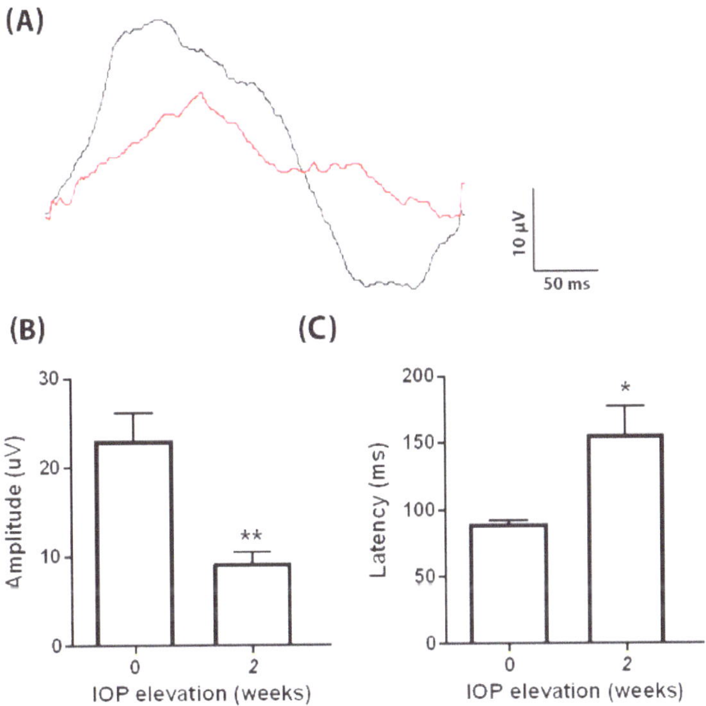

Figure 5. PERG reduction in *egl1* mice. (**A**) Representative PERG waveforms recorded from *egl1* mice before (black) and 2 weeks after (red) IOP elevation. IOP elevation decreased and delayed the PERG response. (**B,C**) Bar graphs show decreased PERG responses (**B**) and increased PERG latencies (**C**) after 2 weeks of IOP elevation. Data are presented as means ± SEM ($n = 5$–6) *: $p < 0.05$, **: $p < 0.01$.

3.6. Glaucomatous ONH Changes in egl1 Mice

Elevated IOP induces glial activation in the ONH and optic nerve axonal degeneration, hallmarks of glaucomatous optic neuropathy. To evaluate these phenotypes in *egl1* mice, we collected eyes from C57BL/6J and *egl1* mice before and 2 and 8 weeks after IOP elevation and performed immunofluorescence staining using specific antibodies against glial fibrillary acidic protein (GFAP), ionized calcium-binding adaptor molecule 1 (Iba-1), and neurofilament (NF) (Figure 6). Eyes of C57BL/6J and *egl1* mice with normal pressure exhibited similar expression levels and patterns of GFAP, Iba-1, and NF. Elevated IOP increased expression levels of GFAP and Iba-1, which peaked at 2 weeks, indicating astrocyte activation and microglia infiltration in the ONH in response to glaucomatous insult. Quantification of GFAP and Iba-1 fluorescence intensity showed a trend of increase at 2 weeks after IOP elevation without statistical significance ($n = 3$–6). The pattern of neurofilaments became disorganized after 2 weeks of IOP elevation. Eyes with extended pressure elevation further exhibited disruption of neurofilaments, suggesting optic nerve axonal damage in the *egl1* mice.

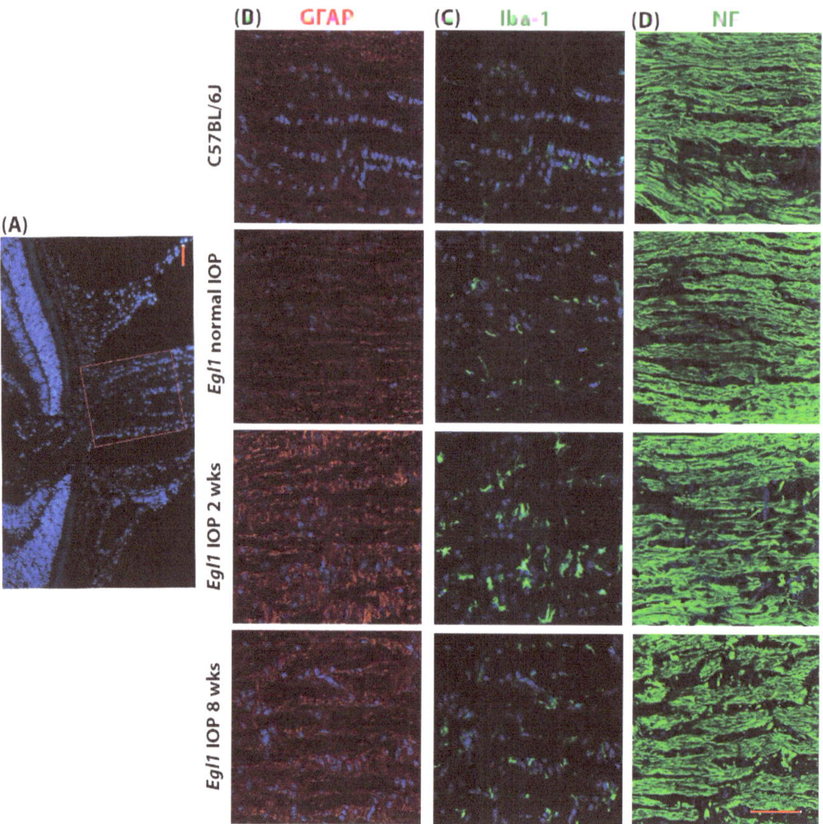

Figure 6. Glaucomatous ONH changes in *egl1* mice. Representative immunofluorescence staining images show increased expression of GFAP and Iba-1 and disorganized NF in the *egl1* mice. Images were taken in the rectangular boxed area in (**A**). (**B–D**) Columns show GFAP (red), Iba-1 (green), and NF (green) staining in C57BL/6J wildtype and *egl1* mice with various IOP levels. Nuclei were labeled with DAPI (blue). Scale bars = 50 μm.

4. Discussion

The present study characterized the glaucoma phenotype in a mouse strain carrying a missense mutation of *Pitx2* named the *egl1* mutation. The *egl1* mutant mice developed glaucoma with a significant variation in the age of onset and phenotypic severity within and between animals. The mice with ocular hypertension demonstrated glaucomatous retinopathy and optic nerve neuropathy.

As a powerful tool, mouse models carrying the *Pitx2* mutation or deletion have been developed to investigate the functions of *PITX2* and the pathogenesis of glaucoma. The *egl1* mice have a chemically induced *Pitx2* mutation and develop early-onset glaucoma. The mouse strain is homozygous for the *egl1* mutation, which was mapped to chromosome 3. High-throughput sequencing further identified the mutation as a single G to T transversion in *Pitx2* exon 2, changing amino acid 115 from arginine to leucine [26,27]. In humans, *PITX2* has been identified as a glaucoma-causing gene. Mutations in *PITX2* are associated with Axenfeld–Rieger syndrome, which involves ocular malformations leading to congenital and childhood glaucoma in more than 50% of the affected patients [40]. Similarly, we found that the *egl1* mouse strain developed anterior segment abnormalities, and by the age of 6 weeks, approximately 50% of the animals developed early-onset glaucoma despite having an identical genotype and genetic background. Gender difference in the risk for glaucoma and glaucoma blindness has been well documented [41,42]. In the *egl1* mice, ocular hypertension developed in male and female mice at a similar percentage. Larger-scale studies are needed to reveal potential sex-associated differences in glaucoma development and severity in *egl1* animals. The IOP elevation in *egl1* animals exhibited significant variation and asymmetry. The onset of ocular hypertension was as early as 3 weeks old, and IOP was elevated to a maximum of 20 to 37 mmHg. Individual animals exhibited significant ocular hypertension in one eye and normal IOP in another. The variation and asymmetry in the development of ocular hypertension may require larger animal cohorts in the research design to ensure a sufficient sample size for statistical analysis. On the other hand, the wide range of IOP values offers a spectrum of disease severity and eyes with normal tension, providing an ideal strain-matched control.

Through our longitudinal IOP measurements, we also found that the IOP elevation in the *egl1* mice demonstrated two patterns: sustained mild elevation until the age of 4 months or high elevation followed by normal pressure. Ciliary body atrophy has been described in the $Pitx2^{+/-}$ mice as well as in human chronic cases when left untreated [25]. Similarly, the histologic examination showed ciliary body atrophy in the *egl1* mice with prolonged ocular hypertension. The reduction of IOP following a high pressure elevation in the *egl1* mice is likely due to ciliary body atrophy. Aqueous humor is produced by the epithelium of the ciliary body and drained primarily through the conventional outflow pathway consisting of the trabecular meshwork (TM), Schlemm's canal, the scleral collector channels, and aqueous veins, sequentially. It has been recognized that increased aqueous humor outflow resistance is associated with elevated IOP in glaucoma [43,44]. In addition, studies in humans and monkeys have shown that aqueous outflow facility (the reciprocal of aqueous humor outflow resistance) naturally declines with aging, even in healthy eyes [45,46]. Thus, we measured the AH outflow facility in young and aged animals and assessed the correlation between IOP and AH outflow facility. There was a negative trend of correlation between IOP and AH outflow facility, suggesting a role played by declined outflow facility in IOP elevation in the *egl1* mice. However, we did not detect a statistically significant correlation between IOP and facility.

Pattern ERG is a well-accepted standard for assessing RGC function. It allows non-invasive and longitudinal evaluation of RGC function. It detects RGC functional loss prior to significant morphological damage [47]. Using PERG, we detected RGC functional loss at 2 weeks after IOP elevation when a significant RGC death was undetectable. The results further demonstrate the glaucomatous phenotype in *egl1* animals and support the use of PERG as a sensitive tool to monitor the disease progression in the *egl1* mouse strain.

The quantification of RGCs showed a mild correlation between RGC loss and IOP exposure. In addition, outer retinal dysfunction has been reported in *egl1* mice. At 3 months of age, there is a decreased rod b-wave and a lower cone response, which indicates a possible outer retinal degeneration [27]. A separate study also showed that the deletion of the *Pitx2* gene disrupts the development of the retinal pigment epithelium [23]. It is also possible that outer retinal degeneration may in turn affect RGC health within the inner retina and contribute to the decrease of RGC numbers.

Recent studies have shown that glia play an important role in the pathogenesis of glaucoma [48–51]. Astrocytes, the major glial cell population in the ONH, are considered the mediators of axonal injury in glaucoma. They respond to mechanical compression due to IOP elevation and undergo a number of cellular changes, such as upregulation of GFAP cytoskeleton and neurotrophic or neuroinflammatory factors [49,52]. Similarly, microglia in the ONH become activated and redistributed, express cytokines and other secreted factors, and exert neuroprotective or neurotoxic effects in glaucomatous eyes [51]. In the *egl1* animals, we found upregulated GFAP, and Iba-1 expression also peaked at an early stage of the disease, suggesting the activation of astrocytes and infiltration of microglia following glaucomatous insults. This further validates the glaucoma phenotype in the mutant animals. The results also support the use of *egl1* mice as an in vivo model to investigate mechanisms of glial activation in glaucoma.

In summary, this study demonstrates that the *egl1* mouse strain exhibits key features of early-onset glaucoma and provides an important tool to study glaucomatous neurodegeneration. Additional studies are needed to further characterize glaucomatous neuropathy in *egl1* animals.

Author Contributions: Conceptualization, Y.L.; methodology, B.K., S.A.M., J.C.M. and Y.L.; validation, B.K., S.A.M., J.C.M. and Y.L.; formal analysis, B.K., S.A.M., J.C.M. and Y.L.; investigation, B.K., S.A.M., J.C.M. and Y.L.; resources, J.C.M. and Y.L.; data curation, B.K., S.A.M., J.C.M. and Y.L.; writing—original draft preparation, B.K., S.A.M., J.C.M. and Y.L.; writing—review and editing, B.K., S.A.M., J.C.M. and Y.L.; visualization, B.K., S.A.M., J.C.M. and Y.L.; supervision, Y.L.; project administration, Y.L.; funding acquisition, Y.L. All authors have read and agreed to the published version of the manuscript.

Funding: This research was funded by the Knights Templar Eye Foundation grant number [2021-7], and the APC was funded by Department of Pharmacology and Neuroscience, University of North Texas Health Science Center.

Institutional Review Board Statement: The animal study protocol was approved by the Institutional Animal Care and Use Committee of the University of North Texas Health Science Center (IACUC 2020-0023; date of approval: 26 May 2020).

Acknowledgments: This work was supported by the Knights Templar Eye Foundation.

Conflicts of Interest: The authors declare no conflict of interest.

References

1. Lee, D.A.; Higginbotham, E.J. Glaucoma and its treatment: A review. *Am. J. Health Pharm.* **2005**, *62*, 691–699. [CrossRef] [PubMed]
2. Papadopoulos, M.; Cable, N.; Rahi, J.; Khaw, P.T. The British Infantile and Childhood Glaucoma (BIG) Eye Study. *Investig. Ophthalmol. Vis. Sci.* **2007**, *48*, 4100–4106. [CrossRef] [PubMed]
3. Chan, J.Y.Y.; Choy, B.N.; Ng, A.L.; Shum, J.W. Review on the Management of Primary Congenital Glaucoma. *J. Curr. Glaucoma Pr.* **2015**, *9*, 92–99. [CrossRef] [PubMed]
4. Anderson, D.R. The development of the trabecular meshwork and its abnormality in primary infantile glaucoma. *Trans. Am. Ophthalmol. Soc.* **1981**, *79*, 458–485.
5. Pilat, A.V.; Proudlock, F.A.; Shah, S.; Sheth, V.; Purohit, R.; Abbot, J.; Gottlob, I. Assessment of the anterior segment of patients with primary congenital glaucoma using handheld optical coherence tomography. *Eye* **2019**, *33*, 1232–1239. [CrossRef]
6. Reis, L.M.; Tyler, R.C.; Kloss, B.A.V.; Schilter, K.F.; Levin, A.V.; Lowry, R.B.; Zwijnenburg, P.J.G.; Stroh, E.; Broeckel, U.; Murray, J.C.; et al. PITX2 and FOXC1 spectrum of mutations in ocular syndromes. *Eur. J. Hum. Genet.* **2012**, *20*, 1224–1233. [CrossRef]
7. Protas, M.E.; Weh, E.; Footz, T.; Kasberger, J.; Baraban, S.C.; Levin, A.V.; Katz, L.J.; Ritch, R.; Walter, M.A.; Semina, E.V.; et al. Mutations of conserved non-coding elements of PITX2 in patients with ocular dysgenesis and developmental glaucoma. *Hum. Mol. Genet.* **2017**, *26*, 3630–3638. [CrossRef]
8. Prosser, J. and V. van Heyningen, PAX6 mutations reviewed. *Hum. Mutat.* **1998**, *11*, 93–108. [CrossRef]
9. Riise, R.; Storhaug, K.; Brøndum-Nielsen, K. Rieger syndrome is associated with PAX6 deletion. *Acta Ophthalmol. Scand.* **2001**, *79*, 201–203. [CrossRef]
10. Khan, A.O.; Aldahmesh, M.A.; Al-Amri, A. HeterozygousFOXC1Mutation (M161K) Associated with Congenital Glaucoma and Aniridia in an Infant and a Milder Phenotype in Her Mother. *Ophthalmic Genet.* **2008**, *29*, 67–71. [CrossRef]
11. Sarfarazi, M.; Stoilov, I. Molecular genetics of primary congenital glaucoma. *Eye* **2000**, *14*, 422–428. [CrossRef]
12. Piedra, M.; Icardo, J.M.; Albajar, M.; Rodriguez-Rey, J.C.; Ros, M. Pitx2 Participates in the Late Phase of the Pathway Controlling Left-Right Asymmetry. *Cell* **1998**, *94*, 319–324. [CrossRef]
13. Ryan, A.K.; Blumberg, B.; Rodriguez-Esteban, C.; Yonei-Tamura, S.; Tamura, K.; Tsukui, T.; De La Peña, J.; Sabbagh, W.; Greenwald, J.; Choe, S.; et al. Pitx2 determines left–right asymmetry of internal organs in vertebrates. *Nature* **1998**, *394*, 545–551. [CrossRef] [PubMed]
14. Hjalt, T.A.; Semina, E.V.; Amendt, B.A.; Murray, J.C. The Pitx2 protein in mouse development. *Dev. Dyn.* **2000**, *218*, 195–200. [CrossRef]
15. Holmberg, J.; Liu, C.-Y.; Hjalt, T.A. PITX2 Gain-of-Function in Rieger Syndrome Eye Model. *Am. J. Pathol.* **2004**, *165*, 1633–1641. [CrossRef]

16. Footz, T.; Idrees, F.; Acharya, M.; Kozlowski, K.; Walter, M.A. Analysis of Mutations of thePITX2Transcription Factor Found in Patients with Axenfeld-Rieger Syndrome. *Investig. Ophthalmol. Vis. Sci.* **2009**, *50*, 2599–2606. [CrossRef] [PubMed]
17. Huang, L.; Meng, Y.; Guo, X. Novel PITX2 Mutations including a Mutation Causing an Unusual Ophthalmic Phenotype of Axenfeld-Rieger Syndrome. *J. Ophthalmol.* **2019**, *2019*, 5642126. [CrossRef]
18. Markintantova, I.V.; Firsova, N.V.; Smirnova, I.A.; Panova, I.G.; Sukhikh, G.T.; Zinov'eva, R.D.; Mitashov, V.I. Localization of the PITX2 gene expression in human eye cells in the course of prenatal development. *Izv. Akad. Nauk. Ser. Biol.* **2008**, *2*, 139–145.
19. Houssaye, G.D.L.; Bieche, I.; Roche, O.; Vieira, V.; Laurendeau, I.; Arbogast, L.; Zeghidi, H.; Rapp, P.; Halimi, P.; Vidaud, M.; et al. Identification of the first intragenic deletion of the PITX2 gene causing an Axenfeld-Rieger Syndrome: Case report. *BMC Med. Genet.* **2006**, *7*, 82.
20. Zhang, L.; Peng, Y.; Ouyang, P.; Liang, Y.; Zeng, H.; Wang, N.; Duan, X.; Shi, J. A novel frameshift mutation in the PITX2 gene in a family with Axenfeld-Rieger syndrome using targeted exome sequencing. *BMC Med. Genet.* **2019**, *20*, 105. [CrossRef]
21. E Hendee, K.; A Sorokina, E.; Muheisen, S.S.; Reis, L.M.; Tyler, R.C.; Markovic, V.; Cuturilo, G.; A Link, B.; Semina, E.V. PITX2 deficiency and associated human disease: Insights from the zebrafish model. *Hum. Mol. Genet.* **2018**, *27*, 1675–1695. [CrossRef] [PubMed]
22. Lin, C.R.; Kioussi, C.; O'Connell, S.; Briata, P.; Szeto, D.; Liu, F.; Izpisúa-Belmonte, J.C.; Rosenfeld, M.G. Pitx2 regulates lung asymmetry, cardiac positioning and pituitary and tooth morphogenesis. *Nature* **1999**, *401*, 279–282. [CrossRef] [PubMed]
23. Evans, A.L.; Gage, P. Expression of the homeobox gene Pitx2 in neural crest is required for optic stalk and ocular anterior segment development. *Hum. Mol. Genet.* **2005**, *14*, 3347–3359. [CrossRef] [PubMed]
24. Liu, Y.; Semina, E.V. pitx2 Deficiency Results in Abnormal Ocular and Craniofacial Development in Zebrafish. *PLoS ONE* **2012**, *7*, e30896. [CrossRef] [PubMed]
25. Chen, L.; Gage, P. HeterozygousPitx2Null Mice Accurately Recapitulate the Ocular Features of Axenfeld-Rieger Syndrome and Congenital Glaucoma. *Investig. Ophthalmol. Vis. Sci.* **2016**, *57*, 5023–5030. [CrossRef] [PubMed]
26. Yang, Y.; Li, X.; Wang, J.; Tan, J.; Fitzmaurice, B.; Nishina, P.M.; Sun, K.; Tian, W.; Liu, W.; Liu, X.; et al. A missense mutation in Pitx2 leads to early-onset glaucoma via NRF2-YAP1 axis. *Cell Death Dis.* **2021**, *12*, 1017. [CrossRef] [PubMed]
27. Chang, B.; Wang, J.; FitzMaurice, B.; Nishina, P. A new mouse model of early-onset glaucoma. *IOVS* **2017**, *58*, 2125.
28. Daniel, S.; Meyer, K.; Clark, A.; Anderson, M.; McDowell, C. Effect of ocular hypertension on the pattern of retinal ganglion cell subtype loss in a mouse model of early-onset glaucoma. *Exp. Eye Res.* **2019**, *185*, 107703. [CrossRef]
29. Wang, W.-H.; Millar, J.C.; Pang, I.-H.; Wax, M.B.; Clark, A.F. Noninvasive Measurement of Rodent Intraocular Pressure with a Rebound Tonometer. *Investig. Ophthalmol. Vis. Sci.* **2005**, *46*, 4617–4621. [CrossRef]
30. Murru, S.; Hess, S.; Barth, E.; Almajan, E.R.; Schatton, D.; Hermans, S.; Brodesser, S.; Langer, T.; Kloppenburg, P.; Rugarli, E.I. Astrocyte-specific deletion of the mitochondrial m -AAA protease reveals glial contribution to neurodegeneration. *Glia* **2019**, *67*, 1526–1541. [CrossRef]
31. Silverman, S.M.; Ma, W.; Wang, X.; Zhao, L.; Wong, W.T. C3- and CR3-dependent microglial clearance protects photoreceptors in retinitis pigmentosa. *J. Exp. Med.* **2019**, *216*, 1925–1943. [CrossRef]
32. Manousi, A.; Göttle, P.; Reiche, L.; Cui, Q.-L.; Healy, L.M.; Akkermann, R.; Gruchot, J.; Schira-Heinen, J.; Antel, J.P.; Hartung, H.-P.; et al. Identification of novel myelin repair drugs by modulation of oligodendroglial differentiation competence. *EBioMedicine* **2021**, *65*, 103276. [CrossRef] [PubMed]
33. Shepard, A.R.; Millar, J.C.; Pang, I.-H.; Jacobson, N.; Wang, W.H.; Clark, A.F. Adenoviral gene transfer of active human transforming growth factor-[beta]2 elevates intraocular pressure and reduces outflow facility in rodent eyes. *Investig. Ophthalmol. Vis. Sci.* **2010**, *51*, 2067–2076. [CrossRef]
34. Millar, J.C.; Phan, T.N.; Pang, I.-H. Assessment of Aqueous Humor Dynamics in the Rodent by Constant Flow Infusion. *Programmed Necrosis* **2017**, *1695*, 109–120. [CrossRef]
35. Millar, J.C.; Clark, A.F.; Pang, I.-H. Assessment of Aqueous Humor Dynamics in the Mouse by a Novel Method of Constant-Flow Infusion. *Investig. Ophthalmol. Vis. Sci.* **2011**, *52*, 685–694. [CrossRef] [PubMed]
36. Stankowska, D.L.; Millar, J.C.; Kodati, B.; Behera, S.; Chaphalkar, R.M.; Nguyen, T.; Nguyen, K.T.; Krishnamoorthy, R.R.; Ellis, D.Z.; Acharya, S. Nanoencapsulated hybrid compound SA-2 with long-lasting intraocular pressure–lowering activity in rodent eyes. *Mol. Vis.* **2021**, *27*, 37–49.
37. Chou, T.-H.; Bohorquez, J.; Toft-Nielsen, J.; Ozdamar, O.; Porciatti, V. Robust Mouse Pattern Electroretinograms Derived Simultaneously From Each Eye Using a Common Snout Electrode. *Investig. Ophthalmol. Vis. Sci.* **2014**, *55*, 2469–2475. [CrossRef]
38. Choudhury, S.; Liu, Y.; Clark, A.F.; Pang, I.-H. Caspase-7: A critical mediator of optic nerve injury-induced retinal ganglion cell death. *Mol. Neurodegener.* **2015**, *10*, 40. [CrossRef]
39. Maddineni, P.; Kasetti, R.B.; Patel, P.D.; Millar, J.C.; Kiehlbauch, C.; Clark, A.F.; Zode, G.S. CNS axonal degeneration and transport deficits at the optic nerve head precede structural and functional loss of retinal ganglion cells in a mouse model of glaucoma. *Mol. Neurodegener.* **2020**, *15*, 48. [CrossRef]
40. Semina, E.V.; Reiter, R.; Leysens, N.J.; Alward, W.L.M.; Small, K.W.; Datson, N.; Siegel-Bartelt, J.; Bierke-Nelson, D.; Bitoun, P.; Zabel, B.U.; et al. Cloning and characterization of a novel bicoid-related homeobox transcription factor gene, RIEG, involved in Rieger syndrome. *Nat. Genet.* **1996**, *14*, 392–399. [CrossRef]
41. Vajaranant, T.S.; Nayak, S.; Wilensky, J.T.; Joslin, C.E. Gender and glaucoma: What we know and what we need to know. *Curr. Opin. Ophthalmol.* **2010**, *21*, 91–99. [CrossRef] [PubMed]

42. Gupta, P.; Zhao, D.; Guallar, E.; Ko, F.; Boland, M.; Friedman, D.S. Prevalence of Glaucoma in the United States: The 2005–2008 National Health and Nutrition Examination Survey. *Investig. Ophthalmol. Vis. Sci.* **2016**, *57*, 2905–2913. [CrossRef] [PubMed]
43. Stamer, W.D.; Acott, T.S. Current understanding of conventional outflow dysfunction in glaucoma. *Curr. Opin. Ophthalmol.* **2012**, *23*, 135–143. [CrossRef] [PubMed]
44. Johnson, M. 'What controls aqueous humour outflow resistance? *Exp. Eye Res.* **2006**, *82*, 545–557. [CrossRef]
45. Croft, M.A. Aging Effects on Accommodation and Outflow Facility Responses to Pilocarpine in Humans. *Arch. Ophthalmol.* **1996**, *114*, 586–592. [CrossRef]
46. Croft, M.A.; Lütjen-Drecoll, E.; Kaufman, P.L. Age-related posterior ciliary muscle restriction—A link between trabecular meshwork and optic nerve head pathophysiology. *Exp. Eye Res.* **2017**, *158*, 187–189. [CrossRef]
47. Liu, Y.; McDowell, C.M.; Zhang, Z.; Tebow, H.E.; Wordinger, R.J.; Clark, A.F. Monitoring Retinal Morphologic and Functional Changes in Mice Following Optic Nerve Crush. *Investig. Ophthalmol. Vis. Sci.* **2014**, *55*, 3766–3774. [CrossRef]
48. Varela, H.J.; Hernandez, M.R. Astrocyte Responses in Human Optic Nerve Head With Primary Open-Angle Glaucoma. *J. Glaucoma* **1997**, *6*, 303–313. [CrossRef]
49. Liddelow, S.A.; Guttenplan, K.A.; Clarke, L.E.; Bennett, F.C.; Bohlen, C.J.; Schirmer, L.; Bennett, M.L.; Münch, A.E.; Chung, W.-S.; Peterson, T.C.; et al. Neurotoxic reactive astrocytes are induced by activated microglia. *Nature* **2017**, *541*, 481–487. [CrossRef]
50. Neufeld, A.H. Microglia in the optic nerve head and the region of parapapillary chorioretinal atrophy in glaucoma. *Arch. Ophthalmol.* **1999**, *117*, 1050–1056. [CrossRef]
51. Zhao, X.; Sun, R.; Luo, X.; Wang, F.; Sun, X. The Interaction between Microglia and Macroglia in Glaucoma. *Front. Neurosci.* **2021**, *15*, 610788. [CrossRef] [PubMed]
52. Morgan, J.E. Optic nerve head structure in glaucoma: Astrocytes as mediators of axonal damage. *Eye* **2000**, *14*, 437–444. [CrossRef] [PubMed]

Article

Sex-Related Motor Deficits in the Tau-P301L Mouse Model

Luana Cristina Camargo [1,2], Dominik Honold [1], Robert Bauer [1], N. Jon Shah [3,4,5], Karl-Josef Langen [3,6], Dieter Willbold [1,2], Janine Kutzsche [1], Antje Willuweit [3] and Sarah Schemmert [1,*]

[1] Institute of Biological Information Processing, Structural Biochemistry (IBI-7), Forschungszentrum Jülich, 52425 Jülich, Germany; l.camargo@fz-juelich.de (L.C.C.); d.honold@fz-juelich.de (D.H.); Robert.Bauer@hhu.de (R.B.); d.willbold@fz-juelich.de (D.W.); j.kutzsche@fz-juelich.de (J.K.)

[2] Institut für Physikalische Biologie, Heinrich-Heine-Universität Düsseldorf, 40225 Düsseldorf, Germany

[3] Institute of Neuroscience and Medicine, Medical Imaging Physics (INM-4), Forschungszentrum Jülich, 52425 Jülich, Germany; n.j.shah@fz-juelich.de (N.J.S.); k.j.langen@fz-juelich.de (K.-J.L.); a.willuweit@fz-juelich.de (A.W.)

[4] JARA-Brain-Translational Medicine, JARA Institute Molecular Neuroscience and Neuroimaging, 52062 Aachen, Germany

[5] Department of Neurology, RWTH Aachen University, 52062 Aachen, Germany

[6] Department of Nuclear Medicine, RWTH Aachen University, 52062 Aachen, Germany

* Correspondence: s.schemmert@fz-juelich.de; Tel.: +49-2461-619447

Abstract: The contribution of mouse models for basic and translational research at different levels is important to understand neurodegenerative diseases, including tauopathies, by studying the alterations in the corresponding mouse models in detail. Moreover, several studies demonstrated that pathological as well as behavioral changes are influenced by the sex. For this purpose, we performed an in-depth characterization of the behavioral alterations in the transgenic Tau-P301L mouse model. Sex-matched wild type and homozygous Tau-P301L mice were tested in a battery of behavioral tests at different ages. Tau-P301L male mice showed olfactory and motor deficits as well as increased Tau pathology, which was not observed in Tau-P301L female mice. Both Tau-P301L male and female mice had phenotypic alterations in the SHIRPA test battery and cognitive deficits in the novel object recognition test. This study demonstrated that Tau-P301L mice have phenotypic alterations, which are in line with the histological changes and with a sex-dependent performance in those tests. Summarized, the Tau-P301L mouse model shows phenotypic alterations due to the presence of neurofibrillary tangles in the brain.

Keywords: tauopathy; Tau-P301L mouse models; behavior; phosphorylated Tau; motor deficits; cognitive deficits; sex-related deficits

Citation: Camargo, L.C.; Honold, D.; Bauer, R.; Shah, N.J.; Langen, K.-J.; Willbold, D.; Kutzsche, J.; Willuweit, A.; Schemmert, S. Sex-Related Motor Deficits in the Tau-P301L Mouse Model. *Biomedicines* **2021**, *9*, 1160. https://doi.org/10.3390/biomedicines9091160

Academic Editor: Martina Perše

Received: 29 July 2021
Accepted: 1 September 2021
Published: 4 September 2021

Publisher's Note: MDPI stays neutral with regard to jurisdictional claims in published maps and institutional affiliations.

1. Introduction

Tau protein is a microtubule associated protein, located in the axons, which plays a major role in the stabilization of microtubules [1] and trafficking [2–4]. It is expressed by the microtubule-associated protein Tau (*MAPT*) gene located on the chromosome 17. In total, six isoforms can be produced by the presence/absence of exon 2, 3 (N-terminal) and 10 (microtubule-binding domain). Therefore, the isoform expression varies from 0N3R, which is the shortest form, to 2N4R, which is the longest form. In humans, the 3R is more frequent during the development, while both 3R and 4R, are present in similar amount in the adult brain [5,6]. Phosphorylation of the Tau protein can occur at different sites by different kinases, a process that assists in Tau physiological function. Under pathological conditions, the Tau binding site to the microtubules is hyperphosphorylated and results in loss of its function. Hyperphosphorylated Tau then assembles into paired helical filament (PHF) forming the neurofibrillary tangles (NFTs) in the dendrites [6,7]. Pathological Tau is present in different neurodegenerative diseases called tauopathies.

Tauopathies, in turn, are a heterogeneous class of diseases that can be classified as primary and secondary tauopathies. In secondary tauopathies, the presence of NFTs occurs as a second event probably due to the toxicity downstream of another event, e.g., aggregation of amyloid-β (Aβ) into neuritic plaques in Alzheimer's disease (AD). In primary tauopathies, the presence of NFTs occurs first and is mainly responsible for the arising neurodegeneration, e.g., in frontotemporal dementia (FTD) [8]. In those dementias, the formation of NFTs in a specific region is correlated with progression of the disease and brain atrophy [9,10]. Considering that brain atrophy and cognitive deficits are a consequence of neurodegeneration and synaptic dystrophy, it is postulated that the presence of NFTs induces synaptic deficits and neurodegeneration [11,12]. Besides in dementias, pathological Tau can also be found in patients with epilepsy, chronic traumatic encephalopathy and other neurological disorders [13]. Similar to AD, most of the FTDs and other tauopathies are sporadic and, unlike AD, different mutations can cause the familial FTDs. The mutations in the *MAPT* gene are genetic causes of FTDs with parkinsonism linked to chromosome 17 (FTDP-17) [14,15]. Those mutations prevent Tau from binding to microtubules due to hyperphosphorylation [16].

Many transgenic mouse models have been developed with different Tau mutations. Those models provide a more detailed understanding of how hyperphosphorylated Tau and NFTs affect the pathophysiology, depending on the type of mutation and the isoform. The most common transgenic models of tauopathy are constructed with the human Tau-P301L mutation [17,18]. The Tau-P301L mouse models only include the 4R Tau isoform, since this mutation is located in the exon 10. Terwel and collaborators [19] developed a transgenic mouse model expressing human Tau-P301L (homozygous) under the regulation of a *thy1* gene promoter at moderate levels. This mouse model did not develop severe motor deficits, but a strong paralysis in the limbs, starting at nine months of age. The mice died before the age of 12 months due to respiratory problems [19,20]. Moreover, Tau-P301L mice showed NFTs at nine months of age in the brainstem and cortex [19]. The presence of NFTs in different areas of the brainstem was postulated to be the cause of respiratory deficits and strong moribund conditions [20]. At earlier ages, this mouse model also showed increased long-term potentiation (LTP) in the dentate gyrus (DG) [21].

Nowadays, mouse models are considered a method to represent human diseases and to test newly developed substances as treatment. Mouse models, especially for neurodegenerative diseases, have recently been under some criticism, in part because many clinical trials failed even though the compounds did previously show promising results in animal models. Very often in these cases, however, treatment studies in mice often had an insufficient study design, which does not mimic the human situation very well. It is essential to know your animal model as well as possible, especially concerning the selection of behavioral tests and to characterize them in longitudinal studies, instead of just analyzing deficits at one specific age, as well as doing this in a sex-specific manner. The objective of this study was to carry out a longitudinal and sex-related characterization of the Tau-P301L model to clarify the onset of the disease with a broader behavioral test battery and to have an in-depth understanding about the deficits of the model. As described before, the Tau-P301L model was evaluated in few behavioral experiments (beam walk, rotarod and novel object recognition) and some studies were cross-sectional. A longitudinal study is advantageous since the onsets of each behavioral deficit occur at different time points; therefore, the cross-sectional studies have limited information regarding the course of the disease. Thus, the present study focused on the characterization of general, motor and cognitive alterations induced by pathological Tau in the Tau-P301L mouse model at different ages and sexes.

2. Materials and Methods

2.1. Animals

Tau-P301L mice were first described by Terwel et al. [19] and were backcrossed from a FVB to a C57BL/6J background. Mice were maintained in a homozygous colony. In this study, we compared homozygous Tau-P301L mice with age- and sex-matched wild type (WT) mice from a parallel breeding.

Mice were bred in-house with a 12/12 h light/dark cycle. In each cage, three to five mice were housed and food and water were available *ad libitum*. All behavioral experiments were approved by the responsible authorities (*Landesamt für Natur, Umwelt und Verbraucherschutz (LANUV)*, North Rhine-Westphalia, Germany, number 84-02.04.2014.A362, 81-02.04.2018.A400 and 81-02.04.2019.A304; approval was received on 05/02/2019, 21/02/2019 and 21/01/2019, respectively) and were performed longitudinally at different ages (2, 4, 6 and 8 months). For all behavioral tests, 7 female and 12 male mice of both genotypes were included.

2.2. Behavioral Tests

2.2.1. Habituation/Dishabituation Olfactory Test

Olfactory deficits from Tau-P301L mice were evaluated by performing the habituation/dishabituation olfactory test [22]. Three different aromas (bacon, cheesecake and hazelnut) (Perfumer's Apprentice, Scotts Valley, CA, USA) were sprayed on a cotton pad which was placed into an embedding cassette. The bacon aroma was placed in the cage for 24 h before the test for habituation. Later, the bacon aroma was presented again to the mice for six times for 30 s each. Next, the bacon aroma was replaced by cheesecake and hazelnut aroma once (30 s each). The time the mice sniffed each embedding cassette was recorded for analysis.

2.2.2. Nesting Behavior Test

Nesting behavior was performed as previously described [23]. One hour before the dark cycle of the animal facility, the mice were single caged with new nesting material. The next morning, the built nest was scored from 1 to 5, whereby 1 was no nest and 5 was a fully built nest.

2.2.3. Marble Burying Test

In the marble burying test [24], mice were placed in a cage with 5 cm of bedding material with 12 equally distant marbles for 30 min, which were placed on the top of the bedding material. Later, the mice were placed back in the habituation cage and the number of marbles each mouse had buried was counted for analysis.

2.2.4. SHIRPA Test Battery

To evaluate the phenotypic alterations of Tau-P301L mice in comparison to the WT mice, the SmithKline Beecham Pharmaceuticals; Harwell, MRC Mouse Genome Centre and Mammalian Genetics Unit; Imperial College School of Medicine at St Mary's; Royal London Hospital, St Bartholomew's and the Royal London School of Medicine; Phenotype Assessment (SHIRPA)-test battery was performed (protocol adapted from [25]). In this test, the different parameters described in Table 1 were evaluated in a scoring system from 0 to 3 (0 = no alteration; 1 = slightly altered; 2 = altered; 3 = strongly altered).

Table 1. Evaluated Parameter on the SHIRPA test.

Parameters	Description
Restlessness	Difficulty staying in one body position for an extended period of time
Apathy	Motionless and lowered head
Stereotyped behaviour	
Convulsion	
Abnormal body carriage	Body posture
Alertness	Response to object proximity
Abnormal gait	Uncommon walk, e.g., paddling, waddling, running
Startle response	Response to an acoustic signal
Loss of righting reflex	Time when the mouse return to standing position when turned on its back
Touch response	
Pinna reflex	
Cornea reflex	
Forelimb placing reflex	Response to stretch their front paws when hanged in proximity to the surface
Hanging behaviour	Mouse stays on the rod or falls
Pain response	Response to tail pinch
Grooming	Overall fur condition

2.2.5. Open Field Test

In the open field test, mice were placed in a cubicle arena (40 cm) for 30 min. During this time, mice were allowed to freely explore the arena, imaginarily divided into different zones (border, center, corner). For evaluation, tracking software was used (EthoVision XT15, Noldus Information Technology, Wageningen, The Netherlands). The following parameters were analyzed: velocity, locomotion, exploration time, time spent in center, border and corner zone.

2.2.6. Accelerating Rotarod

The accelerating Rotarod (Ugo Basile, Gemonio, Italy) test consisted of four trials. In the first trial, the mice were placed onto the rod and should stay there for at least 60 s at 10 rpm (habituation to the apparatus). If they fell, the trial was repeated. In the last three trials, the mice should stay on the rod for 300 s at 4 to 40 rpm. For evaluation, the latency time to fall was noted and the mice were placed back into their home cages. Three sessions in each trial with an interval of 15 min were performed [26].

2.2.7. Modified Pole Test

In order to gain a deeper understanding of the developed motor deficits, a modified version of the so-called pole test was performed [27]. For this, mice were placed facing down on the top of a pole and the way they walked down was scored three times. The scoring system was: 0 = running, 1 = partly running, 2 = slipped and 3 = fallen. This procedure was repeated three times with an interval of 15 min between each trial. For the final evaluation, the sum of the three scores was calculated.

2.2.8. Novel Object Recognition Test

For the novel object recognition test (NOR), two identical objects (familiar object) were presented to the mice during 10 min in the same arena used for the open field test. In the inter-trial interval of 20 min, the mice were placed back in their home cages. Afterwards, the mice were placed back into the arena where one familiar object was replaced by a new object (novel object). The time of exploration was evaluated as the time the mouse spent with the nose at least 2 cm from the object. This was analyzed by EthoVision XT15 (Noldus Information Technology, Wageningen, The Netherlands).

For evaluation, the discrimination index was calculated by the following formula:

$$\frac{Tnovel - Tfamiliar}{Tnovel + Tfamiliar} \tag{1}$$

where *Tnovel* was the time the mice explored the novel object and *Tfamiliar* was the time the mice explored the familiar object.

2.2.9. T-Maze Spontaneous Alternation

In the T- maze spontaneous alternation [28], the mice were placed in the starting arm in an arena with three arms (start, left and right arm) (31 cm × 10 cm) in a "T" format. In the first trial, only the left or right arm was free to be explored and the opposite one was closed by a gate. Once the mice came back to the start arm, both arms were free to be explored and the second trial started. The same procedure was performed for 14 trials or a maximum of 15 min. If a mouse did not reach seven trials, it was excluded from the experiment.

The spontaneous alternation was calculated by the following formula:

$$\frac{number\ of\ correct\ choices}{total\ of\ trials} \tag{2}$$

Correct choices are considered as interactions with the arm opposite to the one that the mouse previously entered in the maze.

2.2.10. Fear Conditioning Test

In order to evaluate the associative memory deficits, the cued and contextual fear conditioning was performed [29] starting with 4 months of age. On the habituation day, mice were placed in the apparatus (Ugo Basile, Gemonio, Italy) for 120 s of habituation. Afterwards, a sound (50%; 2000 Hz) was presented for 30 s, and during the last 2 s, a mild shock (0.35 mA) was also given. The mice stayed in the cage for additional 60 s before returning to their home cages.

The next day, the contextual fear conditioning was evaluated. The mice were placed in the same cage for 5 min and neither the shock nor the sound were presented. After 25 min, the cued fear conditioning was evaluated. The walls and floor of the cage were changed and only the sound was presented three times to the mice. The freezing (%) was analyzed with tracking software (EthoVision XT15, Noldus Information Technology, Wageningen, the Netherlands).

2.2.11. Morris Water Maze

The performance of the Morris water maze (MWM) [30] was divided into 3 stages: training, probe and reversal test. For the MWM training, the mice were placed in a pool (diameter of 120 cm × 60 cm height) filled with water divided into 4 quadrants (NE, NW, SE, SW) with a hidden platform (diameter of 10 cm × 31.5 cm height). An opaque, non-toxic liquid was added into the water to prevent the mice from seeing the platform. For a maximum of 60 s, the mice had to find the hidden platform. In case the mice did not find it, they were placed onto the platform for 10 s for acquisition (to orientate themselves). This trial was then repeated four times per mouse. Additionally, at each trial, the mice were placed in a different starting position. These trials were performed for four consecutive days. On the fifth day, the platform was removed and the probe trial was performed. Moreover, the reversal test was also performed, similar to the training, for three consecutive days and the platform was placed in a different position (opposite position). Similar to the previous cognitive tests, the evaluation and tracking was analyzed by tracking software (EthoVision XT15, Noldus Information Technology, Wageningen, The Netherlands). In the training and reversal test, the time the mice needed to find the hidden platform (escape latency) was analyzed. In the probe trial, the time spent in the platform zone was analyzed. The MWM was performed only at 8 months of age. One female mouse developed a forelimb paralysis and was, therefore, excluded from the MWM experiment.

2.3. Histology

After the performance of the last behavioral tests (MWM), mice were deeply anesthetized for tissue collection. The brains were snap frozen and one hemisphere was cut into 20 μm sagittal sections using a Cryotome (Leica Biosystems Nussloch GmbH, Wetzlar, Germany). Before the staining procedure, the brain slices were placed in 4% formalin and washed three times with TBS for 5 min. Antigen retrieval was performed in citrate buffer, pH 6 at 85 °C for 30 min and slides were washed thre times with TBS for 5 min. In order to remove the endogenous peroxidases, the sections were incubated in 0.6% H_2O_2 in methanol for 15 min and washed once with deionized water and two times with TBS for 5 min. Then, the sections were blocked in 10% horse serum for 1 h and incubated overnight with the primary antibody (AT8 (1:500; MN1020, Thermo Fisher scientific, Waltham, MA, USA) or AT100 (1:500; MN1060, Thermo Fisher scientific, Waltham, MA, USA) in 1% horse serum in TBS at 4 °C. On the subsequent day, the sections were washed and incubated with the secondary antibody (biotinylated goat anti-mouse, 1:1000; Extra2, Sigma-Aldrich, Darmstadt, Germany) for 2 h. Afterwards, slides were again washed and incubated with ExtrAvidin® (1:1000; Extra2, Sigma-Aldrich, Darmstadt, Germany) for additional 2 h, followed by a washing step. Finally, the sections were colored with DAB and saturated nickel ammonium sulphate solution, washed, dehydrated in an ascending alcohol series and mounted with DPX (Sigma-Aldrich, Darmstadt, Germany).

To evaluate neurodegeneration and neuroinflammation, the following staining procedure was done. The brain slides were placed in 4% formalin and washed three times with TBS-T (1% triton) for 5 min. Antigen retrieval was performed in 70% formic acid and slides were washed. In order to remove the endogenous peroxidases, the sections were incubated in 3% H_2O_2 in methanol solution for 15 min and washed. Then, the sections were incubated overnight with the primary antibody (NeuN (1:1000; Merck, Darmstadt, Germany) and GFAP (1:1000; MN1060, Thermo Fisher scientific, Waltham, MA, USA) in 3% BSA in TBS-T at 4 °C. The next day, the sections were washed and incubated with the secondary antibody (biotinylated goat anti-rabbit, 1:1000; Thermo Fisher scientific, Dreieich, Germany) for 2 h. Afterwards, the same procedure was performed as described above. For the detection of reactive microglia (CD11b, 1:2000, Abcam, Berlin, Germany), the staining procedure was the same as previously described although the primary antibody was incubated in 1% normal goat serum (NGS) and 1% bovine serum albumin (BSA) at room temperature for 1.5 h and the washing buffer was TBS. Subsequently, the same procedure was performed as described above.

The images were taken with a LMD6000 microscope and a DFC310 FX camera (Leica Biosystems Nussloch GmbH, Wetzlar, Germany) or with a Zeiss SteREO Lumar V12 microscope and the according software (Zeiss AxioVision 6.4 RE). For pathological Tau, the positive signals in the brainstem (hind and midbrain), cerebellum and cortex were counted with ImageJ software (National Institute of Health, Bethesda, MD, USA). For neuronal death, the positive signals in the brainstem (hind and midbrain), cerebellum and cortex were counted with Cell profiler software (Broad Institute, Cambridge, MA, USA). For reactive microglia and reactive astrocyte analysis, the stained areas (percentage) in the brainstem (hind and midbrain), cerebellum and cortex were analyzed with CellProfiler software (Broad Institute, Cambridge, MA, USA). For each staining, eight males and seven females were analyzed and four to eight slides were taken per mouse for analysis.

2.4. Statistical Analysis

The statistical analyses were performed using GraphPad Prism 8.3 (GraphPad Software, San Diego, CA, USA). Two-way ANOVA and Sidak's multiple comparison post hoc were used as statistical analysis to compare the sex-matched WT with Tau-P301L mice at each age in all behavioral tests, except the habituation/dishabituation olfactory test, the NOR and the MWM. In the habituation/dishabituation olfactory test, the two-way ANOVA and Sidak's multiple comparison post hoc was performed to compared the WT with Tau-P301L mice for each odor presented. In the NOR, the discrimination index was compared against the theoretical mean of 0%, which means the exploration of novel and familiar object are similar; therefore, no discrimination is assumed, and the statistical analysis was calculated by the one-sample T-test for each group. In the MWM, the two-way ANOVA analysis was performed to compare the WT males, WT females, Tau-P310L male and female mice during the training days in both, the training test and the reversal trial. In the MWM probe trial, the two-way ANOVA and Tukey's multiple comparison post hoc was performed to analyze the difference in the time spent at each quadrant. To evaluate the differences in the histology, the two-way ANOVA and Sidak's multiple comparison post hoc was used to compare the males to females and to compare the Tau-P301L to the WT.

3. Results

3.1. Tau-P301L Male Mice Show Phenotypic Alterations Beginning at 4 Months of Age

In order to analyze the phenotypic alterations of male and female Tau-P301L mice, several behavioral tests were performed. Due to the fast-phenotypic progression, the Tau-P301L mice were regularly observed (at least once a week) and observations were reported in a score sheet. Regarding the general behavior, neither male nor female Tau-P301L mice showed any abnormalities in their home cages until 7 months of age. From 7 months of age, Tau-P301L mice progressed to a prominent paralysis of the limbs, and consequently, a loss of body weight and reduced movement in the home cages, as Terwel and collaborator [19] described. By comparison of the body weight, Tau-P301L mice had similar weight compared to WT mice, although Tau-P301L female mice had a slightly higher weight compared to WT female mice at 4 months of age (Figure S1). Tau-P301L mice did not display any deficits in the nesting behavior and marble burying compared to WT mice at all analyzed ages (Figure S2). Moreover, WT male mice buried a smaller number of marbles throughout aging. Furthermore, Tau-P301L mice had a non-significant trend of burying less marbles throughout aging (Figure S2). In the habituation/dishabituation olfactory test, Tau-P301L male mice explored the novel aroma less (cheesecake and hazelnut aroma) compared to the male WT mice at six months of age (Figure 1) (two-way ANOVA, $p = 0.0036$, $p = 0.0139$, respectively). Tau-P301L male mice were not able to discriminate the cheesecake and hazelnut from the bacon aroma. At 6 months of age, unlike Tau-P301L male mice, the Tau-P301L female mice did not show any olfactory deficits (Figure 1).

In the SHIRPA test battery, Tau-P301L male and female mice had phenotypic alterations compared to the sex-matched WT mice starting at 4 months of age (Table 2). Tau-P301L mice showed an abnormal gait, as demonstrated by a waddling walk. Moreover, they appeared to be less agile; they were slower than WT mice at 8 months of age. Furthermore, Tau-P301L mice showed an abnormal body carriage (hunched back) compared to WT male mice starting at 4 months of age. When lifted up by the tail, Tau-P301L mice presented clasping of all limbs, especially with increasing age, which can be described as slight paralysis starting at 6 months of age. This paralysis increased dramatically with age. Those findings are in correspondence with those published by Terwel et al. [19]. When placed hanging on a rod, Tau-P301L mice were not able to hold nor hang with both forelimbs starting with 4 months of age, but some WT mice showed similar impairments at 6 months of age. Finally, some Tau-P301L mice showed a mild loss of postural reflex when placed on their back starting with 4 months of age (Table 2; Table S1). Tau-P301L mice had higher SHIRPA scores compared to WT mice from 4 months onward (Figure 2) (two-way ANOVA; 4 months: $p = 0.0008$, 6 months: $p = 0.0009$, 8 months: $p < 0.0001$). Moreover, an age-dependent deterioration of the phenotype was observed starting at 2 months of age (two-way ANOVA; 2 vs. 4: $p < 0.0001$; 2 vs. 6: $p = 0.0003$; 2 vs. 8: $p = 0.0204$). Tau-P301L female mice had a higher score compared to WT female mice at 4 months of age (two-way ANOVA; $p = 0.0022$) as well as an increased score compared to 2 and 6 months (two-way ANOVA; 2 vs. 6: $p = 0.0002$; 6 vs. 8: $p = 0.0030$) (Figure 2).

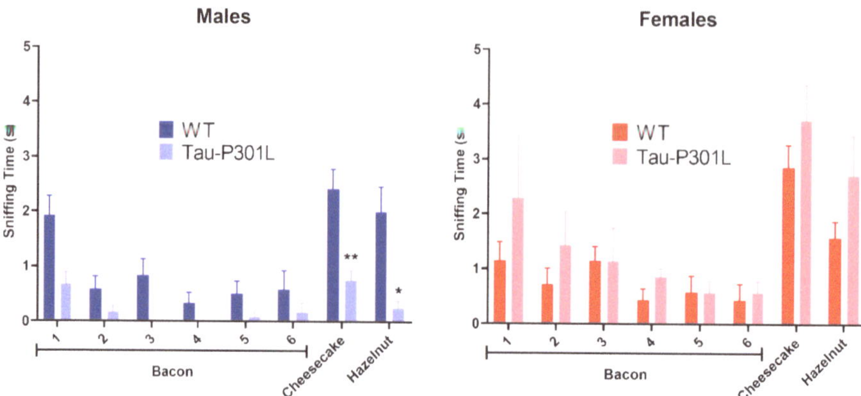

Figure 1. Tau-P301L mice develop olfactory deficits in the habituation/dishabituation olfactory test at 6 months of age. The bacon aroma was presented six times to the mice, the cheesecake and hazelnut aroma were presented afterwards. The sniffing (exploration) time was evaluated as the time the mouse placed the nose on the box with aroma-sprayed cotton. Tau-P301L male mice ($n = 12$) smelled the new aroma less (cheesecake and hazelnut) compared to the age-matched WT male mice ($n = 12$), but this was not observed in the females ($n = 7$). The two-way ANOVA was used as statistical analysis. *: $p < 0.05$ and **: $p < 0.01$ compared to the age-matched WT. Data are given as mean ±SEM.

Table 2. Tau-P301L mice showed phenotypic alterations in different evaluated parameters in the SHIRPA test starting at 4 months of age.

Parameters	Phenotypic Alterations
Restlessness	No alterations
Apathy	No alterations
Stereotyped behavior	No alterations
Convulsion	No alterations
Abnormal body carriage	Hunchback
Alertness	No alterations
Abnormal gait	Waddling walk and slower compared to WT
Startle response	No alterations
Loss of righting reflex	Some Tau-P301L mice have light loss of righting reflex
Touch response	Less responsive to touch than WT
Pinna reflex	No alterations
Cornea reflex	No alterations
Forelimb placing reflex	Paralysis ("Clasping") of the limbs
Hanging behavior	Tau-P301L male mice fall faster from the rod than WT male mice
Pain response	No alterations
Grooming	The Tau-P301L male mice have very good fur condition compared to WT

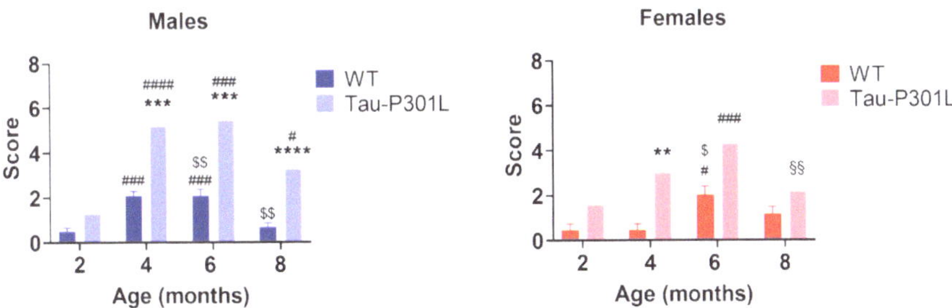

Figure 2. Tau-P301L male mice show phenotypic alteration in the SHIRPA test battery. Both Tau-P301L mice and WT were evaluated at 2, 4, 6 and 8 months of age. At 4, 6 and 8 months of age, Tau-P301L male mice (n = 12) had a higher score compared to the age-matched WT male mice (n = 12). Only at 4 months of age, Tau-P301L female mice (n = 7) had a higher score compared to the age-matched WT female mice (n = 7). Two-way ANOVA was performed. **: $p < 0.01$, ***: $p < 0.001$ and ****: $p < 0.0001$ compared to the age-matched WT. #: $p < 0.05$, ###: $p < 0.001$ and ####: $p < 0.0001$ compared to 2 months, genotype-matched. \$: $p < 0.05$ and \$\$: $p < 0.01$ compared to 4 months, genotype-matched. §§: $p < 0.01$ compared to 6 months, genotype-matched. Data are given as mean ±SEM.

3.2. Tau-P301L Male Mice Display Early Motor Deficits

In the open field test, Tau-P301L male mice had motor deficits from to 2 months of age, since they were slower (two-way ANOVA; 2 months: $p = 0.0068$; 4 months: $p = 0.0059$; 6 months: $p = 0.0015$; 8 months: $p = 0.0298$) (Figure 3A) and travelled less (two-way ANOVA; 2 months: $p = 0.0073$; 4 months: $p = 0.0060$; 6 months: $p = 0.0020$; 8 months: $p = 0.0296$) (Figure 3B). This deficit persisted until 8 months of age and progressed throughout aging (two-way ANOVA; 2 vs. 6: $p = 0.0003$; 2 vs. 8: $p = 0.0003$; 4 vs. 8: $p = 0.0031$). Regarding the exploratory behavior, Tau-P301L male mice also spent less time exploring the arena compared to WT male mice beginning at 2 months of age (two-way ANOVA; 2 months: $p = 0.0216$; 4 months: $p = 0.0198$; 6 months: $p = 0.0011$; 8 months: $p = 0.0454$) (Figure 3C). Mice of both genotypes spent the same amount of time in the corner, border and center zone of the arena, demonstrating that Tau-P301L mice do not have increased anxiety levels compared to WT mice (Figure S4). Similar to the previous data, Tau-P301L female mice did not show any differences compared to the WT female mice, although both Tau-P301L and WT female mice showed a decrease of velocity, distance travelled and active time with aging (Figure 3D–F).

Analysis of the modified pole test revealed that Tau-P301L male mice had higher scores compared to the WT male mice starting at 6 months of age (Figure 4) (two-way ANOVA; 6 months: $p = 0.0365$ and 8 months: $p = 0.0040$). This indicates that Tau-P301L mice developed motor deficits in this test and the deficits progressed throughout aging (two-way ANOVA; 2 vs. 8: $p = 0.0003$; 4 vs. 8: $p = 0.0031$) (Figure 4). Tau-P301L female mice had similar performance as the WT female mice, indicating no motor deficits in this test.

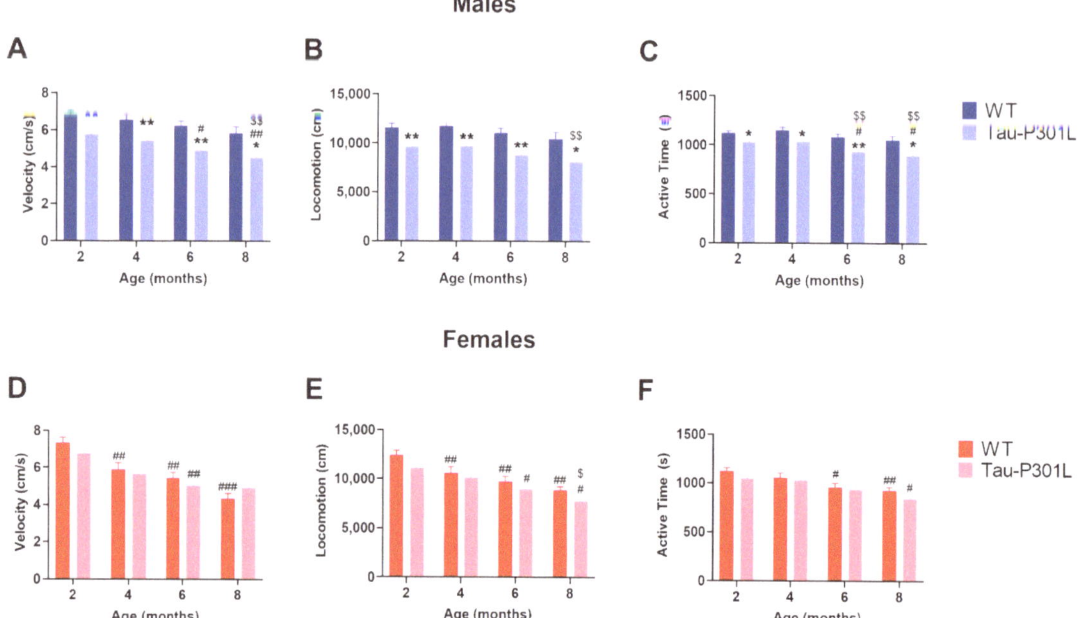

Figure 3. Tau-P301L male mice develop motor deficits in the open field test. Both Tau-P301L mice and WT were evaluated at 2, 4, 6 and 8 months of age. At all ages, Tau-P301L male mice (n = 12) were slower (**A**), travelled less (**B**) and were less active (**C**) compared to the age-matched WT males (n = 12). Tau-P301L female mice (n = 7) had similar velocity (**D**), locomotion (**E**) and active time (**F**) compared to the age-matched WT females (n = 7). Two-way ANOVA was performed. *: $p < 0.05$ and **: $p < 0.01$ compared to the age-matched WT. #: $p < 0.05$, ##: $p < 0.01$ and ###: $p < 0.001$ compared to 2 months genotype-matched. \$: $p < 0.05$ and \$\$: $p < 0.01$ compared to 4 months genotype-matched. Data are given as mean ±SEM.

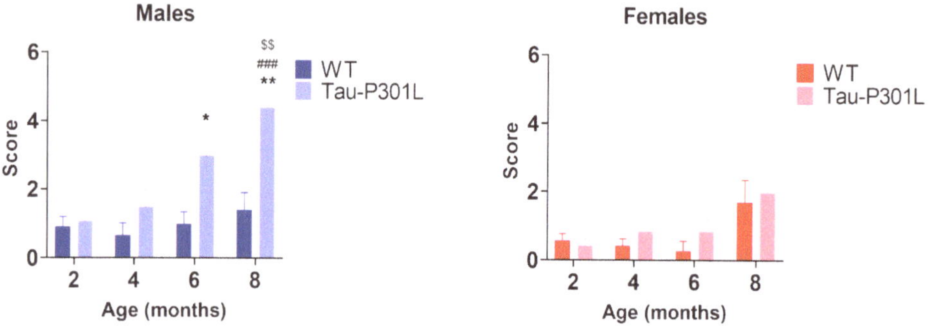

Figure 4. Tau-P301L male mice displayed motor deficits in the modified pole test. Both Tau-P301L mice and WT mice were evaluated at 2, 4, 6 and 8 months of age. The test was performed three times with a 15 min intertrial interval and the sum of the three trials was used for analysis. At 6 and 8 months of age, Tau-P301L male mice (n = 12) had a higher score compared to the age-matched WT male mice (n = 12), but this was not observed in the females (n = 7). Two-way ANOVA was performed. *: $p < 0.05$ and **: $p < 0.01$ compared to the age-matched WT. ###: $p < 0.001$ compared to 2 months genotype-matched. \$\$: $p < 0.01$ compared to 4 months genotype-matched. Data are given as mean ±SEM.

Analysis of the Rotarod performance of both Tau-P301L male and female mice did not show any motor alteration in the accelerating Rotarod (Figure S3). Similar to the previously described paralysis, these results are in correspondence with those published by Terwel et al. [19].

3.3. Tau-P301L Show Mild Cognitive Deficits in the NOR

In order to analyze the development of possible cognitive deficits, several behavioral tests were performed. In the NOR, Tau-P301L mice were not able to discriminate between the novel and familiar object beginning at 6 months of age (Figure 5) (one sample *t*-test against 0%, 2 months: $p = 0.0001$; 4 months: $p = 0.0045$; 6 months: $p = 0.2078$ and 8 months: $p = 0.1365$). WT male mice were able to discriminate the novel from the familiar object at all analyzed ages significantly (one sample *t*-test against 0%, 2 months: $p = 0.0001$; 4 months: $p < 0.0001$; 6 months: $p = 0.0157$ and 8 months: $p = 0.0153$). Tau-P301L female mice did not discriminate the novel from the familiar object at 4 months of age (Figure 6) (one sample *t*-test against, 2 months: $p = 0.0266$; 4 months: $p = 0.1221$; 6 months: $p = 0.1945$ and 8 months: $p = 0.1293$) unlike the WT female mice (one sample *t*-test against, 2 months: $p = 0.0037$; 4 months: $p = 0.0009$; 6 months: $p = 0.0562$ and 8 months: $p = 0.0187$). In summary, since Tau-P301L mice did not significantly explore the novel object more, they had deficits in the recognition memory beginning at 6 (males) and 4 (females) months of age.

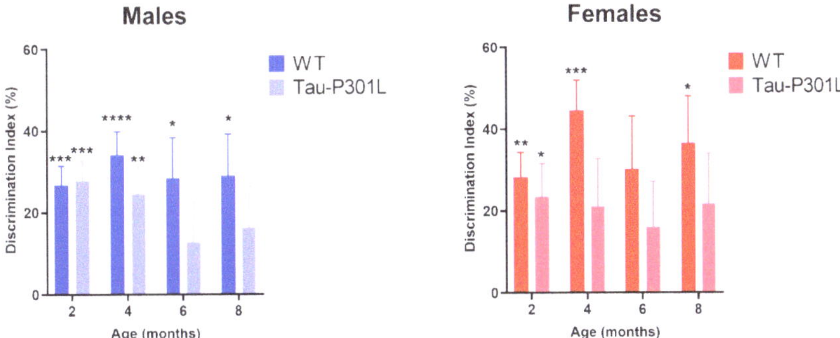

Figure 5. Deficits in recognition memory in Tau-P301L mice in the novel object recognition test (NOR). Both Tau-P301L mice and WT were evaluated at 2, 4, 6 and 8 months of age. Tau-P301L male mice ($n = 12$) were not able to discriminate the novel from the familiar object at 6 months of age and Tau-P301L female mice ($n = 7$) at 4 months of age. Both WT male ($n = 12$) and female mice ($n = 7$) were able to discriminate the novel object. The one sample *t*-test against 0% was used to evaluate the missing discrimination from the novel object. *: $p < 0.05$, **: $p < 0.01$, ***: $p < 0.001$ and ****: $p < 0.0001$. Data are given as mean ±SEM.

No cognitive deficits were detectable, neither in the T-maze spontaneous alternation (Figure S5), nor in the contextual and cued fear conditioning within the here-analyzed ages (Figure S6). Furthermore, no differences could be detected between Tau-P301L and WT mice in the MWM. During the four days of training, all tested mice showed similar escape latencies (Figure 6). In the probe trial, all genotypes spent a similar amount of time in the target quadrant (NW). Moreover, no difference was detectable between Tau-P301L mice, neither between males nor between females. During the reversal trial, Tau-P301L mice and non-transgenic mice spent a similar amount of time to find the platform. Overall, Tau-P301L mice did not have any cognitive deficits in the MWM at the age of 8 months.

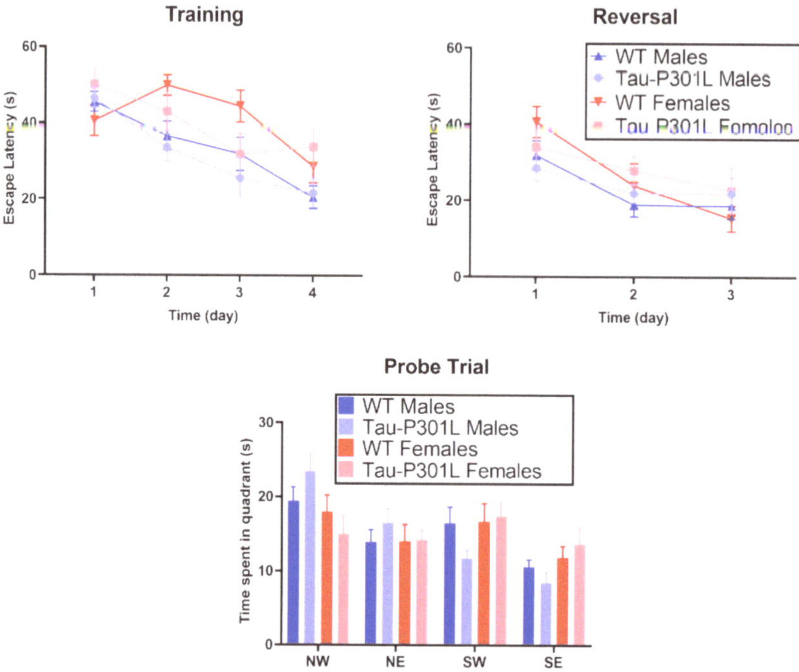

Figure 6. Tau-P301L mice did not have any deficits in the Morris Water Maze (MWM). Tau-P301L and WT mice were evaluated at 8 months of age. In the training and reversal test, both Tau-P301L mice (males: $n = 12$; females: $n = 7$) and WT mice (males: $n = 12$; females: $n = 7$) spent a similar amount of time to find the platform throughout the days. In the probe trial, both Tau-P301L mice and WT male mice explored the target quadrant similarly (NW). Mixed effect and two-way ANOVA were used for analysis, respectively. Data are given as mean ±SEM.

3.4. Tau-P301L Showed Distinct Tau Pathology in the Brain at 8 Months of Age

After performance of the MWM, the brains from all mice were collected. Regarding the histopathology, an AT8-positive signal was found at a significantly higher number in the brains of Tau-P301L male mice compared to WT male mice. AT8 antibody binds to pSer202 and pThr204 and the phosphorylation of this site increases with age [19]. Therefore, those phosphorylated sites occur mainly in PHF [31,32]. In Tau-P301L male mice, more pathological Tau is found compared to WT male mice in the hindbrain (two-way ANOVA; males: $p = 0.0131$), the midbrain (two-way ANOVA; males: $p = 0.0032$), the cortex (two-way ANOVA; males: $p = 0.0318$) and the cerebellum (two-way ANOVA; males: $p = 0.0009$) (Figure 7). Moreover, Tau-P301L male mice ($n = 8$) had more AT-8 positive signals than Tau-P30L female mice ($n = 7$) in the midbrain (two-way ANOVA; $p = 0.0452$) and the cerebellum (two-way ANOVA; $p = 0.0412$). Finally, Tau-P301L female mice did not have more pathological Tau compared to WT male mice in any analyzed brain region.

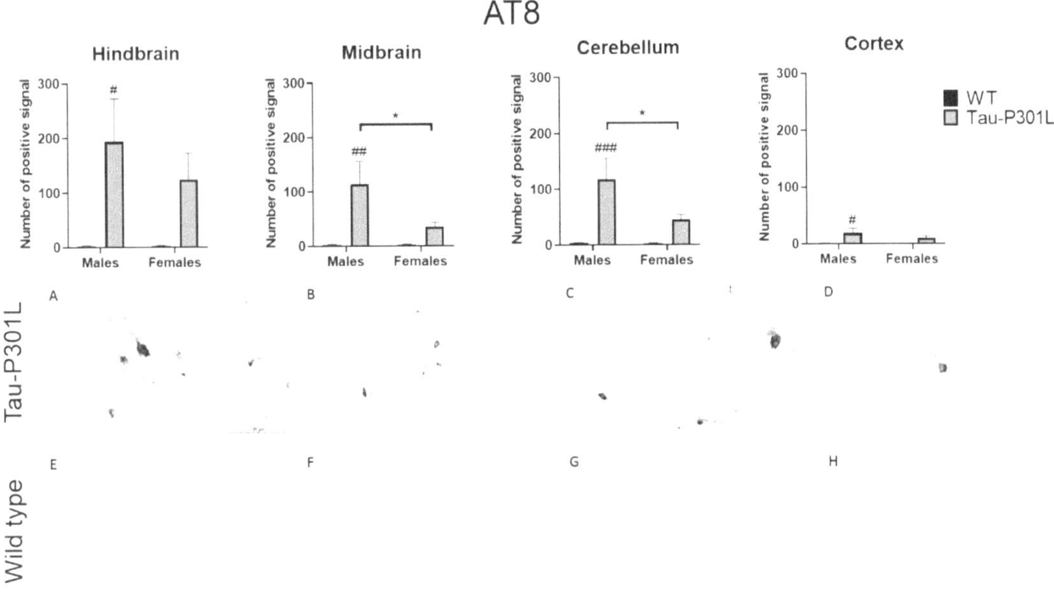

Figure 7. Tau-P301L mice show pathological Tau in different areas of the brain at 8 months of age. The phosphorylated Tau was detected by AT8 antibody. The hindbrain (**A,E**), cortex (**D,H**), midbrain (**B,F**) and cerebellum (**C,G**) from Tau-P301L (**A–D**) and wild type (WT) male mice (**E–H**) were analyzed. The positive signal was counted at different regions of the brain using ImageJ software. A two-way ANOVA was used for analysis. *: $p < 0.05$; #: $p < 0.05$, ##: $p < 0.01$ and ###: $p < 0.001$ compared to sex-matched WT. Scale bar is 125 µm.

Using the AT100 antibody that recognizes pSer214 and pThr212, which are only present in PHF [33] (Figure 8), it was found that Tau-P301L male mice had increased AT100 positive signal in the midbrain (two-way ANOVA; males: $p = 0.0004$), the hindbrain (two-way ANOVA; males: $p = 0.0449$) and the cerebellum (two-way ANOVA; males: $p = 0.0024$) compared to WT male mice. However, in the cortex, the number of positive signals was not significantly different from WT male mice. Regarding the sex, Tau-P301L male mice had an increased amount of AT100 positive signal only in the midbrain compared to Tau-P301L female mice (two-way ANOVA; $p = 0.0083$). Taken together, those regions with both AT100 and AT8 positive signal are mainly responsible for the motor coordination response and this could be an explanation for the motor deficits observed in the Tau-P301L male mice and not in the female mice.

More specifically, pathological Tau is present throughout different nuclei in the hindbrain, especially in the locus coeruleus (LC), pontine reticular nuclei, vestibular nucleus (medial and spinal) and reticular nuclei (parvicellular and intermediate). In the midbrain, the nuclei with Tau pathology were found in the vestibular tegmental area, substantia nigra reticular, periaqueductal gray (PAG), midbrain reticular nuclei and superior colliculus. In the cerebellum, the main region where pathological Tau is present is the interposed nucleus. The pathological Tau observed in those regions were AT8- and AT100-positive, but as expected, more AT8 signal was observed compared to AT100. In the striatum, olfactory bulb and hippocampus, neither AT8 nor AT100 signal was detected; therefore, there is no pathological Tau in those regions.

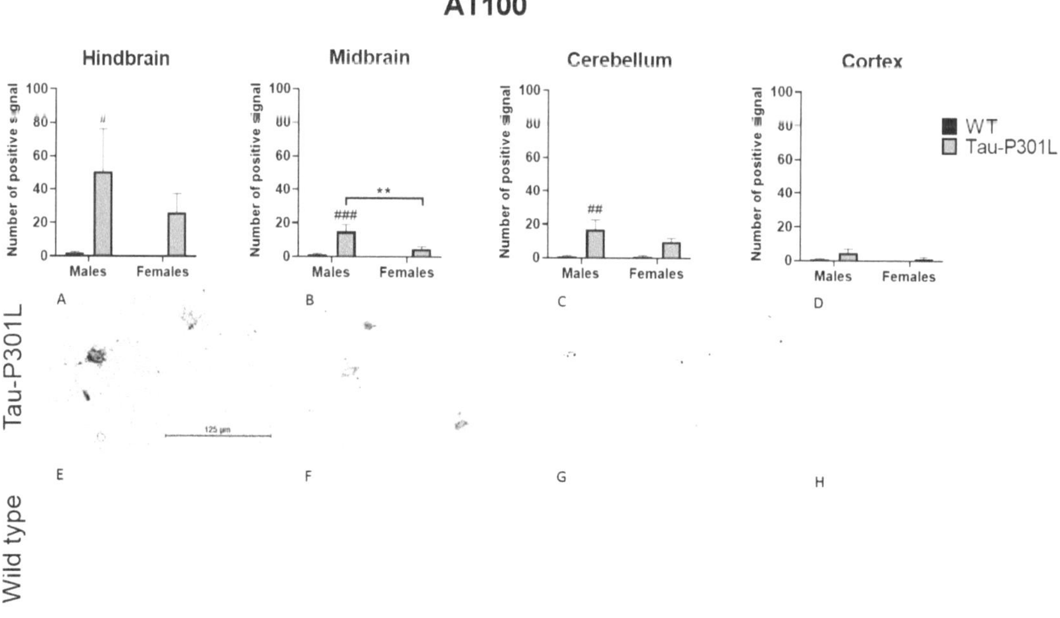

Figure 8. Tau-P301L mice had phosphorylated Tau in different areas of the brain at 8 months. The phosphorylated Tau was detected by AT100 antibody. The hindbrain (**A,E**), cortex (**D,H**), midbrain (**B,F**) and cerebellum (**C,G**) from Tau-P301L male (**A–D**) and wild type (WT) male mice (**E–H**) were analyzed. The positive signal was counted in different regions of the brain using ImageJ software. Two-way ANOVA and Multiple *t*-test were used for analysis. **: $p < 0.01$; #: $p < 0.05$, ##: $p < 0.01$ and ###: $p < 0.001$ compared to sex-matched WT. Scale bar is 125 µm.

Regarding the correlation between individuals, there is a clear relationship between the presence of Tau phosphorylation and the outcome of the behavioral test. Unfortunately, it was not possible to observe a statistically significant correlation between the results of the behavioral tests and the AT8/100 staining (Table S2).

Regarding neuronal loss, Tau-P301L male mice had fewer neurons in the hindbrain compared to WT (Table 3). The neurodegeneration was detected in the same region where the presence of AT8-positive signals but not AT100-positive signals was abundant. One could speculate that, since NFTs are mainly present intracellularly, the neuronal death in the hindbrain is inversely correlated with AT100 positive signal. Therefore, the increase of neuronal death would explain the low amount of AT100 positive signal in the hindbrain. No decrease of neurons was observed in Tau-P301L female mice in any region. In contrast, the reactive astrocytes and microglia were not increased in any brain region of Tau-P301L mice.

Table 3. Neuronal loss and gliosis in Tau-P301L mice in different brain regions.

Staining	Brain Region	WT		Tau-P301L		Significance
		Males	Females	Males	Females	
NeuN (Spot Count)	Hindbrain	1434.7 ± 111.3	1307.6 ± 208.9	817.1 ± 181.7	1286.0 ± 246.4	WT males vs. Tau-P301L males ($p = 0.012$)
	Midbrain	1376.8 ± 184.5	1487.8 ± 108.9	1191.9± 139.9	1474.3 ± 237.6	n.s
	Cortex	3545.1 ± 197.0	2879.1 ± 262.8	4141.1 ± 210.7	3651.3 ± 285.7	n.s
	Cerebellum	1834.0 ± 83.6	1259.0 ± 123.5	2060.1 ± 158.2	1461.1 ± 187.0	n.s
GFAP (Stained Area)	Hindbrain	27.0 ± 1.0	29.3 ± 1.5	31.5 ± 3.6	33.4 ± 1.1	n.s
	Midbrain	18.6 ± 2.8	21.0 ± 3.3	19.9 ± 2.8	31.2 ± 2.1	n.s
	Cortex	15.7 ± 2.7	23.4 ± 3.1	21.7 ± 4.4	31.8 ± 1.7	n.s
	Cerebellum	8.5 ± 1.3	11.0 ± 1.8	11.8 ± 1.6	15.4 ± 1.0	n.s
CD11b (Stained Area)	Hindbrain	6.7 ± 0.9	5.6 ± 0.4	6.5 ± 0.6	5.7 ± 0.8	n.s
	Midbrain	6.0 ± 0.4	5.2 ± 0.6	5.5 ± 0.7	5.1 ± 0.5	n.s
	Cortex	6.0 ± 0.5	5.9 ± 0.6	6.9 ± 0.9	6.0 ± 0.7	n.s
	Cerebellum	6.5 ± 0.4	4.9 ± 0.7	6.6 ± 0.3	5.8 ± 0.3	n.s

Quantification of activated astrocytes (GFAP), reactive microglia (CD11b) and neuronal nuclei (NeuN) of 8-month-old Tau-P301L (Tau-P301L) and wild type (WT) mice. The spot count analysis per selected area analysis was done in different brain regions (cortex, cerebellum, midbrain and hindbrain), resulting in a significant decrease of neurons in the Tau-P301L males' hindbrain. Analysis of gliosis, evaluated as stained area, revealed no differences between groups in different regions. Not statistically significant is represented by n.s.

4. Discussion

Translational research is essential to understand the mechanisms of diseases and mouse models play an important role in this context. Even though mouse models have several limitations, they are still the most complete option to be used in basic and preclinical research of neurodegenerative diseases [34]. For this reason, it is essential to characterize different mouse models down to the smallest details in order to obtain the most accurate translation and correlation to the corresponding human disease. In most characterization studies, only a few aspects of the phenotype are investigated and often only single ages are analyzed, e.g., when the first phenotypic differences are detectable, which might give limited information regarding the model. In this study, we focused on a longitudinal characterization study of the Tau-P301L mouse model, which was first described by Terwel and colleagues [19].

In summary, we showed that Tau-P301L mice had behavioral alterations in different behavioral tests probably due to the presence of pathological Tau in different brain regions. In the habituation/dishabituation olfactory test, Tau-P301L male mice spent less time smelling the newly presented aromas compared to WT male mice at 6 months of age. In the SHIRPA test, Tau-P301L mice had phenotypic alterations starting at 4 months of age. Moreover, the males had more prominent deficits compared to the females, especially regarding the motor alterations. In the modified pole test, Tau-P301L male mice had motor deficits demonstrated by a higher score compared to WT mice starting at 6 months of age. In the open field test, Tau-P301L male mice also had motor deficits, since they were slower, travelled less distance and explored less in an age-dependent manner compared to WT mice starting at 2 months of age. The Tau-P301L female mice did not show any of those alterations; therefore, one can assume they did not develop any motor deficits. Regarding the cognitive deficits, Tau-P301L male mice were not able to discriminate the novel from the familiar object in the NOR from 6 months of age. Moreover, Tau-P301L female mice did not discriminate the novel object from the familiar object at 4 months of age. Therefore, the Tau-P301L mouse model also displayed cognitive deficits. Those alterations can be explained by the presence of pathological Tau (AT8 and AT100 positive signal) in the hindbrain, cerebellum and midbrain, in which the latter ones are more pronounced in the Tau-P301L male mice than in female mice. The presence of pathological Tau induced neurodegeneration in the hindbrain in Tau-P301L male mice. Interestingly, the decrease of neurons seems to occur after the increase of AT8 positive signaling the hindbrain, but an increase of AT100 positive signal was not observed. Since AT100 antibody detects later stages of pathological Tau, one might speculate that the lack of increase of AT100 positive signal in the hindbrain might be due to the neuronal death of those neurons, which had pathological Tau. Finally, no increased activated astrocytes and microglia were observed in this study (Table 3); therefore, pathological Tau does not seem to induce the activation of astrocytes and microglia in Tau-P301L mice brain.

In the present study, Tau-P301L mice did not show alterations on the rotarod at any analyzed ages, similar to the results published in previous studies [19,35]. In one study [35], it was described that Tau-P301L male mice on a C57BL/6J background showed phenotypic alterations at early ages (2 to 5 months of age) in some behavioral tests. Corroborating to the present study, Tau-P301L male mice did not develop any deficits at 4 months of age in the nesting and marble burying test. Moreover, we were able to demonstrate that Tau-P301L mice do not develop any deficits as late as 8 months of age in those tests. In the open field test, Tau-P301L mice travelled less as early as 2 months of age. This effect can be observed up to the age of 8 months. Again, these results agree with the one published by Samaey et al. [35]. In contrast to the study published by Samaey et al. [35], we were not able to observe any difference between Tau-P301L and WT mice in the amount of time they explored the different zones (border, center and corner).

Described for the first time, Tau-P301L male mice had phenotypic alterations in the SHIRPA test battery. Starting at 4 months of age, Tau-P301L mice developed postural changes described by a hunched back, mild deficits in the hanging behavior and some mice developed a loss of the postural reflex. Then, beginning at 6 months of age, Tau-P301L mice started to display clasping of the limbs, which can be considered a paralysis, as well as an abnormal gait described as a waddling walk. Those alterations progressed with age. This result contrasts with those shown by other groups, analyzing other mouse models of tauopathy, since they did not describe any differences in the SHIRPA compared to the WT [36,37]. Only one study described a similar result regarding the hanging behavior. In this study, it was shown that the motor skills of Tau58-2/B mice (Tau-P301L mutation) were so limited regarding this specific subtest that the mice could not perform the test adequately [38]. In the present study, we also conducted a modified pole test. We were able to show that the mice exhibit deficits in this test that become more pronounced with increasing age. The motor deficits in the modified pole test were also described for another mouse model of tauopathy, called SJLB mouse model [39]. Moreover, Tau-P301L mice had olfactory deficits in the habituation/dishabituation olfactory test at 6 months of age. This deficit was also observed in another Tau-P301L mouse model [40]. Regarding the histopathology, neither AT8- nor AT100-positive signals were detected in the olfactory bulb/cortex. Therefore, another pathophysiological mechanism might play a role in the olfactory deficits observed in this study. Another explanation for this alteration is that the olfactory deficits bear on cognitive deficits, so one might speculate that Tau-P301L male mice were not able to recognize the new aroma.

The pathology described for this Tau-P301L mouse model is similar to the same and other mouse models with this specific mutation [19–21,41,42]. Therefore, the strain background and the used promoter do not seem to have any influence on the appearance of NFTs in the brains. However, the presence of NFTs can occur in different brain regions since the pR5 mouse model also shows tauopathy in the hippocampus [41], which is not observed in the mouse model from this study. The Tau pathology in Tau-P301L mice mainly occurs in the brainstem and might be an explanation for the behavioral deficits. The presence of pathological Tau in the brainstem, especially in SNr [43–47] and superior colliculus [48,49], can be related to motor deficits in Tau-P301L male mice. Moreover, the lack of NFTs in Tau-P301L female mice in those regions can be also related to the lack of motor deficits. Sex dimorphism is observed in other transgenic mouse models, but the results are contradictory [50]. Therefore, more experiments are needed in order to understand these sex-related differences more precisely.

Regarding the cognitive deficits, both Tau-P301L female and male mice developed deficits in the NOR that are in line with the Tau aggregation in the LC, since it also plays a role in cognition, and NFTs in this region induce cognitive deficits [51,52]. Tau-P301L mice did not have any deficits in the MWM, T-Maze and contextual fear conditioning, probably due to the lack of NFTs in the hippocampus, which play a main role in processing of the spatial memory [53]. Tau-P301L mice also did not show cognitive deficits in the cued fear conditioning, probably also due to the lack of NFTs in the amygdala, which is the region that processes fear memory [54].

Sex differences in neurodegenerative diseases are observed in both animals and humans. In humans, women have a higher probability to develop AD than men as well as developing a more severe pathology [55,56]. In AD mice, Aβ levels are also higher and cognitive deficits are more prominent in females. Regarding Tau pathology, not much information is available about transgenic models. In the present study, Tau-P301L male mice had motor deficits compared to females, even though both sexes had cognitive deficits. Another study also demonstrated motor deficits in the Tau-P301S males and later cognitive deficits compared to females, but a similar tau pathology in the brain [57]. In a triple transgenic mouse model, which develops both Aβ plaques and NFTs, 3xTg-AD females had a higher amount of Tau pathology and cognitive deficits compared to males. This discrepancy might be due to the age of the tested mice, since AD mice develop the alterations later than Tau mice [58]. Additionally, female reproductive senescence is reached at 12 months of age, when estrogen levels are decreased [59]. Estrogen is known to have neuroprotective effects and its decrease might explain the severity of tau pathology in the 3xTg-AD females [60–63]. In this study, reproductive active females were evaluated and the estrogen levels might explain the milder Tau pathology in females. Still, it is important to highlight that the comparison of different models must be done with caution, since each model has different behavioral and physiopathological outcomes. Moreover, more studies are needed to further explain the remarkable sex differences in the Tau-P301L mouse model, as observed in this study.

In conclusion, this longitudinal study demonstrates that Tau-P301L mice have alterations due to the presence of pathological Tau in the brain that agree with age and are sex-dependent. Tau-P301L male mice had olfactory deficits, motor deficits and increased Tau pathology in the brain. None of those alterations were observed in Tau-P310L female mice. Both sexes, however, had phenotypic alterations in the SHIRPA test battery and cognitive deficits in the NOR. It is possible to determine that the disease onset in the males occurs as early as 2 months of age regarding the motor deficits and 6 months of age regarding the cognitive deficits.

Supplementary Materials: The following are available online at https://www.mdpi.com/article/10.3390/biomedicines9091160/s1, Figure S1: Tau-P301L mice had similar weight compared to WT mice; Figure S2: Tau-P301L mice had similar performance in the nesting and marble burying test compared to WT mice; Table S1: Tau-P301L mice (Tau) had increased scores in different parameters compared to WT starting with 4 months of age in the SHIRPA test battery; Figure S3: Tau-P301L mice had similar performance in the Rotarod test compared to WT mice; Figure S4: Tau-P301L mice spent similar amount of time in the border and center of the open field compared to WT mice; Figure S5: Tau-P301L mice had similar performance in the T-maze spontaneous alternation compared to WT mice; Figure S6: Tau-P301L mice froze similarly compared to WT mice in the cued and contextual fear conditioning; Table S2: Correlation between behavioral tests and AT8 as well as AT100 staining.

Author Contributions: Conceptualization, S.S., L.C.C. and A.W.; experiments, L.C.C., D.H. and R.B.; data analysis, L.C.C., D.H. and S.S.; writing—original draft preparation, L.C.C.; writing—review and editing, S.S., J.K., A.W., K.-J.L., N.J.S. and D.W.; funding acquisition, D.W. and K.-J.L. All authors have read and agreed to the published version of the manuscript.

Funding: D.W. was supported by grants from the Russian Science Foundation (RSF) (project no. 20-64-46027) and by the Technology Transfer Fund of the Forschungszentrum Jülich. K.-J.L. and

D.W. were supported by "Portfolio Drug Research" of the "Impuls und Vernetzungs-Fonds der Helmholtzgemeinschaft".

Institutional Review Board Statement: This study was conducted according to the German Law on the protection of animals (TierSchG §§ 7–9) and was approved by the local ethics committee before start of the experiments (Landesamt fur Natur, Umwelt und Verbraucherschutz, North Rhine-Westphalia, Germany, numbers 84-02.04.2011.A359, 84-02.04.2014.A362, 81-02.04.2018.A400 and 81-02.04.2019.A304, which were approved on 9 December 2014, 5 February 2019, 21 February 2019 and 21 January 2019, respectively).

Informed Consent Statement: Not applicable.

Data Availability Statement: All data from the study are available in this manuscript.

Conflicts of Interest: The authors declare no conflict of interest.

References

1. Jean, D.C.; Baas, P.W. It cuts two ways: Microtubule loss during Alzheimer disease. *Embo J.* **2013**, *32*, 2900–2902. [CrossRef]
2. Dixit, R.; Ross, J.L.; Goldman, Y.E.; Holzbaur, E.L.F. Differential regulation of dynein and kinesin motor proteins by tau. *Science* **2008**, *319*, 1086–1089. [CrossRef] [PubMed]
3. Combs, B.; Mueller, R.L.; Morfini, G.; Brady, S.T.; Kanaan, N.M. Tau and Axonal Transport. Misregulation in Tauopathies. *Adv. Exp. Med. Biol.* **2019**, *1184*, 81–95. [CrossRef]
4. Barbier, P.; Zejneli, O.; Martinho, M.; Lasorsa, A.; Belle, V.; Smet-Nocca, C.; Tsvetkov, P.O.; Devred, F.; Landrieu, I. Role of Tau as a Microtubule-Associated Protein: Structural and Functional Aspects. *Front. Aging Neurosci.* **2019**, *11*, 204. [CrossRef]
5. Tapia-Rojas, C.; Cabezas-Opazo, F.; Deaton, C.A.; Vergara, E.H.; Johnson, G.V.W.; Quintanilla, R.A. It's all about tau. *Prog. Neurobiol.* **2019**, *175*, 54–76. [CrossRef]
6. Fischer, I.; Baas, P.W. Resurrecting the Mysteries of Big Tau. *Trends Neurosci.* **2020**, *43*, 493–504. [CrossRef]
7. Mandelkow, E.M.; Biernat, J.; Drewes, G.; Gustke, N.; Trinczek, B.; Mandelkow, E. Tau domains, phosphorylation, and interactions with microtubules. *Neurobiol. Aging* **1995**, *16*, 355–362. [CrossRef]
8. Götz, J.; Halliday, G.; Nisbet, R.M. Molecular Pathogenesis of the Tauopathies. *Annu. Rev. Pathol. Mech. Dis.* **2019**, *14*, 239–261. [CrossRef]
9. Quintas-Neves, M.; Teylan, M.A.; Besser, L.; Soares-Fernandes, J.; Mock, C.N.; Kukull, W.A.; Crary, J.F.; Oliveira, T.G. Magnetic resonance imaging brain atrophy assessment in primary age-related tauopathy (PART). *Acta Neuropathol. Commun.* **2019**, *7*, 204. [CrossRef]
10. Braak, H.; Alafuzoff, I.; Arzberger, T.; Kretzschmar, H.; Del Tredici, K. Staging of Alzheimer disease-associated neurofibrillary pathology using paraffin sections and immunocytochemistry. *Acta Neuropathol.* **2006**, *112*, 389–404. [CrossRef]
11. Liao, D.; Miller, E.C.; Teravskis, P.J. Tau acts as a mediator for Alzheimer's disease-related synaptic deficits. *Eur. J. Neurosci.* **2014**, *39*, 1202–1213. [CrossRef]
12. Jadhav, S.; Cubinkova, V.; Zimova, I.; Brezovakova, V.; Madari, A.; Cigankova, V.; Zilka, N. Tau-mediated synaptic damage in Alzheimer's disease. *Transl. Neurosci.* **2015**, *6*, 214–226. [CrossRef]
13. Puvenna, V.; Engeler, M.; Banjara, M.; Brennan, C.; Schreiber, P.; Dadas, A.; Bahrami, A.; Solanki, J.; Bandyopadhyay, A.; Morris, J.K.; et al. Is phosphorylated tau unique to chronic traumatic encephalopathy? Phosphorylated tau in epileptic brain and chronic traumatic encephalopathy. *Brain Res.* **2016**, *1630*, 225–240. [CrossRef]
14. Poorkaj, P.; Grossman, M.; Steinbart, E.; Payami, H.; Sadovnick, A.; Nochlin, D.; Tabira, T.; Trojanowski, J.Q.; Borson, S.; Galasko, D.; et al. Frequency of tau gene mutations in familial and sporadic cases of non-Alzheimer dementia. *Arch. Neurol.* **2001**, *58*, 383–387. [CrossRef]
15. Forrest, S.L.; Kril, J.J.; Stevens, C.H.; Kwok, J.B.; Hallupp, M.; Kim, W.S.; Huang, Y.; McGinley, C.V.; Werka, H.; Kiernan, M.C.; et al. Retiring the term FTDP-17 as MAPT mutations are genetic forms of sporadic frontotemporal tauopathies. *Brain* **2017**, *141*, 521–534. [CrossRef]
16. Hasegawa, M.; Smith, M.J.; Goedert, M. Tau proteins with FTDP-17 mutations have a reduced ability to promote microtubule assembly. *FEBS Lett.* **1998**, *437*, 207–210. [CrossRef]
17. Roberson, E.D. Mouse models of frontotemporal dementia. *Ann. Neurol.* **2012**, *72*, 837–849. [CrossRef] [PubMed]
18. Kitazawa, M.; Medeiros, R.; Laferla, F.M. Transgenic mouse models of Alzheimer disease: Developing a better model as a tool for therapeutic interventions. *Curr. Pharm. Des.* **2012**, *18*, 1131–1147. [CrossRef]
19. Terwel, D.; Lasrado, R.; Snauwaert, J.; Vandeweert, E.; Van Haesendonck, C.; Borghgraef, P.; Van Leuven, F. Changed Conformation of Mutant Tau-P301L Underlies the Moribund Tauopathy, Absent in Progressive, Nonlethal Axonopathy of Tau-4R/2N Transgenic Mice*. *J. Biol. Chem.* **2005**, *280*, 3963–3973. [CrossRef]
20. Dutschmann, M.; Menuet, C.; Stettner, G.M.; Gestreau, C.; Borghgraef, P.; Devijver, H.; Gielis, L.; Hilaire, G.; Van Leuven, F. Upper airway dysfunction of Tau-P301L mice correlates with tauopathy in midbrain and ponto-medullary brainstem nuclei. *J. Neurosci.* **2010**, *30*, 1810–1821. [CrossRef] [PubMed]

21. Boekhoorn, K.; Terwel, D.; Biemans, B.; Borghgraef, P.; Wiegert, O.; Ramakers, G.J.; de Vos, K.; Krugers, H.; Tomiyama, T.; Mori, H.; et al. Improved long-term potentiation and memory in young tau-P301L transgenic mice before onset of hyperphosphorylation and tauopathy. *J. Neurosci.* **2006**, *26*, 3514–3523. [CrossRef] [PubMed]

22. Lehmkuhl, A.M.; Dirr, E.R.; Fleming, S.M. Olfactory assays for mouse models of neurodegenerative disease. *J. Vis. Exp. JoVE* **2014**, e51804. [CrossRef] [PubMed]

23. Deacon, R. Assessing burrowing, nest construction, and hoarding in mice. *J. Vis. Exp. JoVE* **2012**, e2607. [CrossRef]

24. Deacon, R.M. Digging and marble burying in mice: Simple methods for in vivo identification of biological impacts. *Nat. Protoc.* **2006**, *1*, 122–124. [CrossRef]

25. Rogers, D.C.; Fisher, E.M.; Brown, S.D.; Peters, J.; Hunter, A.J.; Martin, J.E. Behavioral and functional analysis of mouse phenotype: SHIRPA, a proposed protocol for comprehensive phenotype assessment. *Mamm. Genome Off. J. Int. Mamm. Genome Soc.* **1997**, *8*, 711–713. [CrossRef]

26. Alexandru, A.; Jagla, W.; Graubner, S.; Becker, A.; Bäuscher, C.; Kohlmann, S.; Sedlmeier, R.; Raber, K.A.; Cynis, H.; Rönicke, R.; et al. Selective hippocampal neurodegeneration in transgenic mice expressing small amounts of truncated Aβ is induced by pyroglutamate-Aβ formation. *J. Neurosci.* **2011**, *31*, 12790–12801. [CrossRef]

27. Dunkelmann, T.; Schemmert, S.; Honold, D.; Teichmann, K.; Butzküven, E.; Demuth, H.-U.; Shah, N.J.; Langen, K.-J.; Kutzsche, J.; Willbold, D.; et al. Comprehensive Characterization of the Pyroglutamate Amyloid-β Induced Motor Neurodegenerative Phenotype of TBA2.1 Mice. *J. Alzheimer's Dis.* **2018**, *63*, 115–130. [CrossRef]

28. Spowart-Manning, L.; Van der Staay, F. The T-maze continuous alternation task for assessing the effects of putative cognition enhancers in the mouse. *Behav. Brain Res.* **2004**, *151*, 37–46. [CrossRef]

29. Curzon, P.; Rustay, N.R.; Browman, K.E. Frontiers in Neuroscience. Cued and Contextual Fear Conditioning for Rodents. In *Methods of Behavior Analysis in Neuroscience*; Buccafusco, J.J., Ed.; CRC Press/Taylor & Francis. Taylor & Francis Group, LLC.: Boca Raton, FL, USA, 2009.

30. Morris, R. Developments of a water-maze procedure for studying spatial learning in the rat. *J. Neurosci. Methods* **1984**, *11*, 47–60. [CrossRef]

31. Augustinack, J.C.; Schneider, A.; Mandelkow, E.-M.; Hyman, B.T. Specific tau phosphorylation sites correlate with severity of neuronal cytopathology in Alzheimer's disease. *Acta Neuropathol.* **2002**, *103*, 26–35. [CrossRef]

32. Goedert, M.; Jakes, R.; Vanmechelen, E. Monoclonal antibody AT8 recognises tau protein phosphorylated at both serine 202 and threonine 205. *Neurosci. Lett.* **1995**, *189*, 167–170. [CrossRef]

33. Zheng-Fischhöfer, Q.; Biernat, J.; Mandelkow, E.-M.; Illenberger, S.; Godemann, R.; Mandelkow, E. Sequential phosphorylation of Tau by glycogen synthase kinase-3β and protein kinase A at Thr212 and Ser214 generates the Alzheimer-specific epitope of antibody AT100 and requires a paired-helical-filament-like conformation. *Eur. J. Biochem.* **1998**, *252*, 542–552. [CrossRef]

34. Scearce-Levie, K.; Sanchez, P.E.; Lewcock, J.W. Leveraging preclinical models for the development of Alzheimer disease therapeutics. *Nat. Rev. Drug Discov.* **2020**, *19*, 447–462. [CrossRef]

35. Samaey, C.; Schreurs, A.; Stroobants, S.; Balschun, D. Early Cognitive and Behavioral Deficits in Mouse Models for Tauopathy and Alzheimer's Disease. *Front. Aging Neurosci.* **2019**, *11*, 335. [CrossRef]

36. Polydoro, M.; Acker, C.M.; Duff, K.; Castillo, P.E.; Davies, P. Age-Dependent Impairment of Cognitive and Synaptic Function in the htau Mouse Model of Tau Pathology. *J. Neurosci.* **2009**, *29*, 10741–10749. [CrossRef]

37. Bodea, L.-G.; Evans, H.T.; Van der Jeugd, A.; Ittner, L.M.; Delerue, F.; Kril, J.; Halliday, G.; Hodges, J.; Kiernan, M.C.; Götz, J. Accelerated aging exacerbates a pre-existing pathology in a tau transgenic mouse model. *Aging Cell* **2017**, *16*, 377–386. [CrossRef]

38. Van der Jeugd, A.; Vermaercke, B.; Halliday, G.M.; Staufenbiel, M.; Götz, J. Impulsivity, decreased social exploration, and executive dysfunction in a mouse model of frontotemporal dementia. *Neurobiol. Learn. Mem.* **2016**, *130*, 34–43. [CrossRef]

39. Takenokuchi, M.; Kadoyama, K.; Chiba, S.; Sumida, M.; Matsuyama, S.; Saigo, K.; Taniguchi, T. SJLB mice develop tauopathy-induced parkinsonism. *Neurosci. Lett.* **2010**, *473*, 182–185. [CrossRef]

40. Hu, Y.; Ding, W.; Zhu, X.; Chen, R.; Wang, X. Olfactory Dysfunctions and Decreased Nitric Oxide Production in the Brain of Human P301L Tau Transgenic Mice. *Neurochem. Res.* **2016**, *41*, 722–730. [CrossRef] [PubMed]

41. Lewis, J.; McGowan, E.; Rockwood, J.; Melrose, H.; Nacharaju, P.; Van Slegtenhorst, M.; Gwinn-Hardy, K.; Murphy, M.P.; Baker, M.; Yu, X.; et al. Neurofibrillary tangles, amyotrophy and progressive motor disturbance in mice expressing mutant (P301L) tau protein. *Nat. Genet.* **2000**, *25*, 402–405. [CrossRef]

42. Götz, J.; Chen, F.; Barmettler, R.; Nitsch, R.M. Tau Filament Formation in Transgenic Mice Expressing P301L Tau*. *J. Biol. Chem.* **2001**, *276*, 529–534. [CrossRef]

43. You, Y.; Botros, M.B.; Enoo, A.A.V.; Bockmiller, A.; Herron, S.; Delpech, J.C.; Ikezu, T. Cre-inducible Adeno Associated Virus-mediated Expression of P301L Mutant Tau Causes Motor Deficits and Neuronal Degeneration in the Substantia Nigra. *Neuroscience* **2019**, *422*, 65–74. [CrossRef]

44. Mirra, S.S.; Murrell, J.R.; Gearing, M.; Spillantini, M.G.; Goedert, M.; Crowther, R.A.; Levey, A.I.; Jones, R.; Green, J.; Shoffner, J.M.; et al. Tau Pathology in a Family with Dementia and a P301L Mutation in Tau. *J. Neuropathol. Exp. Neurol.* **1999**, *58*, 335–345. [CrossRef] [PubMed]

45. Klein, R.L.; Dayton, R.D.; Lin, W.L.; Dickson, D.W. Tau gene transfer, but not alpha-synuclein, induces both progressive dopamine neuron degeneration and rotational behavior in the rat. *Neurobiol. Dis.* **2005**, *20*, 64–73. [CrossRef] [PubMed]

46. Kovacs, G.G. Invited review: Neuropathology of tauopathies: Principles and practice. *Neuropathol. Appl. Neurobiol.* **2015**, *41*, 3–23. [CrossRef]

47. Dickson, D.W.; Ahmed, Z.; Algom, A.A.; Tsuboi, Y.; Josephs, K.A. Neuropathology of variants of progressive supranuclear palsy. *Curr. Opin. Neurol.* **2010**, *23*, 394–400. [CrossRef]

48. Armstrong, R.A.; McKee, A.C.; Cairns, N.J. Pathology of the Superior Colliculus in Chronic Traumatic Encephalopathy. *Optom. Vis. Sci. Off. Publ. Am. Acad. Optom.* **2017**, *94*, 33–42. [CrossRef] [PubMed]

49. Dugger, B.N.; Tu, M.; Murray, M.E.; Dickson, D.W. Disease specificity and pathologic progression of tau pathology in brainstem nuclei of Alzheimer's disease and progressive supranuclear palsy. *Neurosci. Lett.* **2011**, *491*, 122–126. [CrossRef]

50. Dennison, J.L.; Ricciardi, N.R.; Lohse, I.; Volmar, C.-H.; Wahlestedt, C. Sexual Dimorphism in the 3xTg-AD Mouse Model and Its Impact on Pre-Clinical Research. *J. Alzheimer's Dis.* **2021**, *80*, 41–52. [CrossRef]

51. Grudzien, A.; Shaw, P.; Weintraub, S.; Bigio, E.; Mash, D.C.; Mesulam, M.M. Locus coeruleus neurofibrillary degeneration in aging, mild cognitive impairment and early Alzheimer's disease. *Neurobiol Aging* **2007**, *28*, 327–335. [CrossRef]

52. Kelly, S.C.; He, B.; Perez, S.E.; Ginsberg, S.D.; Mufson, E.J.; Counts, S.E. Locus coeruleus cellular and molecular pathology during the progression of Alzheimer's disease. *Acta Neuropathol. Commun.* **2017**, *5*, 8. [CrossRef]

53. Tanila, H. Wading pools, fading memories—Place navigation in transgenic mouse models of Alzheimer's disease. *Front. Aging Neurosci.* **2012**, *4*, 11. [CrossRef]

54. Izquierdo, I.; Furini, C.R.; Myskiw, J.C. Fear Memory. *Physiol. Rev.* **2016**, *96*, 695–750. [CrossRef]

55. Dye, R.V.; Miller, K.J.; Singer, E.J.; Levine, A.J. Hormone replacement therapy and risk for neurodegenerative diseases. *Int. J. Alzheimers Dis.* **2012**, *2012*, 258454. [CrossRef]

56. Phung, T.K.; Waltoft, B.L.; Laursen, T.M.; Settnes, A.; Kessing, L.V.; Mortensen, P.B.; Waldemar, G. Hysterectomy, oophorectomy and risk of dementia: A nationwide historical cohort study. *Dement. Geriatr. Cogn. Disord.* **2010**, *30*, 43–50. [CrossRef]

57. Sun, Y.; Guo, Y.; Feng, X.; Jia, M.; Ai, N.; Dong, Y.; Zheng, Y.; Fu, L.; Yu, B.; Zhang, H.; et al. The behavioural and neuropathologic sexual dimorphism and absence of MIP-3α in tau P301S mouse model of Alzheimer's disease. *J. Neuroinflamm.* **2020**, *17*, 72. [CrossRef]

58. Yang, J.-T.; Wang, Z.-J.; Cai, H.-Y.; Yuan, L.; Hu, M.-M.; Wu, M.-N.; Qi, J.-S. Sex. Differences in Neuropathology and Cognitive Behavior in APP/PS1/tau Triple-Transgenic Mouse Model of Alzheimer's Disease. *Neurosci. Bull.* **2018**, *34*, 736–746. [CrossRef]

59. Clinton, L.K.; Billings, L.M.; Green, K.N.; Caccamo, A.; Ngo, J.; Oddo, S.; McGaugh, J.L.; LaFerla, F.M. Age-dependent sexual dimorphism in cognition and stress response in the 3xTg-AD mice. *Neurobiol. Dis.* **2007**, *28*, 76–82. [CrossRef]

60. Goodman, Y.; Bruce, A.J.; Cheng, B.; Mattson, M.P. Estrogens attenuate and corticosterone exacerbates excitotoxicity, oxidative injury, and amyloid beta-peptide toxicity in hippocampal neurons. *J. Neurochem.* **1996**, *66*, 1836–1844. [CrossRef]

61. Luine, V.N. Estradiol increases choline acetyltransferase activity in specific basal forebrain nuclei and projection areas of female rats. *Exp. Neurol.* **1985**, *89*, 484–490. [CrossRef]

62. Behl, C.; Widmann, M.; Trapp, T.; Holsboer, F. 17-beta estradiol protects neurons from oxidative stress-induced cell death in vitro. *Biochem. Biophys. Res. Commun.* **1995**, *216*, 473–482. [CrossRef]

63. Bruce-Keller, A.J.; Keeling, J.L.; Keller, J.N.; Huang, F.F.; Camondola, S.; Mattson, M.P. Antiinflammatory effects of estrogen on microglial activation. *Endocrinology* **2000**, *141*, 3646–3656. [CrossRef] [PubMed]

Article

Dexamethasone-Induced Adipose Tissue Redistribution and Metabolic Changes: Is Gene Expression the Main Factor? An Animal Model of Chronic Hypercortisolism

Flaviane de Fatima Silva [1,*,†], Ayumi Cristina Medeiros Komino [1,*,†], Sandra Andreotti [1], Gabriela Boltes Reis [1], Rennan Oliveira Caminhotto [1], Richardt Gama Landgraf [2], Gabriel Orefice de Souza [1], Rogerio Antonio Laurato Sertié [3], Sheila Collins [4,5], Jose Donato, Jr. [1] and Fabio Bessa Lima [1]

[1] Department of Physiology and Biophysics, Institute of Biomedical Sciences, University of Sao Paulo, Sao Paulo 05508-000, Brazil
[2] Department of Pharmaceutical Sciences, Federal University of Sao Paulo, Diadema 09913-030, Brazil
[3] Department of Nutrition Sciences, University of Alabama at Birmingham, Birmingham, AL 35294, USA
[4] Division of Cardiovascular Medicine, Department of Medicine, Vanderbilt University Medical Center, Nashville, TN 37232, USA
[5] Department of Molecular Physiology and Biophysics, Vanderbilt University Medical Center, Nashville, TN 37232, USA
* Correspondence: silva.flavianef@gmail.com (F.d.F.S.); ayumikomino@gmail.com (A.C.M.K.)
† These authors contributed equally to this work.

Citation: de Fatima Silva, F.; Komino, A.C.M.; Andreotti, S.; Boltes Reis, G.; Caminhotto, R.O.; Landgraf, R.G.; de Souza, G.O.; Sertié, R.A.L.; Collins, S.; Donato, J., Jr.; et al. Dexamethasone-Induced Adipose Tissue Redistribution and Metabolic Changes: Is Gene Expression the Main Factor? An Animal Model of Chronic Hypercortisolism. *Biomedicines* **2022**, *10*, 2328. https://doi.org/10.3390/biomedicines10092328

Academic Editor: Martina Perše

Received: 1 July 2022
Accepted: 6 September 2022
Published: 19 September 2022

Abstract: Chronic hypercortisolism has been associated with the development of several metabolic alterations, mostly caused by the effects of chronic glucocorticoid (GC) exposure over gene expression. The metabolic changes can be partially explained by the GC actions on different adipose tissues (ATs), leading to central obesity. In this regard, we aimed to characterize an experimental model of iatrogenic hypercortisolism in rats with significant AT redistribution. Male Wistar rats were distributed into control (CT) and GC-treated, which received dexamethasone sodium phosphate (0.5 mg/kg/day) by an osmotic minipump, for 4 weeks. GC-treated rats reproduced several characteristics observed in human hypercortisolism/Cushing's syndrome, such as HPA axis inhibition, glucose intolerance, insulin resistance, dyslipidemia, hepatic lipid accumulation, and AT redistribution. There was an increase in the mesenteric (meWAT), perirenal (prWAT), and interscapular brown (BAT) ATs mass, but a reduction of the retroperitoneal (rpWAT) mass compared to CT rats. Overexpressed lipolytic and lipogenic gene profiles were observed in white adipose tissue (WAT) of GC rats as BAT dysfunction and whitening. The AT remodeling in response to GC excess showed more importance than the increase of AT mass per se, and it cannot be explained just by GC regulation of gene transcription.

Keywords: Cushing's syndrome; adipose plasticity; visceral obesity; glucocorticoids; dexamethasone; gene expression; lipolysis; lipogenesis; BAT dysfunction; BAT whitening

1. Introduction

Glucocorticoids (GCs) are steroid hormones synthesized and released by cells of the fasciculate zone of the adrenal gland, stimulated by the hypothalamic-hypophyseal-adrenal (HPA) axis in response to several physiologic, environmental, psychological, and stressing stimuli [1]. Cortisol is the main GC in humans and is secreted in a rhythmical circadian pattern of oscillation with more intense peaks of release in the early morning hours [2]. In rodents, the main GC is corticosterone, but with peaks of release at night, due to the nocturnal behavior [3]. This rhythmical synthesis, release, and action are of utmost importance and affect many physiological processes, including regulatory actions on metabolism, growth, development, and inflammatory response [2,4]. The loss of the circadian rhythm and the inhibition of the HPA axis in response to GC excess are the main factors that define Cushing's syndrome, iatrogenic being the most common [5,6].

Synthetic GC such as dexamethasone is used worldwide as a treatment for several diseases due to its anti-inflammatory and immunosuppressive effects [7,8], with more relevance in the last two years due to massive use as part of COVID-19 treatment, mainly in severe cases, which the patients need hospitalization for long periods [9–11]. Compared to other synthetic GCs, dexamethasone has the advantage of low affinity for mineralocorticoid receptors (MR) [12], which could decrease the chances of side effects, since most are a result of the GC-MR binding [13,14]. However, dexamethasone has a higher affinity to the GC receptor (GR) and does not need 11-βHSD1 for its activation, making this GC more potent and increasing the risk of side effects by the GC-GR interaction [12,15].

The classic effects of GC are genomic through GR. The dimeric complex GC-GR is translocated to the nucleus, where the complex can bind the GC responsive elements (GREs) in the regulatory regions of target genes, or act directly by inhibiting or activating other transcription factors, in a GREs-independent pathway [2,16,17]. These bindings can induce or repress gene expression and, consequently, the synthesis of proteins/enzymes that regulate several physiological processes such as glucose homeostasis and lipid metabolism [2,8]. However, these genomic effects require hours or days to happen [18], and in the case of chronic GC exposure, mainly in excess, can result in side effects in many tissues and organs, as observed in Cushing syndrome. In that regard, if there were a way to dissociate these gene expression side effects in other tissues from the anti-inflammatory properties, the risk of complications due to long-term GC exposure would decrease [19].

One of the major characteristics of Cushing's syndrome is the specific fat accumulation in the central region of the body, such as the abdomen, chest, head, and neck [20,21]. This centripetal fat distribution seems to be due to the hyperplasia and hypertrophy of visceral adipocytes and the differentiation of preadipocytes [22–24]. Previous results of our research group did not observe central fat distribution [25]. Other studies using GC excess focused on one or two specific AT territories, or with more emphasis on muscle or liver effects [26–29]. Considering that central obesity contributes to the development of several metabolic complications [30], and each AT deposit is composed of a distinct subpopulation of adipocytes [31,32] and has a different vascularization and inter-organ drainage [33,34], it becomes important to investigate how GC excess affects the AT in different anatomic locations.

In this work, we present a model of chronic iatrogenic hypercortisolism in young adult rats [35], which can be useful to assess and understand how the GCs excess impacts AT redistribution and plasticity.

2. Materials and Methods

2.1. Ethics Approval and Animals

All protocols described were approved by the Committee of Ethics in the Use of Animals of the Institute of Biomedical Sciences, University of Sao Paulo (CEUA-ICB/USP #89/2016; #26/2017; 9535190219). For the experiments, SPF-certified 8 weeks old male Wistar Hannover rats obtained from the Institute of Biomedical Sciences Animal Facility were used, since sex could be an important variable in adipose tissue analyzes [36]. Each rat was housed in an individual open polycarbonate cage (model 1291H, $425 \times 266 \times 185$ mm and 800 cm^2 of floor area, Tecniplast$^®$, Buguggiate, VA, Itália Italy) in our laboratory private room at Animal Care of the Department of Physiology and Biophysics for 4 weeks of acclimatization to the new space, light cycle, and human–rat bond before the treatment protocol starts. The room was equipped with a controlled air system (20 air renewal/h and 45–55% humidity), temperature (22 ± 2 °C), and light (12 h light/12 h dark, with lights on at 22h00–reverse cycle, with appropriated infrared light). The rats received standard rodents chow Nuvilab$^®$ CR-1 (Nuvital, Colombo, PR, Brazil) and filtered water ad libitum, and these parameters were monitored and replaced twice a week, as well as the cage's bedding (GOOD LIFE PINUS RG, Granja R.G, Suzano, SP, Brazil) and the enrichment-autoclaved handcrafted cardboard rolls. The cages were always paired in the rack, ensuring the rats could have visual/olfactory contact with the rat(s) in the cage(s) beside.

Only the same two researchers handled the rats during the acclimatization and treatment period for daily monitoring, cleaning, anesthesia protocols, pre and postoperative care, surgery, oGTT, and euthanasia. No external people were allowed in the room to avoid possible external stress, except the Animal Care veterinarian when necessary. The treatment protocol started when rats were young adults (12 weeks old). For treatment, 36 rats were randomly divided into two groups: control (CT, n = 16) and treated with glucocorticoid (GC; n = 20) using the Random Sequence feature (https://www.random.org/sequences/, accessed on 10 July 2017).

2.2. Chronic GC Treatment

To ensure continuous and invariable GC administration, an osmotic pump (model 2004, ALZET®, Cupertino, CA, USA) was surgically implanted in the dorsal subcutaneous region. The surgeries were realized in an appropriate experimental room at Animal Care. Each rat was weighed and anesthetized with ketamine hydrochloride (100 mg/kg; i.p., Dopalen, Ceva, Sao Paulo, Brazil) and xylazine hydrochloride (5 mg/kg; i.p., Anasedan, Ceva, Sao Paulo, Brazil) and monitored until complete sedation and anesthesia. At this point, we performed the first nasal–anal measure (Day 0). For eyes hydration during the surgery time (~10 min), we used sterile NaCl 0.9% solution. The rats were submitted to trichotomy of the under scapulae region, local asepsis with 10% polyvidone iodine, and placed in the sterile surgical field. The 1.0 cm horizontal incision was performed with a scalpel, a subcutaneous space was opened with a Blunt scissor, and the minipump containing the GC dexamethasone sodium phosphate (sc-204715, Santa Cruz Biotechnology, Dallas, TX, USA) was allocated according to the manufacturer's instructions. The incision was closed with a continuous suture (4-0, Seda-Silk, Ethicon* by Johnson & Johnson, Sao Jose dos Campos, SP, Brazil), the local was cleaned with 10% polyvidone iodine, and a topic antibiotic ointment was applied (250 U Bacitracin + 3.5 mg Neomycin, Nebacetin®, Takeda Pharma, Jaguariuna, SP, Brazil). The rats remained on a tissue-covered 37 °C heated surface (EFF421, INSIGHT®, Ribeirao Preto, SP, Brazil), being monitored for vital parameters and ocular hydration until complete recovery from anesthesia. The wound recovery and the rats' behavior were monitored daily. The antibiotic ointment was applied once a day in the first week, and the suture fell off about 10 days post-surgery. No oral analgesics were administrated as postoperative. CT rats were submitted to the sham surgery. The GC was diluted in sterile 0.9% NaCl solution (6.25 mg in 200 µL). Considering 375 g as the mean initial body weight and the daily amount released by the pump (6 µL), the corresponding daily dose was 0.5 mg/kg per day of GC for 4 weeks.

The body weight was evaluated on day 0 (mini-pump implantation) and day 28 (euthanasia), and also weekly. The food intake was measured weekly through the difference between the chow offered and the leftover (in grams), expressed as daily food intake per week and as the average of the 28-days daily food intake, both relatives to 100 g of body weight (g/100 g b.w.)

2.3. Oral Glucose Tolerance Test (oGTT)

In the last week of treatment and after 8 h of food deprivation, the rats of both groups were submitted to the oral glucose tolerance test (in the nocturnal phase of the light cycle at ZT [zeitgebber time] 22). After topic anesthesia (5% xylocaine gel), blood samples were collected from a 1.0 mm tail tip cut at time 0 (basal glucose and insulin assessment), followed by oral (gavage) administration of glucose 75 mg/100 g of body weight. New blood collections for glucose and insulin measurements were performed at times 5, 10, 15, 30, 45, 60, and 90 min after glucose administration. The blood glucose was assessed in a glucometer (One Touch Ultra®, Johnson & Johnson, Sao Jose dos Campos, SP, Brazil). For insulin analysis, the blood collected in heparinized capillary tubes was diluted (1:1; NaCl 0.9% plus heparin 10 IU/mL), centrifuged, and the plasma fraction was utilized for insulin ELISA measurement (#A05105, Bertin Bioreagent, Montigny-le-Bretonneux, France).

For the statistical analysis, the areas under curves (AUCs) for blood glucose and insulin were calculated.

2.4. Euthanasia

The endpoint was the 28th day of treatment for all animals. Due to reasons beyond our control, the euthanasia could not be performed on the 28th day in two occasions, and six rats (3 from the CT group and 3 from the GC group) were not considered in this study. The final number of rats/samples was 13 for the CT group and 17 for the GC group.

The rats from both groups were submitted to 12 h of food deprivation and euthanized during the first hour after lights off (1000–1100 h, 12 < ZTs < 13) to encompass the peak of corticosterone release by CT rats [37] and determine the inhibition of the HPA axis in GC group. The rats were anesthetized with thiopental (40 mg/kg; i.p.), weighted, and had their nasal–anal length measured (Day 28). The euthanasia was performed by decapitation for truncal blood collection, and serum was stored at $-80\,^{\circ}$C for hormonal and biochemical analysis. Afterward, the following tissues were dissected, weighed, processed as described below, or stored at $-80\,^{\circ}$C for later analysis: adrenal glands (right and left); mesenteric (meWAT), perirenal (prWAT), retroperitoneal (rpWAT), epididymal (epWAT), subcutaneous inguinal (scWAT), and interscapular brown (BAT) ATs; gastrocnemius (G), soleus (S), and extensor digitorum longus (E) muscles, tibia bone, hypothalamus, and the liver. The experiment was not blinded, the researches involved in the sample collection knew the identification of each rat. Each step was done by the same person to avoid possible differences in the dissection delimitation or weighting criteria.

The data about food intake, adrenal glands, ATs, muscles, and liver mass were expressed in grams (g) or milligrams (mg) of tissue, or relative to 100 g of body weight (g or mg/100 g b.w.). The change in body weight was calculated by the difference between the initial and final body weight in grams. The WAT mass was calculated as the sum of the five WAT depots analyzed in this study and presented as WAT or WAT + BAT, relative to 100 g of body weight (g/100 g b.w.). The feed efficiency was calculated by the ratio between weight change (g) and food intake (g). Additionally, 6 and 9 samples from CT and GC groups, respectively, were randomly selected previously to euthanasia to assess the G, S, and E dry muscle weight, and the remaining muscle samples were frozen for further analysis not presented in this study. For dry weight measurement, muscles were placed at $37\,^{\circ}$C for 3 days, and then weighted.

2.5. Hormonal and Biochemical Analysis

The blood serum collected at euthanasia was used for the measurements of corticosterone (#501320, Cayman Chemical, Ann Arbor, MI, USA, glucose (#133, Labtest, Lagoa Santa, MG, Brazil), lactate (#138, Labtest, Lagoa Santa, MG, Brazil), insulin (#A05105, Bertin Bioreagent, Montigny-le-Bretonneux, France), leptin and adiponectin (#RADPCMAG-82K-07, Millipore, Darmstadt, Germany), non-esterified fatty acids-NEFA (NEFA-HR (2), Wako, Neuss, Germany), triglycerides (#87, Labtest, Lagoa Santa, MG, Brazil), total cholesterol (#76, Labtest, Lagoa Santa, MG, Brazil), HDL-cholesterol (#13, Labtest, Lagoa Santa, MG, Brazil), alanine aminotransferase-ALT (#1008, Labtest, Lagoa Santa, MG, Brazil), and aspartate aminotransferase-AST (#109, Labtest, Lagoa Santa, MG, Brazil). The serum concentration of LDL cholesterol was estimated by the Friedewald equation [38], as well as VLDL cholesterol. The HOMA-IR index was calculated as previously described [39] but using the average of fasting blood glucose and insulin product from CT rats as a correction factor, instead of the 22.5 used for humans.

For liver triglycerides assessment, lipids were extracted from liver samples with chloroform-methanol by the Folch method [40], and the triglycerides in the lipid extract were determined by enzymatic assay (#87, Labtest, Lagoa Santa, MG, Brazil) and expressed by milligrams of triglycerides by 100 milligrams of the liver.

2.6. Histological Analysis

One of the adrenal glands (sagittal section), a section of the liver left medial lobe, and sections of the ATs were fixed in a solution of 10% formaldehyde in PBS (phosphate buffered saline; pH 7.4) for 24 h and submitted to histological processing as previously described [41]. The sections were stained with hematoxylin and eosin (HE) and captured at $100\times$, $200\times$ or $400\times$ magnification (DS-Ri2 microscope, Nikon, Tokyo, Japan), as indicated in the Figures captions.

The adipocyte volume for each WAT deposit in isolated adipose cells was also evaluated. For this purpose, adipocytes of meWAT, prWAT, rpWAT, epWAT, and scWAT were isolated [42], fixed in a solution of 4% formaldehyde in PBS. For the adipocytes image capture, the cells suspension was added to a glass slide, and the pictures were clicked at $100\times$ magnification. The adipocyte volume in picoliters(pL) was calculated as previously described [43] and was presented as frequency distribution for each group and WAT territory. All morphological analyses were performed using Motic Image Plus 3.0 software (Motic® Instruments, Schertz, TX, USA).

BAT slides were also immunostained with UCP1 antibody, performed in Leica Bond Max IHC (Leica Biosystems, Nussloch, Germany). The slides were submitted to deparaffinization and antigen recovery by heat (#AR9640, Leica Biosystems, Nussloch, Germany) for 20 min. Slides were then placed in a protein block (#x0909, DAKO, Carpinteria, CA, USA) for 10 min, followed by anti-UCP1 (#ab10983, 1:1000, Abcam, Waltham, MA, USA) incubation for 60 min. The detection was performed using Bond Refine Polymer (#DS9800, Leica Biosystems, Nussloch, Germany). The slides were dehydrated, cleaned, and covered. Images were captured using the Ks 300 Imaging System 3.0 (Carl Zeiss Vision GmbH, Aalen, Germany) at $100\times$ magnification.

2.7. RNA Isolation and Gene Expression Analysis

For the gene expression, 6 samples (hypothalamus) and 7–12 samples (adipose tissue) for each group were randomly selected. The remaining samples were frozen for further analysis not presented in this study. The final number of samples in gene expression for CT and GT groups were, respectively: 12-12 (meWAT), 12-11 (prWAT), 8-11 (rpWAT), 11-11 (epWAT), 7-11 (scWAT), and 8-8 (BAT).

Hypothalamus and ATs RNA were extracted following the TRIzol Reagent® (#15596026, Invitrogen, Waltham, MA, USA) and purified by PureLinkTM RNA Mini kit (#121830-18A, Ambion by Life Technologies, Carlsbad, CA, USA). Assessment of RNA quantity and quality was performed with Epoch Microplate Spectrophotometer (Biotek, Winooski, VT, USA). All samples were treated with RNase-free DNase I (#18068, Invitrogen, Waltham, MA, USA) before cDNA synthesis. For reverse transcription, 2 µg of hypothalamus RNA was used with SuperScript® II Reverse Transcriptase (#18064, Invitrogen, Waltham, MA, USA) and random primers p(dN)6 (Sigma-Aldrich, San Louis, MO, USA), while in ATs, RNA was used with SuperScript® III Reverse Transcriptase (#18080, Invitrogen, Waltham, MA, USA) and random primer (#48190, Invitrogen, Waltham, MA, USA).

The qPCR of hypothalamus samples was performed using the 7500 Fast Real-Time PCR System (Applied Biosystems, Waltham, MA, USA) with SYBR Green PCR Master Mix (Applied Biosystems, Waltham, MA, USA). The primers used are described in Table 1.

The qPCR reactions of ATs samples were performed using TaqMan® Gene Expression (Applied Biosystems, Waltham, MA, USA) for each interest gene: *Abhd5* (Rn01446981_m1); *Acaca* (Rn00573474_m1); *Acly* (Rn00566411_m1); *Actb* (Rn00667869_m1); *Adrb1* (Rn00824536_s1); *Adrb2* (Rn00560650_s1); *Adrb3* (Rn00565393_m1); *Agpat1* (Rn01525981_g1); *Agpat2* (Rn01438505_m1); *Akt1* (Rn00583646_m1); *Aqp7* (Rn00569727_m1); *B2m* (Rn00560865_m1); *Cd36* (Rn01442639_m1); *Cidea* (Rn04181355_m1); *Dgat1* (Rn00584870_m1); *Dgat2* (Rn01506787_m1); *Dio2* (Rn00581867_m1); *Fabp4* (Rn00670361_m1); *Fasn* (Rn00569117_m1); *G0s2* (Rn01412529_g1); *G6pd* (Rn01529640_g1); *Gk* (Rn00577740_m1); *Gpam* (Rn00568620_m1); *Gpd1* (Rn00573596_m1); *Hsd11b1* (Rn00567167_m1); *Insr* (Rn00690703_m1); *Irs1* (Rn02132493_s1); *Irs2* (Rn01482270_s1); *Ldhb* (Rn00754925_m1); *Lipe* (Rn00689222_m1); *Lpl* (Rn00561482_m1); *Me1* (Rn00561502_m1); *Mgll* (Rn00593297_m1);

P2rx5 (Rn00589966_m1); *Pck1* (Rn01529014_m1); *Pde3b* (Rn00568191_m1); *Pik3cg* (Rn01769524_m1); *Pik3r1** (Rn01644964_m1); *Plin1* (Rn00558672_m1); *Pnpla2* (Rn01479969_m1); *Pparg* (Rn00440945_m1); *Ppargc1a* (Rn00580241_m1); *Prdm16* (Rn01516224_m1); *Prkaca* (Rn01432300_g1); *Prkacb* (Rn01748540_g1); *Rpl37a* (Rn02114291_s1); *Scd* (Rn06152614_s1); *Scd2* (Rn00821391_g1); *Slc16a1* (Rn00562332_m1); *Slc16a7* (Rn00560072_m1), *Slc2a1* (Rn01417099_m1); *Slc2a4* (Rn00562597_m1); *Slc3a2/Pat2* (Rn00595142_m1); *Tfrc* (Rn01474695_m1); *Tmem26* (Rn01428021_m1); *Ucp1* (Rn00562126_m1); *Zic1* (Rn00575376_m1). The reactions were performed in the StepOnePlusTM Real-Time PCR System (Applied Biosystems, Waltham, MA, USA), using TaqMan® Universal PCR Master Mix (#4304437, Applied Biosystems, Waltham, MA, USA).

The gene expression analysis was performed by relative quantification ($2^{-\Delta\Delta CT}$), and the housekeeping gene was used according to tests performed on each tissue. For the hypothalamus, we used the geometric mean of *Actb*, *Gapdh*, and *Ppia*. For the ATs samples, five different genes were tested as housekeeping-*Actb*, *B2m*, *Hprt1*, *Rpl37a*, and *Tfrc*; and the gene without variation between the CT and GC groups was choose to the normalizations, as follow: *Actb* on BAT, prWAT, and rpWAT; *B2m* on epWAT; *Rpl37a* on meWAT; and *Tfrc* on scWAT samples.

Table 1. Primers used on the hypothalamus qPCR.

Gene	Primer Sequences (5′-3′)	
	Forward	**Reverse**
Actb	AGCCTGGATGGCTACGTACA	CCTCTGAACCCTAAGGCCAA
Agrp	AGGACTCGTGCAGCCTTACAC	GCAGAGGTGCTAGATCCACAGAA
Cartpt	CCGCCTTGGCAGCTCCTT	CCGAGCCCTGGACATCTACT
Crh	CCGATAATCTCCATCAGTTTCCTG	TGGATCTCACCTTCCACCTTCTG
Gapdh	CCGTTCAGCTCTGGGATGAC	GGGCAGCCCAGAACATCAT
Hcrp	AGGGAGAGGCAATCCGGAGAG	GCGGCCTCAGACTCCT
Npy	CCCTCAGCCAGAATGCCCAA	CCGCCCGCCATGATGCTAGGTA
Lepr	CCAGAAGAAGAGGACCAAATATCAC	ACTTAATTTCCAAAAGCCTGAAACA
Ppia	TATCTGCACTGCCAAGACTGAGT	CTTCTTGCTGGTCTTGCCATTCC
Pmch	CTTCTACGTTCCTGATGGACTT	ATGCTGGCCTTTTCTTTGTTT
Pomc	GCAAGCCAGCAGGTTGCT	ATAGACGTGTGGAGCTGGTGC
Tnf	GGTTGTCTTTGAGATCCATGC	TCTCAAAACTCGAGTGACAAGC

2.8. BAT Oxidative and Lipogenic Capacity

The citrate synthase maximal activity in BAT samples was evaluated as previously described [44]. Samples were diluted (1:100) and the linear variation of absorbance in the function of time was used to calculate the maximal activity of the enzyme, and the results were normalized by µg of protein. Basal oxidative capacity to D-[U-^{14}C]-glucose (#CFB96, Amersham Biosciences, Buckinghamshire, UK) and [1-^{14}C]-palmitic acid (#CFA23, Amersham Biosciences, Buckinghamshire, UK) in BAT samples (50 mg) was performed according to Sertié et al. [45], modified to tissue, and expressed as ^{14}CO$_2$ nmol/100 mg of BAT. The D-[U-^{14}C]-glucose incorporated into lipids was performed to assess the lipogenic capacity as previously described [46], modified to BAT samples, and the results are expressed as nmol/100 mg of BAT.

2.9. In Vivo Body Temperature Analysis

Corporal (dorsal and ventral) thermal images were captured from anesthetized CT and GC rats on days 0 and 28th, between 12 < ZT < 13, using the FLIR® E53 camera (Teledyne FLIR, Wilsonville, OR, USA). The images were analyzed by the FLIR® Thermal

Studio software version 1.9.23 (Teledyne FLIR, Wilsonville, OR, USA), and the results were expressed as the mean temperature (°C) of the corporal and BAT area (interscapular region).

2.10. Statistical Analysis

All data were tested to distribution (Shapiro–Wilk test), and homogeneity of variances (F test). For comparisons between CT and GC groups, unpaired Student t-test or Mann–Whitney test were used, as indicated in the figure captions. When the F test was significant, Welch correction was performed after the Student t-test. For nasal–anal length comparison on days 0 and 28th, a Two-Way ANOVA was used, with Bonferroni posthoc due to significant interaction. For correlation analyses between plasma leptin levels and adipocytes volume, after the data showed normal distribution, Pearson's test was used. When there was a correlation, linear regression was used to draw the line that represents the correlation. The level of significance assumed was 5% ($p < 0.05$).

3. Results

3.1. Inhibition of the HPA Axis

Chronic GC administration for 4 weeks promoted HPA axis inhibition, as indicated by a 94% reduction in serum corticosterone (Figure 1a), and a 59% reduction of adrenal gland mass (Figure 1b) in GC rats. The reduction of adrenal mass, mainly due to fasciculate (F) and reticular (R) zone atrophy, was confirmed by the histological analysis (Figure 1c). The treatment also promoted a reduction in spleen mass (Figure 1d). The expression of *Hsd11b1* was increased in all WATs, probably due to the absence of corticosterone inhibition/regulation (Figure 1e).

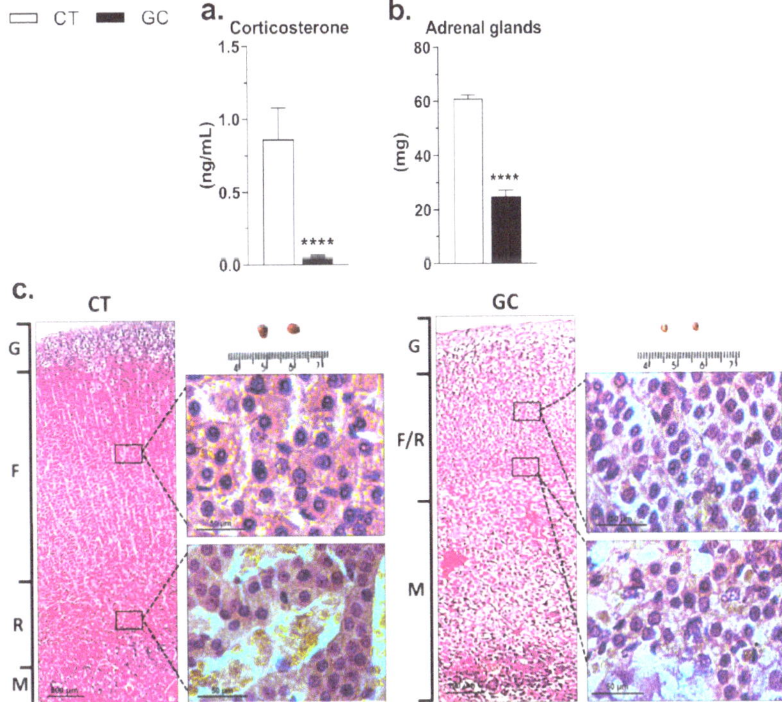

Figure 1. *Cont.*

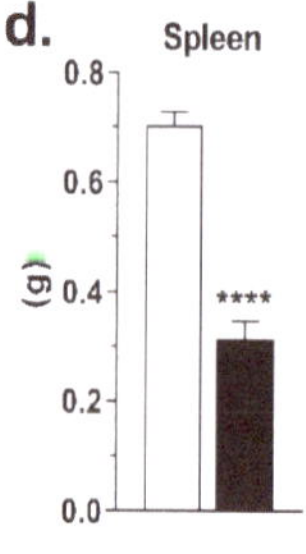
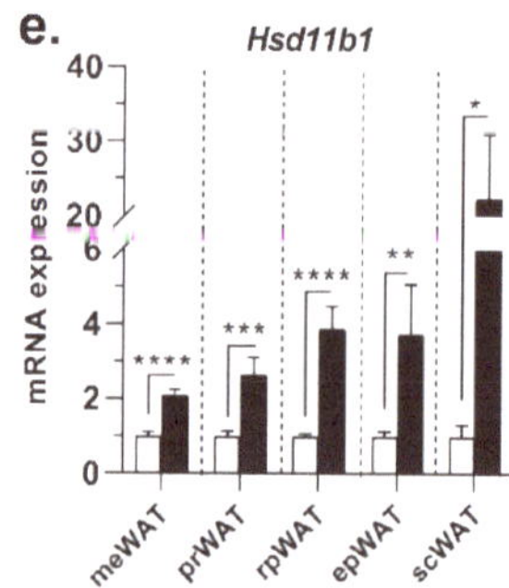

Figure 1. **Glucocorticoid continuous administration promoted inhibition of the HPA axis.** (**a**) Serum corticosterone, (**b**) adrenal gland mass, and (**c**) histological sections; (**d**) spleen mass, and (**e**) *Hsd11b1* gene expression in mesenteric—meWAT, perirenal—prWAT, retroperitoneal—rpWAT, epididymal—epWAT, and subcutaneous inguinal—scWAT of control (CT), and glucocorticoid-treated rats (GC). Data are mean ± SEM of 13 (CT) and 17 (GC) rats (**a,b,d**), and 7–12 (CT) 11-12 (GC) samples (**e**). * $p < 0.05$; ** $p < 0.01$; *** $p < 0.001$ and **** $p < 0.0001$ vs. CT (Unpaired Student *t*-test; Mann–Whitney test in **a** and **d**). (**c**) HE staining—100× and 400× magnification. G: Glomerulosa zone; F: Fasciculata zone; R: Reticularis zone, and M: Medulla.

3.2. Hypercortisolism and Metabolic Changes

The chronic GC exposure promoted glucose intolerance, as evidenced by the incremental area under the curve (AUC) comparison (Figure 2a). We also evaluated the insulinemia during oGTT and, as shown in Figure 2b, the GC-treated rats appeared to have an initial delay in insulin secretion (5 to 15 min), and a hyperinsulinemic profile was observed during the test, as confirmed by AUC analysis (Figure 2b). No difference was observed in 12 hours-fasting glycemia, but GC rats showed hyperinsulinemia in this condition (Figure 2c), resulting in increased HOMA-IR (Figure 2d).

The GC excess did not change serum adiponectin levels (Figure 2e) but promoted the increase in leptin (Figure 2f), NEFA, triglycerides, cholesterol (Figure 2g), and lactate (Figure 2h) compared to the CT group. Additionally, the treatment increased serum levels of ALT but did not change AST (Figure 2i). GC-treated rats also showed an increase in liver mass, with changes in hepatocytes morphology (Figure 2j) and increase in liver triglycerides content (Figure 2k).

The food intake changed along with the treatment in GC rats. While food intake was reduced in the first week, in the last week GC rats exhibited increased food intake compared to the CT group (Figure 3a). When the entire period is analyzed, the average food intake did not change in response to GC treatment (Figure 3b), but feed efficiency was negative (Figure 3c). We also evaluated the hypothalamic gene expression of *Npy, Agrp, Pomc, Cartpt*, and Lepr, in which only *Npy* and *Lepr* were increased by GC administration (Figure 3d). Other genes were analyzed in the hypothalamus, but they did not differ between CT and GC groups (Figure S1a).

GC rats showed accentuated body weight loss (Figures 3e and S1c). To better understand it, we also evaluated the nasal–anal length as three different types of skeletal muscle and the tibia bone, and GC rats showed a reduction of CNA at the end of the treatment (Figure 3f), as well as the gastrocnemius and EDL muscle mass, while soleus muscle showed no difference in mass (Figures 3g and S1b). The treatment also promoted a decrease in the tibia bone weight and length but did not change tibia circumference (Figure 3h).

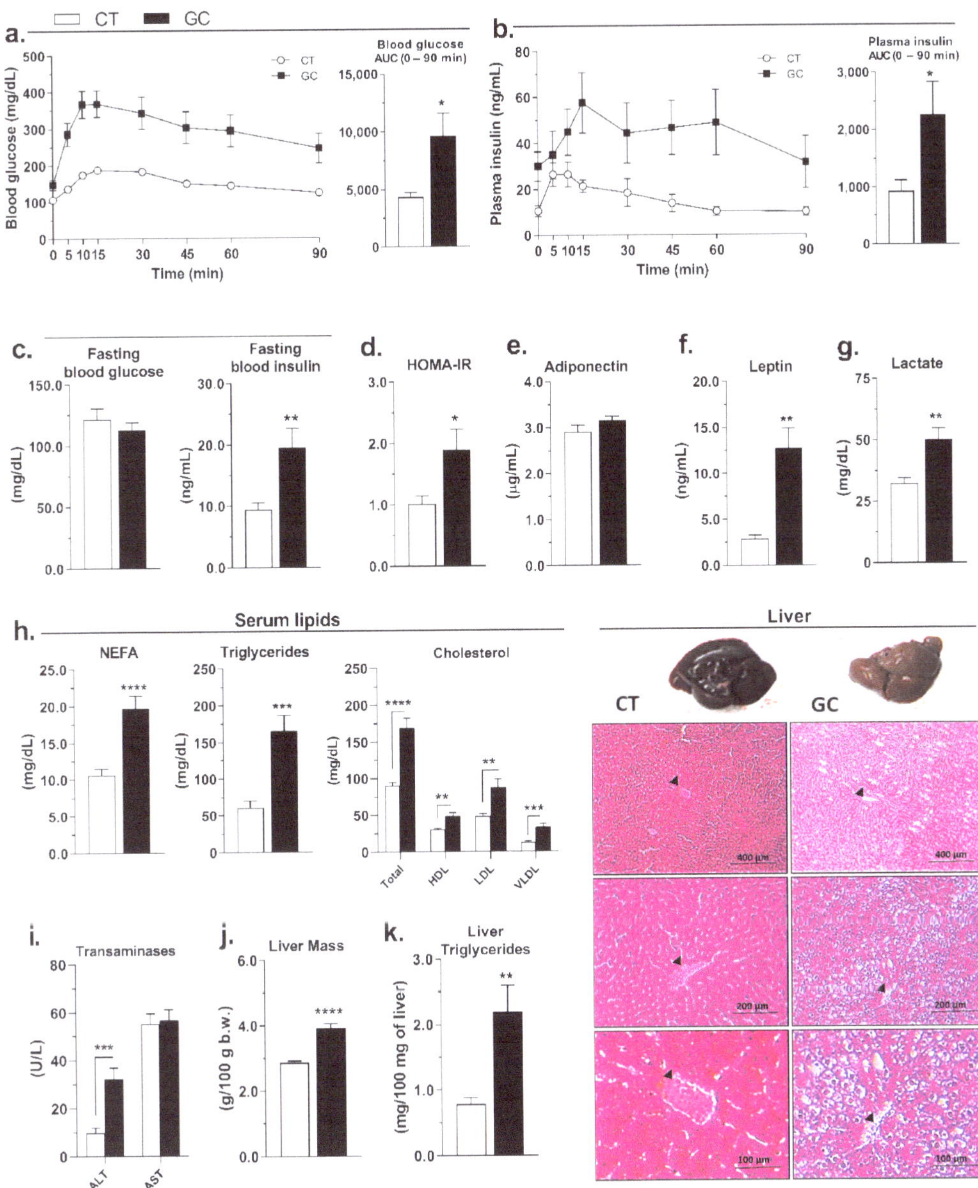

Figure 2. Chronic iatrogenic hypercortisolism promotes glucose intolerance, insulin resistance, dyslipidemia, and increase of hepatic lipid content. (**a**) Blood glucose and (**b**) plasma insulin during oGTT test; (**c**) 12h fasted blood glucose and insulin; (**d**) HOMA-IR; serum (**e**) adiponectin, (**f**) leptin; (**g**) lipids-NEFA, triglycerides, total cholesterol, HDL, LDL, and VLDL-cholesterol fractions; (**h**) lactate, and the (**i**) transaminases ALT and AST; (**j**) liver mass, (**k**) triglycerides content, and histological sections of control (CT), and glucocorticoid-treated (GC) rats. HE staining; $100\times$, $200\times$, and $400\times$ magnification; the black arrows indicates the central vein. Data are mean $\pm$ SEM of 13 (CT) and 17 (GC) rats/samples. * $p < 0.05$; ** $p < 0.01$; *** $p < 0.001$, and **** $p < 0.0001$ vs. CT (unpaired Student *t*-test, Mann–Whitney test in **b**-AUC, and **c**-fasting glucose).

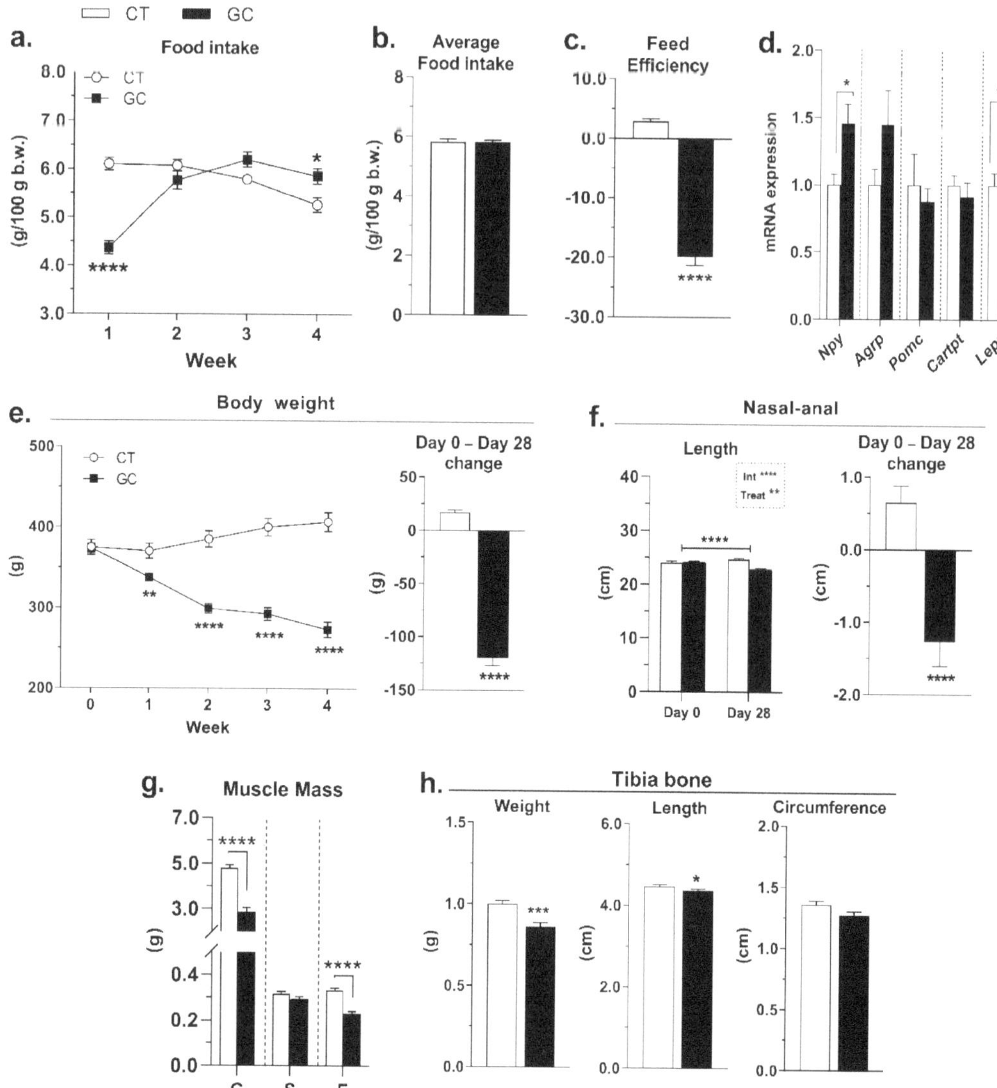

Figure 3. Food intake and body changes in response to GC excess. (**a**) Daily food intake per week; (**b**) average 28 days daily food intake; (**c**) feed efficiency; (**d**) hypothalamic gene expression of *Npy*, *Agrp*, *Pomc*, *Cartpt*, and *Lepr*; (**e**) body weight change; (**f**) nasal–anal length; (**g**) gastrocnemius—G, soleus—S, and extensor digitorum longus—E muscle mass; and (**h**) tibia bone weight, length, and circumference of control (CT), and glucocorticoid-treated (GC) rats. Data are mean ± SEM of 13 (CT) and 17 (GC) rats/samples (**a–c,e–h**), and 6 (CT) and 6 (GC) samples (**d**). * $p < 0.05$; ** $p < 0.01$; *** $p < 0.001$ and **** $p < 0.0001$ vs. CT (unpaired Student *t*-test; Mann–Whitney test in **h**-length). (**f**) **** $p < 0.0001$ (Repeated Measures Two-Way ANOVA with Bonferroni's post hoc test).

3.3. GC-Induced Adipose Tissue Redistribution

GC-treated rats showed AT redistribution: an increase in two visceral depots—meWAT and prWAT, and expressive reduction of rpWAT—but no significant difference was observed in epWAT and scWAT (Figure 4a). The interscapular BAT of GC rats' mass was 3.4 fold bigger than the CT group, and an unilocular phenotype was also observed

(Figure 4b). GC treatment did not change the WAT mass sum of all five different depots analyzed in this study (Figure 4c), as no difference was observed when considering WAT + BAT mass (Figure 4d).

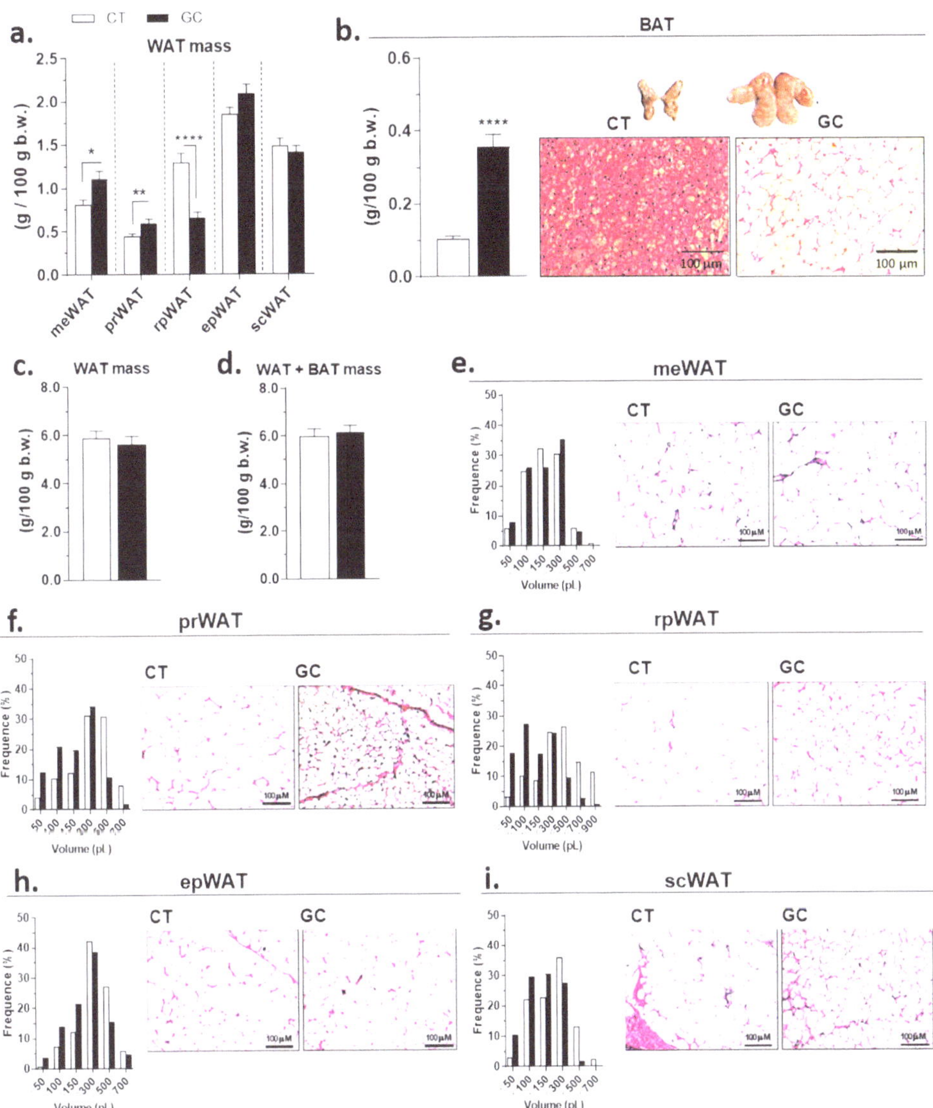

Figure 4. Adipose tissue redistribution after chronic glucocorticoid treatment. (**a**) adipose mass of mesenteric—meWAT, perirenal—prWAT, retroperitoneal—rpWAT, epididymal—epWAT, and subcutaneous inguinal—scWAT; (**b**) interscapular brown adipose tissue—BAT mass and histological sections; (**c**) WAT mass; (**d**) WAT + BAT mass; frequency distribution of adipocyte volume (pL) and histological sections of (**e**) meWAT, (**f**) prWAT, (**g**) rpWAT, (**h**) epWAT, and (**i**) scWAT of control (CT), and glucocorticoid-treated (GC) rats. HE staining; $100\times$ magnification. Data are mean $\pm$ SEM of 13 (CT) and 17 (GC) rats/samples. * $p < 0.05$; ** $p < 0.01$, and **** $p < 0.0001$ vs. CT (unpaired Student *t*-test).

Despite the increase in meWAT mass, there were no changes in the adipocytes volume frequency (Figure 4e). An increase in the frequency of adipocytes with smaller volumes was observed in prWAT (Figure 4f), rpWAT (Figure 4g), epWAT (Figure 4h), and scWAT (Figure 4i), even though the difference in mass was among the fat pads. In the CT group, there was a correlation between the serum leptin and adipocyte volume in all WAT deposits, but this effect was abolished by GC treatment (Figure S2).

3.4. Direct and Permissive Actions by GCs in Gene Expression of WAT Lipolysis and Lipogenesis Pathways

The beta-adrenergic receptors' gene expression diverged among the analyzed WAT (Figure 5). The meWAT (Figure 5a), prWAT (Figure 5b), and epWAT (Figure 5d) did not show any difference between the groups, while in rpWAT (Figure 5c), *Adrb1* decreased expression in GC rats, and *Adrb2* and *Adrb3* increased in scWAT (Figure 5e). *Prkaca* increased in meWAT (Figure 5a), epWAT (Figure 5d), and scWAT (Figure 5e), whereas *Prkacb* only increased in meWAT (Figure 5a) of GC rats. *Plin1* is overexpressed in scWAT (Figure 5e) and the other lipolytic genes—*Abdh5*, *G0s2*, *Fabp4*, *Pnpla2*, *Lipe*, *Mgll*, *Aqp7*, and *Cd36*—showed higher mRNA expression in GC rats, except for *Pnpla2* in prWAT (Figure 5b), and *Lipe* and *Aqp7* in rpWAT (Figure 5c).

The same overexpressed profile was observed concerning lipogenic genes (Figure 6). All WATs increased the expression of *Gpam*, *Agpat1*, *Agpat2*, *Dgat1*, *G6pd*, *Acly*, *Acaca*, *Slc16a1*, *Ldhb*, *Pck1*, and *Gpd1*, but some genes showed different patterns in the WATs. In the meWAT, there was no change in *Lpl*, *Scd*, and *Scd2*, but the expression of *Dgat2*, *Me1*, *Fasn*, *Slc16a7*, *Slc2a1*, and *Gk* were increased in GC group (Figure 6a). Additionally, the treatment promoted the reduction of *Lpl*, and *Scd* expression in prWAT and rpWAT, but did not change the expression of *Me1*, *Fasn*, *Scd2*, *Slc16a7*, and *Slc2a1* in these deposits (Figure 6b,c). Therefore, in the prWAT the GC excess did not change *Gk*, but increased *Dgat2* expression (Figure 6b), while in rpWAT there was no change in *Dgat2*, but increased *Gk* expression (Figure 6c). No differences were observed in the expression of *Lpl*, *Dgat2*, *Me1*, *Fasn*, and *Scd2* in the epWAT (Figure 6d), also as *Scd* and *Scd2*, in the scWAT (Figure 6e), but the expression of *Slc16a7*, *Slc2a1*, and *Gk* increased in both deposits in the GC group. In addition, *Scd* expression was higher in epWAT, and *Lpl*, *Dgat2*, *Me1*, and *Fasn* expression were increased in scWAT (Figure 6d,e).

Regarding the insulin pathway, the only gene that decreased expression was *Pik3cg* in prWAT (Figure S3b) and rpWAT (Figure S3c). *Insr*, *Irs1*, *Irs2*, *Pik3r1*, *Akt1*, *Pde3b*, and *Slc2a4* showed elevated mRNA expression in GCs rats or did not change (Figure S3). *Gpr81* expression was also increased in meWAT, epWAT, and scWAT, but no difference was observed in prWAT and rpWAT (Figure S3g).

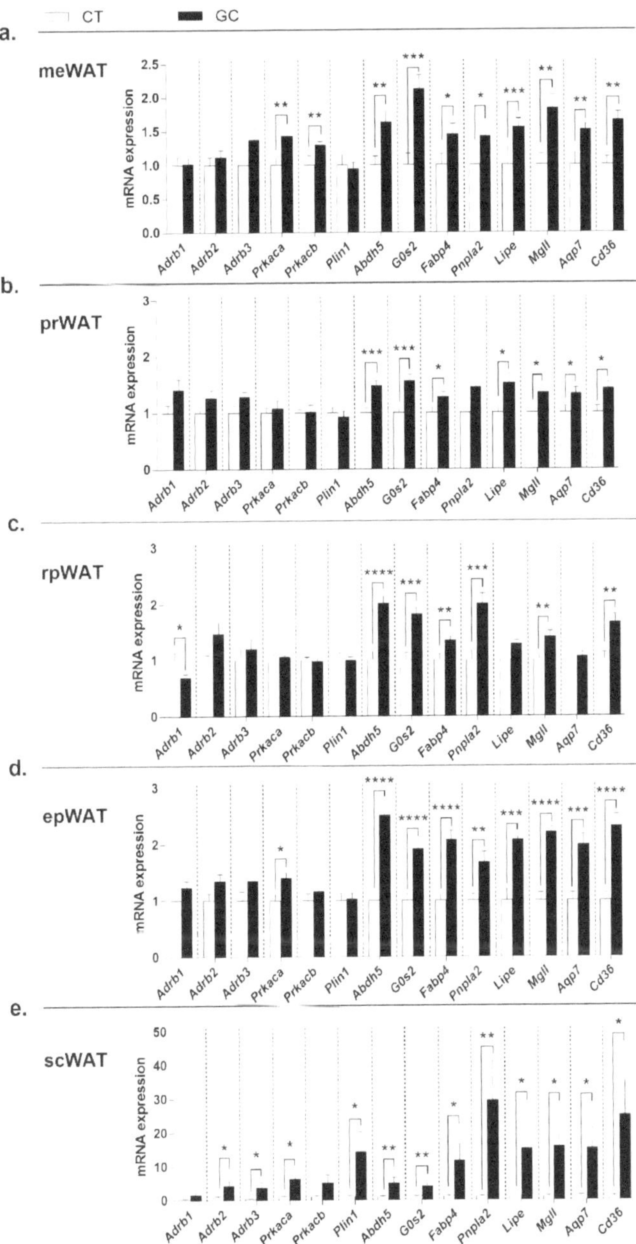

Figure 5. Gene expression of the lipolytic pathway in white adipose tissue. (**a**) mesenteric—meWAT; (**b**) perirenal—prWAT; (**c**) retroperitoneal—rpWAT; (**d**) epididymal—epWAT; and (**e**) subcutaneous inguinal—scWAT. Genes of beta-adrenergic receptors—*Adrb1*, *Adrb2*, and *Adbr3*; catalytic subunits of PKA—*Prkaca* and *Prkacb*; perilipin in white mature adipocytes—*Plin1*; lipase cofactors—*Abdh5*, *G0s2*, and *Fabp4*; lipases—*Pnpla2*, *Lipe*, and *Mgll*; lipolytic products channel and transporter—*Aqp7* and *Cd36*—of control (CT; n = 7–12), and glucocorticoid-treated (GC; n = 11–12) rats. Data are mean ± SEM. * $p < 0.05$; ** $p < 0.01$, *** $p < 0.001$, and **** $p < 0.0001$ vs. CT (unpaired Student *t*-test; Mann–Whitney test in (**a**) (*Adrb2*, *Adrb3*, *Fabp4*, *Abdh5*, *Mgll*, and *Aqp7*), (**b**) (*Adrb3* and *Fabp4*), (**c**) (*Adrb2* and *Adrb3*), (**d**) (*Adrb2*, *Adrb3*, and *Prkacb*), and **e** (all, except *G0s2*)).

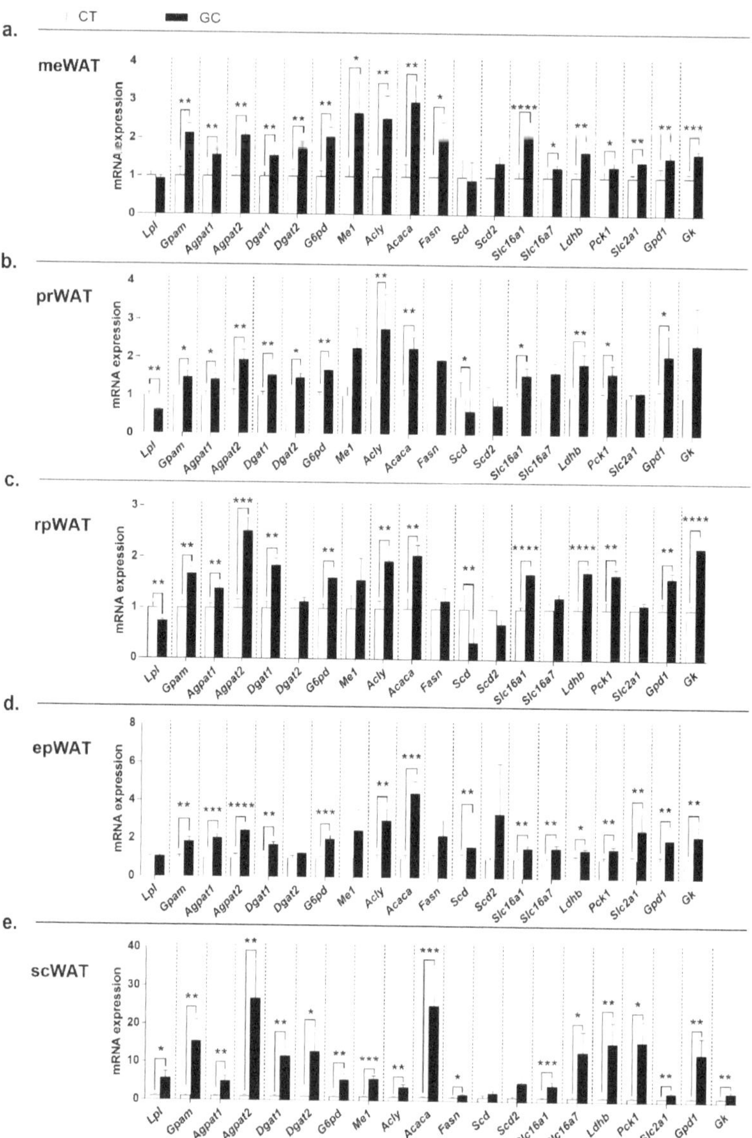

Figure 6. Gene expression of the lipogenic pathway in white adipose tissue. (**a**) mesenteric—meWAT; (**b**) perirenal—prWAT; (**c**) retroperitoneal—rpWAT; (**d**) epididymal—epWAT; and (**e**) subcutaneous inguinal. Genes of lipoprotein lipase—*Lpl*; enzymes of fatty acids esterification—*Gpam, Agpat1, Agpat2, Dgat1,* and *Dgat2*; cytosolic NADPH recyclers—*G6pd* and *Me1*; enzymes of lipogenesis de novo—*Acly, Acaca,* and *Fasn*; fatty acid desaturases—*Scd* and *Scd2*; monocarboxylate transporters—*Slc16a1* and *Slc16a7*; lactate dehydrogenase—*Ldhb*; the key enzyme of glyceroneogenesis—*Pck1*; glucose transporter 1—*Slc2a1*; the last enzyme of glycerol-3-phosphate generation glycolytic pathway and glyceroneogenesis—*Gpd1*; and the glycerol kinase—*Gk*—of control (CT; n = 7–12), and glucocorticoid-treated (GC; n = 11–12) rats. Data are mean ± SEM. * $p < 0.05$; ** $p < 0.01$; *** $p < 0.001$, and **** $p < 0.0001$ vs. CT (unpaired Student t-test; Mann–Whitney test in (**a**) (*Gpam, G6pd, Me1, Acly, Acaca, Fasn, Scd, Scd2* and *Gpd1*), (**b**) (*Agpat1, Dgat2, G6pd, Me1, Acly, Acaca, Scd, Slc16a7,* and *Gk*), (**c**) (*Me1, Fasn, Scd,* and *Slc2a1*), (**d**) (*Agpat2, G6pd, Acly,* and *Ldhb*), and (**e**) (*Lpl, Agpat2, Dgat1, Dgat2, Acaca, Scd2, Slc16a1, Slc16a7, Pck1, Slc2a1,* and *Gpd1*)).

3.5. BAT Whitening in Response to GC Excess

In addition to the increase in mass and unilocular phenotype (Figure 3b), the BAT of GC rats showed a decrease in the *Ucp1* gene (Figure 7a) and protein, as observed in the UCP1 immunostaining (Figure 7b). The gene expression of the pathway *Adrb3-Prdm16-Ppargc1a-Pparg*, which regulates *Ucp1* expression, was also reduced in GC rats BAT (Figure 7c), as well as the expression of the BAT markers *Zic1, Tmem26, Pat2, Cidea*, and *Dio2* (Figure 7d). The activity of citrate synthase (Figure 7e) and the oxidative capacity for palmitic acid (Figure 7f) and glucose (Figure 7g) were also reduced in the BAT of GC-treated rats.

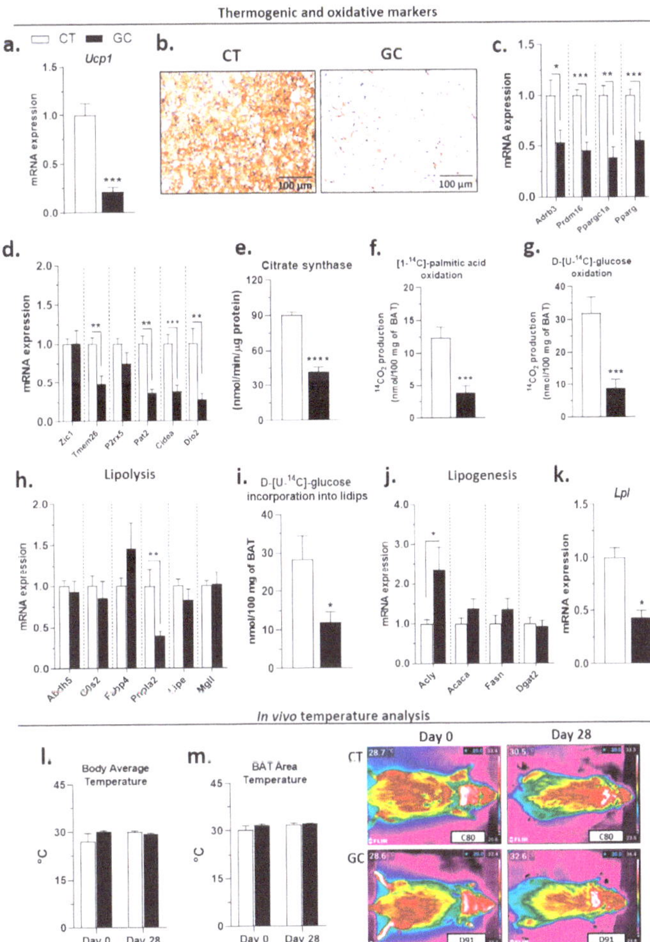

Figure 7. BAT gene and functional changes after chronic GC exposure. (**a**) *Ucp1* gene expression; (**b**) UCP1 immunohistochemistry; (**c**) *Ucp1* expression regulators—*Adrb3, Prdm16, Ppargc1a*, and *Pparg*; (**d**) brown adipocytes markers—*Zic1, Tmem26, Pat2, P2rx5, Cidea*, and *Dio2*; (**e**) citrate synthase maximal activity; oxidation of (**f**) [1-^{14}C]-palmitic acid and (**g**) D-[U-^{14}C]- glucose; (**h**) lipolysis pathway genes—*Abhd5, G0s2, Fabp4, Pnpla2, Lipe*, and Mgll; (**i**) incorporation of D-[U-C^{14}]- glucose into lipids; (**j**) lipogenic pathway genes—*Acly, Acaca, Fasn*, and *Dgat2*; (**k**) *Lpl* expression; *in vivo* temperature analysis of (**l**) body and (**m**) BAT area on days 0 and 28th of control (CT), and glucocorticoid-treated (GC) rats. (**b**) UCP1 staining: 100× magnification. Data are mean ± SEM of 8 (CT) and 8 (GC) samples (**a,c,d,h,j,k**), 8 (CT) and 9 (GC) samples (**e,f,g,i**), and 10 (CT) and 11 (GC) rats (**l,m**). * $p < 0.05$, ** $p < 0.01$, *** $p < 0.001$, and **** $p < 0.0001$ vs. CT (unpaired Student *t*-test; Mann–Whitney test in (**c**)-*Adrb3* and *Ppargc1a*).

In the lipolytic pathway, GC treatment decreased the *Pnpla2* expression but did not change *Abdh5, G0s2, Lipe, Fabp4*, and *Mgll* in the BAT (Figure 7h). The lipogenic capacity was measured by the incorporation of D-[U-^{14}C]-glucose into the lipids, which was also reduced in the BAT of the GC group (Figure 7i), even the increase in *Acly* (Figure 7j). Therefore, no significant changes were observed in the other lipogenic genes—*Acaca, Fasn*, and *Dgat2* in the GC group (Figure 7j)—but the treatment reduced the *Lpl* expression (Figure 7k). GC administration did not change the insulin pathway genes evaluated—*Insr, Pik3r1, Akt1, Slc2a4*, and *Slc2a1*, except *Pik3cg*, which was reduced in GC-rats BAT (Figure S3f). The *Gpr81* expression was reduced in BAT of GC-rats (Figure S3g).

Despite these changes in BAT, known for its thermogenic effects, there was no difference between CT and GC groups in body average temperature (Figure 7l) and BAT area temperature (Figure 7m) on days 0 or 28 of treatment.

4. Discussion

Hypercortisolism models are common, but few are focused on the GC effects in several AT deposits. Additionally, some methods can be more stressful to the animals (daily administration or manipulation), which can lead to results interferences, or even not assure the exact daily dose, as in the case of treatments by drinking water. The GCs effects depend on the dose, time of use, and individual parameters, with pediatric/young patients being more susceptible to severe side effects [47]. For this reason, we build an experimental design less stressful for the animals, which supports continuous and invariable delivery of GC in a dose used in previous animal studies [29] and also pharmacologically in humans [48]. We chose young adult rats (12 weeks old) [35] to avoid more possible interferences in another hormonal axis (e.g., sexual hormones in the pubertal phase, which can also affect the AT distribution).

The most important characteristic of hypercortisolism/Cushing's syndrome is the disruption of HPA axis control and loss of the circadian rhythm of GC production and release [5]. The intense reduction of corticosterone levels and adrenal cortical mass found in GC-treated rats due to the atrophy of the fasciculate and reticular zones of the adrenal cortex (Figure 1b,c) confirms the HPA axis inhibition and adrenal insufficiency [49]. The immunosuppressive effect of treatment was confirmed by the GC rats' spleen mass (Figure 1d) [50].

GC excess induces several metabolic alterations, mainly in glucose and lipid metabolism [51,52]. One factor that could lead to insulin resistance, as observed in the present model (Figure 2a–d), is the increase in circulating NEFA, as shown in Figure 2g. This effect is probably by the permissive role of GC to the actions of the lipolytic hormones as catecholamines, which contributes to an exacerbated mobilization of NEFA from adipose depots and allows their ectopic accumulation in non-specialized tissues as the liver (Figure 2k), favoring the local and systemic insulin resistance [33]. The increased serum ALT corroborates with the increased liver mass, the morphological alterations, and with the increase of triglycerides content in this tissue (Figure 2i–k), suggesting damage to this vital organ in GC rats. Moreover, GCs also have direct hepatic actions, promoting the increase in the expression of the de novo lipogenesis key enzymes such as acetyl-CoA carboxylase and fatty acid synthase [53], and increasing the synthesis and secretion of VLDLs [54]. Together, these effects contribute significantly to the establishment of dyslipidemia, another characteristic of human Cushing syndrome reproduced in our animal model (Figure 2g).

These metabolic changes could also be a result of GC effects on food intake. Even with the increased serum leptin and hypothalamic *Lepr* expression (Figures 2f and 3d), GC rats showed higher food intake in the last week of treatment (Figure 3a). Indeed, GC rats exhibited upregulation of *Npy* expression in the hypothalamus (Figure 3d). These findings corroborate with previous studies, showing that GCs can increase the hypothalamic expression of orexigenic neuropeptides such as NPY and AgRP, promoting increased food intake [55,56]. However, the average food intake did not differ between the groups when the entire period is analyzed (Figure 3b) and, due to the intense loss of body weight (Figure 3e), the GC rats had negative feed efficiency (Figure 3c). In this sense, we hypothesize that, even with the same average food intake during the 28 days, the nutrients may have a different metabolic fate in the GC-treated rats.

In humans, weight gain and an increase in central adiposity are the most visible features of Cushing syndrome [6,57,58]. However, a previous study with excess of synthetic GC induced weight loss in rats [29]. In concordance, our GC-treated rats were visibly "smaller and skinny", and showed a decrease in nasal–anal length compared to CT rats (Figure 3f). GC-rats showed loss of muscle mass in two of the three muscles analyzed (Figures 3g and S1b) and reduced tibia mass (g) and length (cm) compared to CT rats. It highlights the difference in the body composition of these animals, which are caused by GC catabolic effects [59,60]. For this reason, we consider normalizing the AT tissue weights by 100g of body weight, aiming to get more accurate comparisons between the groups.

WAT is the most prevalent AT in humans and rodents, and depending on the local accumulation, there is a risk to develop diseases [61,62]. The increase of visceral adipose tissues (VAT) is more related to metabolic changes than subcutaneous adipose tissue (SAT), even SAT being the predominant adipose site in the body [61,63]. Humans facing Cushing syndrome develop a central fat accumulation pattern, with an increase in VAT and loss of peripheral SAT [60,64]. A previous study of our group did not observe significant AT redistribution typical of Cushing's syndrome [25]. In our model, despite the accentuated overall weight loss (Figure 3e), there was a notable visceral/BAT fat redistribution without changes in the relative amount of WAT/WAT+BAT in treated rats compared to the CT group (Figure 4a–d). Additionally, GC rats display a divergent frequency of adipocyte size (Figure 4e–i), with smaller adipocytes compared to CT rats, which could lead to an improvement in the metabolic condition of GC rats [65,66], but it did not occur.

Small adipocytes present increased secretion of adiponectin, an insulin-sensitizing hormone [67], an effect not observed in our model (Figure 2e). Another adipokine, leptin, has the opposite pattern—its secretion is related to hypertrophic adipocytes or increased adipose mass [68,69]. Our model showed increased leptin levels even with smaller adipocytes and no difference in total adipose tissue mass (Figures 2f, 4c and S2). All these data confirmed an AT remodeling in our model and reinforce the evidence of AT's role or its dysfunction in Cushing's syndrome pathology.

GCs are known as lipolytic hormones in AT [1,4,22,70,71], and the increased serum concentrations of NEFA in GC-treated rats suggest an increase of lipolysis. Since the majority of adverse effects provoked by long-term GC use are genomic [19], we analyzed gene transcription of factors that could interfere directly with the lipolysis pathway, such as lipases (*Pnpla2*, *Lipe*, and *Mgll*) and their cofactors (*Abdh5*, *G0s2*, and *Fabp4*), or indirectly, like beta catecholamines receptors (*Adrb1*, *Adrb2*, and *Adrb3*), catalytic subunits of PKA (*Prkaca* and *Prkacb*), perilipin A (*Plin1*), and transporter/channel for products (*Cd36* and *Aqp7*). None of the changes in the expression of these genes could explain the WAT redistribution observed, since the pattern of expression was similar among the different deposits (Figure 5).

Another phenomenon that could explain the AT redistribution is lipogenesis. This term includes many pathways, such as the esterification of fatty acids in glycerol-3-phosphate (G3P). Proteins involved are encoded by *Gpam*, *Agpat1*, *Agpat2*, *Dgat1*, and *Dgat2* gene in WAT [72], and formation de novo of fatty acids (*Acly*, *Acaca*, and *Fasn* are the genes of key-enzymes [73,74], *G6pd* and *Me1* are genes of enzymes that recycle NADPH [73]). Additionally, *Scd* and *Scd2* are genes that encode proteins not directly involved with the de novo lipogenesis pathway, signaling its occurrence [75], and G3P generation pathways, *Gk*, *Pepckc*, and *Gpd1*, encode key-enzymes, and the other genes, *Slc16a1*, *Slc16a7*, *Ldhb*, and *Slc2a1*, are involved with proteins that manage substrates for these pathways [26,73,76,77]. Again, there was a pattern of overexpression for most of lipogenic genes in the WAT of GC-treated rats (Figure 6), and the same pattern were observed in the genes of the insulin cascade (Figure S3a–e)—the major lipogenic hormone [78,79]. Despite increased HOMA-IR, at the genomic level there is no proof of insulin resistance in the WAT of GC-treated rats [80]. Regardless of metabolic reprogramming occasioned by GC treatment and all the depots of WAT respond to the inhibition of the HPA axis, increasing the expression of *Hsd11b1* (Figure 1e), their actions are promiscuous and wide, not justifying the AT redistribution that appears to be one of the main causes of metabolic alterations.

In the present experimental model, the BAT of GC-treated rats completely loses its well-known characteristics: there was an increase in mass, changes in morphology to an unilocular phenotype, reduced expression of genes related to thermogenesis/brown adipocytes markers, and most importantly, the decreased oxidative capacity (Figure 7a–g), which characterizes a whitening process [81,82]. Our results corroborate with some studies that have shown that GC excess leads to changes in the morphology and/or thermogenic function [83–87]. However, it is still unclear whether BAT dysfunctions are a cause or a consequence of the AT redistribution/obesity that occurs in Cushing's syndrome, but the BAT whitening may also explain some metabolic changes in this model.

Among the factors that can promote BAT dysfunction in obesity is the reduction of vascularization and, consequently, the β-adrenergic activation pathway that regulates the BAT activity, favoring *Ucp1* expression [88–90], and stimulates the lipolytic pathway to mobilize fatty acids for oxidation [91]. Along with the decreased *Ucp1*, the BAT of the GC rats showed reduced expression of *Adrb3* and *Atgl*, which encodes the first enzyme of the lipolytic pathway (Figure 7c,h). Fatty acids are the primary energy substrate for thermogenesis in brown adipocytes [91], and the intracellular triacylglycerol stock is replenished mainly by uptake of circulating fatty acids derived from lipolysis [92], and in second place, by the de novo lipogenesis from glucose uptake [93]. However, GC rats also present reduced *Lpl* expression (Figure 7k), which may indicate reduced fatty acids uptake. This result, added to the decreased oxidative capacity and possible inhibition of the lipolytic pathway, certainly plays a role in the dyslipidemia presented by the GC rats (Figure 2g).

The unilocular phenotype of brown adipocytes of GC-treated rats reveals a significant accumulation of triacylglycerol, which could be a combination of the factors of decreased fatty acids oxidation/thermogenesis/lipolysis plus an increased lipogenic capacity in this fat. Although the *Acly* expression was increased, the de novo lipogenesis, measured by the glucose incorporation into lipids, was reduced in the BAT of GC rats (Figure 7i,j). Since there was no difference in the expressions of *Slc2a1* and *Slc2a4* between CT and GC groups (Figure S3f), is not possible to suppose that the GC excess did not affect the BAT glucose uptake. The glucose may have different fates in BAT than oxidation to CO_2, as shown by our results. For example, a previous study showed that brown adipocytes convert a large amount of glucose to lactate [94], which is mostly exported to the circulation but can also serve as a substrate for lipogenesis [76,95], or regulate pathways such as lipolysis through the activation of GPR81 [96]. Such pathways require further studies in our model since GC rats have higher serum lactate compared to CT rats (Figure 2h), but decreased *Gpr81* in BAT, and only rpWAT, which decreased in mass, showed no alteration in the *Gpr81* expression (Figure S3g).

Besides the significant changes in BAT, GC treatment did not promote significant changes in body and BAT area temperature by the thermal imaging method we use in this study (Figure 7l,m). Given these results, we hypothesize that the BAT dysfunction is being compensated by the browning of WAT territories in GC rats, which is actually under investigation by our research group.

In conclusion, this experimental model of Cushing's syndrome in rats reproduced several characteristics of this syndrome observed in humans, such as HPA axis inhibition, AT redistribution with an increase in VAT (meWAT and prWAT), and metabolic changes such as glucose intolerance, insulin resistance, dyslipidemia, and hepatic lipid accumulation. The AT redistribution and remodeling in response to GC excess showed more importance than the increase of AT mass per se. This pattern observed in AT cannot be explained just by GC regulation of gene transcription, and other mechanisms should be explored.

Supplementary Materials: The following supporting information can be downloaded at: https://www.mdpi.com/article/10.3390/biomedicines10092328/s1, Figure S1. Hypothalamic gene expression of *Pmch*, *Hcrp*, *Crh*, and *Tnf*; and representative body images of CT and GC rats; Figure S2. Correlation between plasma leptin levels and adipocytes volume of meWAT; prWAT; rpWAT; epWAT; scWAT, or WAT and WAT + BAT mass of CT and GC rats; Figure S3. Expression of classic insulin pathway genes in meWAT; prWAT; rpWAT; epWAT; scWAT, and BAT; and *Gpr81* in all fat deposits of CT and GC rats.

Author Contributions: Conceptualization, F.d.F.S., A.C.M.K., and F.B.L.; Data curation, F.d.F.S., A.C.M.K., S.A., G.B.R., R.O.C., R.G.L., G.O.d.S., R.A.L.S. and J.D.J.; Formal analysis, F.d.F.S. and A.C.M.K.; Funding acquisition, F.B.L.; Investigation, F.d.F.S. and A.C.M.K.; Methodology, F.d.F.S., A.C.M.K., R.G.L., Sheila Collins, and F.B.L.; Project administration, F.B.L.; Supervision, S.C., J.D.J. and F.B.L.; Writing–original draft, F.d.F.S. and A.C.M.K.; Writing–review & editing, F.d.F.S., A.C.M.K., S.A., G.B.R., R.O.C., R.G.L., G.O.d.S., R.A.L.S., S.C., J.D.J., and F.B.L. All authors have read and agreed to the published version of the manuscript.

Funding: Sao Paulo Research Foundation (FAPESP), Sao Paulo, Brazil: grants #2016/25129-4, #2018/11145-3, and #2018/16856-5.

Institutional Review Board Statement: The study was conducted according to the guidelines of the Declaration of Helsinki, and all the protocols were approved by the Committee of Ethics in the Use of Animals of the Institute of Biomedical Sciences, University of Sao Paulo (CEUA-ICB/USP) protocols #89/2016 (approved in 09/19/2016), #26/2017 (approved in 03/21/2017), and 9535190219 (approved on 11/19/2019).

Informed Consent Statement: Not applicable.

Data Availability Statement: The datasets generated during and/or analyzed during the current study are available from the corresponding author on reasonable request.

Acknowledgments: The authors thank Larissa de Sá Lima for the technical support and to Cristoforo Scavone for the laboratory equipment availability.

Conflicts of Interest: The authors declare no conflict of interest.

References

1. Peckett, A.J.; Wright, D.C.; Riddell, M.C. The Effects of Glucocorticoids on Adipose Tissue Lipid Metabolism. *Metabolism* **2011**, *60*, 1500–1510. [CrossRef] [PubMed]
2. Ramamoorthy, S.; Cidlowski, J.A. Corticosteroids. Mechanisms of Action in Health and Disease. *Rheum. Dis. Clin.* **2016**, *42*, 15–31. [CrossRef] [PubMed]
3. Mitsugi, N.; Kimura, F. Simultaneous Determination of Blood Levels of Corticosterone and Growth Hormone in the Male Rat: Relation to Sleep-Wakefulness Cycle1 *Neuroendocrinology* **1985**, *41*, 125–130. [CrossRef] [PubMed]
4. Macfarlane, D.P.; Forbes, S.; Walker, B.R. Glucocorticoids and Fatty Acid Metabolism in Humans: Fuelling Fat Redistribution in the Metabolic Syndrome. *J. Endocrinol.* **2008**, *197*, 189–204. [CrossRef]
5. Pivonello, R.; De Martino, M.C.; De Leo, M.; Simeoli, C.; Colao, A. Cushing's Disease: The Burden of Illness. *Endocrine* **2017**, *56*, 10–18. [CrossRef]
6. Nieman, L.K. Diagnosis of Cushing's Syndrome in the Modern Era. *Endocrinol. Metab. Clin.* **2018**, *47*, 259–273. [CrossRef]
7. Buttgereit, F.; Spies, C.M.; Bijlsma, J.W.J. Novel Glucocorticoids: Where Are We Now and Where Do We Want to Go? *Clin. Exp. Rheumatol.* **2015**, *33*, 29–33.
8. Schoneveld, O.J.L.M.; Gaemers, I.C.; Lamers, W.H. Mechanisms of Glucocorticoid Signalling. *Biochim. Biophys. Acta Gene Struct. Expr.* **2004**, *1680*, 114–128. [CrossRef]
9. Ahmed, M.H.; Hassan, A. Dexamethasone for the Treatment of Coronavirus Disease (COVID-19): A Review. *SN Compr. Clin. Med.* **2020**, *2*, 2637–2646. [CrossRef]
10. Lester, M.; Sahin, A.; Pasyar, A. The Use of Dexamethasone in the Treatment of COVID-19. *Ann. Med. Surg.* **2020**, *56*, 218–219. [CrossRef]
11. Chatterjee, S.; Ghosh, R.; Vardhan, B.; Ojha, U.K.; Kalra, S. An Epidemic of Iatrogenic Cushing's Syndrome in Anticipation in Post-COVID Era. *J. Fam. Med. Prim. Care* **2022**, *11*, 412. [CrossRef]
12. Rebuffat, A.G.; Tam, S.; Nawrocki, A.R.; Baker, M.E.; Frey, B.M.; Frey, F.J.; Odermatt, A. The 11-Ketosteroid 11-Ketodexamethasone Is a Glucocorticoid Receptor Agonist. *Mol. Cell. Endocrinol.* **2004**, *214*, 27–37. [CrossRef]
13. Caprio, M.; Fève, B.; Claës, A.; Viengchareun, S.; Lombès, M.; Zennaro, M.-C. Pivotal Role of the Mineralocorticoid Receptor in Corticosteroid-Induced Adipogenesis. *FASEB J.* **2007**, *21*, 2185–2194. [CrossRef]

14. Beaupere, C.; Liboz, A.; Fève, B.; Blondeau, B.; Guillemain, G. Molecular Mechanisms of Glucocorticoid-Induced Insulin Resistance. *Int. J. Mol. Sci.* **2021**, *22*, 623. [CrossRef]

15. He, Y.; Yi, W.; Suino-Powell, K.; Zhou, X.E.; Tolbert, W.D.; Tang, X.; Yang, J.; Yang, H.; Shi, J.; Hou, L.; et al. Structures and Mechanism for the Design of Highly Potent Glucocorticoids. *Cell Res.* **2014**, *24*, 713. [CrossRef]

16. Heitzer, M.D.; Wolf, I.M.; Sanchez, E.R.; Witchel, S.F.; DeFranco, D.B. Glucocorticoid Receptor Physiology. *Rev. Endocr. Metab. Disord.* **2007**, *8*, 321–330. [CrossRef]

17. Kadmiel, M.; Cidlowski, J.A. Glucocorticoid Receptor Signaling in Health and Disease. *Trends Pharmacol. Sci.* **2013**, *34*, 518. [CrossRef]

18. Lecoq, L.; Vincent, P.; Lavoie-Lamoureux, A.; Lavoie, J.P. Genomic and Non-Genomic Effects of Dexamethasone on Equine Peripheral Blood Neutrophils. *Vet. Immunol. Immunopathol.* **2009**, *128*, 126–131. [CrossRef]

19. Clark, A.R. Anti-Inflammatory Functions of Glucocorticoid-Induced Genes. *Mol. Cell. Endocrinol.* **2007**, *275*, 79–97. [CrossRef]

20. Feelders, R.A.; Pulgar, S.J.; Kempel, A.; Pereira, A.M. The Burden of Cushing's Disease: Clinical and Health-Related Quality of Life Aspects. *Eur. J. Endocrinol.* **2012**, *167*, 311–326. [CrossRef]

21. Nieman, L.K. Cushing's Syndrome: Update on Signs, Symptoms and Biochemical Screening. *Eur. J. Endocrinol.* **2015**, *173*, M33–M38. [CrossRef] [PubMed]

22. Campbell, J.E.; Peckett, A.J.; D'Souza, A.M.; Hawke, T.J.; Riddell, M.C. Adipogenic and Lipolytic Effects of Chronic Glucocorticoid Exposure. *Am. J. Physiol. Cell Physiol.* **2011**, *300*, C198–C209. [CrossRef] [PubMed]

23. Halvorsen, Y.D.C.; Bond, A.; Sen, A.; Franklin, D.M.; Lea-Currie Renee, Y.; Sujkowski, D.; Ellis, P.N.; Wilkison, W.O.; Gimble, J.M. Thiazolidinediones and Glucocorticoids Synergistically Induce Differentiation of Human Adipose Tissue Stromal Cells: Biochemical, Cellular, and Molecular Analysis. *Metab. -Clin. Exp.* **2001**, *50*, 407–413. [CrossRef] [PubMed]

24. Lee, M.J. Hormonal Regulation of Adipogenesis. *Compr. Physiol.* **2017**, *7*, 1151–1195. [CrossRef]

25. Chimin, P.; Farias, T.D.S.M.; Torres-Leal, F.L.; Bolsoni-Lopes, A.; Campaña, A.B.; Andreotti, S.; Lima, F.B. Chronic Glucocorticoid Treatment Enhances Lipogenic Activity in Visceral Adipocytes of Male Wistar Rats. *Acta Physiol.* **2014**, *211*, 409–420. [CrossRef]

26. Ferreira, G.N.; Rossi-Valentim, R.; Buzelle, S.L.; Paula-Gomes, S.; Zanon, N.M.; Garófalo, M.A.R.; Frasson, D.; Navegantes, L.C.C.; Chaves, V.E.; Kettelhut, I.C. Differential Regulation of Glyceroneogenesis by Glucocorticoids in Epididymal and Retroperitoneal White Adipose Tissue from Rats. *Endocrine* **2017**, *57*, 287–297. [CrossRef]

27. Minetto, M.A.; Qaisar, R.; Agoni, V.; Motta, G.; Longa, E.; Miotti, D.; Pellegrino, M.A.; Bottinelli, R. Quantitative and Qualitative Adaptations of Muscle Fibers to Glucocorticoids. *Muscle Nerve* **2015**, *52*, 631–639. [CrossRef]

28. Alev, K.; Vain, A.; Aru, M.; Pehme, A.; Purge, P.; Kaasik, P.; Seene, T. Glucocorticoid-Induced Changes in Rat Skeletal Muscle Biomechanical and Viscoelastic Properties: Aspects of Aging. *J. Manipulative Physiol. Ther.* **2018**, *41*, 19–24. [CrossRef]

29. Wu, T.; Jiang, J.; Yang, L.; Li, H.; Zhang, W.; Chen, Y.; Zhao, B.; Kong, B.; Lu, P.; Zhao, Z.; et al. Timing of Glucocorticoid Administration Determines Severity of Lipid Metabolism and Behavioral Effects in Rats. *Chronobiol. Int.* **2017**, *34*, 78–92. [CrossRef]

30. Chanson, P.; Salenave, S. Metabolic Syndrome in Cushing's Syndrome. *Neuroendocrinology* **2010**, *92*, 96–101. [CrossRef]

31. Pellegrinelli, V.; Carobbio, S.; Vidal-Puig, A. Adipose Tissue Plasticity: How Fat Depots Respond Differently to Pathophysiological Cues. *Diabetologia* **2016**, *59*, 1075–1088. [CrossRef]

32. Lynes, M.D.; Tseng, Y.H. Deciphering Adipose Tissue Heterogeneity. *Ann. N. Y. Acad. Sci.* **2018**, *1411*, 5–20. [CrossRef]

33. Choe, S.S.; Huh, J.Y.; Hwang, I.J.; Kim, J.I.; Kim, J.B. Adipose Tissue Remodeling: Its Role in Energy Metabolism and Metabolic Disorders. *Front. Endocrinol.* **2016**, *7*, 30. [CrossRef]

34. Chusyd, D.E.; Wang, D.; Huffman, D.M.; Nagy, T.R. Relationships between Rodent White Adipose Fat Pads and Human White Adipose Fat Depots. *Front. Nutr.* **2016**, *3*, 10. [CrossRef]

35. Sengupta, P. The Laboratory Rat: Relating Its Age With Human's. *Int. J. Prev. Med.* **2013**, *4*, 624.

36. Swarbrick, M.; Zhou, H.; Seibel, M. Local and Systemic Effects of Glucocorticoids on Metabolism: New Lessons from Animal Models. *Eur. J. Endocrinol.* **2021**, *185*, R113–R129. [CrossRef]

37. Velasco, A.; Huerta, I.; G-Granda, T.; Cachero, T.G.; Menéndez, E.; Marin, B. Circadian Rhythms of Plasma Corticosterone at Different Times after Induction of Diabetes. Responses to Corticoadrenal Stimulation in Light and Dark Phases. *Life Sci.* **1993**, *52*, 965–974. [CrossRef]

38. Friedewald, W.T.; Levy, R.I.; Fredrickson, D.S. Estimation of the Concentration of Low-Density Lipoprotein Cholesterol in Plasma, without Use of the Preparative Ultracentrifuge. *Clin. Chem.* **1972**, *18*, 499–502. [CrossRef]

39. Wallace, T.M.; Levy, J.C.; Matthews, D.R. Use and Abuse of HOMA Modeling. *Diabetes Care* **2004**, *27*, 1487–1495. [CrossRef]

40. Folch, J.; Lees, M.; Sloane Stanley, G.H. A Simple Method for the Isolation and Purification of Total Lipides from Animal Tissues. *J. Biol. Chem.* **1957**, *226*, 497–509. [CrossRef]

41. Batista, J.L.; Neves, R.X.; Peres, S.B.; Yamashita, A.S.; Shida, C.S.; Farmer, S.R.; Seelaender, M. Heterogeneous Time-Dependent Response of Adipose Tissue during the Development of Cancer Cachexia. *J. Endocrinol.* **2012**, *215*, 363–373. [CrossRef]

42. Rodbell, M. Metabolism of Isolated Fat Cells. I. Effects of Hormones on Glucose. *J. Biol. Chem.* **1964**, *239*, 375–380. [CrossRef]

43. Di Girolamo, M.; Mendlinger, S.; Fertig, J.W. A Simple Method to Determine Fat Cell Size and Number in Four Mammalian Species. *Am. J. Physiol.* **1971**, *221*, 850–858. [CrossRef]

44. De Oliveira Caminhotto, R.; Andreotti, S.; Komino, A.C.M.; de Fatima Silva, F.; Sertié, R.A.L.; Christoffolete, M.A.; Reis, G.B.; Lima, F.B. Physiological Concentrations of β-Hydroxybutyrate Do Not Promote Adipocyte Browning. *Life Sci.* **2019**, *232*, 116683. [CrossRef]

45. Sertié, R.A.L.; Andreotti, S.; Proença, A.R.G.; Campaña, A.B.; Lima, F.B. Fat Gain with Physical Detraining Is Correlated with Increased Glucose Transport and Oxidation in Periepididymal White Adipose Tissue in Rats. *Braz. J. Med. Biol. Res.* **2015**, *48*, 650–653. [CrossRef]

46. Sertie, R.A.L.; Curi, R.; Oliveira, A.C.; Andreotti, S.; Caminhotto, R.O.; de Lima, T.M.; Proença, A.R.G.; Reis, G.B.; Lima, F.B. The Mechanisms Involved in the Increased Adiposity Induced by Interruption of Regular Physical Exercise Practice. *Life Sci.* **2019**, *222*, 103–111. [CrossRef]

47. Liu, D.; Ahmet, A.; Ward, L.; Krishnamoorthy, P.; Mandelcorn, E.D.; Leigh, R.; Brown, J.P.; Cohen, A.; Kim, H. A Practical Guide to the Monitoring and Management of the Complications of Systemic Corticosteroid Therapy. *Allergy Asthma Clin. Immunol.* **2013**, *9*, 30. [CrossRef]

48. Fadel, R.; Morrison, A.R.; Vahia, A.; Smith, Z.R.; Chaudhry, Z.; Bhargava, P.; Miller, J.; Kenney, R.M.; Alangaden, G.; Ramesh, M.S.; et al. Early Short-Course Corticosteroids in Hospitalized Patients with COVID-19. *Clin. Infect. Dis.* **2020**, *71*, 2114–2120. [CrossRef] [PubMed]

49. Paragliola, R.M.; Papi, G.; Pontecorvi, A.; Corsello, S.M. Treatment with Synthetic Glucocorticoids and the Hypothalamus-Pituitary-Adrenal Axis. *Int. J. Mol. Sci.* **2017**, *18*, 2201. [CrossRef] [PubMed]

50. McKay, L.I.; Cidlowski, J.A. Corticosteroids. In *Holland-Frei Cancer Medicine*, 5th ed.; Bast, R.J., Kufe, D., Pollock, R., Eds.; Decker: Hamilton, ON, Canada, 2000; Chapter 54.

51. Scaroni, C.; Zilio, M.; Foti, M.; Boscaro, M. Glucose Metabolism Abnormalities in Cushing Syndrome: From Molecular Basis to Clinical Management. *Endocr. Rev.* **2017**, *38*, 189–219. [CrossRef] [PubMed]

52. Mir, N.; Chin, S.A.; Riddell, M.C.; Beaudry, J.L. Genomic and Non-Genomic Actions of Glucocorticoids on Adipose Tissue Lipid Metabolism. *Int. J. Mol. Sci.* **2021**, *22*, 8503. [CrossRef]

53. Zhao, L.F.; Iwasaki, Y.; Zhe, W.; Nishiyama, M.; Taguchi, T.; Tsugita, M.; Kambayashi, M.; Hashimoto, K.; Terada, Y. Hormonal Regulation of Acetyl-CoA Carboxylase Isoenzyme Gene Transcription. *Endocr. J.* **2010**, *57*, 317–324. [CrossRef]

54. Dolinsky, V.W.; Douglas, D.N.; Lehner, R.; Vance, D.E. Regulation of the Enzymes of Hepatic Microsomal Triacylglycerol Lipolysis and Re-Esterification by the Glucocorticoid Dexamethasone. *Biochem. J.* **2004**, *378*, 967–974. [CrossRef]

55. Bowles, N.P.; Karatsoreos, I.N.; Li, X.; Vemuri, V.K.; Wood, J.A.; Li, Z.; Tamashiro, K.L.K.; Schwartz, G.J.; Makriyannis, A.M.; Kunos, G.; et al. A Peripheral Endocannabinoid Mechanism Contributes to Glucocorticoid-Mediated Metabolic Syndrome. *Proc. Natl. Acad. Sci. USA* **2015**, *112*, 285–290. [CrossRef]

56. Christ-Crain, M.; Kola, B.; Lolli, F.; Fekete, C.; Seboek, D.; Wittmann, G.; Feltrin, D.; Igreja, S.C.; Ajodha, S.; Harvey-White, J.; et al. AMP-Activated Protein Kinase Mediates Glucocorticoid-Induced Metabolic Changes: A Novel Mechanism in Cushing's Syndrome. *FASEB J.* **2008**, *22*, 1672–1683. [CrossRef]

57. Lee, S.W.; Son, J.Y.; Kim, J.M.; Hwang, S.S.; Han, J.S.; Heo, N.J. Body Fat Distribution Is More Predictive of All-Cause Mortality than Overall Adiposity. *Diabetes Obes. Metab.* **2018**, *20*, 141–147. [CrossRef]

58. Schoettl, T.; Fischer, I.P.; Ussar, S. Heterogeneity of Adipose Tissue in Development and Metabolic Function. *J. Exp. Biol.* **2018**, *221*, jeb162958. [CrossRef]

59. Pleasure, D.E.; Walsh, G.O.; Engel, W.K. Atrophy of Skeletal Muscle in Patients With Cushing's Syndrome. *Arch. Neurol.* **1970**, *22*, 118–125. [CrossRef]

60. Ferrau, F.; Korbonits, M. Metabolic Comorbidities in Cushing's Syndrome. *Eur. J. Endocrinol.* **2015**, *173*, M133–M157. [CrossRef]

61. Fox, C.S.; Massaro, J.M.; Hoffmann, U.; Pou, K.M.; Maurovich-Horvat, P.; Liu, C.Y.; Vasan, R.S.; Murabito, J.M.; Meigs, J.B.; Cupples, L.A.; et al. Abdominal Visceral and Subcutaneous Adipose Tissue Compartments. *Circulation* **2007**, *116*, 39–48. [CrossRef]

62. Palmer, B.F.; Clegg, D.J. The Sexual Dimorphism of Obesity. *Mol. Cell. Endocrinol.* **2015**, *402*, 113–119. [CrossRef]

63. Smith, S.R.; Lovejoy, J.C.; Greenway, F.; Ryan, D.; De Jonge, L.; De La Bretonne, J.; Volafova, J.; Bray, G.A. Contributions of Total Body Fat, Abdominal Subcutaneous Adipose Tissue Compartments, and Visceral Adipose Tissue to the Metabolic Complications of Obesity. *Metab. -Clin. Exp.* **2001**, *50*, 425–435. [CrossRef]

64. Lee, M.J.; Pramyothin, P.; Karastergiou, K.; Fried, S.K. Deconstructing the Roles of Glucocorticoids in Adipose Tissue Biology and the Development of Central Obesity. *Biochim. Biophys. Acta -Mol. Basis Dis.* **2014**, *1842*, 473–481. [CrossRef]

65. Andersson, D.P.; Hogling, D.E.; Thorell, A.; Toft, E.; Qvisth, V.; Näslund, E.; Thörne, A.; Wirén, M.; Löfgren, P.; Hoffstedt, J.; et al. Changes in Subcutaneous Fat Cell Volume and Insulin Sensitivity after Weight Loss. *Diabetes Care* **2014**, *37*, 1831–1836. [CrossRef]

66. Hoffstedt, J.; Arner, E.; Wahrenberg, H.; Andersson, D.P.; Qvisth, V.; Löfgren, P.; Rydén, M.; Thörne, A.; Wirén, M.; Palmér, M.; et al. Regional Impact of Adipose Tissue Morphology on the Metabolic Profile in Morbid Obesity. *Diabetologia* **2010**, *53*, 2496–2503. [CrossRef]

67. Bahceci, M.; Gokalp, D.; Bahceci, S.; Tuzcu, A.; Atmaca, S.; Arikan, S. The Correlation between Adiposity and Adiponectin, TNF-Alpha, IL-6 and High Sensitivity CRP Protein Levels. *J. Endocrinol. Investig.* **2007**, *30*, 210–214. [CrossRef]

68. Skurk, T.; Alberti-Huber, C.; Herder, C.; Hauner, H. Relationship between Adipocyte Size and Adipokine Expression and Secretion. *J. Clin. Endocrinol. Metab.* **2007**, *92*, 1023–1033. [CrossRef] [PubMed]

69. Considine, R.V.; Sinha, M.K.; Heiman, M.L.; Kriauciunas, A.; Stephens, T.W.; Nyce, M.R.; Hannesian, J.P.O.; Marco, C.C.; Mckee, L.J.; Bauer, T.L.; et al. Serum Immunoreactive-Leptin Concentrations in Normal-Weight and Obese Humans. *N. Engl. J. Med.* **1996**, *334*, 292–295. [CrossRef] [PubMed]

70. Xu, C.; He, J.; Jiang, H.; Zu, L.; Zhai, W.; Pu, S.; Xu, G. Direct Effect of Glucocorticoids on Lipolysis in Adipocytes. *Mol. Endocrinol.* **2009**, *23*, 1161–1170. [CrossRef] [PubMed]

71. Arnaldi, G.; Scandali, V.M.; Trementino, L.; Cardinaletti, M.; Appolloni, G.; Boscaro, M. Pathophysiology of Dyslipidemia in Cushing's Syndrome. *Neuroendocrinology* **2010**, *92*, 86–90. [CrossRef] [PubMed]

72. Coleman, R.A.; Lee, D.P. Enzymes of Triacylglycerol Synthesis and Their Regulation. *Prog. Lipid Res.* **2004**, *43*, 134–176. [CrossRef]

73. Proença, A.R.G.; Sertié, R.A.L.; Oliveira, A.C.; Campaña, A.B.; Caminhotto, R.O.; Chimin, P.; Lima, F.B. New Concepts in White Adipose Tissue Physiology. *Braz. J. Med. Biol. Res.* **2014**, *47*, 192–205. [CrossRef]

74. Lafontan, M. Advances in Adipose Tissue Metabolism. *Int. J. Obes.* **2008**, *32*, S39–S51. [CrossRef]

75. Saponaro, C.; Gaggini, M.; Carli, F.; Gastaldelli, A. The Subtle Balance between Lipolysis and Lipogenesis: A Critical Point in Metabolic Homeostasis. *Nutrients* **2015**, *7*, 9453–9474. [CrossRef]

76. Nye, C.; Kim, J.; Kalhan, S.C.; Hanson, R.W. Reassessing Triglyceride Synthesis in Adipose Tissue. *Trends Endocrinol. Metab.* **2008**, *19*, 356–361. [CrossRef]

77. Nye, C.K.; Hanson, R.W.; Kalhan, S.C. Glyceroneogenesis Is the Dominant Pathway for Triglyceride Glycerol Synthesis in Vivo in the Rat. *J. Biol. Chem.* **2008**, *283*, 27565–27574. [CrossRef]

78. Newsholme, E.A.; Dimitriadis, G. Integration of Biochemical and Physiologic Effects of Insulin on Glucose Metabolism. *Exp. Clin. Endocrinol. Diabetes* **2001**, *109*, 122–134. [CrossRef]

79. Dimitriadis, G.; Mitrou, P.; Lambadiari, V.; Maratou, E. Insulin Effects in Muscle and Adipose Tissue. *Diabetes Res. Clin. Pract.* **2011**, *93*, 52–59. [CrossRef]

80. Kobayashi, N.; Ueki, K.; Okazaki, Y.; Iwane, A.; Kubota, N.; Ohsugi, M. Blockade of Class IB Phosphoinositide-3 Kinase Ameliorates Obesity-Induced in Fl Ammation and Insulin Resistance. *Proc. Natl. Acad. Sci. USA* **2011**, *108*, 5753–5758. [CrossRef]

81. Shimizu, I.; Walsh, K. The Whitening of Brown Fat and Its Implications for Weight Management in Obesity. *Curr. Obes. Rep.* **2015**, *4*, 224–229. [CrossRef]

82. Hill, B.G. Insights into an Adipocyte Whitening Program. *Adipocyte* **2015**, *4*, 75–80. [CrossRef]

83. Bel, J.S.; Tai, T.C.; Khaper, N.; Lees, S.J. Chronic Glucocorticoid Exposure Causes Brown Adipose Tissue Whitening, Alters Whole-Body Glucose Metabolism and Increases Tissue Uncoupling Protein-1. *Physiol. Rep.* **2022**, *10*, e15292. [CrossRef]

84. Mousovich-Neto, F.; Matos, M.S.; Costa, A.C.R.; de Melo Reis, R.A.; Atella, G.C.; Miranda-Alves, L.; Carvalho, D.P.; Ketzer, L.A.; Corrêa da Costa, V.M. Brown Adipose Tissue Remodelling Induced by Corticosterone in Male Wistar Rats. *Exp. Physiol.* **2019**, *104*, 514–528. [CrossRef]

85. Poggioli, R.; Ueta, C.B.; Drigo, R.A.E.; Castillo, M.; Fonseca, T.L.; Bianco, A.C. Dexamethasone Reduces Energy Expenditure and Increases Susceptibility to Diet-Induced Obesity in Mice. *Obesity* **2013**, *21*, 415–420. [CrossRef]

86. Scotney, H.; Symonds, M.E.; Law, J.; Budge, H.; Sharkey, D.; Manolopoulos, K.N. Glucocorticoids Modulate Human Brown Adipose Tissue Thermogenesis in Vivo. *Metabolism* **2017**, *70*, 125–132. [CrossRef]

87. Van Den Beukel, J.C.; Boon, M.R.; Steenbergen, J.; Rensen, P.C.N.; Meijer, O.C.; Themmen, A.P.N.; Grefhorst, A. Cold Exposure Partially Corrects Disturbances in Lipid Metabolism in a Male Mouse Model of Glucocorticoid Excess. *Endocrinology* **2015**, *156*, 4115–4128. [CrossRef]

88. Cannon, B.; Nedergaard, J. The Biochemistry of an Inefficient Tissue: Brown Adipose Tissue. *Essays Biochem.* **1985**, *20*, 110–164.

89. Murano, I.; Barbatelli, G.; Giordano, A.; Cinti, S. Noradrenergic Parenchymal Nerve Fiber Branching after Cold Acclimatisation Correlates with Brown Adipocyte Density in Mouse Adipose Organ. *J. Anat.* **2009**, *214*, 171–178. [CrossRef] [PubMed]

90. Soumano, K.; Desbiens, S.; Rabelo, R.; Bakopanos, E.; Camirand, A.; Silva, J.E. Glucocorticoids Inhibit the Transcriptional Response of the Uncoupling Protein-1 Gene to Adrenergic Stimulation in a Brown Adipose Cell Line. *Mol. Cell. Endocrinol.* **2000**, *165*, 7–15. [CrossRef]

91. McNeill, B.T.; Morton, N.M.; Stimson, R.H. Substrate Utilization by Brown Adipose Tissue: What's Hot and What's Not? *Front. Endocrinol.* **2020**, *11*, 571659. [CrossRef] [PubMed]

92. Berbeé, J.F.P.; Boon, M.R.; Khedoe, P.P.S.J.; Bartelt, A.; Schlein, C.; Worthmann, A.; Kooijman, S.; Hoeke, G.; Mol, I.M.; John, C.; et al. Brown Fat Activation Reduces Hypercholesterolaemia and Protects from Atherosclerosis Development. *Nat. Commun.* **2015**, *6*, 6356. [CrossRef]

93. Sanchez-Gurmaches, J.; Tang, Y.; Jespersen, N.Z.; Wallace, M.; Martinez Calejman, C.; Gujja, S.; Li, H.; Edwards, Y.J.K.; Wolfrum, C.; Metallo, C.M.; et al. Brown Fat AKT2 Is a Cold-Induced Kinase That Stimulates ChREBP-Mediated De Novo Lipogenesis to Optimize Fuel Storage and Thermogenesis. *Cell Metab.* **2018**, *27*, 195–209.e6. [CrossRef]

94. Petersen, C.; Nielsen, M.D.; Andersen, E.S.; Basse, A.L.; Isidor, M.S.; Markussen, L.K.; Viuff, B.M.; Lambert, I.H.; Hansen, J.B.; Pedersen, S.F. MCT1 and MCT4 Expression and Lactate Flux Activity Increase during White and Brown Adipogenesis and Impact Adipocyte Metabolism. *Sci. Rep.* **2017**, *7*, 13101. [CrossRef]

95. Saggerson, E.D.; McAllister, T.W.J.; Baht, H.S. Lipogenesis in Rat Brown Adipocytes. Effects of Insulin and Noradrenaline, Contributions from Glucose and Lactate as Precursors and Comparisons with White Adipocytes. *Biochem. J.* **1988**, *251*, 701–709. [CrossRef]
96. Liu, C.; Wu, J.; Zhu, J.; Kuei, C.; Yu, J.; Shelton, J.; Sutton, S.W.; Li, X.; Su, J.Y.; Mirzadegan, T.; et al. Lactate Inhibits Lipolysis in Fat Cells through Activation of an Orphan G-Protein-Coupled Receptor, GPR81. *J. Biol. Chem.* **2009**, *284*, 2811–2822. [CrossRef]

biomedicines

Article

Taste-Driven Responsiveness to Fat and Sweet Stimuli in Mouse Models of Bariatric Surgery

Aurélie Dastugue [1], Cédric Le May [2], Séverine Ledoux [3], Cindy Le Bourgot [4], Pascaline Delaby [5], Arnaud Bernard [1] and Philippe Besnard [1,6,*]

1. UMR 1231, INSERM/University of Bourgogne-Franche-Comté/AgroSup Dijon, 21000 Dijon, France; aurelie.dastugue@agrosupdijon.fr (A.D.); arnaud.bernard@u-bourgogne.fr (A.B.)
2. Institut du Thorax, CNRS, INSERM, University of Nantes, 44000 Nantes, France; cedric.lemay@univ-nantes.fr
3. Explorations Fonctionnelles, Hôpital Louis Mourier (APHP), 92700 Colombe, France; severine.ledoux@aphp.fr
4. Tereos, 77230 Moussy-le-Vieux, France; cindy.lebourgot@tereos.com
5. Lesieur, 92600 Asnières/Seine, France; pdelaby@lesieur.fr
6. Physiologie de la Nutrition, AgroSup Dijon, 21000 Dijon, France
* Correspondence: pbesnard@u-bourgogne.fr; Tel.: +33-380-774-091

Abstract: A preferential consumption of healthier foods, low in fat and sugar, is often reported after bariatric surgery, suggesting a switch of taste-guided food choices. To further explore this hypothesis in well-standardized conditions, analysis of licking behavior in response to oily and sweet solutions has been realized in rats that have undergone a Roux-en-Y bypass (RYGB). Unfortunately, these studies have produced conflicting data mainly due to methodological differences. Paradoxically, whereas the vertical sleeve gastrectomy (VSG) becomes the most commonly performed bariatric surgery worldwide and is easier to perform and standardize in small animals, its putative impacts on the orosensory perception of energy-dense nutrients remains unknown. Using brief-access licking tests in VSG or RYGB mice, we found that (i) VSG induces a significant reduction in the fat mass in diet-induced obese (DIO) mice, (ii) VSG partially corrects the licking responses to lipid and sucrose stimuli which are degraded in sham-operated DIO mice, (iii) VSG improves the willingness to lick oily and sucrose solutions in DIO mice and (iv) RYGB leads to close outcomes. Altogether, these data strongly suggest that VSG, as RYGB, can counteract the deleterious effect of obesity on the orosensory perception of energy-dense nutrients in mice.

Keywords: obesity; bariatric surgery; gustation; brief-access licking tests; fat and sugar

Citation: Dastugue, A.; Le May, C.; Ledoux, S.; Le Bourgot, C.; Delaby, P.; Bernard, A.; Besnard, P. Taste-Driven Responsiveness to Fat and Sweet Stimuli in Mouse Models of Bariatric Surgery. *Biomedicines* **2022**, *10*, 741. https://doi.org/10.3390/biomedicines10040741

Academic Editor: Martina Perše

Received: 28 January 2022
Accepted: 18 March 2022
Published: 22 March 2022

Publisher's Note: MDPI stays neutral with regard to jurisdictional claims in published maps and institutional affiliations.

1. Introduction

The prevalence of morbid obesity (BMI $\geq$ 40 Kg/m^2) has increased dramatically in the past decades worldwide, becoming a major public health priority. Due to multiple associated co-morbidities, such as type 2 diabetes, hypertension, cancer, heart and neurodegenerative diseases [1], severe obesity is at the origin of a reduction in the life expectancy ranging from 5 to 20 years [2,3], this trend being all the more dramatic when obesity is precocious [4].

Among the treatments of obesity, bariatric surgery appears to be currently the most efficient method to reduce body weight and correct the associated metabolic disorders in patients with morbid obesity. The different surgery procedures developed have been frequently divided into two categories: the restrictive methods that encompass the adjustable gastric banding, the vertical banded gastroplasty and the vertical sleeve gastrectomy (VSG), and the malabsorptive approaches including the biliopancreatic diversion, the biliointestinal bypass and the Roux-en-Y gastric bypass (RYGB). However, this simplistic distinction proved insufficient to explain the strong variability in the long-term benefits of these surgery procedures, the mechanisms of action being multiple and complex.

Two surgical methods predominate in clinical practice: the VSG and the RYGB. By reason of its long-term effectiveness, RYGB has long been regarded as the bariatric gold standard [5]. This procedure consists of the creation of a small gastric pouch with a new alimentary limb ensured by a Roux-en-Y gastro-jejunostomy, the biliary and pancreatic juice transit being maintained by a jejuno-jejunostomy (biliopancreatic limb), allowing their flow into the remaining small intestine (common channel) [6]. Despite its efficiency, RYGB is gradually being supplanted as the first surgical intervention by VSG that consists of realizing a tube-like structure by removing 75% to 80% of the stomach. This less invasive and challenging procedure combines metabolic benefits and efficient, but sometimes transient, weight loss with a low postoperative complication rate [7]. In 2019, The American Society for Metabolic and Bariatric Surgery (ASMBS) estimated that nearly two-thirds of obesity surgery realized in the USA was VSG.

Paradoxically, how RYGB and VSG work is not fully understood, likely due to the complexity of homeostatic induced changes and the difficulty of linking mechanistic modifications to functional consequences in humans. To overcome these limitations and clarify this question, rat and mouse models of RYGB and VSG have been adapted and generated successfully. The adaptation of these surgeries in mice requires great expertise, high surgical dexterity due to the small size of animals, and adapted specific pre and postoperative protocols of anesthesia/analgesia/nutritional care [8]. From a surgical point of view, the VSG procedure is relatively similar between mice and humans, is well tolerated and the postsurgical survival rate is relatively good in most published studies (reviewed in [9]. However, due to differences in stomach musculature, RYGB is more difficult to transpose from humans to mice and several different models have been described [10]. The mortality rate of the different surgical setups is only reported in 50% of published studies and the average is 29% (reviewed in [9]).

Despite anatomical and physiological differences between rodents and humans (e.g., lack of gallbladder in rats and of the forestomach in humans, species specificities of bile acid metabolism), important functional postoperative similarities were found between these animal models and patients, including not only positive outcomes as fat mass loss, induction of gut satiety hormone secretion, improvement of gut microbiota composition and of insulin sensitivity, but also negative impacts, such as post-RYGB vitamin and iron deficiencies, bone demineralization and alcohol addiction (for review see [9,11]). Although the translational jump from rodent data to human physiology requires caution, these rodent models of bariatric surgery constitute a useful tool in providing mechanistic hypothesis essential to better understand how bariatric surgery corrects homeostatic dysfunctions induced by obesity.

Among the multiple postoperative changes, a preferential consumption for low-caloric foods is often reported by patients who have undergone RYGB or VSG (for reviews, see [12–14]). This healthier food selection, which might play a significant role in the success of long-term weight loss, raises the possibility that obesity surgery might also affect the orosensory perception of energy nutrients (i.e., sugar and fat). However, the physiological relevance of this observation is not yet clearly established. Indeed, most of the published clinical data were generated using indirect approaches, such as self-reported questionnaires [14], which can often generate biased responses, especially in patients with obesity [15]. Moreover, the analysis of the taste sensitivity using direct sensory methods, such as alternative forced-choice tests or taste strips, led to discrepant results [14], likely due to the great genetic variability and food habit heterogeneity between patients. To further investigate this question in well-controlled experimental conditions, sweet taste sensitivity has been compared in sham-operated controls and in RYGB rats using brief-access licking tests (e.g., 10 s). The advantage of this behavioral test was to provide information about the taste-driven licking activity (i.e., immediate pleasure or "liking") and motivation (incentive salience or "wanting") in response to an oral stimulus independent of post-ingestive cues. Unfortunately, these studies led to controversial outcomes mainly due to methodological differences (Table 1), with some of them reporting a positive correlation between the licking rate and the sucrose concentration, suggesting that RYGB improves sweetness sensitivity [16], while others did not see any change [17–19] or even found the opposite result [20,21]. Similarly, the relationship between RYGB and an orosensory perception of dietary lipids also remains unclear, with studies being scarce and conflicting.

Table 1. Impact of RYGB on the taste-driven responses to sweet and fat stimuli in rats.

Strains	Obesity	Surgery	Methods	Findings	References
Male Sprague rats Sham = 11, Surg = 11, Lean = 7	Yes (HFD for 14–16 weeks)	RYGB 20% gastr pouch Alim limb ± 15 cm Bilio-pancr lim ± 40 cm Comm limb ± 25 cm	Licking tests (Davis MS160, trial = 10 s, session = 30 min, 4-6 months after RYGB). Sucrose or corn oil stimuli in random order. Fed animals	RYGB rats display licking profiles close to lean controls for low concentrations of sucrose or corn oil	[16]
Male Sprague rats, Sham = 7, Surg = 7	No (Std diet)	RYGB Gastr pouch remnant Alim limb ± 50 cm Bilio-pancr lim ± 20 cm Comm limb ± 25–30 cm	Licking tests (Davis MS160, trial = 10 s, session = 30 min, 1–1.5 months after RYGB). Sucrose stimuli in random order. Water restriction	No ≠ on the sucrose lick scores between Sham and RYGB groups with or without 23 h fasting	[18]
Female Sprague rats Sham = 7, Surg = 7	No (Std diet)	RYGB Gastr pouch remnant Alim limb ± 50 cm Bilio-pancr lim ± 20 cm Comm limb ± 25–30 cm	Licking tests (Davis MS160 trial = 10 s, session = 30 min, 2–3 months after RYGB). Sucrose stimuli in random order. Water restriction	No attenuation of licking response to a concentration series of sucrose solutions	[19]
Male Long-Evans rats Oletf-RYGB = 6 Pair-fed Oletf-Sham = 4 Leto-RYGB = 6 Pair-fed Leto-Sham = 4	Yes (Genetic CCK-R$^{-/-}$)	RYGB 20% gastr pouch Alim limb ± 10 cm Bilio-pancr lim ± 15 cm Comm limb ± 25–30 cm	Licking tests (Davis MS160 trial = 10 s, session = 20 min, 1 months after RYGB). Sucrose stimuli in random order. Water restriction	RYGB reduces the licking response to high sucrose in OLEF (= obese) rats in contrast to LETO (= lean) rats	[20]
Male Sprague rats Sham = 11, Surg = 9	No (Std diet)	RYGB 10% gastr pouch Alim limb ± 15 cm, Bilio-pancr lim ± 20–25 cm, Comm limb ± 50 cm	Licking tests (Davis MS160 trial = 10 s, session = 30 min, 1 months after RYGB). Sucrose stimuli in random order. Water restriction	RYGB decreases the number of licks for highest sucrose concentrations	[21]
Male Wistar rats Sham = 8, Surg = 8	No (Std diet)	RYGB Gastr pouch remnant Alim limb ± 50 cm Bilio-pancr lim ± 20 cm Comm limb ± 25–30 cm	Licking tests (Davis MS160 trial = 10 s, session = 30 min, 5 months after RYGB). Intralipid stimuli in random order. Water restriction	No ≠ on the Intralipid lick scores between Sham and RYGB groups with or without water restriction	[22]

Sham—sham-operated controls; Std—standard laboratory chow; HFD—high fat diet; RYGB— Roux-en-Y gastric bypass; CCK-1 R$^{-/-}$—cholecystokinin-1 receptor null-mice; Gastr—gastric; Alim—alimentary; Bilio-pancr—bilio-pancreatic; Comm—common.

Le Roux et al. found no difference in the Intralipid^{TM} lick scores between RYGB rats and sham-operated controls [22], while Shin et al. reported an improvement in the licking responses to corn oil after RYGB [16]. Finally, although the VSG is also associated with healthier changes in the food selection in rats [23–25], the potential impact of this bariatric procedure on the orosensory perception of fat and sweet stimuli is unknown.

The purpose of the present study was to evaluate the respective effects of VSG on the short-term licking responses to fat and sweet stimuli. Brief-access tests were also performed after RYGB in reference to the previously published studies. We have chosen to conduct experiments in the mouse whose multiple transgenic models make possible further studies on the potential mechanisms linking obesity, bariatric surgery and orosensory perception of energy-dense foods. Therefore, the present study was designed according to two complementary goals: explore, for the first time, the effects of VSG on the licking behavior in response to fat and sweet stimuli in obese animals and re-study the role of RYGB on this behavioral parameter, previous data being conflicting. Two distinct cohorts of mice with their own lean and obese controls were used. Bariatric surgery (VSG or RYGB) was performed in diet-induced obese (DIO) mice (VSG-DIO and RYGB-DIO). Since lipid content of diets affects the taste perception independently of body weight changes in rodents [26], mice that underwent VSG or RYGB were maintained on HFD after surgery and were compared to the Sham DIO controls in order to explore whether the bariatric surgery per se may modify the taste perception.

2. Materials and Methods

2.1. Animals

Experiments were performed in accordance with the European guidelines for the care and use of laboratory animals. Protocols were approved by the French National Animal Ethics Committee (APAFIS#12728-201712151028893 and APAFIS#19877-2019032109244847). For the estrous cycle affecting the taste reactivity response [27], experiments were only conducted in male mice. Six-week-old C57Bl/6J mice (Charles River Laboratories, Eculie, France) were housed individually in a conventional cage (Eurostandard type II, floor area = 530 cm^2, Tecniplast, France; bedding material, Scobis Uno, Mucedola SRL, Italia; enrichment: Rodis Uno, Mucedola SRL, Settimo Milanese, Italy) in a controlled environment with a constant temperature (23 $\pm$ 1 °C), humidity (60 $\pm$ 5%), a dark period from 7 p.m. to 7 a.m., with free access to tap water and chow. Mice were fed either a standard laboratory chow (Sdt-4RF21 autoclaved, 315.0 Kcal/100 g, Mucedola SRL, Settimo Milanese, Italy) or an obesogenic high fat diet (HFD-4RF25 autoclaved, Mucedola SRL, Settimo Milanese, Italy, + 31.8% palm oil, wt/wt, 505.4 Kcal/100 g). At the end of the experiments, the fat mass was determined by molecular resonance imaging (EchoMRI-Echo Medical Systems, Houston, TX, USA, Figure 1A).

2.2. Surgical Procedures and Postoperative Care

The surgical methods used have been fully described and video-captured elsewhere [8]. Prior surgery, all mice received analgesics (buprenorphine 0.1 mg/kg), antibiotics (marbofloxacin 10 mg/kg) and pro-kinetics (metoclopramide 1 mg/kg) via subcutaneous (sc) injections. Anesthesia was induced in a chamber with 5% isoflurane, (air flow of 0.4 L/min and O$_2$ flow 0.4 L/min). The VSG procedure consists of removing a large part of the stomach by discarding the forestomach and a part of the corpus (See Figure 1A). Briefly, after performing a median laparotomy, we sutured the pylorus vessels along the stomach's greater curvature with 8.0 Prolene single sutures to avoid future bleeding. We next performed a resection of the cardiac (forestomach) and pyloric region (corpus) to discard approximately 80% of the stomach. The incision site was sutured with 8.0 Prolene running suture from the gastro-oesophagal origin to the end of the incision. For the RYGB procedure (Figure 1B), after externalization of the first small intestinal loop (duodenum), a double ligature with 5.0 Prolene was realized 4 cm after the end of the duodenum in order to transect the proximal jejunum between the two ligatures. The distal part of the proximal jejunum was moved upon realizing an anastomosis with the stomach which constitutes the alimentary limb. The proximal part of the jejunum was used to perform a side-to-side anastomosis with the median jejunal segment. This Y-shaped anatomical rearrangement leads to a biliopancreatic limb, the remained jejunum constituting the common limb (Figure 1B).

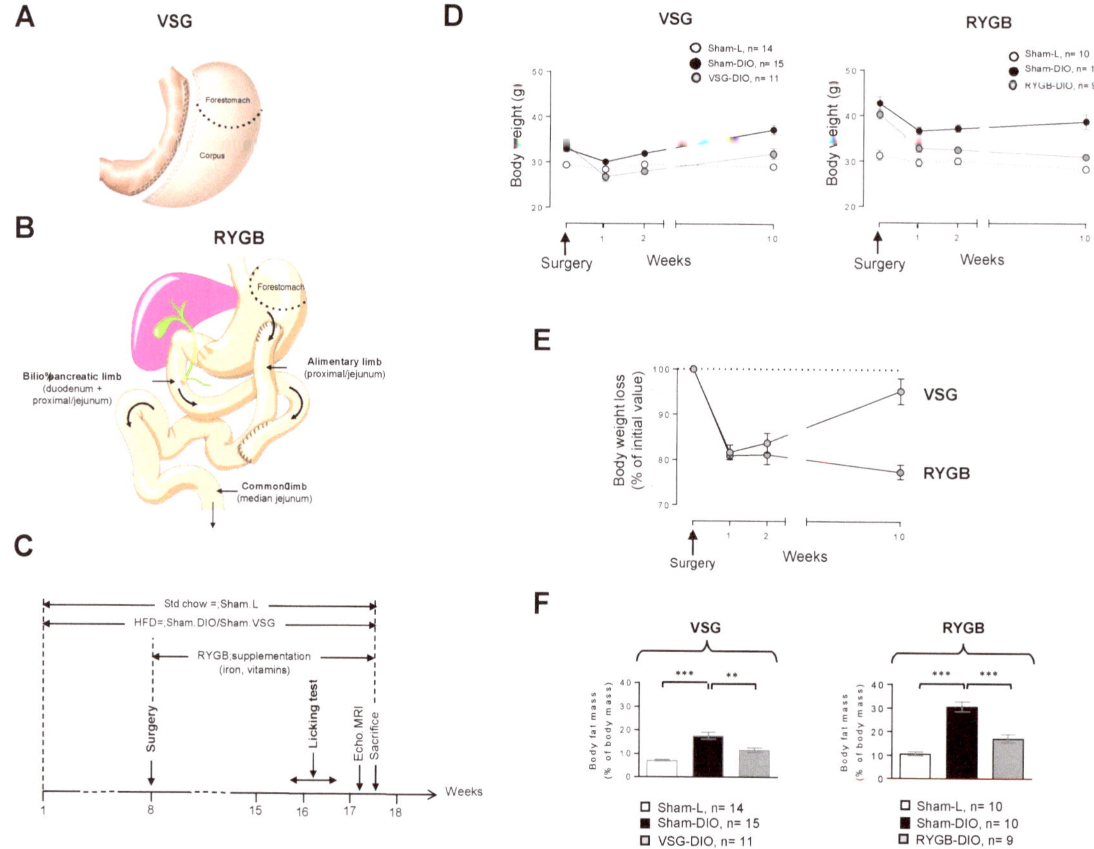

Figure 1. (**A**) Vertical sleeve gastrectomy (VSG) procedure used. (**B**) Roux-en-Y gastric bypass (RYGB) procedure used. (**C**) Time-course of the experiment. (**D**) Evolution of body weight after surgery. (**E**) Evolution of body weight loss after surgery. (**F**) Fat mass at the end of the experiment. Sham-operated lean controls (Sham-L); sham-operated obese controls (Sham-DIO); obese mice that underwent a vertical sleeve gastrectomy (VSG-DIO) or a Roux-en-Y gastric bypass (RYGB-DIO). Mean ± SEM, Wilcoxon and Mann–Whitney tests: ** $p < 0.01$, *** $p < 0.001$. Std—standard laboratory chow; HFD—high fat diet; DIO—diet-induced obesity.

During these two surgery procedures, special care has been taken to avoid any damage to neural and vascular systems, especially at the level of esophagogastric and pyloric areas. The mean operative time for the VSG procedure was 33.5 ± 1.4 min and 60.33 ± 0.88 min for RYGB. Lean and obese controls were sham-operated, taking into account the operative time for each of the surgeries. In our hands, while we have rarely observed deaths following a VSG procedure, the mortality rate in our modified RYGB mouse model varies from 0% to 50%, depending on the training and dexterity of the surgeon. Importantly, all the deaths observed following RYGB occurred in the 5 days after surgery and were mostly due to anastomotic leakage intestinal stenosis. After surgery, all mice received analgesics (buprenorphine 0.1 mg/kg twice daily, from day 0 to day 3 after surgery), antibiotics (marbofloxacin 10 mg/kg from day 0 to day 3 after surgery), and pro-kinetics (metoclopramide 1 mg/kg from day 0 to day 5 after surgery) via sc injections. For gastric bypass inducing vitamin malabsorption and anemia [9], RYGB mice were daily supplemented with poly-vitamins (800 mg/180 mL in drinking water) and iron (0.5 mg/kg/day by sc injection, Figure 1C). Sham-controls received the vehicle alone. For five days after surgery, mice were housed in a 30 °C room and received a gel diet food (Geldiet Highfat, Safe laboratories; lard 10%, liquid sugar 10%, water 57%). Solid food was reintroduced 3 days after surgery.

2.3. Brief-Access Licking Tests

Licking tests were performed using an original octagonal-shaped gustometer with eight bottles, each being equipped with a lickometer and a computer-controlled shutter allowing random access during a short time (10 s in the present study) to bottles filled with a specific solution [28]. By forcing the animal to move to have access to the drinking source, this system allows the analysis of licking behavior which mirrors the immediate pleasure gained from the consumption of a rewarding stimulus (i.e., "liking") and brings information on the motivation to drink (i.e., "wanting"). The concept and procedures are detailed elsewhere [28]. In brief, each of the mice was individually subjected to two training sessions before the taste-testing sessions. The first training is a time of habituation to the apparatus environment with free access to the eight bottles filled with water. Analysis of the number of licks on each bottle provided information about the exploration capacity (curiosity, stress) of the animal, the existence of a bottle location preference, and the rhythmic oromotor activity. During the second training, the mouse learns to drink with random access to the eight bottles as during the taste-testing sessions, with the difference that the only solution available is water. The animal has access to a first bottle for 10 s after the first lick. After this first trial, all shutters remain closed for 10 s before another one is opened in a randomized manner among the seven remaining shutters. The time taken by the mouse to perform the first lick was defined as the latency. Training and taste-testing sessions were computer-controlled to start at the first lick and stop 30 min later. The mouse can initiate as many trials as it wants during this session time. When the mouse has licked the eight bottles (eight trials), it has realized a block. At the end of one block, another block started, so that the number of blocks, and thus of trials, mirrors the motivation for the stimulus. To avoid a memory bias, the shutter opening was designed according to a computer random draw for each block. An animal is excluded from the data if it is unable to realize at least one block in 30 min. To promote licking, mice were overnight water-restricted before each session according to previously published data [18,21,22,29], the experiments being performed in the morning (Figure 1A). To avoid the satiety signals influencing the initial rate of licking [29,30], mice were also food deprived during the dark period before the taste-testing sessions. The restriction led to body mass variations being too limited ($-7.7 \pm 0.3\%$ of initial body weight on average) to induce physiological disturbances usually associated with dehydration [31]. This body mass change was followed by a nearly full weight regain during the resting time between two successive sessions. No noticeable distress signs were observed in any of the mice. Taste testing sessions were performed using sucrose (seven concentrations from 0.01 to 0.6 M in water), then rapeseed oil (seven concentrations from 0.03% to 4% in water plus 0.3 xanthan gum to facilitate oil solubilization and minimize textural cues). For each of these sessions, the control solution was the vehicle alone (water alone and water + 0.3% xanthan gum for sucrose and rapeseed oil, respectively).

2.4. Statistical Analysis

The statistical tests were performed using R software (v 3.4.4, the R Foundation, Vienna, Austria) with an α level of 0.05. Data not having a normal distribution and variances being not equal, means the difference between groups were analyzed using non-parametric tests (Wilcoxon–Mann–Whitney). The statistical analysis was performed using the obese mice (Sham-DIO) as a reference group, thus providing information on both the efficiency of the obesogenic diet (Sham-L vs. Sham-DIO) and bariatric surgery (Sham-DIO vs. VSG-DIO or RYGB-DIO) on studied parameters. Therefore, lean controls (Sham-L) and obese mice that underwent VSG or RYGB have not been compared since these groups cumulate two types of differences (diet and surgery). A principal component analysis (PCA), normalized and centered, was done with the R-commander package (v2.4.4) using all studied parameters.

3. Results

3.1. VSG Reduces the Fat Mass despite a Transient Weight Loss

The time-course of the experiment is shown in Figure 1C. Animals were maintained on the same diet throughout the experiment, standard chow for the Sham-L groups and HFD for Sham-DIO, RYGB-DIO, and VSG-DIO animals. Surgery was performed 2 months after the beginning of the protocol. The evolution of body weights after surgery is shown in Figure 1D. By contrast to RYGB, the weight loss after VSG was transient, with mice tending to return to their preoperative weight 10 weeks after surgery (Figure 1E). Despite this difference, VSG as RYGB deeply affected the body composition by reducing the fat mass as compared to their respective sham-DIO controls (Figure 1F).

3.2. VSG Improves the Licking Responses to Oily and Sucrose Solutions, as RYGB

In order to study the taste-driven responses to lipid and sweet stimuli, brief-access licking tests (licks/10 s for 30 min) were carried out using an octagonal-shaped gustometer (Figure 2A) according to the protocol shown in Figure 2B.

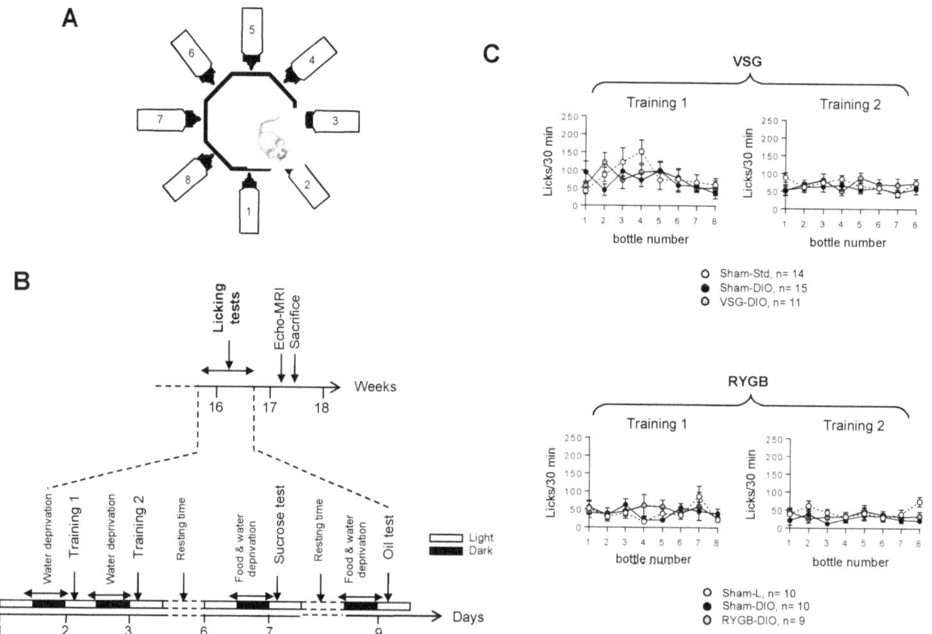

Figure 2. Analysis of the licking behavior in sham-operated lean (Sham-L) and obese (Sham-DIO) controls and in obese mice that underwent a vertical sleeve gastrectomy (VSG-DIO) or a Roux-en-Y gastric bypass (RYGB-DIO) during the training sessions. (**A**) Design of the gustometer. (**B**) Time-course of licking tests session. (**C**) Analysis number of licks per bottle during the training sessions (30 min). Training 1: mice have free access to the eight bottles filled with water. Training 2: mice are subjected to random and intermittent access (10 s) to each of the eight bottles filled with water. Mean ± SEM, Wilcoxon and Mann–Whitney tests. DIO—diet-induced obesity.

The training sessions failed to reveal substantial behavioral differences among the three groups of mice whatever the surgical procedures used, showing that the experimental design did not affect the animals' adaptability to a new environment (training 1) nor their ability to learn how the device works (training 2, Figure 2C). They also suggest that all the mice displayed similar oromotor and mobility activities.

In Sham-L (VSG or RYGB) mice, the licking activity showed a typical dose–response curve to growing concentrations of oily emulsion (Figure 3A) and sucrose solution (Figure 3B).

As previously observed [16], the licking rates decreased dramatically in Sham-DIO mice (VSG or RYGB) as compared to the lean controls (Sham-L), confirming that obese animals are unable to perceive properly fat and sweet stimuli. By contrast, DIO mice that have undergone VSG or RYGB displayed a greater licking activity than their respective obese controls (Sham-DIO). This sensory improvement was substantial, as pointed out by the comparison of both the area under curves (AUC, inserts Figure 3) and the total licks number during the 30 min session. However, this correction was partial, as the licking rates of VSG and RYGB mice remained lower than those found in their respective lean controls (Sham-L).

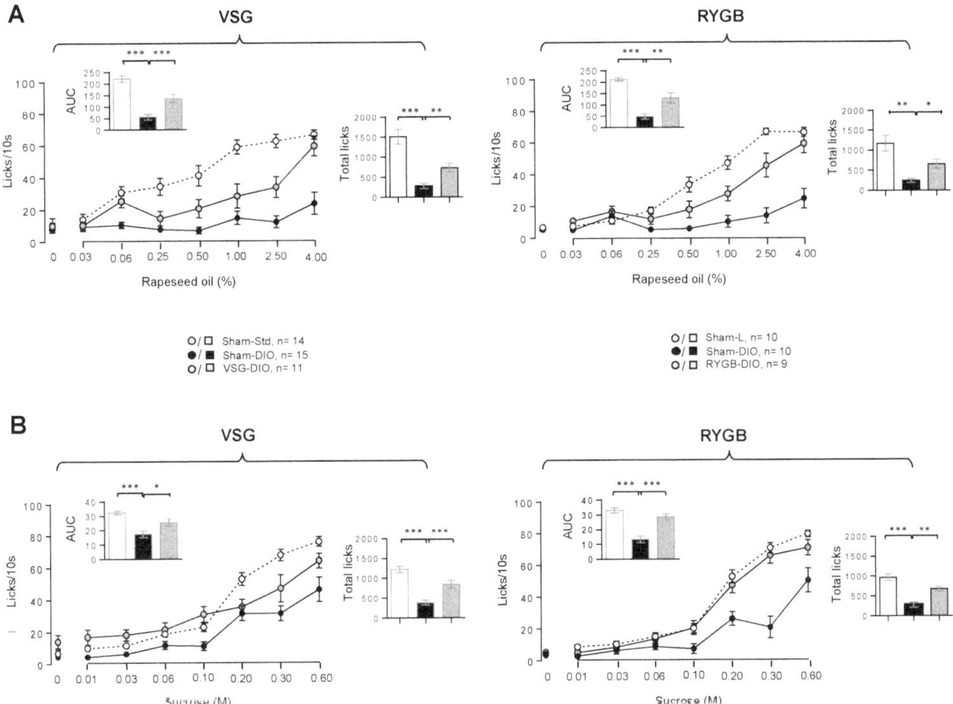

Figure 3. Taste-driven responsiveness to rapeseed oil (**A**) and sucrose (**B**) in sham-operated lean (Sham-L) and obese (Sham-DIO) controls and in obese mice that underwent a vertical sleeve gastrectomy (VSG-DIO) or a Roux-en-Y gastric bypass (RYGB-DIO). Brief-access taste-testing responses (licks/10 s) to various concentrations of the oily emulsion or sucrose solution were presented randomly. Zero on the X-axis represents the licking rate in response to the control solution without oil or sucrose. Analysis of both areas under curves (AUC, Inserts) and total licks during the 30 min of taste-testing sessions were performed using the Sham-DIO group as reference. Mean ± SEM. * $p < 0.05$, ** $p < 0.01$, *** $p < 0.001$. DIO—diet-induced obesity.

3.3. VSG Improves the Willingness to Lick Oily and Sucrose Solutions in DIO Mice

To explore the motivational component of the taste-driven responsiveness to fat and sweet stimuli, the latency (i.e., the time to initiate the first lick following the access to a solution), and the number of trials (the number of different solutions licked in 30 min) was determined. Latency revealed motivational divergences between groups (Figure 4): the time-response to oily stimuli being greater in obese mice than in lean controls, VSG (and RYGB) animals being in an intermediate position (Figure 4A). While the latency was significantly reduced in the VSG group, only a downward trend was found in the RYGB group (Figure 4A, inserts). Similar changes were also found in response to various sucrose concentrations (Figure 4B). Sham-DIO mice also initiated a fewer number of trials during the 30 min taste-testing session than Sham-L mice, both in response to oily (Figure 4A) or sucrose (Figure 4B) solutions. This relative disinterest was partially reversed in the VSG group, whatever the stimuli tested, but was significant only for sucrose in the RYGB group (Figure 4).

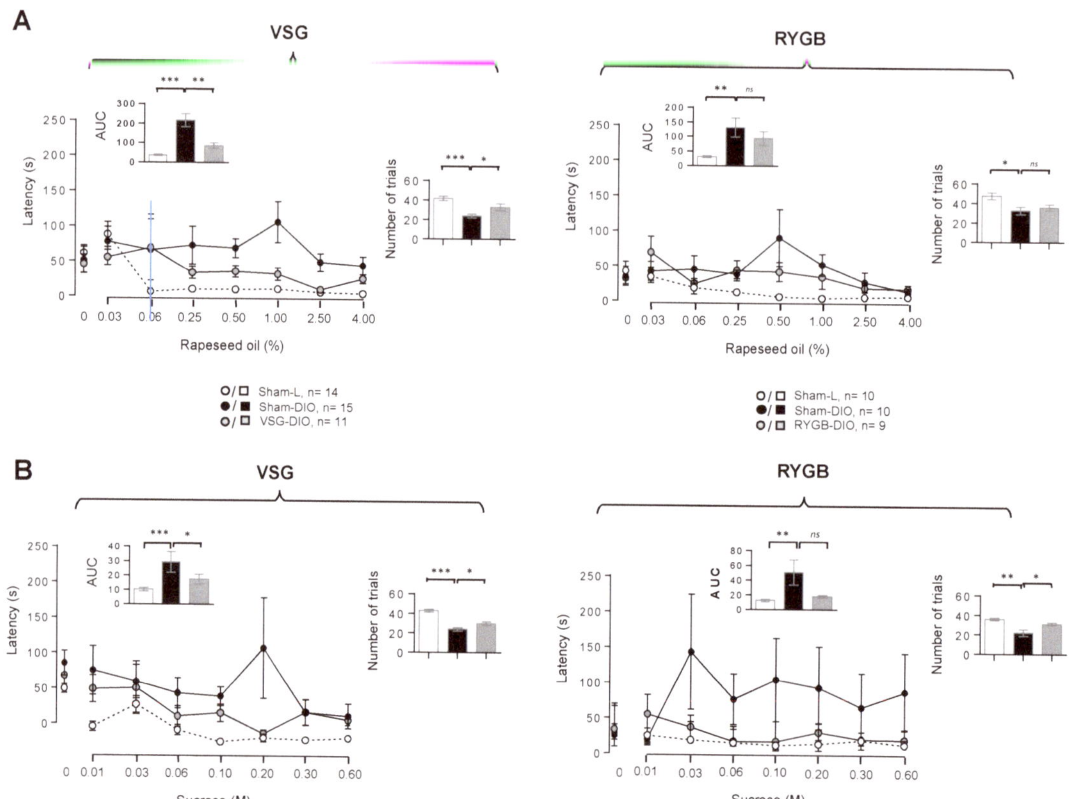

Figure 4. Motivational components of the taste-driven responses to oily emulsion and sucrose solution in sham-operated lean (Sham-L) and obese (Sham-DIO) controls and in obese mice that underwent a vertical sleeve gastrectomy (VSG-DIO) or a Roux-en-Y gastric bypass (RYGB-DIO). Latency is the time between the presentation of a bottle and the initiation of the first lick and the number of trials is the number of solutions licked during the 30 min taste-testing session. (**A**) Oily solutions. (**B**) Sucrose solutions. Zero on the X-axis represents the licking rate in response to the control solution without oil or sucrose. Analysis of both areas under curves (AUC, Inserts) and a number of trials during the 30 min of taste-testing sessions was performed using the Sham-DIO group as reference. Mean ± SEM. *ns*—non-significant, * $p < 0.05$, ** $p < 0.01$, *** $p < 0.001$. DIO—diet-induced obesity.

3.4. VSG and RYGB Bring Mice Subjected to Obesogenic Diet Closer to Lean Controls

To gain insight into the impacts of the VSG and RYGB procedures in this animal model, a multivariate analysis (principal component analysis—PCA) was performed. The values on the x- and y-axes, describing, respectively, the component score for the dimension 1 and 2, accounted for 68.6 and 11.2% (total = 79.8%) of inertia for the VSG procedure and 64.6 and 14.7% (total, 79.3%) for the RYGB (Figure 5A) demonstrating the existence of strong relationships between studied variables. The confidence ellipse analysis highlighted that obese mice (Sham-DIO) were clearly distinct from the lean controls (Sham-L), whereas the VSG-DIO and RYGB-DIO mice were found in an intermediate position suggesting an improvement of the physiological parameters studied in animals that underwent a bariatric surgery whatever the procedure used. This phenomenon takes place in spite of VSG and

RYGB mice being maintained on the obesogenic diet during the course of the study. It is noteworthy that all the mice groups were mainly defined on dimension 1 (Figure 5B).

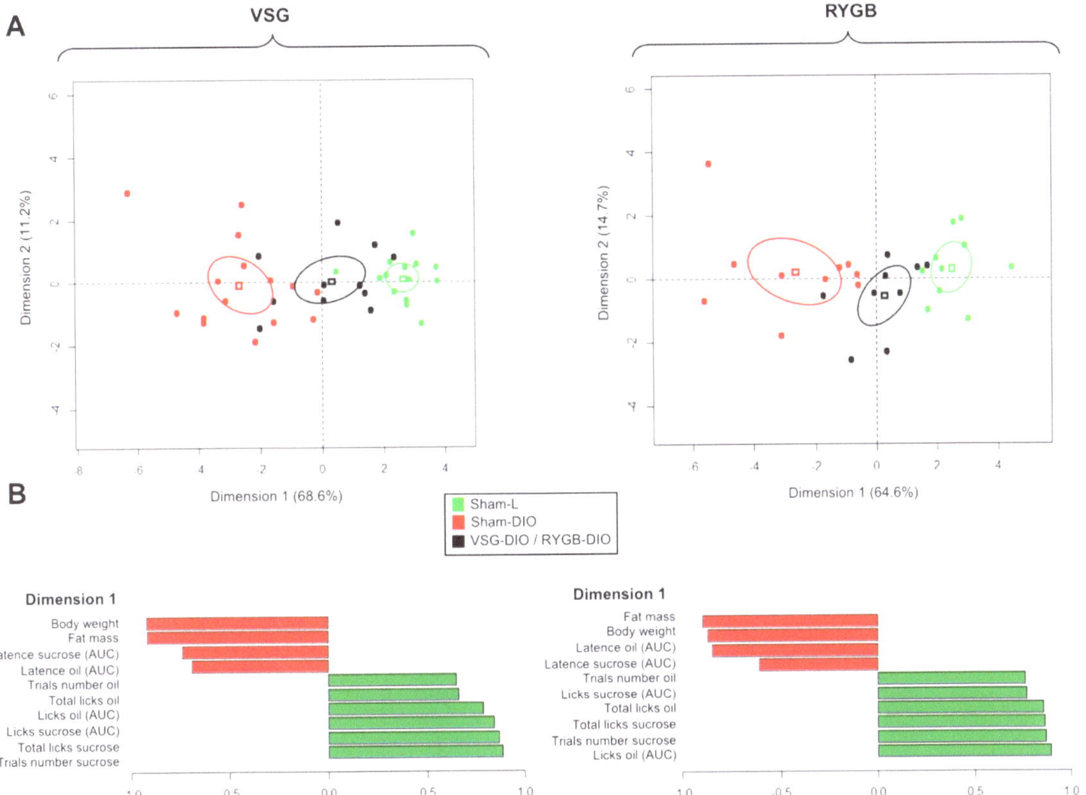

Figure 5. Principal component analysis (PCA) was performed using the studied variables in sham-operated lean (Sham-L) and obese (Sham-DIO) controls and in obese mice that underwent a vertical sleeve gastrectomy (VSG-DIO) or a Roux-en-Y gastric bypass (RYGB-DIO). (**A**) Confidence ellipse analysis with the cluster distribution along the dimension 1 & 2, each dot representing a mouse. (**B**) Main variables characterizing each group of mice.

4. Discussion

Most of the patients that have undergone an RYGB, or a VSG reports a change in their orosensory perception of tasty energy-dense foods, a phenomenon usually associated with a preferential consumption of healthier foods, low in sugar and fat (for reviews see, [13,14]). Although these observations suggest a switch in the fat and sweet taste sensitivity, this assumption remains poorly substantiated in humans. To date, the exploration of this issue using rodent models was limited to RYGB rats and has led to discrepant data (Table 1). The present study shows that VSG in mice is a suitable model to better understand how this surgery, commonly performed in humans, can counteract the deleterious effect of obesity on the orosensory perception of energy-dense nutrients observed in rodents [16,32] and re-explores the role of RYGB on this sensory parameter.

Consistent with previous rodent studies [16,28], we found that the licking responses to fat and sweet stimuli were systematically lower in obese mice (Sham-DIO) than in lean controls (Sham-L). Interestingly, this behavioral change was corrected, at least partially, in VSG mice despite they were maintained on HFD after the intervention, showing that this obesity-associated dysfunction is a reversible phenomenon, as previously suggested, using food-restricted DIO mice [32]. For the lipids, our data are congruent with those obtained in RYGB rats by Shin et al. [16] for the same oil concentration range (i.e., from 0.03% to 4%, *w/w*). By contrast, they differ from Le Roux et al.'s data [22], showing a positive correlation between the licking rate and Intralipid^{TM} concentrations, both in RYGB and sham-operated rats. The nature of lipids used as a stimulus might partly explain this discrepancy. In contrast to the oil solution used herein and by Shin and co-workers [16], triglycerides (TG) are encapsulated by egg phospholipids in Intralipid^{TM}, limiting their hydrolysis by the lingual TG-lipase and, thus, the release of long-chain fatty acids (LCFA). This step is essential for the generation of a lipid signal by the taste bud cells [33]. Therefore, it is likely that the Intralipid^{TM} emulsion is orally perceived, rather through its textural cues via the trigeminal system, than by the LCFA receptors lining the taste bud cells, in contrast to the oily solutions. An improvement of the orosensory perception of sucrose was also observed in VSG mice.

Our data also strongly suggest that VSG modifies the two main components of the orosensory perception of foods, i.e., the hedonic response ("liking") and the incentive salience ("wanting") [34]. Indeed, the licking rate, in response to fat and sweet stimuli, which mirrors the pleasure gained from the consumption of a rewarding stimulus, was higher in VSG-DIO mice than in Sham-DIO controls, suggesting again, the orosensory acuity. Moreover, VSG is also associated with a significant change in two indicators of motivation to lick: the time to initiate the first lick during a trial or latency [35] and the number of trials initiated during a session [19]. Indeed, reduced latency and a greater number of trials were found in VSG mice, suggesting a greater motivation to lick oil and sucrose solutions as compared to obese controls. Whether VSG can "reset" the obesity-mediated dysfunction in brain regions processing reward-related behavior, as reported in RYGB rats [36], is not yet known. This point is significant because the reward deficiency observed in DIO animals is thought to be involved in their tendency to overeat high-rewarding fatty foods [16], probably to gain the desired hedonic response [37]. Therefore, the positive outcome of VSG on the orosensory perception of energy-dense nutrients likely contributes to the healthier food choices found in operated mice rats [23–25].

In our hands, similar licking changes were also found after RYGB. This last observation substantiates Shin's data showing that RYGB rats display licking profiles close to lean controls for low concentrations of corn oil and sucrose [16]. However, post-surgery evolution of the body weight is the main difference between the two types of bariatric procedures studied. In contrast to RYGB animals, but as reported in most of the rodent studies [9], VSG mice tend to return to their preoperative weight 10 weeks after surgery, despite a persistent reduction in fat mass. Therefore, the orosensory benefits of VSG in mice might be dependent on the body composition rather than on the long-term weight loss. By releasing pro-inflammatory signaling factors, the white adipose tissue greatly contributes to the obesity-mediated homeostatic disruption, leading to multiple negative functional consequences (e.g., insulin resistance). Using the same cohort of VSG mice, we have recently found that the sham-DIO controls were characterized by a plasmatic neurotoxic signature elicited by the inflammation-mediated overactivation of the tryptophan (Trp) catabolism along the kynurenine (Kyn) pathway. This metabolic change was associated with a reduction in lingual fungiform density and a disturbance of responses to the oily solution during two-bottle preference tests [38]. Interestingly, these metabolic and functional troubles were widely corrected in VSG-DIO mice, suggesting that the Trp/Kyn pathway is one of the factors implicated in the cross-talk between nutritional obesity, bariatric surgery and orosensory perception of lipids. The present data strengthens this previous finding by demonstrating the positive impact of VSG on the licking behavior in response to fat (and sweet) stimuli. Nevertheless, the experimental design used herein has some limits. It does not allow us to determine whether the licks changes found after VSG (and RYGB) results from the surgery per se or is a postoperative consequence (e.g., fat mass loss). In the same way, the fact that VSG and RYGB experiments were performed using two distinct cohorts with their own controls precludes a direct comparison. However, despite significant anatomical and functional species specificities (e.g., lack of forestomach in humans), the main postoperative benefits observed in clinical studies, including the reduction in both fat mass (present study) and insulin resistance [33] were reproduced in VSG mice. Moreover, the plausible involvement of the Trp/Kyn metabolism in the degradation of fatty taste perception identified in the mouse was also observed in a sub-group of patients displaying a significant post-VSG reduction in its inflammatory status [33]. Overall, these data point out that VSG in mice is a suitable model to further explore the molecular mechanisms and functional consequences linking obesity and taste-driven food choices.

In conclusion, this study brings the first demonstration that VSG can counteract, at least partially, the deleterious effects of obesity on the orosensory perception of fat and sweet stimuli and supports the existence of similar action for RYGB. A better understanding of complex mechanisms at the origin of surgery-induced food changes might lead in the future to new strategies targeting the gusto-olfactory system facilitating the long-term compliance with healthy dietary recommendations in patients undergoing bariatric surgery.

Author Contributions: Conceptualization, P.B., S.L., C.L.M., A.B., P.D. and C.L.B.; methodology, A.D., C.L.M., A.B. and P.B.; formal analysis, A.D., A.B. and P.B.; investigation, A.D., A.B. and P.B.; writing—original draft preparation, P.B.; writing—review and editing, P.B., A.D., A.B. and S.L.; supervision, P.B. and A.B.; project administration, A.B.; funding acquisition, P.B. All authors have read and agreed to the published version of the manuscript.

Funding: The HumanFATaste2 program from which this study is derived is supported by a grant from the French National Research Agency under the program "Investissements d'Avenir" (ANR-11-LABX-0021-LipSTIC) and by financial supports from the Lesieur/Avril group, Tereos company, and Benjamin Delessert Institute.

Institutional Review Board Statement: French National Animal Ethics Committee (APAFIS#12728-201712151028893 and APAFIS#19877-2019032109244847.

Informed Consent Statement: Not applicable.

Data Availability Statement: Not applicable.

Acknowledgments: The authors are indebted to Audrey Ayer and Claire Blanchard (Univ et CHU de Nantes, CNRS, INSERM, Institut du Thorax) for the surgical interventions and are grateful to Christophe Blanchard (AgroSupDijon) for animal accommodations. We thank Xavier Collet (UMR 1048 INSERM/Univ Toulouse 2), from the HumanFATaste2 consortium, and Bertrand Cariou (Univ/CHU Nantes, Institut du thorax) for fruitful discussions.

Conflicts of Interest: No conflict to be disclosed. P.D. and C.L.B. are employed by Lesieur and Tereos companies, respectively.

Abbreviations

DIO—diet-induced obesity; HFD—high-fat diet; RYGB— Roux-en-Y gastric by-pass; VSG—vertical sleeve gastrectomy.

References

1. Guh, D.P.; Zhang, W.; Bansback, N.; Amarsi, Z.; Birmingham, C.L.; Anis, A.H. The incidence of co-morbidities related to obesity and overweight: A systematic review and meta-analysis. *BMC Public Health* **2009**, *9*, 88. [CrossRef] [PubMed]
2. Fontaine, K.R.; Redden, D.T.; Wang, C.; Westfall, A.O.; Allison, D.B. Years of life lost due to obesity. *JAMA* **2003**, *289*, 187–193. [CrossRef] [PubMed]
3. Bhaskaran, K.; Dos-Santos-Silva, I.; Leon, D.A.; Douglas, I.J.; Smeeth, L. Association of BMI with overall and cause-specific mortality: A population-based cohort study of 3.6 million adults in the UK. *Lancet Diabetes Endocrinol.* **2018**, *6*, 944–953. [CrossRef]
4. Lindberg, L.; Danielsson, P.; Persson, M.; Marcus, C.; Hagman, E. Association of childhood obesity with risk of early all-cause and cause-specific mortality: A Swedish prospective cohort study. *PLoS Med.* **2020**, *17*, e1003078. [CrossRef] [PubMed]
5. Sharples, A.J.; Mahawar, K. Systematic Review and Meta-Analysis of Randomised Controlled Trials Comparing Long-Term Outcomes of Roux-En-Y Gastric Bypass and Sleeve Gastrectomy. *Obes. Surg.* **2020**, *30*, 664–672. [CrossRef] [PubMed]
6. Elder, K.A.; Wolfe, B.M. Bariatric surgery: A review of procedures and outcomes. *Gastroenterology* **2007**, *132*, 2253–2271. [CrossRef]

7. Rondelli, F.; Bugiantella, W.; Vedovati, M.C.; Mariani, E.; Balzarotti Canger, R.C.; Federici, S.; Guerra, A.; Boni, M. Laparoscopic gastric bypass versus laparoscopic sleeve gastrectomy: A retrospective multicenter comparison between early and long-term post-operative outcomes. *Int. J. Surg.* **2017**, *37*, 36–41. [CrossRef]

8. Ayer, A.; Borel, F.; Moreau, F.; Prieur, X.; Neunlist, M.; Cariou, B.; Blanchard, C.; Le May, C. Techniques of Sleeve Gastrectomy and Modified Roux-en-Y Gastric Bypass in Mice. *J. Vis. Exp. JoVE* **2017**. [CrossRef]

9. Stevenson, M.; Lee, J.; Lau, R.G.; Brathwaite, C.E.M.; Ragolia, L. Surgical Mouse Models of Vertical Sleeve Gastrectomy and Roux-en Y Gastric Bypass: A Review. *Obes. Surg.* **2019**, *29*, 4084–4094. [CrossRef]

10. Yin, D.P.; Gao, Q.; Ma, L.L.; Yan, W.; Williams, P.E.; McGuinness, O.P.; Wasserman, D.H.; Abumrad, N.N. Assessment of different bariatric surgeries in the treatment of obesity and insulin resistance in mice. *Ann. Surg.* **2011**, *254*, 73–82. [CrossRef]

11. Lutz, T.A.; Bueter, M. The Use of Rat and Mouse Models in Bariatric Surgery Experiments. *Front. Nutr.* **2016**, *3*, 25. [CrossRef] [PubMed]

12. Miras, A.D.; le Roux, C.W. Bariatric surgery and taste: Novel mechanisms of weight loss. *Curr. Opin. Gastroenterol.* **2010**, *26*, 140–145. [CrossRef] [PubMed]

13. Ahmed, K.; Penney, N.; Darzi, A.; Purkayastha, S. Taste Changes after Bariatric Surgery: A Systematic Review. *Obes. Surg.* **2018**, *28*, 3321–3332. [CrossRef] [PubMed]

14. Nance, K.; Acevedo, M.B.; Pepino, M.Y. Changes in taste function and ingestive behavior following bariatric surgery. *Appetite* **2020**, *146*, 104423. [CrossRef]

15. Ravelli, M.N.; Ramirez, Y.P.G.; Quesada, K.; Rasera, I.; de Oliveira, M.R.M. Underreporting of Energy Intake and Bariatric Surgery. *Metab. Pathophysiol. Bariatr. Surg.* **2016**, 429–437. [CrossRef]

16. Shin, A.C.; Zheng, H.; Pistell, P.J.; Berthoud, H.R. Roux-en-Y gastric bypass surgery changes food reward in rats. *Int. J. Obes.* **2011**, *35*, 642–651. [CrossRef]

17. Weineland, S.M.; Arvidsson, D.; Kakoulidis, T.; Dahl, J. Acceptance and commitment therapy for bariatric surgery patients, a pilot rct. *Obes. Surg.* **2011**, *21*, 956–1156. [CrossRef]

18. Mathes, C.M.; Bueter, M.; Smith, K.R.; Lutz, T.A.; le Roux, C.W.; Spector, A.C. Roux-en-Y gastric bypass in rats increases sucrose taste-related motivated behavior independent of pharmacological GLP-1-receptor modulation. *Am. J. Physiol. Regul. Integr. Comp. Physiol.* **2012**, *302*, R751–R767. [CrossRef]

19. Hyde, K.M.; Blonde, G.D.; Bueter, M.; le Roux, C.W.; Spector, A.C. Gastric bypass in female rats lowers concentrated sugar solution intake and preference without affecting brief-access licking after long-term sugar exposure. *Am. J. Physiol. Regul. Integr. Comp. Physiol.* **2020**, *318*, R870–R885. [CrossRef]

20. Hajnal, A.; Kovacs, P.; Ahmed, T.; Meirelles, K.; Lynch, C.J.; Cooney, R.N. Gastric bypass surgery alters behavioral and neural taste functions for sweet taste in obese rats. *Am. J. Physiol. Gastrointest. Liver Physiol.* **2010**, *299*, G967–G979. [CrossRef]

21. Tichansky, D.S.; Glatt, A.R.; Madan, A.K.; Harper, J.; Tokita, K.; Boughter, J.D. Decrease in sweet taste in rats after gastric bypass surgery. *Surg. Endosc.* **2011**, *25*, 1176–1181. [CrossRef] [PubMed]

22. Le Roux, C.W.; Bueter, M.; Theis, N.; Werling, M.; Ashrafian, H.; Lowenstein, C.; Athanasiou, T.; Bloom, S.R.; Spector, A.C.; Olbers, T.; et al. Gastric bypass reduces fat intake and preference. *Am. J. Physiol. Regul. Integr. Comp. Physiol.* **2011**, *301*, R1057–R1066. [CrossRef] [PubMed]

23. Wilson-Perez, H.E.; Chambers, A.P.; Sandoval, D.A.; Stefater, M.A.; Woods, S.C.; Benoit, S.C.; Seeley, R.J. The effect of vertical sleeve gastrectomy on food choice in rats. *Int. J. Obes.* **2013**, *37*, 288–295. [CrossRef] [PubMed]

24. Chambers, A.P.; Wilson-Perez, H.E.; McGrath, S.; Grayson, B.E.; Ryan, K.K.; D'Alessio, D.A.; Woods, S.C.; Sandoval, D.A.; Seeley, R.J. Effect of vertical sleeve gastrectomy on food selection and satiation in rats. *Am. J. Physiol. Endocrinol. Metab.* **2012**, *303*, E1076–E1084. [CrossRef] [PubMed]

25. Saeidi, N.; Nestoridi, E.; Kucharczyk, J.; Uygun, M.K.; Yarmush, M.L.; Stylopoulos, N. Sleeve gastrectomy and Roux-en-Y gastric bypass exhibit differential effects on food preferences, nutrient absorption and energy expenditure in obese rats. *Int. J. Obes.* **2012**, *36*, 1396–1402. [CrossRef]

26. Ahart, Z.C.; Martin, L.E.; Kemp, B.R.; Dutta Banik, D.; Roberts, S.G.E.; Torregrossa, A.M.; Medler, K.F. Differential Effects of Diet and Weight on Taste Responses in Diet-Induced Obese Mice. *Obesity* **2020**, *28*, 284–292. [CrossRef]

27. Clarke, S.N.; Ossenkopp, K.P. Taste reactivity responses in rats: Influence of sex and the estrous cycle. *Am. J. Physiol.* **1998**, *274*, R718–R724. [CrossRef]

28. Dastugue, A.; Merlin, J.F.; Maquart, G.; Bernard, A.; Besnard, P. A New Method for Studying Licking Behavior Determinants in Rodents: Application to Diet-Induced Obese Mice. *Obesity* **2018**, *26*, 1905–1914. [CrossRef]

29. Davis, J.D. Do we really need new concepts to explain satiety? I think not. *Appetite* **1988**, *11*, 42–47. [CrossRef]

30. Davis, J.D.; Perez, M.C. Food deprivation- and palatability-induced microstructural changes in ingestive behavior. *Am. J. Physiol.* **1993**, *264*, R97–R103. [CrossRef]

31. Bekkevold, C.M.; Robertson, K.L.; Reinhard, M.K.; Battles, A.H.; Rowland, N.E. Dehydration parameters and standards for laboratory mice. *J. Am. Assoc. Lab. Anim.Sci. JAALAS* **2013**, *52*, 233–239. [PubMed]

32. Chevrot, M.; Bernard, A.; Ancel, D.; Buttet, M.; Martin, C.; Abdoul-Azize, S.; Merlin, J.F.; Poirier, H.; Niot, I.; Khan, N.A.; et al. Obesity alters the gustatory perception of lipids in the mouse: Plausible involvement of lingual CD36. *J. Lipid Res.* **2013**, *54*, 2485–2494. [CrossRef]

33. Kawai, T.; Fushiki, T. Importance of lipolysis in oral cavity for orosensory detection of fat. *Am. J. Physiol. Regul. Integr. Comp. Physiol.* **2003**, *285*, R447–R454. [CrossRef]
34. Berridge, K.C.; Robinson, T.E.; Aldridge, J.W. Dissecting components of reward: 'liking', 'wanting', and learning. *Curr. Opin. Pharmacol.* **2009**, *9*, 65–73. [CrossRef] [PubMed]
35. D'Aquila, P.S.; Galistu, A. Within-session decrement of the emission of licking bursts following reward devaluation in rats licking for sucrose. *PLoS ONE* **2017**, *12*, e0177705. [CrossRef]
36. Thanos, P.K.; Michaelides, M.; Subrize, M.; Miller, M.L.; Bellezza, R.; Cooney, R.N.; Leggio, L.; Wang, G.J.; Rogers, A.M.; Volkow, N.D.; et al. Roux-en-Y Gastric Bypass Alters Brain Activity in Regions that Underlie Reward and Taste Perception. *PLoS ONE* **2015**, *10*, e0125570. [CrossRef]
37. Johnson, P.M.; Kenny, P.J. Dopamine D2 receptors in addiction-like reward dysfunction and compulsive eating in obese rats. *Nat. Neurosci.* **2010**, *13*, 635–641. [CrossRef]
38. Bernard, A.; Le May, C.; Dastugue, A.; Ayer, A.; Blanchard, C.; Martin, J.C.; Pais de Barros, J.P.; Delaby, P.; Le Bourgot, C.; Ledoux, S.; et al. The Tryptophan/Kynurenine Pathway: A Novel Cross-Talk between Nutritional Obesity, Bariatric Surgery and Taste of Fat. *Nutrients* **2021**, *13*, 1366. [CrossRef] [PubMed]

Article

Towards Standardisation of a Diffuse Midline Glioma Patient-Derived Xenograft Mouse Model Based on Suspension Matrices for Preclinical Research

Elvin 't Hart [1], John Bianco [1], Helena C. Besse [2], Lois A. Chin Joe Kie [1], Lesley Cornet [1], Kimberly L. Eikelenboom [1], Thijs J.M. van den Broek [1], Marc Derieppe [1], Yan Su [1], Eelco W. Hoving [1], Mario G. Ries [2,†] and Dannis G. van Vuurden [1,*,†]

[1] Princess Máxima Center for Pediatric Oncology, Heidelberglaan 25, 3584 CS Utrecht, The Netherlands
[2] Center for Imaging Sciences, University Medical Center Utrecht, Heidelberglaan 100, 3584 CX Utrecht, The Netherlands
* Correspondence: d.g.vanvuurden@prinsesmaximacentrum.nl; Tel.:+31-(06)-50006665
† These authors contributed equally to this work.

Abstract: Diffuse midline glioma (DMG) is an aggressive brain tumour with high mortality and limited clinical therapeutic options. Although in vitro research has shown the effectiveness of medication, successful translation to the clinic remains elusive. A literature search highlighted the high variability and lack of standardisation in protocols applied for establishing the commonly used HSJD-DIPG-007 patient-derived xenograft (PDX) model, based on animal host, injection location, number of cells inoculated, volume, and suspension matrices. This study evaluated the HSJD-DIPG-007 PDX model with respect to its ability to mimic human disease progression for therapeutic testing in vivo. The mice received intracranial injections of HSJD-DIPG-007 cells suspended in either PBS or Matrigel. Survival, tumour growth, and metastases were assessed to evaluate differences in the suspension matrix used. After cell implantation, no severe side effects were observed. Additionally, no differences were detected in terms of survival or tumour growth between the two suspension groups. We observed delayed metastases in the Matrigel group, with a significant difference compared to mice with PBS-suspended cells. In conclusion, using Matrigel as a suspension matrix is a reliable method for establishing a DMG PDX mouse model, with delayed metastases formation and is a step forward to obtaining a standardised in vivo PDX model.

Keywords: HSJD-DIPG-007; diffuse midline glioma; PDX model; Matrigel; metastases

Citation: 't Hart, E.; Bianco, J.; Besse, H.C.; Chin Joe Kie, L.A.; Cornet, L.; Eikelenboom, K.L.; van den Broek, T.J.; Derieppe, M.; Su, Y.; Hoving, E.W.; et al. Towards Standardisation of a Diffuse Midline Glioma Patient-Derived Xenograft Mouse Model Based on Suspension Matrices for Preclinical Research. *Biomedicines* **2023**, *11*, 527. https://doi.org/10.3390/biomedicines11020527

Academic Editors: Martina Perše and Shaker A. Mousa

Received: 15 December 2022
Revised: 31 January 2023
Accepted: 9 February 2023
Published: 11 February 2023

1. Introduction

Paediatric high-grade gliomas (pHGGs) are malignant brain tumours found in the hemispheres and midline structures of the brain and account for 10% of all central nervous system (CNS) tumours in children, while being responsible for 40% of all fatal cases. Diffuse intrinsic pontine glioma (DIPG) is a particularly aggressive and invasive pHGG subtype arising in the brainstem (pons) and has been recognised as a distinct type within the paediatric diffuse high-grade glioma family in the 5th edition of the WHO Classification of Tumours of the Central Nervous System [1], and as such, these tumours have been reclassified to 'diffuse midline glioma, H3K27-altered' (DMG). Alterations in H3K27 in DMG include point mutations at the histone H3K27M, predominantly with H3.3 expression and to a lesser degree H3.1 with up to 80% of tumours harbouring one of these mutations [2]. H3.3 mutations cause trimethylation loss of the chromatin with altered manifestations of the oncogenes and tumour-suppressor genes [3]. Loss of H3K27 trimethylation by overexpression of EZHIP has been observed in H3K27 wildtype DMG [4]. In addition, DMGs are also commonly associated with mutations in the TP53 gene (up to 60%) and to a lesser extent with mutations in PPM1D (up to 30%) [5]. Combined, these mutations increase the aggressiveness of DMGs and are associated with a poor overall prognosis. Genomic analyses have revealed that DMGs are molecularly complex, also harbouring mutations in ACVR1, ATRX, H3F3A, HIST1H3B/c, MYC, PDGFRA, PIK3CA, PTEN, and RB1 that can cooperate with mutated TP53 and PPM1D to promote tumour formation [2,6–9]. In addition to pontine localization, DMGs can occur in other midline structures, such as the thalamus and spinal cord.

Pontine DMGs are mainly diagnosed in children between 6 and 9 years of age. Rapid progression of this disease results in a median survival of 11 months and a 95% fatality rate within 2 years after diagnosis [10,11]. DMG is commonly found in the brainstem, a delicate brain region responsible for the execution of vital functions [1,12]. Clinical symptoms are caused by pressure of the tumour and dysfunction of the brainstem, resulting in cranial nerve deficits, such as facial and abducens nerve palsy, multiple cranial neuropathies, and long-tract and cerebellar signs, such as paresis and ataxia [13]. Because of the delicate location and invasive nature of DMG, radical surgery is impossible, while chemotherapy is complicated by the presence of the blood–brain barrier (BBB), preventing 98% and 100% of small and large molecules from entering the brain [14,15]. While tumour progression can cause BBB disruption and subsequently increased BBB permeability, most of the BBB in DMG remains intact over the course of the disease. Even when BBB disruption is observed, this occurs mostly at the core of the tumour lesion after onset of local tissue necrosis [16]. Therefore, the current standard of care of DMG is fractionated radiotherapy of 1.8–2 Gy daily cumulating to a total dose of 54–60 Gy, with concurrent temozolomide causing temporal tumour growth delay, but also inevitable recurrence [17]. Metastasis along the neuroaxis is rarely seen at diagnosis (2%) but can increase to up to 17.3% at disease progression [10,18], where an under-recognised pattern of subventricular spread was observed in the majority of investigated cases, with infiltration of the subventricular zone as well as tumour nodules in the frontal horns of the lateral ventricles [19].

Although intensive research has been conducted for the treatment of DMG, little clinical progress has been made to date [20]. Even though in vitro drug screening has evidenced several promising chemotherapeutic candidates for DMG treatment, the successful translation to preclinical in vivo studies has demonstrated to be challenging [21–25]. Additionally, therapeutic translational complexity is added due to the biological differences between patients and animal models of the disease [26]. Although small animals do not develop DMG spontaneously, in vivo studies are made possible by establishing genetically engineered mouse models (GEMMs) or patient-derived xenografts (PDXs) [27]. GEMM models have an altered genomic profile to mimic the human disease allowing genetic/fundamental research to be conducted, while PDX models use orthotopic injection of human primary DMG cells in (partially) immune-deficient animals. The role of these models in preclinical research is to facilitate recapitulation of human malignancies and the associated disease progression, allowing validation of therapeutic agents or interventional techniques before clinical trials [28,29].

HSJD-DIPG-007 is an established DMG cell line from the Sant Joan de Déu Hospital in Barcelona, derived from the autopsy of a radiotherapy-naive, 6-year-old male that died one month after diagnosis and received one course of chemotherapy (cisplatin and irinotecan). These HSJD-DIPG-007 tumour cells harbour mutations in H3F3A K27M, ACVR1 R206H, PPM1Dp.P428fs, and PIK3CAp.H1047R [5,30]. In recent years, HSJD-DIPG-007 has increasingly been used as a cell line for DMG PDX mouse models [31]. This model displays an intact BBB as well as an invasive growth pattern that mimics human pathology for a large part of disease progression, rendering it appealing for evaluating therapeutic response and efficiency [32]. However, a standardised method in establishing orthotopic in vivo models using HSJD-DIPG-007 cells has not yet been developed. Current protocols using this cell type vary between studies on several levels, such as use of cell suspension matrix, site of implantation, and volume/number of tumour cells inoculated. The lack of a standardised approach also complicates comparison while potentiating different experimental outcomes. Finally, the development time and extent of diffuse, infiltrative growth, and metastasis make these models difficult to compare to human disease progression. Because metastases at diagnosis is a relatively rare occurrence in DMG patients, optimising the cell implantation procedure in a standardised manner could better mimic tumour growth progression in vivo, with a greater correlation with human disease progression.

We postulated that using Matrigel instead of phosphate-buffered saline (PBS) as a cell suspension matrix for tumour cell inoculation in preclinical models would prevent premature cell dissemination. Local confinement of the tumour is particularly relevant for locoregional treatment paradigms, such as convection-enhanced and focused ultrasound-mediated drug delivery to the brainstem in preclinical research. The aim of this study is to provide a literature overview of the HSJD-DIPG-007 DMG PDX model, to extract common features, and to investigate the impact of dissimilarities. For the latter, the presented work focuses on the comparison of the extent of infiltrative and metastatic growth patterns of the model with cells inoculated with either PBS or Matrigel as the suspension substrate in athymic nude mice and the relevance of the time delay between inoculation and the onset of therapy.

2. Materials and Methods

2.1. Literature Search

A literature search was performed to identify publications using the HSJD-DIPG-007 cell line to create an overview of preclinical DMG tumour models using this cell line without any exclusion criteria. Upon study inclusion, data were classified based on (1) animal host and age, (2) location of injection, (3) injected volume and cell concentration, (4) cell suspension matrix, and (5) treatment and follow-up. The age of the mice was categorised based on their postnatal ($\leq$3 weeks), adolescent (3–9 weeks), and adult (>9 weeks) phase [33].

2.2. Animals

All experiments were conducted on 6–8-week-old male athymic nude Foxn1$^{-/-}$ mice (Code 069, Envigo, Horst, The Netherlands) in accordance with guidelines of the Dutch Ethical Committee and the Animal Welfare Body of Utrecht University (AVD3990020209445, approval date: 11/02/2020). A total of 34 mice were used for the study, consisting of 15 for DMG PDX /tumour growth and survival validation and 19 that were sacrificed at designated timepoints for histological analysis. Mice were housed under specific pathogen-free conditions in separately ventilated cages, at up to four animals/cage, and allowed to acclimatise for 2 weeks before experimental procedures. Mice were kept on regular laboratory food and water ad libitum, with a fixed 12 h light/dark cycle in accordance with ARRIVE guidelines [2]. Measurable outcomes in PDX models of DMG are not influenced by gender, and as such gender dimension was not relevant for this study [34]. A detailed description of housing conditions of animals is available as Supplementary Material.

2.3. Cells

HSJD-DIPG-007 cells are patient-derived from the autopsy of a pontine tumour in a 9-year-old male, kindly provided by Dr. Ángel Montero Carcaboso (Sant Joan de Déu Barcelona Hospital). Genetic information is archived in the Cellosaurus Database (CVCL_VU70, www.cellosaurus.org). Cells were grown and maintained in 1:1 Neurobasal-A and Advanced DMEM/F-12 medium containing 10 mM HEPES buffer, 1$\times$ MEM nonessential amino acids, 1% GlutaMAX, 1 mM Sodium pyruvate, 1$\times$ B-27 minus vitamin-A (ThermoFisher, Waltham, MA, USA), 10 ng/mL PDGF-AA, 10 ng/mL PDGF-BB, 20 ng/mL bFGF, 20 ng/mL EGF (Peprotech, London, UK), 2 µg/mL heparin (Leo Pharma, Amsterdam, The Netherlands), and 1 mg/mL primocin (InvivoGen, San Diego, CA, USA). Medium was refreshed every 3–4 days. Single-cell suspensions were obtained using Accutase (ThermoFisher, Waltham, MA, USA). Cells were cultured at 37 °C, 5% CO_2, and 95% humidity. For in vivo tumour growth monitoring by bioluminescence imaging (BLI), HSJD-DIPG-007 cells were transduced to express firefly luciferase as previously described [35].

Following infection, eGFP-lucF-gene-positive HSJD-DIPG-007 cells were selected using a Sony SH800 Cell Sorter (Sony, Tokyo, Japan). Before cell implantation, HSJD-DIPG-007 cells were suspended in 1× PBS (pH 7.4) or Matrigel (50% *v/v*, in PBS, Corning, Corning, NY, USA) and kept on ice until used.

2.4. Drugs

Pre- and postsurgical analgesia was managed with 67 μg/mL carprofen (Faculty of Veterinary Medicine pharmacy, Utrecht, The Netherlands) per os (p.o.) in drinking water with an additional subcutaneous (s.c.) injection of 5 mg/kg before surgery. Further pain suppression was performed by s.c. injection of 0.5% lidocaine (B. Braun, Melsungen, Germany) during surgery. Anaesthesia was maintained with isoflurane (Zoetis, Capelle aan den IJssel, The Netherlands), mixed with air, 3% induction, 1.8% maintenance. BLI signal of engrafted cells was monitored by intraperitoneal (i.p.) injection of 150 mg/kg D-luciferin (Cayman Chemical, Uden, The Netherlands) in PBS. Euthanasia was performed via i.p. injection of a mix of 7.14 mg/mL ketamine (Alfasan, Woerden, The Netherlands) and 0.714 mg/mL sedazine (AST Farma, Oudewater, The Netherlands) in PBS.

2.5. Tumour Cell Implantation

Twenty-four hours before and after orthotopic intracranial injection with eGFP-lucF-gene-positive HSJD-DIPG-007 cells, mice received carprofen p.o. in drinking water. Thirty minutes before surgery, carprofen was administrated s.c. for local pain management. After anaesthesia with isoflurane, mice were fixed on a stereotactic frame with bite and ear bars. Eye cream was applied to prevent eye damage, while the mice were kept warm during the procedure. After incision of the skin, a drop of lidocaine was added before removal of the facia on the skull. Using a high-speed drill, a burr hole was made in the skull 0.8 mm posterior and 1.0 mm lateral to the lambda. At a depth of 4.5 mm in the pontine region, a total of 5×10^5 HSJD-DIPG-007 cells suspended in 4.3 μL of PBS or Matrigel were injected at a rate of 2 μL/min using a 5 μL Hamilton syringe fitted with a 26-gauge needle. After injection, the needle remained in place for 7 min before being slowly retracted to prevent cell accumulation in the needle tract. Wound closure was performed by applying topical skin adhesive Histoacryl (B. Braun, Melsungen, Germany) before placing the mice under a heating lamp until awake. Possible signs of stress and postoperative complications (lack of food/water intake, antisocial behaviour, and motor deficits) were carefully monitored.

2.6. Tumour Growth Assessment with Bioluminescence

Mice were weighed three times a week, while their tumour growth was monitored twice a week by measuring the BLI signal of engrafted eGFP-lucF-gene-positive HSJD-DIPG-007 cells using the MILABS U-OI camera (MILABS, Houten, The Netherlands). For signal measurement, mice were anaesthetised with isoflurane and injected 5 min later with D-luciferin before positioning in the camera. BLI images were taken under anaesthesia from 5 to 30 min after D-luciferin injection with a 60 s exposure. Signal intensity was quantified within the region of interest (ROI) of the whole animal head by using ImageJ software (v1.53t, National Institute of Health, Bethesda, USA) [36]. Mice were sacrificed with ketamine/sedazine after reaching their scientific or humane endpoints. Humane endpoints were defined based on 20% weight loss from cell implantation, 15% weight loss within two days, or development of neurological deficiencies, such as circling, hyperexcitability, convulsions, or ataxia.

2.7. Histological Analysis

Histopathological elements, tumour size, location, and proliferation were determined by human vimentin and haematoxylin and eosin (H & E) staining. After euthanasia, mice were transcardially perfused with PBS followed by 10% formalin, after which the brain was excised and postfixed in 10% formalin for 48 h before paraffin embedding. Sagittal sections of 4μm were made using a microtome (Leica Biosystems, Wetzlar, Germany) and mounted on Superfrost® Plus microscope slides. Before staining, sections were deparaffinized and subjected to antigen retrieval with sodium citrate buffer (10 mM, pH 6, 95–100 °C, and 30 min). Endogenous peroxidase activity was reduced by incubation in 3% hydrogen peroxidase for 20 min, after which sections were washed twice with deionized water and once with 1× Tris-buffered saline containing 0.1% Tween (TBST). Sections were blocked for 1 h at room temperature with antibody diluent clear (VWRKBD09-125, VWR, Radnor, USA) before overnight incubation at 4 °C with rabbit antihuman vimentin [SP20] (1:5–1:8, ab27608, Abcam, Cambridge, UK) followed by washing with TBST. Sections were then incubated for 2 h at room temperature with biotinylated affinity-purified goat antirabbit secondary antibody (1:500, BA-1000, and IgG (H + L), Vector Laboratories, Newark, NJ, USA) before washing with TBST. VECTASTAIN® Elite ABC-HRP Peroxidase (PK-6100, Vector Laboratories, Newark, USA) was applied for 1 h at room temperature followed by a 3–4 min incubation in 3,3′-diaminobenzidine (DAB, K346711-2, Agilent Dako, Amstelveen, The Netherlands) before counterstaining with haematoxylin (Epredia, Breda, The Netherlands).

2.8. Data and Statistical Analysis

Weight and tumour growth measured by BLI signal were analysed using an independent t-test. Survival was analysed using a Kaplan–Meier plot and Log-rank test. Metastases formation in olfactory bulb and spinal cord were analysed by a nonparametric Kolmogorov–Smirnov test to compare cumulative distributions. A p-value of ≤ 0.05 was considered statistically significant. Statistical analyses were performed using GraphPad Prism (v9, GraphPad Software, LLC, Boston, MA, USA). Photographic and electronic images were obtained on a Leica DMi8 and processed using Adobe Photoshop 21 (Adobe Inc., San Jose, CA, USA).

3. Results

3.1. HSJD-DIPG-007 PDX Model in the Literature

A total of 20 articles were published between 2016 and 2022 using the HSJD-DIPG-007 cell line for establishing a DMG PDX mouse model [5,25,31,32,37–52]. An overview of these studies is given in Table 1. For the orthotopic generation of DMG, 65% of the studies described injection in the pontine/brainstem region, 20% in the 4th ventricle, and 15% in a combination of both 4th ventricle/pons. In 75% of the studies, adolescent mice were used for establishing the tumour model, 15% used early postnatal mice, and 10% did not define the age. Athymic nude, nude BALB/c, NOD-SCID, and NOD-SCID gamma (NSG) nude mice were used as host animals in 30%, 20%, 30%, and 15% of the cases, respectively, with one study (representing 5%) using an athymic nude rat. The injection volume ranged between 1 µL and 5 µL, with 45% of the cases injecting 5×10^5 HSJD-DIPG-007 cells. Only one study used 7.5×10^5 cells suspended in 7.5 µL for establishing the PDX model using athymic nude rat as the host. PBS, Matrigel, or medium were used as suspension matrices in 20%, 40%, and 10% of the studies reported, respectively. In 20% of the studies, an undefined suspension matrix was used, while the remaining 10% used combinations or other matrices. The treatment applications ranged from day 0 up to day 80 after cell inoculation. Despite the high variety of treatments performed in these studies, prolonged survival or delayed tumour growth was observed in 82% of cases reporting treatment outcomes.

3.2. Well-Being and Weight Profiles upon Implantation Procedure

Orthotopic injections of HSJD-DIPG-007 cells suspended in PBS (n = 9) or Matrigel (n − 6) did not give rise to deleterious neurological complications following implantation. The time frame of the surgery and anaesthesia affected the wakefulness of the mice afterwards, where extended procedures resulted in lengthier recovery times until the mice were fully active and mobile (observation), even though mouse core temperature was monitored and maintained throughout the procedure. Following cell implantation, mice initially lost weight, but gained on average 16% of their initial weight by day 37 for PBS and 15% by day

30 for Matrigel-injected mice. No significant differences in overall weight gain or loss were measured between the PBS and Matrigel groups (Figure 1). An early and aggressive tumour onset can explain the severe weight loss in one PBS mouse (Supplementary Figure S1).

Table 1. Summary of studies using HSJD-DIPG-007 for establishing a DMG PDX model from 2016 to 2022.

Animal Host	Age (Weeks)	Location	Total Cells Inoculated	Volume	Suspension Matrix	Days Before Treatment	Treatment Strategy	Treatment Efficiency	Reference
Athymic nude	n.d.	Brainstem	5×10^5	n.d.	n.d.	21	RG7388	Enhanced survival	[5]
NOD-SCID	7	Pons	3×10^5	2 µL	Matrigel	28	Panobinostat	No	[25]
NOD-SCID	5	Pons	2×10^5	5 µL	Matrigel	None	None	Not assessed	[31]
Athymic nude	6	Pons	5×10^5	5 µL	PBS	37	Doxorubicin and FUS	No	[32]
Athymic nude	n.d.	Pons	5×10^5	5 µL	n.d.	14	BGB324, Panobinostat and CED	Enhanced survival combined with CED	[37]
NOD-SCID	6–8	4th Ventricle	5×10^5	4 µL	Matrigel	28	OKN-007 and LDN-193189	Reduced cellular activity	[38]
NOD-SCID gamma (NSG)	8–10	Pons	4×10^5	2 µL	Medium: Matrigel (1:1)	21	2-DG and IDH1 inhibitor	Enhanced survival and decreased growth	[39]
Athymic nude rat	4	4th Ventricle	7.5×10^5	7.5 µL	n.d.	28	SN-38	Not assessed	[40]
Nude BALB/c	8	4th Ventricle/Pons	2×10^5	3 µL	n.d.	21	DCA, Metformin and RT	Enhanced survival combined with RT	[41]
NOD-SCID	7–8	4th Ventricle/Pons	2×10^5	2 µL	Matrigel	28	CBL0137 and Panobinostat	Enhanced survival in combination	[42]
Nude BALB/c	6	Pons	1×10^4	1 µL	HBSS	21	CRAd.S.pK7	Enhanced survival	[43]
NOD-SCID gamma (NSG)	5–7	Pons	1×10^5	2 µL	Serum free media	80	ALDH+/− and GDC-0084	Enhanced survival	[44]
Nude BALB/c	5–7	Brainstem	2×10^5	2 µL	Matrigel	30	DMFO & AMXT	Enhanced survival and decreased growth	[45]
Athymic nude	3	4th Ventricle	5×10^5	5 µL	Matrigel	25	Vandetanib and Everolimus	Enhanced survival in combination	[46]
Athymic nude	7–9	Striatum/Pons	5×10^5	5 µL	PBS	75	Bevacizumab	Not assessed	[47]
Athymic nude	6–8	Pons	5×10^5	5 µL	PBS	7–8	Doxorubicin and CED	No	[48]
NOD-SCID	4–5	4th Ventricle	5×10^5	2 µL	PBS	0	HSV1716	Reduced cellular growth	[49]
Nude BALB/c	5–6	4th Ventricle/Pons	2×10^5	2 µL	Matrigel	28–35	Temozolomide and RT	RT enhanced survival	[50]
NOD-SCID gamma (NSG)	0–2 days	Brainstem	1×10^3	n.d.	Tumour stem medium	0	GSK2830371	Enhanced survival	[51]
NOD-SCID	3	Pons	5×10^5	5 µL	Matrigel	21–28	LDN-193189 and LDN-214117	Enhanced survival	[52]

n.d.: Not defined.

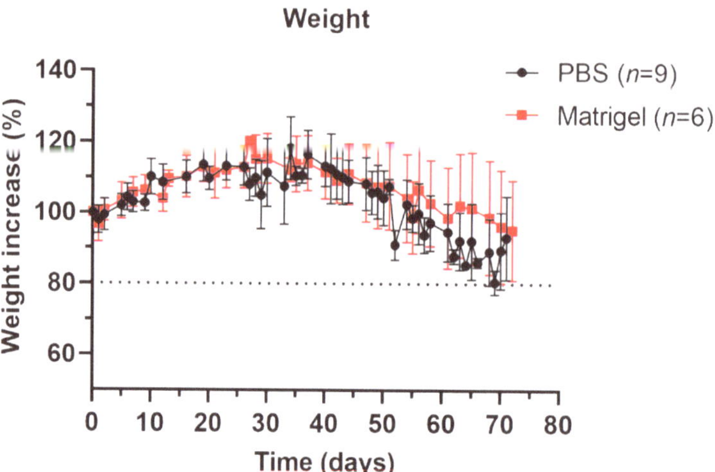

Figure 1. Weight profiles of mice inoculated with HSJD-DIPG-007 cells suspended either in PBS or Matrigel. Changes in weight after intracranial implantation of HSJD-DIPG-007 cells suspended in either PBS or Matrigel were monitored. Weight increased consistently up to day 37 for PBS and day 30 for Matrigel suspension groups, after which weight loss set in, lasting until terminal endpoint. No significant differences between PBS and Matrigel groups were observed. Dotted line represents the weight threshold of the humane endpoint. Data points are expressed as mean weight ± SD.

3.3. Survival and Tumour Growth Using PBS or Matrigel as Suspension Matrices

Despite different suspension matrices being used, no significant differences in survival between the PBS and Matrigel groups were observed. Mice with PBS or Matrigel survived up to 90 and 100 days and with a median overall survival of 70 and 75 days, respectively (Figure 2A). PBS mice were sacrificed in 2/9 cases based on neurological symptoms of motor functions, such as tremors and paralysis, 6/9 based on weight loss, and 1/9 for both conditions. Matrigel mice were sacrificed in 2/6 cases based on neurological symptoms, 3/6 based on weight loss, and in 1/6 case, the animal passed away during BLI. The BLI signal confirmed successful cell implantation in all animals of both treatment groups. Steady exponential tumour growth was observed up to day 30 post implantation, after which the growth increase exceeded around day 40 1.8 AU/day for both (Figure 2B,C and Supplementary Figure S2). The increased exponential growth could be indicative of locoregional metastasis formation with the tumour spreading outside the injection location of the pons. No significant differences in tumour growth were observed between the PBS and Matrigel groups.

3.4. Metastases Occurrence in Olfactory Bulb and Spinal Cord

Because DMGs are tumours beginning in the pontine region and spreading on mid-disease in the majority of cases to adjacent areas, the HSJD-DIPG-007 PDX model should preferably recapitulate this growth pattern, in particular for the evaluation of locoregional treatment at the initial stage of disease [53]. Based on the BLI signal, the first onset of metastases in the frontal lobe (olfactory bulb) was observed at day 26 after inoculation in the PBS group and at day 44 after inoculation in the Matrigel group. A median metastasis-free survival (MMFS) in the olfactory bulb of 33 vs. 58 days was found for PBS and Matrigel mice, with a significant difference between the groups (Figure 3A). Metastatic formations in the spinal cord were first observed at day 37 post inoculation in the PBS group and at day 44 post inoculation in the Matrigel group, with an MMFS of 47 and 68 days, respectively (Figure 3B). At time of death, two of the nine mice in the PBS group had not developed metastases, while only one had metastasis in the olfactory bulb. Of the six Matrigel mice, two did not develop metastases, and one developed an olfactory bulb metastasis.

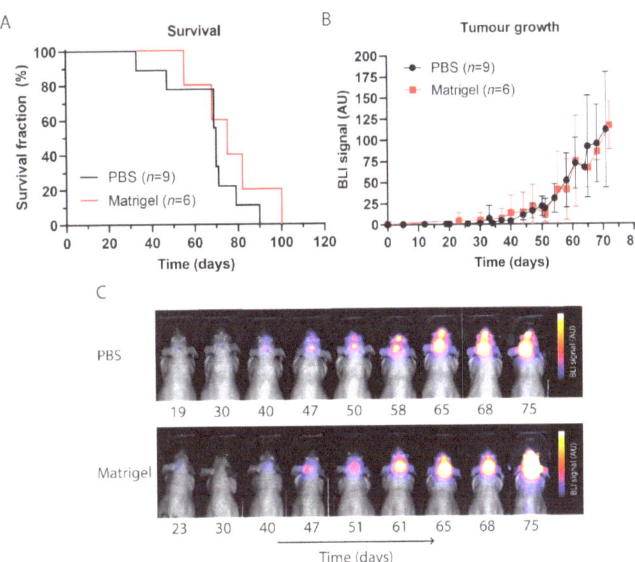

Figure 2. Survival and tumour growth following inoculation with HSJD-DIPG-007 cells suspended in PBS or Matrigel. (**A**) Kaplan–Meier curve showing survival following cell inoculation and tumour progression. No significant difference between PBS and Matrigel suspension groups was observed. (**B**) Tumour volume over time in both PBS and Matrigel suspension groups, showing that tumour growth was comparable up to 75 days. Data points are expressed as mean signal intensity $\pm$ SD. (**C**) BLI signal showing tumour growth over time. Tumour development within the pontine region, as well as metastatic development in the olfactory bulb region, can be seen in both PBS and Matrigel groups, progressing with time. AU = arbitrary unit.

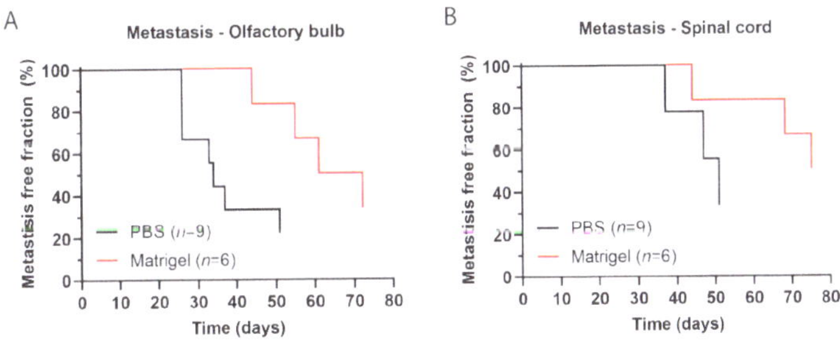

Figure 3. Distant metastatic formations over time monitored through BLI signal following HSJD-DIPG-007 cell inoculation into the pontine region. (**A**) Metastasis in the olfactory bulb in PBS and Matrigel suspension groups with a median onset of 33 and 53 days, respectively. A significant difference between the two groups was found ($p < 0.05$). (**B**) Metastasis in the spinal cord in PBS and Matrigel suspension groups with a median onset of 47 and 68 days, respectively, but without a significant difference ($p > 0.05$).

Mice with cells suspended in PBS showed local growth up to day 31, with subsequent locoregional progression as well as metastatic formations in the mid cerebrum/lateral ventricle, with eventual spreading into the cerebellum and olfactory bulb. Mice with cells suspended in Matrigel showed local growth up to day 31 and the presence of tumour cells in the lateral ventricles, with delayed locoregional progression and distal striatal infiltration with inevitable invasion of the whole brain at day 55 (Figure 4). No histopathological or morphological differences were observed in the mice of both groups by H & E staining (Supplementary Figure S3). Antihuman vimentin staining confirmed the local injection of

HSJD-DIPG-007 in the pontine region of the mice and showcased that contamination of the cerebrospinal fluid (CSF) and dissemination to other brain structures in proximity to the injection site through the perivascular system (PVS) can occur (Figure 5).

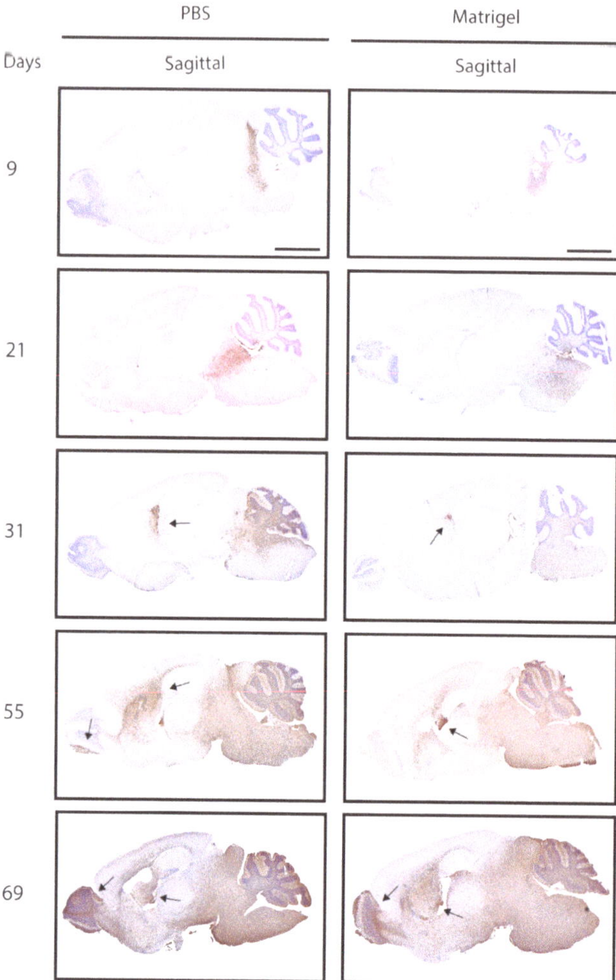

Figure 4. Antihuman vimentin staining of mouse brains showing tumour progression over time following inoculation with HSJD-DIPG-007 cells suspended in PBS or Matrigel. Tumour progression over time can be seen in both PBS and Matrigel groups through the accumulation and spread of human vimentin-positive cells (brown staining) within the pons and other, more distant brain regions. Metastases can be observed from day 31 in both PBS and Matrigel suspension groups (black arrows). Whole brain invasion of tumour cells can be observed at day 55 in both groups. Mouse brains in both groups were counterstained with haematoxylin. $n = 9$ for PBS group and $n = 10$ for Matrigel group. Scale bar = 2 mm.

A comparative histopathological analysis of clinical autopsy-derived DMG with the orthotopic E98 DIPG mouse model was previously performed by Caretti and colleagues [53]. To determine the clinical relevance of the HSJD-DIPG-007 PDX model, we used the histology panel of DMG patient tissue of Carreti et al. for a comparative assessment of disease progression (Figure 6). Perivascular tumour dissemination in the HSJD-DIPG-007 PDX model was seen to be like that observed in the DMG patient (Figure 6C,D). Similarities were also observed in brain parenchyma invasion in the HSJD-DIPG-007 model and clinical DMG (Figure 6G,H), as well as vascular proliferation (Figure 6K,L).

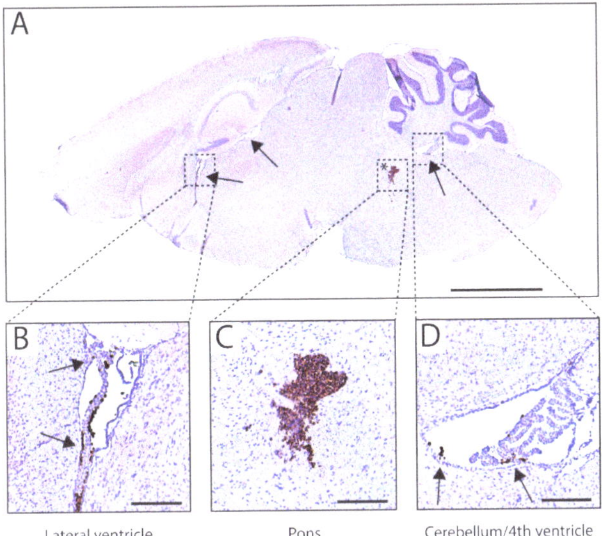

Figure 5. HSJD-DIPG-007 cells within the pons and disseminated throughout the brain immediately following inoculation. (**A**) Matrigel-suspended HSJD-DIPG-007 cells within the pons area (asterisk) as well as in distant brain structures (arrows) at time zero, identified via antihuman vimentin staining. Magnification of HSJD-DIPG-007 cells present in the choroid plexus of the lateral ventricle (**B**), pons (**C**), and choroid plexus of the cerebellum/4th ventricle (**D**). Counterstaining is with haematoxylin. Scale bar = 2 mm for **A**, 100 μm for (**B–D**).

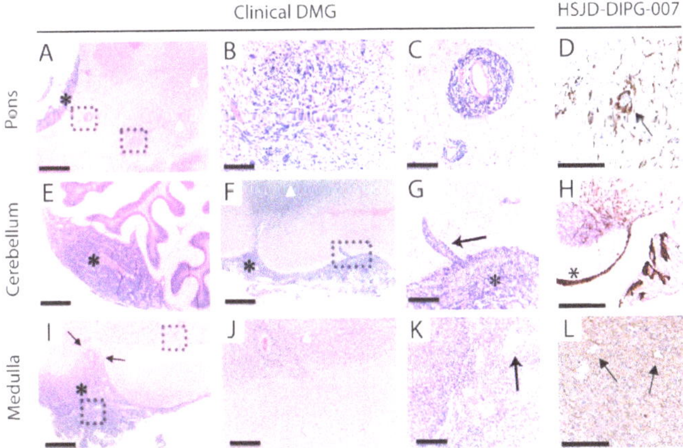

Figure 6. Comparative assessment of clinical DMG and HSJD-DIPG-007 PDX model histopathology. The comparative clinical DMG panel (panes **A–C,E–G,I–K**) was adapted and reproduced with permission from Caretti et al. [53]. (**A,E,F,I**) H & E staining of clinical DMG at the pons, cerebellum, and medulla. (**B,C,G,J,K**) Magnifications of H & E staining indicated by the dotted squares. (**D,H,L**) Antihuman vimentin staining of HSJD-DIPG-007 PDX model at the pons, cerebellum, and medulla. (**A,E–I**) Asterisks indicate leptomeningeal growth. (**D,G**) Arrows indicate perivascular growth, and (**I,K,L**) represent blood vessels in dense tumour areas. Scale bars = 250 μm (**A,F,I**), 62.5 μm (**B,C,G**), 125 μm (**J,K**), 500 μm (**E**), and 50 μm (**D,H,L**).

4. Discussion

DMG is an invasive paediatric brain tumour with a high mortality rate, and because of the location and infiltrative nature of the disease, radiotherapy is the only effective palliative treatment option currently available [10,17]. As DMG rarely develops naturally in animals, PDX animal models are important for preclinical in vivo therapy efficacy validation. However, due to translational complexities between preclinical research and clinical applicability, there is a demand for PDX models emulating human disease progression. Metastases and immediate organ-specific proliferation infrequently occurs in patients in early stages of disease progression but can be seen in late/end stages of DMG [19]. Ideally DMG models would reproduce this form of early disease progression, which is observed in most of the patients as an initial diffuse local tumour proliferation in the pontine area, with subsequent expansion through the medulla, the cerebellum, and the thalamic areas. Because DMG patients suffer from a rapid progression of disease in the vital pontine area leading to a poor prognosis, late-stage disease beyond this point is rarely observed.

The HSJD-DIPG-007 cell line, derived from the autopsy of a 6-year-old, is widely used for establishing DMG PDX models, but the high heterogeneity among protocols indicates that a universal procedure is yet to be developed. With the aim of facilitating a standardised inoculation method, this study investigated and optimised the growth pattern of the HSJD-DIPG-007 PDX model for local or metastatic phenotype treatment based on two different suspension matrices.

Studies in other cancer models, such as pancreatic cancer, have shown the importance of the suspension matrix when tumour cells are injected locally and the issues, such as leakage, low tumour formation, and development of metastases, that can arise [54]. The local introduction of cells into the brain is a delicate matter, requiring precision in location, as well as in the injection procedure to avoid positive and negative pressure build-up that could dissipate the cells in an unfavourable manner. A possible alternative to common suspension matrices, such as growth medium and PBS, could be the basement membrane Matrigel, due to its composition resembling the extracellular matrix of many tissues, as well as its favourable viscoelastic properties where it remains liquid at low temperatures but polymerises to a dense matrix at temperatures above 10 °C [55]. No standardised procedure in establishing the HSJD-DIPG-007 PDX model could be discerned from the studies outlined in Table 1. Protocols differed considerably in all the reported parameters, as well as in the stated level of detail provided, with the most noteworthy differences being in the suspension matrix used, day at which treatment was initiated, and treatment modality or efficacy. In our study comparing PBS and Matrigel, we selected to use animals at 6–8 weeks of age for inoculation of the HSJD-DIPG-007 cell line, corresponding to the age range used by 50% of the studies reported in Table 1 and commonly used for in vivo studies. To ensure adequate cell grafting, we also opted for 5×10^5 total cell inoculation for both the PBS and Matrigel suspension groups.

The initial observation made in comparing the PBS and Matrigel groups was in weight stability following HSJD-DIPG-007 cell implantation. Substantial fluctuations, including rapid gains/losses in a short period of time are reliable indications of health in in vivo animal models. In our study, no significant differences in weight were observed between the PBS and Matrigel groups. The weight gained in the first 4–5 weeks after cell implantation could be due to the young 6–8-week age of the mice, in which they were still in their adolescent and body growth phase [33].

As DMG progresses quite rapidly in children, symptoms are typically not evident for 4 to 6 weeks before diagnosis [56]. Patients present to the clinic when disease progression is relatively advanced, with a triad of symptoms consisting of cranial neuropathy, long tract signs, and cerebellar signs [57]. By this stage, DMG may have been developing for 12 months or more. In our study, mice in both groups were asymptomatic and gained weight for 30 days, as seen in Figure 1, after which tumour presence could be verified and followed by BLI, and weight loss began to occur. The subsequent weight loss could be attributed to the progression of the tumour in the pontine region, the consequence of which could be diminished appetite. Figure 2 shows how the BLI signal intensified rapidly post day 30, with a strong signal being observed in brain regions outside the graft area, suggesting the presence of metastatic formations. However, this increase in the BLI signal did not correspond with overall survival, as no significant differences were observed irrespective of whether cells were inoculated using PBS or Matrigel. Both groups also had comparable tumour growth rates during the steady growth phase of the first 4 weeks post inoculation as well as in the exponential growth phase thereafter, further supporting the similar survival times observed and confirming that the suspension matrix on its own does not influence local tumour growth or survival.

Strikingly, the analysis of the BLI signal did show differences between the PBS and Matrigel groups in terms of metastatic formations within the olfactory bulbs, as seen in Figure 3, suggesting that Matrigel does significantly delay the onset of metastases by an average of approximately 3 weeks. Delays in the spinal cord were also observed in the Matrigel group, and even though these were not found to be significant, the suggestion that Matrigel influences metastasis formation is present. The polymerisation of the cell-loaded Matrigel upon injection into the pons could have contributed to reduced cellular leakage into the brain parenchyma or into the needle tract produced during the inoculation procedure, without altering the tumour growth rate. The observation that the use of one suspension matrix significantly delays metastases when compared to another further emphasises the need for a standardised protocol in establishing DMG PDX models through orthotopic injection of cells into the pons. This was confirmed, as seen in Figure 4, through antihuman vimentin staining of HSJD-DIPG-007 cell-inoculated mouse brains at various stages of tumour development. The staining confirmed that the pons was accurately targeted and that the cells successfully engrafted and were able to induce local tumour formation. Tumour growth rapidly progressed with time, while advanced metastatic formations were observed by 31 days within the midbrain and by 55 days within the olfactory bulbs of the PBS suspension group and not in the Matrigel group.

The PVS and cerebrospinal fluid hydrodynamics are important factors that must also be considered, especially when DMG PDX models are established. It has previously been shown that connections between the CSF and nasal lymphatic vessels in mammals, including humans and rodents, share common characteristics [58]. Metastases can initially be seen in the location of the lateral ventricle, followed by formations in the olfactory bulbs, which coincides with the direction of the CSF flow in both humans and rodents. In humans, CFS circulates in a caudal-directed manner through the ventricles to the subarachnoid space, resulting in an exchange of various substances in a to-and-from manner between the CSF and interstitial compartments [59].

A proportion of the CSF drains into the cribriform place, while the rest is recycled into the brain parenchyma through perivascular spaces surrounding blood vessels. Perivascular space connections penetrating deep into the brainstem and 4th ventricle have also been observed. New PVS connections between ventricles and different parts of the brain parenchyma have been revealed, suggesting a possible role for the ventricles as a source or sink for solutes in the brain [60].

These observations further demonstrate that Matrigel, as a suspension matrix, is more favourable in supporting local tumour growth at the site of inoculation and delays the onset of metastases, especially to the olfactory bulbs, in a significant manner, further supporting its use as a more suitable suspension matrix than PBS. It may be that Matrigel supresses perivascular proliferation of inoculated cells, resulting in a model of disease progression that more closely resembles that seen in patients as described by Caretti and colleagues [19].

Although Matrigel delays metastases overall, the inoculation procedure itself is not infallible. When injecting a substance 4–5 mm deep within the mouse brain, the needle passes through several structures and does cause a degree of disruption to adjacent tissues, while also disturbing the CSF present in the brain. As shown in Figure 5, rogue cells can be seen already circulating within the brain outside the pontine region immediately following inoculation. The circulating CSF could potentially distribute these cells through the lateral ventricles and on to the olfactory bulbs, which also act as a CSF sink and outflow to the nasal lymphatics [60], where they can give rise to metastatic formations developing very early following initial inoculation. Therefore, it is imperative that any residual cells that may remain on the outside of the needle while filling with cell-containing suspension matrix be removed thoroughly before injection into the brain.

When the observations of tumour volume and BLI signal of Figure 3 are compared with the immunohistochemical images of the tumour progression in Figure 4, we can see that although the BLI signal is not detected before 40 days post inoculation of the HSJD-DIPG-007 cells, tumour progression with extensive infiltration of the brain parenchyma by tumour cells is already present by day 21. This suggests that the BLI signal alone is not reliable in determining early and local tumour formations and thus should not be used as a measure to determine the onset of treatment as tumour size and burden, including the presence of metastases could be underestimated. Such an underestimation could render treatment regimens unsuccessful because of a too large tumour burden rather than treatment inefficacy, leading to false negatives and the ultimate rejection of suitable drug or treatment candidates. From the immunohistological data obtained, in addition to using Matrigel as a cell suspension, we would suggest that treatment initiation be performed between 7 and 14 days post cell inoculation. This timeframe would allow for cells to engraft and tumour formation to occur to a point where the burden is not too high to render treatment ineffective and not too low to result in false positives. It is noteworthy that of the studies listed in Table 1, only 1/5 initiated treatment within this timeline, while the majority started therapy three or more weeks following cell implantation. For locoregional therapy approaches, treating within two weeks would also ensure that the entire tumour within the pons is targeted, and not later-occurring metastatic formations within other brain regions, which are missed, especially in the distant olfactory bulbs.

In summary, the HSJD-DIPG-007 PDX mouse model is one that has gained interest in DMG research as it does resemble human disease progression in a clinically relevant manner, as Figure 6 shows. Vital elements, such as perivascular tumour dissemination, invasion of the parenchyma, and vascular proliferation, are well emulated in the HSJD-DIPG-007 PDX model. As Caretti [53] showed in both the clinical and E98 DMG tumours and observations in the HSJD-DIPG-007 model used in this study, perivascular migration appears to be a route by which invasion of the brain parenchyma can occur by tumour cells located in the subarachnoid space. However, for this model to be optimally utilised in preclinical research, standardisation of its establishment needs to be achieved.

Based on our results, we propose a standardised method of using Matrigel as a suspension matrix to inoculate cells within the pons to delay metastases to other brain regions. We also suggest treatment initiation be within 1–2 weeks of grafting to ensure an adequate but not overbearing tumour burden for assessment of treatment strategies. Further standardisation of this model assessing the animal host used, total cells inoculated, injection volume, and graft location is needed so that a reliable and reproducible model that recapitulates the histological characteristics of DMG can be established.

Supplementary Materials: The following supporting information can be downloaded at: https://www.mdpi.com/article/10.3390/biomedicines11020527/s1, Figure S1: Weight profiles of individual mice after inoculation with HSJD-DIPG-007 cells suspended in PBS or Matrigel; Figure S2: Tumour growth profiles of individual mice following inoculation with HSJD-DIPG-007 cells suspended in PBS or Matrigel; Figure S3: Haematoxylin & Eosin staining of mouse brains with tumour progression over time following inoculation with HSJD-DIPG-007 cells suspended in PBS or Matrigel Table S1: List of conditions, materials, and manufacturers related to the in vivo experiments including living conditions, cage enrichments, food, water, and health monitoring of the animals by pathogen detection in the research facility.

Author Contributions: Conceptualization, J.B., H.C.B., M.G.R. and D.G.v.V.; Methodology, E.'t.H., J.B., H.C.B., Y.S.; Validation, E.'t.H., J.B., L.A.C.J.K., L.C., K.L.E., T.J.M.v.d.B. and M.D.; Formal analysis, E.'t.H. and H.C.B.; Investigation, E.'t.H., J.B., H.C.B., L.A.C.J.K., L.C., K.L.E., T.J.M.v.d.B. and M.D.; Resources, E.'t.H., H.C.B. and Y.S.; Writing–original draft, E.'t.H. and J.B.; Writing–review & editing,

H.C.B., E.W.H., M.G.R. and D.G.v.V.; Visualization, E.'t.H. and J.B.; Supervision, J.B., E.W.H., M.G.R. and D.G.v.V.; Project administration, E.'t.H., J.B., M.G.R. and H.C.B.; Funding acquisition, D.G.v.V. All authors have read and agreed to the published version of the manuscript.

Funding: The Dutch Cancer Society (KWF) grant number 10911.

Institutional Review Board Statement: This study was conducted in accordance with the guidelines of the Dutch Ethical Committee and the Animal Welfare Body of Utrecht University (AVD3990020209445, approval date: 11 February 2020).

Informed Consent Statement: Not applicable.

Data Availability Statement: No new data sets were created or analysed in this study therefore data sharing is not applicable for this article.

Acknowledgments: This project was supported by KWF Young Investigator Award (KWF 10911, D.G. van Vuurden). The authors also thank Ángel Montero Carcaboso (Sant Joan de Déu Barcelona Hospital) for kindly providing the HSJD-DIPG-007 patient-derived cell line used in this study. Graphical abstract created with BioRender.com.

Conflicts of Interest: The authors declare no conflict of interest.

References

1. Louis, D.N.; Perry, A.; Wesseling, P.; Brat, D.J.; Cree, I.A.; Figarella-Branger, D.; Hawkins, C.; Ng, H.K.; Pfister, S.M.; Reifenberger, G.; et al. The 2021 WHO Classification of Tumors of the Central Nervous System: A Summary. *Neuro. Oncol.* **2021**, *23*, 1231–1251. [CrossRef]
2. Khuong-Quang, D.-A.; Buczkowicz, P.; Rakopoulos, P.; Liu, X.-Y.; Fontebasso, A.M.; Bouffet, E.; Bartels, U.; Albrecht, S.; Schwartzentruber, J.; Letourneau, L.; et al. K27M Mutation in Histone H3.3 Defines Clinically and Biologically Distinct Subgroups of Pediatric Diffuse Intrinsic Pontine Gliomas. *Acta Neuropathol.* **2012**, *124*, 439–447. [CrossRef]
3. Chan, K.-M.; Fang, D.; Gan, H.; Hashizume, R.; Yu, C.; Schroeder, M.; Gupta, N.; Mueller, S.; James, C.D.; Jenkins, R.; et al. The Histone H3.3K27M Mutation in Pediatric Glioma Reprograms H3K27 Methylation and Gene Expression. *Genes Dev.* **2013**, *27*, 985–990. [CrossRef]
4. Castel, D.; Kergrohen, T.; Tauziède-Espariat, A.; Mackay, A.; Ghermaoui, S.; Lechapt, E.; Pfister, S.M.; Kramm, C.M.; Boddaert, N.; Blauwblomme, T.; et al. Histone H3 Wild-Type DIPG/DMG Overexpressing EZHIP Extend the Spectrum Diffuse Midline Gliomas with PRC2 Inhibition beyond H3-K27M Mutation. *Acta Neuropathol.* **2020**, *139*, 1109–1113. [CrossRef]
5. Xu, C.; Liu, H.; Pirozzi, C.J.; Chen, L.H.; Greer, P.K.; Diplas, B.H.; Zhang, L.; Waitkus, M.S.; He, Y.; Yan, H. TP53 Wild-Type/PPM1D Mutant Diffuse Intrinsic Pontine Gliomas Are Sensitive to a MDM2 Antagonist. *Acta Neuropathol. Commun.* **2021**, *9*, 178. [CrossRef]
6. Zarghooni, M.; Bartels, U.; Lee, E.; Buczkowicz, P.; Morrison, A.; Huang, A.; Bouffet, E.; Hawkins, C. Whole-Genome Profiling of Pediatric Diffuse Intrinsic Pontine Gliomas Highlights Platelet-Derived Growth Factor Receptor Alpha and Poly (ADP-Ribose) Polymerase as Potential Therapeutic Targets. *J. Clin. Oncol.* **2010**, *28*, 1337–1344. [CrossRef]
7. Paugh, B.S.; Broniscer, A.; Qu, C.; Miller, C.P.; Zhang, J.; Tatevossian, R.G.; Olson, J.M.; Geyer, J.R.; Chi, S.N.; da Silva, N.S.; et al. Genome-Wide Analyses Identify Recurrent Amplifications of Receptor Tyrosine Kinases and Cell-Cycle Regulatory Genes in Diffuse Intrinsic Pontine Glioma. *J. Clin. Oncol.* **2011**, *29*, 3999–4006. [CrossRef]
8. Buczkowicz, P.; Hoeman, C.; Rakopoulos, P.; Pajovic, S.; Letourneau, L.; Dzamba, M.; Morrison, A.; Lewis, P.; Bouffet, E.; Bartels, U.; et al. Genomic Analysis of Diffuse Intrinsic Pontine Gliomas Identifies Three Molecular Subgroups and Recurrent Activating ACVR1 Mutations. *Nat. Genet.* **2014**, *46*, 451–456. [CrossRef]
9. Zhang, L.; Chen, L.H.; Wan, H.; Yang, R.; Wang, Z.; Feng, J.; Yang, S.; Jones, S.; Wang, S.; Zhou, W.; et al. Exome Sequencing Identifies Somatic Gain-of-Function PPM1D Mutations in Brainstem Gliomas. *Nat. Genet.* **2014**, *46*, 726–730. [CrossRef]
10. Hoffman, L.M.; Veldhuijzen van Zanten, S.E.M.; Colditz, N.; Baugh, J.; Chaney, B.; Hoffmann, M.; Lane, A.; Fuller, C.; Miles, L.; Hawkins, C.; et al. Clinical, Radiologic, Pathologic, and Molecular Characteristics of Long-Term Survivors of Diffuse Intrinsic Pontine Glioma (DIPG): A Collaborative Report from the International and European Society for Pediatric Oncology DIPG Registries. *J. Clin. Oncol.* **2018**, *36*, 1963–1972. [CrossRef]
11. Jansen, M.H.; Veldhuijzen van Zanten, S.E.; Sanchez Aliaga, E.; Heymans, M.W.; Warmuth-Metz, M.; Hargrave, D.; van der Hoeven, E.J.; Gidding, C.E.; de Bont, E.S.; Eshghi, O.S.; et al. Survival Prediction Model of Children with Diffuse Intrinsic Pontine Glioma Based on Clinical and Radiological Criteria. *Neuro. Oncol.* **2015**, *17*, 160–166. [CrossRef] [PubMed]
12. Tate, M.C.; Lindquist, R.A.; Nguyen, T.; Sanai, N.; Barkovich, A.J.; Huang, E.J.; Rowitch, D.H.; Alvarez-Buylla, A. Postnatal Growth of the Human Pons: A Morphometric and Immunohistochemical Analysis. *J. Comp. Neurol.* **2015**, *523*, 449–462. [CrossRef]
13. Fisher, P.G.; Breiter, S.N.; Carson, B.S.; Wharam, M.D.; Williams, J.A.; Weingart, J.D.; Foer, D.R.; Goldthwaite, P.T.; Tihan, T.; Burger, P.C. A Clinicopathologic Reappraisal of Brain Stem Tumor Classification. Identification of Pilocystic Astrocytoma and Fibrillary Astrocytoma as Distinct Entities. *Cancer* **2000**, *89*, 1569–1576. [CrossRef] [PubMed]
14. Pardridge, W.M. Blood-Brain Barrier Delivery. *Drug Discov. Today* **2007**, *12*, 54–61. [CrossRef]

15. Stupp, R.; Hegi, M.E.; Mason, W.P.; van den Bent, M.J.; Taphoorn, M.J.B.; Janzer, R.C.; Ludwin, S.K.; Allgeier, A.; Fisher, B.; Belanger, K.; et al. Effects of Radiotherapy with Concomitant and Adjuvant Temozolomide versus Radiotherapy Alone on Survival in Glioblastoma in a Randomised Phase III Study: 5-Year Analysis of the EORTC-NCIC Trial. *Lancet Oncol.* **2009**, *10*, 459–466. [CrossRef]

16. Warren, K.E. Beyond the Blood:Brain Barrier: The Importance of Central Nervous System (CNS) Pharmacokinetics for the Treatment of CNS Tumors, Including Diffuse Intrinsic Pontine Glioma. *Front. Oncol.* **2018**, *8*, 239. [CrossRef] [PubMed]

17. Stupp, R.; Mason, W.P.; van den Bent, M.J.; Weller, M.; Fisher, B.; Taphoorn, M.J.B.; Belanger, K.; Brandes, A.A.; Marosi, C.; Bogdahn, U.; et al. Radiotherapy plus Concomitant and Adjuvant Temozolomide for Glioblastoma. *N. Engl. J. Med.* **2005**, *352*, 987–996. [CrossRef]

18. Gururangan, S.; McLaughlin, C.A.; Brashears, J.; Watral, M.A.; Provenzale, J.; Coleman, R.E.; Halperin, E.C.; Quinn, J.; Reardon, D.; Vredenburgh, J.; et al. Incidence and Patterns of Neuraxis Metastases in Children with Diffuse Pontine Glioma. *J. Neurooncol.* **2006**, *77*, 207–212. [CrossRef]

19. Caretti, V.; Bugiani, M.; Freret, M.; Schellen, P.; Jansen, M.; van Vuurden, D.; Kaspers, G.; Fisher, P.G.; Hulleman, E.; Wesseling, P.; et al. Subventricular Spread of Diffuse Intrinsic Pontine Glioma. *Acta Neuropathol.* **2014**, *128*, 605–607. [CrossRef] [PubMed]

20. Hargrave, D.; Bartels, U.; Bouffet, E. Diffuse Brainstem Glioma in Children: Critical Review of Clinical Trials. *Lancet Oncol.* **2006**, *7*, 241–248. [CrossRef]

21. Veringa, S.J.E.; Biesmans, D.; van Vuurden, D.G.; Jansen, M.H.A.; Wedekind, L.E.; Horsman, I.; Wesseling, P.; Vandertop, W.P.; Noske, D.P.; Kaspers, G.J.L.; et al. In Vitro Drug Response and Efflux Transporters Associated with Drug Resistance in Pediatric High Grade Glioma and Diffuse Intrinsic Pontine Glioma. *PLoS ONE* **2013**, *8*, e61512. [CrossRef] [PubMed]

22. Grasso, C.S.; Tang, Y.; Truffaux, N.; Berlow, N.E.; Liu, L.; Debily, M.-A.; Quist, M.J.; Davis, L.E.; Huang, E.C.; Woo, P.J.; et al. Functionally Defined Therapeutic Targets in Diffuse Intrinsic Pontine Glioma. *Nat. Med.* **2015**, *21*, 555–559. [CrossRef]

23. Mueller, S.; Hashizume, R.; Yang, X.; Kolkowitz, I.; Olow, A.K.; Phillips, J.; Smirnov, I.; Tom, M.W.; Prados, M.D.; James, C.D.; et al. Targeting Wee1 for the Treatment of Pediatric High-Grade Gliomas. *Neuro. Oncol.* **2014**, *16*, 352–360. [CrossRef] [PubMed]

24. Caretti, V.; Hiddingh, L.; Lagerweij, T.; Schellen, P.; Koken, P.W.; Hulleman, E.; van Vuurden, D.G.; Vandertop, W.P.; Kaspers, G.J.L.; Noske, D.P.; et al. WEE1 Kinase Inhibition Enhances the Radiation Response of Diffuse Intrinsic Pontine Gliomas. *Mol. Cancer Ther.* **2013**, *12*, 141–150. [CrossRef]

25. Hennika, T.; Hu, G.; Olaciregui, N.G.; Barton, K.L.; Ehteda, A.; Chitranjan, A.; Chang, C.; Gifford, A.J.; Tsoli, M.; Ziegler, D.S.; et al. Pre-Clinical Study of Panobinostat in Xenograft and Genetically Engineered Murine Diffuse Intrinsic Pontine Glioma Models. *PLoS ONE* **2017**, *12*, e0169485. [CrossRef] [PubMed]

26. Perlman, R.L. Mouse Models of Human Disease: An Evolutionary Perspective. *Evol. Med. Public Health* **2016**, *2016*, 170–176. [CrossRef]

27. Chen, Z.; Peng, P.; Zhang, X.; Mania-Farnell, B.; Xi, G.; Wan, F. Advanced Pediatric Diffuse Pontine Glioma Murine Models Pave the Way towards Precision Medicine. *Cancers* **2021**, *13*, 1114. [CrossRef]

28. Funato, K.; Major, T.; Lewis, P.W.; Allis, C.D.; Tabar, V. Use of Human Embryonic Stem Cells to Model Pediatric Gliomas with H3.3K27M Histone Mutation. *Science* **2014**, *346*, 1529–1533. [CrossRef]

29. Monje, M.; Mitra, S.S.; Freret, M.E.; Raveh, T.B.; Kim, J.; Masek, M.; Attema, J.L.; Li, G.; Haddix, T.; Edwards, M.S.B.; et al. Hedgehog-Responsive Candidate Cell of Origin for Diffuse Intrinsic Pontine Glioma. *Proc. Natl. Acad. Sci. USA* **2011**, *108*, 4453–4458. [CrossRef]

30. Taylor, K.R.; Mackay, A.; Truffaux, N.; Butterfield, Y.; Morozova, O.; Philippe, C.; Castel, D.; Grasso, C.S.; Vinci, M.; Carvalho, D.; et al. Recurrent Activating ACVR1 Mutations in Diffuse Intrinsic Pontine Glioma. *Nat. Genet.* **2014**, *46*, 457–461. [CrossRef]

31. Vinci, M.; Burford, A.; Molinari, V.; Kessler, K.; Popov, S.; Clarke, M.; Taylor, K.R.; Pemberton, H.N.; Lord, C.J.; Gutteridge, A.; et al. Functional Diversity and Cooperativity between Subclonal Populations of Pediatric Glioblastoma and Diffuse Intrinsic Pontine Glioma Cells. *Nat. Med.* **2018**, *24*, 1204–1215. [CrossRef]

32. Haumann, R.; Bianco, J.I.; Waranecki, P.M.; Gaillard, P.J.; Storm, G.; Ries, M.; van Vuurden, D.G.; Kaspers, G.J.L.; Hulleman, E. Imaged-Guided Focused Ultrasound in Combination with Various Formulations of Doxorubicin for the Treatment of Diffuse Intrinsic Pontine Glioma. *Transl. Med. Commun.* **2022**, *7*, 8. [CrossRef]

33. Brust, V.; Schindler, P.M.; Lewejohann, L. Lifetime Development of Behavioural Phenotype in the House Mouse (Mus Musculus). *Front. Zool.* **2015**, *12*, S17. [CrossRef]

34. Qi, L.; Kogiso, M.; Du, Y.; Zhang, H.; Braun, F.K.; Huang, Y.; Teo, W.-Y.; Lindsay, H.; Zhao, S.; Baxter, P.; et al. Impact of SCID Mouse Gender on Tumorigenicity, Xenograft Growth and Drug-Response in a Large Panel of Orthotopic PDX Models of Pediatric Brain Tumors. *Cancer Lett.* **2020**, *493*, 197–206. [CrossRef] [PubMed]

35. Kholosy, W.M.; Derieppe, M.; van den Ham, F.; Ober, K.; Su, Y.; Custers, L.; Schild, L.; van Zogchel, L.; Wellens, L.; Ariese, H.; et al. Neuroblastoma and DIPG Organoid Coculture System for Personalized Assessment of Novel Anticancer Immunotherapies. *J. Pers. Med.* **2021**, *11*, 869. [CrossRef]

36. Schneider, C.A.; Rasband, W.S.; Eliceiri, K.W. NIH Image to ImageJ: 25 Years of Image Analysis. *Nat. Methods* **2012**, *9*, 671–675. [CrossRef]

37. Meel, M.H.; de Gooijer, M.C.; Metselaar, D.S.; Sewing, A.C.P.; Zwaan, K.; Waranecki, P.; Breur, M.; Buil, L.C.M.; Lagerweij, T.; Wedekind, L.E.; et al. Combined Therapy of AXL and HDAC Inhibition Reverses Mesenchymal Transition in Diffuse Intrinsic Pontine Glioma. *Clin. Cancer Res.* **2020**, *26*, 3319–3332. [CrossRef] [PubMed]

38. Thomas, L.; Smith, N.; Saunders, D.; Zalles, M.; Gulej, R.; Lerner, M.; Fung, K.-M.; Carcaboso, A.M.; Towner, R.A. OKlahoma Nitrone-007: Novel Treatment for Diffuse Intrinsic Pontine Glioma. *J. Transl. Med.* **2020**, *18*, 424. [CrossRef] [PubMed]
39. Chung, C.; Sweha, S.R.; Pratt, D.; Tamrazi, B.; Panwalkar, P.; Banda, A.; Bayliss, J.; Hawes, D.; Yang, F.; Lee, H.-J.; et al. Integrated Metabolic and Epigenomic Reprograming by H3K27M Mutations in Diffuse Intrinsic Pontine Gliomas. *Cancer Cell* **2020**, *38*, 334–349. [CrossRef]
40. Chaves, C.; Declèves, X.; Taghi, M.; Menet, M.-C.; Lacombe, J.; Varlet, P.; Olaciregui, N.G.; Carcaboso, A.M.; Cisternino, S. Characterization of the Blood-Brain Barrier Integrity and the Brain Transport of SN-38 in an Orthotopic Xenograft Rat Model of Diffuse Intrinsic Pontine Glioma. *Pharmaceutics* **2020**, *12*, 399. [CrossRef]
41. Shen, H.; Yu, M.; Tsoli, M.; Chang, C.; Joshi, S.; Liu, J.; Ryall, S.; Chornenkyy, Y.; Siddaway, R.; Hawkins, C.; et al. Targeting Reduced Mitochondrial DNA Quantity as a Therapeutic Approach in Pediatric High-Grade Gliomas. *Neuro Oncol.* **2020**, *22*, 139–151. [CrossRef]
42. Ehteda, A.; Simon, S.; Franshaw, L.; Giorgi, F.M.; Liu, J.; Joshi, S.; Rouaen, J.R.C.; Pang, C.N.I.; Pandher, R.; Mayoh, C.; et al. Dual Targeting of the Epigenome via FACT Complex and Histone Deacetylase Is a Potent Treatment Strategy for DIPG. *Cell Rep.* **2021**, *35*, 108994. [CrossRef] [PubMed]
43. Chastkofsky, M.I.; Pituch, K.C.; Katagi, H.; Zannikou, M.; Ilut, L.; Xiao, T.; Han, Y.; Sonabend, A.M.; Curiel, D.T.; Bonner, E.R.; et al. Mesenchymal Stem Cells Successfully Deliver Oncolytic Virotherapy to Diffuse Intrinsic Pontine Glioma. *Clin. Cancer Res.* **2021**, *27*, 1766–1777. [CrossRef]
44. Surowiec, R.K.; Ferris, S.F.; Apfelbaum, A.; Espinoza, C.; Mehta, R.K.; Monchamp, K.; Sirihorachai, V.R.; Bedi, K.; Ljungman, M.; Galban, S. Transcriptomic Analysis of Diffuse Intrinsic Pontine Glioma (DIPG) Identifies a Targetable ALDH-Positive Subset of Highly Tumorigenic Cancer Stem-like Cells. *Mol. Cancer Res.* **2021**, *19*, 223–239. [CrossRef] [PubMed]
45. Khan, A.; Gamble, L.D.; Upton, D.H.; Ung, C.; Yu, D.M.T.; Ehteda, A.; Pandher, R.; Mayoh, C.; Hébert, S.; Jabado, N.; et al. Dual Targeting of Polyamine Synthesis and Uptake in Diffuse Intrinsic Pontine Gliomas. *Nat. Commun.* **2021**, *12*, 971. [CrossRef] [PubMed]
46. Carvalho, D.M.; Richardson, P.J.; Olaciregui, N.; Stankunaite, R.; Lavarino, C.; Molinari, V.; Corley, E.A.; Smith, D.P.; Ruddle, R.; Donovan, A.; et al. Repurposing Vandetanib plus Everolimus for the Treatment of ACVR1-Mutant Diffuse Intrinsic Pontine Glioma. *Cancer Discov.* **2022**, *12*, 416–431. [CrossRef]
47. Jansen, M.H.A.; Lagerweij, T.; Sewing, A.C.P.; Vugts, D.J.; van Vuurden, D.G.; Molthoff, C.F.M.; Caretti, V.; Veringa, S.J.E.; Petersen, N.; Carcaboso, A.M.; et al. Bevacizumab Targeting Diffuse Intrinsic Pontine Glioma: Results of 89Zr-Bevacizumab PET Imaging in Brain Tumor Models. *Mol. Cancer Ther.* **2016**, *15*, 2166–2174. [CrossRef]
48. Sewing, A.C.P.; Lagerweij, T.; van Vuurden, D.G.; Meel, M.H.; Veringa, S.J.E.; Carcaboso, A.M.; Gaillard, P.J.; Peter Vandertop, W.; Wesseling, P.; Noske, D.; et al. Preclinical Evaluation of Convection-Enhanced Delivery of Liposomal Doxorubicin to Treat Pediatric Diffuse Intrinsic Pontine Glioma and Thalamic High-Grade Glioma. *J. Neurosurg. Pediatr.* **2017**, *19*, 518–530. [CrossRef] [PubMed]
49. Cockle, J.V.; Brüning-Richardson, A.; Scott, K.J.; Thompson, J.; Kottke, T.; Morrison, E.; Ismail, A.; Carcaboso, A.M.; Rose, A.; Selby, P.; et al. Oncolytic Herpes Simplex Virus Inhibits Pediatric Brain Tumor Migration and Invasion. *Mol. Ther. Oncolytics* **2017**, *5*, 75–86. [CrossRef]
50. Tsoli, M.; Shen, H.; Mayoh, C.; Franshaw, L.; Ehteda, A.; Upton, D.; Carvalho, D.; Vinci, M.; Meel, M.H.; van Vuurden, D.; et al. International Experience in the Development of Patient-Derived Xenograft Models of Diffuse Intrinsic Pontine Glioma. *J. Neurooncol.* **2019**, *141*, 253–263. [CrossRef]
51. Akamandisa, M.P.; Nie, K.; Nahta, R.; Hambardzumyan, D.; Castellino, R.C. Inhibition of Mutant PPM1D Enhances DNA Damage Response and Growth Suppressive Effects of Ionizing Radiation in Diffuse Intrinsic Pontine Glioma. *Neuro Oncol.* **2019**, *21*, 786–799. [CrossRef]
52. Carvalho, D.; Taylor, K.R.; Olaciregui, N.G.; Molinari, V.; Clarke, M.; Mackay, A.; Ruddle, R.; Henley, A.; Valenti, M.; Hayes, A.; et al. ALK2 Inhibitors Display Beneficial Effects in Preclinical Models of ACVR1 Mutant Diffuse Intrinsic Pontine Glioma. *Commun. Biol.* **2019**, *2*, 156. [CrossRef] [PubMed]
53. Caretti, V.; Zondervan, I.; Meijer, D.H.; Idema, S.; Vos, W.; Hamans, B.; Bugiani, M.; Hulleman, E.; Wesseling, P.; Vandertop, W.P.; et al. Monitoring of Tumor Growth and Post-Irradiation Recurrence in a Diffuse Intrinsic Pontine Glioma Mouse Model. *Brain Pathol.* **2011**, *21*, 441–451. [CrossRef] [PubMed]
54. Jiang, Y.-J.; Lee, C.-L.; Wang, Q.; Zhou, Z.-W.; Yang, F.; Jin, C.; Fu, D.-L. Establishment of an Orthotopic Pancreatic Cancer Mouse Model: Cells Suspended and Injected in Matrigel. *World J. Gastroenterol.* **2014**, *20*, 9476–9485. [CrossRef]
55. Kleinman, H.K.; McGarvey, M.L.; Liotta, L.A.; Robey, P.G.; Tryggvason, K.; Martin, G.R. Isolation and Characterization of Type IV Procollagen, Laminin, and Heparan Sulfate Proteoglycan from the EHS Sarcoma. *Biochemistry* **1982**, *21*, 6188–6193. [CrossRef]
56. Johung, T.B.; Monje, M. Diffuse Intrinsic Pontine Glioma: New Pathophysiological Insights and Emerging Therapeutic Targets. *Curr. Neuropharmacol.* **2017**, *15*, 88–97. [CrossRef]
57. Donaldson, S.S.; Laningham, F.; Fisher, P.G. Advances toward an Understanding of Brainstem Gliomas. *J. Clin. Oncol. Off. J. Am. Soc. Clin. Oncol.* **2006**, *24*, 1266–1272. [CrossRef]
58. Johnston, M.; Zakharov, A.; Papaiconomou, C.; Salmasi, G.; Armstrong, D. Evidence of Connections between Cerebrospinal Fluid and Nasal Lymphatic Vessels in Humans, Non-Human Primates and Other Mammalian Species. *Cereb. Fluid Res.* **2004**, *1*, 2. [CrossRef] [PubMed]

59. Orešković, D.; Klarica, M. Development of Hydrocephalus and Classical Hypothesis of Cerebrospinal Fluid Hydrodynamics: Facts and Illusions. *Prog. Neurobiol.* **2011**, *94*, 238–258. [CrossRef] [PubMed]
60. Magdoom, K.N.; Brown, A.; Rey, J.; Mareci, T.H.; King, M.A.; Sarntinoranont, M. MRI of Whole Rat Brain Perivascular Network Reveals Role for Ventricles in Brain Waste Clearance. *Sci. Rep.* **2019**, *9*, 11480. [CrossRef]

Article

Cardiac Cell Exposure to Electromagnetic Fields: Focus on Oxdative Stress and Apoptosis

Ilenia Martinelli [1,2,†], Mathieu Cinato [1,2,†], Sokhna Keita [1,2], Dimitri Marsal [1,2], Valentin Antoszewski [2], Junwu Tao [3,‡] and Oksana Kunduzova [1,2,*,‡]

[1] National Institute of Health and Medical Research (INSERM) U1297, CEDEX 4, 31432 Toulouse, France; ilenia.martinelli@unicam.it (I.M.); cinato.mathieu@aol.fr (M.C.); sokhna.keita@inserm.fr (S.K.); dimitri.marsal@inserm.fr (D.M.)

[2] Unité Mixte de Recherche (UMR) 1297, Université Toulouse III Paul-Sabatier, 31062 Toulouse, France; valentin.antoszewski@univ-tlse3.fr

[3] Laboratoire Plasma et Conversion d'Energie (LAPLACE), National Polytechnic Institute of Toulouse, The Ecole Nationale Supérieure d'Electrotechnique, d'Electronique, d'Informatique, d'Hydraulique et des Télécommunications (ENSEEIHT), Toulouse University III, 31071 Toulouse, France; tao@laplace.univ-tlse.fr

* Correspondence: oxana.koundouzova@inserm.fr
† These authors contributed equally to this work.
‡ These authors contributed equally to this work.

Citation: Martinelli, I.; Cinato, M.; Keita, S.; Marsal, D.; Antoszewski, V.; Tao, J.; Kunduzova, O. Cardiac Cell Exposure to Electromagnetic Fields: Focus on Oxdative Stress and Apoptosis. *Biomedicines* **2022**, *10*, 929. https://doi.org/10.3390/biomedicines10050929

Academic Editor: Igor Belyaev

Received: 14 March 2022
Accepted: 13 April 2022
Published: 19 April 2022

Publisher's Note: MDPI stays neutral with regard to jurisdictional claims in published maps and institutional affiliations.

Abstract: Exposure to electromagnetic fields (EMFs) is a sensitive research topic. Despite extensive research, to date there is no evidence to conclude that exposure to EMFs influences the cardiovascular system. In the present study, we examined whether 915 MHz EMF exposure affects myocardial antioxidative and apoptotic status in vitro and in vivo. No statistically significant difference in the apoptotic cell profile and antioxidant capacity was observed between controls and short-term EMF-exposed mouse cardiomyocytes and H9C2 cardiomyoblasts. Compared with sham-exposed controls, mice subjected to a 915 MHz EMF for 48 h and 72 h had no significant effect on structural tissue integrity and myocardial expression of apoptosis and antioxidant genes. Therefore, these results indicate that short-term exposure to EMF in cardiac cells and tissues did not translate into a significant effect on the myocardial antioxidant defense system and apoptotic cell death.

Keywords: electromagnetic fields; cardiomyoblasts; oxidative stress; apoptosis

1. Introduction

Human exposure to electric and magnetic fields (EMFs) is an integral part of modern societies. Sources of exposure to EMFs are largely varied and increasingly prevalent. The EMFs generate the electromagnetic waves of different frequencies from natural environments, such as the solar energy and geomagnetic field, or from man-made sources, including transmission towers, telecommunications, home appliances, mobile phones, Wi-Fi, and base stations [1,2]. A number of studies have reported that EMF exposure of living systems can affect vital cellular processes [3–5]. However, the massive findings originated from in vitro and in vivo studies remain controversial. The lack of sufficient scientific data on EMF-exposed animal models has led to diverse apprehensions [6–9].

EMFs coupled with biological systems depend on the frequency ranges of the employed signals, which are classified as extremely low frequency field(s) (ELF, between 1 and 100 kHz) and high frequency (HF) fields, in the band of the radio frequency fields (RF, 100 kHz–3 GHz), and of the microwaves (MW, above 3 GHz) [10,11]. Numerous in vitro and in vivo studies have been performed in order to dissect the impact of EMFs generated by mobile phones and their base station in the range of 890–960 MHz. Several reports suggest that exposure to 900 MHz EMFs emitted by a mobile phone may cause cell apoptosis and oxidative stress in kidney, endometrium, and brain tissue [12–16]. Multiple molecular mechanisms have been proposed to explain the observed cellular effects of EMFs in the biological system. It has been claimed that possible effects could be triggered by generation of reactive oxygen species (ROS), disturbance of DNA-repair processes, and ROS-associated signaling pathways. In this scenario, the balance between pro- and antioxidant enzymes can be upset by an increase in ROS or by a decrease in antioxidant enzymes, including superoxide dismutase (SOD), catalase, and glutathione peroxidase (GSH-Px) [1,17].

Cardiac cells are particularly sensitive to changes in antioxidant capacity and ROS status due to the high energy metabolism in mitochondria [18]. Mitochondria, the main ROS generators, are abundant in heart tissue, constituting 40% of the cell volume of adult cardiac cells [19]. ROS are generated at an accelerated rate in the failing heart, suggesting that oxidative stress is common in cardiac cell dysfunction. Excessive ROS formation in cells has been shown to trigger the intrinsic apoptotic pathways via multiple mechanisms, including Bcl-2-associated X protein (Bax) translocation to the mitochondria, release of cytochrome c, and caspase activation [18]. Due to the impairment of transcriptional responses to oxidative stress and the decreased expression of antioxidant defense enzymes, oxidative stress was demonstrated to increase in damaged heart tissue [18]. While excessive accumulation of ROS-mediated damage is widely accepted as one of the primary causes of cardiac cell dysfunction, ROS also act in signaling pathways [19].

Given the ubiquitous nature of EMFs and their widespread applications, conclusive investigations into the potential effects of EMFs on cardiac cells are critical. However, there are many unknowns related to the influence of EMFs on antioxidant potential and cell death status in the heart. In this study, we examined the short-term exposure to 915 MHz EMFs on myocardial antioxidant capacity and apoptotic cell death in vitro and in vivo. Since morphological and biochemical analyses are the most applied methods to detect apoptotic cell death after EMF exposure [20], here, we examined the messenger ribonucleic acid (mRNA) levels of key apoptosis-related genes *Bcl2*, *Bax*, *caspase-3*, and *caspase-8* [21]. Among endogenous defense mechanisms responsible for the EMF-induced effects, catalase and SOD are crucial antioxidative enzymes that have been implicated in human diseases [22]. We also determined the expression levels of antioxidant defenses, including *catalase* and mitochondrial *SOD* in cardiac cells and tissues exposed to EMFs.

2. Materials and Methods

2.1. In Vitro Studies

Rat cardiomyoblasts (H9C2 cell line) were maintained in a growth medium with Dulbecco's Modified Eagle Medium (DMEM) Glutamax supplemented with 10% fetal bovine serum and 1% penicillin–streptomycin at 37 °C in 5% CO_2. The cultured H9C2 cells were grown to 80–90% in a Petri dish and were exposed to EMFs for 24 h, 48 h, and 72 h. Cardiac myocytes were isolated from male 10-week-old C57BL6 mice using the method described by Xu et al. [23]. Briefly, hearts were perfused by the Langendorff method with HEPES-buffered Earle's balanced salt solution (GIBCO-BRL) supplemented with 6 mM glucose, amino acids and vitamins (buffer A), and then with buffer A containing 0.8 mg/mL collagenase B (Boehringer-Mannheim, Mannheim, Germany) and 10 μM $CaCl_2$. The enzyme solutions were filtered (2-μm pores) and recirculated through the heart until the flow rate doubled (12–20 min); the left ventricle was then removed and minced in collagenase containing buffer A. Thereafter, the tissue pieces were transferred to fresh (enzyme free) buffer A supplemented with 1.25 mg/mL taurine, 5 mg/mL BSA (Sigma-Aldrich, St. Louis, MO, USA), and 150 μM $CaCl_2$ and mechanically dissociated by gentle trituration. The resulting suspension was filtered and isolated cells were obtained by sedimentation. The cultured primary cardiomyocytes were exposed to 915 MHz EMFs for 1 h, 3 h, and 24 h.

2.1.1. Radiofrequency Equipment to In Vitro Studies

To evaluate the biological effects of radiofrequency EMF on living cardiac cells, we used a transverse electromagnetic (TEM) cell (FCC-TEM-JM1 from Fischer Custom Communications, Torrance, CA, USA). A solid state radiofrequency generator (2022D from

Marconi Instruments Limited, Saint Albans, UK) was used to power the TEM cell. In order to define dielectric property of the biological medium, the technique of measuring samples reported on a microstrip transmission line was used [24]. The TEM cell used in an in vitro experiment was modeled on HFSS, as described previously [24].

2.1.2. In Vitro EMF Exposition Parameters

In vitro cell culture experiments were conducted in a standard cell culture incubator at 37 °C, 5% CO_2, and at 21% O_2 levels. Control cells were treated in the same manner as the exposed ones but were not subjected to EMFs at any point. Using Marconi 2022D (Marconi Instruments Limited, Saint Albans, UK) tune without modulation (for the continuous wave mode), the 20 mW power level led to a maximum local specific absorption rate (SAR) of ~0.08 W/kg in cultured cells. The simulation was performed using high frequency structure simulator (HFSS) software version 15. The 20 mW corresponded to the highest output power of the generator. The TEM cell was placed inside a cell culture incubator. The power supply of the TEM cell was carried out by a coaxial cable through an orifice for the passage of pipes.

2.2. In Vivo Studies

2.2.1. Animals

The investigation conforms to the Guide for the Care and Use of Laboratory Animals published by the US National Institutes of Health (NIH Publication no. 85-23, revised 1985) and was performed in accordance with the recommendations of the French Accreditation of the Laboratory Animal Care (approved by the local Centre National de la Recherche Scientifique ethics committee). The RjOrl:Swiss female mice at 6 months of age were used for these investigations (Envigo RMS, Gannat, France). After arrival, animals were randomized and housed in groups of four in polycarbonate cages, enriched with paper. Mice were allowed free access to standard food pellets and tap water. Temperature was controlled at 21 ± 2 °C. Two days prior to EMF exposure, mice were adapted to new environmental conditions in a room in which, subsequently, the whole experiment was performed. The light was on a 12 h light–12 h dark cycle, with light on at 8 am. Mice were then randomly segregated into two groups (control, n = 6 and EMFs, n = 10) and exposed to a 915 MHz EMF for 48 h and 72 h. Control animals were treated in the same manner but were not subjected to EMFs at any point.

2.2.2. Radiofrequency Equipment to In Vivo Studies

In vivo experiments were performed in a Giga-TEM (GTEM) cell to accommodate cages (up to 4) in which the animals exposed to EMFs. The model of the GTEM cell is shown in Figure 1. Twenty-four hours before the experiment, the mice were placed in the GTEM to adaptation.

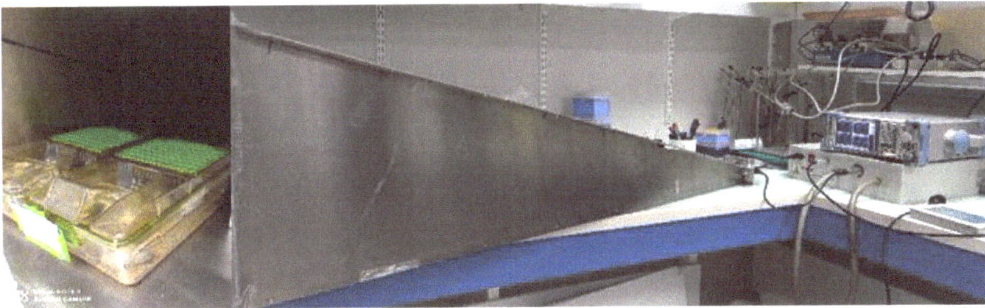

Figure 1. GTEM cell to explore in vivo studies.

We used a solid-state radiofrequency generator of a fixed 915 MHz frequency (WSPS-915–1000) (Chengdu Wattsine Electronics Technology, Chengdu, Sichuan, China). A full-wave electromagnetic simulation was taken to verify if matched load conditions with one or several cages could be obtained by center conductor tuning [25]. In these simulations, the webster was modeled by the sphere of equivalent volume. Three spheres were placed along the wave propagation axis at the end of the GTEM cell. By using a relative permittivity of 55 and equivalent conductivity of 3 S/m, the estimated SAR values reached ~40 W/kg. The calculation of whole body SAR was taken by volume integration of absorption power using MATLAB codes. The field data were taken from HFSS simulation exportation. The input RF power was 4 W in the HFSS simulation and in the experimentation.

2.3. Quantitative RT–PCR Analysis

The expression of genes was assessed using quantitative polymerase chain reaction (qRT-PCR). Total RNAs were isolated from the rat cardiomyoblast cell line H9C2, primary cardiomyocytes, and cardiac tissue using the RNeasy mini kit (Qiagen, Hilden, Germany). Total RNAs (300 ng) were reverse transcribed as previously described [26]. Primer sequences were detailed in Supplementary Table S1. The expression of target mRNA was normalized to GAPDH and RPLP0 mRNA expression.

2.4. Morphology

Heart tissue integrity was assessed by hematoxylin and eosin (H&E) stain, according to standard methods. In both the experimental group and the control group, mice were sacrificed to collect hearts at 48 h and 72 h. Tissues were collected under a protocol approved by the Animal Care and Use Committee of INSERM. Tissues were snap frozen for further analyses. In cardiac tissue, the H&E stain was performed to evaluate morphological features, including damaged myofibers, collagen accumulation, necrotic myofibers, and myocyte size.

2.5. TUNEL Assay

The apoptosis level was assessed using the DeadEndFluorometric TUNEL system according to the manufacturer's instructions (Promega, Madison, WI, USA). TUNEL staining were performed in heart cryosections to detect nuclear DNA fragmentation during apoptosis.

2.6. Western Blot

Proteins from cardiac tissues were extracted using the RIPA buffer and quantified using the Bio-Rad Protein Assay (Bio-Rad, Hercules, CA, USA). Proteins were resolved by SDS-PAGE and western blotting. Immunoreactive bands were detected by chemiluminescence with the Clarity Western ECL Substrate (Bio-Rad, Hercules, CA, USA) on a ChemiDoc MP Acquisition system (Bio-Rad, Hercules, CA, USA), as previously described [26]. Antibodies used in this study were SOD2 and tubulin; both were purchased from Santa Cruz Biotechnology (Santa Cruz, CA, USA). Tubulin was used as a loading control.

2.7. Statistical Analysis

Data are expressed as mean $\pm$ SEM. A statistical comparison between the two groups was performed by Student's *t*-test and the ANOVA test was used to assess the statistical significance of the difference between more than two study group means. The Kruskal–Wallis test was used to assess the statistical significance of the difference of a non-parametric variable between more than the two study groups using GraphPad Prism version 5.00 (GraphPad Software, Inc., San Diego, CA, USA). p-values < 0.05 were considered to indicate statistically significant differences.

3. Results

In order to determine whether radiofrequency EMFs affect the antioxidant defense system and apoptosis at the cellular level, adult mouse cardiomyocytes were exposed to 915 MHz EMF at 13 dBm. Gene expression profiles of apoptosis-related factors, including *Bax*, *Bcl2*, *caspase 3*, and *caspase 8*, were determined in cultured cardiomyocytes exposed to EMFs for 1 h, 3 h, and 24 h. As shown in Figure 2, exposure of myocytes to a 915 MHz did not significantly change the mRNA expression of *Bax*, *Bcl2*, *caspase 3*, and *caspase 8* at 1 h and 3 h as compared to control conditions. In exposed cells for 24 h, *Bax* and *caspase 3* mRNA levels showed a tendency to increase, but this did not reach statistical significance (Figure 2B,D).

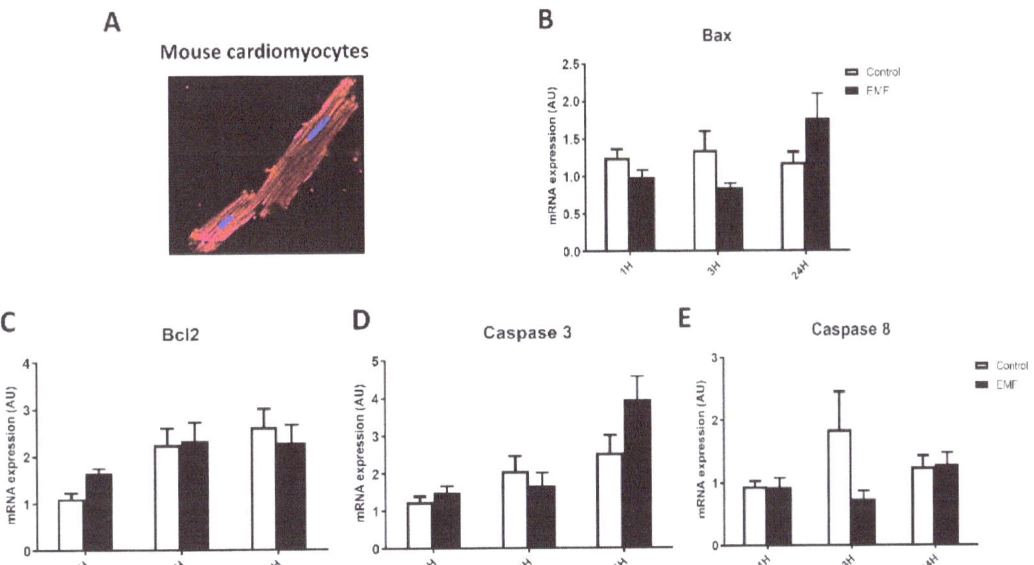

Figure 2. Apoptosis-related gene expression in mouse cardiomyocytes exposed to EMFs. Representative pictures of cultured mouse cardiomyocytes (**A**) and mRNA expression levels of *Bax* (**B**), *Bcl2* (**C**), *caspase 3* (**D**), and *caspase 8* (**E**) in cardiomyocytes exposed to 915 MHz EMFs for 1 h, 3 h, and 24 h. The results present the mean ± SEM of three independent experiments.

To evaluate the level of antioxidant capacity in cardiac cells in conditions of the electromagnetic field, we next examined the expression of *catalase* and *SOD2* genes in cardiomyocytes subjected to a 915 MHz frequency EMF for 1 h, 3 h, and 24 h. As shown in Figure 3, real-time qPCR analysis revealed no differences for expression levels of *catalase* and *SOD2* between control and EMF-exposed cardiomyocytes.

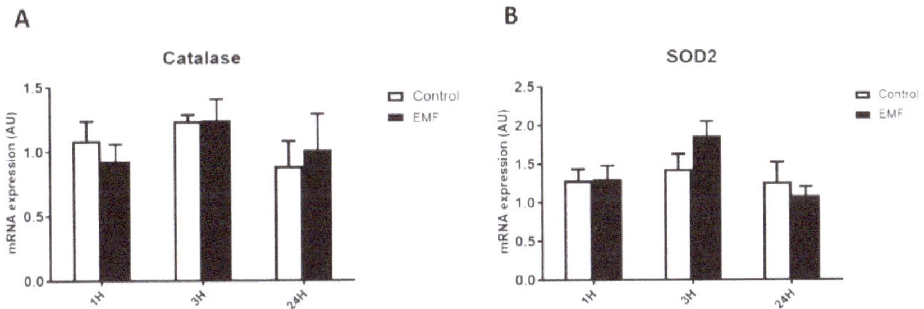

Figure 3. Effects of 915 MHz EMFs on antioxidant enzyme gene expression in cultured cardiomyocytes. Quantitative analysis of the mRNA expression levels of *catalase* (**A**) and *SOD2* (**B**) in cardiomyocytes exposed to EMFs for 1 h, 3 h, and 24 h. The results present the mean ± SEM of three independent experiments.

These results suggest an adaptive response of cells subjected to a 915 MHz continuous wave radiofrequency exposure for 1 h, 3 h, and 24 h.

Interestingly, similar results were obtained in H9C2 cardiomyoblasts. As shown in Figure 4, no significant differences were observed in the expression levels of apoptosis-related factors and antioxidant enzyme *catalase* in cells subjected to EMFs for 24 h, 48 h, and 72 h. These data suggest that short-term exposition to EMFs in cardiac cells did not translate into a significant effect on antioxidant enzymes and apoptotic cell death.

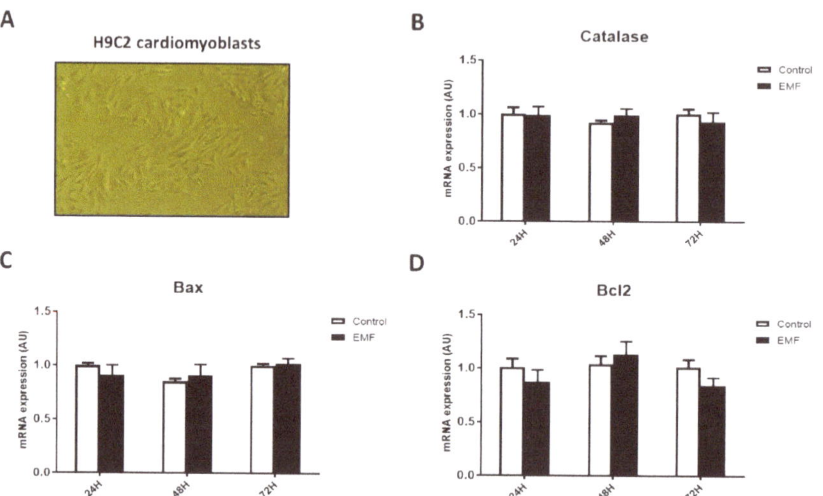

Figure 4. Gene expression profile of antioxidant enzyme *catalase* and pro- and anti-apoptotic factors in H9C2 cells exposed to radiofrequency EMFs at 915 MHz for 24 h, 48 h, and 72 h. Representative pictures of H9C2 cardiomyoblasts (**A**) and mRNA expression levels of *catalase* (**B**), *Bax* (**C**), *Bcl2* (**D**). The results present the mean ± SEM of three independent experiments.

In order to evaluate in vivo effects of EMFs in mice, we next examined whether exposition of mice to EMFs affected structural integrity, apoptosis, and antioxidant capacity in cardiac tissues. As shown in Figure 5A, no significant differences between EMF and control groups were observed at the structural level in terms of collagen accumulation, necrotic myofibers, and myocyte size. We next examined whether electromagnetic stress could cause cardiac apoptosis after mice were exposed to 915 MHz for 48 h and 72 h. As shown in Figure 5B,C, both control mice and EMF-exposed mice exhibited minimal apoptosis assessed by TUNEL staining in the left ventricles.

Furthermore, no significant differences were found in the expression of genes involved in the antioxidant system and apoptosis regulation in cardiac tissue from mice exposed to 915 MHz EMFs and control animals (Figure 6).

Western blot analysis further confirmed no significant changes in the expression of antioxidant enzyme SOD2 in cardiac tissues from mice subjected to 915 MHz for 48 h and 72 h as compared to control mice (Figure 7).

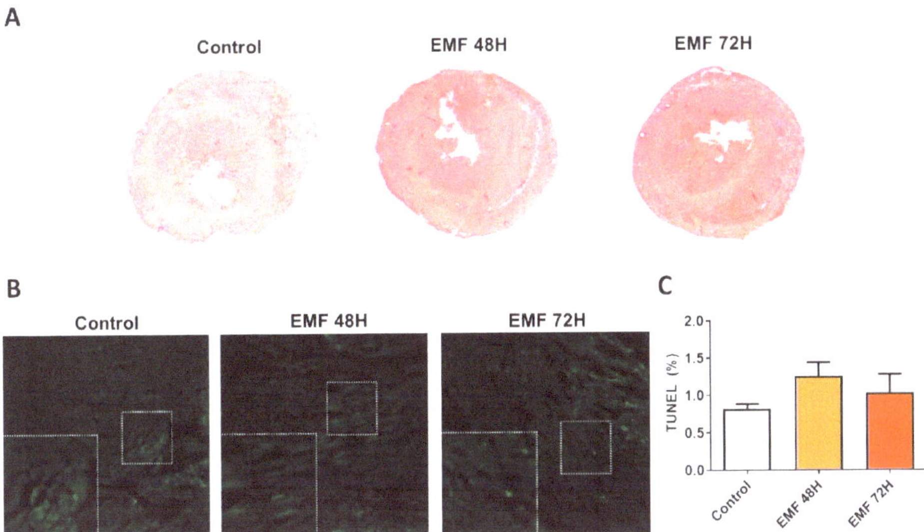

Figure 5. Effects of 915 MHz EMFs on cardiac structure and apoptosis in mice exposed for 48 h and 72 h. Representative images of cardiac sections stained with H&E (**A**). Scale bar: 2.5 mm. Representative images of cardiac sections stained with TUNEL (**B**) and quantification (**C**). Scale bar: 100 μm. The results present the mean ± SEM, $n = 3$.

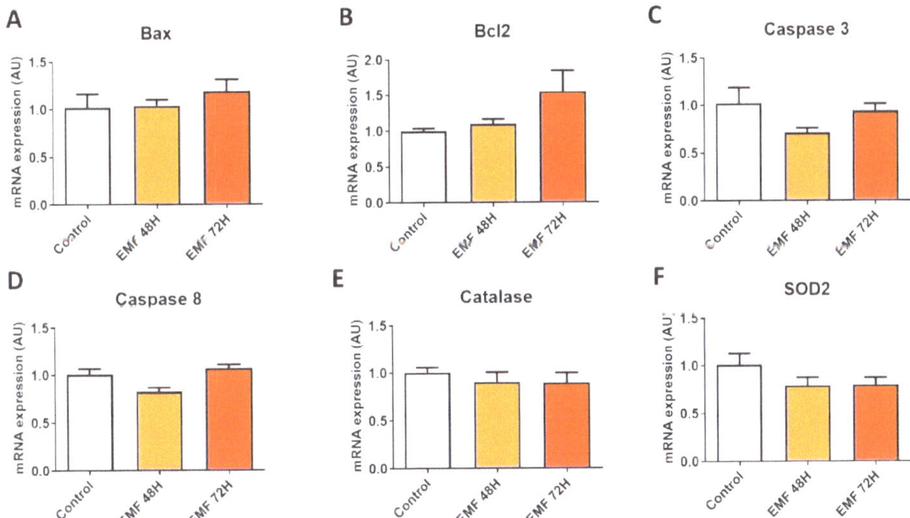

Figure 6. Gene expression profile of antioxidant enzymes and apoptosis-related factors in mice subjected to EMFs. Quantitative analysis of the mRNA expression levels of *Bax* (**A**), *Bcl2* (**B**), *caspase 3* (**C**), *caspase 8* (**D**), *catalase* (**E**), and *SOD2* (**F**) in mice exposed to 915 MHz EMFs for 48 h and 72 h and control animals. The results present the mean ± SEM, $n = 6–10$.

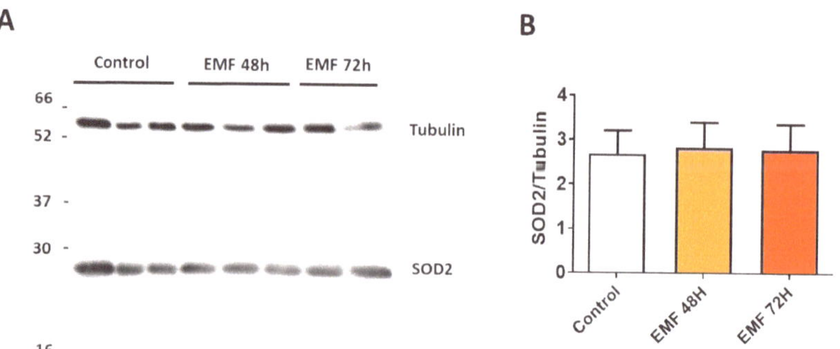

Figure 7. Myocardial expression of SOD2 in mice exposed to 915 MHz EMFs. Representative western blot image (**A**) and quantification (**B**) of SOD2 protein expression levels in mice exposed to 915 MHz EMFs for 48 h and 72 h and control animals The results present the mean $\pm$ SEM, $n = 6$–10.

4. Discussion

The use of mobile phones and base stations are increasing worldwide [27]; however, the biological effects of EMF exposure on cardiac cells and tissues remain unclear. Cardiomyocytes are sensitive to a wide range of internal and external stimuli linked to ROS formation [18]. Oxidative stress and cell apoptotic reprogramming are largely involved in the pathophysiology of human diseases, where both survival and cell death signaling cascades have been reported to be modulated by ROS [28]. The present study was designed to test the hypothesis that 915 MHz EMFs affect myocardial antioxidant capacity and apoptotic cell death in vitro and in vivo.

Potential harmful effects of EMF on biological systems have been considered controversial. Currently, there are debates about the influence of EMFs on DNA damage or oxidative stress. Indeed, some data have described damaging activities, conversely others have described no evident influence [29]. The observed bio-effects are dependent on various conditions, such as cells oxidative status, level of anti-oxidative enzymes, cell type, and parameters of the applied EMF [29]. Studies in vitro and in vivo revealed that EMFs at very low frequencies induced an increased production of free radicals, which can determine DNA double-strand breaks [29]. Interestingly, in vitro, in primary rat neocortical astroglial cell culture, 900 MHz induces ROS production and DNA damage [30]. Moreover, a range of studies have demonstrated that various cell lines do respond differently to EMF exposure, as, in vitro, six types of different cell lines exposed to EMF at 1800 MHz for 1 h and 24 h showed DNA damage in a cell type-dependent manner [31], while human glioblastoma cell lines exposed to 1950 MHz of EMF did not cause chromosomal damage after 50, 100, 150, and 200 h [32]. In addition, no significant ROS generation was measured in human cell lines exposed to 1800 MHz [33]. In accordance, our study showed no remarkably differences in oxidative status. Indeed, the short-term EMF-exposure of myocytes to 915 MHz did not significantly change the gene expression of antioxidant enzymes as compared to control conditions. Moreover, in vivo, the 1950 MHz EMF exposure did not affect oxidative stress in the aging brain [34]. On the contrary, a previous study showed that EMF induced oxidative damage and it was associated with neurodegenerative diseases with deleterious effects on brain tissue [35]. Another study reported that RF exposure at 1800 MHz slightly elevated the concentration of lipid peroxidation markers in the in brain, blood, liver, and kidney [36].

The activation of apoptosis is also considered to be involved in possible damage induced by EMF. An in vitro study reported that 1950 MHz signals induced apoptosis in astrocytes with the involvement of *Bax* and *Bcl2* [37]. In the present study, to elucidate whether EMF affected cell death in cardiac cells, we evaluated mRNA expression of *Bax* (pro-apoptotic factor) and *Bcl2* (anti-apoptotic factor) in H9C2 cells subjected to EMF 915

MHz for 24 h, 48 h, and 72 h. We did not observe significant differences in *Bax*, *Bcl2*, and *catalase* levels after short-term EMF exposure.

Exposure to EMFs can damage biological tissues by inducing changes in structure and function [1,4]. We provide in vivo evidence that 915 MHz of EMF exposure did not induce alterations in myocardial structure and apoptosis in mice at 48 h and 72 h. Furthermore, an analysis of antioxidant system capacity EMF-exposed mice demonstrated no significant alterations in the expression of antioxidant enzymes as compared to control mice. Thus, an in vivo study does not support the hypothesis that short-term exposure to 915 MHz promotes changes in the heart. However, we cannot rule out that long-term exposure to electromagnetic stress may be risk factors for myocardial damage.

5. Conclusions

Both in vitro and in vivo short-term studies found no evidence that EMFs affect antioxidant and apoptosis status in cardiac cells and tissues. Further studies examining dynamic changes in oxidative stress and apoptosis after long-term cardiac cell exposure to EMFs are warranted. These data provide an important reference in relation to the cellular antioxidant defense system and programmed cell death in response to electromagnetic stress.

Supplementary Materials: The following supporting information can be downloaded at: https://www.mdpi.com/article/10.3390/biomedicines10050929/s1, Table S1: List of primer sequences.

Author Contributions: Conceptualization, O.K. and J.T.; methodology, O.K. and J.T.; software, J.T.; validation, I.M. and M.C.; formal analysis, S.K.; investigation, I.M., M.C., S.K., D.M. and V.A.; resources, O.K. and J.T.; data curation, I.M. and M.C.; writing—original draft preparation, I.M. and M.C.; writing—review and editing, I.M.; supervision, O.K. and J.T.; project administration, O.K.; funding acquisition, O.K. All authors have read and agreed to the published version of the manuscript.

Funding: This research was funded by ANSES, grant number EST-2016 RF-01 and by INSERM.

Institutional Review Board Statement: All experiments were conducted in accordance with the guidelines established by the European Communities Council Directive (2010/63/EU Council Directive Decree) and approved by the local ethical committee at Institute of Metabolic and Cardiovascular diseases (project N 20/1048/14/02, national agreement 2020072717253128).

Data Availability Statement: The data presented in this study are available on request from the corresponding author.

Acknowledgments: We are grateful to Angelo Parini and Cecile Gautier for the administrative.

Conflicts of Interest: The authors declare no conflict of interest.

References

1. Kıvrak, E.G.; Yurt, K.K.; Kaplan, A.A.; Alkan, I.; Altun, G. Effects of electromagnetic fields exposure on the antioxidant defense system. *J. Microsc. Ultrastruct.* **2017**, *5*, 167–176. [CrossRef] [PubMed]
2. Dauda Usman, J.; Isyaku, U.M.; Magaji, R.A.; Fasanmade, A.A. Assessment of electromagnetic fields, vibration and sound exposure effects from multiple transceiver mobile phones on oxidative stress levels in serum, brain and heart tissue. *Sci. Afr.* **2020**, *7*, e00271. [CrossRef]
3. Singh, S.; Kapoor, N. Health implications of electromagnetic fields, mechanisms of action, and research needs. *Adv. Biol.* **2014**, *2014*, 198609. [CrossRef]
4. Kaszuba-Zwoińska, J.; Gremba, J.; Gałdzińska-Calik, B.; Wójcik-Piotrowicz, K.; Thor, P.J. Electromagnetic field induced biological effects in humans. *Przegl. Lek.* **2015**, *72*, 636–641.
5. Schuermann, D.; Mevissen, M. Manmade Electromagnetic fields and oxidative stress-biological effects and consequences for health. *Int. J. Mol. Sci.* **2021**, *22*, 3772. [CrossRef]
6. Kaprana, A.E.; Karatzanis, A.D.; Prokopakis, E.P.; Panagiotaki, I.E.; Vardiambasis, I.O.; Adamidis, G.; Christodoulou, P.; Velegrakis, G.A. Studying the effects of mobile phone use on the auditory system and the central nervous system: A review of the literature and future directions. *Eur. Arch. Otorhinolaryngol.* **2008**, *265*, 1011–1019. [CrossRef]
7. Regel, S.J.; Achermann, P. Cognitive performance measures in bioelectromagnetic research—Critical evaluation and recommendations. *Environ. Health* **2011**, *10*, 10. [CrossRef]
8. Genuis, S.J.; Lipp, C.T. Electromagnetic hypersensitivity: Fact or fiction? *Sci. Total Environ.* **2012**, *414*, 103–112. [CrossRef]

9. De Luca, C.; Thai, J.C.; Raskovic, D.; Cesareo, E.; Caccamo, D.; Trukhanov, A.; Korkina, L. Metabolic and genetic screening of electromagnetic hypersensitive subjects as a feasible tool for diagnostics and intervention. *Mediat. Inflamm.* **2014**, *2014*, 924184. [CrossRef]

10. Ahlbom, A.; Bergqvist, U.; Bernhardt, J.H.; Cesarini, J.P.; Court, L.A.; Grandolfo, M.; Hietanen, M.; McKinlay, A.F.; Repacholi, M.H.; Sliney, D.H.; et al. Guidelines for limiting exposure to time-varying electric, magnetic, and electromagnetic fields (up to 300 GHz). *Health Phys.* **1998**, *74*, 494–521.

11. International Commission on Non-Ionizing Radiation Protection. Guidelines for limiting exposure to time-varying electric and magnetic fields (1 Hz to 100 kHz). *Health Phys.* **2010**, *99*, 818–836. [CrossRef] [PubMed]

12. Özorak, A.; Nazıroğlu, M.; Çelik, Ö.; Yüksel, M.; Özçelik, D.; Özkaya, M.O.; Çetin, H.; Kahya, M.C.; Kose, S.A. Wi-Fi (2.45 GHz)- and mobile phone (900 and 1800 MHz)-induced risks on oxidative stress and elements in kidney and testis of rats during pregnancy and the development of offspring. *Biol. Trace Elem. Res.* **2013**, *156*, 221–229. [CrossRef] [PubMed]

13. Ulubay, M.; Yahyazadeh, A.; Deniz, Ö.G.; Kıvrak, E.G.; Altunkaynak, B.Z.; Erdem, G.; Kaplan, S. Effects of prenatal 900 MHz electromagnetic field exposures on the histology of rat kidney. *Int. J. Radiat. Biol.* **2015**, *91*, 35–41. [CrossRef] [PubMed]

14. Deniz, Ö.G.; Kıvrak, E.G.; Kaplan, A.A.; Altunkaynak, B.Z. Effects of folic acid on rat kidney exposed to 900 MHz electromagnetic radiation. *J. Microsc. Ultrastruct.* **2017**, *5*, 198–205. [CrossRef]

15. Oral, B.; Guney, M.; Ozguner, F.; Karahan, N.; Mungan, T.; Comlekci, S.; Cesur, G. Endometrial apoptosis induced by a 900-MHz mobile phone: Preventive effects of vitamins E and C. *Adv. Ther.* **2006**, *23*, 957–973. [CrossRef]

16. Meral, I.; Mert, H.; Mert, N.; Deger, Y.; Yoruk, I.; Yetkin, A.; Keskin, S. Effects of 900-MHz electromagnetic field emitted from cellular phone on brain oxidative stress and some vitamin levels of guinea pigs. *Brain Res.* **2007**, *1169*, 120–124. [CrossRef]

17. Calcabrini, C.; Mancini, U.; De Bellis, R.; Diaz, A.R.; Martinelli, M.; Cucchiarini, L.; Sestili, P.; Stocchi, V.; Potenza, L. Effect of extremely low-frequency electromagnetic fields on antioxidant activity in the human keratinocyte cell line NCTC 2544. *Biotechnol. Appl. Biochem.* **2017**, *64*, 415–422. [CrossRef]

18. Peoples, J.N.; Saraf, A.; Ghazal, N.; Pham, T.T.; Kwong, J.Q. Mitochondrial dysfunction and oxidative stress in heart disease. *Exp. Mol. Med.* **2019**, *51*, 1–13. [CrossRef]

19. Doenst, T.; Nguyen, T.D.; Abel, E.D. Cardiac metabolism in heart failure: Implications beyond ATP production. *Circ. Res.* **2013**, *113*, 709–724. [CrossRef]

20. Romeo, S.; Zeni, O.; Scarfi, M.R.; Poeta, L.; Lioi, M.B.; Sannino, A. Radiofrequency electromagnetic field exposure and apoptosis: A scoping review of in vitro studies on mammalian cells. *Int. J. Mol. Sci.* **2022**, *23*, 2322. [CrossRef]

21. Elmore, S. Apoptosis: A review of programmed cell death. *Toxicol. Pathol.* **2007**, *35*, 495–516. [CrossRef] [PubMed]

22. Cichon, N.; Bijak, M.; Synowiec, E.; Miller, E.; Sliwinski, T.; Saluk-Bijak, J. Modulation of antioxidant enzyme gene expression by extremely low frequency electromagnetic field in post-stroke patients. *Scand. J. Clin. Lab. Investig.* **2018**, *78*, 626–631. [CrossRef] [PubMed]

23. Xu, H.; Guo, W.; Nerbonne, J.M. Four kinetically distinct depolarization-activated K+ currents in adult mouse ventricular myocytes. *J. Gen. Physiol.* **1999**, *113*, 661–678. [CrossRef] [PubMed]

24. Cinato, M.; Tao, J.; Keita, S.; Lefeuvre, S.; Kunduzova, O. Investigation of Electromagnetic Fields on Cardiac Cells. In Proceedings of the 17th International Conference on Microwave and High Frequency Heating (AMPERE 2019), Valence, Spain, 9–12 September 2019; pp. 1–7.

25. Tao, J.; Li, Z.-G.; Vuong, T.-H.; Kunduzova, O. GTEM cell design for controlled exposure to electromagnetic fields in biological system. *bioRxiv* **2022**. [CrossRef]

26. Martinelli, I.; Timotin, A.; Moreno-Corchado, P.; Marsal, D.; Kramar, S.; Loy, H.; Joffre, C.; Boal, F.; Tronchere, H.; Kunduzova, O. Galanin promotes autophagy and alleviates apoptosis in the hypertrophied heart through FoxO1 pathway. *Redox Biol.* **2021**, *40*, 101866. [CrossRef]

27. Miller, A.B.; Sears, M.E.; Morgan, L.L.; Davis, D.L.; Hardell, L.; Oremus, M.; Soskolne, C.L. Risks to health and well-being from radio-frequency radiation emitted by cell phones and other wireless devices. *Front. Public Health* **2019**, *7*, 223. [CrossRef]

28. Redza-Dutordoir, M.; Averill-Bates, D.A. Activation of apoptosis signalling pathways by reactive oxygen species. *Biochim. Biophys. Acta* **2016**, *1863*, 2977–2992. [CrossRef]

29. Saliev, T.; Begimbetova, D.; Masoud, A.R.; Matkarimov, B. Biological effects of non-ionizing electromagnetic fields: Two sides of a coin. *Prog. Biophys. Mol. Biol.* **2019**, *141*, 25–36. [CrossRef]

30. Campisi, A.; Gulino, M.; Acquaviva, R.; Bellia, P.; Raciti, G.; Grasso, R.; Musumeci, F.; Vanella, A.; Triglia, A. Reactive oxygen species levels and DNA fragmentation on astrocytes in primary culture after acute exposure to low intensity microwave electromagnetic field. *Neurosci. Lett.* **2010**, *473*, 52–55. [CrossRef]

31. Xu, S.; Chen, G.; Chen, C.; Sun, C.; Zhang, D.; Murbach, M.; Kuster, N.; Zeng, Q.; Xu, Z. Cell type-dependent induction of DNA damage by 1800 MHz radiofrequency electromagnetic fields does not result in significant cellular dysfunctions. *PLoS ONE* **2013**, *8*, e54906. [CrossRef]

32. Al-Serori, H.; Kundi, M.; Ferk, F.; Mišík, M.; Nersesyan, A.; Murbach, M.; Lah, T.T.; Knasmüller, S. Evaluation of the potential of mobile phone specific electromagnetic fields (UMTS) to produce micronuclei in human glioblastoma cell lines. *Toxicol. Vitr.* **2017**, *40*, 264–271. [CrossRef] [PubMed]

33. Lantow, M.; Lupke, M.; Frahm, J.; Mattsson, M.O.; Kuster, N.; Simko, M. ROS release and Hsp70 expression after exposure to 1800 MHz radiofrequency electromagnetic fields in primary human monocytes and lymphocytes. *Radiat. Environ. Biophys.* **2006**, *45*, 55–62. [CrossRef] [PubMed]

34. Jeong, Y.J.; Son, Y.; Han, N.K.; Choi, H.D.; Pack, J.K.; Kim, N.; Lee, Y.S.; Lee, H.J. Impact of long-term RF-EMF on oxidative stress and neuroinflammation in aging brains of C57BL/6 mice. *Int. J. Mol. Sci.* **2018**, *19*, 2103. [CrossRef] [PubMed]

35. Consales, C.; Merla, C.; Marino, C.; Benassi, B. Electromagnetic fields, oxidative stress, and neurodegeneration. *Int. J. Cell Biol.* **2012**, *2012*, 683897. [CrossRef]

36. Bodera, P.; Stankiewicz, W.; Antkowiak, B.; Paluch, M.; Kieliszek, J.; Sobiech, J.; Niemcewicz, M. Influence of electromagnetic field (1800 MHz) on lipid peroxidation in brain, blood, liver and kidney in rats. *Int. J. Occup. Med. Environ. Health* **2015**, *28*, 751–759. [CrossRef]

37. Liu, Y.X.; Tai, J.L.; Li, G.Q.; Zhang, Z.W.; Xue, J.H.; Liu, H.S.; Zhu, H.; Cheng, J.D.; Liu, Y.L.; Li, A.M.; et al. Exposure to 1950-MHz TD-SCDMA electromagnetic fields affects the apoptosis of astrocytes via caspase-3-dependent pathway. *PLoS ONE* **2012**, *7*, e42332. [CrossRef]

Article

Gabapentin Increases Intra-Epidermal and Peptidergic Nerve Fibers Density and Alleviates Allodynia and Thermal Hyperalgesia in a Mouse Model of Acute Taxol-Induced Peripheral Neuropathy

Michal Klazas [1], Majdi Saleem Naamneh [2], Wenhua Zheng [3] and Philip Lazarovici [2,*]

[1] Pharmacy Unit, School of Pharmacy Institute for Drug Research, Faculty of Medicine, The Hebrew University of Jerusalem, Jerusalem 9112002, Israel

[2] Pharmacology Unit, School of Pharmacy Institute for Drug Research, Faculty of Medicine, The Hebrew University of Jerusalem, Jerusalem 9112002, Israel

[3] Center of Reproduction, Development and Aging and Institute of Translation Medicine, Faculty of Health Sciences, University of Macau, Taipa, Macau 999078, China

* Correspondence: philipl@ekmd.huji.ac.il; Tel.: +972-2-6758729; Fax: +972-2-6757490

Abstract: The clinical pathology of Taxol-induced peripheral neuropathy (TIPN), characterized by loss of sensory sensitivity and pain, is mirrored in a preclinical pharmacological mice model in which Gabapentin, produced anti-thermal hyperalgesia and anti-allodynia effects. The study aimed to investigate the hypothesis that gabapentin may protect against Taxol-induced neuropathic pain in association with an effect on intra-epidermal nerve fibers density in the TIPN mice model. A TIPN study schedule was induced in mice by daily injection of Taxol during the first week of the experiment. Gabapentin therapy was performed during the 2nd and 3rd weeks. The neuropathic pain was evaluated during the whole experiment by the Von Frey, tail flick, and hot plate tests. Intra-epidermal nerve fibers (IENF) density in skin biopsies was measured at the end of the experiment by immunohistochemistry of ubiquitin carboxyl-terminal hydrolase PGP9.5 pan-neuronal and calcitonin gene-related (CGRP) peptides-I/II- peptidergic markers. Taxol-induced neuropathy was expressed by 80% and 73% reduction in the paw density of IENFs and CGPR, and gabapentin treatment corrected by 83% and 46% this reduction, respectively. Gabapentin-induced increase in the IENF and CGRP nerve fibers density, thus proposing these evaluations as an additional objective end-point tool in TIPN model studies using gabapentin as a reference compound.

Keywords: Taxol-induced peripheral neuropathy (TIPN); intra-epidermal nerve fibers (IENF); calcitonin gene-related peptides (CGRP); gabapentin (Neurontin); neuropathic pain; analgesia; mice model

Citation: Klazas, M.; Naamneh, M.S.; Zheng, W.; Lazarovici, P. Gabapentin Increases Intra-Epidermal and Peptidergic Nerve Fibers Density and Alleviates Allodynia and Thermal Hyperalgesia in a Mouse Model of Acute Taxol-Induced Peripheral Neuropathy. *Biomedicines* 2022, 10, 3190. https://doi.org/10.3390/biomedicines10123190

Academic Editor: Martina Perše

Received: 3 November 2022
Accepted: 5 December 2022
Published: 8 December 2022

Publisher's Note: MDPI stays neutral with regard to jurisdictional claims in published maps and institutional affiliations.

1. Introduction

Chemotherapy-induced peripheral neuropathy represents a common, dose-limiting, adverse effect of cytotoxic anticancer drug chemotherapy [1]. Taxol (paclitaxel) is one of the most commonly used taxane drugs that specifically inhibit the function of microtubules and, hence, the formation of the mitotic spindle, therefore blocking tumor cell proliferation [2] and angiogenesis [3]. For this reason, Taxol represents a first-line chemotherapy drug in oncology, commonly used to inhibit the progression of ovarian, breast, cervical, endometrial, carcinoma tumors, etc. [4]. Unfortunately, Taxol-induces peripheral neuropathy (TIPN) is an adverse effect, with an incidence of 30 to 50% [5]. It is manifested by loss of cutaneous sensation followed by persistent pain, such as mechanical allodynia (an exaggerated painful response to mechanical stimulation due to axonal degeneration and neurotoxicity to sensory neurons) and thermal hyperalgesia (painful perception of temperature), thus leading to the dose reduction or discontinuation of therapy. Therefore, there is an unmet clinical need to find safe and efficacious drugs for the therapy of TIPN.

To identify possible methods of prevention and treatment for TIPN, it is important to use a validated, preclinical animal model. While several models have been developed, only a few have proven translatable and effective [6]. C57BL/6J mice are commonly used for TIPN research, proving to be most efficacious, quantitatively defined as being reproducible and producing significant peripheral polyneuropathy in at least 90% of the experiments and at least two neurobehavioral outcomes such as thermal hyper- and hypo-algesia and mechanical allodynia [6]. Moreover, C57BL/6J mice are widely used and commercially available, and, in addition to their inbred nature, allow for transgenic and knockout models. The removal of genetic variability minimizes phenotypic or trait variability, thereby enhancing the reproducibility of the model [7]. In the majority of TIPN experiments, the age of the mice ranged from four weeks up to adulthood [8] and TIPN was induced by cumulative doses of Taxol ranging from 4 mg/kg up to 180 mg/kg, delivered in the majority of preclinical studies by the intraperitoneal (i.p.) route of administration, causing a significant peripheral polyneuropathy in 91% of the experiments using males [6]. Mice receiving a Taxol regimen mirror common clinical syndrome of hyperalgesia, allodynia, and numbness as early as the first or third day after receiving their first dose [9]. Low treatment doses guarantee mice welfare, which allows the consistent reproducibility of some clinical manifestations of peripheral neuropathy and their measurement through behavioral assays. Higher doses of paclitaxel result in severe motor deficits and weight loss, obscuring sensory dysfunction and quantification. Therefore, Taxol intermittent low-dose regimen represents the most translatable preclinical dosing protocol in the TIPN mice model. However, it is important to realize that no animal model will represent the full clinical situation perfectly. Nevertheless, the results with mice TIPN models treated with Taxol show high efficacy in causing TIPN (also across sex and various strains used) [6], and this is in line with the findings that Taxol is a chemotherapeutic drug causing one of the highest TIPN rates in patients [10]. With this background, in the present study, we generated the TIPN model in young C57BL/6J male mice by injection of Taxol (6 mg/kg), administered i.p., once daily, for a total of eight injections, to a cumulative intermittent low-dose of 48 mg/kg, to induce prolonged mechanical hypersensitivity and thermal hypoalgesia, mimicking one cycle of chemotherapy treatment in cancer patients [8]. Theoretically, it is preferable to use animal models that replicate all symptoms observed in humans. This remains, however and until today, very challenging. Measures such as numbness, tingling, and ongoing pain rely on verbal reports from the patient, often occur spontaneously, and, therefore, are very difficult to replicate in mice models. Fortunately, the investigations into novel measures of ongoing pain in rodents is an emerging area for research, but, for now, developing animal models of TIPN, which replicate all the symptoms that patients report, remains very challenging, and we, therefore, focused in this study on allodynia/hyperalgesia and intra-epidermal nerve fiber (IENF) density in skin biopsies. Another important issue regarding the clinical relevance of the TIPN model is that the present mice model is tumor-free, whereas, in the clinical situation, most TIPN patients have or experienced previous cancer, which may confound the results related to the use of animal models. We assume that, by using an effective and robust TIPN mice model that mimics the neurobehavioral and pathological clinical situation as much as possible, the translation to the clinical situation will improve in identifying future promising therapies for TIPN.

A recent update of the American Society of Clinical Oncology guideline on the recommended prevention and treatment approaches in the management of chemotherapy-induced peripheral neuropathy in adult cancer survivors reconfirmed that no agents are recommended and suggested that gabapentinoids might be helpful and worth trying for chemotherapy-induced neuropathy [11]. The U.S. Food and Drug Administration (FDA) approve gabapentin (brand name Neurontin) for adjunctive therapy in the treatment of partial seizures and postherpetic neuralgia and various "off-label" (unapproved) uses, including treatment of neuropathic pain caused by diabetic neuropathy, central pain, and TIPN [12], but without solid preclinical research in animal models and clinical trial support [13,14]. In preclinical rodent studies of neuropathic pain, gabapentin was generally used as a positive control to assess dose–response analgesic effects, with a duration of several weeks after frequent administration, and with an effective dose (50%) (ED50) of 27.8 mg/kg [15] with repeated dosing [16]. For example, the analgesic effects of Gabapentin after the induction of chemotherapy-induced neuropathy were observed at a low dose in cisplatin-evoked pain-like behavior, with reduction of mechanical allodynia in mice of both genders that received six i.p. injections of cisplatin (2.3 mg/kg/day) every other day, over the course of two weeks [17]. It has been reported that daily administration of Gabapentin 30 or 100 mg/kg, p.o.(per os, orally), for 14 days [18], 300 mg/kg i.p., twice weekly for 18 days [19], and 100 mg/kg, i.p., representing four daily injections after the peak symptoms [16] suppressed paclitaxel-induced peripheral neuropathy, mechanical hyperalgesia, and allodynia in rat models, respectively. Similarly, a few studies indicated that gabapentin reversed, in C57BL/6 mice, paclitaxel-induced allodynia, with an ED_{50} of 67.4 mg/kg i.p. 0.5–4 h after paclitaxel injection [20] and prevented TIPN at 100 mg/kg i.p., daily injected for eight days in ICR mice [21] and, at 100 mg/kg p.o, injected daily for 14 days in BALB/c mice [18]. However, the effect of gabapentin on intra-epidermal nerve fibers in these rodents' pharmacological TIPN model using skin biopsies was not yet investigated, representing a significant shortcoming in terms of the relevance of use of Gabapentin as a gold standard compound and of the clinical translational utility of this rodent model.

Intra-epidermal nerve fibers (IENFs) are free nerve endings arising from unmyelinated and thinly myelinated sensory neurons within the skin dermis and are important for the transmission of peripheral pain, allowing the quantification of a small-fiber neuropathy, which may manifest clinically with pain and dysesthesia sensory symptoms, such as burning, stinging and tingling sensations or numbness [22]. Analysis of intra-epidermal nerve fibers (IENFs) in skin biopsy samples has become a standard clinical tool for diagnosing peripheral neuropathies in human patients [23]. A significant reduction in the density of IENFs in the skin contributes to the neuropathic pain in Taxol-induced peripheral neuropathy in rat [24] and mice models [25]. IENF density quantification directly correlates with neurophysiological and neurobehavioral changes and represents, therefore, useful outcome measurement in experimental neuropathy models [26–28]. Hence, treatments that protect against chemotherapeutic-induced reduction in the density of IENFs may reduce the development of neuropathic pain and alleviate dysesthesias. Ongoing clinical trials are assessing pharmacological therapeutic strategies to manage chemotherapy-induced peripheral neuropathy, based on preclinical studies in different rodent models [29,30]. However, the relationship between IENF density and TIPN was not yet investigated in mice models under a therapeutic protocol of gabapentin-induced analgesia. Moreover, although the loss of skin innervation in human patients and rodent models of TIPN is now consistently observed, it is yet unknown the extent to which calcitonin gene-related peptides-I/II (CGRP-I/II)- peptidergic nerve fibers were affected during the progressive loss of IENF and if their respective loss is corrected by gabapentin treatment.

Therefore, the present study aimed to assess whether treatment with gabapentin can affect IENFs and CGRP density in the mice' hind paws at a time point that significantly confers an analgesic effect in an experimental TIPN mice model that closely mimics the course of peripheral neuropathy in human patients. To the best of our knowledge, this is the first study showing the decreased density of CGRP and PGP 9.5-positive nerve fibers innervating the paw skin in the TIPN model and their increase after 21 days of gabapentin treatment, justifying gabapentin use as a gold standard reference compound in the preclinical TIPN mouse model [31–33], as found in a randomized, placebo-controlled, clinical trial [34]. Present findings refine the TIPN mice model and further support the claim that skin biopsy with quantification of the density of IENF and CGRP, using generally agreed-upon counting rules, is a reliable, efficient, and objective technique, complementary to analgesia, to assess gabapentin effects on peripheral neuropathy both in patients and in animal models of peripheral neuropathy.

2. Materials and Methods

2.1. Animals

Young male mice C57BL/6JOlaHsd (four-weeks-old, 18.52 ± 0.32 g at study initiation) were purchased from Envigo RMS Ltd. (Rehovot, Israel) and served as subjects in these experiments, after acclimation for five to seven days to laboratory conditions. During acclimation and throughout the entire study duration, animals were housed within a limited access in a specific pathogen free rodent facility and kept in groups with a maximum of 8–10 mice per cage made of polypropylene (Euro standard type IV, floor area of the cage: $425 \times 266 \times 185$ mm (800 cm^2)), fitted with solid bottoms and a static cage filter top (Tecniplast Co., Buguggiate, Italy), filled with 7090 Teklad sani-chips animal bedding (Envigo RMS Ltd., Rehovot, Israel), and having two plastic tubes in each cage as enrichment material. During housing, animals were monitored twice daily for health status. No adverse events were observed. During the acclimation period, the mice were assigned to experimental groups in order to reduce possible litter effects. This was conducted using an online random number generator (https://www.graphpad.com/quickcalcs/randomize1/ (accessed on 10 June 2021)) assigning 10 mice subjects to a group. The experiment, including three groups (a. Healthy-Saline, Control; b. Taxol treated-Diseases and c. Taxol-induced disease treated with Gabapentin), was repeated twice (two blocks) during a 12-month period of time. Cages were likewise allocated to treatment groups with randomly generated numbers. Each experimental group was kept in a separate cage to avoid cross-contamination, which can occur through the consumption of fecal material. Mice were provided *ad libitum* with a commercial Teklad 2018S global 18% protein rodent diet (Envigo RMS Ltd., Rehovot, Israel) and had free access to drinking acidified water (using HCl to a pH of 3 ± 0.2) that was monitored periodically and supplied to each cage via polyethylene bottles with stainless steel sipper tubes. Mice were housed in automatically controlled environmental conditions, and the temperature was maintained at 17–23 °C, with a relative humidity of 30–70%, on a 12-h light/dark cycle (light cycle: 7 a.m.–7 p.m; dark cycle: 7 p.m.–7 a.m.) and 15–30 air changes/h in the study room. Random neurobehavioral measurements of 30 mice were blindly conducted between 10 a.m. and 4 p.m. by two female experimenters. If an injection was administered on the same day as the behavior tests, it was administered after all testing had been completed (4–6 p.m.), with the exception of gabapentin and vehicle (saline), which were applied 2 h prior to the neurobehavioral testing. At the end of the study, 54 animals were euthanized by intraperitoneal (i.p.) injection using an overdose of 1000 mg/kg of sodium pentobarbital. The study was twice conducted using 30 mice in each experiment [n = 10 control (healthy, saline), versus n = 10 Taxol (disease), and n = 10 Taxol and gabapentin (disease-gabapentin)]. Two mice before, and four mice during the TIPN study, were excluded due to paw skin injuries that may affect paw biopsies, and, therefore, the neurobehavioral data of these mice were omitted from the data analysis resulting with n = 18 for each experimental group. Two female experimenters were blinded to the allocation of the mice and conduct of the experiments, and another female investigator was blinded for the outcome assessments and data analysis. All procedures were carried out and strictly adhered to the guidelines of the Committee for Research and Ethical Issues of the International Association for the Study of Pain and were approved by an application form (MD-1-1-1027-3117) submitted to the Hebrew University Committee for Ethical Conduct in the Care and Use of Laboratory Animals that approved on 1 September 2020 that the study complies with the rules and regulations set forth. Animal studies (Supplementary Materials-Arrive authors check list) are reported in compliance with the ARRIVE guidelines [35].

2.2. Chemicals, Drugs, and Administration

Taxol (paclitaxel) was purchased from both Sigma-Aldrich-Merck (33069-62-4, Saint-Louis, MO, USA), and Tocris (33069-62-4, Minneapolis, MN, USA), and gabapentin was acquired from Thermo Fischer Co. (60142-96-3, Waltham, MA, USA) and Cayman (60142-96-3, Ann Arbor, MI, USA). Sodium chloride physiological solution (saline, 0.9% NaCl) and pentobarbital sodium were obtained from Sigma-Aldrich (57-33-0, Saint Louis, MO, USA). Taxol was dissolved in a solution composed of 50% Cremophor EL and 50% absolute ethanol to a concentration of 25 mg/mL and stored protected from the light at $-20\,^{\circ}$C for about 10 days, and then diluted in normal saline (NaCl 0.9%) to a final concentration of 1.2 mg/mL just before administration. Animals in the control healthy group received an injection of vehicle (Cremophor EL: Ethanol, 1:1) diluted in sterile 0.9% NaCl solution to a final concentration of 33.3%. Gabapentin was dissolved in water to a stock solution of 100 mg/mL and, before use, further diluted in saline to a concentration of 30 mg/mL. Animals in the control healthy group received a vehicle saline injection. Drugs and vehicles after filtration on MF-Millipore™ membrane filter (0.22 µm pore size) were administered via the intraperitoneal route (i.p.) in an injection volume of 0.1 mL (about 5 mL/kg body weight). The affinity-purified anti-ubiquitin carboxyl-terminal hydrolase PGP9.5, C-terminal polyclonal antibody produced in rabbits was purchased from Sigma-Aldrich-Merck (SAB1306173, Saint Louis, MO, USA), and recombinant anti-calcitonin gene-related peptides-I/II (CGRP-I/II) antibody [EPR23804-95] (ab283568) and DAPI staining solution (ab228549) were purchased from Abcam, Cambridge, UK). Alexa Fluor 594 affinity pure donkey (red, 112-585-167) and Alexa Fluor 488 goat (green 111-545-144) anti-rabbit IgG-conjugated secondary polyclonal antibody were acquired from Jackson ImmunoResearch Laboratories (West Grove, PA, USA).

2.3. Taxol-Induced Peripheral Neuropathy (TIPN)

The present acute TIPN study schedule (Figure 1) was induced in mice by injection of Taxol (6 mg/kg) administered i.p., once daily, from day 0 to 7 (1st week, disease induction), for a total of eight injections, to a cumulative dose of 48 mg/kg. This protocol has been well characterized to produce allodynia [28,36]. Gabapentin and vehicle (saline) were injected 2 h before the Von Frey test on study days 8, 10, 12, 14, 16, and 20 (2nd and 3rd week, therapeutic effect). The neuropathic pain was evaluated by measuring mechanical allodynia using the von Frey test, which was performed in the healthy mice (two days before "−2" for mice adaptation, and one day before "−1" the first Taxol injection, for baseline response) and on the study on days 4, 8, 10, 12, 14, 16, 20, and 21. Mechanical allodynia appeared on the first measurement after Taxol injection (day 4) and persisted for the whole duration of the experiment. The tail flick-test was performed on study day "−1" (baseline, healthy mice) and on days 10, 14, and 21. The hot plate test was performed on study day "−1" (baseline, healthy mice) and days 8, 13, and 20. The body weight of animals was measured three times a week throughout the three weeks study (Figure 1).

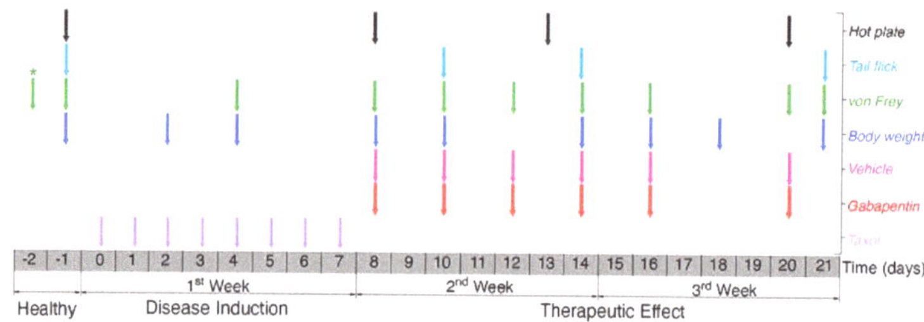

Figure 1. The schematic presentation of the TIPN study protocol. A TIPN was induced in mice (n = 40) by daily injection of Taxol from day 0 to 7 (1st week, disease induction) to a cumulative dose of 48 mg/kg. Gabapentin and vehicle saline were injected during the 2nd and 3rd week (therapeutic effect). The neuropathic pain was evaluated by measuring mechanical allodynia using the von Frey test in the healthy mice (two days (*) and one day before the first Taxol injection), for baseline response and during the disease and therapy period, and thermal hypersensitivity was measured four times each, with the tail-flick and the hot plate tests during the entire protocol. The body weight of animals was measured three times a week throughout the entire protocol. Healthy represents the mice cohort before TIPN. Colored arrows represent the day of drug delivery and performance of the neurobehavioral test.

2.4. Evaluation of Mechanical Allodynia (Von Frey Touch Test)

Allodynia response to tactile stimulation was assessed using the von Frey apparatus (World Precision Instruments, Sarasota, FL, USA) [37] and Electronic Von Frey–e-VF Hand-held (product code: 38450, Ugo Basile Co., Gemonio, Italy) [38]. For acclimation to the test environment, the mice were positioned on a metal mesh surface and allowed to move freely. The animals' cabins were covered with red cellophane to diminish environmental disturbances. The test started after 30 min (cessation of exploratory behavior). The set of von Frey monofilaments provided an approximate logarithmic scale of actual force and a linear scale of perceived intensity. Briefly, von Frey filaments, with approximately equal logarithmic incremental bending forces, were chosen (von Frey numbers equivalent to 0.008, 0.02, 0.04, 0.07, 0.16, 0.40, 0.60, 1.00, 1.40, 2.00, 4.00, 6.00, 8.00, 10.0, 15.0, 26.0 and 60.0 g). When the tip of a fiber of a given length and diameter was pressed against the skin at a right angle and was randomly applied to the left and right plantar surface of the hind paw for 3 s, the force of application increased until the fiber bent. Thereafter, the probe continued to advance, causing the fiber to bend more, but without additional force being applied. The mice exhibited a paw withdrawal reflex of withdrawal, lifting, licking, or shaking the paw, considered a positive response. The minimal force needed to elevate the withdrawal reflex was considered as the value of reference. Decreases in force needed to induce withdrawal were indicative of allodynia, as the force applied is a non-painful stimulus under normal conditions. Evaluations were based on the means ± SEM of mechanical allodynia data. Each treatment group was compared to the vehicle group using statistical analysis.

2.5. Evaluation of Thermal Allodynia

2.5.1. Tail-Flick Test

Tail-flick test is a nociceptive essay based on the measurement of the latency of the avoidance response to thermal stimulus in mice [37]. The thermal stimulus was applied on the tail using the Tail Flick Unit–Thermal stimulation (D'Amour and Smith method, product code: 37360, Ugo Basile Co., Gemonio, Italy). When the animal felt discomfort, it reacted with a sudden tail movement. The tail flick or twitch reaction time was then measured and used as an index of animal pain sensitivity. The test was repeated several times on the same animal (with about 5 min resting time between each evaluation), before and after gabapentin administration. In this test, the animals were quiet and immobile during the measurement, which was performed without any holder or restrainer. Evaluations were based on the means ± SEM of thermal allodynia data. Each treatment group was compared to the vehicle group using statistical analysis.

2.5.2. Hot Plate Test

The mice were individually placed on a hot plate (product code: 35300, Ugo Basile Co., Gemonio, Italy), with the temperature adjusted to $55 \pm 1\,°C$. The latency time to withdrawal, shaking, licking the paws, flinching, or jumping to avoid the heat was recorded, and the animal was immediately removed from the hot plate. The time that the mice were left on the plate was limited to 25 s to avoid tissue damage (in case of reduction in the sensory threshold of the animal's experience) [39]. Evaluations were based on the means ± SEM of thermal allodynia data. Each treatment group was compared to the vehicle group using statistical analysis.

2.6. Immunohistochemically (IHC) Staining and Intra-Epidermal Nerve Fiber Counting

Upon experiment termination, 3 mm-diameter punch biopsies were harvested from the hind paw. The tissue was fixed in parafomaldehyde 4%, at $4\,°C$ for 24 hrs., and washed three times with 0.1 M of phosphate-buffered saline (PBS), and stored at $4\,°C$. Thereafter, it was transferred to sucrose 15% and sucrose 30% in PBS at $4\,°C$ for 24 h until freeze embedding. The samples were freeze-embedded in a perpendicular plane in OCT (optimal cutting temperature compound) medium without decalcifying agents. The frozen tissues were sectioned with 20 μm thickness and stained. After three washings for 5 min in

0.1% Triton-PBS, immunofluorescence staining was performed with primary antibodies at two non-consecutive, free-floating sections using the pan-axonal marker anti-PGP 9.5 and two non-consecutive sections with the peptidergic nerve fiber marker anti-CGRP-I/II. After three washings for 15 min in 0.1% Triton PBS, sections were stained with Alexa Fluor 594 or Alexa Fluor 488 conjugated secondary antibodies, respectively. Finally, after three additional washings for 15 min in 0.1% Triton PBS, the sections were mounted with Fluoromount-G containing DAPI and stored at 4 °C, in the dark, until evaluation. Care was taken to orientate skin sections such that the surface of the epidermis was parallel to the upper limit of the photographed field. The intra-epidermal nerve fiber (IENF) density in the skin was counted using a fluorescent microscope (Olympus BX43, software-cellSens Standard v1, Olympus, Tokyo, Japan), according to published guidelines [26]. Briefly, nerves that branch after crossing the basement membrane were counted as a single unit, and nerves that split or branched below the basement membrane were counted as two units. Nerves that crossed the basement membrane were counted. However, nerve fibers that approached the basement membrane and nerve fragments in the epidermis that did not cross the basement membrane were not counted. IENF number was counted blindly in two sections per 9 samples (n = 9). The IENF density was calculated using the software ImageJ (ImageJ for Mac OS X, version 1.51) as the total number of fibers per unit length of the epidermis (IENF number/tissue in mm), excluding folds, tears, or hair follicles. In addition, peptidergic CGRP positive neuronal fibers were counted in two sections per 4 samples (n = 4).

2.7. Statistics

The software GraphPad Prism version 8.0.2 (GraphPad Software Inc., San Diego, CA, USA) was used for plotting graphs, data, and statistical analyses. Statistical analyses were first performed using Shapiro-Wilk for normality test. For multiple comparisons with parametric datasets, the one-way analysis of variance (ANOVA) was performed, and for non-parametric datasets, the Kruskal-Wallis test was performed to test for independence. Dunn's multiple comparisons post-hoc test was used to analyze differences between specific groups. The results in the figures are expressed as scatter plots with mean $\pm$ SEM. The differences were considered significant at * $p \leq 0.05$, ** $p \leq 0.01$, *** $p \leq 0.001$, and **** $p \leq 0.0001$. Based on prior experience with TIPN assay, and according to a power analysis (JMP, t-test comparison of the mean), in which we have equal size control, and samples with the variables of 26% expected difference in the thermal and mechanical allodynia, 20% expected standard deviation of the data, a desired power of 0.8, and alpha of 0.05, the program predicts a need for ten mice per treatment group.

3. Results

3.1. Gabapentin Inhibited Mechanical Allodynia and Thermal Hyperalgesia in Taxol-Induced Peripheral Neuropathic Pain

A TIPN study schedule was induced in mice by injection of Taxol administered i.p., once daily, from day 0 to 7 (1st week, disease induction), for a total of eight injections (Figure 1), to a cumulative dose of 48 mg/kg. This treatment was well tolerated, with the exception that mice receiving Taxol (Figure 2A-disease group) were slower to gain weight from days 18 and 21 compared to healthy control mice. However, in the Gabapentin disease-treated group, the mice did not lose the weight (Figure 2A).

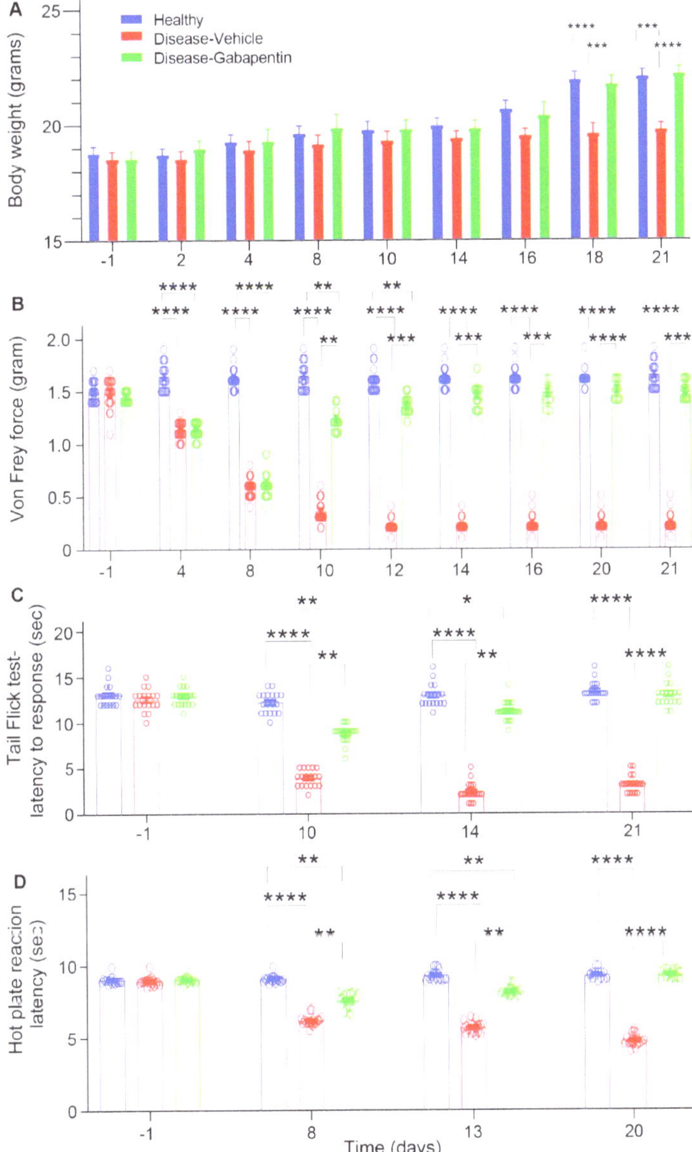

Figure 2. The inhibitory effects of gabapentin on the body weight and on Taxol-induced mechanical and thermal allodynia in mice. (**A**) Taxol-treated mice (disease-vehicle) gained less weight than controls (healthy) from day 18–21, but their weight normalized in gabapentin-treated mice (disease-gabapentin). (**B**) Taxol-induced hypersensitive response to von Frey filament stimulations test (disease-vehicle) compared to control (healthy). Taxol-induced mechanical allodynia was significantly improved during two weeks of gabapentin treatment (disease-gabapentin). (**C**) Tail-flick test latency to response was significantly decreased in Taxol-treated mice (disease-vehicle) compared to control (healthy), but it was normalized in gabapentin-treated mice (disease-gabapentin). (**D**) The reaction latency times measured using the hot plate thermal hyperalgesia test were significantly decreased in Taxol-treated mice (disease-vehicle) compared to control (healthy), but they were normalized in gabapentin-treated mice (disease-gabapentin). The results are means $\pm$ SEM (n = 18); The level of significance (*) assumed was 1% (** $p \leq 0.01$); 0.1% (*** $p \leq 0.001$); 0.01% (**** $p \leq 0.0001$); one-way ANOVA, followed by Dunn's post-test.

Taxol-induced persistent mechanical allodynia was measured up to three weeks after the 1st Taxol injection (Figure 2B). Gabapentin was administered i.p., at a dose of 150 mg/kg, beginning 24 h after the last Taxol injection (Figure 1, 2nd and 3rd week, therapeutic effect), at a cumulative dose of 900 mg/kg, that is, in the range of the doses used to treat chemotherapy-induced, peripheral neuropathy in mice [16,18], reported efficacious for the management of pain in patients [40–42], and lower than the maximally tolerated dose in mice of 2000 mg/kg/day [43]. At baseline (day-1, Figure 2B), there were no significant differences between the healthy, disease-vehicle-treated, and disease-gabapentin-treated mice groups. The von Frey mechanical hyperalgesia tests revealed that, compared with the healthy group, the disease-vehicle-treated group showed a very significant decrease in the force needed to induce paw withdrawal reflex, from the 4th to 10th day (from 1.13 ± 0.02 g on the 4th day to 0.35 ± 0.03 g on the 10th day, $p \leq 0.0001$ vs. healthy, n = 18 in each group), that persisted until 21st day, indicative of peripheral nerve degeneration, neuropathic effect. By contrast, the paw withdrawal reflex threshold was gradually increased in the mice treated with gabapentin from the 10th to 21st day (from 1.23 ± 0.02 g on the 10th day to 1.48 ± 0.03 g on the 21st day, $p \leq 0.01$–0.001 vs. disease-vehicle, n = 17–18 in each group) (Supplementary Materials, Data Set S1 Von Frey for all end points) which was interpreted as an analgesic effect.

Using the tail-flick test, at baseline (day-1, Figure 2C), the tail withdrawal latency in the healthy, disease-vehicle treated, and diseased-gabapentin-treated mice groups were similar, with values of 13.05 ± 0.26 sec, 12.50 ± 0.28 sec and 12.83 ± 0.23 sec (n = 18), respectively, thus showing good baseline comparability. Taxol treatment decreased the tail withdrawal latency by about 70% from day 10 to 21 (disease-vehicle group, 3.83 ± 0.22 sec on day 10, $p \leq 0.0001$, Figure 2C). After administration of gabapentin in the disease-treated group, withdrawal latencies were significantly increased ($p \leq 0.01$–0.0001) to values of 8.72 ± 0.24 sec, 11.11 ± 0.25 sec, and 12.83 ± 0.31 sec (n = 18), measured on days 10, 14 and 21, respectively (Supplementary Materials, Data Set S2 tail flick for all end points). This anti-hyperalgesia effect of gabapentin was also investigated using the hot plate test (Figure 2D). In the hot plate test, the reaction latency of the Taxol treatment group (disease-vehicle), compared to the healthy group, was significantly reduced by 31–48 % (n = 18) on day eight (6.2 ± 0.08 sec; $p \leq 0.0001$), 13 (5.7 ± 0.09 sec; $p \leq 0.0001$) and 20 (4.7 ± 0.1 sec; $p \leq 0.0001$). After administration of gabapentin in the disease-treated group (diseased-gabapentin), withdrawal latencies were significantly increased to values of 7.5 ± 0.15 sec ($p \leq 0.01$), 8.1 ± 0.09 sec. ($p \leq 0.01$), and 9.2 ± 0.08 sec ($p \leq 0.0001$) (n = 18), measured on days 8, 13, and 20, respectively (Figure 2D) (Supplementary Materials, Data Set S3 hot plate for all end points). These findings indicate that gabapentin significantly inhibited the Taxol-induced thermal allodynia in the acute thermal pain models of the hot plate and tail-flick tests.

3.2. Gabapentin Inhibited Taxol-Induced IENF Loss

IENF loss had been reported in human patients and rodent models of chemotherapy-induced neuropathy [23]. We analyzed the effect of gabapentin treatment on the IENF density by visualization with PGP9.5 immunostaining in the hind paw of Taxol-treated mice three weeks after the first Taxol treatment. Control, healthy mice showed an abundant distribution of nerve fibers entering the epidermis. However, as expected, the IENF density was significantly decreased by about 80% in response to Taxol treatment (disease-vehicle group, Figures 3 and 4). Gabapentin significantly conferred recovery from Taxol-induced IENF loss by increasing intra-epidermal nerve fiber density to about 83% of the healthy mice (Figures 3 and 4) (Supplementary Materials, Data Set S4 IENF and GCPR number for all end points).

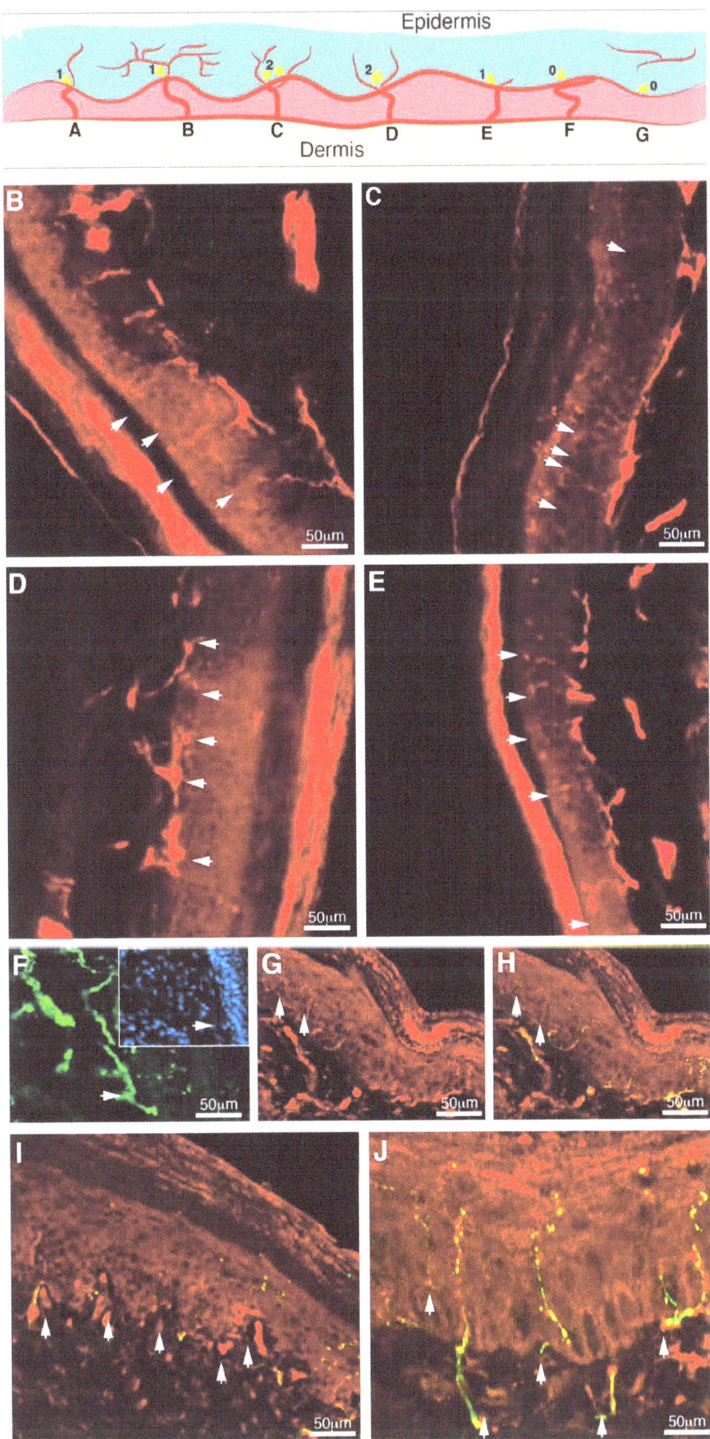

Figure 3. A schematic diagram depicting the intra-epidermal nerve fiber (IENF) counting rules and immunohistochemistry photos of punch biopsies of the different experimental groups. (**A**) Diagram

of skin innervation with IENF estimation rules that are widely used in human studies: nerves (red lines), basement membrane (pink area), dermis (yellow), and epidermis (blue-gray shade). (A) Count as a single unit nerve (score 1) that crosses the basement membrane of the epidermis; (B) Nerves that branch after crossing the basement membrane are also counted as a single unit (score 1); (C) Nerves that split below the basement membrane are counted as two units (score 2); (D) Nerves that appear to branch within the basement membrane are counted as two units (score 2); (E) Nerve fragments that do cross the basement membrane are counted as one unit (score 1) (F) Nerve fibers that approach the basement membrane but do not cross it are not counted (score 0); (G) Nerve fragments in the epidermis that do not cross the basement membrane in the section are not counted (score 0); (B) Healthy mice—white arrows indicate nerves labeled with anti-PGP 9.5 antibody (red, n = 9); (C) Taxol-treated mice (disease-vehicle)—white arrows indicate nerves (n = 9); (D,E) Gabapentin-treated mice (disease-vehicle)—white arrows indicate nerves (n = 9); (F) CGRP-positive nerves (green, n = 4) represent peptidergic IENFs labeled with white arrows; Insert: DAPI labeled nuclei in blue; (G) IENFs immunostained for PGP 9.5 (red), representing all epidermal nerve fibers labeled with white arrows; (H) Merged F and G view; (I). Taxol-treated mice (disease-vehicle; n = 4)—white arrows indicate peptidergic IENFs; (J) Gabapentin-treated mice (disease-vehicle; n = 4)—white arrows indicate peptidergic IENFs.

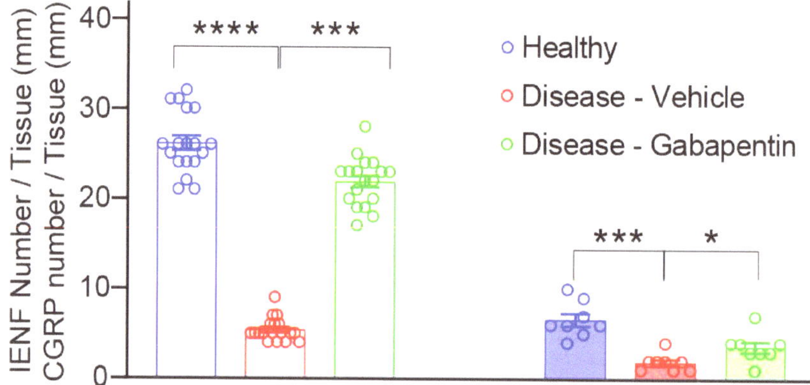

Figure 4. Quantification of immunohistochemically staining of IENF (left) and CGRP-I/II (right) density in skin biopsies of healthy, Taxol-treated (disease-vehicle) and gabapentin-treated mice (disease-gabapentin). The epidermal total (IENF) and CGRP-peptidergic nerve fiber innervation density (mean +/− SEM) on day 21 were measured and expressed as the number of IENF (left bars) and CGRP (right bars) per length of the section (IENF/mm, n = 9; CGRP/mm, n = 4); The level of significance (*) assumed was 5% (* $p \leq 0.05$); 0.1% (*** $p \leq 0.001$); 0.01% (**** $p \leq 0.0001$); one-way ANOVA followed by Dunn's post-test.

Traditionally, the quantification of IENF has relied on immunostaining nerves with the pan-axonal marker, protein gene product 9.5 (PGP 9.5), to measure the density of nerve fibers crossing the basement membrane into the epidermis as a marker of nerve degeneration. No previous studies have quantified peptidergic nerve fiber marker anti-CGRP-I/II in the TIPN mice model to allow the distinction between peptidergic and total IENF concerning neuropathy and gabapentin analgesic effects. Figure 4 indicates that about 25% of IENFs express CGRP-I/II peptidergic marker. CGRP-I/II peptidergic nerve fiber density was significantly decreased by about 73% in response to Taxol treatment (disease-vehicle group, Figures 3 and 4). Gabapentin significantly conferred recovery from Taxol-induced IENF loss by increasing CGRP-I/II to about 46% of the level in the healthy mice (Figures 3 and 4) (Supplementary Materials, Data Set S4 IENF and GCPR number for all end points).

4. Discussion

TIPN reflects a serious form of neuropathic pain that correlates with a reduction in the density of IENFs in the skin of cancer patients, which adversely influences treatment. Because of the need for effective pharmacotherapies to treat TIPN that are developed in rodent pharmacological models with well characterized gold-standard drugs, the present study examined whether treatment with gabapentin would affect total (IENF) and peptidergic (CGRP-I/II) nerve fiber density at a time point where it alleviates allodynia and thermal hyperalgesia in a mouse model of acute Taxol-induced peripheral neuropathy.

The findings of this study indicate that gabapentin significantly treated Taxol-induced allodynia and thermal hyperalgesia. Similar, Taxol treatment regimens (multiple injections) and doses (10–50 mg/kg), as used in this study, have been reported to produce painful neuropathy in mice that manifested as thermal hyperalgesia and mechanical allodynia [9,20,21,44]. Present findings of TIPN have been found in other acute and chronic models of peripheral neuropathy, such as mouse models of diabetes [45], HIV-associated distal sensory polyneuropathy [46], anti-retroviral drugs-induced peripheral neuropathy [47], and drug-induced neuropathy [48].

Administration of gabapentin [18,20,49] and other anticonvulsant drugs [50,51] to TIPN rodent models alleviated the development of Taxol-induced allodynia and/or thermal hyperalgesia. In line with these studies, in the present experimental paradigm, gabapentin treatment progressively alleviated Taxol-induced mechanical allodynia and thermal hyperalgesia, as evidenced by increased von Frey force and tail flick and hot plate reaction latencies after two days from the first injection, and for 11–12 days of successive injections (Figure 2), when compared with the healthy and disease-vehicle treated group. These findings are also supported by a study indicating that gabapentin induced significant anti-nociceptive effects compared to other anti-epileptics such as lamotrigine and topiramate [52]. Pregabalin (Lyrica), another γ-aminobutyric acid (GABA) analog, such as gabapentin, also showed a significant increase in tail withdrawal latency and significantly inhibited thermal hyperalgesia in mice [53] and rat [54] TIPN models.

Taxol-induced peripheral neuropathy is associated with degeneration or loss of intraepidermal nerve fibers (IENFs) that have been associated with thermal hyperalgesia and mechanical allodynia [55–57]. It is plausible that this reduction in the density of IENFs in the skin contributes to neuropathic pain in TIPN mice models [25]. IENF density quantification directly correlates with neurophysiological and neurobehavioral changes and represents, therefore, useful outcome measurement in experimental neuropathy models [26–28]. The principal finding in the current study is that, in an acute TIPN mice model, there is a significantly decreased proportion of IENFs, including those immunolabelled for CGRP-I/II. To the best of our knowledge, this is the first study to identify quantitative changes in the CGRP-I/II nerve fibers, in addition to IENF, that might be important in the development of Taxol-induced neuropathy and Gabapentin-induced analgesic effect. The decrease in the number of CGRP-immunolabelled nerve fibers could be the result of enhanced CGRP release or a loss of nerve fibers. The decrease in the PGP 9.5-immunoreactive and the double-labeled CGRP and PGP 9.5 nerve fibers in the same anatomical area may indicate a degenerative loss of nerve fibers in the skin and their regeneration after Gabapentin treatment. The IENFs are considered to relay pain information to the central nervous system and have been divided into two distinct groups based on neurochemical criteria: peptidergic fibers containing one or more neuropeptides, such as CGRP, substance P, somatostatin, and the remaining nerve fibers are non-peptidergic. The question of whether the density of the non-peptidergic nerve fibers is also affected by gabapentin deserves further investigation. Present results suggest that CGRP-I/II-peptidergic nerve fibers as total IENF are significantly lost in the mouse epidermis after TIPN, and increased after 21 days of Gabapentin treatment, in parallel to behavioral deficits, suggesting that the neurobehavioral symptoms and their alleviation may be tied to damage within the peptidergic and non-peptidergic nerve fibers. Quantitation of IENF is proposed as a more objective tool for the evaluation of peripheral neuropathy in animal models of TIPN and other peripheral neuropathies.

The precise molecular mechanism of mechanical allodynia and sensitivity to heat observed upon Taxol treatment are not clearly understood. The present working hypothesis claims that Taxol probably causes dysfunctional microtubules in dorsal root ganglia, axons, and Schwann cells, resulting in dysfunction in calcium signaling, neuropeptide, growth factor release, mitochondrial damage, reactive oxygen species formation, and ion channels, cumulatively contributing to the neuropathic pain [36]. Previous experimental studies had shown that Gabapentin binds with a high affinity to the $\alpha 2\delta 1$-subunit of the different types of voltage-dependent calcium channels [58], known to interact with N-methyl-d-aspartate-sensitive glutamate receptors, neurexin-1α, thrombospondins (adhesion molecules), and other presynaptic proteins [59], thereby reducing the release of both excitatory and inhibitory neurotransmitters from the rat spinal cord dorsal horn [60]. This possible mechanism could account in part for the anti-nociceptive effect of gabapentin. The ability of gabapentin to restore IENF and CGRP I/II-peptidergic nerve fibers density in the TIPN model maybe be explained by an increased level of skin neurotrophins, such as nerve growth factor (NGF), known to induce sprouting of cutaneous sensory nerve fibers [61]. Indeed, several studies have implicated NGF in gabapentinoids' analgesic and IENF effects in rat diabetic skin [61,62], a possibility deserving further investigation. Moreover, gabapentinoid treatment promoted corticospinal axon regrowth ability following murine spinal cord injury [63], indicating an ability to induce nerve fiber sprouting.

In conclusion, the results from this study confirm that, in the acute TIPN mice model, Taxol-induced peripheral neuropathy is expressed by allodynia and thermal hyperalgesia. Gabapentin alleviated these effects, in parallel to the increase in IENF and CGRP I/II-peptidergic nerve fibers density, refining the TIPN model and indicating that it can be used as a gold standard reference in the TIPN mice model for preclinical research and development of drugs. We assume that further quantification of the different types of IENF innervations in mice footpads in studies with the TIPN mice model will become a useful popular tool and will offer objective quantification metrics in addition to the animal neurobehavioral outcomes related to pain. Skin biopsies might be useful in detecting early changes of IENF density, which predict the progression of neuropathy and assess degeneration and regeneration of IENF. However, further TIPN studies using mice of different gender, age, and genetic backgrounds are warranted to confirm the potential usefulness of skin biopsy with the measurement of IENF/ CGRP I/II nerve fiber density as an outcome diagnostic and/or therapeutic measure in preclinical, pharmacological rodent models research.

Supplementary Materials: The following supporting information can be downloaded at: https://www.mdpi.com/article/10.3390/biomedicines10123190/s1; Arrive authors check list; Data Set S1 Von Frey for all end points; Data Set S2 Tail Flick for all end points; Data Set S3 Hot plate for all end points; Data Set S4 IENF and GCPR number for all end points.

Author Contributions: M.K. made substantial contributions to the acquisition, analysis of data, statistics, and references' organization. M.S.N. contributed to the drug preparation and analyses and acquisition of IHC data. P.L. and W.Z. supervised the study and made substantial contributions to the conception and design of the experiments, analysis, and interpretation of data, and they also revised the article and approved the final version. All authors have read and agreed to the published version of the manuscript.

Funding: The study was a part of the research project funded by the Bioconvergence grant 73184 from the Israel Innovation Authority for 2020–2021.

Institutional Review Board Statement: The animal study protocol was approved by the Hebrew University Committee for Ethical Conduct in the Care and Use of Laboratory Animals (MD-1-1-1027-3117, 1 September 2020).

Informed Consent Statement: Not applicable.

Data Availability Statement: The data presented in this study are available in the article and Supplementary Material.

Acknowledgments: PL holds the Jacob Gitlin Chair in Physiology and is affiliated and supported by David R. Bloom Center for Pharmacy, Adolf and Klara Brettler Center for Research in Molecular Pharmacology and Therapeutics, and the Grass Center for Drug Design and Synthesis of Novel Therapeutics at the Hebrew University of Jerusalem, Israel. The authors acknowledge partial technical assistance received from MD Biosciences, Rehovot, Israel. Zehava Cohen is acknowledged for the professional help with the graphic.

Conflicts of Interest: The authors declare no conflict of interest. The funders had no role in the design of the study; in the collection, analyses, or interpretation of data; in the writing of the manuscript; or in the decision to publish the results.

References

1. Li, T.; Mizrahi, D.; Goldstein, D.; Kiernan, M.C.; Park, S.B. Chemotherapy and peripheral neuropathy. *Neurol. Sci.* **2021**, *42*, 4109–4121. [CrossRef] [PubMed]
2. Cheng, Z.; Lu, X.; Feng, B. A review of research progress of antitumor drugs based on tubulin targets. *Transl. Cancer Res.* **2020**, *9*, 4020–4027. [CrossRef] [PubMed]
3. Yu, D.-L.; Lou, Z.-P.; Ma, F.-Y.; Najafi, M. The interactions of paclitaxel with tumour microenvironment. *Int. Immunopharmacol.* **2022**, *105*, 108555. [CrossRef] [PubMed]
4. Wiseman, L.R.; Spencer, C.M. Paclitaxel. An update of its use in the treatment of metastatic breast cancer and ovarian and other gynaecological cancers. *Drugs Aging* **1998**, *12*, 305–334. [CrossRef] [PubMed]
5. Farquhar-Smith, P. Chemotherapy-induced neuropathic pain. *Curr. Opin. Support. Palliat. Care* **2011**, *5*, 1–7. [CrossRef] [PubMed]
6. Gadgil, S.; Ergün, M.; Heuvel, S.A.V.D.; Van Der Wal, S.E.; Scheffer, G.J.; Hooijmans, C.R. A systematic summary and comparison of animal models for chemotherapy induced (peripheral) neuropathy (CIPN). *PLoS ONE* **2019**, *14*, e0221787. [CrossRef] [PubMed]
7. Tuttle, A.H.; Philip, V.; Chesler, E.J.; Mogil, J.S. Comparing phenotypic variation between inbred and outbred mice. *Nat. Methods* **2018**, *15*, 994–996. [CrossRef] [PubMed]
8. Pennypacker, S.D.; Fonseca, M.M.; Morgan, J.W.; Dougherty, P.M.; Cubillos-Ruiz, J.R.; Strowd, R.E.; Romero-Sandoval, E.A. Methods and protocols for chemotherapy-induced peripheral neuropathy (CIPN) mouse models using paclitaxel. *Methods Cell Biol.* **2022**, *168*, 277–298. [CrossRef] [PubMed]
9. Toma, W.; Kyte, S.L.; Bagdas, D.; Alkhlaif, Y.; Alsharari, S.D.; Lichtman, A.H.; Chen, Z.-J.; Del Fabbro, E.; Bigbee, J.W.; Gewirtz, D.A.; et al. Effects of paclitaxel on the development of neuropathy and affective behaviors in the mouse. *Neuropharmacology* **2017**, *117*, 305–315. [CrossRef] [PubMed]
10. Molassiotis, A.; Cheng, H.L.; Lopez, V.; Au, J.S.K.; Chan, A.; Bandla, A.; Leung, K.T.; Li, Y.C.; Wong, K.H.; Suen, L.K.P.; et al. Are we mis-estimating chemotherapy-induced peripheral neuropathy? Analysis of assessment methodologies from a prospective, multinational, longitudinal cohort study of patients receiving neurotoxic chemotherapy. *BMC Cancer* **2019**, *19*, 132. [CrossRef] [PubMed]
11. Loprinzi, C.L.; Lacchetti, C.; Bleeker, J.; Cavaletti, G.; Chauhan, C.; Hertz, D.L.; Kelley, M.R.; Lavino, A.; Lustberg, M.B.; Paice, J.A.; et al. Prevention and management of chemotherapy-induced peripheral neuropathy in survivors of adult cancers: ASCO guideline update. *J. Clin. Oncol.* **2020**, *38*, 3325–3348. [CrossRef] [PubMed]
12. Moore, J.; Gaines, C. Gabapentin for chronic neuropathic pain in adults. *Br. J. Community Nurs.* **2019**, *24*, 608–609. [CrossRef] [PubMed]
13. Goodman, C.W.; Brett, A.S. A clinical overview of Off-label use of gabapentinoid drugs. *JAMA Intern. Med.* **2019**, *179*, 695–701. [CrossRef] [PubMed]
14. Mack, A. Examination of the evidence for Off-label use of gabapentin. *J. Manag. Care Pharm.* **2003**, *9*, 559–568. [CrossRef] [PubMed]
15. Garrone, B.; Di Matteo, A.; Amato, A.; Pistillo, L.; Durando, L.; Milanese, C.; Di Giorgio, F.P.; Tongiani, S. Synergistic interaction between trazodone and gabapentin in rodent models of neuropathic pain. *PLoS ONE* **2021**, *16*, e0244649. [CrossRef] [PubMed]
16. Xiao, W.; Boroujerdi, A.; Bennett, G.; Luo, Z. Chemotherapy-evoked painful peripheral neuropathy: Analgesic effects of gabapentin and effects on expression of the alpha-2-delta type-1 calcium channel subunit. *Neuroscience* **2007**, *144*, 714–720. [CrossRef] [PubMed]
17. Woller, S.; Corr, M.; Yaksh, T. Differences in cisplatin-induced mechanical allodynia in male and female mice. *Eur. J. Pain* **2015**, *19*, 1476–1485. [CrossRef] [PubMed]
18. Kato, N.; Tateishi, K.; Tsubaki, M.; Takeda, T.; Matsumoto, M.; Tsurushima, K.; Ishizaka, T.; Nishida, S. Gabapentin and duloxetine prevent oxaliplatin- and paclitaxel-induced peripheral neuropathy by inhibiting extracellular signal-regulated kinase 1/2 (ERK1/2) phosphorylation in spinal cords of mice. *Pharmaceuticals* **2020**, *14*, 30. [CrossRef] [PubMed]
19. Zbârcea, C.E.; Ciotu, I.C.; Bild, V.; Chiriță, C.; Tănase, A.M.; Șeremet, O.C.; Ștefănescu, E.; Arsene, A.L.; Bastian, A.E.; Ionică, F.E.; et al. Therapeutic potential of certain drug combinations on paclitaxel-induced peripheral neuropathy in rats. *Rom. J. Morphol. Embryol.* **2017**, *58*, 507–516. [PubMed]

20. Donvito, G.; Wilkerson, J.L.; Damaj, M.I.; Lichtman, A.H. Palmitoylethanolamide reverses paclitaxel-induced allodynia in mice. *J. Pharmacol. Exp. Ther.* **2016**, *359*, 310–318. [CrossRef] [PubMed]

21. Huehnchen, P.; Boehmerle, W.; Endres, M. Assessment of paclitaxel induced sensory polyneuropathy with "Catwalk" automated gait analysis in mice. *PLoS ONE* **2013**, *8*, e76772. [CrossRef] [PubMed]

22. Pereira, M.P.; Mühl, S.; Pogatzki-Zahn, E.M.; Agelopoulos, K.; Ständer, S. Intraepidermal nerve fiber density: Diagnostic and therapeutic relevance in the management of chronic pruritus: A review. *Dermatol. Ther.* **2016**, *6*, 509–517. [CrossRef] [PubMed]

23. Mangus, L.M.; Rao, D.B.; Ebenezer, G.J. Intraepidermal nerve fiber analysis in human patients and animal models of peripheral neuropathy: A comparative review. *Toxicol. Pathol.* **2019**, *48*, 59–70. [CrossRef] [PubMed]

24. Ko, M.-H.; Hu, M.-E.; Hsieh, Y.-L.; Lan, C.-T.; Tseng, T.-J. Peptidergic intraepidermal nerve fibers in the skin contribute to the neuropathic pain in paclitaxel-induced peripheral neuropathy. *Neuropeptides* **2014**, *48*, 109–117. [CrossRef] [PubMed]

25. Ferrari, G.; Nallasamy, N.; Downs, H.; Dana, R.; Oaklander, A.L. Corneal innervation as a window to peripheral neuropathies. *Exp. Eye Res.* **2013**, *113*, 148–150. [CrossRef] [PubMed]

26. Lauria, G.; Lombardi, R.; Borgna, M.; Penza, P.; Bianchi, R.; Savino, C.; Canta, A.R.; Nicolini, G.; Marmiroli, P.; Cavaletti, G. Intraepidermal nerve fiber density in rat foot pad: Neuropathologic-neurophysiologic correlation. *J. Peripher. Nerv. Syst.* **2005**, *10*, 202–208. [CrossRef] [PubMed]

27. Wozniak, K.M.; Vornov, J.J.; Wu, Y.; Liu, Y.; Carozzi, V.A.; Rodriguez-Menendez, V.; Ballarini, E.; Alberti, P.; Pozzi, E.; Semperboni, S.; et al. Peripheral neuropathy induced by microtubule-targeted chemotherapies: Insights into acute injury and long-term recovery. *Cancer Res* **2018**, *78*, 817–829. [CrossRef] [PubMed]

28. Höke, A.; Ray, M. Rodent models of chemotherapy-induced peripheral neuropathy. *ILAR J.* **2014**, *54*, 273–281. [CrossRef] [PubMed]

29. Bouchenaki, H.; Danigo, A.; Sturtz, F.; Hajj, R.; Magy, L.; Demiot, C. An overview of ongoing clinical trials assessing pharmacological therapeutic strategies to manage chemotherapy-induced peripheral neuropathy, based on preclinical studies in rodent models. *Fundam. Clin. Pharmacol.* **2020**, *35*, 506–523. [CrossRef] [PubMed]

30. HHopkins, H.L.; Duggett, N.A.; Flatters, S.J.L. Chemotherapy-induced painful neuropathy: Pain-like behaviours in rodent models and their response to commonly used analgesics. *Curr. Opin. Support. Palliat. Care* **2016**, *10*, 119–128. [CrossRef] [PubMed]

31. Yamamoto, S.; Egashira, N. Drug repositioning for the prevention and treatment of chemotherapy-induced peripheral neuropathy: A mechanism- and screening-based strategy. *Front. Pharmacol.* **2021**, *11*, 607780. [CrossRef] [PubMed]

32. Wodarski, R.; Clark, A.; Grist, J.; Marchand, F.; Malcangio, M. Gabapentin reverses microglial activation in the spinal cord of streptozotocin-induced diabetic rats. *Eur. J. Pain* **2009**, *13*, 807–811. [CrossRef] [PubMed]

33. Moreira, D.R.M.; Santos, D.S.; Santo, R.F.D.E.; dos Santos, F.E.; Filho, G.B.D.O.; Leite, A.C.L.; Soares, M.B.P.; Villarreal, C.F. Structural improvement of new thiazolidinones compounds with antinociceptive activity in experimental chemotherapy-induced painful neuropathy. *Chem. Biol. Drug Des.* **2017**, *90*, 297–307. [CrossRef] [PubMed]

34. Aghili, M.; Zare, M.; Mousavi, N.; Ghalehtaki, R.; Sotoudeh, S.; Kalaghchi, B.; Akrami, S.; Esmati, E. Efficacy of gabapentin for the prevention of paclitaxel induced peripheral neuropathy: A randomized placebo controlled clinical trial. *Breast J.* **2019**, *25*, 226–231. [CrossRef] [PubMed]

35. Lilley, E.; Stanford, S.; Kendall, D.E.; Alexander, S.P.; Cirino, G.; Docherty, J.R.; George, C.H.; Insel, P.A.; Izzo, A.A.; Ji, Y.; et al. ARRIVE 2.0 and the British Journal of Pharmacology: Updated guidance for 2020. *Br. J. Pharmacol.* **2020**, *177*, 3611–3616. [CrossRef] [PubMed]

36. Staff, N.P.; Fehrenbacher, J.C.; Caillaud, M.; Damaj, M.I.; Segal, R.A.; Rieger, S. Pathogenesis of paclitaxel-induced peripheral neuropathy: A current review of in vitro and in vivo findings using rodent and human model systems. *Exp. Neurol.* **2019**, *324*, 113121. [CrossRef] [PubMed]

37. Deuis, J.R.; Dvorakova, L.S.; Vetter, I. Methods used to evaluate pain behaviors in rodents. *Front. Mol. Neurosci.* **2017**, *10*, 284. [CrossRef] [PubMed]

38. Castel, D.; Sabbag, I.; Brenner, O.; Meilin, S. Peripheral neuritis trauma in pigs: A neuropathic pain model. *J. Pain* **2015**, *17*, 36–49. [CrossRef] [PubMed]

39. Taliyan, R.; Sharma, P.L. Possible mechanism of protective effect of thalidomide in STZ-induced-neuropathic pain behavior in rats. *Inflammopharmacology* **2011**, *20*, 89–97. [CrossRef] [PubMed]

40. Baos, S.; Rogers, C.A.; Abbadi, R.; Alzetani, A.; Casali, G.; Chauhan, N.; Collett, L.; Culliford, L.; de Jesus, S.E.; Edwards, M.; et al. Effectiveness, cost-effectiveness and safety of gabapentin versus placebo as an adjunct to multimodal pain regimens in surgical patients: Protocol of a placebo controlled randomised controlled trial with blinding (GAP study). *BMJ Open* **2020**, *10*, e041176. [CrossRef] [PubMed]

41. Rullán, M.; PHN group; Bulilete, O.; Leiva, A.; Soler, A.; Roca, A.; González-Bals, M.J.; Lorente, P.; Llobera, J. Efficacy of gabapentin for prevention of postherpetic neuralgia: Study protocol for a randomized controlled clinical trial. *Trials* **2017**, *18*, 24. [CrossRef] [PubMed]

42. Tassone, D.M.; Boyce, E.; Guyer, J.; Nuzum, D. Pregabalin: A novel γ-aminobutyric acid analogue in the treatment of neuropathic pain, partial-onset seizures, and anxiety disorders. *Clin. Ther.* **2007**, *29*, 26–48. [CrossRef] [PubMed]

43. Available online: https://www.medicines.org.uk/emc/product/4636/smpc#PRECLINICAL_SAFETY (accessed on 3 October 2022).

44. Nieto, F.R.; Entrena, J.M.; Cendán, C.M.; Del Pozo, E.; Vela, J.M.; Baeyens, J.M. Tetrodotoxin inhibits the development and expression of neuropathic pain induced by paclitaxel in mice. *Pain* **2008**, *137*, 520–531. [CrossRef] [PubMed]

45. Jolivalt, C.G.; Frizzi, K.E.; Marquez, A.; Ochoa, J.; Calcutt, N.A. Phenotyping peripheral neuropathy in mouse models. *Curr. Protoc. Mouse Biol.* **2017**, *6*, 223–255. [CrossRef] [PubMed]

46. Han, M.M.; Frizzi, K.E.; Ellis, R.J.; Calcutt, N.A.; Fields, J.A. Prevention of HIV-1 TAT protein-induced peripheral neuropathy and mitochondrial disruption by the antimuscarinic pirenzepine. *Front. Neurol.* **2021**, *12*, 663373. [CrossRef] [PubMed]

47. Dalakas, M.C. Peripheral neuropathy and antiretroviral drugs. *J. Peripher. Nerv. Syst.* **2001**, *6*, 14–20. [CrossRef] [PubMed]

48. Manji, H. Drug-induced neuropathies. *Handb. Clin. Neurol.* **2013**, *115*, 729–742. [CrossRef] [PubMed]

49. Gauchan, P.; Andoh, T.; Ikeda, K.; Fujita, M.; Sasaki, A.; Kato, A.; Kuraishi, Y. Mechanical allodynia induced by paclitaxel, oxaliplatin and vincristine: Different effectiveness of gabapentin and different expression of voltagedependent calcium channel.alpha.2.delta.-1 subunit. *Biol. Pharm. Bull.* **2009**, *32*, 732–734. [CrossRef] [PubMed]

50. Thangamani, D.; Edafiogho, I.O.; Masocha, W. The anticonvulsant enaminone E139 attenuates paclitaxel-induced neuropathic pain in rodents. *Sci. World J.* **2013**, *2013*, 240508. [CrossRef] [PubMed]

51. Flatters, S.J.; Bennett, G.J. Ethosuximide reverses paclitaxel- and vincristine-induced painful peripheral neuropathy. *Pain* **2004**, *109*, 150–161. [CrossRef] [PubMed]

52. Paudel, K.R.; Bhattacharya, S.; Rauniar, G.; Das, B. Comparison of antinociceptive effect of the antiepileptic drug gabapentin to that of various dosage combinations of gabapentin with lamotrigine and topiramate in mice and rats. *J. Neurosci. Rural Pract.* **2011**, *2*, 130–136. [CrossRef] [PubMed]

53. Sałat, K.; Cios, A.; Wyska, E.; Sałat, R.; Mogilski, S.; Filipek, B.; Więckowski, K.; Malawska, B. Antiallodynic and antihyperalgesic activity of 3-[4-(3-trifluoromethyl-phenyl)-piperazin-1-yl]-dihydrofuran-2-one compared to pregabalin in chemotherapy-induced neuropathic pain in mice. *Pharmacol. Biochem. Behav.* **2014**, *122*, 173–181. [CrossRef] [PubMed]

54. Park, H.J.; Joo, H.S.; Chang, H.W.; Lee, J.Y.; Hong, S.H.; Lee, Y.; Moon, D.E. Attenuation of neuropathy-induced allodynia following intraplantar injection of pregabalin. *Can. J. Anaesth.* **2010**, *57*, 664–671. [CrossRef] [PubMed]

55. Boyette-Davis, J.; Xin, W.; Zhang, H.; Dougherty, P.M. Intraepidermal nerve fiber loss corresponds to the development of Taxol-induced hyperalgesia and can be prevented by treatment with minocycline. *Pain* **2011**, *152*, 308–313. [CrossRef] [PubMed]

56. Siau, C.; Xiao, W.; Bennett, G.J. Paclitaxel- and vincristine-evoked painful peripheral neuropathies: Loss of epidermal innervation and activation of Langerhans cells. *Exp. Neurol.* **2006**, *201*, 507–514. [CrossRef] [PubMed]

57. Jin, H.W.; Flatters, S.J.L.; Xiao, W.H.; Mulhern, H.L.; Bennett, G.J. Prevention of paclitaxel-evoked painful peripheral neuropathy by acetyl-l-carnitine: Effects on axonal mitochondria, sensory nerve fiber terminal arbors, and cutaneous Langerhans cells. *Exp. Neurol.* **2008**, *210*, 229–237. [CrossRef] [PubMed]

58. Calandre, E.P.; Rico-Villademoros, F.; Slim, M. Alpha$_2$delta ligands, gabapentin, pregabalin and mirogabalin: A review of their clinical pharmacology and therapeutic use. *Expert Rev. Neurother.* **2016**, *16*, 1263–1277. [CrossRef] [PubMed]

59. Taylor, C.P.; Harris, E.W. Analgesia with gabapentin and pregabalin may involve N-Methyl-d-Aspartate receptors, neurexins, and thrombospondins. *J. Pharmacol. Exp. Ther.* **2020**, *374*, 161–174. [CrossRef] [PubMed]

60. Yamaoka, J.; Di, Z.H.; Sun, W.; Kawana, S. Changes in cutaneous sensory nerve fibers induced by skin-scratching in mice. *J. Dermatol. Sci.* **2007**, *46*, 41–51. [CrossRef] [PubMed]

61. Al-Massri, K.F.; Ahmed, L.A.; El-Abhar, H.S. Pregabalin and lacosamide ameliorate paclitaxel-induced peripheral neuropathy via inhibition of JAK/STAT signaling pathway and Notch-1 receptor. *Neurochem. Int.* **2018**, *120*, 164–171. [CrossRef] [PubMed]

62. Evans, L.; Andrew, D.; Robinson, P.; Boissonade, F.; Loescher, A. Increased cutaneous NGF and CGRP-labelled trkA-positive intra-epidermal nerve fibres in rat diabetic skin. *Neurosci. Lett.* **2012**, *506*, 59–63. [CrossRef] [PubMed]

63. Sun, W.; Larson, M.J.; Kiyoshi, C.M.; Annett, A.J.; Stalker, W.A.; Peng, J.; Tedeschi, A. Gabapentinoid treatment promotes corticospinal plasticity and regeneration following murine spinal cord injury. *J. Clin. Investig.* **2019**, *130*, 315–358. [CrossRef] [PubMed]

Article

Low-Dose Metformin as a Monotherapy Does Not Reduce Non-Small-Cell Lung Cancer Tumor Burden in Mice

Nicole L. Stott Bond [1,2], Didier Dréau [3], Ian Marriott [3], Jeanette M. Bennett [4], Michael J. Turner [2], Susan T. Arthur [2] and Joseph S. Marino [2,*]

[1] Distance Education, Technology and Integration, University of North Georgia, Dahlonega, GA 30597, USA; nicole.bond@ung.edu

[2] Laboratory of Systems Physiology, Department of Applied Physiology, Health, and Clinical Sciences, University of North Carolina at Charlotte, Charlotte, NC 28223, USA; miturner@uncc.edu (M.J.T.); sarthur8@uncc.edu (S.T.A.)

[3] Department of Biological Sciences, University of North Carolina at Charlotte, Charlotte, NC 28223, USA; ddreau@uncc.edu (D.D.); imarriot@uncc.edu (I.M.)

[4] Department of Psychological Science, University of North Carolina at Charlotte, Charlotte, NC 28223, USA; jbenne70@uncc.edu

* Correspondence: jmarin10@uncc.edu

Abstract: Non-small-cell lung cancer (NSCLC) makes up 80–85% of lung cancer diagnoses. Lung cancer patients undergo surgical procedures, chemotherapy, and/or radiation. Chemotherapy and radiation can induce deleterious systemic side effects, particularly within skeletal muscle. To determine whether metformin reduces NSCLC tumor burden while maintaining skeletal muscle health, C57BL/6J mice were injected with Lewis lung cancer (LL/2), containing a bioluminescent reporter for in vivo tracking, into the left lung. Control and metformin (250 mg/kg) groups received treatments twice weekly. Skeletal muscle was analyzed for changes in genes and proteins related to inflammation, muscle mass, and metabolism. The LL/2 model effectively mimics lung cancer growth and tumor burden. The in vivo data indicate that metformin as administered was not associated with significant improvement in tumor burden in this immunocompetent NSCLC model. Additionally, metformin was not associated with significant changes in key tumor cell division and inflammation markers, or improved skeletal muscle health. Metformin treatment, while exhibiting anti-neoplastic characteristics in many cancers, appears not to be an appropriate monotherapy for NSCLC tumor growth in vivo. Future studies should pursue co-treatment modalities, with metformin as a potentially supportive drug rather than a monotherapy to mitigate cancer progression.

Keywords: Lewis lung model; lung cancer; skeletal muscle; cachexia

Citation: Stott Bond, N.L.; Dréau, D.; Marriott, I.; Bennett, J.M.; Turner, M.J.; Arthur, S.T.; Marino, J.S. Low-Dose Metformin as a Monotherapy Does Not Reduce Non-Small-Cell Lung Cancer Tumor Burden in Mice. *Biomedicines* **2021**, *9*, 1685. https://doi.org/10.3390/biomedicines9111685

Academic Editor: Martina Perše

Received: 21 October 2021
Accepted: 11 November 2021
Published: 14 November 2021

1. Introduction

Lung cancer is the second most common cancer and represents ~13% of all new cancer cases in the United States (SEER, National Cancer Institute). Lung cancer contributed to ~145,000 fatalities in 2019 [1], with the yearly diagnoses expected to reach 225,000 in 2030, in the United States alone [2]. Cigarette smoke is one of the largest contributors to lung cancer diagnoses, but now it has now been established that a combination of lifestyle, genetic, and environmental components contributes to an individual's risk and development of lung cancer [3]. Specifically, factors that put individuals at a greater risk for lung cancer include cigarette smoke, environmental pollutants, alcohol consumption, adverse dietary consideration, physical inactivity, and hereditary markers [3]. Lung cancer patients have a 5 year relative survival rate of only 19% (16% for men and 22% for women), making it one of the lower survival rates among cancers [1]. While treatments continue to improve, the prevalence and severity of lung cancer necessitates more refinement of treatment modalities.

Continuous advances are bringing new insight into oncology therapeutics [4], especially through drug repositioning [5]. This is an attractive tactic since new drug characterization and approval requires an extensive investment in time and money [6]. Observational studies, pre-clinical trials, and clinical trials have provided insights into the efficacy of drug repositioning for cancer prevention and cancer therapy [7].

Metformin canonically facilitates improved insulin sensitivity and overall glucose uptake for type 2 diabetes (T2D) patients, but recent studies show the potential of repositioning metformin due to its anti-cancer properties [8–12]. Importantly, the literature suggests that metformin decreases lung cancer risk for T2D patients and increases survival for lung cancer patients with co-morbid T2D [13–16]. Whether this is due to normalization of glycemia and insulinemia, or results from a direct effect on tumor burden, remains to be determined.

Metformin elicits anti-tumorigenic effects in many cancers, including prostate, colon, skin, and obesity-activated thyroid cancer [10,12,17,18]. In cancers, many signaling pathways components, including AMPK, mTOR, MAPK, and insulin-like growth factors contribute to the anti-tumorigenic effects of metformin [19]. In particular, metformin activates AMPK inhibiting cell mitosis and proliferation, particularly via protein p53 activation [5]. While metformin demonstrates anti-neoplastic effects via cell cycle arrest, the efficacy of metformin and the mechanism underlying this agent's action on non-small-cell lung cancer (NSCLC) tumor development remains unclear. Filling this knowledge gap is crucial to the successful repositioning of metformin as an anti-cancer therapeutic. Utilizing metformin independently or in conjugation with other treatment modalities could mitigate the side effects many cancer patients experience while receiving more potent oncology therapeutics.

Following diagnosis, lung cancer patients often undergo surgical procedures, chemotherapy, or radiation, but these can drive systemic complications, negatively affecting patient welfare and recovery timelines. One of the most common systemic effects of conventional cancer treatment is cachexia, the rapid loss of skeletal muscle and adipose tissue [20,21]. Cachexia occurs in more than 50% of lung cancer patients undergoing chemotherapy, radiotherapy, or a combination of both [22,23], and more than 60% of patients with advanced NSCLC present respiratory complications and increased rates of cachexia [24]. Furthermore, patients with cancer-induced cachexia often exhibit a lower tolerance and responsiveness to chemotherapy, shortened survival times, far greater symptom burdens, and systemic inflammation [25,26]. Higher morbidity and mortality rates also correlate with the degree of weight loss and rapid decreases in BMI, both of which are independent prognostic factors for cancer patients, with or without cachexia [24,27].

Few treatment options are available for cachexia and these effects are irreversible even during remission, making such repercussions even more debilitating [28,29]. Metformin may be an attractive target to manage cancer-induced metabolic dysfunction and cachexia. Within skeletal muscle, which is the largest insulin-sensitive tissue in the body, metformin increases peroxisome proliferator-activated receptor-coactivator-1α (PGC-1α) protein expression, a transcriptional co-activator involved in mitochondrial biogenesis, glucose metabolism, and muscle fiber type differentiation [30]. PGC-1α increases the expression of genes involved in energy metabolism, which is thought to protect skeletal muscle from atrophy, and suppresses forkhead box O3 (FoxO3), a transcription factor that induces the expression of ubiquitin-ligases involved in atrophy [31]. Metformin also preserves the satellite cell pool in a lower metabolic state which sustains quiescence and delays satellite cell activation [32]. Maintenance of the stem cell population is crucial for preservation of skeletal muscle mass, repair, and function [33].

Although metformin has been used as an anti-cancer therapy in clinical trials, its efficacy against NSCLC remains understudied. Furthermore, it is currently unknown how the combination of metformin treatment and NSCLC directly influences skeletal muscle health and metabolism. In the present study, we have investigated whether metformin treatment suppresses tumor growth in C57BL/6J mice with NSCLC, and we have investigated the effects of NSCLC tumor progression on skeletal muscle health. Importantly, we have employed a mouse model in the present study where the animals are neither obese nor diabetic and this has allowed us to investigate the direct effects of metformin on tumor burden. Determining the efficacy of metformin therapy against NSCLC could provide new treatment options for cancer patients and provide valuable insights into the physiological disparities that underlie NSCLC progression.

2. Materials and Methods

2.1. Experimental Animals

Six-week-old male ($n = 12$) and female ($n = 12$) C57BL/6J mice (Jackson Laboratory, Bar Harbor, ME, USA) were randomly assigned (manually) into a control group (lung cancer without metformin treatment) ($n = 12$; 6 males, 6 females) and a metformin treatment group (lung cancer with metformin treatment) ($n = 12$; 6 males, 6 females). All animals were housed individually in cages with filter lids and placed in rooms with a 12:12 h light:dark cycle. Mice were housed in cages measuring 7.5 inches in width, 11.5 inches in length and 5 inches in height (Allentown Inc. and Ancare, Bellmore, NY, USA). The floor surface area was 86.25 square inches. Teklad corncob bedding was used throughout the study (7092A; Envigo, Cumberland, VA, USA). For enrichment, all cages included a small plastic hide (Bio-care, Flemington, NJ, USA) and a Nestlet 2 inch square for nestling (Ancare, Bellmore, NY, USA). The animal housing facility was equipped with 24-h temperature monitoring and alarms to ensure a constant ambient temperature of 65–75 °F and 20–60% humidity (depending on season). Animals were acclimated for 5 days prior to use. When an animal exhibited signs of distress (>20% reduction in body weight), the animal was immediately captured by daily weigh-ins and euthanized. Control ($n = 7$) and metformin ($n = 9$) animals completed the study and were used in statistical calculations. Some control mice ($n = 5$) and metformin-treated mice ($n = 3$) mice presented extreme tumor burdens and did not survive for the full length of study and were excluded from statistical calculations (Table 1). Figure 1 outlines the study progression.

Table 1. C57Bl/6J mice survival and metastases following injection with LL/2 cells.

Group	Mice Began Study, *n*	Mice Survived, *n*	Mice with Signal, *n*	Mice with Metastases, *n*
Males	6	3	3	2
Females	6	4	4	1
Total Control	12	7	7	3
Males	6	3	3	0
Females	6	6	6	3
Total Metformin	12	9	9	3

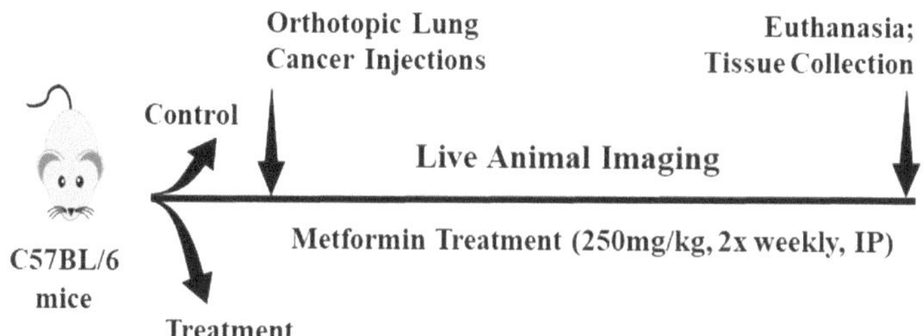

Figure 1. Experimental timeline for Lewis lung carcinoma development in an immunocompetent mouse model. Male ($n = 12$) and female ($n = 12$) C57BL/6J mice were implanted with 1000 Lewis lung carcinoma cells harboring luciferase reporter expression. Live animal imaging was continuous for the duration of the study. Once a bioluminescent signal was detected, vehicle or metformin treatment (250 mg/kg, 2× weekly, intraperitoneal injection) began.

All mice were provided with ad libitum access to water and standard rodent chow (Teklad Diets 2919; Envigo, Cumberland, VA, USA). Food mass was measured weekly and the total amount of food consumed over the study was used to determine total caloric intake. The energy density of the standard rodent chow was 3.3 kcal/g. Male and female C57BL/6J mice were used to address metformin's efficacy on reducing lung tumor burden in immunocompetent mice. The Lewis lung carcinoma immunocompetent mouse model mimics lung tumor development including the immune system modulations. The non-small-cell lung carcinoma (NSCLC) Lewis lung carcinoma (LL/2) cells are syngeneic with C57BL/6J mice and stably and constitutively expresses a luciferase reporter (Imanis Life Sciences, Rochester, MN, USA), allowing tumor growth monitoring over time with a live animal imaging system. The LL/2 orthotopic model effectively mimics lung cancer growth and tumor burden in accordance with other murine Lewis lung cancer models [34–36]. All aspects of this study were approved by the Institutional Animal Care and Use Committee at The University of North Carolina at Charlotte.

2.2. Culturing Non-Small-Cell Lung Cancer Cells

NSCLC cells (Imanis Life Sciences, Rochester, MN, USA) were grown in standard growth media (Dulbecco's Modified Eagle Medium) with 10% fetal bovine serum and 1% penicillin-streptomycin. Cells were passaged with 2 µg/mL puromycin to maintain high luciferase fluorescence expression. Cells were maintained at 37 °C for 48 h or until predetermined time points.

2.3. Orthotopic Injection

Animal hair was removed from the ventral and left thoracic regions and were then aseptically prepared. Prior to receiving an LL/2 cancer injection, all animals were imaged and baseline images acquired using an in vivo imaging system (IVIS). Under anesthesia (1–3% isoflurane), mice received one orthotopic lung injection of LL/2 cells into the left lung. LL/2 cells (1.0×10^3) were administered in PBS and Matrigel® (10 µg; Dulbecco's Modified Eagle's Medium with 50 ug/mL gentamycin phenol red free, Corning, Glendale, AZ, USA). Matrigel® facilitated both tumor cell growth and homing within the lung tissue [36]. A small incision (3–5 mm) was made to expose the area surrounding the seventh and eighth ribs. Cells were injected orthotopically into the lung using a sterile 29-gauge syringe and the incision was closed with a wound clip. Following surgery, all mice were individually housed and allowed to recover for one week. Animal weights were recorded weekly throughout the study. Any mouse showing signs of distress or exceeding 20% body mass loss was euthanized in accordance with approved IACUC guidelines.

2.4. In Vivo Imaging

Tumor growth in all animals was initially monitored weekly using bioluminescent imaging via IVIS. Cell visualization in vivo occurred by giving all animals D-luciferin (150 mg/kg) 15 min prior to imaging. The area to be imaged was shaved and cleaned to remove any hair that could interfere with the bioluminescent signal detected. All images were captured within a 30 min window following D-luciferin injection. All mice were imaged weekly until a bioluminescent signal was detected. Following detection, each mouse was imagined bi-weekly and treatment commenced.

2.5. Metformin Treatment

Control and metformin-treated mice were injected intraperitoneally (i.p.) with saline (PBS, 1×) and metformin (250 mg/kg, twice weekly). Metformin hydrochloride (1084; Sigma Aldrich, St. Louis, MO, USA) was dissolved in 1× PBS and sterile filtered (0.2 µm) for a final dose of 250 mg/kg. Metformin preparations were cultured on nutrient agar plates to ensure sterility. This metformin dose is commonly used in many mouse cancer studies [37,38]. While metformin dosing is typically daily, the mice used in this study also received injections for bioluminescent imaging so we minimized administrations to twice weekly. Control mice received a placebo of $1\times$ PBS solution via an i.p. injection twice a week. At 5 weeks post-tumor implantation, mice were euthanized (>4% isoflurane), and tissue was collected, snap frozen on liquid nitrogen, and stored at $-80\,^{\circ}$C.

2.6. Tumor Burden

Tumor burden was assessed with the Living Image analysis (Version 4.5.5, Perkin Elmer, USA). The region of interest (ROI) was determined by outlining the tumor bioluminescent signal with minimum detection parameters set to 5%. Brightness, contrast, and opacity were maintained between all images regardless of time point. A separate ROI was drawn on each mouse to determine background signal. Mice with metastases were identified as having more ROIs at a single time point. Each bioluminescent signal was first normalized to the background signal for the same image and all animals were normalized to the baseline image of the same mouse. Total signal counts for animals with multiple detectable bioluminescent signals were added together to determine total tumor burden for a single mouse at a single timepoint. Mice with a saturated signal were excluded from analyses.

2.7. Tumor Tissue and Gastrocnemius Muscle Homogenization and mRNA Extraction

Tumor tissue ($\leq$30 mg) was placed into a microcentrifuge tube with beads in ~300 μL (or sufficient volume not exceeding 10% of tissue mass) QIAzol lysis reagent (79306; Qiagen, Germantown, MD, USA). Tissue was disrupted with a bead blaster homogenizer (BeadBlasterTM 24 Microtube, Sigma, St. Louis, MO, USA) with 2 separate rounds of 2–30 s intervals at 619 m/s followed by 1 min of rest. Following lysis, tumor mRNA was extracted utilizing a RNeasy Lipid Tissue Mini Kit (74804; Qiagen, Germantown, MD, USA). Following the addition of chloroform, the upper aqueous phase was removed and placed into a clean tube and washed multiple times. mRNA from homogenized tissue was eluted using RNAse-free water though a RNeasy column.

The left gastrocnemius muscle was homogenized using $\leq$30 mg of tissue in 300 μL of buffer RLT supplemented with 1% β-mercaptoethanol. Tissue was disrupted with a bead blaster homogenizer (BeadBlasterTM 24 Microtube, Sigma, St. Louis, MO, USA) with 2 separate rounds of 2–30 s intervals at 619 m/s followed by 1 min of rest. Following lysis, mRNA was extracted utilizing an RNeasy Fibrous Tissue kit (74704; Qiagen, Germantown, MD, USA). Proteinase K and RNase-free water were added to each sample, allowed to incubate at 55 $^{\circ}$C for 10 min, and centrifuged at 10,000$\times$ g for 3 min. Supernatant was transferred to a clean tube. Following the addition of ethanol, the upper aqueous phase was removed and placed into a clean tube and washed multiple times. mRNA was eluted using RNAse-free water though an RNeasy column.

The quality and quantity of mRNA was assessed using a NanoDrop 1000. Briefly, 2 μL of RNAse-free water was used to blank the NanoDrop and 2 μL of sample was loaded onto the pedestal and quantified. The quality of mRNA was determined according to the 260/280 and 260/230 ratios.

2.8. cDNA and Real-Time PCR

mRNA (1 μg of RNA/reaction) was reverse transcribed to cDNA using Applied Biosystems cDNA synthesis kit (4368814; Fisher Scientific, Suwanee, GA, USA). Real-time polymerase chain reaction (qPRC) was used to evaluate gene expression targets involved in cell cycle regulation, tumor suppression, skeletal muscle mass, metabolism, and inflammation. Regulators of the cell cycle included cyclin D kinase 4 (CDK4) and protein 27 (p27). Tumor suppression targets included protein 21 (p21). F4/80, a macrophage marker, and hairy and enhancer of Split-1 (HES1), a downstream target gene involved in cellular determination and fate, were also included in our analyses. Genes involved in inflammatory responses included F4/80 and tumor necrosis alpha (TNF-α). Phosphatase and tensin homolog (PTEN), an atrophy-associated gene, and peroxisome proliferator-activated receptor-γ coactivator 1 alpha (PGC-1α), a gene involved in skeletal muscle metabolism were also assessed. Table 2 shows all primers used for gene expression analyses. Briefly, Radiant Green HI-ROX SYBR Green was utilized for all qPCR reactions. Glyceraldehyde 3-phosphate (GAPDH) was the housekeeping gene for all qPCR experiments. SYBR green ROX cycling occurred under the following conditions: cDNA was activated at 95 °C for 2 min followed by 20 cycles of 95 °C for 5 s (denaturation) and 60 °C for 20 s (annealing/extension).

Table 2. Primers Used for Gene Expression Analyses.

Primer		Sequence
p27	Forward	TCTCTTCGGCCCGGTCAAT
	Reverse	AAATTCCACTTGCGCTGACTC
F4/80	Forward	CTTTGGCTATGGGCTTCCAGTC
	Reverse	GCAAGGAGGACAGAGTTTATCGTG
CDK4	Forward	ATGGCTGCCACTCGATATGAA
	Reverse	TCCTCCATTAGGAACTCTCACAC
IL-6	Forward	CTGCAAGAGCTTCCATCCAGTT
	Reverse	GAAGTAGGGAAGGCCGTGG
Hes1	Forward	GGTCCTGGAATAGTGCTACCG
	Reverse	CACCGGGGAGGAGGAATTTTT
TNF-α	Forward	CCAGACCCTCACACTCAGATC
	Reverse	CACTTGGTGGTTTGCTACGAC
PGC-1α	Forward	TGATGTGAATGACTTGGATACAGACA
	Reverse	GCTCATTGTTGTACTGGTTGGATATG
MAFbx	Forward	CCAGGATCCGCAGCCCTCCA
	Reverse	ATGCGGCGCGTTGGGAAGAT
GAPDH	Forward	ATGTTTGTGATGGGTGTGAA
	Reverse	ATGCCAAAGTTGTCATGGAT

p27: cyclin-dependent kinase inhibitor protein 27; CDK4: cyclin-dependent kinase 4; IL-6: interleukin 6; Hes1: hairy and enhancer split protein; TNF-α: tumor necrosis factor alpha; PGC-1α: peroxisome proliferator-activated receptor-γ coactivator 1 alpha; MAFbx: muscle-specific ubiquitin ligases muscle atrophy F-box; GAPDH: glyceraldehyde 3-phosphate dehydrogenase.

2.9. Gastrocnemius Tissue Protein Isolation and Quantification

Upon sacrifice, the skeletal muscle tissue was harvested, and muscle weights were taken for the gastrocnemius muscle. Gastrocnemius tissue ($\leq$30 mg) was placed into a microcentrifuge tube with beads in cell lysis buffer (30 μL/mg tissue) containing ice cold radioimmunoprecipitation assay (RIPA) buffer (sc-24948; Santa Cruz, Dallas, TX, USA), supplemented with 10% sodium dodecyl sulfate (SDS), 1% Triton X-100, protease cocktail inhibitor. Tissue was disrupted with a bead blaster homogenizer (BeadBlasterTM 24 Microtube; Sigma, St. Louis, MO, USA) with 2 separate rounds of 2–30 s intervals at 619 m/s followed by 1 min of rest. Samples were placed on ice for 5 min on ice between the 2 separate rounds. Following lysis, protein underwent centrifugation at $10,000 \times g$ (rcf) for 10 min at 4 °C. Protein supernatant concentrations were quantified using a Pierce BCA protein kit (23225; Thermo Fisher, Allentown, PA, USA).

2.10. Western Blotting

Western blotting was used to assess the expression level of proteins regulating skeletal muscle metabolism. Protein samples prepared in 1× loading buffer, supplemented with 10% β-mercaptoethanol, were denatured at 95 °C for 3 min and then immediately placed on ice for 5 min. Protein samples (30 μg/well) were loaded onto 10% SDS-page gels and were run at 225 V for 40 min in 1× running buffer. Following electrophoresis, the gel was placed into 1× Towbin's transfer buffer, supplemented with 20% methanol, for 15 min. Proteins were transferred onto a 0.45 μm Polyvinylidene difluoride (PVDF-FL) membrane at 100 V for 90 min in 4 °C. Following transfer, membranes were washed once in 1× Tris-buffered saline (TBS) for 5 min. Next, the membrane underwent blocking in Odyssey Blocking Buffer and TBS (1:1) for 1 h at room temp. After blocking, the primary antibodies were added overnight (16 h). Primary antibodies were directed against the following: pAMPK (1:500; CS, #4188), AMPK (1:500; CS, #2532), pSTAT3 Ser 727 (1:500, CS, #9134), STAT3 (1:500; CS, #4904), REDD1 (1:1000; FS, 3PIPA520495), and GAPDH (1:5000; CS, #MAB473). Following removal of the primary antibodies, the membrane underwent 3×5 min washes in $1\times$ Tris-buffered saline with Tween 20 (TBST). Secondary antibodies (1:10,0000 in TBST) were targeted to primary antibodies and incubated at room temp for 2 h. Next, membranes were washed twice in $1\times$ TBST and twice in $1\times$ TBS. Membranes were imaged using the Odyssey® Licor CLx System.

Using the Odyssey® Licor CLx System, bands were quantified and expressed using arbitrary units as a measure of integrated optical density. Phosphorylated proteins (pSTAT3) were normalized to total (STAT3) protein expression. Total protein expressions (STAT3, AMPK, REDD1) were normalized to glyceraldehyde 3-phosphate dehydrogenase levels (GAPDH).

2.11. Statistical Analyses

An unpaired Student's *t*-test was used to assess baseline body mass between all control and metformin mice. A mixed-effects model (time $\times$ treatment) was use to assess normalized body mass between treatment groups and food consumption for the duration of the study. An unpaired Student's *t*-test was used to identify any differences in time to caloric intake, signal detection, and length of treatment. Overall survival was determined by a Logrank test. An unpaired Student's *t*-test was used to compare differences in gene expression, except where variances significantly differed ($p < 0.05$). In those cases, a Welch's *t*-test was used to compare differences in gene expression between control and treatment animals. Outliers were identified using a Grubb's test. Significance was established with an a priori alpha value of 0.05. All statistics were completed in GraphPad Prism (Version 9.1, GraphPad Software, San Diego, CA, USA).

3. Results

3.1. Body Mass in C57BL/6J Mice with NSCLC

There was no differences in baseline body mass between control and metformin-treated mice (unpaired Student's *t*-test, $p = 0.774$). As expected, body mass of control and metformin animals increased through the duration of the study (Figure 2). Mixed modeling (time x treatment) from all mice with a detectable bioluminescent signal indicated significant increases in body mass with time [$F(2.320, 27.85) = 8.788$, $p < 0.001$] but not treatment [$F(1, 14) = 4.510$, $p = 0.0520$] or an interaction (time x treatment) effect [$F(5, 60) = 1.943$, $p = 0.1005$]. There were no differences detected between male and female cohorts, supporting the comparison of treatment cohorts pooling both sexes. Since body mass was not significantly different at baseline body mass was represented as a fold change from baseline.

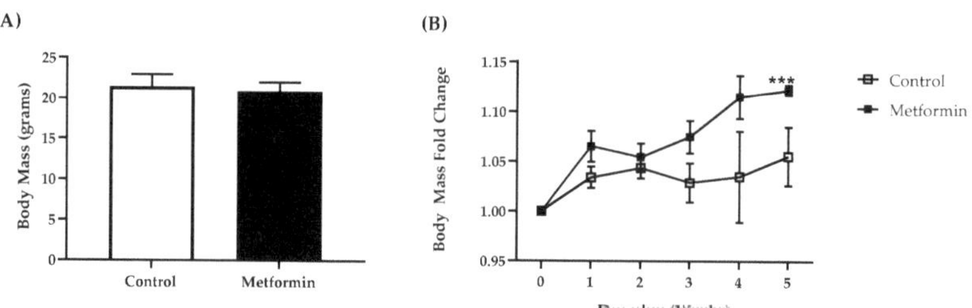

Figure 2. Body weight in C57BL/6J male and female mice following orthotopic LL/2 non-small lung cancer cell implantation. (**A**) Body mass (grams) between control and treatment mice. (**B**) Body mass fold change between control and metformin- (250 mg/kg) treated mice following orthotopic LL/2 cancer cell implantation. Data were analyzed using mixed modeling (time $\times$ treatment). *** $p < 0.001$, main effect for time. Control, $n = 7$; metformin, $n = 9$. Data shown as mean $\pm$ SEM.

3.2. Food Consumption in C57BL/6J Mice with NSCLC

Control and metformin-treated animals continued consuming food for the duration of the study (Figure 3). Overall, most animals maintained a healthy body mass, good ambulatory movement, and an appetite even with tumor burden. Mixed modeling (time × treatment) indicated significant increases in food consumption with time [F (2.149, 24.35) = 5.566, p = 0.009], independent of treatment. Control mice had significantly lower (p = 0.018) total caloric consumption compared to metformin-treated animals (Figure 3).

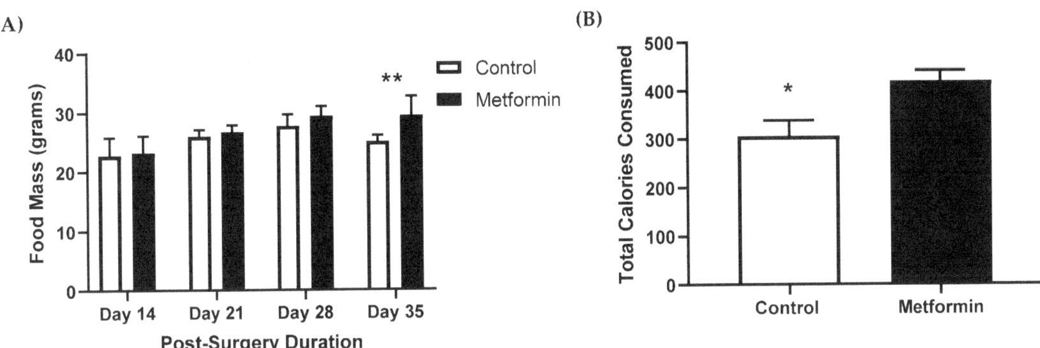

Figure 3. Total food mass and caloric consumption in C57BL/6J male and female mice following orthotopic LL/2 non-small lung cancer cell implantation. (**A**) Food consumption (grams) between control and treatment mice following tumor injection. Data were analyzed using mixed modeling (time × treatment). ** p = 0.009, main effect for time compared to Day 14. Control, n = 7; metformin, n = 8. Data shown as mean ± SEM. (**B**) Caloric Consumption between control and metformin- (250 mg/kg) treated mice following orthotopic LL/2 implantation. Data were analyzed using an unpaired t-test. * p = 0.018 compared to metformin animals. Control, n = 7; metformin, n = 8. Data shown as mean ± SEM.

3.3. Time to Tumor Detection and Length of Treatment

There was no differences in overall time to detectable bioluminescent signal between control and metformin animals (unpaired Student's t-test, p = 0.790) (Figure 4). The treatment timeline between cohorts remained similar, irrespective of treatment (p = 0.753) (Figure 4).

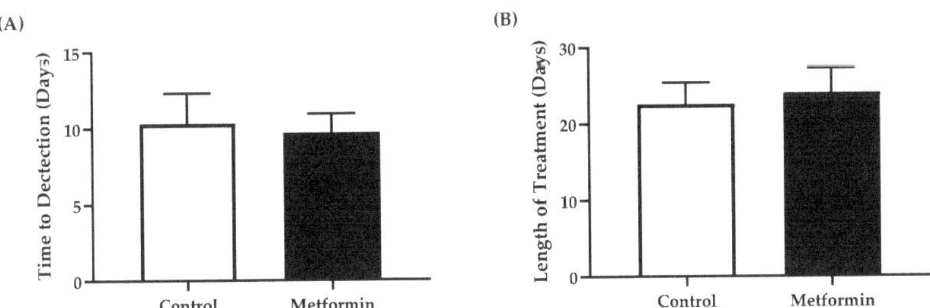

Figure 4. Time to tumor detection and length of treatment in C57BL/6J male and female mice following orthotopic LL/2 non-small lung cancer cell implantation. (**A**) Number of days to a discernable bioluminescent signal following orthotopic injection of LL/2 cells into C57BL/6J mice. Data were analyzed using an unpaired t-test. Control, n = 7; metformin, n = 9. Data shown as mean ± SEM. (**B**) Number of days C57BL/6J mice with NSCLC underwent treatment with control or metformin (250 mg/kg). Data were analyzed using an unpaired t-test. Control, n = 7; metformin, n = 8. Data shown as mean ± SEM.

3.4. NSCLC Tumor Burden and Animal Survival

The mean survival times for control (37 ± 5.6 days) and metformin treatment (40 ± 1.4 days) groups were not statistically significant (Figure 5). There were no significant differences in mean survival time between groups (Welch's *t*-test, $p = 0.412$). Similarly, no difference was detected in overall survival between control or metformin-treated mice with a detectable bioluminescent signal (log rank test, $p = 0.827$) (Figure 5) nor was there any significant trends ($p = 0.0515$). Some mice developed metastases (control ($n = 3$); metformin ($n = 3$)), which led to an increased tumor burden in those animals. However, mice with a saturated signal were excluded from tumor burden analysis due to limitations within the Living Imaging software (Version 4.5.5, Perkin Elmer, USA). Mice with evidence of metastasis did not exhibit overt indications of declining health compared to mice without metastasis. Moreover, similar tumor burdens were recorded in the groups tested, irrespective of treatment (unpaired Student's *t*-test, $p = 0.615$) (Figure 5). LL/2 tumor signals representative of observations made in a female control and a male metformin-treated mice were similar (Figure 5).

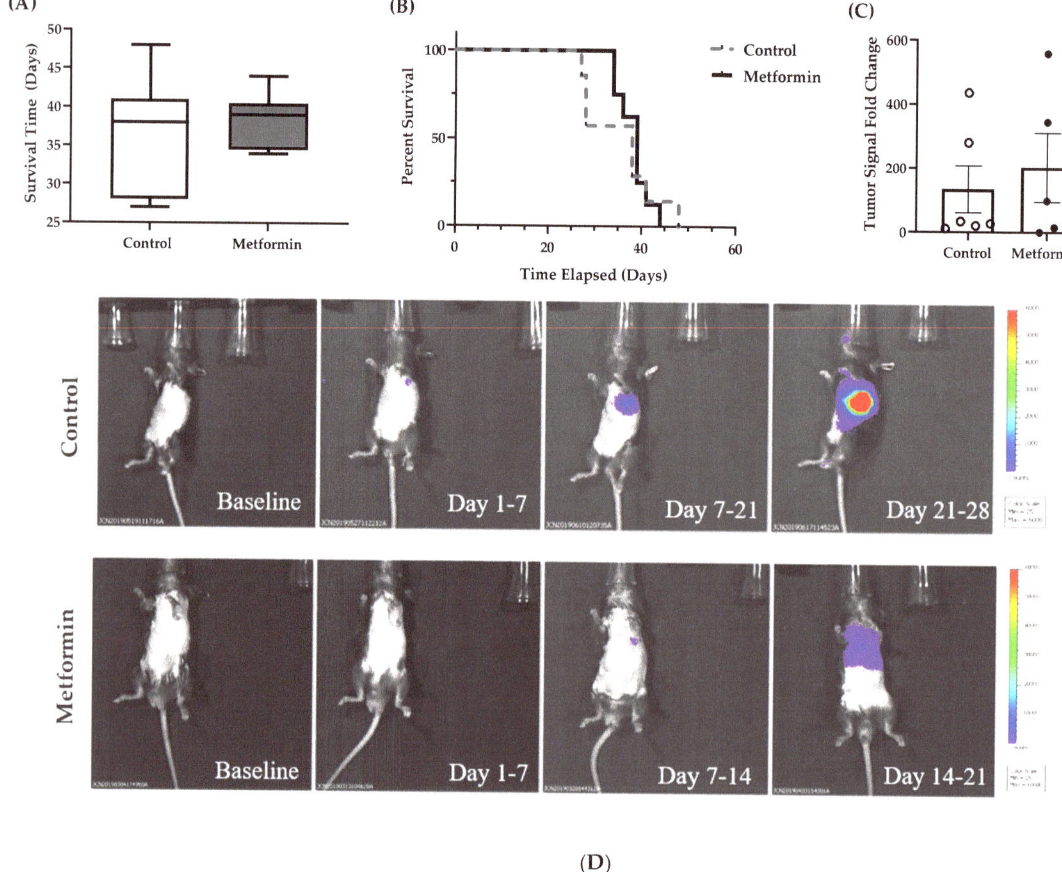

Figure 5. Survival time, tumor burden, and non-small-cell lung tumor growth in C57BL/6J male and female mice following orthotopic LL/2 non-small lung cancer cell implantation. (**A**) Survival duration (days) between control and metformin- (250 mg/kg) treated mice with a detectable bioluminescent signal. Data were analyzed with a Welch's *t*-test. (**B**) Percent survival of both control and metformin- (250 mg/kg) treated mice following detection of a bioluminescent signal. Data were analyzed with a Log-rank test. Control, $n = 7$; metformin, $n = 8$. (**C**) NSCLC tumor burden fold change between control and metformin- (250 mg/kg) treated mice with a detectable bioluminescent signal. Mice with a saturated signal were removed

from analyses. Data were analyzed with a Welch's *t*-test. Control, *n* = 6; metformin, *n* = 5. Data shown as mean ± SEM. (**D**) Male and female C57BL/6J mice orthotopically administered 1000 Lewis lung carcinoma cells harboring luciferase reporter expression. Bioluminescent signals were tracked throughout the duration of the study via an in vivo imaging system. Once a bioluminescent signal was detected, vehicle or metformin treatment (250 mg/kg, 2× weekly, intraperitoneal injection) began. Top row: Control C57BL/6J female mouse; Bottom row: metformin-treated C57BL/6J male mouse. Scale represents increased tumor burden as signal intensity increases from blue to red.

3.5. NSCLC Tumor Gene Expression

No significant differences were detected in gene expression from NSCLC tumors collected from C57BL/6J mice (Unpaired Student's *t*-tests, $p > 0.05$). p27, CDK4, F480, IL-6 or Hes1 gene expression were similar between tumors collected from control and metformin-treated mice (p27, $p = 0.639$; CDK4, $p = 0.973$; F480, $p = 0.488$; IL-6, $p = 0.203$; Hes1, $p = 0.118$) (Figure 6). However, Hes1 expression showed a modestly significant effect for sex when males and females are separated within each treatment group [$F(1, 8) = 6.828$; $p = 0.031$].

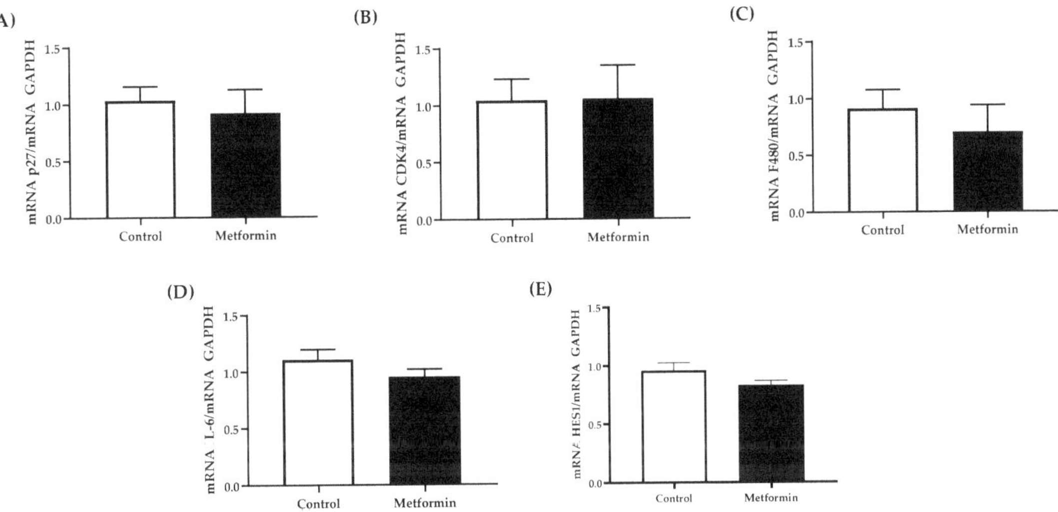

Figure 6. Gene expression in tumor mass collected from C57BL/6J male and female mice following orthotopic LL/2 non-small lung cancer cell implantation. (**A**) mRNA cyclin-dependent kinase inhibitor (p27)/GAPDH; (**B**) mRNA cyclin-dependent kinase (CDK4)/GAPDH; (**C**) mRNA F480/GAPDH; (**D**) mRNA interleukin-6/GAPDH; (**E**) mRNA hairy and enhancer of split 1 (Hes1)/GAPDH. Tumor mRNA expression from C57BL/6J mice with NSCLC concomitant with or without metformin (250 mg/kg) treatment. Data were analyzed using an unpaired *t*-test. Sample size: Control, *n* = 6; metformin, *n* = 6. Data shown as mean ± SEM.

3.6. Maintenance of Skeletal Muscle Mass

Skeletal muscle mass was maintained in all mice, irrespective of treatment (Table 3). Gastrocnemius muscle mass between control (Left: 100.0 ± 7.6 mg; Right: 102.0 ± 8.6 mg) and metformin- (Left: 102.5 ± 6.2 mg; Right: 97.5 ± 6.2 mg) treated mice did not significantly differ (Left: $p = 0.731$; Right: $p = 0.776$).

Table 3. Gastrocnemius Muscle Mass in C57BL/6J mice orthotopically implanted with LL/2 non-small lung cancer cells.

	Left Gastrocnemius (mg)	Right Gastrocnemius (mg)
Control	100.0 ± 7.6	102.0 ± 8.6
Metformin	102.5 ± 6.2	97.5 ± 6.2

Data shown as mean ± SEM.

3.7. Gastrocnemius Gene Expression

Genes involved in maintaining skeletal muscle mass and inflammatory signaling were not significantly different with regard to treatment (Figure 7). Metformin did not alter skeletal muscle PGC1-α mRNA (p = 0.816), MAFBx mRNA levels (p = 0.325), TNF-α mRNA levels (p = 0.111) or F480 mRNA levels (p = 0.076) expression. Two outliers were removed from PGC1-α mRNA expression data. Separation via treatment and sex revealed no significant differences in gene expression.

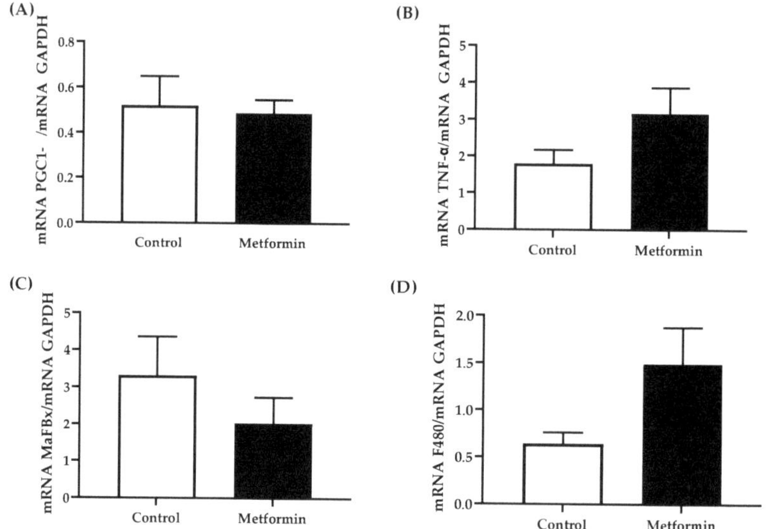

Figure 7. Gene expression in gastrocnemius muscle from C57BL/6J mice with orthotopically implanted LL/2 non-small lung cancer cells. (**A**) mRNA Peroxisome proliferator-activated receptor-gamma coactivator-1alpha (PGC1-α)/GAPDH; (**B**) mRNA tumor necrosis factor-alpha (TNF-α)/GAPDH; (**C**) mRNA muscle atrophy F-box (MAFbx)/GAPDH; (**D**) mRNA F480/GAPDH. Gastrocnemius mRNA expression from C57BL/6J mice with NSCLC concomitant with or without metformin (250 mg/kg) treatment. Data were analyzed using an unpaired *t*-test. Sample size: Control, n = 7; metformin, n = 7. Data shown as mean ± SEM.

3.8. Gastrocnemius Protein Expression

Skeletal muscle proteins that promote atrophy and regulate metabolism did not reveal detectable differences between control and metformin-treated groups (pSTAT3, p = 0.5889; STAT3, p = 0.6534; AMPK, p = 0.6387; REDD1, p = 0.6998) (Figure 8).

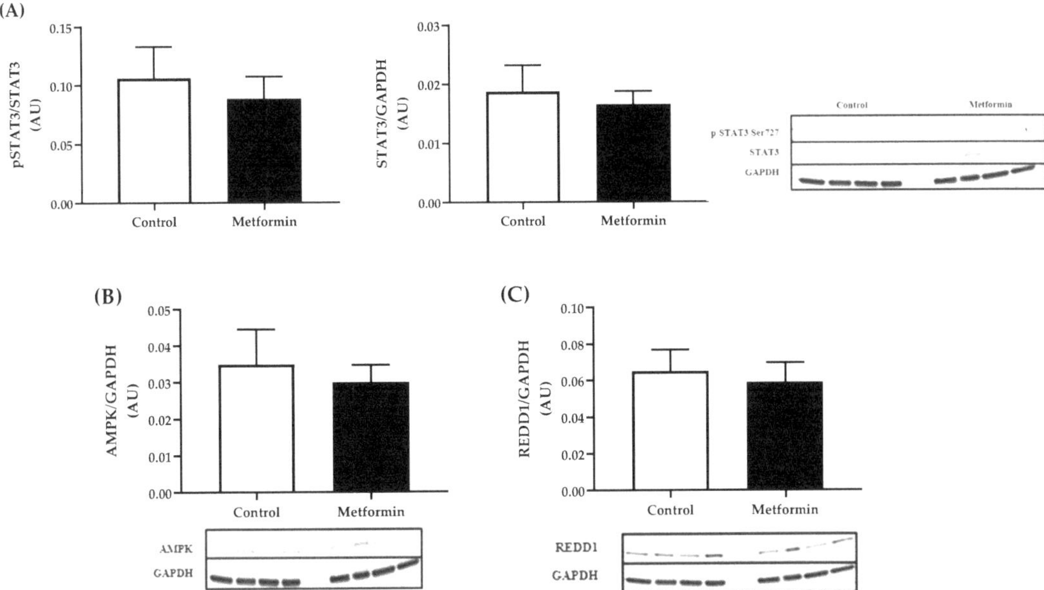

Figure 8. Protein expression in gastrocnemius muscle from C57BL/6J mice with orthotopically implanted with LL/2 non-small lung cancer cells. (**A**) STAT3 expression in gastrocnemius muscle from C57BL/6J mice with NSCLC. Phospho (p)-Signal transducer and activator of transcription 3 (STAT3) Ser727/Total STAT3 and STAT3/GAPDH expression (arbitrary units, AU) in gastrocnemius muscle from C57BL/6J mice with NSCLC concomitant with or without metformin (250 mg/kg) treatment. Data were analyzed using an unpaired *t*-test. (**B**) AMPK expression in gastrocnemius muscle from C57BL/6J mice with NSCLC. Total adenosine monophosphate-activated protein kinase (AMPK)/GAPDH expression (arbitrary units, AU) in gastrocnemius muscle from C57BL/6J mice with NSCLC concomitant with or without metformin (250 mg/kg) treatment. Data were analyzed using an unpaired *t*-test. (**C**) REDD1 expression in gastrocnemius muscle from C57BL/6J mice with NSCLC. Regulated in development and DNA damage responses 1 (REDD1)/GAPDH expression (arbitrary units, AU) in gastrocnemius muscle from C57BL/6J mice with NSCLC concomitant with or without metformin (250 mg/kg) treatment. Data were analyzed using an unpaired *t*-test. Sample size: Control, *n* = 7; metformin, *n* = 8. Data shown as mean ± SEM.

4. Discussion

The present study aimed to assess the effects of metformin as a stand-alone, i.e., monotherapy treatment in altering LL/2 non small lung tumor progression and its ability to support skeletal muscle health during LL/2 tumor progression in C57BL/6J mice. Our data indicate that metformin administered at a dose of 250 mg/kg, twice weekly, via i.p. in an immunocompetent model of NSCLC, was not associated with significant improvements in tumor burden. Moreover, there were no marked differences in gene expression of key tumor cell division (p27, CDK4 and Hes1) and inflammation markers (F4/80 and IL-6) following metformin treatment. Similarly, metformin was not associated with significant improvement in skeletal muscle health. Of note, as no control cohort without LL/2 cells was available, whether skeletal muscles became unhealthy is unknown.

In the conditions tested, no significant differences in tumor fold change or cell cycle regulatory genes (p27 and CDK4) were identified between the control and metformin-treated mice, the possibility exists that our dosing frequency was insufficient to exert effects. Indeed, metformin has a relatively short half-life, a high rate of absorption in the small intestine, and a nearly complete clearance via the kidneys, supporting that in our conditions the bioavailability of metformin is limited and delivery to the tumor site is inadequate [39,40]. Furthermore, it is important to acknowledge that metformin has a hormetic response such that the concentration of metformin within a target tissue influences the mechanism of action, which was elegantly reviewed by Panfoli et al. [41]. Therefore, the tissue concentration supporting the classical effects of metformin as an anti-diabetic drug, may differ from that necessary to alter the cellular and molecular signaling within a tumor. Additionally, appropriate delivery of the anti-cancer therapeutic is of the utmost importance. Oral gavage or via drinking water may prove to be a better route of administration. Indeed, delivering medicines through drinking water results in more consistent drug levels in the plasma when compared to drug delivery via i.p. injections [42,43]. Mice treated with metformin through drinking water, rather than i.p. injections, had an average blood plasma concentration of 32 µM (range of 9.1–55.7 µM), that could allow more consistent drug delivery to the tumor site [42].

Notably, the application of nanoparticle technology has provided an advantageous approach to more innovative cancer treatments. Specifically in NSCLC lines, nanoparticle carriers encapsulated biomolecules and successfully reached target tissues, resulting in either silencing or knockdown of genes to attenuate tumor cell growth [44,45]. Nanocarriers also possess many unique characteristics, making them excellent vehicles for drug delivery with the potential to better regulate pharmacokinetic effects [46]. This could lead to improved uptake of a nanoparticle into a target cell, resulting in increased drug bioavailability such as metformin, more controlled release of a therapeutic, increased drug stability, and reduced side effects from more conventional cancer treatments [47].

Although there was no significant reduction in F4/80 or IL-6 gene expression, animals receiving metformin treatment showed a trend for lower IL-6 gene expression. IL-6 is a multifaceted cytokine that acts as a key mediator of inflammation. High serum concentrations of IL-6 are associated with tumor progression, metastases, and poor clinical outcomes, especially for colorectal cancer patients [48]. In lung cancer patients, metformin has also been shown to reduce IL-6 driven epithelial-mesenchymal transitions, which plays an important role in tumorigenesis [49]. Together these findings suggest that metformin might mitigate tumor migration via effects on IL-6 production.

Metformin has been shown to reduce infiltration of tumor-associated inflammatory macrophages [50]. A previous study indicated that metformin (0.5–2.0 mM in vitro; 100 mg/kg/daily, i.p. in vivo) blocked alternatively activated (M2) macrophage polarization, which is often associated with tumor-driven angiogenesis, tumor migration and invasion, and suppression of anti-tumor immune responses [50]. However, it should be noted that this study differed in terms of metformin dosing strategy administered daily (100 mg/kg, i.p.) versus our twice weekly (250 mg/kg, i.p.). Interestingly, metformin reduced Lewis lung cancer metastases without affecting tumor growth in vivo [50]. This suggests that while metformin may not be directly targeting tumor growth, it is affecting the tumor microenvironment and possibly mitigating metastases. Low-dose metformin (50 mg/kg/day) administration in esophageal squamous cell carcinoma has been previously shown to not affect proliferation or apoptosis of cancer cells, but did increase the formation of tumor-suppressing macrophages in vitro [51]. Similarly, low-dose metformin treatment (250 mg/day) leads to a reprogramming of the tumor immune microenvironment in humans with esophageal cancer [51]. In contrast to the present study, metformin was administered daily, rather than twice weekly, which leads to differing bioavailability of metformin in the tumor microenvironment. As such, metformin may play a significant role in modulation of the tumor microenvironment rather than having a direct anti-tumorigenic impact on the tumor cells, particularly for prostate cancer cells [52].

Key skeletal muscle markers of muscle metabolism (PGC1-α 1) and atrophy (MAFbx) were used to assess overall skeletal muscle health, but exhibited no marked differences in gene expression. However, separating expression based on sex reveals some variation within each treatment, suggesting a potential source of noise evidenced in graphs grouped by treatment (Figure 7). PGC1-α 1 is a transcriptional co-activator critical for regulating energy metabolism and mitochondrial biogenesis [53]. Higher expression of PGC1-α suppresses atrophy-associated genes (muscle RING finger 1 and muscle atrophy F-box (MAFbx)/atrogin-1) and lower expressions of PGC1-α can be associated with rapid muscle atrophy such as cancer cachexia [31]. Metformin has also been previously shown to increase levels of PGC1-α in skeletal muscle via AMPK phosphorylation [30]. Here, neither MAFbx nor PGC1-α showed marked changes during NSCLC cancer development or in response to metformin treatment, suggesting that conditions in this study were not sufficient to induce rapid atrophy (<6 weeks).

In the present study, body mass and gastrocnemius muscle mass were also maintained, indicating that weight loss was probably not an indicative marker of cancer-induced cachexia. Since muscle mass was not significantly affected in this immunocompetent model of LL/2, it is likely that the balance between protein synthesis and degradation was maintained, suggesting that cancer-induced cachexia was not achieved in this study possibly because the endpoint of this study preceded the development of cachexia.

Although there were no significant differences in gene expression markers or correlations between tumor burden or inflammatory markers, a modest inflammatory response occurred within skeletal muscle. Indeed, metformin-treated mice showed a non-significant trend for elevated gene expression levels of markers of inflammation, specifically F4/80, that suggested greater macrophage infiltration and/or activation and the inflammatory cytokine TNF-α. Importantly, infiltration of pro-inflammatory F4/80 positive macrophages has been shown to be linked to obesity, insulin resistance, and cancer cachexia [54–56]. Low-grade inflammation coincides with the onset of insulin resistance, which can be indicative of declining skeletal muscle health and reduced glucose disposal. Elevated TNF-α levels are also associated with increased catabolic activity in skeletal muscle, such as protein degradation, insulin resistance, impaired myogenesis and contractile dysfunction [57,58].

Signal transducer and activator of transcription 3 (STAT3), a cytokine transcription factor, has been linked with systemic inflammation in cancer cachexia [59]. Importantly STAT3 is a critical regulator of satellite cell self-renewal and this signaling component plays an important role in muscle wasting, including cachexia [60]. Findings from the present study revealed no phosphorylation of pSTAT3 Ser727 or change in total STAT3 protein expression, suggesting that skeletal muscle wasting, if present, did not occur via this signaling pathway. Because the orthotopic injection mimics tumor development in the lungs, it is possible that a longer timeline or a combination of treatment modalities with irradiation or chemotherapeutics could better mimic the onset of muscle wasting.

Previous studies employing the Lewis lung carcinoma mouse model have shown an attenuation in the expression of fundamental genes involved in the phosphatidylinositol 3-kinase (PI3K)-protein kinase B (Akt) pathway have been observed [61]. The PI3K/AKT pathway, which is often constitutively active in tumor cells, plays an important role in cellular proliferation, growth, metabolism, and protein synthesis [62]. Reduced expression of regulatory genes in the PI3K/AKT pathway could lead to mitochondrial dysfunction and skeletal muscle wasting [61]. Importantly, metformin treatment in tumor bearing rats has been reported to decrease skeletal muscle wasting and improve protein metabolism, attenuating cancer-induced cachexia [63].

Regulated in development and DNA damage response (REDD1) is a ubiquitous protein that is a well-known endogenous inhibitor of the AKT/mTOR pathway [64]. Not surprisingly, this means that REDD1 plays a role in regulating cell growth, mitochondrial function, oxidative stress, and apoptosis [65]. Recent studies have highlighted the importance of REDD1 in maintaining skeletal muscle mass [66]. The present study revealed no differences in REDD1 expression in control or metformin-treated animals. In contrast, a murine model of Lewis lung carcinoma has shown skeletal muscle mass loss between 28–35 days post-tumor development concomitant with increased REDD1 gene expression. The increased REDD1 expression was also associated with lower mTOR expression, suggesting that REDD1 may curb mTOR signaling during later stages of cachexia development [67]. Variations in REDD1 expression in the present study compared to previous investigations [67,68] may be attributed to the variations in lung cancers cell implantation approach.

In vitro incubation of cancer cells with metformin suggested anti-neoplastic potential, although these effects were not supported by our in vivo findings. Futures studies should consider including more frequent metformin dosing in combination with standard chemotherapeutics known to induce deleterious effects to skeletal muscles. In addition, metformin's potential as a tumor suppressor maybe be supportive in adjuvant therapies or in combination with other cancer treatments. A formative study by Della Corte et al. [69] demonstrated that metformin enhanced the anti-tumor properties of the MEK inhibitor, selumetinib, during in vitro and in vivo treatments. Specifically, the combination of metformin and selumetinib nearly doubled the reduction in proliferation of several human lung cancer cell lines and significantly mitigated tumor growth in mice [69]. Furthermore, human clinical trials demonstrated high safety when combining metformin with erlotinib, a tyrosine kinases inhibitor of the epidermal growth factor receptor, in non-diabetic NSCLC patients as a second-line therapy [70]. Therefore, metformin combination therapies may work synergistically to manage tumor growth by mitigating activity of the PI3K/Akt and MAPK pathways [69,70].

Author Contributions: Conceptualization, N.L.S.B., D.D. and J.S.M.; Data curation, N.L.S.B. and J.S.M.; Formal analysis, N.L.S.B., D.D., I.M., J.M.B., M.J.T., S.T.A. and J.S.M.; Funding acquisition, J.S.M.; Investigation, N.L.S.B.; Methodology, N.L.S.B., D.D., I.M., J.M.B., M.J.T. and S.T.A.; Writing—original draft, N.L.S.B.; Writing—review and editing, D.D., I.M., J.M.B., M.J.T., S.T.A. and J.S.M. All authors have read and agreed to the published version of the manuscript.

Funding: This research was funded by the Targeted Research Internal Seed Program at The University of North Carolina at Charlotte.

Institutional Review Board Statement: All aspects of this study were approved by the Institutional Animal Care and Use Committee at The University of North Carolina at Charlotte (Protocol #18-005; approved on 29 July 2018).

Informed Consent Statement: Not applicable.

Data Availability Statement: Data available upon request. Please email jmarin10@uncc.edu for access to data.

Acknowledgments: The authors would like to thank Chandra Williams for her direction and assistance with the planning and early stages of this project, including her chief involvement with the lung cancer injection surgeries.

Conflicts of Interest: The authors declare no conflict of interest.

References

1. American Cancer Society. *Cancer Facts and Figures 2019*; American Cancer Society: Atlanta, GA, USA, 2019; Available online: https://www.cancer.org/content/dam/cancer-org/research/cancer-facts-and-statistics/annual-cancer-facts-and-figures/2019/cancer-facts-and-figures-2019.pdf (accessed on 5 December 2020).
2. Rahib, L.; Smith, B.D.; Aizenberg, R.; Rosenzweig, A.B.; Fleshman, J.M.; Matrisian, L.M. Projecting cancer incidence and deaths to 2030: The unexpected burden of thyroid, liver, and pancreas cancers in the United States. *Cancer Res.* **2014**, *74*, 2913–2921. [CrossRef]
3. Molina, J.R.; Yang, P.; Cassivi, S.D.; Schild, S.E.; Adjei, A.A. Non-small cell lung cancer: Epidemiology, risk factors, treatment, and survivorship. *Mayo Clin. Proc.* **2008**, *83*, 584–594. [CrossRef]
4. Hirsch, F.R.; Suda, K.; Wiens, J.; Bunn, P.A. New and emerging targeted treatments in advanced non-small-cell lung cancer. *Lancet* **2016**, *388*, 1012–1024. [CrossRef]
5. Irie, H.; Banno, K.; Yanokura, M.; Iida, M.; Adachi, M.; Nakamura, K.; Umene, K.; Nogami, Y.; Masuda, K.; Kobayashi, Y.; et al. Metformin: A candidate for the treatment of gynecological tumors based on drug repositioning. *Oncol. Lett.* **2016**, *11*, 1287–1293. [CrossRef] [PubMed]
6. Williams, C.T. Food and Drug Administration Drug Approval Process: A History and Overview. *Nurs. Clin. N. Am.* **2016**, *51*, 1–11. [CrossRef] [PubMed]
7. Sleire, L.; Førde, H.E.; Netland, I.A.; Leiss, L.; Skeie, B.S.; Enger, P.Ø. Drug repurposing in cancer. *Pharmacol. Res.* **2017**, *124*, 74–91. [CrossRef]

8. Queiroz, E.A.I.F.; Puukila, S.; Eichler, R.; Sampaio, S.C.; Forsyth, H.L.; Lees, S.J.; Barbosa, A.M.; Dekker, R.F.H.; Fortes, Z.B.; Khaper, N. Metformin Induces Apoptosis and Cell Cycle Arrest Mediated by Oxidative Stress, AMPK and FOXO3a in MCF-7 Breast Cancer Cells. *PLoS ONE* **2014**, *9*, e98207. [CrossRef]

9. Zakikhani, M.; Dowling, R.; Fantus, I.G.; Sonenberg, N.; Pollak, M. Metformin Is an AMP Kinase–Dependent Growth Inhibitor for Breast Cancer Cells. *Cancer Res.* **2006**, *66*, 10269–10273. [CrossRef]

10. Sarmento-Cabral, A.; L-López, F.; Gahete, M.D.; Castaño, J.P.; Luque, R.M. Metformin Reduces Prostate Tumor Growth, in a Diet-Dependent Manner, by Modulating Multiple Signaling Pathways. *Mol. Cancer Res.* **2017**, *15*, 862–874. [CrossRef] [PubMed]

11. Park, J.W.; Lee, J.H.; Park, Y.H.; Park, S.J.; Cheon, J.H.; Kim, W.H.; Kim, T.I. Sex-dependent difference in the effect of metformin on colorectal cancer-specific mortality of diabetic colorectal cancer patients. *World J. Gastroenterol.* **2017**, *23*, 5196–5205. [CrossRef] [PubMed]

12. Checkley, L.A.; Rho, O.; Angel, J.M.; Cho, J.; Blando, J.; Beltran, L.; Hursting, S.D.; DiGiovanni, J. Metformin Inhibits Skin Tumor Promotion in Overweight and Obese Mice. *Cancer Prev. Res.* **2014**, *7*, 54–64. [CrossRef] [PubMed]

13. Zhu, N.; Zhang, Y.; Gong, Y.I.; He, J.; Chen, X. Metformin and lung cancer risk of patients with type 2 diabetes mellitus: A meta-analysis. *Biomed. Rep.* **2015**, *3*, 235–241. [CrossRef]

14. Tseng, C.-H. Metformin and lung cancer risk in patients with type 2 diabetes mellitus. *Oncotarget* **2017**, *8*, 41132–41142. [CrossRef] [PubMed]

15. Hung, M.-S.; Chuang, M.-C.; Chen, Y.-C.; Lee, C.-P.; Yang, T.-M.; Chen, P.-C.; Tsai, Y.-H.; Yang, Y.-H. Metformin Prolongs Survival in Type 2 Diabetes Lung Cancer Patients With EGFR-TKIs. *Integr. Cancer Ther.* **2019**, *18*, 1534735419869491. [CrossRef]

16. Tsai, M.-J.; Yang, C.-J.; Kung, Y.-T.; Sheu, C.-C.; Shen, Y.-T.; Chang, P.-Y.; Huang, M.-S.; Chiu, H.-C. Metformin decreases lung cancer risk in diabetic patients in a dose-dependent manner. *Lung Cancer* **2014**, *86*, 137–143. [CrossRef]

17. Algire, C.; Amrein, L.; Zakikhani, M.; Panasci, L.; Pollak, M. Metformin blocks the stimulative effect of a high-energy diet on colon carcinoma growth in vivo and is associated with reduced expression of fatty acid synthase. *Endocr. Relat. Cancer* **2010**, *17*, 351–360. [CrossRef] [PubMed]

18. Park, J.; Kim, W.G.; Zhao, L.; Enomoto, K.; Willingham, M.; Cheng, S.-Y. Metformin blocks progression of obesity-activated thyroid cancer in a mouse model. *Oncotarget* **2016**, *7*, 34832–34844. [CrossRef] [PubMed]

19. Lei, Y.; Yi, Y.; Liu, Y.; Liu, X.; Keller, E.T.; Qian, C.-N.; Zhang, J.; Lu, Y. Metformin targets multiple signaling pathways in cancer. *Chin. J. Cancer* **2017**, *36*, 17. [CrossRef]

20. Baracos, V.E.; Martin, L.; Korc, M.; Guttridge, D.C.; Fearon, K.C.H. Cancer-associated cachexia. *Nat. Rev. Dis. Prim.* **2018**, *4*, 17105. [CrossRef]

21. Evans, W.J. Skeletal muscle loss: Cachexia, sarcopenia, and inactivity. *Am. J. Clin. Nutr.* **2010**, *91*, 1123S–1127S. [CrossRef] [PubMed]

22. Sorensen, J. Lung Cancer Cachexia: Can Molecular Understanding Guide Clinical Management? *Integr. Cancer Ther.* **2018**, *17*, 1000–1008. [CrossRef]

23. Aniort, J.; Stella, A.; Philipponnet, C.; Poyet, A.; Polge, C.; Claustre, A.; Combaret, L.; Béchet, D.; Attaix, D.; Boisgard, S.; et al. Muscle wasting in patients with end-stage renal disease or early-stage lung cancer: Common mechanisms at work. *J. Cachexia Sarcopenia Muscle* **2019**, *10*, 323–337. [CrossRef]

24. Currow, D.C.; Maddocks, M.; Cella, D.; Muscaritoli, M. Efficacy of Anamorelin, a Novel Non-Peptide Ghrelin Analogue, in Patients with Advanced Non-Small Cell Lung Cancer (NSCLC) and Cachexia—Review and Expert Opinion. *Int. J. Mol. Sci.* **2018**, *19*, 3471. [CrossRef]

25. LeBlanc, T.; Nipp, R.D.; Rushing, C.N.; Samsa, G.P.; Locke, S.; Kamal, A.H.; Cella, D.F.; Abernethy, A.P. Correlation Between the International Consensus Definition of the Cancer Anorexia-Cachexia Syndrome (CACS) and Patient-Centered Outcomes in Advanced Non-Small Cell Lung Cancer. *J. Pain Symptom Manag.* **2015**, *49*, 680–689. [CrossRef]

26. Aoyagi, T.; Terracina, K.P.; Raza, A.; Matsubara, H.; Takabe, K. Cancer cachexia, mechanism and treatment. *World J. Gastrointest. Oncol.* **2015**, *7*, 17–29. [CrossRef]

27. Fearon, K.C.; Voss, A.C.; Hustead, D.S. Definition of cancer cachexia: Effect of weight loss, reduced food intake, and systemic inflammation on functional status and prognosis. *Am. J. Clin. Nutr.* **2006**, *83*, 1345–1350. [CrossRef] [PubMed]

28. Burckart, K.; Beca, S.; Urban, R.J.; Sheffield-Moore, M. Pathogenesis of muscle wasting in cancer cachexia: Targeted anabolic and anticatabolic therapies. *Curr. Opin. Clin. Nutr. Metab. Care* **2010**, *13*, 410–416. [CrossRef]

29. Tisdale, M.J. Cancer cachexia. *Curr. Opin. Gastroenterol.* **2010**, *26*, 146–151. [CrossRef]

30. Suwa, M.; Egashira, T.; Nakano, H.; Sasaki, H.; Kumagai, S. Metformin increases the PGC-1α protein and oxidative enzyme activities possibly via AMPK phosphorylation in skeletal muscle in vivo. *J. Appl. Physiol.* **2006**, *101*, 1685–1692. [CrossRef] [PubMed]

31. Sandri, M.; Lin, J.; Handschin, C.; Yang, W.; Arany, Z.P.; Lecker, S.H.; Goldberg, A.L.; Spiegelman, B.M. PGC-1alpha protects skeletal muscle from atrophy by suppressing FoxO3 action and atrophy-specific gene transcription. *Proc. Natl. Acad. Sci. USA* **2006**, *103*, 16260–16265. [CrossRef] [PubMed]

32. Pavlidou, T.; Marinkovic, M.; Rosina, M.; Fuoco, C.; Vumbaca, S.; Gargioli, C.; Castagnoli, L.; Cesareni, G. Metformin Delays Satellite Cell Activation and Maintains Quiescence. *Stem Cells Int.* **2019**, *2019*, 5980465. [CrossRef]

33. Yin, H.; Price, F.; Rudnicki, M.A. Satellite Cells and the Muscle Stem Cell Niche. *Physiol. Rev.* **2013**, *93*, 23–67. [CrossRef]

34. Latteyer, S.; Christoph, S.; Theurer, S.; Hönes, G.S.; Schmid, K.W.; Führer, D.; Moeller, L.C. Thyroxine promotes lung cancer growth in an orthotopic mouse model. *Endocr. Relat. Cancer* **2019**, *26*, 565–574. [CrossRef]

35. Mordant, P.; Loriot, Y.; Lahon, B.; Castier, Y.; Lesèche, G.; Soria, J.-C.; Vozenin, M.-C.; Decraene, C.; Deutsch, E. Bioluminescent Orthotopic Mouse Models of Human Localized Non Small Cell Lung Cancer. Feasibility and Identification of Circulating Tumour Cells. *PLoS ONE* **2011**, *6*, e26073. [CrossRef]

36. Ogawa, F.; Amano, H.; Ito, Y.; Matsui, Y.; Hosono, K.; Kitasato, H.; Satoh, Y.; Majima, M. Aspirin reduces lung cancer metastasis to regional lymph nodes. *Biomed. Pharmacother.* **2014**, *68*, 79–86. [CrossRef] [PubMed]

37. Lengyel, E.; Litchfield, L.M.; Mitra, A.K.; Nieman, K.; Mukherjee, A.; Zhang, Y.; Johnson, A.; Bradaric, M.; Lee, W.; Romero, I.L. Metformin inhibits ovarian cancer growth and increases sensitivity to paclitaxel in mouse models. *Am. J. Obstet. Gynecol.* **2015**, *212*, 479.e1–479.e10. [CrossRef]

38. Memmott, R.M.; Mercado, J.R.; Maier, C.R.; Kawabata, S.; Fox, S.D.; Dennis, P.A. Metformin Prevents Tobacco Carcinogen–Induced Lung Tumorigenesis. *Cancer Prev. Res.* **2010**, *3*, 1066–1076. [CrossRef]

39. Pernicova, I.; Korbonits, M. Metformin—Mode of action and clinical implications for diabetes and cancer. *Nat. Rev. Endocrinol.* **2014**, *10*, 143–156. [CrossRef] [PubMed]

40. Bailey, C.J.; Turner, R.C. Metformin. *N. Engl. J. Med.* **1996**, *334*, 574–579. [CrossRef] [PubMed]

41. Panfoli, I.; Puddu, A.; Bertola, N.; Ravera, S.; Maggi, D. The Hormetic Effect of Metformin: "Less Is More"? *Int. J. Mol. Sci.* **2021**, *22*, 6297. [CrossRef] [PubMed]

42. Dowling, R.J.; Lam, S.; Bassi, C.; Mouaaz, S.; Aman, A.; Kiyota, T.; Al-Awar, R.; Goodwin, P.; Stambolic, V. Metformin Pharmacokinetics in Mouse Tumors: Implications for Human Therapy. *Cell Metab.* **2016**, *23*, 567–568. [CrossRef]

43. Heinig, K.; Bucheli, F. Fast liquid chromatographic-tandem mass spectrometric (LC–MS–MS) determination of metformin in plasma samples. *J. Pharm. Biomed. Anal.* **2004**, *34*, 1005–1011. [CrossRef] [PubMed]

44. Bai, J.; Duan, J.; Liu, R.; Du, Y.; Luo, Q.; Cui, Y.; Su, Z.; Xu, J.; Xie, Y.; Lu, W. Engineered targeting tLyp-1 exosomes as gene therapy vectors for efficient delivery of siRNA into lung cancer cells. *Asian J. Pharm. Sci.* **2020**, *15*, 461–471. [CrossRef]

45. Mehta, A.; Dalle Vedove, E.; Isert, L.; Merkel, O.M. Targeting KRAS Mutant Lung Cancer Cells with siRNA-Loaded Bovine Serum Albumin Nanoparticles. *Pharm. Res.* **2019**, *36*, 133. [CrossRef]

46. Duncan, R.; Gaspar, R. Nanomedicine(s) under the microscope. *Mol. Pharm.* **2011**, *8*, 2101–2141. [CrossRef]

47. Chaudhary, S.; Singh, A.; Kumar, P.; Kaushik, M. Strategic targeting of non-small-cell lung cancer utilizing genetic material-based delivery platforms of nanotechnology. *J. Biochem. Mol. Toxicol.* **2021**, *35*, e22784. [CrossRef]

48. Knüpfer, H.; Preiss, R. Serum interleukin-6 levels in colorectal cancer patients—A summary of published results. *Int. J. Color. Dis.* **2009**, *25*, 135–140. [CrossRef]

49. Zhao, Z.; Cheng, X.; Wang, Y.; Han, R.; Li, L.; Xiang, T.; He, L.; Long, H.; Zhu, B.; He, Y. Metformin Inhibits the IL-6-Induced Epithelial-Mesenchymal Transition and Lung Adenocarcinoma Growth and Metastasis. *PLoS ONE* **2014**, *9*, e95884. [CrossRef] [PubMed]

50. Ding, L.; Liang, G.; Yao, Z.; Zhang, J.; Liu, R.; Chen, H.; Zhou, Y.; Wu, H.; Ruiyang, L.; He, Q. Metformin prevents cancer metastasis by inhibiting M2-like polarization of tumor associated macrophages. *Oncotarget* **2015**, *6*, 36441–36455. [CrossRef] [PubMed]

51. Wang, S.; Lin, Y.; Xiong, X.; Wang, L.; Guo, Y.; Chen, Y.; Chen, S.; Wang, G.; Lin, P.; Chen, H.; et al. Low-Dose Metformin Reprograms the Tumor Immune Microenvironment in Human Esophageal Cancer: Results of a Phase II Clinical Trial. *Clin. Cancer Res.* **2020**, *26*, 4921–4932. [CrossRef] [PubMed]

52. Liu, Q.; Tong, D.; Liu, G.; Gao, J.; Wang, L.-A.; Xu, J.; Yang, X.; Xie, Q.; Huang, Y.; Pang, J.; et al. Metformin Inhibits Prostate Cancer Progression by Targeting Tumor-Associated Inflammatory Infiltration. *Clin. Cancer Res.* **2018**, *24*, 5622–5634. [CrossRef] [PubMed]

53. Brown, J.L.; Rosa-Caldwell, M.E.; Lee, D.E.; Blackwell, T.A.; Brown, L.A.; Perry, R.A.; Haynie, W.S.; Hardee, J.P.; Carson, J.; Wiggs, M.P.; et al. Mitochondrial degeneration precedes the development of muscle atrophy in progression of cancer cachexia in tumour-bearing mice. *J. Cachexia Sarcopenia Muscle* **2017**, *8*, 926–938. [CrossRef]

54. Fink, L.N.; Costford, S.R.; Lee, Y.S.; Jensen, T.E.; Bilan, P.J.; Oberbach, A.; Blüher, M.; Olefsky, J.M.; Sams, A.; Klip, A. Pro-Inflammatory macrophages increase in skeletal muscle of high fat-Fed mice and correlate with metabolic risk markers in humans. *Obesity* **2014**, *22*, 747–757. [CrossRef] [PubMed]

55. Ham, D.; Murphy, K.; Chee, A.; Lynch, G.; Koopman, R. Glycine administration attenuates skeletal muscle wasting in a mouse model of cancer cachexia. *Clin. Nutr.* **2014**, *33*, 448–458. [CrossRef]

56. Batista, M.L.; Neves, R.X.; Peres, S.B.; Yamashita, A.S.; Shida, C.S.; Farmer, S.R.; Seelaender, M. Heterogeneous time-dependent response of adipose tissue during the development of cancer cachexia. *J. Endocrinol.* **2012**, *215*, 363–373. [CrossRef] [PubMed]

57. Li, Y.-P.; Reid, M.B. Effect of tumor necrosis factor-α on skeletal muscle metabolism. *Curr. Opin. Rheumatol.* **2001**, *13*, 483–487. [CrossRef]

58. Thoma, A.; Lightfoot, A.P. NF-kB and Inflammatory Cytokine Signalling: Role in Skeletal Muscle Atrophy. *Adv. Exp. Med. Biol.* **2018**, *1088*, 267–279.

59. Zimmers, T.A.; Fishel, M.L.; Bonetto, A. STAT3 in the systemic inflammation of cancer cachexia. *Semin. Cell Dev. Biol.* **2016**, *54*, 28–41. [CrossRef]

60. Guadagnin, E.; Mázala, D.; Chen, Y.-W. STAT3 in Skeletal Muscle Function and Disorders. *Int. J. Mol. Sci.* **2018**, *19*, 2265. [CrossRef]
61. Constantinou, C.; De Oliveira, C.C.F.; Mintzopoulos, D.; Busquets, S.; He, J.; Kesarwani, M.; Mindrinos, M.; Rahme, L.G.; Argiles, J.M.; Tzika, A.A. Nuclear magnetic resonance in conjunction with functional genomics suggests mitochondrial dysfunction in a murine model of cancer cachexia. *Int. J. Mol. Med.* **2011**, *27*, 15–24. [CrossRef]
62. Hemmings, B.A.; Restuccia, D.F. PI3K-PKB/Akt Pathway. *Cold Spring Harb. Perspect. Biol.* **2012**, *4*, a011189. [CrossRef]
63. Oliveira, A.G.; Gomes-Marcondes, M.C.C. Metformin treatment modulates the tumour-induced wasting effects in muscle protein metabolism minimising the cachexia in tumour-bearing rats. *BMC Cancer* **2016**, *16*, 418. [CrossRef] [PubMed]
64. Brugarolas, J.; Lei, K.; Hurley, R.L.; Manning, B.D.; Reiling, J.H.; Hafen, E.; Kaelin, W.G. Regulation of mTOR function in response to hypoxia by REDD1 and the TSC1/TSC2 tumor suppressor complex. *Genes Dev.* **2004**, *18*, 2893–2904. [CrossRef]
65. Britto, F.A.; Dumas, K.; Giorgetti-Peraldi, S.; Ollendorff, V.; Favier, F.B. Is REDD1 a metabolic double agent? Lessons from physiology and pathology. *Am. J. Physiol. Physiol.* **2020**, *319*, C807–C824. [CrossRef] [PubMed]
66. Gordon, B.S.; Steiner, J.L.; Williamson, D.L.; Lang, C.H.; Kimball, S.R. Emerging role for regulated in development and DNA damage 1 (REDD1) in the regulation of skeletal muscle metabolism. *Am. J. Physiol. Metab.* **2016**, *311*, E157–E174. [CrossRef]
67. Puppa, M.J.; Gao, S.; Narsale, A.A.; Carson, J.A. Skeletal muscle glycoprotein 130's role in Lewis lung carcinoma–induced cachexia. *FASEB J.* **2014**, *28*, 998–1009. [CrossRef] [PubMed]
68. Niu, M.; Li, L.; Su, Z.; Wei, L.; Pu, W.; Zhao, C.; Ding, Y.; Wazir, J.; Cao, W.; Song, S.; et al. An integrative transcriptome study reveals Ddit4/Redd1 as a key regulator of cancer cachexia in rodent models. *Cell Death Dis.* **2021**, *12*, 652. [CrossRef]
69. Della Corte, C.M.; Ciaramella, V.; Di Mauro, C.; Castellone, M.; Papaccio, F.; Fasano, M.; Sasso, F.C.; Martinelli, E.; Troiani, T.; De Vita, F.; et al. Metformin increases antitumor activity of MEK inhibitors through GLI1 downregulation in LKB1 positive human NSCLC cancer cells. *Oncotarget* **2016**, *7*, 4265–4278. [CrossRef]
70. Morgillo, F.; Fasano, M.; Della Corte, C.M.; Sasso, F.C.; Papaccio, F.; Viscardi, G.; Esposito, G.; DI Liello, R.; Normanno, N.; Capuano, A.; et al. Results of the safety run-in part of the METAL (METformin in Advanced Lung cancer) study: A multicentre, open-label phase I–II study of metformin with erlotinib in second-line therapy of patients with stage IV non-small-cell lung cancer. *ESMO Open* **2017**, *2*, e000132. [CrossRef]

Article

Probing Skin Barrier Recovery on Molecular Level Following Acute Wounds: An In Vivo/Ex Vivo Study on Pigs

Enamul Haque Mojumdar [1,2,*], Lone Bruhn Madsen [3], Henri Hansson [4], Ida Taavoniku [3], Klaus Kristensen [3], Christina Persson [5], Anna Karin Morén [4], Rajmund Mokso [6], Artur Schmidtchen [7,8], Tautgirdas Ruzgas [1,2] and Johan Engblom [1,2]

[1] Department of Biomedical Science, Faculty of Health and Society, Malmö University, SE-205 06 Malmö, Sweden; tautgirdas.ruzgas@mau.se (T.R.); johan.engblom@mau.se (J.E.)

[2] Biofilms-Research Center for Biointerfaces (BRCB), Malmö University, SE-205 06 Malmö, Sweden

[3] Timeline Bioresearch AB, Scheelevägen 2, SE-223 63 Lund, Sweden; lbm@timelinebioresearch.se (L.B.M.); it@timelinebioresearch.se (I.T.); kk@timelinebioresearch.se (K.K.)

[4] Galenica AB, Medeon Science Park, SE-205 12 Malmö, Sweden; Henri.Hansson@galenica.se (H.H.); AnnaKarin.Moren@galenica.se (A.K.M.)

[5] Department of Occupational and Environmental Dermatology, Lund University, Skåne University Hospital, SE-205 02 Malmö, Sweden; Christina.ML.Persson@skane.se

[6] Department of Solid Mechanics & MAX IV Laboratory, Lund University, SE-221 00 Lund, Sweden; rajmund.mokso@maxiv.lu.se

[7] Division of Dermatology and Venereology, Department of Clinical Sciences, Lund University, SE-221 84 Lund, Sweden; artur.schmidtchen@med.lu.se

[8] Copenhagen Wound Healing Center, Bispebjerg Hospital, Department of Biomedical Sciences, University of Copenhagen, DK-2400 Copenhagen, Denmark

* Correspondence: enamul.mojumdar@gmail.com; Tel.: +46-40-665-74-69

Citation: Mojumdar, E.H.; Madsen, L.B.; Hansson, H.; Taavoniku, I.; Kristensen, K.; Persson, C.; Morén, A.K.; Mokso, R.; Schmidtchen, A.; Ruzgas, T.; et al. Probing Skin Barrier Recovery on Molecular Level Following Acute Wounds: An In Vivo/Ex Vivo Study on Pigs. *Biomedicines* **2021**, *9*, 360. https://doi.org/10.3390/biomedicines9040360

Academic Editor: Martina Perše

Received: 10 March 2021
Accepted: 26 March 2021
Published: 31 March 2021

Publisher's Note: MDPI stays neutral with regard to jurisdictional claims in published maps and institutional affiliations.

Abstract: Proper skin barrier function is paramount for our survival, and, suffering injury, there is an acute need to restore the lost barrier and prevent development of a chronic wound. We hypothesize that rapid wound closure is more important than immediate perfection of the barrier, whereas specific treatment may facilitate perfection. The aim of the current project was therefore to evaluate the quality of restored tissue down to the molecular level. We used Göttingen minipigs with a multi-technique approach correlating wound healing progression in vivo over three weeks, monitored by classical methods (e.g., histology, trans-epidermal water loss (TEWL), pH) and subsequent physicochemical characterization of barrier recovery (i.e., small and wide-angle X-ray diffraction (SWAXD), polarization transfer solid-state NMR (PTssNMR), dynamic vapor sorption (DVS), Fourier transform infrared (FTIR)), providing a unique insight into molecular aspects of healing. We conclude that although acute wounds sealed within two weeks as expected, molecular investigation of stratum corneum (SC) revealed a poorly developed keratin organization and deviations in lipid lamellae formation. A higher lipid fluidity was also observed in regenerated tissue. This may have been due to incomplete lipid conversion during barrier recovery as glycosphingolipids, normally not present in SC, were indicated by infrared FTIR spectroscopy. Evidently, a molecular approach to skin barrier recovery could be a valuable tool in future development of products targeting wound healing.

Keywords: skin barrier; stratum corneum; lipid; acute wound; in vivo/ex vivo; trans-epidermal water loss (TEWL); pH; histology; polarization transfer solid state NMR (PTssNMR); small and wide-angle X-ray diffraction (SWAXD)

1. Introduction

Skin is one of the largest organs in terrestrial life, comprising an essential barrier towards the external harsh environment and being paramount for our survival. It has numerous vital functions, where perhaps the most important is to maintain body homeostasis, and among others preventing excessive water loss [1–3]. Mammalian skin consists of three layers: the epidermis, dermis, and hypodermis (Figure 1) [4]. The epidermis can be further subdivided into four distinct layers: stratum basale (SB), stratum spinosum (SS), stratum granulosum (SG), and stratum corneum (SC). Skin barrier function is assured by the outermost layer of the epidermis—the SC [1,5]. Formation of the barrier begins in the deepest layer of the epidermis where keratinocytes in SB proliferate and gradually migrate through SS and SG [6,7]. During their migration, the cells flatten, anucleate, and adopt the typical size and shape of corneocytes when they reach SC, and finally they are expelled from the skin surface by desquamation. In normal skin, the epidermis undergoes constant renewal with an average turnover of 20 to 30 days [8].

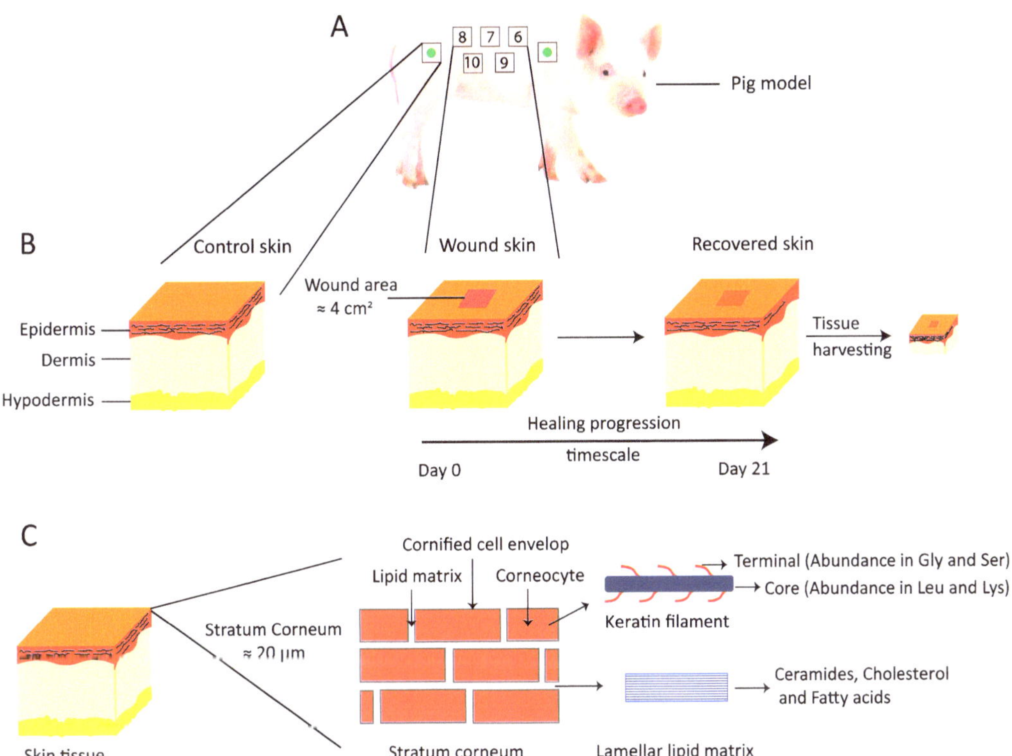

Figure 1. (**A**) A schematic of the wound healing study design. The study was performed on two non-naïve Göttingen minipigs. A total of 10 wounds were generated on each pig, five on the left side of the pigs back and five on the right side, as shown here with numbers from 6 to 10. The green filled circle inside the square indicates regions from where the control skin tissues were harvested at the end of the in vivo study. (**B**) A simplified drawing showing three major layers of the skin: the epidermis, the dermis, and the hypodermis. The wound incision area, which is approximately 2 × 2 cm², is shown, along with the time scale for healing progression over 21 days. After the in vivo study, the recovered skin was harvested for further evaluation ex vivo. (**C**) A cartoon presenting the thin top layer of the epidermis, the stratum corneum, which can be illustrated with bricks and mortar. The bricks in the cartoon represent anucleated corneocytes, mainly composed of keratin filaments. The corneocytes are embedded in a lipid lamellar matrix constituting the mortar. These lipid lamellae mainly comprise ceramides, cholesterol, and free fatty acids.

The SC is generally about 10–15 μm thick and consists of a lipid–protein complex where 10–15 layers of protein-bound stacked dead cells, corneocytes, are embedded in a multilamellar lipid matrix [9,10]. The corneocytes are filled with keratin intermediate filaments, proteins (enzymes), and compounds constituting the natural moisturizing factor (NMF), and weigh approximately 85% of the dry SC [11,12]. Corneocytes are responsible for mechanical properties such as viscoelasticity and plasticity of the SC [13]. The extracellular lipids in SC are derived from lamellar bodies found in SG as membrane-bound granules and are responsible for skin barrier function. During the transition from SG to SC, the lamellar bodies secrete their lipid content (phospholipids, glucosylceramides, sphingomyelin, and cholesterol) together with hydrolytic enzymes to the extracellular space through exocytosis [9,14]. These lipids are then further enzymatically processed to the barrier constituting lipids (mainly ceramides (Cer), cholesterol (Chol), and free fatty acids (Ffa)), which together form stacks of structured solid lipid lamellae in the extracellular space of SC [15,16]. In healthy human SC, the lipids exhibit both a long (LPP) and a short (SPP) periodicity phase with repeat distances of approximately 13 and 6 nm, respectively, while the lipid chains pack laterally predominantly in an orthorhombic arrangement [17,18]. The mainly solid lipid lamellae also comprise a small fraction of fluid lipids that play an important role for the macroscopic barrier properties [19,20]. In several skin diseases, e.g., atopic dermatitis and psoriasis, the barrier function of the skin is compromised [21–24]. The impaired barrier function is often attributed to alterations in SC lipid composition and their lamellar assembly [25]. The fraction of long acyl Cers is, e.g., significantly reduced in the diseased skin, and these specific lipids are key to the formation of LPP, which plays a crucial role in the SC barrier function [26,27]. It has furthermore been reported that lipids in diseased skin preferentially pack in a less dense hexagonal instead of orthorhombic lateral arrangement [26,28].

The SC is often conceptualized by the "brick and mortar" model in which the corneocytes are the bricks and the intercellular lipids serve as the mortar [29]. Unlike cell/plasma membranes, the lipids in SC are mainly solid at ambient temperature and thus form a robust barrier [29,30]. Nevertheless, despite the solid nature of the extracellular lipid matrix and the very low permeability of SC, the barrier is not completely sealed. A minute imperceptible fraction of water (about 300–400 mL/day) evaporates continuously from the skin surface, commonly referred to as trans-epidermal water loss (TEWL) [31]. Injury or other damage to the skin perturb the barrier functionality and require immediate action for rapid wound closure, healing, and restoration. The wound healing process is highly complex and involves a series of subsequent overlapping phases: coagulation, inflammation, proliferation, and remodeling [32]. In acute wounds, the healing normally progresses through these different phases, but the cascade pathway is not always smooth and can come to a halt, as seen with chronic wounds [33,34]. Lack of proper diagnostics and mistreatment often results in scar formation, delayed healing, and recurrent wounds.

There are several factors that may promote wound healing, where hydration and pH in particular are claimed to play a crucial role [35–41]. Superfluous or reduced hydration may impede healing by, e.g., precluding formation of low molecular weight anti-inflammatory compounds, essential for the process [42]. One example is amino acids in the skin, sourced through proteolytic degradation of filaggrin, which requires hydrated conditions [43]. It has furthermore been demonstrated that wounds with alkaline pH have a lower healing progression rate [44,45]. Under normal circumstances, the skin surface is acidic as a result of, e.g., secreted Ffas from sebaceous glands, NMFs, and other filaggrin degradation products from the keratinocytes [46–53]. Skin's acidic nature and associated pH gradient is disturbed in wounds, and the more alkaline pH favors protease activity, destroying the cellular matrix and the growth factors that are essential for wound healing [40,54]. Therefore, restoring an acidic milieu in the wound bed would effectively control infection, protease activity, oxygen circulation, etc., which are essential for faster healing rate [55,56]. Several studies have indeed shown the benefits of acidic supplementation in treating burn wounds and skin infections, as well as reducing bacterial colony formation on the skin surface [40,57,58].

Fully recovered skin with no loss of skin appendages should possess a barrier functionality close to that of healthy control skin. However, this recovery may not be straightforward when substantial tissue damage occurs, leading to wide open wounds. Our body's natural response to such acute trauma would be to reestablish the skin barrier function and maintain body homeostasis, i.e., prevent excessive water loss [59,60]. We hypothesize that rapid wound closure is more important than immediate perfection of the barrier, although it is reasonable to assume that various types of treatment may also facilitate perfection and thereby minimize, e.g., degree of scar formation. As a first step we address this question by monitoring the natural wound healing progression and skin barrier recovery on pigs in vivo. Determining the quality of the restored cutaneous tissue down to molecular level will provide further understanding on how successful the healing has been and serve as a control in future testing of potential treatment to improve healing rate and barrier perfection. We decided to use minipigs as the test model since pig skin is anatomically and physiologically very similar to human skin [61]. Both humans and pigs share similar epidermal turnover time, heal through physiologically similar processes, and also have almost identical skin barrier constitution. These similarities make pigs popular test subjects when it comes to probe pharmacokinetics, skin barrier properties, wound healing progression, etc., and serve as a strong argument for our selection here.

More specifically, the aim of the current paper was to gain further understanding on the mechanism of wound healing on molecular level by studying the natural healing progression of acute wounds in vivo and then evaluate the quality of the restored tissue in comparison to healthy control skin from the same individuals. Two pigs were employed in parallel, and wound healing was followed over three weeks by visual inspection, monitoring degree of wound closure, pH at wound site, skin surface temperature, and TEWL. The recovered tissue was then excised from the wound sites together with adjacent healthy control skin and subjected to classical histology to examine potential morphological differences. The harvested tissue was also analyzed with physicochemical methods to gain further details on molecular structure and dynamics of the SC and its constituents. Using small and wide-angle X-ray diffraction (SWAXD) and Fourier transform infrared (FTIR) spectroscopy, we obtained information regarding lipid lamellae and protein organization in the SC. Dynamic vapor sorption (DVS) provided water uptake capacity of SC with respect to ambient humidity, and polarization transfer solid state NMR (PTssNMR) provided further details on the dynamics of fluid and solid material in the SC that could be linked to the SC organization probed by SWAXD. This multi-technique approach correlating wound healing progression monitored by classical established methods and subsequent physicochemical characterization of barrier recovery provides a unique insight into molecular aspects of healing. We are confident that it would serve as a very valuable support for future considerations in the development of new supplements that could further speed up and improve the healing by, e.g., reducing scar formation.

2. Materials and Methods

2.1. Materials

Bovine pancreas trypsin and ethanol were purchased from Sigma-Aldrich Chemie GmbH, Schnelldorf, Germany. K_2SO_4, which was used to prepare saturated salt solution to maintain constant humidity in the desiccator, was obtained from Merck, Darmstadt, Germany. Water used to prepare salt solutions was of Millipore quality produced by MilliQ water filtration system from Merck, Darmstadt, Germany with a resistivity of 18 MΩ.cm at 25 °C.

2.2. In Vivo Study

2.2.1. Animal Model

The in vivo wound healing study was performed on 2 castrated male Göttingen minipigs from Ellegaard Göttingen Minipigs, Denmark. The minipigs were non-naïve and aged between 9 and 10 months.

2.2.2. Ethical Permission

The in vivo wound healing study was conducted according to the Swedish ethical conduct of animal experiments with license number M131–16 and extension 974/2019.

2.2.3. Wound Induction

On day 0, the pigs were anesthetized with an intramuscular injection of Zoletile mixture (Tiletamine, Zolazepam, Butorphanol, Ketamine, Xylazine) 1 mL/10 kg. When needed, additional Zoletil mixture was given in a reduced dosage (0.5 mL/10 kg). The pigs were also continuously given oxygen via a face mask. Eye ointment (Viscotears 2 mg/g, Bausch + Lomb GmbH, Berlin, Germany) was applied over the cornea to avoid desiccation. Heart rate and oxygen saturation of blood (SpO_2) were continuously monitored with a pulse oximeter. The back of the pigs was clipped with an electric clipper, washed with a sponge containing antimicrobial skin cleaner based on chlorhexidine gluconate (Mediscrub, Rovers medical devices B.V., Lekstraat, the Netherlands), shaved with a disposable razorblade, and wiped with chlorhexidine (Klorhexidinsprit 5 mg/mL, Fresenius Kabi AB, Uppsala, Sweden). The wound areas were marked out with a marker-pen and subsequently disinfected with 70% ethanol (Solveco, Ref 1393, Stockholm, Sweden).

Five full thickness wounds were induced on each side of both pigs, yielding a total of 10 wounds per pig. Each wound was approximately 2×2 cm^2, and there was 4 cm in between each wound. The wounds were induced with a sterile disposable scalpel (Paragon, size 22, Ref P508, Swann Morton Ltd., Sheffield, United Kingdom) in a depth of approximately 4–8 mm. The skin in the wounded area was fixed with forceps and removed from the subcutaneous fat with scissors.

In order to avoid the wounds drying out, we used a sterile gauze moist with 0.9% NaCl to cover the wounds until all wounds had been induced.

2.2.4. Pain Treatment

During anesthesia on day 0, the pigs were given an intramuscular injection with buprenorphine (Temgesic 0.3 mg/mL, Indivior UK Limited, East Yorkshire, UK) as well as a fentanyl plaster (Fentanyl ratiopharm 50 µg/h, Ratiopharm GmbH, Ulm, Germany) attached to the abdominal skin. The fentanyl plaster was estimated to continuously provide analgesia for approximately 4 days. Additionally, pigs were administered oral paracetamol (Alvedon 500 mg, GlaxoSmithKline Consumer Healthcare AB, Brondby, Denmark) the first 2 days after wound induction. At the dressing change on day 4, the fentanyl dose was reduced by changing to another fentanyl plaster (Fentanyl Sandoz 25 µg/h, Sandoz A/S, Copenhagen, Denmark), and following this period, no further treatment was considered necessary.

2.2.5. Dressing Change

Primary layer: Each wound was individually covered with an absorbent foam dressing (Mepilex Transfer, ref 294800, Mölnlycke Health Care, Sweden) cut into pieces of 3×3 cm^2, and an 5×5 cm^2 adhesive film (Tegaderm Roll, Ref 16002, 3 M Health Care, Dusseldorf, Germany) to avoid leakage.

Secondary layer: Sterile gauze (Cutisoft Cotton, ref 71738-17, BSN medical, Hamburg, Germany) attached with sports tape (Strappal, Ref 71490-00, BSN Medical, Vibraye, France) was used to cover the wounded area and the primary dressing.

Tertiary layer: An elastic, water-repellent bandage (Vet–Flex, Kruuse, Langeskov, Denmark) was bandaged around the pigs neck, back, and belly.

2.2.6. Study Termination

Study termination was conducted at study day 21.

The pigs were anesthetized (Zoletil mixture). Blood samples, measurements, and photographs were taken before the pig was euthanized with an intravenous overdose of pentobarbital (Exagon vet. 400 mg/mL, Richter Pharma AG, Wels, Austria). Following euthanasia, the wounds were sampled and frozen for subsequent analysis.

2.3. Trans-Epidermal Water Loss (TEWL) Measurement

TEWL was evaluated using a Delfin VapoMeter wireless with serial number SWL4001. The VapoMeter is equipped with a closed cylindrical chamber that contains sensors for humidity and temperature. The TEWL is calculated from the linearly increased humidity in the chamber after placing the device in contact with the skin. The evaporation rate follows Fick's Law of diffusion, indicating the quantity being transported per a defined area and period of time. The results were expressed in grams per square meter per hour ($g/m^2 \cdot h$). All the TEWL values were recorded at ambient surrounding conditions, which were controlled externally to approximately 50% relative humidity (RH) and at 23 °C.

2.4. Measurement of pH and Skin Temperature

The pH of the wounds and control skin as well as skin surface temperature were measured using the SOFT PLUS (Callegari S.r.l., Parma, Italy) instrument equipped with pH electrode and temperature probe. The pH was recorded in selected wound and control positions on every occasion of dressing change during the healing cycle. The temperature was recorded only on the control skin surface at those occasions when we also measured pH and TEWL.

2.5. Wound Closure Monitoring and Photographic Documentation

After each removal of the primary dressings, photographs were taken with a digital camera (Nikon D50, lens: Sigma 50 mm 1:2.8 DG Macro D) of each wound position to document wound closure. Each photograph included a label of pig number and wound number, as well as a ruler for size determination. The wound areas were measured using the program Image J Fiji.

2.6. Histology

Frozen skin tissues from control and healed skin at position 10 were used to prepare histology slides. Classical hematoxylin and eosin (H&E) staining was performed, followed by epoxy embedding using the standard protocol as described by Arcega et al. [62]. The stained tissues were cut to a thickness of 5 µm slices and imaged with a light microscope (Olympus BX 43 with 40× objective lens) using the software ImageView version x64.

2.7. Sample Treatment Ex Vivo

Skin for ex vivo experiments were harvested from all the healed skin positions (approximately 3×3 cm²) as well as from the control skin sites (approximately 5×5 cm²) close to the head and tail region of the pigs' backs (see Figure 1A) after the in vivo session and stored at −80 °C until further used. For histology preparations, we used the excised tissue without further modifications. For other experiments where isolated SC was required, we carefully removed SC from the remaining skin according to the protocol described below.

2.8. Isolation of Stratum Corneum (SC)

Prior to SC separation, the excised skin was thawed and carefully rinsed under cold tap water with the epidermis side upwards to avoid any contact of subcutaneous fat with the epidermis. Then, the hair was removed from the skin surface with a trimmer. The control skin was dermatomed (Dermatome, Integra LifeSciences, Plainsboro, NJ, USA) to a thickness of ≈500 µm, while the smaller pieces of healed skin had to be dissected using a scalpel. The skin was then placed on a filter paper soaked in trypsin solution (0.2 wt % trypsin in MilliQ) and kept at 4 °C for about 20 h. The SC sheets were peeled off from the viable epidermis using forceps and rinsed in excess MilliQ water 5 times to get rid of all the trypsin. The SC sheets were dried under vacuum in a desiccator and stored at −20 °C for subsequent use.

2.9. Sample Preparation Method for SWAXD, NMR, DVS, and FTIR Studies

To prepare samples for NMR experiments, we used approximately 20 mg of dry SC, which was then equilibrated for 48 h at 32 °C either at dry conditions or at 97% RH, provided by a saturated K_2SO_4 salt solution in the desiccator. After equilibration, the samples were quickly transferred to NMR inserts (Bruker) with screw caps to avoid dehydration/hydration and subsequently placed in the NMR rotors (Bruker) with caps on it. The same method but with less sample (3–5 mg) was used to prepare samples for SWAXD experiments. After equilibration, the samples were quickly transferred to screwtight sandwich cells with polyethylene film as support windows for SWAXD measurements. The DVS and the attenuated total reflectance (ATR)–FTIR measurements were performed at dry conditions only, using approximately 5–7 mg and 1–3 mg of dry SC, respectively.

2.10. Small and Wide Angle X-ray Diffraction (SWAXD)

SWAXD studies were performed using a SAXSLab Ganesha 300XL instrument (SAXS-LAB ApS, Skovlunde, Denmark), equipped with 2D 300K Pilatus detector (Dectris Ltd., Baden, Switzerland). The scattering intensity (I) was recorded as a function of the scattering vector q in reciprocal Ångström (Å^{-1}), where q is defined as $q = \frac{4\pi sin\theta}{\lambda}$. Here, θ is the scattering angle and λ is the wavelength of the incident X-ray beam, where $\lambda(Cu–K_\alpha) = 1.54$ Å. The d-spacing was calculated from the peak position of q by using equation $d = \frac{2\pi}{q}$. The two-dimensional (2D) scattering pattern recorded by the detector was radially averaged using the software SAXSGui to obtain 1D I vs. q data. The exposure time varied from 15 to 120 min and the temperature was controlled to 32 °C using an external circulating water bath.

2.11. Polarization Transfer Solid State Nuclear Magnetic Resonance (PTssNMR)

PTssNMR is a combination of three different experiments that are performed on the same sample in a sequential manner, i.e., direct polarization (DP), cross polarization (CP), and insensitive nuclei enhanced by polarization transfer (INEPT) [63,64]. The experiments were carried out on a Bruker Avance AVII 500 NMR spectrometer equipped with a 4 mm CP/magic angle spinning (MAS) probe, operated at 5 kHz frequency, with ^{1}H and ^{13}C resonance frequencies of 500 and 125 MHz, respectively. The temperature calibration was performed with methanol and set to 32 °C in all experiments [65]. A spectral width of 250 parts per million (ppm) was used and the number of scans per experiment was 2048 with an acquisition time and a recycle delay of 0.05 and 5 s, respectively. This gave a total estimated time of approximately 9 h for all 3 experiments. Data processing was done using a line broadening of 20 Hz, zero filling from 1597 to 8192 time domain points, Fourier transformation, phase correction, and baseline correction using an in-house Matlab code partially derived from matNMR [66,67].

2.12. Dynamic Vapor Sorption (DVS) Measurements

Water sorption measurements were performed using the TA Instruments (New Castle, DE, USA) Q5000 SA Dynamic vapor sorption microbalance (DVS). To compare the sorption isotherms, we ran both control and healed SC samples in parallel in a single experiment. Dry samples were placed on 2 separate pans hanging on a balance in the DVS and exposed to a stream of N_2 with controlled RH at 32 °C. The water sorption of the samples was continuously recorded by the microbalance.

2.13. Fourier Transform Infrared (FTIR) Spectroscopy

ATR–FTIR spectra of control and healed SC samples were recorded using a Thermo Nicolet Nexus 6700 instrument from Thermo Scientific (Waltham, MA, USA). All the experiments were performed in the attenuated total reflectance (ATR) mode in dry conditions and at ambient temperature.

3. Results

In the present work we employed Göttingen minipigs aged 9 to 10 months to study healing progression of acute wounds in vivo and to compare the quality of the restored tissue down to a molecular level with that of control healthy skin excised from the same individuals. A comparative analysis in terms of skin barrier recovery at the molecular level concerning SC molecular dynamics, lipid composition, and lipid–protein structural organization was made with the long-term aim of obtaining a better understanding of the mechanism behind wound healing and how it can be affected to promote and perfectionate tissue recovery. Ten full thickness wounds, five on each side of the pigs back, were incised on each pig on day 0 when initiating the in vivo study (Figure 1A). The study was carried out over 21 days, which was a fairly reasonable time to recover the skin tissue with all its major layers (Figure 1B). During the in vivo cycle, we measured a set of physical parameters, i.e., TEWL, pH, and skin surface temperature, in order to evaluate the healing progression in real time and compared with control skin. The wounds were also photographed and monitored visually for any sort of inflammation that might delay the healing process. When the in vivo part was terminated on day 21, healed skin was harvested from all wound sites on each pig together with adjacent control healthy skin. The different layers in the healed tissue were visualized along with that of control skin using hematoxylin and eosin. The molecular details regarding SC lipid and protein organization (Figure 1C) were obtained with SWAXD, PTssNMR, and FTIR. All results are presented and discussed below. First, we present the data obtained in vivo in real time during wound healing progression, and then the data obtained on ex vivo post-healing excised tissue are described.

3.1. Wound Healing Progression In Vivo

3.1.1. Visualizing Wound Closure and Healing Cycle In Vivo

Images were captured for all wound positions on both pigs throughout the healing cycle from day 0 of wound incision until day 21 when the in vivo part of the study was terminated. Figure 2A displays representative images for pig 1 on wound position 1 and provides a qualitative overview of the healing progression. Two more representative image series of the healing cycle on the two pigs are provided in Figures S1 and S2. The images clearly demonstrate an overall good skin recovery at the end of the healing cycle on day 21. When we examined more closely, formation of granulation tissue on the wound bed on day 7 was noticeable, and on day 10, further contraction from the side was also evident. This indicates that both primary and secondary wound healing took place and full wound closure was observed on day 14. A similar wound closure on day 14 was observed for all the wounds on both pigs on day 14. No sign of infection was observed during the healing process when the dressings were changed in either of the pigs. From day 14 and further until day 21, no visual differences were noticed.

When comparing the healing progression cycle on pig 1 and pig 2, we noticed a few differences. At the beginning, the overall healing moved faster on pig 2. However, at the later part of the cycle, from day 14 onwards, the healing process was faster on pig 1. The wounds were narrower in pig 1 compared to pig 2 on day 14, which indicates faster tissue regeneration and contraction from the sides in pig 1.

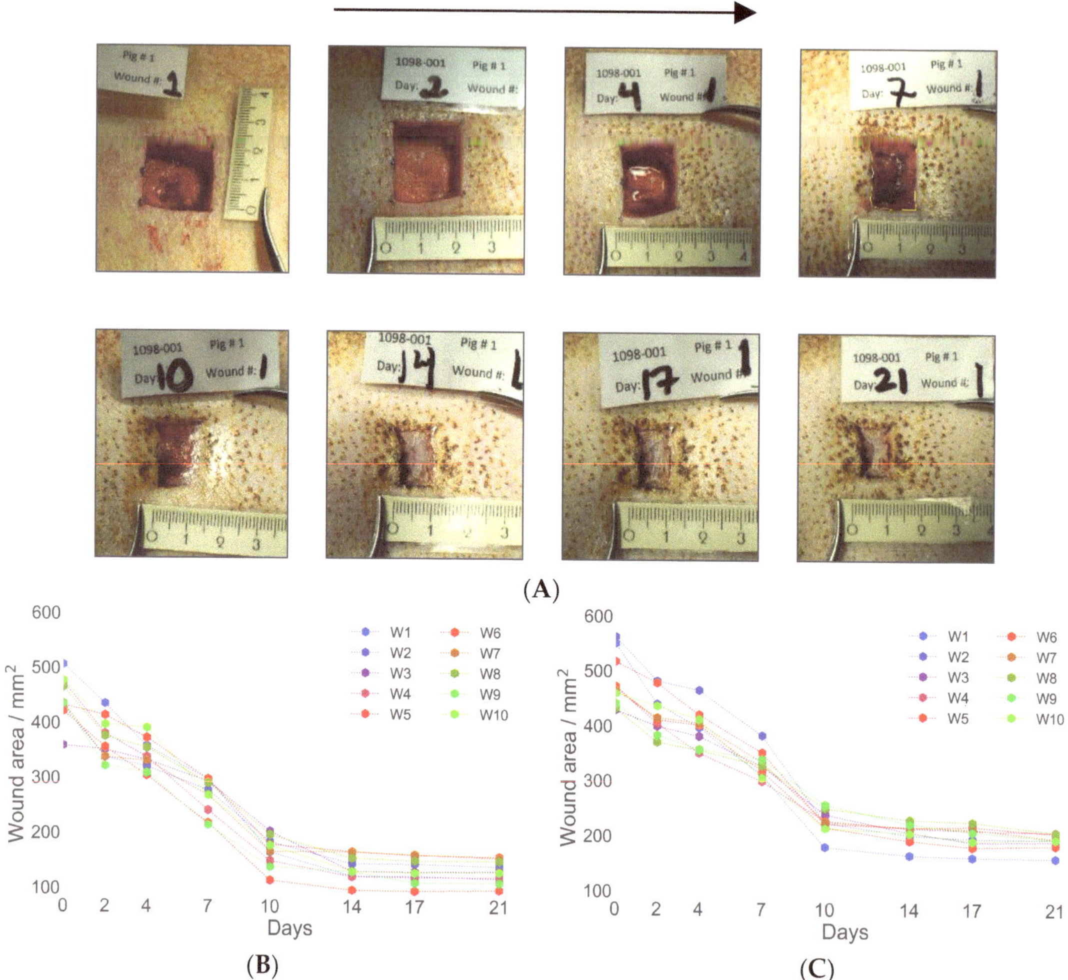

Figure 2. (**A**) Representative wound images on pig 1 at various time points during the healing progression cycle. The arrow on top of the figure indicates ascending date for healing cycle and the readers are encouraged to follow the images in the arrow direction. The images presented origin from position 1 and were captured on the days when the dressing was changed. The name plates on top of the images indicate the in vivo wound study number, pig number, study days when the wound dressing was changed, and the wound position number. The ruler on the bottom of the images was used as a guide to set the scale when calculating the wound marked area, e.g., see marking on day 7. Calculated wound area was presented as a function of days during the healing cycle for pig 1 (**B**) and pig 2 (**C**). The labelling W1, W2 and so on in the figure legends indicate the various wound positions on each pig.

We further calculated wound area closure of each wound on both pigs during the healing cycle at various time points. Table 1 tabulated wound area values calculated for wound position 1. For all other wound positions, the calculated area values are provided in Table S1. The detailed procedure of area calculation is given in the supplementary text. Figure 2B,C shows the calculated wound area for all wounds plotted as a function of time during the healing cycle for pig 1 and pig 2. The overall trend looked similar in both pigs, with the area gradually decreasing during the first 10 days and plateauing after day 10 until the end of the healing cycle on day 21. This is expected when examining wound images and it can be seen that wounds were almost sealed on day 10. The remaining minor sealing happened before day 14, for which we saw minor decrease in the closure of the wounds. On day 17 post-wound induction, the wounds were completely closed. The wound area calculations therefore provided the values of the wound scar lines and indicated no change in those markings during the healing process. Comparing pig 1 and pig 2, the area revealing wound closure with marked scar formation was higher in pig 1 compare to pig 2 (Figure 2B,C).

Table 1. Summary of calculated wound area, trans-epidermal water loss (TEWL), pH, and skin surface temperature measurements performed during the in vivo wound healing. The measurements were recorded at the time points when the dressings were changed during the healing cycle. * For wound area calculations and TEWL measurements, the values corresponding to wound position 1 are provided. ** For pH, the readings provided correspond to wound position 1 for pig 1 and position 5 for pig 2. The skin surface temperature values presented here originate from control measurements on the left side of the pigs' backs. The "–" sign indicates that no measurement was performed at that particular occasion.

	Wound Area * (mm^2)	TEWL * (g/m^2·h)		pH **		Temperature (°C)
Pig 1/Days	**Wound**	**Control**	**Wound**	**Control**	**Wound**	**Control**
0	505.6	14.4	14.4	–	–	–
2	435.1	11.9	160	5	8.4	34.2
4	358.0	12.8	188	5.3	8.2	34.2
7	297.6	14.7	191	5.3	8.1	33.7
10	165.6	12.1	165	5.2	7.3	33.5
14	122.0	12.9	30.5	5.1	5.3	33.5
17	122.5	12.2	13.5	5.2	5.2	33.3
21	117.2	12.3	7.9	5.2	5.1	33.6
Pig 2/Days						
0	549.3	15.3	11.9	–	–	–
2	439.3	13.7	144	5.3	8.5	34.7
4	397.1	11.8	195	5.2	8.3	34.4
7	310.2	13.6	183	5.4	–	34.3
10	180.2	13.7	167	5.2	7.8	34.1
14	165.4	11.0	25.2	5.3	5.8	33.4
17	161.3	12.4	12.6	5.4	5.5	33.7
21	159.9	12.6	7.2	5.1	5.1	33.1

3.1.2. TEWL and pH as Barrier Indicator of Healed Skin

There always exists a water gradient in the skin with lower water activity on the skin surface, regulated by the water activity of the ambient, and a higher water activity in the deeper layers of the dermis. This gradient in water activity gives rise to diffusional transport of water from the inside of the body towards the ambient and subsequent evaporation from the skin surface. This continuous imperceptible water loss is normally referred to as TEWL and is widely measured and used as an indicator to assess the skin barrier function. In general, a higher water loss from the skin surface would indicate a compromised/reduced skin barrier and vice versa. We here recorded TEWL values on various wounds in both pigs and also on the control site at various time points during the wound healing study in order to assess skin barrier recovery. Figure 3A presents bar plots of the recorded TEWL values (average with standard deviation) against days for control and healed skin sites at various time points of the healing cycle for pig 1. Additionally, the individual values for several wounds on pig 1 are plotted as a function of time in Figure 3B. The corresponding curves for pig 2 are provided in Figure S3. Table 1 also compiles representative TEWL values for wound position 1 in both pigs along with control measurement values. All the recorded TEWL values from the various wound positions in both pigs at different days of the wound healing as well as all the control measurements during the in vivo study are tabulated in Table S2.

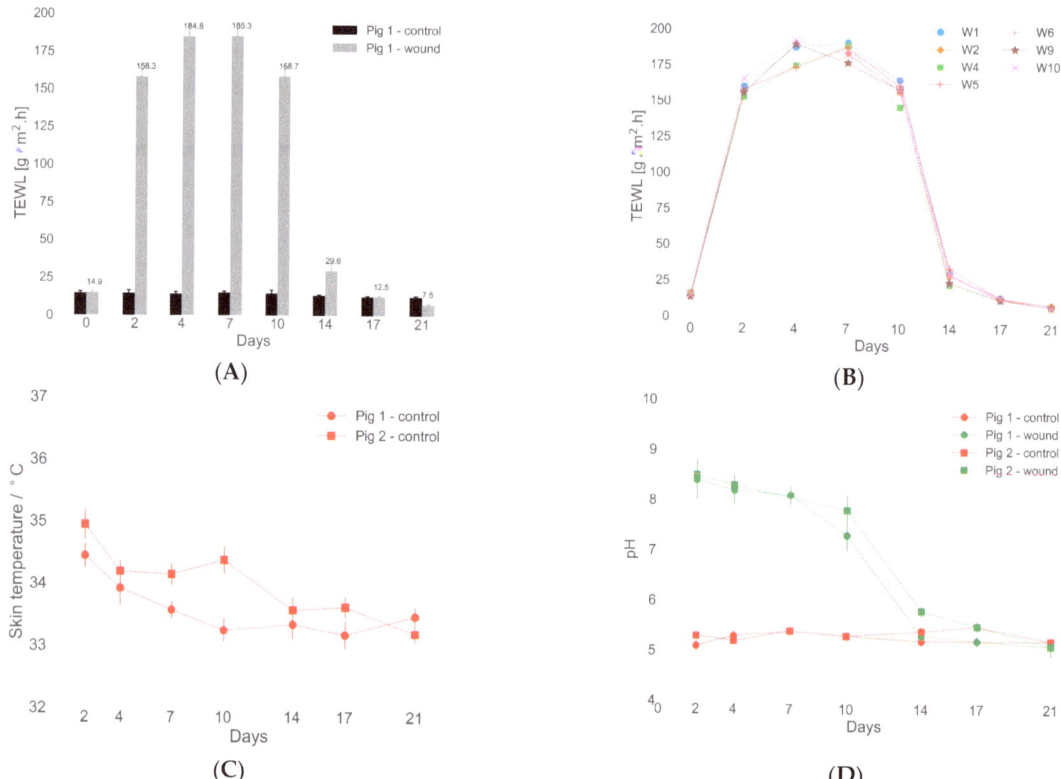

Figure 3. Variations in TEWL, skin temperature, and pH during healing. (**A**) Measured TEWL vs. time for control skin and healing skin for pig 1, position 1. On day 0, the readings were recorded before the wound incision. (**B**) TEWL values vs. time for individual wounds obtained from pig 1 when the wound dressings were changed. (**C**) Skin surface temperature vs. time measured on both pigs. (**D**) pH measurements on the wounds (position 1, pig 1, and position 5, pig 2) and corresponding control sites were plotted vs. time for both pigs. All the TEWL and pH values were recorded at ambient controlled conditions, i.e., approximately 50% relative humidity (RH) and 23 °C. Error bars depict standard deviation.

An overall similar trend for the recorded TEWL values was observed for both control and healing skin in both pigs as shown in Figure 3A,B. The average TEWL value recorded on day 0 for control skin was approximately 15 ± 0.2 for pig 1 and 14 ± 0.1 g/m^2·h for pig 2. TEWL values were also recorded on day 0 for the particular skin sites before wound incision. On every occasion, when the wound dressings were changed and TEWL was recorded for healing skin, we also measured TEWL for control skin. The recorded control values were very similar for both pigs throughout the study period. When TEWL was measured for healing skin on day 2, the recorded values were, as expected due to the open wound sites and moist environment, high, i.e., 149.2 ± 1.3 and 153.2 ± 1.5 g/m^2·h for pig 1 and pig 2, respectively. Higher TEWL values were also recorded on day 4 in both pigs and decreased only slightly on day 7 and further on day 10. The highest TEWL value was recorded for pig 2 on day 4 and was 191 ± 1.6 g/m^2·h. On day 14, a dramatic decrease in TEWL was observed in both pigs with recorded values of 32 ± 0.3 and 34 ± 0.4 g/m^2·h for pig 1 and pig 2, respectively. The timing of the sharp decrease in TEWL around day 14 was in accordance with wound closure that was also observed on day 14 (Figure 2A, Figures S1 and S2) and established the connection that the measured TEWL values were reduced due to sealing of the wound surface. The TEWL values continued to decrease further and returned to comparable control values on day 17. At this point, the TEWL values recorded were 13 ± 0.4 and 14 ± 0.5 g/m^2·h for pig 1 and pig 2, respectively.

Surprisingly, the TEWL values further reduced on day 21 down to 7.8 ± 0.3 g/m^2·h in pig 2, which was below the normal values recorded for control skin.

In addition to the determination of TEWL, skin surface temperature and pH of the wounds were also measured during the in vivo study at each occasion where the wound dressings were changed. The recorded skin surface temperature on both pigs at control positions were plotted as a function of study days and presented in Figure 3C, and the values are listed in Table 1. A full list of temperatures recorded at various control positions is revealed in Table S3. The skin surface temperature on both pigs showed similar trends and remained steady with little fluctuations of 1–2 degrees throughout the study.

The pH values measured for two randomly chosen wounds and corresponding control sites are presented in Figure 3D. Hence, pH was determined for wound position 1 for pig 1 and position 5 for pig 2, and the values are revealed in Table 1. All pH values measured for the individual wounds during the healing cycle are provided in Table S4. The average measured pH on the control skin site was 5.3 ± 0.1 for both pigs and was steady during the whole cycle. When the pH values were recorded for the different wounds, a slightly basic pH of around 8.0 was observed in both pigs on day 2 and remained high until day 7. On day 2, the pH values measured were 8.4 ± 0.4 and 8.5 ± 0.3 for pig 1 and pig 2 in wound positions 1 and 5, respectively. The pH of the wounds started decreasing after day 7, and on day 10, the recorded values were 7.3 ± 0.3 and 7.8 ± 0.6 for pig 1 and pig 2 on wound positions 1 and 5, respectively. On day 17, when the wounds were almost dry, the measured pH returned to around 5.5 in both pigs and did not change further during the remaining time of the in vivo cycle, hence resembling the values for the control measurements.

3.2. Ex Vivo Characterization of Recovered Tissue after Healing Compared to Control Skin

3.2.1. Morphology of Healed vs. Control Skin

To evaluate the quality of recovered tissue after healing, we visualized various skin layers with light microscopy using hematoxylin and eosin (H&E) staining. Figure 4 displays images of histological preparations from excised skin of pig 2 at wound position 10 on day 0 (Figure 4A,B) as well as skin harvested after in vivo termination on day 21 (Figure 4C,D). Additional images of control and healed skin tissue are provided in Figure S4. The images captured from skin harvested on day 0 could serve as a control since the healed skin tissue was harvested from the same wound position after terminating the in vivo study.

Figure 4A clearly demonstrates the dermis and the individual layers of the epidermis for control skin as labelled in the image. The corresponding layers were likewise observed in the healed skin, as shown in Figure 4C. However, further details in various layers of the epidermis revealed differences between control and healed skin. In control skin, the dermal epidermal ridges developed from the basal cells were quite prominent, while the healed skin did not present such ridges (Figure 4A,C). The cells in the basal layer of the epidermis in control skin were columnar shaped, ordered, and densely packed (Figure 4A). In the healed skin, on the other hand, these cells were flat, distorted, and disorganized (Figure 4C). As the basal cells undergo progressive maturation during migration towards stratum spinosum and granulosum layers, referred to as keratinization, they became flatter in shape. The healed skin revealed less densely packed cells in epidermis, and the cells were also highly stretched and flattened. The surface layer of the epidermis, the stratum corneum (SC), appeared to have a similar thickness in both control and healed skin (Figure 4A,C). Further morphological differences in the maturation of corneocyte cells in the healed SC could not be discerned from these light microscopy images due to limitations in resolution. Figure 4B,D displays a closeup on the dermis region for both control and healed skin. From these images, we clearly noticed more epithelization in the control compared to healed skin, indicating less developed dermis in recovered skin.

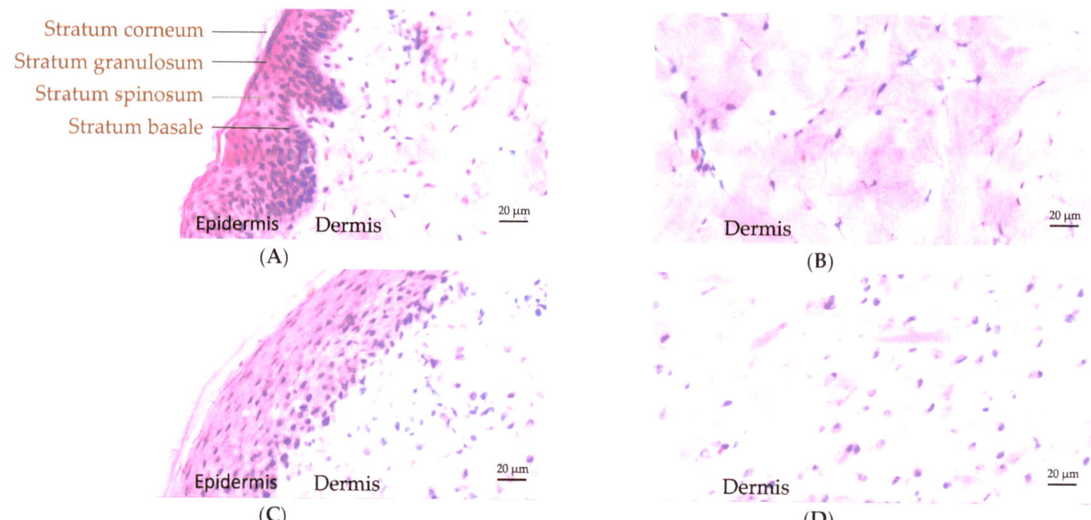

Figure 4. Classical hematoxylin and eosin (H&E) staining of control (**A**,**B**) and healed (**C**,**D**) pig skin. The skin surface is directed towards the top/left. The individual epidermal layers can be identified in the control skin labelled (**A**). Magnifications of the dermis part for control and healed skin are provided in (**B**,**D**).

3.2.2. SC Lipid and Protein Molecular Organization Investigated Using SWAXD

SWAXD can provide detailed information on molecular ordering (within approximately 3–150 Å) in the complex architecture of the SC, particularly regarding variations in lipid lamellar organization and lateral hydrocarbon chain packing, protein secondary structure, and keratin filament organization. Figure 5 displays the SAXD and WAXD spectra of the control and healed SC for pig 1 measured at dry conditions and at 97% RH. The corresponding spectra obtained for pig 2 are provided in Figure S5. All the SWAXD peaks detected in SC from the two pigs at dry and wet conditions are compiled in Table 2 along with peaks reported in the literature for pig SC. The SWAXD pattern of both control and healed SC revealed several peaks corresponding to structures formed by SC lipids and protein (Figure 5). A few of these peaks, highlighted in gray in Table 2, showed differences between control and healed SC, while the remaining peaks did not change. Below, we present the origin of specific peaks and then describe the differences we observed between control and healed SC.

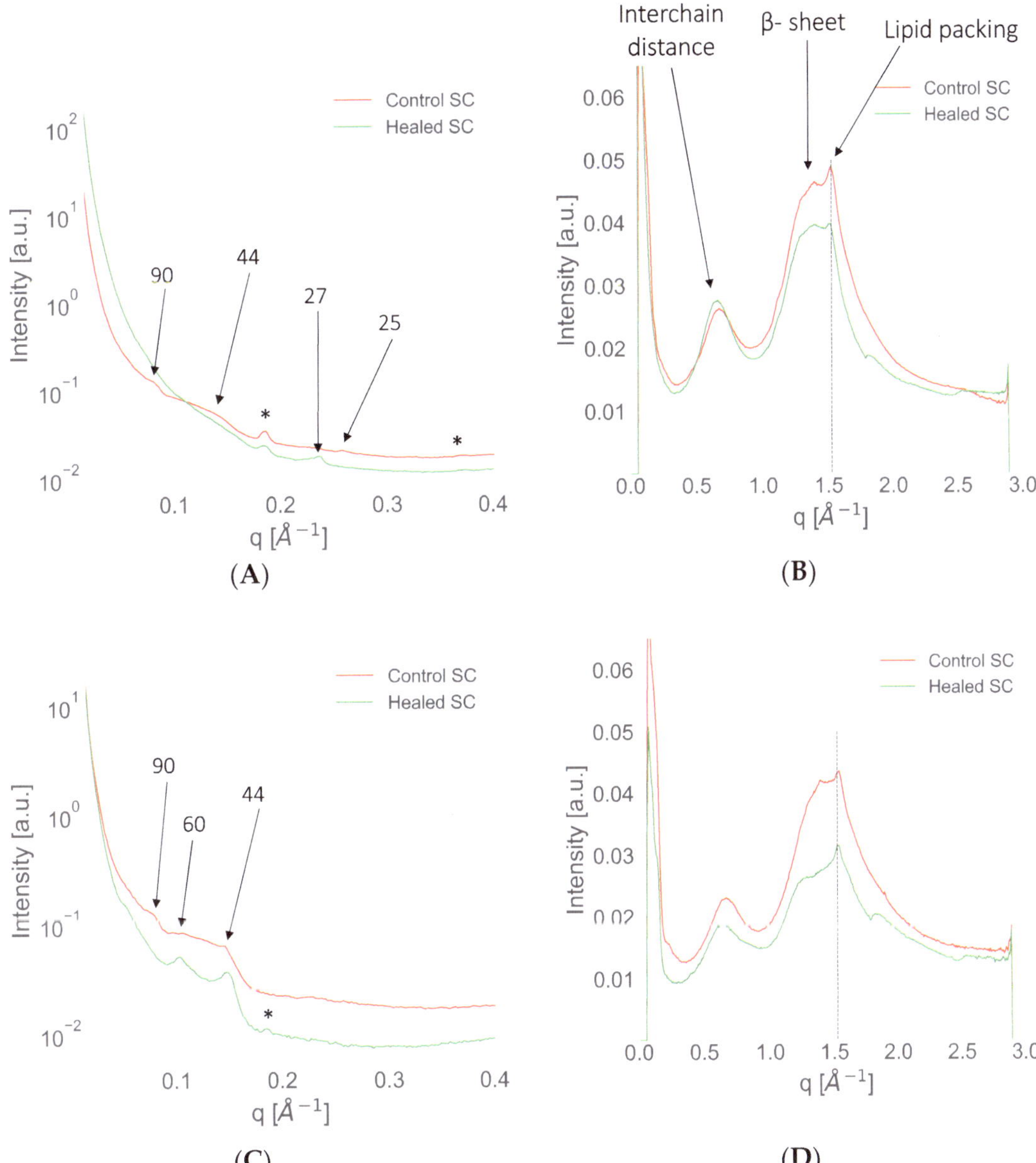

Figure 5. Small and wide-angle X-ray diffraction (SWAXD) spectra of control and healed stratum corneum (SC) excised from pig 1 and recorded at dry (**A**,**B**) and wet (97% RH) conditions (**C**,**D**). The SAXD spectra are provided on the left column whereas the WAXD spectra are on the right. In the SAXD region, several peaks attributed to lipid lamellar ordering and keratin packing are identified. The numbers with arrows indicate the *d*-spacing in Å for various peaks. Phase-separated crystalline Chol was also detected in some spectra and then indicated by an asterisk (*) sign. In the WAXD region, the shaded lines indicate peaks originating from the keratin interchain distance and show a change in peak position following hydration of the sample. The lipid acyl chain ordering indicated by dotted lines do not show a shift in peak position when comparing dry and hydrated conditions. The secondary β-sheet structure of keratin could also be detected and is marked in the plot. All SWAXD measurements were performed at 32 °C, which represents average physiological skin temperature.

SAXD pattern from control SC at dry condition showed a weak peak at $q \approx 0.07$ Å^{-1} which corresponded to a d-spacing of approximately 90 Å (Figure 5A). The origin of this peak has been interpreted as being due to the keratin filament rod diameter or due to the thickness of corneocyte bound lipids [15,68]. A broad hump centered around $q \approx 0.14$ Å^{-1} that corresponded to a d-spacing of approximately 44 Å was also observed in control SC at dry conditions (Figure 5A). This hump may be the second order of the 90 Å peak and/or the third order of the LPP, which has also been reported in the literature [15,68–70]. Additionally, another peak at $q \approx 0.25$ Å^{-1}, which corresponded to a d-spacing of approximately 25 Å, was detected and could be the second order of a SPP with shorter d-spacing (Figure 5A). The dry SC of control skin also revealed a clear peak at $q \approx 0.18$ Å^{-1}, which corresponded to a d-spacing of approximately 33 Å (Figure 5A). This peak originated from crystalline Chol present in the samples [15]. When hydrating the control sample at 97% RH, the peaks at $q \approx 0.07$ and 0.14 Å^{-1}, which corresponded to d-spacings of 90 and 44 Å, respectively, could still be detected along with the Chol peak at $q \approx 0.18$ Å^{-1} (Figure 5C). In addition, in the hydrated state, another peak at $q \approx 0.1$ appeared, which corresponded to a d-spacing of approximately 60 Å (Figure 5C). This peak could be attributed to the first order of the SPP in accordance with the literature [15]. When examining the SAXD pattern of the healed SC at dry conditions, we observed a peak at $q \approx 0.23$ Å^{-1} corresponding to a d-spacing of approximately 27 Å (Figure 5A). No other peaks were observed at dry conditions except the Chol peak at $q \approx 0.18$ Å^{-1}. Healed SC at wet conditions revealed two peaks at $q \approx 0.1$ and 0.14 Å^{-1}, which corresponded to d-spacings of approximately 60 and 44 Å (Figure 5C). The 60 Å peak could be attributed to the first order of the SPP, whereas the 44 Å peak might have been due to keratin rod diameter or third order of LPP, as described for control SC. The SAXD pattern for pig 2 showed similar trends as observed for pig 1 in both control and healed SC (Figure S5A,C).

The WAXD spectrum of control SC at dry conditions revealed several prominent peaks. The peak at $q \approx 0.67$ Å^{-1}, which corresponded to a d-spacing of approximately 9.3 Å is attributed to the keratin interchain distance of two-polypeptide chains (Figure 5B) [15,68,71,72]. Hydration at 97% RH led to a shift in this peak position to $q \approx 0.65$ Å^{-1}, which corresponded to an increase in d-spacing to approximately 9.7 Å (Figure 5D). At dry conditions, the control SC also exhibited a peak at approximately 1.39 Å^{-1}, which corresponded to a d-spacing of 4.5 Å (Figure 5B). Previously, this peak was attributed to the β-sheet secondary structure of keratin [73,74]. The control SC furthermore showed a prominent peak at $q \approx 1.52$ Å^{-1} at dry conditions, which corresponded to a d-spacing of approximately 4.1 Å (Figure 5B). This d-spacing is typical of hexagonally packed lipid acyl chains [15]. This peak did not shift on hydration. The WAXD spectrum of the healed SC exhibited similar features as the control SC. The keratin interchain distance in healed SC showed a shift in peak position from $q \approx 0.67$ Å^{-1} in dry state to 0.64 Å^{-1} in wet state, which corresponded to a change in d-spacing from approximately 9.3 to 9.8 Å (Figure 5D). The protein secondary structure was less pronounced in healed SC at wet conditions and the lipid chain packing was not affected by the hydration. The WAXD pattern in pig 2 showed similar trends as observed for pig 1 at dry and wet conditions for both control and healed SC (Figure S5B,D).

In summary, using SWAXD, we identified differences in the protein molecular structure between control and healed SC (keratin filament rod diameter and interchain distances). Only small differences were resolved for the lipid organization, which might have been due to the relatively low flux in the inhouse X-ray source.

3.2.3. Water Uptake in SC Examined by DVS and ^{1}H NMR

Water sorption isotherms provided a relation between the water content of a sample and the ambient relative humidity (RH). Figure 6A depicts the sorption isotherm of control and healed SC. When sorption measurements were performed with DVS to examine water uptake in SC from both control and healed SC, the spectra were almost identical in the low humidity regime (<50% RH, Figure 6A). In the higher humidity regime (>50% RH), the water uptake was slightly higher in healed SC compared to control SC. At 96% RH, the water uptake in healed SC was approximately 40 wt %, whereas control SC took around 33 wt %. Further evaluation of water uptake in these samples were performed with ^{1}H NMR, and Figure 6B shows NMR spectra of control and healed SC equilibrated at 97% RH. Both ^{1}H NMR spectra for the respective control and healed SC showed a pronounced peak at approximately 4.75 ppm, indicative of water present in the samples.

Table 2. Compilation of SWAXD peaks detected in control and healed SC spectra for both pigs at dry and wet conditions. The peaks are presented with their Q positions (Å^{-1}) and corresponding d-spacings (Å) and compared with d-spacings reported in the literature. The differences observed between control and healed SC at dry and wet conditions are highlighted in gray. (*)—indicates peaks for which higher diffraction orders are reported in the literature; NR—not resolved; RT—room temperature; Chol—cholesterol. All the SWAXD measurements were carried out at 32 °C.

	Dry			Wet			d (Å) in Literature for Pig	Remarks
	Pig 1 Q (Å^{-1})	Pig 2 Q (Å^{-1})	d (Å)	Pig 1 Q (Å^{-1})	Pig 2 Q (Å^{-1})	d (Å)		
Control SC	–	–	–	–	–	–	120–132 * [15,69]	LPP with higher orders. Observed at RT and recrystallization from 120 °C.
	0.07	–	90	0.07	0.07	90	90 [15]	Keratin rod diameter/thickness of corneocyte bound lipids.
	–	–	–	0.1; 0.2	–	60	60 [15]	SPP. Detected at RT and skin temperature.
	0.25	–	25	–	0.12; 0.25	50–52; 25	– / –	SPP first order with shorter d-spacing. SPP second order with shorter d-spacing.
	0.14	0.14	44	0.14	0.14	44	45 [15,69]	Possibly second order of 90 Å phase/higher order of LPP/other phase.
	0.18	0.18	33	0.18	0.18	33	34 [15,68]	Anhydrous Chol crystals.
	1.52	1.52	4.1	1.52	1.52	4.1	4.1 [15,69,75]	Crystalline hexagonal packing of hydrocarbon lipid tails.
	0.67	0.67	9.3	0.65	0.65	9.7	9.0–10.4 * [15,69,75]	Keratin interchain distance. Swelling upon hydration.
	NR	NR	–	NR	NR	–	5.1–5.4 * [15,69,75]	α-Helical secondary structure of keratin protein.
	1.39	1.39	4.5	1.39	NR	4.5	4.6–4.9 * [15,75]	β-Sheet secondary structure of keratin protein.
Healed SC	–	–	–	–	–	–	120–132 * [15,69]	LPP with higher orders. Observed at RT and recrystallization from 120 °C.
	–	–	–	–	–	–	90 [15]	Keratin rod diameter/thickness of corneocyte bound lipids.
	–	–	–	0.1	–	60	60 [15]	SPP with higher orders. Detected at RT and skin temperature.
	0.23	0.12; 0.23	52–54, 27	–	0.12	52	– / –	SPP first order with shorter d-spacing. SPP second order with shorter d-spacing.
	–	–	–	0.14	0.14	44	45 [15,69]	Possibly second order of 90 Å phase/higher order of LPP/other phase.
	0.18	0.18	33	0.18	–	33	34 [15,69]	Anhydrous Chol crystals.
	1.52	1.52	4.1	1.52	1.52	4.1	4.1 [15,69,75]	Crystalline hexagonal packing of hydrocarbon lipid tails.
	0.67	0.65	9.3–9.7	0.64	0.60	9.8–10.4	9.0–10.4 * [15,69,75]	Keratin interchain distance. Swelling upon hydration. More pronounced in healed SC compared to control SC.
	NR	NR	–	NR	NR	–	5.1–5.4 * [15,69,75]	α-Helical secondary structure of keratin protein.
	NR	1.39	4.5	NR	NR	–	4.6–4.9 * [15,75]	β-Sheet secondary structure of keratin protein (not observed in wet condition).

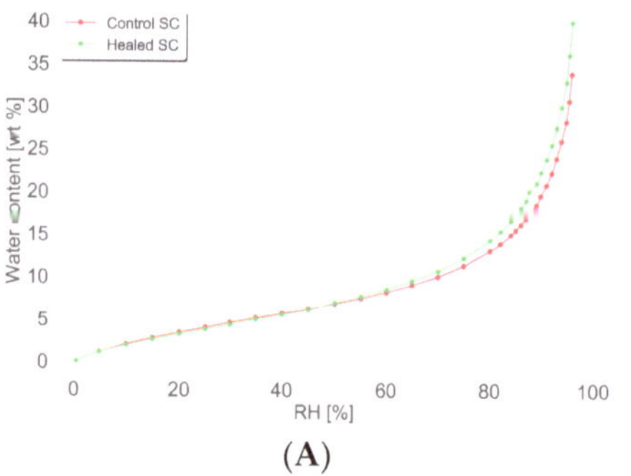 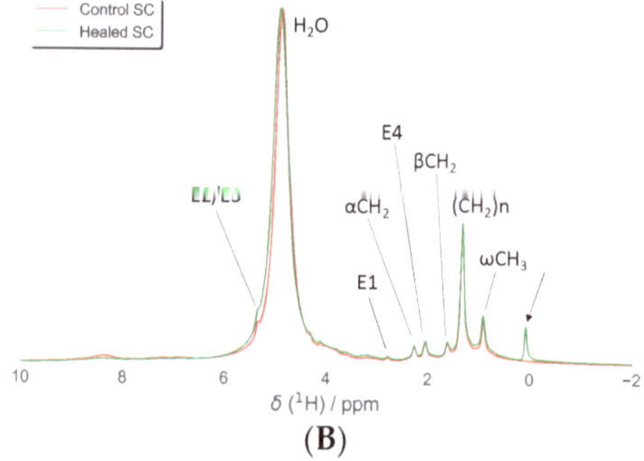

Figure 6. (**A**) Water sorption measurements for control and healed SC at 32 °C, expressed as water content, (in wt % with respect to dry SC) plotted as a function of RH. (**B**) ^{1}H NMR spectra recorded on control and healed SC at 97% RH. The peak close to 5 ppm in both spectra was due to water in the sample from hydration at 97% RH. Several sharp peaks attributed to lipid molecular segments (assignments in Figure 7E) were also detected, indicative of fluid lipids present in both samples. The arrow close to 0 ppm in the healed SC was due to the silicon present in the wound dressing.

3.2.4. Molecular Structure and Dynamics of SC Constituents Examined by PTssNMR

PTssNMR relies on the natural abundance of ^{13}C present in the sample and provides details on the SC lipid and protein mobility with close to atomic resolution. A detailed description of the method and its applicability is given elsewhere [11,76–79]. Briefly, the PTssNMR method comprises three individual experiments (i.e., DP, CP, and INEPT) performed on the very same sample. It relies on the natural abundance of ^{13}C in the sample. The DP experiment utilizes a direct pulse and acquires all ^{13}C signals present in the sample and can be used as a reference. The CP and INEPT experiments involve polarization transfer from ^{1}H→^{13}C, where CP boosts signals for solid/rigid carbons and INEPT does so for mobile/fluid carbons present in the sample [63,64]. When overlaying the DP, CP, and INEPT spectra in one plot, one can unveil a detailed molecular picture regarding the SC molecular environment.

Here, PTssNMR spectra were recorded for control and healed SC equilibrated at both dry conditions and at 97% RH. The spectra for the lower chemical shift region (10–95 ppm) are shown in Figure 7, while the corresponding spectra for the higher ppm range (100–150 ppm) are provided in Figure S7. The forest of peaks in all these NMR spectra revealed a complex molecular architecture in the SC. The first impression when looking at the spectra is the dominant CP signal (blue), which demonstrates that the main part of SC is in solid/rigid state in both control and healed SC, independent on the degree of hydration. However, the presence of an INEPT (red) signal is also evident, indicating that a fraction of the SC constituents is in mobile state. The notable lipid peaks detected in all these spectra originated from the lipid acyl chain terminals (ωCH$_3$ at $\approx$14.6 ppm, (ω-1)CH$_2$ at $\approx$23.3 ppm, and (ω-2)CH$_2$ at $\approx$32.5 ppm), lipid main chain methylene (–CH$_2$–)$_n$ all-trans (AT), and trans/gauche (TG) isomers at $\approx$33 and 31 ppm. We further noticed molecular mobility in the unsaturated lipid acyl chains. The E1 and E4 carbon segments showed mobility at approximately 25.7 and 27.6 ppm, respectively (assignments in Figure 7E), while the E2 and E3 segments showed mobility in the higher ppm range, at approximately 128 and 131 ppm, respectively (Figure S7). Signals from the lipid headgroups were also observed at approximately 34.8, 56.3, and 62.3 ppm for αCH$_2$, Cer C1, and for Cer C2, respectively (assignments in Figure 7E).

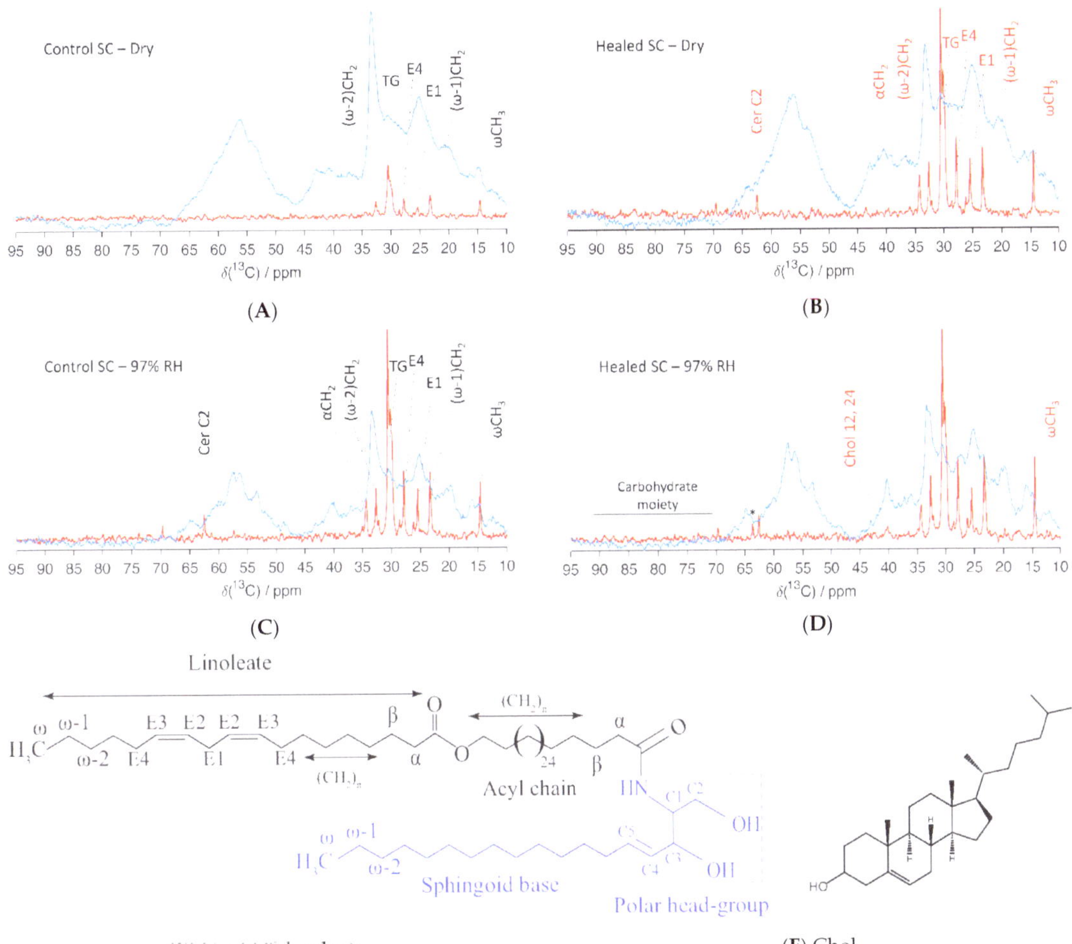

Figure 7. Natural abundance ¹³C polarization transfer solid state NMR (PTssNMR) study on control (**A,C**) and healed (**B,D**) SC samples in dry (**A,B**) and wet (**C,D**) conditions. The individual direct polarization (DP) (grey), cross polarization (CP) (blue), and insensitive nuclei enhanced by polarization transfer (INEPT) (red) spectra were overlaid in all experiments for the purpose of comparison. The resonance lines originating from various SC lipid molecular segments along with Chol are labelled in the spectra. The labelling is provided in black for control and red in healed SC when changes in the INEPT signals were observed compared to control SC. The peak labeled with (*) in the healed SC at 97% RH might indicate the presence of glycosphingolipid. Assignments of various lipid molecular segments with their numberings are depicted with a representative ceramide (Cer) structure—Cer EOS linoleate (**E**), as well as for Chol (**F**).

In dry state, both control and healed SC showed mobility in the terminal and central parts of the lipid acyl chains. However, the mobility was much higher in healed SC compared to control SC. The unsaturated carbon segments in the chains also showed mobility in both control and healed SC (Figure 7 and Figure S7), and the mobility of E1 and E4 carbon segments was higher in healed SC compared to control SC. Similar higher mobility of E2 and E3 segments were noticed in the high ppm regime for healed SC (Figure S7). Furthermore, the healed SC showed mobility in the headgroup region for αCH₂ and Cer2 carbon segments, which was not seen with control SC. No molecular mobility could be detected for Chol carbons and protein components for either control or healed SC at dry conditions. When hydrating SC at 97% RH, lipid mobility was increased in all the carbon segments for which we detected signals in dry conditions. We also observe additional mobility in the lipid head-groups at αCH_2 for control SC, which was not seen in dry state. No differences in mobility between control and healed SC could be detected at hydrated conditions, except for a slightly higher mobility of the lipid chain terminal ωCH_3 and Chol in healed SC. For healed SC, we observed an additional signal at 63.5 ppm and a few weak signals between 65 and 100 ppm, which could be attributed to the presence of a carbohydrate moiety from glycosphingolipids present in the sample [80,81]. Note, we did not notice any mobility in the protein segments even at 97% RH in either control or healed SC, which differed from previous observations from samples of pig ear SC [75]. The weak signal at approximately 40 ppm could be attributed to the Chol 12,14 carbon segments.

4. Discussion

Cutaneous wound healing is a complex process and is not yet fully understood. Hence, in order to investigate the wound healing mechanism and evaluate post-wound healing skin barrier recovery in comparison with control skin barrier, we conducted an in vivo study in Göttingen minipigs. In general, Göttingen minipigs are well-established animal models for human wound healing and therefore form the model of choice for the current study. The length of the in vivo study was 21 days, which was an appropriate time for wound closure. No visual sign of inflammation was observed in any of the induced wound sites during the study, which indicated a normal healing process with no complications involved. The wound healing progression was examined in real time during the in vivo period using traditional approaches, e.g., monitoring wound closure and measuring wound pH and TEWL, which allowed for the comparison of these parameters to those previously reported in literature on wound healing. Histological examination on excised healed skin tissue was also performed to evaluate the recovered dermis and epidermis and correlate with the literature. As our main interest lies in the healed skin barrier recovery in comparison to the control skin barrier, we performed further ex vivo tissue characterization with several physicochemical methods, including SWAXD, FTIR, water uptake capacity, and PTssNMR characterization of the SC, which revealed molecular details and provided insight to what extent the wound has been recovered when compared with control SC. The physicochemical characterization was then further discussed in relation to the traditional measures of wound healing progression performed during the in vivo session.

In control healthy skin, the TEWL, which is commonly used as a measure of the skin barrier efficiency, has been reported to be 10–15 $g/m^2 \cdot h$ [82–84]. In the current study, the obtained TEWL value was around 15 $g/m^2 \cdot h$ for the control skin. Differences in measured TEWL could be due to instrumental aspects, and variations in ambient conditions such as humidity, temperature, etc. [85]. When TEWL was measured repeatedly on control skin, the values were very similar in both pigs. The measured TEWL on the wound sites were significantly higher than control at the beginning of the wound healing cycle and dropped abruptly on day 14, which was due to sealing of the wound opening as visualized in the wound images. This shows an excellent correlation between sharply decreased TEWL and wound sealing on day 14. The TEWL of the wounds reached the control values on day 17 of the healing cycle. Interestingly, the TEWL decreased below control values when measured on day 21. This reduction was not due to fluctuations in humidity and temperature, which were controlled and stable during the whole in vivo cycle. The low TEWL in healed skin might instead have been due to possible loss of skin appendages, thicker skin due to scar formation, differences in SC molecular composition, etc.

An important aspect of assessing wound healing progression is to evaluate the pH of the wound fluid. The wound healing proceeds more rapidly at an acidic pH compared to alkaline or neutral pH [44,86]. This has been explained by the fact that protease inhibitors and several other enzymes essential for wound healing work best at slightly acidic conditions [40,56]. When pH was measured on the wound sites, an alkaline pH around 8.0 was recorded at the beginning of the wound healing cycle. Similar high pH values on the wound sites have also previously been reported in the literature and could have several reasons, including exposure of wound tissue to the body's physiological pH, release of molecules on the wound site as a result of acute wounds, bacterial colonization, or the production of ammonia generated by urease with urea as substrate [56,87]. As the healing progresses and the amount of wound fluid gradually decreases, the wound pH also shows a declining trend. This is indicative of wound healing progression without visual inflammation. It is also consistent with literature reports showing that the wound environment generally progresses from an initial alkaline condition through a neutral and then acidic state during the healing cycle [45,54,88–90]. The pH of the control skin was also recorded and shown to be slightly acidic, i.e., around 5.0. We here acknowledge the fundamental problem of interpreting the measured pH reading obtained by placing an electrode on a dry skin surface. In the present wound healing study, the wounds were moist with substantial wound fluid at the beginning of the healing cycle. This leads to a reliable measurement of wound pH until day 10 when the wounds were still moist. On day 14 and onwards, due to the sealing of wounded area, the wound surfaces were dry, and interpreting the pH reading again posed a challenge. However, a pH value around 5.1 was recorded on day 14 and onwards when the wound surfaces were dry, which was in agreement with skin surface pH values reported in the literature [47,48].

Restoration of barrier function to prevent further damage or infection of the wounds is an important aspect of the healing process [41,59]. It is plausible that there is a clear advantage to the wounded individual with respect to accelerate the sealing of the open wound while the remaining healing underneath may proceed at a different rate to perfectionate the barrier. Indeed, histological observations of the wound tissues in both samples revealed several morphological differences in various layers of the epidermis and dermis. In the healed skin tissue, the cells in various layers of the epidermis were still deformed, not densely packed, and organized, which indicated incomplete healing in those layers when compared with control skin. The dermis was furthermore less epithelialized in healed skin when compared to control skin. These observations clearly indicate that after three weeks of wound healing in vivo, cell morphology and tissue rebuilding in the deeper layers had not reached the same level as in the control skin. Due to lack of resolution in light microscopy, details in the superficial thin layer of the skin, the SC, could not be resolved, and instead physicochemical methods were used to detect molecular differences between control and healed skin three weeks post-wound induction in the Göttingen minipigs.

As the transport properties and mechanical strength of SC largely depends on SC hydration, we investigated the water uptake of SC isolated from the excised skin at varying hydrating conditions. When we compared the water uptake of healed and control SC, the sorption profiles looked very similar with only a slight difference at higher RH. The water uptake at 97% RH in both healed and control SC was measured to be between 35 and 40 wt %, which is in line with previous reports [75]. Further examination of SC with SWAXD did, however, reveal differences in protein keratin organization. The control SC exhibited a peak with d-spacing 90 Å, which was attributed to the keratin filament rod diameter and/or the thickness of corneocyte-bound lipids. The corresponding peak was absent in the healed SC, which might indicate that the corneocytes did not fully mature and/or that the keratin organization was less developed. Furthermore, in the healed SC, the keratin interchain distance showed more pronounced swelling upon hydration compared to control SC, which could be another indication of less developed keratin. The structural information obtained from SWAXD can be compared to the information on molecular mobility in the keratin filament amino acids. The PTssNMR data showed no mobility in any of the carbon segments of the keratin filament at the humidities investigated (Figure 7). This contrasts from previous reports on SC from pig ears showing that the protein segments become mobile at RH values above a threshold hydration corresponding to ≈85% RH [75]. It is possible that the sample treatment and the SC extraction and washing procedures may lead to excessive loss of polar NMF molecules. Lack of NMF and other small molecules could lead to reduced water uptake in the corneocytes and subsequent reduction in molecular mobility [91]. It is also possible that the SC molecular properties vary slightly between different sites on the body, which also may explain some of the observed differences compared to previous studies.

The molecular organization of SC lipids have been shown to play an important role in the skin barrier function. When we examined lipid lamellar organization using SWAXD and PTssNMR, both control and healed SC revealed a short lamellar phase, i.e., SPP, phase-separated Chol, along with hexagonal lateral packing of the lipid acyl chains. This is in accordance with literature reports for pig SC, although orthorhombic packing of the lipid chains has also been reported in pig SC on the basis of FTIR data [15,92]. No long lamellar phase, i.e., LPP, was detected in either of the samples. However, the $q \approx 1.4$ Å^{-1} peak, which corresponds to a d-spacing of 45 Å, has been suggested to originate from a third-order refection of the LPP [15,69]. This peak is present in the control SC while absent in the healed SC at dry conditions. Apart from that, no other differences could be observed in lipid organization between control and healed SC due to the lack of higher flux in the in-house X-ray source, and more detailed investigations would require a synchrotron X-ray source.

While SWAXD provides detailed information on ordered structures, it is less powerful to detect small changes in fluid structures. Changes in molecular mobility of SC components are, however, considered crucial to the material properties of SC, including barrier function [93]. The ^{13}C PTssNMR spectra from both control and healed SC show mobility in several lipid segments with increasing mobility at increasing hydration. This is also evident from the ^{1}H NMR analysis of control and healed SC at dry conditions (Figure S6) where the peaks originating from lipid acyl chain terminals, main chain $(CH_2)_n$, lipid head-groups, and double bond unsaturation are sharp and clearly visible. The higher mobility in the healed SC compared to control SC at dry conditions could indicate differences in lipid composition. The healed SC also reveals several other interesting peaks in the region between 55 and 100 ppm, which may come from, e.g., saccharides and phospholipid head-groups [94]. This could indicate presence of glycosphingolipids in the healed SC, a precursor molecule for barrier lipid synthesis that is normally not present in the SC [95]. Incomplete lipid conversion/production during the three weeks we allowed for skin barrier recovery in the current study could indeed affect the SC lipid organization, and our FTIR data also confirm potential presence of glycosphingolipids in the healed SC (Figure S8). A detailed lipid composition analysis of both healed and control SC would help to resolve this issue and will be the subject for a future study.

The healing progression follows the same trends in both pigs when evaluated on the basis of TEWL, pH, and wound closure. TEWL and pH largely returned to control values within three weeks, indicating good healing progression. However, the deeper layers of epidermis and dermis did not mature to resemble control skin within the three-week period. The extracellular lipids of SC did not produce the LPP, and samples from both pigs also revealed the presence of more fluid lipids. Taken together, this indicates that there was a difference in SC lipid composition between the healed skin and the control skin. These phenomena supported our hypothesis that rapid wound closure is crucial for restoring skin barrier, minimizing fluid loss, and preventing further infection while perfecting the lipid composition considered vital for optimal barrier function and restoring the viable parts of the skin underneath may proceed over a longer time scale. The current study may serve as a baseline, as well as a template, for evaluating future formulations, ability to facilitate wound healing time and quality, and possibly reduce scar formation.

5. Conclusions

In the current work, we conducted a study on Göttingen minipigs to gain further understanding on the mechanism of wound healing on molecular level by studying the natural healing progression of acute wounds in vivo and then evaluate the quality and the molecular properties of the excised restored tissue in comparison to healthy control skin from the same individuals. The main findings are as follows:

i.   Acute open wounds sealed within two weeks after incision. Wound closure monitored visually reflected good regeneration of the tissue at the end of the 21 days in vivo session. Histological evaluation of excised healed skin did, however, reveal morphological differences in various layers of the epidermis and less epithelialization in the dermis when compared with control skin.

ii.   The TEWL of the healing skin reached the value of control skin ($\approx$15 g/m^2·h) within three weeks from wound incision. The pH was alkaline ($\approx$8.0) at the beginning of the healing cycle and became slightly acidic ($\approx$5.1) at the end of the session. Both TEWL and pH provided an excellent correlation to visually determined wound closure.

iii.   SC water uptake capacity was similar in both control and healed SC with slightly higher water uptake in healed SC compared to control SC at higher humidity.

iv.   SWAXD studies on excised SC showed poorly developed keratin organization in healed SC. The extracellular lipid organization in healed SC also exhibited deviations in lamellar structure compared to control SC.

v.   PTssNMR revealed the presence of more fluid lipids in healed SC compared to control SC. This could be a result of incomplete lipid conversion during barrier recovery as glycosphingolipids, which were normally not present in SC, were detected by FTIR.

We conclude that although the wounds healed within three weeks, barrier recovery on the molecular level was not complete. Topical supplementation comprising compounds that facilitate lipid maturation and affect subsequent organization in SC may help skin barrier recovery and improve healing with, e.g., reduced scar formation as a result. Evidently a molecular approach to wound healing could be a valuable tool in future product development.

Supplementary Materials: The following are available online at https://www.mdpi.com/article/10.3390/biomedicines9040360/s1: Supplementary text with figure: cutaneous wound area calculation, Table S1: compilation of wound area calculation, Table S2: compilation of measured TEWL values, Table S3: compilation of recorded skin surface temperature, Table S4: compilation of measured pH values, Figure S1: Representative wound images on pig 1, position 4 during the healing cycle, Figure S2: Representative wound images on pig 2, position 10 during the healing cycle, Figure S3: variation in TEWL readings during the healing cycle, Figure S4: Histology images on control and healed skin, Figure S5: SWAXD spectra of control and healed SC excised from pig 2 in dry and wet condition, Figure S6: ^{1}H NMR measurements on control and healed SC in dry condition, Figure S7: ^{13}C PTssNMR spectra on control and healed SC in dry and wet conditions for high (100–150) ppm region, Figure S8: FTIR spectra of control and healed SC in dry condition

Author Contributions: Conceptualization and funding acquisition, J.E., A.S., E.H.M., H.H., K.K., T.R., and R.M.; methodology and experiments, E.H.M., A.K.M., A.S., C.P., I.T., J.E., K.K., and L.B.M.; data evaluation and processing, E.H.M., A.K.M., A.S., C.P., H.H., I.T., J.E., K.K., L.B.M., R.M., and T.R.; writing and reviewing the manuscript, E.H.M., A.S., H.H., J.E., K.K., L.B.M., and T.R. All authors have read and agreed to the published version of the manuscript.

Funding: This work was financially supported by the Swedish Knowledge Foundation (Dnr 20180108), Gustaf Th Ohlsson Foundation and Biofilms—Research Center for Biointerfaces at Malmö University.

Institutional Review Board Statement: The animal experiments were performed according to Swedish Animal Welfare Act SFS 1988:534 and were approved by the Animal Ethics Committee of Malmö/Lund, Sweden (permit numbers, M131-16, including extension 5.8.18-02900/2018 and 5.8.18-00974/2019).

Informed Consent Statement: Not applicable.

Data Availability Statement: Not applicable.

Acknowledgments: We are grateful to Emma Sparr for fruitful discussions and suggestions on the manuscript. We also thank Assoc. Ola Bergendorff for providing and assisting with the TEWL device and Elisabeth Nyborg for her help during the in vivo session.

Conflicts of Interest: The authors declare no conflict of interest.

References

1. Madison, K.C. Barrier function of the skin: "La raison d'etre" of the epidermis. *J. Investig. Dermytol.* **2003**, *121*, 231–241. [CrossRef]
2. Proksch, E.; Brandner, J.M.; Jensen, J.-M. The skin: An indispensable barrier. *Exp. Dermatol.* **2008**, *17*, 1063–1072. [CrossRef]
3. Blatteis, C.M. Age-dependent changes in temperature regulation. A mini review. *Gerontology* **2012**, *58*, 289–295. [CrossRef]
4. Schaefer, H.; Redelmeier, T.E. *Skin Barrier: Principles of Percutaneous Absorption*; Karger: Basel, Switzerland, 1996.
5. Scheuplein, R.J. Permeability of the skin: A review of major concepts and some new developments. *J. Investig. Dermatol.* **1976**, *67*, 672–676. [CrossRef]
6. Eckert, R.L. Structure, function, and differentiation of the keratinocyte. *Physiol. Rev.* **1989**, *69*, 1316–1346. [CrossRef]
7. Watt, F.M. Terminal differentiation of epidermal keratinocytes. *Curr. Opin. Cell Biol.* **1989**, *1*, 1107–1115. [CrossRef]
8. Grove, G.L.; Kligman, A.M. Age-associated changes in human epidermal cell renewal. *J. Gerontol.* **1983**, *38*, 137–142. [CrossRef]
9. Candi, E.; Schmidt, R.; Melino, G. The cornified envelope: A model of cell death in the skin. *Nat. Rev. Mol. Cell Biol.* **2005**, *6*, 328–340. [CrossRef]
10. Holbrook, K.A.; Odland, G.F. Regional differences in the thickness (cell layers) of the human stratum corneum: An ultrastructural analysis. *J. Investig. Dermatol.* **1974**, *62*, 415–422. [CrossRef] [PubMed]
11. Pham, Q.D.; Carlström, G.; Lafon, O.; Sparr, E.; Topgaard, D. Quantification of the amount of mobile components in intact stratum corneum with natural-abundance 13C solid-state NMR. *Phys. Chem. Chem. Phys.* **2020**, *22*, 6572–6583. [CrossRef]
12. Schaefer, H.; Redelmeier, T.E. Structure and dynamics of the skin barrier. In *Skin Barrier: Principles of Percutaneous Absorption*; Krager: Basel, Switzerland, 1996; pp. 1–42.
13. Papir, Y.S.; Hsu, K.; Wildnauer, R.H. The mechanical properties of stratum corneum. *Biochim. Biophys. Acta* **1975**, *399*, 170–180. [CrossRef]
14. Breiden, B.; Sandhoff, K. The role of sphingolipid metabolism in cutaneous permeability barrier formation. *Biochim. Biophys. Acta* **2014**, *1841*, 441–452. [CrossRef]
15. Bouwstra, J.A.; Gooris, G.S.; Bras, W.; Downing, D.T. Lipid organization in pig stratum corneum. *J. Lipid Res.* **1995**, *36*, 685–695. [CrossRef]
16. Wertz, P.W.; Miethke, M.C.; Long, S.A.; Strauss, J.S.; Downing, D.T. The Composition of the Ceramides from Human Stratum Corneum and from Comedones. *J. Investig. Dermatol.* **1985**, *84*, 410–412. [CrossRef] [PubMed]
17. Bouwstra, J.A.; Gooris, G.S.; Van Der Spek, J.A.; Bras, W. Structural investigations of human stratum corneum by small-angle X-ray scattering. *J. Investig. Dermatol.* **1991**, *97*, 1005–1012. [CrossRef]
18. White, S.H.; Mirejovsky, D.; King, G.I. Structure of lamellar lipid domains and corneocyte envelopes of murine stratum corneum. An x-ray diffraction study. *Biochemistry* **1988**, *27*, 3725–3732. [CrossRef]
19. Silva, C.L.; Topgaard, D.; Kocherbitov, V.; Sousa, J.J.S.; Pais, A.A.C.C.; Sparr, E. Stratum corneum hydration: Phase transformations and mobility in stratum corneum, extracted lipids and isolated corneocytes. *Biochim. Biophys. Acta (BBA)–Biomembr.* **2007**, *1768*, 2647–2659. [CrossRef]
20. Alonso, A.; Meirelles, N.C.; Yushmanov, V.E.; Tabak, M. Water increases the fluidity of intercellular membranes of stratum corneum: Correlation with water permeability, elastic, and electrical resistance properties. *J. Investig. Dermatol.* **1996**, *106*, 1058–1063. [CrossRef]
21. Park, Y.H.; Jang, W.H.; Seo, J.A.; Park, M.; Lee, T.R.; Park, Y.H.; Kim, D.K.; Lim, K.M. Decrease of ceramides with very long-chain fatty acids and downregulation of elongases in a murine atopic dermatitis model. *J. Investig. Dermatol.* **2012**, *132*, 476–479. [CrossRef]
22. Ishikawa, J.; Narita, H.; Kondo, N.; Hotta, M.; Takagi, Y.; Masukawa, Y.; Kitahara, T.; Takema, Y.; Koyano, S.; Yamazaki, S.; et al. Changes in the ceramide profile of atopic dermatitis patients. *J. Investig. Dermatol.* **2010**, *130*, 2511–2514. [CrossRef] [PubMed]
23. Motta, S.; Monti, M.; Sesana, S.; Mellesi, L.; Ghidoni, R.; Caputo, R. Abnormality of Water Barrier Function in Psoriasis: Role of Ceramide Fractions. *Arch. Dermatol.* **1994**, *130*, 452–456. [CrossRef]
24. Imokawa, G.; Abe, A.; Jin, K.; Higaki, Y.; Kawashima, M.; Hidano, A. Decreased level of ceramides in stratum corneum of atopic dermatitis: An etiologic factor in atopic dry skin? *J. Investig. Dermatol.* **1991**, *96*, 523–526. [CrossRef]
25. Janssens, M.; Van Smeden, J.; Gooris, G.S.; Bras, W.; Portale, G.; Caspers, P.J.; Vreeken, R.J.; Kezic, S.; Lavrijsen, A.P.; Bouwstra, J.A. Lamellar lipid organization and ceramide composition in the stratum corneum of patients with atopic eczema. *J. Investig. Dermatol.* **2011**, *131*, 2136–2138. [CrossRef] [PubMed]
26. Janssens, M.; Van Smeden, J.; Gooris, G.S.; Bras, W.; Portale, G.; Caspers, P.J.; Vreeken, R.J.; Hankemeier, T.; Kezic, S.; Wolterbeek, R.; et al. Increase in short-chain ceramides correlates with an altered lipid organization and decreased barrier function in atopic eczema patients. *J. Lipid. Res.* **2012**, *53*, 2755–2766. [CrossRef]
27. Bouwstra, J.; Gooris, G.; Ponec, M. The lipid organisation of the skin barrier: Liquid and crystalline domains coexist in lamellar phases. *J. Biol. Phys.* **2002**, *28*, 211–223. [CrossRef]
28. Van Smeden, J.; Janssens, M.; Gooris, G.S.; Bouwstra, J.A. The important role of stratum corneum lipids for the cutaneous barrier function. *Biochim. Biophys. Acta* **2014**, *1841*, 295–313. [CrossRef]
29. Elias, P.M. Epidermal lipids, barrier function, and desquamation. *J. Investig. Dermatol.* **1983**, *80*, 44–49. [CrossRef] [PubMed]
30. Nemes, Z.; Steinert, P.M. Bricks and mortar of the epidermal barrier. *Exp. Mol. Med.* **1999**, *31*, 5–19. [CrossRef] [PubMed]

31. Honari, G.; Maibach, H. Chapter 1–Skin Structure and Function. In *Applied Dermatotoxicology*; Maibach, H., Honari, G., Eds.; Academic Press: Boston, MA, USA, 2014; pp. 1–10. [CrossRef]
32. Reinke, J.M.; Sorg, H. Wound repair and regeneration. European surgical research. Europaische chirurgische Forschung. *Rech. Chir. Eur.* **2012**, *49*, 35–43. [CrossRef]
33. Landen, N.X.; Li, D.; Stahle, M. Transition from inflammation to proliferation: A critical step during wound healing. *Cell. Mol. Life Sci. CMLS* **2016**, *73*, 3861–3885. [CrossRef]
34. Wilhelm, K.P.; Wilhelm, D.; Bielfeldt, S. Models of wound healing: An emphasis on clinical studies. *Skin Res. Technol.* **2017**, *23*, 3–12. [CrossRef] [PubMed]
35. Atiyeh, B.S.; Ioannovich, J.; Al-Amm, C.A.; El-Musa, K.A. Management of acute and chronic open wounds: The importance of moist environment in optimal wound healing. *Curr. Pharm. Biotechnol.* **2002**, *3*, 179–195. [CrossRef] [PubMed]
36. Dyson, M.; Young, S.; Pendle, C.L.; Webster, D.F.; Lang, S.M. Comparison of the effects of moist and dry conditions on dermal repair. *J. Investig. Dermatol.* **1988**, *91*, 434–439. [CrossRef] [PubMed]
37. Svensjö, T.; Pomahac, B.; Yao, F.; Slama, J.; Eriksson, E. Accelerated Healing of Full-Thickness Skin Wounds in a Wet Environment. *Plast. Reconstr. Surg.* **2000**, *106*, 602–612. [CrossRef] [PubMed]
38. Vogt, P.M.; Andree, C.; Breuing, K.; Liu, P.Y.; Slama, J.; Helo, G.; Eriksson, E. Dry, moist, and wet skin wound repair. *Ann. Plast. Surg.* **1995**, *34*, 493–499. [CrossRef] [PubMed]
39. Atiyeh, B.; Hayek, S. Moisture and wound healing. *J. Plaies Cicatr.* **2005**, *9*, 7–11.
40. Percival, S.L.; McCarty, S.; Hunt, J.A.; Woods, E.J. The effects of pH on wound healing, biofilms, and antimicrobial efficacy. *Wound Repair Regen.* **2014**, *22*, 174–186. [CrossRef]
41. Ousey, K.; Cutting, K.F.; Rogers, A.A.; Rippon, M.G. The importance of hydration in wound healing: Reinvigorating the clinical perspective. *J. Wound Care* **2016**, *25*, 124–130. [CrossRef]
42. Visscher, M.O.; Robinson, M.; Fugit, B.; Rosenberg, R.J.; Hoath, S.B.; Randall Wickett, R. Amputee skin condition: Occlusion, stratum corneum hydration and free amino acid levels. *Arch. Dermatol. Res.* **2011**, *303*, 117–124. [CrossRef]
43. Harding, C.R.; Watkinson, A.; Rawlings, A.V.; Scott, R. Dry skin, moisturization and corneodesmolysis. *Int. J. Cosmet. Sci.* **2000**, *22*, 21–52. [CrossRef]
44. Roberts, G.; Hammad, L.; Creevy, J.; Shearman, C.; European Wound Management Association. Physical changes in dermal tissues around chronic venous ulcers. In Proceedings of the European Conference, 7th Advances in Wound Management, Harrogate, North Yorkshire; 1998; pp. 104–105.
45. Hoffman, R.; Noble, J.; Eagle, M. The use of proteases as prognostic markers for the healing of venous leg ulcers. *J. Wound Care* **1999**, *8*, 273–276. [CrossRef]
46. Schade, H.; Marchionini, A. Der Säuremantel der Haut nach Gaskettenmessngen. *Klin. Wochenschr* **1928**, *7*, 12–14. [CrossRef]
47. Ali, S.M.; Yosipovitch, G. Skin pH: From Basic Science to Basic Skin Care. *Acta Derm. Venereol.* **2013**, *93*, 261–269. [CrossRef]
48. Braun-Falco, O.; Korting, H.C. Normal pH value of human skin. *Der Hautarzt Z. Dermatol. Venerol. Verwandte Geb.* **1986**, *37*, 126–129.
49. Bandier, J.; Johansen, J.D.; Petersen, L.J.; Carlsen, B.C. Skin pH, atopic dermatitis, and filaggrin mutations. *Dermat. Contact Atopic Occup. Drug* **2014**, *25*, 127–129. [CrossRef] [PubMed]
50. Kezic, S.; Kammeyer, A.; Calkoen, F.; Fluhr, J.W.; Bos, J.D. Natural moisturizing factor components in the stratum corneum as biomarkers of filaggrin genotype: Evaluation of minimally invasive methods. *Br. J. Dermatol.* **2009**, *161*, 1098–1104. [CrossRef]
51. Rippke, F.; Schreiner, V.; Schwanitz, H.J. The acidic milieu of the horny layer: New findings on the physiology and pathophysiology of skin pH. *Am. J. Clin. Dermatol.* **2002**, *3*, 261–272. [CrossRef] [PubMed]
52. Scott, I.R.; Harding, C.R. Filaggrin breakdown to water binding compounds during development of the rat stratum corneum is controlled by the water activity of the environment. *Dev. Biol.* **1986**, *115*, 84–92. [CrossRef]
53. Elias, P.M. The how, why and clinical importance of stratum corneum acidification. *Exp. Dermatol.* **2017**, *26*, 999–1003. [CrossRef] [PubMed]
54. Shukla, V.K.; Shukla, D.; Tiwary, S.K.; Agrawal, S.; Rastogi, A. Evaluation of pH measurement as a method of wound assessment. *J. Wound Care* **2007**, *16*, 291–294. [CrossRef] [PubMed]
55. Schneider, L.A.; Korber, A.; Grabbe, S.; Dissemond, J. Influence of pH on wound-healing: A new perspective for wound-therapy? *Arch. Dermatol. Res.* **2007**, *298*, 413–420. [CrossRef]
56. Nagoba, B.; Suryawanshi, N.; Wadher, B.; Selkar, S. Acidic Environment and Wound Healing: A Review. *Wounds Compend. Clin. Res. Pract.* **2015**, *27*, 5–11.
57. Kurabayashi, H.; Tamura, K.; Machida, I.; Kubota, K. Inhibiting bacteria and skin pH in hemiplegia: Effects of washing hands with acidic mineral water. *Am. J. Phys. Med. Rehabil.* **2002**, *81*, 40–46. [CrossRef]
58. Nyman, E.; Henricson, J.; Ghafouri, B.; Anderson, C.D.; Kratz, G. Hyaluronic Acid Accelerates Re-epithelialization and Alters Protein Expression in a Human Wound Model. *Plast. Reconstr. Surg. Glob. Open* **2019**, *7*, e2221. [CrossRef] [PubMed]
59. Sorg, H.; Tilkorn, D.J.; Hager, S.; Hauser, J.; Mirastschijski, U. Skin Wound Healing: An Update on the Current Knowledge and Concepts. *Eur. Surg. Res.* **2017**, *58*, 81–94. [CrossRef]
60. Cañedo-Dorantes, L.; Cañedo-Ayala, M. Skin Acute Wound Healing: A Comprehensive Review. *Int. J. Inflamm.* **2019**, *2019*, 3706315. [CrossRef]

61. Sullivan, T.P.; Eaglstein, W.H.; Davis, S.C.; Mertz, P. The pig as a model for human wound healing. *Wound Repair Regen. Off. Publ. Wound Heal Soc. Eur. Tissue Repair Soc.* **2001**, *9*, 66–76. [CrossRef] [PubMed]
62. Arcega, R.S.; Woo, J.S.; Xu, H. Performing and Cutting Frozen Sections. In *Biobanking: Methods and Protocols*; Yong, W.H., Ed.; Springer: New York, NY, USA, 2019; pp. 279–288. [CrossRef]
63. Morris, G.A.; Freeman, R. Enhancement of nuclear magnetic resonance signals by polarization transfer. *J. Am. Chem. Soc.* **1979**, *101*, 760–762. [CrossRef]
64. Pines, A.; Gibby, M.G.; Waugh, J.S. Proton-enhanced nuclear induction spectroscopy. A method for high resolution NMR of dilute spins in solids. *J. Chem. Phys.* **1972**, *56*, 1776–1777. [CrossRef]
65. Van Geet, A.L. Calibration of methanol nuclear magnetic resonance thermometer at low temperature. *Anal. Chem.* **1970**, *42*, 679–680. [CrossRef]
66. Chen, L.; Weng, Z.; Goh, L.; Garland, M. An efficient algorithm for automatic phase correction of NMR spectra based on entropy minimization. *J. Magn. Reson.* **2002**, *158*, 164–168. [CrossRef]
67. Van Beek, J.D. MatNMR: A flexible toolbox for processing, analyzing and visualizing magnetic resonance data in Matlab®. *J. Magn. Reson.* **2007**, *187*, 19–26. [CrossRef]
68. Doucet, J.; Potter, A.; Baltenneck, C.; Domanov, Y.A. Micron-scale assessment of molecular lipid organization in human stratum corneum using microprobe X-ray diffraction. *J. Lipid Res.* **2014**, *55*, 2380–2388. [CrossRef]
69. Bjorklund, S.; Engblom, J.; Thuresson, K.; Sparr, E. Glycerol and urea can be used to increase skin permeability in reduced hydration conditions. *Eur. J. Pharm. Sci. Off. J. Eur. Fed. Pharm. Sci.* **2013**, *50*, 638–645. [CrossRef]
70. Yagi, N.; Aoyama, K.; Ohta, N. Microbeam X-ray diffraction study of lipid structure in stratum corneum of human skin. *PLoS ONE* **2020**, *15*, e0233131. [CrossRef]
71. Nakazawa, H.; Ohta, N.; Hatta, I. A possible regulation mechanism of water content in human stratum corneum via intercellular lipid matrix. *Chem. Phys. Lipids* **2012**, *165*, 238–243. [CrossRef]
72. Garson, J.-C.; Doucet, J.; Lévêque, J.-L.; Tsoucaris, G. Oriented Structure in Human Stratum Corneum Revealed by X-Ray Diffraction. *J. Investig. Dermatol.* **1991**, *96*, 43–49. [CrossRef]
73. Kreplak, L.; Doucet, J.; Dumas, P.; Briki, F. New aspects of the alpha-helix to beta-sheet transition in stretched hard alpha-keratin fibers. *Biophys. J.* **2004**, *87*, 640–647. [CrossRef]
74. Rodriguez, J.A.; Ivanova, M.I.; Sawaya, M.R.; Cascio, D.; Reyes, F.E.; Shi, D.; Sangwan, S.; Guenther, E.L.; Johnson, L.M.; Zhang, M.; et al. Structure of the toxic core of [agr]-synuclein from invisible crystals. *Nature* **2015**, *525*, 486–490. [CrossRef]
75. Mojumdar, E.H.; Pham, Q.D.; Topgaard, D.; Sparr, E. Skin hydration: Interplay between molecular dynamics, structure and water uptake in the stratum corneum. *Sci. Rep.* **2017**, *7*, 15712. [CrossRef]
76. Nowacka, A.; Bongartz, N.A.; Ollila, O.H.S.; Nylander, T.; Topgaard, D. Signal intensities in 1H–13C CP and INEPT MAS NMR of liquid crystals. *J. Magn. Reson.* **2013**, *230*, 165–175. [CrossRef]
77. Bjorklund, S.; Nowacka, A.; Bouwstra, J.A.; Sparr, E.; Topgaard, D. Characterization of stratum corneum molecular dynamics by natural-abundance (1)(3)C solid-state NMR. *PLoS ONE* **2013**, *8*, e61889. [CrossRef] [PubMed]
78. Nowacka, A.; Mohr, P.C.; Norrman, J.; Martin, R.W.; Topgaard, D. Polarization transfer solid-state NMR for studying surfactant phase behavior. *Langmuir* **2010**, *26*, 16848–16856. [CrossRef]
79. Andersson, J.M.; Grey, C.; Larsson, M.; Ferreira, T.M.; Sparr, E. Effect of cholesterol on the molecular structure and transitions in a clinical-grade lung surfactant extract. *Proc. Natl. Acad. Sci. USA* **2017**, *114*, 3592–3601. [CrossRef]
80. Lu, Y.; Hu, F.; Miyakawa, T.; Tanokura, M. Complex Mixture Analysis of Organic Compounds in Yogurt by NMR Spectroscopy. *Metabolites* **2016**, *6*. [CrossRef]
81. Yazdani, P.; Wang, B.; Rimaz, S.; Kawi, S.; Borgna, A. Glucose hydrogenolysis over Cu-La2O3/Al2O3: Mechanistic insights. *Mol. Catal.* **2019**, *466*, 138–145. [CrossRef]
82. Klang, V.; Schwarz, J.C.; Lenobel, B.; Nadj, M.; Auböck, J.; Wolzt, M.; Valenta, C. In vitro vs. in vivo tape stripping: Validation of the porcine ear model and penetration assessment of novel sucrose stearate emulsions. *Eur. J. Pharm. Biopharm. Off. J. Arb. Pharm. Verfahr. V* **2012**, *80*, 604–614. [CrossRef]
83. Zhai, H.; Maibach, H.I. *Dermatotoxicology*, 6th ed.; CRC Press LCC: Boca Raton, FL, USA, 2004; pp. 938–955.
84. Boer, M.; Duchnik, E.; Maleszka, R.; Marchlewicz, M. Structural and biophysical characteristics of human skin in maintaining proper epidermal barrier function. *Adv. Dermatol. Allergol. Postępy Dermatol. Alergol.* **2016**, *33*, 1–5. [CrossRef]
85. Du Plessis, J.; Stefaniak, A.; Eloff, F.; John, S.; Agner, T.; Chou, T.-C.; Nixon, R.; Steiner, M.; Franken, A.; Kudla, I.; et al. International guidelines for the in vivo assessment of skin properties in non-clinical settings: Part 2. transepidermal water loss and skin hydration. *Skin Res. Technol.* **2013**, *19*, 265–278. [CrossRef]
86. Lengheden, A.; Jansson, L. pH effects on experimental wound healing of human fibroblasts in vitro. *Eur. J. Oral Sci.* **1995**, *103*, 148–155. [CrossRef] [PubMed]
87. Stüttgen, G.; Schaefer, H. Die Hautoberfläche. In *Funktionelle Dermatologie: Grundlagen der Morphokinetik Pathophysiologie, Pharmakoanalyse und Therapie von Dermatosen*; Stüttgen, G., Schaefer, H., Eds.; Springer: Berlin, Germany, 1974; pp. 184–190. [CrossRef]
88. Gethin, G. The significance of surface pH in chronic wounds. *Wounds UK* **2007**, *3*, 52.
89. Greener, B.; Hughes, A.A.; Bannister, N.P.; Douglass, J. Proteases and pH in chronic wounds. *J. Wound Care* **2005**, *14*, 59–61. [CrossRef]

90. Leveen, H.H.; Falk, G.; Borek, B.; Diaz, C.; Lynfield, Y.; Wynkoop, B.J.; Mabunda, G.A.; Rubricius, J.L.; Christoudias, G.C. Chemical acidification of wounds. An adjuvant to healing and the unfavorable action of alkalinity and ammonia. *Ann. Surg.* **1973**, *178*, 745–753. [CrossRef]
91. Björklund, S.; Andersson, J.M.; Pham, Q.D.; Nowacka, A.; Topgaard, D.; Sparr, E. Stratum corneum molecular mobility in the presence of natural moisturizers. *Soft Matter* **2014**, *10*, 4535–4546. [CrossRef]
92. Ongpipattanakul, B.; Francoeur, M.L.; Potts, R.O. Polymorphism in stratum corneum lipids. *Biochim. Biophys. Acta (BBA) Biomembr.* **1994**, *1190*, 115–122. [CrossRef]
93. Pham, Q.D.; Björklund, S.; Engblom, J.; Topgaard, D.; Sparr, E. Chemical penetration enhancers in stratum corneum–Relation between molecular effects and barrier function. *J. Control. Release* **2016**, *232*, 175–187. [CrossRef]
94. Mojumdar, E.H.; Grey, C.; Sparr, E. Self-Assembly in Ganglioside-Phospholipid Systems: The Co-Existence of Vesicles, Micelles, and Discs. *Int. J. Mol. Sci.* **2019**, *21*, 56. [CrossRef] [PubMed]
95. Hamanaka, S.; Suzuki, A.; Hara, M.; Nishio, H.; Otsuka, F.; Uchida, Y. Human Epidermal Glucosylceramides are Major Precursors of Stratum Corneum Ceramides. *J. Investig. Dermatol.* **2002**, *119*, 416–423. [CrossRef]

MDPI AG
Grosspeteranlage 5
4052 Basel
Switzerland
Tel.: +41 61 683 77 34

Biomedicines Editorial Office
E-mail: biomedicines@mdpi.com
www.mdpi.com/journal/biomedicines